Single Crystal Elastic Constants and Calculated Aggregate Properties: A HANDBOOK

Single Crystal Elastic Constants and Calculated Aggregate Properties: A HANDBOOK

Second Edition

Gene Simmons and Herbert Wang

THE M.I.T. PRESS
Cambridge, Massachusetts, and London, England

Printed and bound in the United States of America
by Halliday Lithograph Corp.

ISBN 0 262 19092 3 (hardcover)

Library of Congress catalog card number: 71-148848

Contents

Acknowledgments

We wish to express our appreciation to several individuals who contributed to this tabulation. Joan Jason and Carol Sande compiled most of the data and helped with the extensive checking of data. R.F.S. Hearmon called to our attention errors in the first edition. Financial support was provided by the National Science Foundation Grant GA-579 and by the National Aeronautics and Space Administration Grant NGR-22-009-176. Computation was done at the Massachusetts Institute of Technology.

Introduction

Data on the elastic properties of single crystals have been compiled previously by Huntington (1958), Hearmon (1946, 1956, 1966, 1969), Alexandrov and Ryzhova (1961), and Simmons (1965). Because of the twofold increase in the amount of data during the last five years, it seems desirable at this time however to bring together many of the elastic constants and some pressure and temperature derivatives in a single tabulation, including both Cs and Ss for each material. It is also desirable to calculate the elastic properties of a random, macroscopically isotropic, aggregate of crystals from the single crystal (anisotropic) elastic constants. This is not yet possible, but bounds may be obtained for the aggregate properties from the single crystal constants. Voigt (1928) averaged the elastic stiffnesses (C_{ij}) over all space. Reuss (1929) averaged the elastic compliances (S_{ij}). Hill (1952) showed that these averaged values are the least upper bound and the greatest lower bound, respectively for the aggregate. These averaged values of elastic properties, known as Voigt and Reuss averages, were tabulated by Simmons (1965) for all materials for which single crystal data had been reported in the literature by mid-1964. The Voigt and Reuss averages are given by:

Voigt	Reuss
$K_V = (A + 2B)/3$	$K_R = 1/(3a + 6b)$
$G_V = (A - B + 3C)/5$	$G_R = 5/(4a - 4b + 3c)$

where

$3A = C_{11} + C_{22} + C_{33}$	$3a = S_{11} + S_{22} + S_{33}$
$3B = C_{23} + C_{31} + C_{12}$	$3b = S_{23} + S_{31} + S_{12}$
$3C = C_{44} + C_{55} + C_{66}$	$3c = S_{44} + S_{55} + S_{66}$.

Hashin and Shtrikman (1962a, b) showed that the Voigt and Reuss bounds could be improved, and they developed expressions for the corresponding bounds of aggregates of cubic crystals. As it turns out, the difference between the upper and lower bounds for cubic crystals is improved by at least an order of magnitude for most materials. For a single phase aggregate of a cubic material, the bulk modulus, K, is given unambiguously by

$$K = \frac{1}{3}(C_{11} + 2C_{12})$$

and the rigidity is bounded by

$$G_1^* = G_1 + 3\left(\frac{5}{G_2 - G_1} - 4\beta_1\right)^{-1}$$

and

$$G_2^* = G_2 + 2\left(\frac{5}{G_1 - G_2} - 6\beta_2\right)^{-1}$$

where

$$\beta_1 = -\frac{3(K + 2G_1)}{5G_1(3K + 4G_1)}$$

$$\beta_2 = -\frac{3(K + 2G_2)}{5G_2(3K + 4G_2)}$$

$$G_1 = \frac{1}{2}(C_{11} - C_{12})$$

$$G_2 = C_{44}$$

and the C_{ij} are the usual single crystal elastic stiffnesses. The larger G_1* and G_2* is termed the Hashin rigidity, G_H, the smaller Shtrikman rigidity, G_S. The Hashin and Shtrikman bounds were tabulated by Simmons (1967) for those cubic materials found in his 1965 compilation.

We tabulate all four bounds (Voigt, Hashin, Shtrikman, Reuss) for cubic materials and just Voigt and Reuss bounds for non-cubic materials. The elastic properties (of aggregates) that are in common use include Young's modulus, shear modulus, Poisson's ratio, bulk modulus, compressibility, velocity of compressional waves, and the velocity of shear waves. Although any two of these properties suffice to characterize a given material uniquely, all are tabulated because different sets are best suited for different uses.

The present compilation is uncritical in the sense that data of low precision are included in the tables along with data of very high quality. With few exceptions there are several entries in the tables for materials for which the elastic data are suspect, and the reader will be warned immediately by the disagreement of the data between different samples and different original investigators. The guiding philosophy has been that all of the available data should be tabulated and the reader permitted to choose the set he believes better for his particular application. The several entries for a given material are usually listed by date of publication of the reference from which it was obtained. Thus newest data is presented last.

An internal check on the derived set of constants was made in the computer program by forming the product-matrix of the C-matrix and the S-matrix (before the elements were rounded off). The off-diagonal terms of the product-matrix were in all cases of the order of 10^{-5} or smaller. This procedure shows only that the arithmetic was properly done and fails to detect errors in the input data.

Considerable effort has gone into checking the input data for errors. Undoubtedly some errors still remain and we would greatly appreciate having these brought to our attention.

No distinction is made in the present compilation between adiabatic and isothermal elastic constants. The difference for most materials is of the order of 10^{-3}, or less. Notable exceptions are the alkali halides, aluminum, and periclase (MgO) which have differences of a few percent. Either set of constants may be calculated from the other by means of the relations given by Hearmon (1961, p. 15).

The density listed with an entry is that which converts data on velocity as a function of direction in a single crystal to elastic constant data through the Christoffel relation

$$\begin{vmatrix} \Gamma_{11} - \rho v^2 & \Gamma_{12} & \Gamma_{13} \\ \Gamma_{12} & \Gamma_{22} - \rho v^2 & \Gamma_{23} \\ \Gamma_{13} & \Gamma_{23} & \Gamma_{33} - \rho v^2 \end{vmatrix} = 0$$

where $\Gamma_{ik} = a_j a_l c_{ijkl}$

and a_i = direction cosine of the normal to the wave front. When the density was not actually reported by the original author, we estimated a value for ρ which we believed appropriate.

Alloy and solid solution compositions are given in atomic and mole percent, respectively, unless explicitly noted.

We calculated dS_{ij}/dX (where X = pressure or temperature) from dC_{ij}/dX through the matrix relation

$$\frac{dS}{dX} = -S \frac{dC}{dX} S$$

since $S = C^{-1}$. "Voigt" and "Reuss" averages for the derivatives were then calculated from the same linear combinations as for the elastic constants themselves, but with dC_{ij}/dX or dS_{ij}/dX used in place of the C_{ij} or S_{ij}.

References for Introduction

1. Alexandrov, K.S., and T.V. Ryzhova, The elastic properties of crystals, Sov. Phys.-Cryst., 6, 228-252, 1961.

2. Hashin, Z., and S. Shtrikman, On some variational principles in anisotropic and nonhomogeneous elasticity, J. Mech. and Phys. Solids, 10, 335-342, 1962a.

3. Hashin, Z., and S. Shtrikman, A variational approach to the theory of the elastic behaviour of polycrystals, J. Mech. and Phys. Solids, 10, 343-352, 1962b.

4. Hearmon, R.F.S., The elastic constants of anisotropic materials, Rev. Mod. Phys., 18, 409-440, 1946.

5. Hearmon, R.F.S., The elastic constants of anisotropic materials II, Advan. Phys., 5, 323-382, 1956.

6. Hearmon, R.F.S., An Introduction to Applied Anisotropic Elasticity, 136 pp., Oxford University Press, Oxford, 1961.

7. Hearmon, R.F.S., The elastic constants of non-piezoelectric crystals, in Landolt-Börnstein Tables, Numerical Data and Functional Relationships in Science and Technology, Vol. I, New Series, Group III, edited by K.-H. Hellwege, pp. 1-39, Springer-Verlag New York, Inc., New York, 1966.

8. Hearmon, R.F.S., The elastic constants of non-piezoelectric crystals, in Landolt-Börnstein Tables, Numerical Data and Functional Relationships in Science and Technology, Vol. II, New Series, Group III, edited by K.-H. Hellwege, pp. 1-39, Springer-Verlag New York, Inc., New York, 1969.

9. Hill, R., The elastic behaviour of a crystalline aggregate, Proc. Phys. Soc. London (A), 65, 349-354, 1952.

10. Huntington, H.B., The elastic constants of crystals, in Solid State Physics, Vol. 7, edited by F. Seitz and D. Turnbull, pp. 213-285, Academic Press, New York, 1958.

11. Reuss, A., Berechnung der Fliessgrenze von Mischkristallen auf Grund der Plastizitätsbedingung für Einkristalle, *Z. Angew. Math. Mech.*, *9*, 49-58, 1929.

12. Simmons, G., Single crystal elastic constants and calculated aggregate properties, *J. Graduate Research Center* (Southern Methodist University), *34*, 1-269, 1965.

13. Simmons, G., Hashin bounds for aggregates of cubic crystals, *J. Graduate Research Center* (Southern Methodist University), *36*, 1-87, 1967.

14. Voigt, W., *Lehrbuch der Krystallphysik*, 962 pp., B.G. Teubner, Leipzig, 1928.

Abbreviations

CODE	Specimen number (used for internal accounting purposes).
REF	Reference.
P	Pressure.
TEMP	Temperature (°K or ROOM).
C	Elastic stiffness with subscripts given at top of each column in the tables.
S	Elastic compliance with subscripts given at top of each column in the tables.
E	Young's modulus.
G	Shear modulus; Modulus of rigidity.
SIGMA	Poisson's ratio.
K	Bulk modulus.
BETA	Compressibility.
VS	Velocity of shear waves in an inifinite medium.
VP	Velocity of compressional waves in an infinite medium.
V	Averaged according to Voigt's scheme.
R	Averaged according to Reuss' scheme.
H	Hashin bound.
S	Shtrikman bound.

Units

Symbol	Unit	Pressure Derivative	Temperature Derivative
P	kb	-------	-------
TEMP	°K	-------	-------
C_{ij}	Mb	None	kb/°K
S_{ij}	Mb^{-1}	Mb^{-2}	$10^{-3}Mb^{-1}$/°K
E	Mb	None	kb/°K
G	Mb	None	kb/°K
SIGMA	None	10^{-4}/kb	10^{-4}/°K
K	Mb	None	kb/°K
BETA	Mb^{-1}	Mb^{-2}	$10^{-3}Mb^{-1}$/°K
VS	km/sec	km/sec/Mb	10^{-3}km/sec/°K
VP	km/sec	km/sec/Mb	10^{-3}km/sec/°K

Mb = Megabars = 10^{12}dynes/cm^2.

kb = Kilobars = 10^9 dynes/cm^2.

None = Dimensionless quantity

Single Crystal Constants

NAME / CODE	DENSITY	REF	TEMP(K)		11	12	44
ALUM (AMMONIUM)				C	0.2500	0.1060	0.0800
10001	1.6400	129	ROOM	S	5.3511	-1.5933	12.5000
ALUM (AMMONIUM)				C	0.2510	0.1080	0.0810
10002	1.6400	129	ROOM	S	5.3758	-1.6172	12.3457
ALUM (POTASSIUM)				C	0.2540	0.1070	0.0840
10003	1.7500	129	ROOM	S	5.2474	-1.5553	11.9048
ALUM (POTASSIUM)				C	0.2560	0.1070	0.0850
10004	1.7600	129	ROOM	S	5.1835	-1.5279	11.7647
ALUM (100.0K- 0.0CR)				C	0.2560	0.1070	0.0860
10005	1.7600	129	ROOM	S	5.1835	-1.5279	11.6279
ALUM (86.5K- 13.5CR)				C	0.2520	0.1050	0.0810
10006	1.7720	129	ROOM	S	5.2567	-1.5461	12.3457
ALUM (60.0K- 40.0CR)				C	0.2470	0.1010	0.0780
10007	1.7960	129	ROOM	S	5.3086	-1.5407	12.8205
ALUM (54.5K- 45.5CR)				C	0.2440	0.1000	0.0780
10008	1.8020	129	ROOM	S	5.3804	-1.5641	12.8205
ALUM (0.0K-100.0CR)				C	0.2370	0.0930	0.0770
10009	1.8450	129	ROOM	S	5.4177	-1.5268	12.9870
ALUM (THALLIUM)				C	0.2900	0.1300	0.0870
10010	2.3200	129	ROOM	S	4.7727	-1.4773	11.4943
ALUM (THALLIUM)				C	0.2885	0.1335	0.0870
10011	2.3200	245	ROOM	S	4.9000	-1.5500	11.5000
ALUM (CH3NH3-AL-S)				C	0.2971	0.1732	0.0584
10012	1.5890	4	ROOM	S	5.8987	-2.1723	17.1233
ALUM (CS-AL-S)				C	0.3115	0.1539	0.0839
10013	1.9990	4	ROOM	S	4.7684	-1.5768	11.9190
ALUM (K-AL-S)				C	0.2460	0.1040	0.0855
10014	1.7500	4	ROOM	S	5.4291	-1.6132	11.6959
ALUM (K-AL-S)				C	0.2465	0.1025	0.0865
10015	1.7530	4	ROOM	S	5.3679	-1.5765	11.5607
ALUM (NA-AL-S)				C	0.3514	0.2202	0.0770
10016	1.7700	4	ROOM	S	5.5023	-2.1197	12.9870
ALUM (NH4-AL-S)				C	0.2520	0.1090	0.0811
10017	1.6420	4	ROOM	S	5.3712	-1.6218	12.3305
ALUM (RB-AL-S)				C	0.2535	0.1033	0.0844
10018	1.8840	4	ROOM	S	5.1630	-1.4948	11.8483
ALUM (TL-AL-S)				C	0.2540	0.1130	0.0814
10019	2.3220	4	ROOM	S	5.4226	-1.6696	12.2850
ALUM (CH3NH3-AL-SE)				C	0.2736	0.1008	0.0543
10020	1.8270	4	ROOM	S	4.5595	-1.2276	18.4162

CUBIC

CODE	NAME / DENSITY	REF	TEMP(K)		11	12	44
ALUM (CS-AL-SE)				C	0.2608	0.1178	0.0742
10021	2.2240	4	ROOM	S	5.3335	-1.6595	13.4771
ALUM (K-AL-SE)				C	0.2330	0.0970	0.0775
10022	1.9860	4	ROOM	S	5.6826	-1.6703	12.9032
ALUM (NH4-AL-SE)				C	0.2384	0.1040	0.0752
10023	1.8880	4	ROOM	S	5.7070	-1.7334	13.2979
ALUM (RB-AL-SE)				C	0.2426	0.0996	0.0778
10024	2.1130	4	ROOM	S	5.4165	-1.5765	12.8535
ALUM (CS-FE-S)				C	0.3038	0.1484	0.0841
10025	2.0650	4	ROOM	S	4.8450	-1.5900	11.8906
ALUM (CH3NH3-GA-S)				C	0.2898	0.1686	0.0562
10026	1.7170	4	ROOM	S	6.0322	-2.2186	17.7936
ALUM (CS-GA-S)				C	0.3069	0.1533	0.0816
10027	2.1270	4	ROOM	S	4.8836	-1.6268	12.2549
ALUM (K-GA-S)				C	0.2356	0.0994	0.0849
10028	1.8980	4	ROOM	S	5.6621	-1.6800	11.7786
ALUM (NH4-GA-S)				C	0.2395	0.1029	0.0805
10029	1.7840	4	ROOM	S	5.6290	-1.6917	12.4224
ALUM (RB-GA-S)				C	0.2450	0.0996	0.0853
10030	2.0250	4	ROOM	S	5.3355	-1.5421	11.7233
ALUM (CS-GA-SE)				C	0.2530	0.1130	0.0756
10031	2.3420	4	ROOM	S	5.4578	-1.6851	13.2275
ALUM (CS-IN-S)				C	0.2957	0.1407	0.0816
10032	2.2120	4	ROOM	S	4.8787	-1.5729	12.2549
ALUM (RB-IN-S)				C	0.2366	0.0954	0.0826
10033	2.1070	4	ROOM	S	5.5013	-1.5808	12.1065
ALUMINUM				C	1.0824	0.6216	0.2841
10034	2.7020	128	ROOM	S	1.5900	-0.5800	3.5200
ALUMINUM				C	1.1200	0.6600	0.2790
10035	2.7020	129	293.0	S	1.5859	-0.5880	3.5842
ALUMINUM				C	1.0730	0.6090	0.2830
10036	2.6990	4	ROOM	S	1.5823	-0.5729	3.5336
ALUMINUM (P= 0KB)				C	1.0567	0.5897	0.2854
10037	2.6990	149	298.0	S	1.5766	-0.5647	3.5039
ALUMINUM (P= 2KB)				C	1.0700	0.6159	0.2892
10038	2.7063	149	298.0	S	1.6129	-0.5892	3.4578
ALUMINUM (P= 4KB)				C	1.0859	0.6062	0.2930
10039	2.7134	149	298.0	S	1.5348	-0.5498	3.4130
ALUMINUM (P= 6KB)				C	1.1060	0.5890	0.2973
10040	2.7206	149	298.0	S	1.4354	-0.4988	3.3636

CUBIC

CODE	NAME DENSITY	REF	TEMP(K)		11	12	44
ALUMINUM	(P= 8KB)			C	1.1211	0.5916	0.3032
10041	2.7276	149	298.0	S	1.4037	-0.4849	3.2982
ALUMINUM	(P=10KB)			C	1.1536	0.5851	0.3082
10042	2.7346	149	298.0	S	1.3161	-0.4429	3.2446
ALUMINUM				C	1.2296	0.7080	0.3090
10043	2.7333	249	0.0	S	1.4040	-0.5130	3.2360
ALUMINUM				C	1.2260	0.7063	0.3073
10044	2.7311	249	63.2	S	1.4090	-0.5150	3.2540
ALUMINUM				C	1.2255	0.7065	0.3067
10045	2.7302	249	70.2	S	1.4110	-0.5160	3.2600
ALUMINUM				C	1.2224	0.7046	0.3063
10046	2.7300	249	77.2	S	1.4140	-0.5170	3.2650
ALUMINUM				C	1.2053	0.6995	0.2999
10047	2.7249	249	133.2	S	1.4460	-0.5310	3.3350
ALUMINUM				C	1.2037	0.7002	0.2986
10048	2.7233	249	143.2	S	1.4520	-0.5340	3.3490
ALUMINUM				C	1.1955	0.6942	0.2974
10049	2.7216	249	153.2	S	1.4590	-0.5360	3.3620
ALUMINUM				C	1.1878	0.6915	0.2949
10050	2.7187	249	173.2	S	1.4730	-0.5420	3.3910
ALUMINUM				C	1.1778	0.6866	0.2922
10051	2.7157	249	193.2	S	1.4880	-0.5480	3.4220
ALUMINUM				C	1.1680	0.6818	0.2895
10052	2.7125	249	213.2	S	1.5030	-0.5540	3.4540
ALUMINUM				C	1.1601	0.6794	0.2868
10053	2.7092	249	233.2	S	1.5190	-0.5610	3.4870
ALUMINUM				C	1.1502	0.6749	0.2840
10054	2.7055	249	253.2	S	1.5360	-0.5680	3.5210
ALUMINUM				C	1.1382	0.6685	0.2811
10055	2.7022	249	273.2	S	1.5540	-0.5750	3.5570
ALUMINUM				C	1.1304	0.6663	0.2783
10056	2.6976	249	293.2	S	1.5720	-0.5830	3.5930
ALUMINUM				C	1.1205	0.6622	0.2755
10057	2.6938	249	313.2	S	1.5910	-0.5910	3.6300
ALUMINUM				C	1.1088	0.6563	0.2726
10058	2.6890	249	333.2	S	1.6110	-0.5990	3.6690
ALUMINUM				C	1.0952	0.6486	0.2698
10059	2.6857	249	353.2	S	1.6320	-0.6070	3.7060
ALUMINUM				C	1.0800	0.6393	0.2670
10060	2.6823	249	373.2	S	1.6540	-0.6150	3.7450

CUBIC

CODE	NAME DENSITY	REF	TEMP(K)		11	12	44
ALUMINUM				C	1.0538	0.6252	0.2614
10061	2.6744	249	413.2	S	1.7000	-0.6330	3.8260
ALUMINUM				C	1.0282	0.6120	0.2557
10062	2.6666	249	453.2	S	1.7500	-0.6530	3.9110
ALUMINUM				C	0.9969	0.5932	0.2501
10063	2.6577	249	493.2	S	1.8040	-0.6730	3.9990
ALUMINUM				C	0.9715	0.5804	0.2443
10064	2.6504	249	533.2	S	1.8610	-0.6960	4.0930
ALUMINUM				C	0.9408	0.5626	0.2385
10065	2.6403	249	573.2	S	1.9240	-0.7200	4.1920
ALUMINUM				C	0.9116	0.5462	0.2326
10066	2.6317	249	613.2	S	1.9910	-0.7460	4.2990
ALUMINUM				C	0.8858	0.5337	0.2265
10067	2.6223	249	653.2	S	2.0640	-0.7760	4.4150
ALUMINUM				C	0.8574	0.5183	0.2201
10068	2.6131	249	693.2	S	2.1420	-0.8070	4.5440
ALUMINUM				C	0.8299	0.5042	0.2134
10069	2.6029	249	733.2	S	2.2280	-0.8420	4.6870
ALUMINUM				C	0.8036	0.4911	0.2063
10070	2.5937	249	773.2	S	2.3200	-0.8800	4.8470
ALUMINUM				C	1.1430	0.6192	0.3162
10071	2.7334	140	0.0	S	1.4127	-0.4964	3.1626
ALUMINUM				C	1.1428	0.6190	0.3160
10072	2.7334	140	20.0	S	1.4128	-0.4964	3.1646
ALUMINUM				C	1.1420	0.6197	0.3155
10073	2.7331	140	40.0	S	1.4164	-0.4982	3.1696
ALUMINUM				C	1.1404	0.6194	0.3141
10074	2.7325	140	60.0	S	1.4197	-0.4997	3.1837
ALUMINUM				C	1.1373	0.6191	0.3128
10075	2.7314	140	80.0	S	1.4268	-0.5029	3.1969
ALUMINUM				C	1.1333	0.6185	0.3104
10076	2.7296	140	100.0	S	1.4356	-0.5069	3.2216
ALUMINUM				C	1.1285	0.6179	0.3078
10077	2.7274	140	120.0	S	1.4466	-0.5118	3.2489
ALUMINUM				C	1.1231	0.6169	0.3050
10078	2.7249	140	140.0	S	1.4584	-0.5171	3.2787
ALUMINUM				C	1.1171	0.6159	0.3022
10079	2.7220	140	160.0	S	1.4721	-0.5232	3.3091
ALUMINUM				C	1.1108	0.6142	0.2993
10080	2.7190	140	180.0	S	1.4850	-0.5287	3.3411

CUBIC

CODE	NAME DENSITY	REF	TEMP(K)		11	12	44
ALUMINUM				C	1.1040	0.6130	0.2964
10081	2.7158	140	200.0	S	1.5008	-0.5358	3.3738
ALUMINUM				C	1.0973	0.6119	0.2935
10082	2.7125	140	220.0	S	1.5170	-0.5431	3.4072
ALUMINUM				C	1.0903	0.6109	0.2907
10083	2.7090	140	240.0	S	1.5348	-0.5511	3.4400
ALUMINUM				C	1.0828	0.6096	0.2878
10084	2.7055	140	260.0	S	1.5536	-0.5596	3.4746
ALUMINUM				C	1.0753	0.6085	0.2849
10085	2.7019	140	280.0	S	1.5736	-0.5687	3.5100
ALUMINUM				C	1.0678	0.6074	0.2821
10086	2.6982	140	300.0	S	1.5941	-0.5780	3.5448
ALUMINUM				C	1.1630	0.6480	0.3090
10087	2.7390	261	4.0	S	1.4301	-0.5117	3.2362
ALUMINUM				C	1.1420	0.6310	0.3060
10088	2.7350	261	77.0	S	1.4433	-0.5137	3.2680
ALUMINUM				C	1.0730	0.6080	0.2830
10089	2.7020	261	300.0	S	1.5793	-0.5712	3.5336
ALUMINUM				C	1.0649	0.6039	0.2828
10090	2.6970	109	293.0	S	1.5928	-0.5764	3.5361
ALUMINUM				C	1.0643	0.6035	0.2821
10091	2.6950	109	300.0	S	1.5935	-0.5766	3.5448
ALUMINUM				C	1.0567	0.6012	0.2795
10092	2.6900	109	325.0	S	1.6111	-0.5842	3.5778
ALUMINUM				C	1.0489	0.5992	0.2762
10093	2.6860	109	350.0	S	1.6308	-0.5929	3.6206
ALUMINUM				C	1.0394	0.5971	0.2726
10094	2.6820	109	375.0	S	1.6565	-0.6044	3.6684
ALUMINUM				C	1.0297	0.5950	0.2687
10095	2.6770	109	400.0	S	1.6838	-0.6166	3.7216
ALUMINUM				C	1.0195	0.5930	0.2649
10096	2.6710	109	425.0	S	1.7142	-0.6304	3.7750
ALUMINUM				C	1.0086	0.5905	0.2612
10097	2.6670	109	450.0	S	1.7467	-0.6450	3.8285
ALUMINUM				C	0.9983	0.5883	0.2574
10098	2.6620	109	475.0	S	1.7793	-0.6597	3.8850
ALUMINUM				C	0.9880	0.5862	0.2532
10099	2.6560	109	500.0	S	1.8135	-0.6753	3.9494
ALUMINUM				C	0.9772	0.5839	0.2493
10100	2.6510	109	525.0	S	1.8505	-0.6921	4.0112

CODE	NAME DENSITY	REF	TEMP(K)		11	12	44
ALUMINUM				C	0.9663	0.5808	0.2453
10101	2.6460	109	550.0	S	1.8860	-0.7080	4.0766
ALUMINUM				C	0.9560	0.5782	0.2413
10102	2.6400	109	575.0	S	1.9224	-0.7245	4.1442
ALUMINUM				C	0.9457	0.5758	0.2372
10103	2.6350	109	600.0	S	1.9612	-0.7422	4.2159
ALUMINUM				C	0.9348	0.5724	0.2332
10104	2.6290	109	625.0	S	1.9999	-0.7595	4.2882
ALUMINUM				C	0.9239	0.5700	0.2291
10105	2.6220	109	650.0	S	2.0453	-0.7804	4.3649
ALUMINUM				C	0.9134	0.5676	0.2248
10106	2.6170	109	675.0	S	2.0906	-0.8012	4.4484
ALUMINUM				C	0.9023	0.5642	0.2205
10107	2.6120	109	700.0	S	2.1360	-0.8218	4.5351
ALUMINUM				C	0.8916	0.5615	0.2162
10108	2.6050	109	725.0	S	2.1850	-0.8443	4.6253
ALUMINUM				C	0.8806	0.5586	0.2120
10109	2.5990	109	750.0	S	2.2372	-0.8683	4.7170
ALUMINUM				C	0.8694	0.5560	0.2076
10110	2.5940	109	775.0	S	2.2954	-0.8954	4.8170
ALUMINUM				C	0.8580	0.5527	0.2032
10111	2.5870	109	800.0	S	2.3534	-0.9220	4.9213
ALUMINUM				C	0.8458	0.5494	0.1989
10112	2.5820	109	825.0	S	2.4206	-0.9532	5.0277
ALUMINUM				C	0.8334	0.5460	0.1945
10113	2.5750	109	850.0	S	2.4928	-0.9867	5.1414
ALUMINUM				C	0.8211	0.5427	0.1900
10114	2.5680	109	875.0	S	2.5695	-1.0225	5.2632
ALUMINUM				C	0.8081	0.5392	0.1855
10115	2.5610	109	900.0	S	2.6559	-1.0629	5.3908
ALUMINUM				C	0.7950	0.5359	0.1810
10116	2.5540	109	925.0	S	2.7516	-1.1079	5.5249
ALUMINUM				C	1.0675	0.6041	0.2834
10117	2.6970	254	298.0	S	1.5851	-0.5728	3.5286
ALUMINUM ANTIMONIDE				C	0.8939	0.4427	0.4155
10118	4.3600	42	300.0	S	1.6649	-0.5514	2.4067
AMMONIUM BROMIDE				C	0.2960	0.0590	0.0530
10119	2.4290	129	ROOM	S	3.6181	-0.6013	18.8679
AMMONIUM BROMIDE				C	0.3380	0.0910	0.0685
10120	2.4360	4	ROOM	S	3.3401	-0.7085	14.5985

CUBIC

CODE	NAME DENSITY	REF	TEMP(K)		11	12	44
AMMONIUM BROMIDE				C	0.3264	0.0623	0.0798
10121	2.4497	102	237.0	S	3.2634	-0.5231	12.5360
AMMONIUM BROMIDE				C	0.3286	0.0640	0.0797
10122	2.4489	102	238.0	S	3.2496	-0.5297	12.5471
AMMONIUM BROMIDE				C	0.3319	0.0667	0.0795
10123	2.4482	102	240.0	S	3.2302	-0.5405	12.5818
AMMONIUM BROMIDE				C	0.3369	0.0711	0.0790
10124	2.4475	102	245.0	S	3.2035	-0.5582	12.6630
AMMONIUM BROMIDE				C	0.3394	0.0734	0.0784
10125	2.4460	102	250.0	S	3.1918	-0.5676	12.7519
AMMONIUM BROMIDE				C	0.3421	0.0763	0.0773
10126	2.4431	102	260.0	S	3.1823	-0.5802	12.9433
AMMONIUM BROMIDE				C	0.3429	0.0777	0.0760
10127	2.4402	102	270.0	S	3.1821	-0.5875	13.1492
AMMONIUM BROMIDE				C	0.3429	0.0783	0.0748
10128	2.4373	102	280.0	S	3.1865	-0.5923	13.3726
AMMONIUM BROMIDE				C	0.3424	0.0784	0.0735
10129	2.4343	102	290.0	S	3.1936	-0.5952	13.6073
AMMONIUM BROMIDE				C	0.3414	0.0782	0.0722
10130	2.4314	102	300.0	S	3.2024	-0.5970	13.8543
AMMONIUM BROMIDE				C	0.3403	0.0778	0.0708
10131	2.4285	102	310.0	S	3.2124	-0.5980	14.1183
AMMONIUM BROMIDE				C	0.3389	0.0772	0.0694
10132	2.4256	102	320.0	S	3.2237	-0.5981	14.4009
AMMONIUM BROMIDE (P= 0KB)				C	0.3411	0.0751	0.0778
10133	2.4445	102	255.0	S	3.1847	-0.5747	12.8535
AMMONIUM BROMIDE (P= 2KB)				C	0.3654	0.0938	0.0856
10134	2.4725	102	255.0	S	3.0574	-0.6245	11.6822
AMMONIUM BROMIDE (P= 4KB)				C	0.3843	0.1085	0.0931
10135	2.4982	102	255.0	S	2.9716	-0.6543	10.7411
AMMONIUM BROMIDE (P= 6KB)				C	0.4010	0.1216	0.1003
10136	2.5222	102	255.0	S	2.9035	-0.6756	9.9701
AMMONIUM BROMIDE (P= 8KB)				C	0.4162	0.1335	0.1073
10137	2.5450	102	255.0	S	2.8461	-0.6912	9.3197
AMMONIUM BROMIDE (P=10KB)				C	0.4302	0.1445	0.1140
10138	2.5667	102	255.0	S	2.7969	-0.7032	8.7719
AMMONIUM BROMIDE (P= 0KB)				C	0.3419	0.0783	0.0792
10139	2.4328	102	295.0	S	3.1978	-0.5959	12.6263
AMMONIUM BROMIDE (P= 2KB)				C	0.3615	0.0929	0.0805
10140	2.4607	102	295.0	S	3.0911	-0.6320	12.4224

CUBIC

CODE	NAME / DENSITY	REF	TEMP(K)		11	12	44
AMMONIUM BROMIDE (P= 4KB)				C	0.3789	0.1063	0.0879
10141	2.4866	102	295.0	S	3.0091	-0.6593	11.3766
AMMONIUM BROMIDE (P= 6KB)				C	0.3949	0.1186	0.0951
10142	2.5109	102	295.0	S	2.9402	-0.6791	10.5152
AMMONIUM BROMIDE (P= 8KB)				C	0.4097	0.1301	0.1020
10143	2.5340	102	295.0	S	2.8819	-0.6946	9.8039
AMMONIUM BROMIDE (P=10KB)				C	0.4236	0.1407	0.1091
10144	2.5561	102	295.0	S	2.8294	-0.7055	9.1659
AMMONIUM CHLORIDE				C	0.3900	0.0720	0.0680
10145	1.5270	129	ROOM	S	2.7207	-0.4240	14.7059
AMMONIUM CHLORIDE				C	0.3790	0.0970	0.0830
10146	1.5260	4	ROOM	S	2.9458	-0.6003	12.0482
AMMONIUM CHLORIDE				C	0.3511	0.2055	0.0697
10147	1.4350	124	293.0	S	5.0161	-1.8520	14.3472
AMMONIUM CHLORIDE				C	0.3700	0.0880	0.0860
10148	1.5310	98	300.0	S	2.9746	-0.5715	11.6279
AMMONIUM CHLORIDE				C	0.4640	0.1540	0.1232
10149	1.5743	100	155.0	S	2.5824	-0.6436	8.1195
AMMONIUM CHLORIDE				C	0.4609	0.1516	0.1212
10150	1.5710	100	170.0	S	2.5921	-0.6417	8.2529
AMMONIUM CHLORIDE				C	0.4547	0.1465	0.1183
10151	1.5666	100	190.0	S	2.6087	-0.6357	8.4509
AMMONIUM CHLORIDE				C	0.4453	0.1383	0.1150
10152	1.5621	100	210.0	S	2.6334	-0.6242	8.6964
AMMONIUM CHLORIDE				C	0.4223	0.1172	0.1102
10153	1.5564	100	230.0	S	2.6920	-0.5846	9.0777
AMMONIUM CHLORIDE				C	0.4055	0.1011	0.1080
10154	1.5532	100	236.0	S	2.7388	-0.5468	9.2558
AMMONIUM CHLORIDE				C	0.4008	0.0966	0.1076
10155	1.5524	100	237.0	S	2.7528	-0.5347	9.2954
AMMONIUM CHLORIDE				C	0.3955	0.0915	0.1071
10156	1.5517	100	238.0	S	2.7694	-0.5205	9.3353
AMMONIUM CHLORIDE				C	0.3887	0.0850	0.1066
10157	1.5510	100	239.0	S	2.7918	-0.5010	9.3844
AMMONIUM CHLORIDE				C	0.3813	0.0779	0.1059
10158	1.5501	100	240.0	S	2.8178	-0.4779	9.4411
AMMONIUM CHLORIDE				C	0.3734	0.0704	0.1051
10159	1.5491	100	241.0	S	2.8483	-0.4516	9.5102
AMMONIUM CHLORIDE				C	0.3507	0.0520	0.0983
10160	1.5391	100	242.5	S	2.9648	-0.3826	10.1729

CUBIC

CODE	NAME DENSITY	REF	TEMP(K)		11	12	44
AMMONIUM CHLORIDE				C	0.3538	0.0551	0.0982
10161	1.5389	100	243.0	S	2.9503	-0.3976	10.1854
AMMONIUM CHLORIDE				C	0.3582	0.0596	0.0980
10162	1.5386	100	244.0	S	2.9309	-0.4181	10.2082
AMMONIUM CHLORIDE				C	0.3609	0.0624	0.0978
10163	1.5384	100	245.0	S	2.9197	-0.4304	10.2302
AMMONIUM CHLORIDE				C	0.3689	0.0709	0.0967
10164	1.5371	100	250.0	S	2.8898	-0.4658	10.3423
AMMONIUM CHLORIDE				C	0.3789	0.0825	0.9328
10165	1.5326	100	270.0	S	2.8622	-0.5119	1.0720
AMMONIUM CHLORIDE				C	0.3814	0.0866	0.0903
10166	1.5279	100	290.0	S	2.8627	-0.5299	11.0742
AMMONIUM CHLORIDE				C	0.3812	0.0880	0.0873
10167	1.5228	100	310.0	S	2.8724	-0.5389	11.4574
AMMONIUM CHLORIDE (P= 2KB)				C	0.4206	0.1164	0.1122
10168	1.5606	100	250.7	S	2.7017	-0.5856	8.9134
AMMONIUM CHLORIDE (P= 4KB)				C	0.4547	0.1485	0.1206
10169	1.5769	100	250.7	S	2.6207	-0.6452	8.2884
AMMONIUM CHLORIDE (P= 6KB)				C	0.4740	0.1665	0.1278
10170	1.5892	100	250.7	S	2.5811	-0.6710	7.8235
AMMONIUM CHLORIDE (P= 8KB)				C	0.4896	0.1808	0.1346
10171	1.6019	100	250.7	S	2.5505	-0.6878	7.4311
AMMONIUM CHLORIDE (P=10KB)				C	0.5035	0.1937	0.1412
10172	1.6133	100	250.7	S	2.5261	-0.7018	7.0816
AMMONIUM CHLORIDE (P=12KB)				C	0.5166	0.2058	0.1474
10173	1.6248	100	250.7	S	2.5041	-0.7134	6.7833
AMMONIUM CHLORIDE (P= 2KB)				C	0.4009	0.1003	0.1033
10174	1.5495	100	265.0	S	2.7720	-0.5547	9.6759
AMMONIUM CHLORIDE (P= 4KB)				C	0.4318	0.1269	0.1162
10175	1.5695	100	265.0	S	2.6727	-0.6071	8.6051
AMMONIUM CHLORIDE (P= 6KB)				C	0.4623	0.1556	0.1245
10176	1.5840	100	265.0	S	2.6046	-0.6559	8.0302
AMMONIUM CHLORIDE (P= 8KB)				C	0.4818	0.1738	0.1317
10177	1.5967	100	265.0	S	2.5664	-0.6804	7.5919
AMMONIUM CHLORIDE (P=10KB)				C	0.4972	0.1880	0.1385
10178	1.6081	100	265.0	S	2.5378	-0.6963	7.2192
AMMONIUM CHLORIDE (P=12KB)				C	0.5111	0.2008	0.1451
10179	1.6210	100	265.0	S	2.5137	-0.7090	6.8913
AMMONIUM CHLORIDE (P= 2KB)				C	0.4054	0.1058	0.1004
10180	1.5460	100	280.1	S	2.7654	-0.5723	9.9592

CUBIC

CODE	NAME / DENSITY	REF	TEMP(K)		11	12	44
AMMONIUM CHLORIDE (P= 4KB)				C	0.4227	0.1201	0.1087
10181	1.5604	100	280.1	S	2.7060	-0.5987	9.2022
AMMONIUM CHLORIDE (P= 6KB)				C	0.4394	0.1338	0.1196
10182	1.5776	100	280.1	S	2.6530	-0.6193	8.3605
AMMONIUM CHLORIDE (P= 8KB)				C	0.4692	0.1618	0.1280
10183	1.5917	100	280.1	S	2.5892	-0.6639	7.8125
AMMONIUM CHLORIDE (P=10KB)				C	0.4888	0.1800	0.1352
10184	1.6026	100	280.1	S	2.5516	-0.6867	7.3959
AMMONIUM CHLORIDE (P=12KB)				C	0.5045	0.1946	0.1424
10185	1.6154	100	280.1	S	2.5242	-0.7026	7.0235
AMMONIUM CHLORIDE (P= 2KB)				C	0.4043	0.1059	0.0977
10186	1.5423	100	295.0	S	2.7752	-0.5760	10.2323
AMMONIUM CHLORIDE (P= 4KB)				C	0.4242	0.1228	0.1056
10187	1.5567	100	295.0	S	2.7096	-0.6083	9.4670
AMMONIUM CHLORIDE (P= 6KB)				C	0.4402	0.1364	0.1135
10188	1.5706	100	295.0	S	2.6619	-0.6297	8.8082
AMMONIUM CHLORIDE (P= 8KB)				C	0.4453	0.1390	0.1226
10189	1.5861	100	295.0	S	2.6374	-0.6274	8.1553
AMMONIUM CHLORIDE (P=10KB)				C	0.4765	0.1684	0.1316
10190	1.5978	100	295.0	S	2.5737	-0.6720	7.6017
AMMONIUM CHLORIDE (P=12KB)				C	0.4955	0.1862	0.1396
10191	1.6102	100	295.0	S	2.5395	-0.6936	7.1613
AMMONIUM CHLORIDE (P= 2KB)				C	0.4032	0.1058	0.0957
10192	1.5389	100	308.0	S	2.7838	-0.5786	10.4515
AMMONIUM CHLORIDE (P= 4KB)				C	0.4221	0.1218	0.1035
10193	1.5532	100	308.0	S	2.7207	-0.6093	9.6590
AMMONIUM CHLORIDE (P= 6KB)				C	0.4389	0.1361	0.1112
10194	1.5671	100	308.0	S	2.6704	-0.6321	8.9896
AMMONIUM CHLORIDE (P= 8KB)				C	0.4544	0.1494	0.1187
10195	1.5797	100	308.0	S	2.6283	-0.6503	8.4253
AMMONIUM CHLORIDE (P=10KB)				C	0.4635	0.1566	0.1262
10196	1.5919	100	308.0	S	2.6014	-0.6570	7.9227
AMMONIUM CHLORIDE (P=12KB)				C	0.4836	0.1748	0.1354
10197	1.6066	100	308.0	S	2.5590	-0.6794	7.3877
AMMONIUM CHLOROSTANNATE				C	0.2170	0.0950	0.1060
10198	2.4000	197	292.0	S	6.2835	-1.9132	9.4340
ARGON				C	0.0529	0.0135	0.0159
10199	1.7710	120	4.2	S	21.0924	-4.2884	62.8931
ARGON				C	0.0335	0.0101	0.0093
10200	1.6451	120	76.8	S	34.6973	-8.0377	107.5269

CODE	NAME / DENSITY	REF	TEMP(K)		11	12	44
BARIUM FLUORIDE				C	0.9010	0.4030	0.2490
10201	4.8300	4	ROOM	S	1.5340	-0.4741	4.0161
BARIUM FLUORIDE				C	0.9122	0.4148	0.2551
10202	4.8840	124	293.0	S	1.5317	-0.4788	3.9200
BARIUM FLUORIDE				C	0.8915	0.4002	0.2535
10203	4.8860	103	ROOM	S	1.5540	-0.4815	3.9448
BARIUM FLUORIDE				C	0.9396	0.4282	0.2644
10204	4.9120	278	195.0	S	1.4892	-0.4662	3.7821
BARIUM FLUORIDE				C	0.9199	0.4157	0.2568
10205	4.8860	278	298.0	S	1.5126	-0.4708	3.8941
BARIUM FLUORIDE				C	0.8948	0.3854	0.2495
10206	4.8860	108	295.0	S	1.5089	-0.4542	4.0080
BARIUM NITRATE				C	0.6020	0.1860	0.1210
10207	3.2440	129	ROOM	S	1.9448	-0.4591	8.2645
BARIUM NITRATE				C	0.6040	0.1860	0.1220
10208	3.2440	129	ROOM	S	1.9364	-0.4559	8.1967
BARIUM NITRATE				C	0.5930	0.1890	0.1210
10209	3.2440	129	ROOM	S	1.9935	-0.4818	8.2645
BARIUM NITRATE				C	0.2925	0.2065	0.1277
10210	3.2560	124	293.0	S	8.2244	-3.4035	7.8309
BARIUM TITANATE				C	1.7278	0.8196	1.0823
10211	6.0350	31	423.0	S	0.8330	-0.2680	0.9240
CADMIUM TELLURIDE				C	0.5351	0.3681	0.1994
10212	5.8544	167	298.0	S	4.2542	-1.7338	5.0150
CADMIUM TELLURIDE (CONSTANT E)				C	0.6207	0.4359	0.1957
10213	5.8600	32	77.0	S	3.8300	-1.5800	5.1100
CALCIUM FLUORIDE				C	1.6280	0.4330	0.3340
10214	3.1800	4	ROOM	S	0.6915	-0.1453	2.9940
CALCIUM FLUORIDE				C	1.6760	0.4720	0.3690
10215	3.1800	4	ROOM	S	0.6809	-0.1496	2.7100
CALCIUM FLUORIDE				C	1.6400	0.4500	0.3380
10216	3.1800	4	ROOM	S	0.6915	-0.1489	2.9586
CALCIUM FLUORIDE				C	1.6600	0.4870	0.3580
10217	3.1800	4	ROOM	S	0.6949	-0.1576	2.7933
CALCIUM FLUORIDE				C	1.6440	0.5020	0.3470
10218	3.1800	4	ROOM	S	0.7097	-0.1660	2.8818
CALCIUM FLUORIDE				C	1.7400	0.5600	0.3593
10219	3.2107	137	4.2	S	0.6815	-0.1659	2.7832
CALCIUM FLUORIDE				C	1.7400	0.5400	0.3593
10220	3.2107	137	20.0	S	0.6738	-0.1596	2.7832

CUBIC

CODE	NAME DENSITY	REF	TEMP(K)		11	12	44
CALCIUM FLUORIDE				C	1.7300	0.5500	0.3586
10221	3.2106	137	40.0	S	0.6828	-0.1647	2.7886
CALCIUM FLUORIDE				C	1.7300	0.5300	0.3576
10222	3.2102	137	60.0	S	0.6750	-0.1583	2.7964
CALCIUM FLUORIDE				C	1.7200	0.5300	0.3562
10223	3.2094	137	80.0	S	0.6801	-0.1602	2.8074
CALCIUM FLUORIDE				C	1.7100	0.5400	0.3548
10224	3.2081	137	100.0	S	0.6893	-0.1654	2.8185
CALCIUM FLUORIDE				C	1.7000	0.5300	0.3533
10225	3.2063	137	120.0	S	0.6906	-0.1641	2.8305
CALCIUM FLUORIDE				C	1.6900	0.5500	0.3518
10226	3.2042	137	140.0	S	0.7043	-0.1729	2.8425
CALCIUM FLUORIDE				C	1.6800	0.5400	0.3506
10227	3.2017	137	160.0	S	0.7056	-0.1716	2.8523
CALCIUM FLUORIDE				C	1.6700	0.5500	0.3487
10228	3.1991	137	180.0	S	0.7156	-0.1773	2.8678
CALCIUM FLUORIDE				C	1.6600	0.5700	0.3465
10229	3.1962	137	200.0	S	0.7307	-0.1868	2.8860
CALCIUM FLUORIDE				C	1.6500	0.5600	0.3449
10230	3.1931	137	220.0	S	0.7320	-0.1855	2.8994
CALCIUM FLUORIDE				C	1.6500	0.5600	0.3429
10231	3.1898	137	240.0	S	0.7320	-0.1855	2.9163
CALCIUM FLUORIDE				C	1.6400	0.5700	0.3409
10232	3.1864	137	260.0	S	0.7430	-0.1916	2.9334
CALCIUM FLUORIDE				C	1.6400	0.5400	0.3390
10233	3.1830	137	280.0	S	0.7286	-0.1805	2.9499
CALCIUM FLUORIDE				C	1.6400	0.5300	0.3370
10234	3.1795	137	300.0	S	0.7241	-0.1768	2.9674
CALCIUM FLUORIDE				C	1.6357	0.4401	0.3392
10235	3.1804	124	293.0	S	0.6901	-0.1463	2.9481
CALCIUM FLUORIDE (P=0.001KB)				C	1.7375	0.4733	0.3609
10236	3.2085	134	77.4	S	0.6515	-0.1395	2.7709
CALCIUM FLUORIDE (P=0.158KB)				C	1.7389	0.4738	0.3610
10237	3.2091	134	77.4	S	0.6510	-0.1394	2.7701
CALCIUM FLUORIDE (P=0.272KB)				C	1.7392	0.4740	0.3610
10238	3.2095	134	77.4	S	0.6510	-0.1394	2.7701
CALCIUM FLUORIDE (P=1.072KB)				C	1.7443	0.4777	0.3621
10239	3.2124	134	77.4	S	0.6498	-0.1397	2.7617
CALCIUM FLUORIDE (P=1.538KB)				C	1.7473	0.4797	0.3627
10240	3.2141	134	77.4	S	0.6491	-0.1398	2.7571

CODE	NAME DENSITY	REF	TEMP(K)		11	12	44
CALCIUM FLUORIDE (P=1.848KB)				C	1.7493	0.4811	0.3630
10241	3.2152	134	77.4	S	0.6486	-0.1399	2.7548
CALCIUM FLUORIDE (P=2.381KB)				C	1.7525	0.4835	0.3636
10242	3.2171	134	77.4	S	0.6479	-0.1401	2.7503
CALCIUM FLUORIDE (P=2.936KB)				C	1.7564	0.4854	0.3642
10243	3.2191	134	77.4	S	0.6467	-0.1400	2.7457
CALCIUM FLUORIDE (P=3.409KB)				C	1.7592	0.4877	0.3648
10244	3.2209	134	77.4	S	0.6462	-0.1403	2.7412
CALCIUM FLUORIDE (P=0.001KB)				C	1.6839	0.4555	0.3500
10245	3.1963	134	194.5	S	0.6712	-0.1429	2.8571
CALCIUM FLUORIDE (P=0.241KB)				C	1.6854	0.4567	0.3504
10246	3.1972	134	194.5	S	0.6708	-0.1430	2.8539
CALCIUM FLUORIDE (P=0.574KB)				C	1.6875	0.4579	0.3508
10247	3.1984	134	194.5	S	0.6702	-0.1430	2.8506
CALCIUM FLUORIDE (P=0.924KB)				C	1.6897	0.4592	0.3512
10248	3.1998	134	194.5	S	0.6696	-0.1431	2.8474
CALCIUM FLUORIDE (P=1.273KB)				C	1.6919	0.4606	0.3516
10249	3.2011	134	194.5	S	0.6690	-0.1432	2.8441
CALCIUM FLUORIDE (P=1.615KB)				C	1.6940	0.4619	0.3520
10250	3.2023	134	194.5	S	0.6684	-0.1432	2.8409
CALCIUM FLUORIDE (P=2.129KB)				C	1.6972	0.4638	0.3527
10251	3.2042	134	194.5	S	0.6675	-0.1433	2.8353
CALCIUM FLUORIDE (P=2.677KB)				C	1.7004	0.4662	0.3533
10252	3.2063	134	194.5	S	0.6668	-0.1435	2.8305
CALCIUM FLUORIDE (P=3.189KB)				C	1.7034	0.4683	0.3539
10253	3.2082	134	194.5	S	0.6660	-0.1436	2.8257
CALCIUM FLUORIDE (P=3.732KB)				C	1.7067	0.4704	0.3546
10254	3.2102	134	194.5	S	0.6651	-0.1437	2.8201
CALCIUM FLUORIDE (P=4.164KB)				C	1.7092	0.4720	0.3551
10255	3.2118	134	194.5	S	0.6645	-0.1438	2.8161
CALCIUM FLUORIDE (P=0.001KB)				C	1.6580	0.4496	0.3409
10256	3.1835	134	273.1	S	0.6820	-0.1455	2.9334
CALCIUM FLUORIDE (P=0.265KB)				C	1.6616	0.4484	0.3410
10257	3.1845	134	273.1	S	0.6798	-0.1445	2.9326
CALCIUM FLUORIDE (P=0.578KB)				C	1.6618	0.4521	0.3415
10258	3.1857	134	273.1	S	0.6810	-0.1456	2.9283
CALCIUM FLUORIDE (P=0.911KB)				C	1.6629	0.4549	0.3419
10259	3.1870	134	273.1	S	0.6814	-0.1464	2.9248
CALCIUM FLUORIDE (P=1.281KB)				C	1.6654	0.4572	0.3423
10260	3.1884	134	273.1	S	0.6810	-0.1467	2.9214

CUBIC

CODE	NAME DENSITY	REF	TEMP(K)		11	12	44
CALCIUM FLUORIDE (P=1.741KB)				C	1.6690	0.4594	0.3429
10261	3.1901	134	273.1	S	0.6800	-0.1468	2.9163
CALCIUM FLUORIDE (P=2.026KB)				C	1.6708	0.4610	0.3433
10262	3.1912	134	273.1	S	0.6796	-0.1470	2.9129
CALCIUM FLUORIDE (P=2.520KB)				C	1.6745	0.4633	0.3439
10263	3.1931	134	273.1	S	0.6786	-0.1471	2.9078
CALCIUM FLUORIDE (P=3.030KB)				C	1.6782	0.4659	0.3446
10264	3.1950	134	273.1	S	0.6776	-0.1472	2.9019
CALCIUM FLUORIDE (P=3.505KB)				C	1.6815	0.4685	0.3452
10265	3.1968	134	273.1	S	0.6769	-0.1475	2.8969
CALCIUM FLUORIDE (P=4.028KB)				C	1.6853	0.4714	0.3458
10266	3.1988	134	273.1	S	0.6760	-0.1478	2.8918
CALCIUM FLUORIDE (P=0.001KB)				C	1.6494	0.4462	0.3380
10267	3.1795	134	295.5	S	0.6852	-0.1459	2.9586
CALCIUM FLUORIDE (P=0.264KB)				C	1.6547	0.4450	0.3387
10268	3.1805	134	295.5	S	0.6821	-0.1446	2.9525
CALCIUM FLUORIDE (P=0.582KB)				C	1.6553	0.4482	0.3391
10269	3.1817	134	295.5	S	0.6829	-0.1455	2.9490
CALCIUM FLUORIDE (P=0.950KB)				C	1.6566	0.4513	0.3396
10270	3.1832	134	295.5	S	0.6834	-0.1463	2.9446
CALCIUM FLUORIDE (P=1.269KB)				C	1.6588	0.4534	0.3400
10271	3.1844	134	295.5	S	0.6830	-0.1466	2.9412
CALCIUM FLUORIDE (P=1.669KB)				C	1.6610	0.4557	0.3405
10272	3.1859	134	295.5	S	0.6827	-0.1470	2.9369
CALCIUM FLUORIDE (P=2.126KB)				C	1.6646	0.4583	0.3411
10273	3.1876	134	295.5	S	0.6818	-0.1472	2.9317
CALCIUM FLUORIDE (P=2.629KB)				C	1.6682	0.4610	0.3417
10274	3.1895	134	295.5	S	0.6809	-0.1474	2.9265
CALCIUM FLUORIDE (P=3.139KB)				C	1.6723	0.4639	0.3424
10275	3.1915	134	295.5	S	0.6799	-0.1476	2.9206
CALCIUM FLUORIDE				C	1.6420	0.4398	0.8406
10276	3.1810	279	298.0	S	0.6867	-0.1451	1.1896
CALCIUM FLUORIDE				C	1.7078	0.5159	0.3413
10277	3.1810	193	293.0	S	0.6810	-0.1580	2.9300
CALCIUM FLUORIDE				C	1.7022	0.5188	0.3378
10278	3.1740	193	323.0	S	0.6850	-0.1600	2.9600
CALCIUM FLUORIDE				C	1.6864	0.5155	0.3322
10279	3.1666	193	373.0	S	0.6920	-0.1620	3.0100
CALCIUM FLUORIDE				C	1.6638	0.5050	0.3268
10280	3.1562	193	423.0	S	0.7000	-0.1630	3.0600

NAME CODE DENSITY		REF	TEMP(K)		11	12	44
CALCIUM FLUORIDE				C	1.6519	0.5038	0.3205
10281	3.1457	193	473.0	S	0.7060	-0.1650	3.1200
CALCIUM FLUORIDE				C	1.6333	0.4957	0.3145
10282	3.1352	193	523.0	S	0.7130	-0.1660	3.1800
CALCIUM FLUORIDE				C	1.6055	0.4793	0.3086
10283	3.1251	193	573.0	S	0.7220	-0.1660	3.2400
CALCIUM FLUORIDE				C	1.5825	0.4714	0.3030
10284	3.1129	193	623.0	S	0.7320	-0.1680	3.3000
CALCIUM FLUORIDE				C	1.5578	0.4649	0.2967
10285	3.1001	193	673.0	S	0.7440	-0.1710	3.3700
CALCIUM FLUORIDE				C	1.5662	0.4827	0.2941
10286	3.0955	193	693.0	S	0.7470	-0.1760	3.4000
CALCIUM FLUORIDE				C	1.6750	0.4516	0.3490
10287	3.1980	278	195.0	S	0.6742	-0.1432	2.8653
CALCIUM FLUORIDE				C	1.6420	0.4398	0.3370
10288	3.1810	278	298.0	S	0.6867	-0.1451	2.9674
CESIUM				C	0.0259	0.0217	0.0160
10289	2.0000	146	4.2	S	163.5402	-74.5551	62.5000
CESIUM				C	0.0248	0.0206	0.0151
10290	1.9800	146	63.0	S	163.7808	-74.3146	66.2252
CESIUM				C	0.0247	0.0206	0.0148
10291	1.9800	146	78.0	S	167.6714	-76.2312	67.5676
CESIUM BROMIDE				C	0.3100	0.0840	0.0750
10292	4.4500	125	293.0	S	3.6472	-0.7776	13.3333
CESIUM BROMIDE				C	0.3000	0.0780	0.0756
10293	4.4330	41	300.0	S	3.7340	-0.7705	13.2275
CESIUM BROMIDE				C	0.3350	0.1025	0.1002
10294	4.5992	173	4.2	S	3.4847	-0.8164	9.9800
CESIUM BROMIDE				C	0.3347	0.1022	0.1001
10295	4.5992	173	10.0	S	3.4857	-0.8154	9.9900
CESIUM BROMIDE				C	0.3334	0.1025	0.0997
10296	4.5992	173	20.0	S	3.5064	-0.8245	10.0301
CESIUM BROMIDE				C	0.3323	0.1024	0.0992
10297	4.5962	173	30.0	S	3.5204	-0.8293	10.0806
CESIUM BROMIDE				C	0.3316	0.1015	0.0984
10298	4.5922	173	40.0	S	3.5208	-0.8251	10.1626
CESIUM BROMIDE				C	0.3311	0.1014	0.0976
10299	4.5872	173	50.0	S	3.5267	-0.8268	10.2459
CESIUM BROMIDE				C	0.3137	0.0866	0.0809
10300	4.4820	173	220.0	S	3.6202	-0.7832	12.3609

CUBIC

CODE	NAME DENSITY	REF	TEMP(K)		11	12	44
CESIUM BROMIDE				C	0.3111	0.0852	0.0792
10301	4.4696	173	240.0	S	3.6434	-0.7833	12.6263
CESIUM BROMIDE				C	0.3091	0.0845	0.0775
10302	4.4571	173	260.0	S	3.6654	-0.7869	12.9032
CESIUM BROMIDE				C	0.3071	0.0818	0.0759
10303	4.4446	173	280.0	S	3.6672	-0.7713	13.1752
CESIUM BROMIDE				C	0.3056	0.0776	0.0743
10304	4.4322	173	300.0	S	3.6474	-0.7386	13.4590
CESIUM BROMIDE				C	0.3349	0.1001	0.0962
10305	4.5692	216	80.0	S	3.4621	-0.7968	10.3928
CESIUM BROMIDE				C	0.3327	0.0993	0.0942
10306	4.5568	216	100.0	S	3.4838	-0.8007	10.6146
CESIUM BROMIDE				C	0.3304	0.0984	0.0922
10307	4.5443	216	120.0	S	3.5057	-0.8046	10.8460
CESIUM BROMIDE				C	0.3281	0.0975	0.0902
10308	4.5318	216	140.0	S	3.5287	-0.8086	11.0877
CESIUM BROMIDE				C	0.3257	0.0966	0.0882
10309	4.5194	216	160.0	S	3.5522	-0.8127	11.3404
CESIUM BROMIDE				C	0.3234	0.0958	0.0862
10310	4.5069	216	180.0	S	3.5762	-0.8171	11.6050
CESIUM BROMIDE				C	0.3210	0.0949	0.0842
10311	4.4945	216	200.0	S	3.6003	-0.8214	11.8821
CESIUM BROMIDE				C	0.3187	0.0940	0.0822
10312	4.4820	216	220.0	S	3.6248	-0.8256	12.1655
CESIUM BROMIDE				C	0.3164	0.0932	0.0801
10313	4.4696	216	240.0	S	3.6500	-0.8303	12.4782
CESIUM BROMIDE				C	0.3140	0.0923	0.0781
10314	4.4571	216	260.0	S	3.6757	-0.8349	12.7992
CESIUM BROMIDE				C	0.3118	0.0915	0.0761
10315	4.4446	216	280.0	S	3.6998	-0.8394	13.1406
CESIUM BROMIDE				C	0.3097	0.0903	0.0750
10316	4.4353	216	295.0	S	3.7184	-0.8394	13.3333
CESIUM BROMIDE				C	0.3085	0.0957	0.0758
10317	4.4400	195	293.0	S	3.8000	-0.9000	13.2000
CESIUM BROMIDE				C	0.3085	0.0957	0.0730
10318	4.4170	195	323.0	S	3.8000	-0.9000	13.7000
CESIUM BROMIDE				C	0.2976	0.0893	0.0685
10319	4.3870	195	373.0	S	3.9000	-0.9000	14.6000
CESIUM BROMIDE				C	0.2876	0.0835	0.0637
10320	4.3560	195	423.0	S	4.0000	-0.9000	15.7000

CUBIC

NAME CODE	DENSITY	REF	TEMP(K)		11	12	44
CESIUM BROMIDE				C	0.3000	0.1000	0.0588
10321	4.2960	195	473.0	S	4.0000	-1.0000	17.0000
CESIUM BROMIDE				C	0.2894	0.0934	0.0541
10322	4.2730	195	523.0	S	4.1000	-1.0000	18.5000
CESIUM BROMIDE				C	0.2696	0.0735	0.0493
10323	4.2410	195	573.0	S	4.2000	-0.9000	20.3000
CESIUM BROMIDE				C	0.2540	0.0653	0.0444
10324	4.2240	195	623.0	S	4.4000	-0.9000	22.5000
CESIUM BROMIDE				C	0.2469	0.0617	0.0398
10325	4.1800	195	673.0	S	4.5000	-0.9000	25.1000
CESIUM BROMIDE				C	0.2340	0.0554	0.0350
10326	4.1230	195	723.0	S	4.7000	-0.9000	28.6000
CESIUM BROMIDE				C	0.3437	0.1035	0.0999
10327	4.6550	258	4.2	S	3.3808	-0.7824	10.0100
CESIUM BROMIDE				C	0.3345	0.0966	0.0951
10328	4.6290	258	77.0	S	3.4340	-0.7695	10.5152
CESIUM BROMIDE				C	0.3072	0.0800	0.0725
10329	4.4720	258	300.0	S	3.6477	-0.7537	13.7931
CESIUM BROMIDE				C	0.3063	0.0807	0.0750
10330	4.4560	234	298.0	S	3.6678	-0.7648	13.3333
CESIUM BROMIDE				C	0.3343	0.1019	0.0972
10331	4.6124	213	77.0	S	3.4881	-0.8148	10.2881
CESIUM BROMIDE				C	0.3091	0.0841	0.0747
10332	4.4560	213	300.0	S	3.6613	-0.7831	13.3869
CESIUM BROMIDE				C	0.3077	0.0827	0.0760
10333	4.4560	55	286.0	S	3.6675	-0.7769	13.1579
CESIUM CHLORIDE				C	0.3640	0.0920	0.0800
10334	3.9900	125	293.0	S	3.0593	-0.6172	12.5000
CESIUM CHLORIDE				C	0.3664	0.0882	0.0804
10335	3.9880	234	298.0	S	3.0105	-0.5841	12.4378
CESIUM CHLORIDE				C	0.3683	0.0893	0.0817
10336	3.9880	55	286.0	S	2.9990	-0.5852	12.2399
CESIUM IODIDE				C	0.2450	0.0710	0.0620
10337	4.5200	125	293.0	S	4.6927	-1.0544	16.1290
CESIUM IODIDE				C	0.2460	0.0670	0.0624
10338	4.5090	41	300.0	S	4.6016	-0.9850	16.0256
CESIUM IODIDE				C	0.2660	0.0751	0.0808
10339	4.6525	216	80.0	S	4.2940	-0.9454	12.3716
CESIUM IODIDE				C	0.2640	0.0739	0.0792
10340	4.6393	216	100.0	S	4.3166	-0.9438	12.6279

CODE	NAME DENSITY	REF	TEMP(K)		11	12	44
CESIUM IODIDE				C	0.2619	0.0728	0.0775
10341	4.6261	216	120.0	S	4.3436	-0.9451	12.8999
CESIUM IODIDE				C	0.2598	0.0818	0.0759
10342	4.6129	216	140.0	S	4.5323	-1.0851	13.1822
CESIUM IODIDE				C	0.2576	0.0707	0.0742
10343	4.5997	216	160.0	S	4.4014	-0.9474	13.4753
CESIUM IODIDE				C	0.2556	0.0696	0.0726
10344	4.5864	216	180.0	S	4.4301	-0.9486	13.7817
CESIUM IODIDE				C	0.2534	0.0685	0.0709
10345	4.5732	216	200.0	S	4.4604	-0.9497	14.0984
CESIUM IODIDE				C	0.2513	0.0675	0.0693
10346	4.5600	216	220.0	S	4.4902	-0.9505	14.4342
CESIUM IODIDE				C	0.2492	0.0665	0.0676
10347	4.5468	216	240.0	S	4.5206	-0.9523	14.7863
CESIUM IODIDE				C	0.2470	0.0654	0.0660
10348	4.5336	216	260.0	S	4.5522	-0.9526	15.1538
CESIUM IODIDE				C	0.2450	0.0644	0.0643
10349	4.5204	216	280.0	S	4.5834	-0.9537	15.5424
CESIUM IODIDE				C	0.2434	0.0636	0.0632
10350	4.5155	216	295.0	S	4.6073	-0.9545	15.8328
CESIUM IODIDE				C	0.2737	0.0793	0.0825
10351	4.7120	258	4.2	S	4.2004	-0.9436	12.1212
CESIUM IODIDE				C	0.2678	0.0771	0.0789
10352	4.6930	258	77.0	S	4.2858	-0.9581	12.6743
CESIUM IODIDE				C	0.2457	0.0647	0.0629
10353	4.5370	258	300.0	S	4.5719	-0.9530	15.8982
CESIUM IODIDE				C	0.2446	0.0661	0.0629
10354	4.5250	234	298.0	S	4.6195	-0.9828	15.8982
CESIUM IODIDE				C	0.2462	0.0659	0.0644
10355	4.5250	55	286.0	S	4.5794	-0.9669	15.5279
CESIUM IODIDE				C	0.2725	0.0767	0.0873
10356	4.7120	174	4.2	S	4.1875	-0.9198	11.4548
CHROMITE				C	3.2250	1.4370	1.1670
10357	4.4500	129	ROOM	S	0.4275	-0.1318	0.8569
CHROMIUM				C	3.9100	0.8960	1.0320
10358	7.2230	40	77.0	S	0.2796	-0.0521	0.9690
CHROMIUM				C	3.5000	0.6780	1.0080
10359	7.2000	40	298.0	S	0.3049	-0.0495	0.9921
CHROMIUM				C	3.4600	0.7620	0.9870
10360	7.1630	40	500.0	S	0.3140	-0.0567	1.0132

CUBIC

CODE / NAME	DENSITY	REF	TEMP(K)		11	12	44
CHROMIUM				C	3.3980	0.5860	0.9900
10361	7.2000	247	298.0	S	0.3100	-0.0456	1.0101
COBALT OXIDE				C	1.9474	1.6352	0.8153
10362	6.4400	12	253.0	S	2.1993	-1.0038	1.2265
COBALT OXIDE				C	1.9502	1.6620	0.8112
10363	6.4400	12	263.0	S	2.3764	-1.0934	1.2327
COBALT OXIDE				C	1.9673	1.6729	0.8070
10364	6.4400	12	273.0	S	2.3272	-1.0695	1.2392
COBALT OXIDE				C	1.9937	1.6607	0.8057
10365	6.4400	12	277.0	S	2.0647	-0.9383	1.2412
COBALT OXIDE				C	2.0016	1.6504	0.8052
10366	6.4400	12	278.0	S	1.9611	-0.8863	1.2419
COBALT OXIDE				C	2.0486	1.6326	0.8039
10367	6.4400	12	281.0	S	1.6653	-0.7386	1.2439
COBALT OXIDE				C	2.0968	1.6094	0.8070
10368	6.4400	12	283.0	S	1.4305	-0.6212	1.2392
COBALT OXIDE				C	2.1679	1.5791	0.8157
10369	6.4400	12	285.0	S	1.1948	-0.5035	1.2259
COBALT OXIDE				C	2.3068	1.5484	0.8235
10370	6.4400	12	287.0	S	0.9407	-0.3778	1.2143
COBALT OXIDE				C	2.4876	1.4516	0.8272
10371	6.4400	12	289.0	S	0.7053	-0.2599	1.2089
COBALT OXIDE				C	2.5293	1.5181	0.8277
10372	6.4400	12	290.0	S	0.7192	-0.2697	1.2082
COBALT OXIDE				C	2.5609	1.5009	0.8282
10373	6.4400	12	291.0	S	0.6889	-0.2545	1.2074
COBALT OXIDE				C	2.5911	1.4871	0.8286
10374	6.4400	12	293.0	S	0.6638	-0.2420	1.2069
COBALT OXIDE				C	2.6123	1.4699	0.8300
10375	6.4400	12	298.0	S	0.6436	-0.2317	1.2048
COBALT OXIDE				C	2.6170	1.4500	0.8320
10376	6.4400	12	303.0	S	0.6317	-0.2252	1.2019
COBALT ZINC FERRITE				C	2.6600	1.5300	0.7800
10377	5.4300	129	303.0	S	0.6482	-0.2367	1.2821
COLUMBIUM				C	2.4653	1.3335	0.2837
10378	8.5780	119	298.0	S	0.6540	-0.2296	3.5251
COLUMBIUM				C	2.4645	1.3323	0.2843
10379	8.5780	119	298.0	S	0.6538	-0.2294	3.5173
COPPER				C	1.7017	1.1492	0.6098
10380	8.9200	128	ROOM	S	1.2900	-0.5200	1.6400

CUBIC

CODE	NAME DENSITY	REF	TEMP(K)		11	12	44
COPPER				C	1.7637	1.2920	0.7519
10381	8.9200	128	ROOM	S	1.4900	-0.6300	1.3300
COPPER				C	1.7000	1.2400	0.6450
10382	8.9300	129	ROOM	S	1.5290	-0.6449	1.5504
COPPER				C	1.6840	1.2140	0.7540
10383	8.9600	129	ROOM	S	1.4995	-0.6282	1.3263
COPPER				C	1.6980	1.2260	0.7530
10384	8.9200	4	ROOM	S	1.4928	-0.6259	1.3280
COPPER (P= 0KB)				C	1.7098	1.2399	0.7545
10385	8.9410	149	298.0	S	1.4983	-0.6298	1.3254
COPPER (P= 2KB)				C	1.7202	1.2484	0.7567
10386	8.9542	149	298.0	S	1.4921	-0.6275	1.3215
COPPER (P= 4KB)				C	1.7305	1.2564	0.7588
10387	8.9673	149	298.0	S	1.4847	-0.6245	1.3179
COPPER (P= 6KB)				C	1.7410	1.2644	0.7594
10388	8.9803	149	298.0	S	1.4769	-0.6213	1.3168
COPPER (P= 8KB)				C	1.7451	1.2654	0.7610
10389	8.9933	149	298.0	S	1.4677	-0.6169	1.3141
COPPER (P=10KB)				C	1.7555	1.2728	0.7627
10390	9.0062	149	298.0	S	1.4586	-0.6131	1.3111
COPPER				C	1.7620	1.2494	0.8177
10391	9.0305	199	0.0	S	1.3788	-0.5720	1.2229
COPPER				C	1.7620	1.2494	0.8176
10392	9.0305	199	10.0	S	1.3788	-0.5720	1.2231
COPPER				C	1.7618	1.2492	0.8173
10393	9.0305	199	20.0	S	1.3788	-0.5720	1.2235
COPPER				C	1.7613	1.2488	0.8168
10394	9.0278	199	30.0	S	1.3791	-0.5721	1.2243
COPPER				C	1.7605	1.2482	0.8160
10395	9.0278	199	40.0	S	1.3796	-0.5724	1.2255
COPPER				C	1.7593	1.2475	0.8149
10396	9.0251	199	50.0	S	1.3809	-0.5729	1.2271
COPPER				C	1.7578	1.2467	0.8135
10397	9.0224	199	60.0	S	1.3828	-0.5738	1.2293
COPPER				C	1.7560	1.2457	0.8118
10398	9.0197	199	70.0	S	1.3849	-0.5747	1.2318
COPPER				C	1.7540	1.2447	0.8100
10399	9.0171	199	80.0	S	1.3875	-0.5759	1.2346
COPPER				C	1.7517	1.2437	0.8080
10400	9.0144	199	90.0	S	1.3910	-0.5775	1.2376

CODE	NAME DENSITY	REF	TEMP(K)		11	12	44
COPPER				C	1.7493	1.2425	0.8059
10401	9.0117	199	100.0	S	1.3942	-0.5790	1.2408
COPPER				C	1.7467	1.2414	0.8036
10402	9.0090	199	110.0	S	1.3982	-0.5809	1.2444
COPPER				C	1.7441	1.2402	0.8013
10403	9.0063	199	120.0	S	1.4019	-0.5826	1.2480
COPPER				C	1.7412	1.2389	0.7990
10404	9.0010	199	130.0	S	1.4062	-0.5846	1.2516
COPPER				C	1.7383	1.2377	0.7966
10405	8.9983	199	140.0	S	1.4108	-0.5868	1.2553
COPPER				C	1.7353	1.2364	0.7942
10406	8.9956	199	150.0	S	1.4155	-0.5889	1.2591
COPPER				C	1.7322	1.2350	0.7917
10407	8.9902	199	160.0	S	1.4202	-0.5911	1.2631
COPPER				C	1.7290	1.2337	0.7892
10408	8.9875	199	170.0	S	1.4254	-0.5936	1.2671
COPPER				C	1.7258	1.2323	0.7867
10409	8.9849	199	180.0	S	1.4304	-0.5959	1.2711
COPPER				C	1.7225	1.2309	0.7841
10410	8.9795	199	190.0	S	1.4358	-0.5984	1.2753
COPPER				C	1.7192	1.2295	0.7816
10411	8.9768	199	200.0	S	1.4412	-0.6009	1.2794
COPPER				C	1.7158	1.2280	0.7790
10412	8.9714	199	210.0	S	1.4466	-0.6034	1.2837
COPPER				C	1.7123	1.2266	0.7763
10413	8.9661	199	220.0	S	1.4526	-0.6063	1.2882
COPPER				C	1.7088	1.2251	0.7736
10414	8.9634	199	230.0	S	1.4584	-0.6090	1.2927
COPPER				C	1.7053	1.2236	0.7709
10415	8.9580	199	240.0	S	1.4643	-0.6117	1.2972
COPPER				C	1.7018	1.2220	0.7681
10416	8.9553	199	250.0	S	1.4699	-0.6143	1.3019
COPPER				C	1.6982	1.2205	0.7654
10417	8.9500	199	260.0	S	1.4761	-0.6173	1.3065
COPPER				C	1.6947	1.2189	0.7626
10418	8.9446	199	270.0	S	1.4818	-0.6199	1.3113
COPPER				C	1.6911	1.2174	0.7597
10419	8.9419	199	280.0	S	1.4881	-0.6229	1.3163
COPPER				C	1.6875	1.2158	0.7568
10420	8.9366	199	290.0	S	1.4943	-0.6257	1.3214

CUBIC

NAME CODE	DENSITY	REF	TEMP(K)		11	12	44
COPPER				C	1.6839	1.2142	0.7539
10421	8.9312	199	300.0	S	1.5004	-0.6286	1.3264
COPPER				C	1.6809	1.2145	0.7511
10422	8.9883	226	300.0	S	1.5105	-0.6336	1.3314
COPPER				C	1.6830	1.2110	0.7570
10423	8.9320	85	298.0	S	1.4936	-0.6250	1.3210
COPPER				C	1.6880	1.2170	0.7538
10424	8.9320	251	300.0	S	1.4963	-0.6268	1.3266
COPPER				C	1.7000	1.2250	0.7580
10425	8.9340	51	300.0	S	1.4838	-0.6214	1.3193
COPPER				C	1.6750	1.2150	0.7450
10426	8.9130	51	350.0	S	1.5305	-0.6434	1.3423
COPPER				C	1.6550	1.2050	0.7310
10427	8.8920	51	400.0	S	1.5635	-0.6587	1.3680
COPPER				C	1.6350	1.1950	0.7180
10428	8.8680	51	450.0	S	1.5980	-0.6748	1.3928
COPPER				C	1.6150	1.1850	0.7040
10429	8.8440	51	500.0	S	1.6340	-0.6915	1.4205
COPPER				C	1.5950	1.1750	0.6910
10430	8.8200	51	550.0	S	1.6718	-0.7092	1.4472
COPPER				C	1.5750	1.1650	0.6770
10431	8.8000	51	600.0	S	1.7114	-0.7276	1.4771
COPPER				C	1.5550	1.1550	0.6630
10432	8.7760	51	650.0	S	1.7529	-0.7471	1.5083
COPPER				C	1.5350	1.1450	0.6500
10433	8.7500	51	700.0	S	1.7965	-0.7676	1.5385
COPPER				C	1.5150	1.1350	0.6360
10434	8.7240	51	750.0	S	1.8425	-0.7891	1.5723
COPPER				C	1.4950	1.1250	0.6230
10435	8.7010	51	800.0	S	1.8908	-0.8119	1.6051
COPPER				C	1.6610	1.1990	0.7560
10436	8.9300	133	ROOM	S	1.5251	-0.6394	1.3228
COPPER(5.0)-ALUMINUM(95.0)				C	3.0822	2.6256	0.2703
10437	2.9240	128	ROOM	S	1.5000	-0.6900	3.7000
COPPER(95.19)-ALUMINUM(4.81)				C	1.6580	1.2160	0.7490
10438	8.6479	192	ROOM	S	1.5898	-0.6726	1.3351
COPPER(90.02)-ALUMINUM(9.98)				C	1.5950	1.1770	0.7660
10439	8.2916	192	ROOM	S	1.6793	-0.7130	1.3055
COPPER(98.42)-GALLIUM(1.58)				C	1.6500	1.1920	0.7430
10440	8.9730	192	ROOM	S	1.5382	-0.6452	1.3459

NAME CODE	DENSITY	REF	TEMP(K)		11	12	44
COPPER(95.85)-GALLIUM(4.15)				C	1.6520	1.2100	0.7410
10441	8.9403	192	ROOM	S	1.5902	-0.6723	1.3495
COPPER(98.97)-GERMANIUM(1.03)				C	1.6660	1.2100	0.7500
10442	8.9781	192	ROOM	S	1.5436	-0.6494	1.3333
COPPER(98.29)-GERMANIUM(1.71)				C	1.6310	1.1820	0.7500
10443	8.9700	192	ROOM	S	1.5682	-0.6590	1.3333
COPPER(75.08)-GOLD(24.92)				C	1.9069	1.3830	0.6631
10444	12.2132	232	293.0	S	1.3440	-0.5650	1.5080
COPPER(75.08)-GOLD(24.92)				C	1.8972	1.3878	0.6472
10445	12.1656	232	373.0	S	1.3800	-0.5830	1.5450
COPPER(75.08)-GOLD(24.92)				C	1.8691	1.3803	0.6277
10446	12.1069	232	473.0	S	1.4360	-0.6100	1.5930
COPPER(75.08)-GOLD(24.92)				C	1.8476	1.3698	0.6177
10447	12.0740	232	523.0	S	1.4680	-0.6250	1.6190
COPPER(75.08)-GOLD(24.92)				C	1.8241	1.3607	0.6035
10448	12.0410	232	573.0	S	1.5120	-0.6460	1.6570
COPPER(75.08)-GOLD(24.92)				C	1.8165	1.3645	0.5959
10449	12.0239	232	598.0	S	1.5480	-0.6640	1.6780
COPPER(75.08)-GOLD(24.92)				C	1.7947	1.3549	0.5886
10450	12.0019	232	623.0	S	1.5900	-0.6840	1.6990
COPPER(75.08)-GOLD(24.92)				C	1.7705	1.3370	0.5851
10451	11.9909	232	633.0	S	1.6130	-0.6940	1.7090
COPPER(75.08)-GOLD(24.92)				C	1.7593	1.3327	0.5817
10452	11.9763	232	643.0	S	1.6380	-0.7060	1.7190
COPPER(75.08)-GOLD(24.92)				C	1.7289	1.3095	0.5780
10453	11.9555	232	653.0	S	1.6660	-0.7180	1.7300
COPPER(75.08)-GOLD(24.92)				C	1.7058	1.2930	0.5754
10454	11.9372	232	658.0	S	1.6930	-0.7300	1.7380
COPPER(75.08)-GOLD(24.92)				C	1.6724	1.2646	0.5714
10455	11.9335	232	660.5	S	1.7140	-0.7380	1.7500
COPPER(75.08)-GOLD(24.92)				C	1.6042	1.2471	0.5596
10456	11.9335	232	660.5	S	1.9480	-0.8520	1.7870
COPPER(75.08)-GOLD(24.92)				C	1.5919	1.2366	0.5587
10457	11.9299	232	663.0	S	1.9580	-0.8560	1.7900
COPPER(75.08)-GOLD(24.92)				C	1.6014	1.2485	0.5565
10458	11.9262	232	668.0	S	1.9700	-0.8630	1.7970
COPPER(75.08)-GOLD(24.92)				C	1.5945	1.2435	0.5549
10459	11.9225	232	673.0	S	1.9810	-0.8680	1.8020
COPPER(75.08)-GOLD(24.92)				C	1.5921	1.2448	0.5516
10460	11.9152	232	683.0	S	2.0010	-0.8780	1.8130

CUBIC

NAME CODE	DENSITY	REF	TEMP(K)		11	12	44
COPPER(75.08)-GOLD(24.92)				C	1.5897	1.2460	0.5488
10461	11.9079	232	693.0	S	2.0210	-0.8880	1.8220
COPPER(75.08)-GOLD(24.92)				C	1.5620	1.2270	0.5405
10462	11.8871	232	723.0	S	2.0730	-0.9120	1.8500
COPPER(75.0)-GOLD(25.0) (DISORDERED)				C	1.8700	1.3240	0.7450
10463	12.3310	96	4.2	S	1.2948	-0.5367	1.3423
COPPER(75.0)-GOLD(25.0) (DISORDERED)				C	1.8490	1.3530	0.6870
10464	12.2140	96	300.0	S	1.4173	-0.5989	1.4556
COPPER(75.0)-GOLD(25.0) (ORDERED)				C	1.8930	1.3190	0.7360
10465	12.3310	96	4.2	S	1.2350	-0.5072	1.3587
COPPER(75.0)-GOLD(25.0) (ORDERED)				C	1.8710	1.3490	0.6780
10466	12.2140	96	300.0	S	1.3501	-0.5656	1.4749
COPPER(97.66)-NICKEL(2.34)				C	1.6933	1.2189	0.7631
10467	8.9886	226	300.0	S	1.4860	-0.6220	1.3104
COPPER(96.98)-NICKEL(3.02)				C	1.6937	1.2187	0.7674
10468	8.9892	226	300.0	S	1.4842	-0.6211	1.3031
COPPER(95.51)-NICKEL(4.49)				C	1.7012	1.2194	0.7731
10469	9.0000	226	300.0	S	1.4642	-0.6113	1.2935
COPPER(93.96)-NICKEL(6.04)				C	1.7105	1.2237	0.7812
10470	8.9894	226	300.0	S	1.4497	-0.6046	1.2801
COPPER(90.27)-NICKEL(9.73)				C	1.7239	1.2261	0.7909
10471	8.9949	226	300.0	S	1.4190	-0.5898	1.2644
COPPER(95.83)-SILICON(4.17)				C	1.6780	1.2420	0.7480
10472	8.7600	192	ROOM	S	1.6091	-0.6844	1.3369
COPPER(94.84)-SILICON(5.16)				C	1.6080	1.1880	0.7490
10473	8.7050	192	ROOM	S	1.6710	-0.7100	1.3351
COPPER(92.31)-SILICON(7.69)				C	1.6580	1.2640	0.7410
10474	8.5662	192	ROOM	S	1.7717	-0.7664	1.3495
COPPER(52.8)-ZINC(47.2)				C	0.5203	0.3351	1.7301
10475	8.3500	115	297.0	S	3.8800	-1.5200	0.5780
COPPER(52.8)-ZINC(47.2)				C	0.5654	0.3666	1.7241
10476	8.2500	115	468.0	S	3.6100	-1.4200	0.5800
COPPER(52.8)-ZINC(47.2)				C	0.4496	0.2606	1.7007
10477	8.1940	115	566.0	S	3.8700	-1.4200	0.5880
COPPER(52.8)-ZINC(47.2)				C	0.4232	0.2471	1.6026
10478	8.1380	115	662.0	S	4.1500	-1.5300	0.6240
COPPER(52.8)-ZINC(47.2)				C	0.4240	0.2688	1.4599
10479	8.0800	115	721.0	S	4.6400	-1.8000	0.6850
COPPER(52.8)-ZINC(47.2)				C	0.4294	0.2853	1.3793
10480	8.0710	115	739.0	S	4.9600	-1.9800	0.7250

CODE / NAME	DENSITY	REF	TEMP(K)		11	12	44
COPPER(52.8)-ZINC(47.2)				C	0.4137	0.2694	1.3755
10481	8.0680	115	741.0	S	4.9700	-1.9600	0.7270
COPPER(52.8)-ZINC(47.2)				C	0.4032	0.2614	1.3532
10482	8.0440	115	779.0	S	5.0600	-1.9900	0.7390
COPPER(51.06WT%)-ZINC(48.94WT%) (P= 0KB)				C	1.2904	1.0956	0.8245
10483	8.3040	149	298.0	S	3.5181	-1.6154	1.2129
COPPER(51.06WT%)-ZINC(48.94WT%) (P= 2KB)				C	1.2999	1.1044	0.8328
10484	8.3190	149	298.0	S	3.5051	-1.6100	1.2008
COPPER(51.06WT%)-ZINC(48.94WT%) (P= 4KB)				C	1.3082	1.1113	0.8380
10485	8.3339	149	298.0	S	3.4802	-1.5985	1.1933
COPPER(51.06WT%)-ZINC(48.94WT%) (P= 6KB)				C	1.3191	1.1211	0.8443
10486	8.3487	149	298.0	S	3.4606	-1.5899	1.1844
COPPER(51.06WT%)-ZINC(48.94WT%) (P= 8KB)				C	1.3281	1.1294	0.8495
10487	8.3634	149	298.0	S	3.4481	-1.5846	1.1772
COPPER(51.06WT%)-ZINC(48.94WT%) (P=10KB)				C	1.3432	1.1437	0.8536
10488	8.3780	149	298.0	S	3.4335	-1.5790	1.1715
COPPER(95.41)-ZINC(4.59)				C	1.6340	1.1920	0.7420
10489	8.9346	192	ROOM	S	1.5913	-0.6712	1.3477
COPPER(100.0)				C	1.6840	1.2140	0.7550
10490	8.9900	208	300.0	S	1.4995	-0.6282	1.3245
COPPER(95.9)-ZINC(4.1)				C	1.6320	1.1760	0.7440
10491	8.9410	208	300.0	S	1.5457	-0.6473	1.3441
COPPER(90.9)-ZINC(9.1)				C	1.5710	1.1370	0.7230
10492	8.8760	208	300.0	S	1.6228	-0.6814	1.3831
COPPER(82.9)-ZINC(17.1)				C	1.4990	1.0970	0.7150
10493	8.7640	208	300.0	S	1.7486	-0.7389	1.3986
COPPER(77.3)-ZINC(22.7)				C	1.4470	1.0710	0.7130
10494	8.6860	208	300.0	S	1.8659	-0.7936	1.4025
DIAMOND				C	9.3202	4.1118	4.1667
10495	3.5110	128	ROOM	S	0.1470	-0.0450	0.2400
DIAMOND				C	9.5000	3.9000	4.3000
10496	3.5110	129	ROOM	S	0.1383	-0.0403	0.2326
DIAMOND				C	11.0000	3.3000	4.4000
10497	3.5030	129	300.0	S	0.1055	-0.0244	0.2273
DIAMOND				C	9.4900	1.5100	5.2100
10498	3.5110	4	ROOM	S	0.1102	-0.0151	0.1919
DIAMOND				C	10.7600	1.2500	5.7580
10499	3.5120	165	300.0	S	0.0952	-0.0099	0.1737
GALENA				C	1.0200	0.3800	0.2500
10500	7.5640	129	ROOM	S	1.2289	-0.3336	4.0000

CUBIC

CODE	NAME DENSITY	REF	TEMP(K)		11	12	44
GALENA				C	1.2700	0.2980	0.2480
10501	7.5640	129	ROOM	S	0.8645	-0.1643	4.0323
GALLIUM ANTIMONIDE				C	0.8850	0.4040	0.4330
10502	5.6190	27	298.0	S	1.5829	-0.4961	2.3095
GALLIUM ANTIMONIDE				C	0.8970	0.4120	0.4280
10503	5.6190	83	302.0	S	1.5683	-0.4936	2.3364
GALLIUM ANTIMONIDE				C	0.8839	0.4033	0.4316
10504	5.6137	170	298.0	S	1.5843	-0.4964	2.3170
GALLIUM ARSENIDE				C	1.1880	0.5380	0.5940
10505	5.3070	27	298.0	S	1.1729	-0.3656	1.6835
GALLIUM ARSENIDE				C	1.1904	0.5384	0.5952
10506	5.3160	78	ROOM	S	1.1695	-0.3642	1.6801
GALLIUM ARSENIDE				C	1.1877	0.5372	0.5944
10507	5.3169	169	298.0	S	1.1722	-0.3651	1.6824
GALLIUM ARSENIDE				C	1.2210	0.5660	0.5990
10508	5.3070	99	77.4	S	1.1595	-0.3672	1.6694
GALLIUM ARSENIDE				C	1.1810	0.5320	0.5940
10509	5.3070	99	300.0	S	1.1757	-0.3651	1.6835
GALLIUM PHOSPHIDE				C	1.4120	0.6253	0.7047
10510	4.1297	276	300.0	S	0.9726	-0.2985	1.4190
GARNET (21.8%FERROUS OXIDE)				C	1.9700	0.9000	0.5700
10511	3.7590	129	ROOM	S	0.7115	-0.2231	1.7544
GARNET (22.7%FERROUS OXIDE)				C	1.9200	0.9900	0.5900
10512	3.6730	129	ROOM	S	0.8023	-0.2730	1.6949
GARNET (23.6%FERROUS OXIDE)				C	2.1000	1.0300	0.6700
10513	3.6300	129	ROOM	S	0.7032	-0.2314	1.4925
GARNET (23.0%FERROUS OXIDE)				C	2.2200	1.0400	0.7000
10514	3.6700	129	ROOM	S	0.6425	-0.2050	1.4286
GARNET (26.2%FERROUS OXIDE)				C	2.2600	1.2600	0.6200
10515	3.7500	129	ROOM	S	0.7364	-0.2636	1.6129
GARNET (28.7%FERROUS OXIDE)				C	2.7300	1.5700	0.6800
10516	4.1300	129	ROOM	S	0.6315	-0.2306	1.4706
GARNET (33.5%FERROUS OXIDE)				C	3.2700	1.2400	0.8900
10517	4.3200	129	ROOM	S	0.3864	-0.1062	1.1236
GARNET (ALMANDINE-PYROPE)				C	2.6760	0.9450	0.8650
10518	3.6580	224	ROOM	S	0.4581	-0.1196	1.1561
GARNET (ALMANDINE-PYROPE)				C	2.6700	0.8960	0.8910
10519	4.0070	224	ROOM	S	0.4505	-0.1132	1.1223
GARNET (ALMANDINE-PYROPE)				C	2.9160	0.9730	0.9460
10520	4.0600	224	ROOM	S	0.4117	-0.1030	1.0571

CODE	NAME / DENSITY	REF	TEMP(K)		11	12	44
GARNET (ALMANDINE-PYROPE)				C	2.8500	0.9680	0.8820
10521	4.1520	224	ROOM	S	0.4239	-0.1075	1.1338
GARNET (ALMANDINE-PYROPE)				C	2.6470	0.9380	0.8280
10522	4.0600	224	ROOM	S	0.4638	-0.1213	1.2077
GARNET (ALMANDITE)				C	3.0480	1.1230	0.9440
10523	4.1830	262	ROOM	S	0.4093	-0.1102	1.0593
GARNET (ALMANDITE-PYROPE)				C	3.0619	1.1246	0.9266
10524	4.1602	242	298.0	S	0.4069	-0.1093	1.0792
GARNET (EUROPIUM IRON)				C	2.5100	1.0700	0.7620
10525	6.2800	25	298.0	S	0.5346	-0.1598	1.3123
GARNET (GADOLINIUM GALLIUM)				C	2.8590	1.1510	0.9030
10526	7.0900	117	298.0	S	0.4549	-0.1306	1.1074
GARNET (GROSSULAR)				C	2.7980	0.8570	0.8960
10527	3.5980	224	ROOM	S	0.4173	-0.0979	1.1161
GARNET (SPESSARTITE-ALMANDITE)				C	3.0730	1.0970	0.9520
10528	4.2470	262	ROOM	S	0.4007	-0.1054	1.0504
GARNET (YTTRIUM IRON)				C	2.6800	1.1060	0.7660
10529	5.1700	25	298.0	S	0.4917	-0.1436	1.3055
GARNET (YTTRIUM IRON)				C	2.6900	1.0770	0.7640
10530	5.1700	58	296.0	S	0.4821	-0.1378	1.3089
GERMANIUM				C	1.3160	0.5090	0.6690
10531	5.3500	129	298.0	S	0.9689	-0.2702	1.4948
GERMANIUM				C	1.2920	0.4790	0.6700
10532	5.3500	162	ROOM	S	0.9682	-0.2619	1.4925
GERMANIUM				C	1.3150	0.4948	0.6840
10533	5.3390	161	73.0	S	0.9574	-0.2618	1.4620
GERMANIUM				C	1.3115	0.4922	0.6819
10534	5.3360	161	123.0	S	0.9589	-0.2617	1.4665
GERMANIUM				C	1.3050	0.4900	0.6788
10535	5.3310	161	173.0	S	0.9639	-0.2631	1.4732
GERMANIUM				C	1.2990	0.4871	0.6756
10536	5.3280	161	223.0	S	0.9678	-0.2639	1.4802
GERMANIUM				C	1.2920	0.4845	0.6725
10537	5.3240	161	273.0	S	0.9730	-0.2654	1.4870
GERMANIUM				C	1.2890	0.4830	0.6710
10538	5.3230	161	298.0	S	0.9750	-0.2657	1.4903
GERMANIUM				C	1.2850	0.4819	0.6690
10539	5.3210	161	323.0	S	0.9783	-0.2668	1.4948
GERMANIUM				C	1.2780	0.4797	0.6655
10540	5.3150	161	373.0	S	0.9841	-0.2686	1.5026

CUBIC

CODE	NAME DENSITY	REF	TEMP(K)		11	12	44
GERMANIUM				C	1.2720	0.4776	0.6620
10541	5.3120	161	423.0	S	0.9889	-0.2699	1.5106
GERMANIUM				C	1.2650	0.4756	0.6580
10542	5.3070	161	473.0	S	0.9949	-0.2719	1.5198
GERMANIUM				C	1.2575	0.4738	0.6540
10543	5.3020	161	523.0	S	1.0018	-0.2742	1.5291
GERMANIUM				C	1.2510	0.4721	0.6509
10544	5.2970	161	573.0	S	1.0078	-0.2761	1.5363
GERMANIUM				C	1.2835	0.4823	0.6666
10545	5.3130	36	298.0	S	0.9804	-0.2678	1.5001
GERMANIUM				C	1.2950	0.4830	0.6720
10546	5.3250	251	300.0	S	0.9684	-0.2631	1.4881
GERMANIUM (P=0)				C	1.3111	0.4923	0.6817
10547	5.3380	163	77.2	S	0.9593	-0.2619	1.4669
GERMANIUM (P=30.000PSI)				C	1.3217	0.5013	0.6843
10548	5.3524	163	77.2	S	0.9560	-0.2629	1.4613
GERMANIUM (P=0)				C	1.2853	0.4826	0.6680
10549	5.3230	163	298.0	S	0.9787	-0.2672	1.4970
GERMANIUM (P=30.000PSI)				C	1.2957	0.4915	0.6709
10550	5.3377	163	298.0	S	0.9753	-0.2682	1.4905
GOLD				C	1.7893	1.4863	0.4367
10551	19.3000	128	ROOM	S	2.2700	-1.0300	2.2900
GOLD				C	1.9505	1.6563	0.4202
10552	19.3000	128	ROOM	S	2.3300	-1.0700	2.3800
GOLD				C	1.7875	1.4872	0.4274
10553	19.3000	128	ROOM	S	2.2900	-1.0400	2.3400
GOLD				C	1.9400	1.6600	0.4000
10554	19.3000	4	ROOM	S	2.4443	-1.1271	2.5000
GOLD				C	1.8700	1.5500	0.4290
10555	19.3000	4	ROOM	S	2.1504	-0.9746	2.3310
GOLD				C	2.0163	1.6967	0.4544
10556	19.4920	191	0.0	S	2.1476	-0.9814	2.2007
GOLD				C	2.0161	1.6967	0.4542
10557	19.4920	191	10.0	S	2.1489	-0.9820	2.2017
GOLD				C	2.0151	1.6961	0.4533
10558	19.4890	191	25.0	S	2.1515	-0.9833	2.2060
GOLD				C	2.0107	1.6937	0.4510
10559	19.4860	191	50.0	S	2.1648	-0.9898	2.2173
GOLD				C	2.0030	1.6886	0.4477
10560	19.4690	191	75.0	S	2.1824	-0.9983	2.2336

CUBIC

CODE	NAME DENSITY	REF	TEMP(K)		11	12	44
GOLD				C	1.9940	1.6822	0.4446
10561	19.4570	191	100.0	S	2.2003	-1.0069	2.2492
GOLD				C	1.9853	1.6759	0.4415
10562	19.4390	191	125.0	S	2.2172	-1.0149	2.2650
GOLD				C	1.9763	1.6693	0.4384
10563	19.4160	191	150.0	S	2.2343	-1.0231	2.2810
GOLD				C	1.9677	1.6633	0.4352
10564	19.3990	191	175.0	S	2.2531	-1.0321	2.2978
GOLD				C	1.9587	1.6567	0.4320
10565	19.3810	191	200.0	S	2.2707	-1.0405	2.3148
GOLD				C	1.9497	1.6503	0.4290
10566	19.3580	191	225.0	S	2.2902	-1.0498	2.3310
GOLD				C	1.9410	1.6440	0.4257
10567	19.3400	191	250.0	S	2.3084	-1.0586	2.3491
GOLD				C	1.9322	1.6378	0.4226
10568	19.3170	191	275.0	S	2.3285	-1.0682	2.3663
GOLD				C	1.9234	1.6314	0.4195
10569	19.3000	191	300.0	S	2.3474	-1.0773	2.3838
GOLD				C	1.9250	1.6300	0.4240
10570	19.3000	51	300.0	S	2.3242	-1.0657	2.3585
GOLD				C	1.9100	1.6150	0.4170
10571	19.2590	51	350.0	S	2.3247	-1.0651	2.3981
GOLD				C	1.8900	1.6050	0.4100
10572	19.2190	51	400.0	S	2.4045	-1.1042	2.4390
GOLD				C	1.8750	1.5900	0.4030
10573	19.1790	51	450.0	S	2.4051	-1.1037	2.4814
GOLD				C	1.8500	1.5800	0.3970
10574	19.1340	51	500.0	S	2.5357	-1.1680	2.5189
GOLD				C	1.8400	1.5650	0.3900
10575	19.0880	51	550.0	S	2.4913	-1.1450	2.5641
GOLD				C	1.8200	1.5550	0.3830
10576	19.0430	51	600.0	S	2.5833	-1.1903	2.6110
GOLD				C	1.8050	1.5400	0.3760
10577	18.9960	51	650.0	S	2.5840	-1.1896	2.6596
GOLD				C	1.7850	1.5300	0.3690
10578	18.9560	51	700.0	S	2.6832	-1.2384	2.7100
GOLD				C	1.7700	1.5200	0.3620
10579	18.9120	51	750.0	S	2.7360	-1.2640	2.7624
GOLD				C	1.7550	1.5050	0.3550
10580	18.8660	51	800.0	S	2.7366	-1.2634	2.8169

CODE	NAME DENSITY	REF	TEMP(K)		11	12	44
GOLD				C	1.9290	1.6380	0.4150
10581	19.3000	133	ROOM	S	2.3550	-1.0814	2.4096
GOLD(100.0)				C	1.9244	1.6298	0.4200
10582	19.2830	114	296.5	S	2.3273	-1.0672	2.3810
GOLD(97.05)-NICKEL(2.95)				C	1.9259	1.6261	0.4300
10583	19.0600	114	296.5	S	2.2881	-1.0475	2.3256
GOLD(90.28)-NICKEL(9.72)				C	1.9231	1.6095	0.4497
10584	18.5330	114	296.5	S	2.1907	-0.9981	2.2237
GOLD(75.80)-NICKEL(24.20)				C	1.9573	1.5819	0.5098
10585	17.3410	114	296.5	S	1.8410	-0.8229	1.9616
GOLD(57.58)-NICKEL(42.42)				C	1.9950	1.5600	0.6186
10586	15.7120	114	296.5	S	1.5977	-0.7011	1.6166
GOLD(75.0)-SILVER(25.0)				C	1.6621	1.3241	0.4854
10587	17.2440	128	ROOM	S	2.0500	-0.9090	2.0600
GOLD(50.0)-SILVER(50.0)				C	1.4894	1.1350	0.5076
10588	15.0300	128	ROOM	S	1.9700	-0.8520	1.9700
GOLD(25.0)-SILVER(75.0)				C	1.3826	1.0448	0.4878
10589	12.7990	128	ROOM	S	2.0700	-0.8910	2.0500
HAFNIUM-COBALT INTERMETALLIC				C	2.6818	1.2998	0.9192
10590	11.6280	229	0.0	S	0.5455	-0.1781	1.0879
HAFNIUM-COBALT INTERMETALLIC				C	2.6818	1.2998	0.9192
10591	11.6280	229	77.0	S	0.5455	-0.1781	1.0879
HAFNIUM-COBALT INTERMETALLIC				C	2.6799	1.2979	0.9191
10592	11.6280	229	100.0	S	0.5456	-0.1780	1.0880
HAFNIUM-COBALT INTERMETALLIC				C	2.6722	1.3002	0.9188
10593	11.6280	229	150.0	S	0.5491	-0.1797	1.0884
HAFNIUM-COBALT INTERMETALLIC				C	2.6348	1.2788	0.9141
10594	11.6280	229	200.0	S	0.5558	-0.1816	1.0940
HAFNIUM-COBALT INTERMETALLIC				C	2.5959	1.2565	0.9080
10595	11.6280	229	250.0	S	0.5630	-0.1836	1.1013
HAFNIUM-COBALT INTERMETALLIC				C	2.5554	1.2330	0.9030
10596	11.6280	229	300.0	S	0.5705	-0.1857	1.1074
HEXAMETHYLENETETRAMINE				C	0.1500	0.0300	0.0700
10597	1.3390	129	ROOM	S	7.1429	-1.1905	14.2857
HEXAMETHYLENETETRAMINE				C	0.1643	0.0433	0.0515
10598	1.3390	4	ROOM	S	6.8382	-1.4263	19.4175
INDIUM ANTIMONIDE				C	0.6660	0.3350	0.3140
10599	5.7907	203	0.0	S	2.2636	-0.7575	3.1847
INDIUM ANTIMONIDE				C	0.6649	0.3351	0.3137
10600	5.7907	203	100.0	S	2.2711	-0.7610	3.1878

CODE	NAME DENSITY	REF	TEMP(K)		11	12	44
INDIUM ANTIMONIDE				C	0.6607	0.3345	0.3107
10601	5.7835	203	200.0	S	2.2944	-0.7712	3.2185
INDIUM ANTIMONIDE				C	0.6472	0.3265	0.3071
10602	5.7751	203	300.0	S	2.3352	-0.7830	3.2563
INDIUM ANTIMONIDE				C	0.6312	0.3166	0.3034
10603	5.7664	203	400.0	S	2.3827	-0.7959	3.2960
INDIUM ANTIMONIDE				C	0.6131	0.3039	0.2998
10604	5.7577	203	500.0	S	2.4291	-0.8050	3.3356
INDIUM ANTIMONIDE				C	0.5906	0.2888	0.2958
10605	5.7490	203	600.0	S	2.4943	-0.8191	3.3807
INDIUM ANTIMONIDE				C	0.6918	0.3788	0.3132
10606	5.7907	235	0.0	S	2.3599	-0.8350	3.1928
INDIUM ANTIMONIDE				C	0.6899	0.3775	0.3123
10607	5.7907	235	50.0	S	2.3647	-0.8363	3.2020
INDIUM ANTIMONIDE				C	0.6872	0.3753	0.3117
10608	5.7907	235	77.6	S	2.3693	-0.8369	3.2082
INDIUM ANTIMONIDE				C	0.6847	0.3735	0.3111
10609	5.7907	235	100.0	S	2.3751	-0.8383	3.2144
INDIUM ANTIMONIDE				C	0.6794	0.3698	0.3096
10610	5.7871	235	150.0	S	2.3882	-0.8418	3.2300
INDIUM ANTIMONIDE				C	0.6744	0.3670	0.3076
10611	5.7835	235	200.0	S	2.4054	-0.8477	3.2510
INDIUM ANTIMONIDE				C	0.6702	0.3652	0.3051
10612	5.7784	235	250.0	S	2.4238	-0.8549	3.2776
INDIUM ANTIMONIDE				C	0.6669	0.3645	0.3020
10613	5.7751	235	300.0	S	2.4434	-0.8635	3.3113
INDIUM ANTIMONIDE				C	0.6720	0.3670	0.3020
10614	5.7890	27	298.0	S	2.4229	-0.8558	3.3113
INDIUM ANTIMONIDE				C	0.6700	0.3649	0.3019
10615	5.7747	77	ROOM	S	2.4232	-0.8544	3.3124
INDIUM ARSENIDE				C	0.8329	0.4526	0.3959
10616	5.6720	104	ROOM	S	1.9448	-0.6847	2.5259
INDIUM ARSENIDE				C	0.8980	0.5025	0.3924
10617	5.6720	106	4.2	S	1.8608	-0.6677	2.5484
INDIUM ARSENIDE				C	0.8465	0.5001	0.3968
10618	5.6720	106	77.4	S	2.1051	-0.7818	2.5202
INDIUM ARSENIDE				C	0.8329	0.4526	0.3959
10619	5.6720	106	293.0	S	1.9448	-0.6847	2.5259
INDIUM ARSENIDE				C	0.8998	0.5064	0.4040
10620	5.6670	215	4.2	S	1.8689	-0.6730	2.4752

CUBIC

NAME / CODE	DENSITY	REF	TEMP(K)		11	12	44
INDIUM ARSENIDE				C	0.8955	0.5025	0.4031
10621	5.6670	215	77.0	S	1.8717	-0.6728	2.4808
INDIUM ARSENIDE				C	0.8821	0.4933	0.3995
10622	5.6670	215	195.0	S	1.8931	-0.6790	2.5031
INDIUM ARSENIDE				C	0.8677	0.4857	0.3956
10623	5.6670	215	300.0	S	1.9264	-0.6914	2.5278
INDIUM PHOSPHIDE				C	1.0220	0.5760	0.4600
10624	4.7800	131	ROOM	S	1.6481	-0.5941	2.1739
IRIDIUM				C	5.9947	2.5582	2.6882
10625	22.5200	206	294.0	S	0.2240	-0.0670	0.3720
IRIDIUM				C	5.8000	2.4200	2.5600
10626	22.5200	159	300.0	S	0.2286	-0.0673	0.3906
IRON				C	2.2809	1.3348	1.1086
10627	7.8600	128	ROOM	S	0.7720	-0.2850	0.9020
IRON				C	2.0936	1.1366	1.1136
10628	7.8600	128	ROOM	S	0.7730	-0.2720	0.8980
IRON				C	2.3688	1.4063	1.1601
10629	7.8600	128	ROOM	S	0.7570	-0.2820	0.8620
IRON				C	2.4200	1.4650	1.1200
10630	7.8600	4	ROOM	S	0.7604	-0.2867	0.8929
IRON				C	2.3700	1.4100	1.1600
10631	7.8600	4	ROOM	S	0.7587	-0.2830	0.8621
IRON				C	2.4310	1.3810	1.2190
10632	7.9240	211	4.2	S	0.6991	-0.2533	0.8203
IRON				C	2.4310	1.3810	1.2190
10633	7.9238	211	20.0	S	0.6991	-0.2533	0.8203
IRON				C	2.4280	1.3800	1.2183
10634	7.9238	211	40.0	S	0.7004	-0.2538	0.8208
IRON				C	2.4240	1.3780	1.2168
10635	7.9226	211	60.0	S	0.7017	-0.2543	0.8218
IRON				C	2.4160	1.3740	1.2148
10636	7.9214	211	80.0	S	0.7043	-0.2553	0.8232
IRON				C	2.4080	1.3700	1.2125
10637	7.9190	211	100.0	S	0.7070	-0.2564	0.8247
IRON				C	2.4000	1.3660	1.2100
10638	7.9166	211	120.0	S	0.7097	-0.2574	0.8264
IRON				C	2.3920	1.3640	1.2070
10639	7.9143	211	140.0	S	0.7136	-0.2592	0.8285
IRON				C	2.3840	1.3620	1.2040
10640	7.9095	211	160.0	S	0.7176	-0.2609	0.8306

CUBIC

CODE	NAME DENSITY	REF	TEMP(K)		11	12	44
IRON				C	2.3760	1.3600	1.2005
10641	7.9047	211	180.0	S	0.7216	-0.2627	0.8330
IRON				C	2.3680	1.3580	1.1970
10642	7.8999	211	200.0	S	0.7256	-0.2645	0.8354
IRON				C	2.3610	1.3570	1.1938
10643	7.8952	211	220.0	S	0.7297	-0.2663	0.8377
IRON				C	2.3540	1.3558	1.1900
10644	7.8906	211	240.0	S	0.7337	-0.2681	0.8403
IRON				C	2.3460	1.3555	1.1862
10645	7.8858	211	260.0	S	0.7390	-0.2706	0.8430
IRON				C	2.3380	1.3547	1.1825
10646	7.8786	211	280.0	S	0.7440	-0.2730	0.8457
IRON				C	2.3310	1.3544	1.1783
10647	7.8740	211	300.0	S	0.7488	-0.2752	0.8487
IRON				C	2.3600	1.3400	1.1900
10648	7.9123	155	77.0	S	0.7197	-0.2607	0.8403
IRON				C	2.2800	1.3200	1.1650
10649	7.8672	155	300.0	S	0.7622	-0.2795	0.8584
IRON (P= 0KB)				C	2.3140	1.3470	1.1640
10650	7.8720	220	300.0	S	0.7560	-0.2781	0.8591
IRON (P= 2KB)				C	2.3290	1.3570	1.1690
10651	7.8810	220	300.0	S	0.7520	-0.2768	0.8554
IRON (P= 4KB)				C	2.3440	1.3670	1.1740
10652	7.8910	220	300.0	S	0.7480	-0.2755	0.8518
IRON (P= 0KB)				C	2.3010	1.3460	1.1660
10653	7.8672	121	300.0	S	0.7648	-0.2823	0.8576
IRON (P= 2KB)				C	2.3140	1.3550	1.1710
10654	7.8766	121	300.0	S	0.7615	-0.2812	0.8540
IRON (P= 4KB)				C	2.3280	1.3640	1.1760
10655	7.8861	121	300.0	S	0.7575	-0.2799	0.8503
IRON (P= 6KB)				C	2.3410	1.3730	1.1820
10656	7.8956	121	300.0	S	0.7542	-0.2788	0.8460
IRON (P= 8KB)				C	2.3550	1.3830	1.1870
10657	7.9051	121	300.0	S	0.7510	-0.2778	0.8425
IRON (P=10KB)				C	2.3680	1.3920	1.1920
10658	7.9154	121	300.0	S	0.7478	-0.2768	0.8389
IRON				C	2.2600	1.4000	1.1600
10659	7.8672	151	298.0	S	0.8411	-0.3217	0.8621
IRON				C	2.2200	1.3200	1.1400
10660	7.8458	151	373.0	S	0.8093	-0.3018	0.8772

CUBIC

CODE	NAME DENSITY	REF	TEMP(K)		11	12	44
IRON				C	2.1900	1.3100	1.1300
10661	7.8318	151	423.0	S	0.8269	-0.3095	0.8850
IRON				C	2.1700	1.2700	1.1200
10662	7.8154	151	473.0	S	0.8115	-0.2996	0.8929
IRON				C	2.1600	1.3000	1.1100
10663	7.7990	151	523.0	S	0.8452	-0.3176	0.9009
IRON				C	2.1100	1.3100	1.1000
10664	7.7804	151	573.0	S	0.9038	-0.3462	0.9091
IRON				C	2.0800	1.2600	1.0900
10665	7.7618	151	623.0	S	0.8855	-0.3340	0.9174
IRON				C	2.0600	1.2200	1.0800
10666	7.7433	151	673.0	S	0.8677	-0.3228	0.9259
IRON				C	2.0300	1.2100	1.0700
10667	7.7249	151	723.0	S	0.8879	-0.3316	0.9346
IRON				C	2.0000	1.2200	1.0500
10668	7.7042	151	773.0	S	0.9298	-0.3523	0.9524
IRON(96.03)-ALUMINUM(3.97)				C	2.2083	1.3249	1.2022
10669	7.6617	150	300.0	S	0.8233	-0.3087	0.8318
IRON(90.35)-ALUMINUM(9.65)				C	2.0489	1.2697	1.2262
10670	7.3619	150	300.0	S	0.9282	-0.3551	0.8155
IRON(85.50)-ALUMINUM(14.50)				C	1.9371	1.2493	1.2434
10671	7.1097	150	300.0	S	1.0444	-0.4095	0.8042
IRON(82.15)-ALUMINUM(17.85)				C	1.8535	1.2465	1.2520
10672	6.9422	150	300.0	S	1.1750	-0.4725	0.7987
IRON(80.17)-ALUMINUM(19.83)				C	1.7942	1.2482	1.2554
10673	6.8459	150	300.0	S	1.2987	-0.5328	0.7966
IRON(77.55)-ALUMINUM(22.45)				C	1.7430	1.2544	1.2624
10674	6.7441	150	300.0	S	1.4428	-0.6038	0.7921
IRON(76.43)-ALUMINUM(23.57)				C	1.7481	1.2993	1.2824
10675	6.7020	150	300.0	S	1.5621	-0.6660	0.7798
IRON(74.95)-ALUMINUM(25.05)				C	1.7099	1.3061	1.3170
10676	6.6436	150	300.0	S	1.7281	-0.7484	0.7593
IRON(73.03)-ALUMINUM(26.97)				C	1.6640	1.2502	1.3064
10677	6.5645	150	300.0	S	1.6911	-0.7255	0.7655
IRON(71.92)-ALUMINUM(28.08)				C	1.6636	1.2258	1.3097
10678	6.5228	150	300.0	S	1.6038	-0.6804	0.7635
IRON(66.00)-ALUMINUM(34.00)				C	1.7167	1.1355	1.2954
10679	6.3023	150	300.0	S	1.2306	-0.4899	0.7720
IRON(59.89)-ALUMINUM(40.11)				C	1.8105	1.1369	1.2709
10680	6.0420	150	300.0	S	1.0713	-0.4132	0.7868

CUBIC

CODE	NAME DENSITY	REF	TEMP(K)		11	12	44
LEAD				C	0.4765	0.4028	0.1441
10681	11.3400	128	ROOM	S	9.3000	-4.2600	6.9400
LEAD				C	0.4660	0.3920	0.1441
10682	11.3437	128	ROOM	S	9.2800	-4.2400	6.9400
LEAD				C	0.5554	0.4542	0.1942
10683	11.5900	270	0.0	S	6.8153	-3.0661	5.1493
LEAD				C	0.5549	0.4541	0.1930
10684	11.5860	270	20.0	S	6.8416	-3.0791	5.1813
LEAD				C	0.5516	0.4520	0.1899
10685	11.5760	270	40.0	S	6.9224	-3.1177	5.2659
LEAD				C	0.5473	0.4504	0.1867
10686	11.5580	270	60.0	S	7.1101	-3.2098	5.3562
LEAD				C	0.5429	0.4483	0.1836
10687	11.5440	270	80.0	S	7.2788	-3.2920	5.4466
LEAD				C	0.5386	0.4460	0.1804
10688	11.5270	270	100.0	S	7.4324	-3.3667	5.5432
LEAD				C	0.5343	0.4437	0.1772
10689	11.5100	270	120.0	S	7.5928	-3.4447	5.6433
LEAD				C	0.5299	0.4413	0.1741
10690	11.4890	270	140.0	S	7.7604	-3.5262	5.7438
LEAD				C	0.5261	0.4395	0.1709
10691	11.4720	270	160.0	S	7.9355	-3.6119	5.8514
LEAD				C	0.5213	0.4367	0.1678
10692	11.4540	270	180.0	S	8.1192	-3.7011	5.9595
LEAD				C	0.5169	0.4343	0.1646
10693	11.4340	270	200.0	S	8.3116	-3.7949	6.0753
LEAD				C	0.5125	0.4321	0.1615
10694	11.4170	270	220.0	S	8.5340	-3.9038	6.1920
LEAD				C	0.5083	0.4297	0.1583
10695	11.3960	270	240.0	S	8.7255	-3.9972	6.3171
LEAD				C	0.5038	0.4274	0.1552
10696	11.3790	270	260.0	S	8.9714	-4.1177	6.4433
LEAD				C	0.4996	0.4254	0.1521
10697	11.3580	270	280.0	S	9.2316	-4.2455	6.5746
LEAD				C	0.4953	0.4229	0.1490
10698	11.3380	270	300.0	S	9.4567	-4.3555	6.7114
LEAD				C	5.5540	4.5420	1.9420
10699	11.5880	15	4.2	S	0.6815	-0.3066	0.5149
LEAD				C	5.4360	4.4860	1.8400
10700	11.5480	15	77.0	S	0.7249	-0.3277	0.5435

CUBIC

CODE	NAME DENSITY	REF	TEMP(K)		11	12	44
LEAD				C	4.9620	4.2340	1.4960
10701	11.3450	15	296.0	S	0.9406	-0.4331	0.6684
LEAD				C	5.1843	4.3513	1.6588
10702	11.4420	184	195.0	S	0.8243	-0.3762	0.6028
LEAD				C	4.9661	4.2309	1.4977
10703	11.3440	184	296.0	S	0.9316	-0.4286	0.6677
LEAD FLUORIDE				C	0.8880	0.4720	0.2454
10704	7.7900	274	300.0	S	1.7845	-0.6193	4.0750
LEAD NITRATE				C	0.4560	0.3090	0.1370
10705	4.5300	34	ROOM	S	4.8455	-1.9572	7.2993
LEAD NITRATE				C	0.3729	0.2765	0.1347
10706	4.5470	124	293.0	S	7.2756	-3.0978	7.4239
LEAD NITRATE				C	0.3760	0.2800	0.1350
10707	4.5300	182	ROOM	S	7.3006	-3.1161	7.4074
LEAD TELLURIDE				C	1.0720	0.0768	0.1300
10708	8.1600	83	302.0	S	0.9419	-0.0630	7.6923
LEAD TELLURIDE				C	1.2798	0.0435	0.1514
10709	8.3672	136	0.0	S	0.7831	-0.0257	6.6050
LEAD TELLURIDE				C	1.2772	0.0445	0.1511
10710	8.3672	136	10.0	S	0.7848	-0.0264	6.6181
LEAD TELLURIDE				C	1.2733	0.0448	0.1509
10711	8.3672	136	20.0	S	0.7872	-0.0268	6.6269
LEAD TELLURIDE				C	1.2683	0.0454	0.1506
10712	8.3646	136	30.0	S	0.7904	-0.0273	6.6401
LEAD TELLURIDE				C	1.2581	0.0455	0.1496
10713	8.3571	136	50.0	S	0.7969	-0.0278	6.6845
LEAD TELLURIDE				C	1.2277	0.0481	0.1465
10714	8.3345	136	100.0	S	0.8169	-0.0308	6.8259
LEAD TELLURIDE				C	1.1952	0.0528	0.1435
10715	8.3121	136	150.0	S	0.8398	-0.0355	6.9686
LEAD TELLURIDE				C	1.1573	0.0609	0.1404
10716	8.2872	136	200.0	S	0.8687	-0.0434	7.1225
LEAD TELLURIDE				C	1.1207	0.0676	0.1374
10717	8.2625	136	250.0	S	0.8985	-0.0511	7.2780
LEAD TELLURIDE				C	1.0795	0.0764	0.1343
10718	8.2379	136	303.2	S	0.9351	-0.0618	7.4460
LEAD (0.17WT.%THALLIUM)				C	0.5582	0.4564	0.1935
10719	11.5880	15	4.2	S	6.7754	-3.0478	5.1680
LEAD (0.17WT.%THALLIUM)				C	0.5425	0.4477	0.1823
10720	11.5480	15	77.0	S	7.2642	-3.2844	5.4855

CUBIC

NAME / CODE	DENSITY	REF	TEMP(K)		11	12	44
LEAD (1.04WT.%THALLIUM)				C	0.5594	0.4576	0.1899
10721	11.5910	15	4.2	S	6.7748	-3.0483	5.2659
LEAD (1.04WT.%THALLIUM)				C	0.5405	0.4459	0.1814
10722	11.5510	15	77.0	S	7.2799	-3.2909	5.5127
LEAD (1.04WT.%THALLIUM)				C	0.4940	0.4214	0.1472
10723	11.3480	15	296.0	S	9.4321	-4.3420	6.7935
LEAD (1.75WT.%THALLIUM)				C	0.5523	0.4531	0.1896
10724	11.5930	15	4.2	S	6.9490	-3.1317	5.2743
LEAD (1.75WT.%THALLIUM)				C	0.5412	0.4474	0.1801
10725	11.5530	15	77.0	S	7.3394	-3.3215	5.5525
LEAD (1.75WT.%THALLIUM)				C	0.4933	0.4213	0.1471
10726	11.3500	15	296.0	S	9.5088	-4.3801	6.7981
LEAD (2.32WT.%THALLIUM)				C	0.5524	0.4526	0.1892
10727	11.5950	15	4.2	S	6.9087	-3.1113	5.2854
LEAD (2.32WT.%THALLIUM)				C	0.5395	0.4461	0.1803
10728	11.5560	15	77.0	S	7.3706	-3.3361	5.5463
LEAD (2.32WT.%THALLIUM)				C	0.4958	0.4226	0.1471
10729	11.3520	15	296.0	S	9.3560	-4.3052	6.7981
LEAD (3.45WT.%THALLIUM)				C	0.5476	0.4492	0.1868
10730	11.5990	15	4.2	S	7.0056	-3.1570	5.3533
LEAD (3.45WT.%THALLIUM)				C	0.5362	0.4432	0.1773
10731	11.5590	15	77.0	S	7.4028	-3.3499	5.6402
LEAD (3.45WT.%THALLIUM)				C	0.4906	0.4190	0.1450
10732	11.3560	15	296.0	S	9.5619	-4.4046	6.8966
LEAD (6.02WT.%THALLIUM)				C	0.5455	0.4487	0.1853
10733	11.6070	15	4.2	S	7.1181	-3.2125	5.3967
LEAD (6.02WT.%THALLIUM)				C	0.5329	0.4421	0.1759
10734	11.5670	15	77.0	S	7.5774	-3.4359	5.6850
LEAD (6.02WT.%THALLIUM)				C	0.4857	0.4151	0.1440
10735	11.3640	15	296.0	S	9.6962	-4.4681	6.9444
LEAD (14.7WT.%THALLIUM)				C	0.5216	0.4324	0.1830
10736	11.6350	15	4.2	S	7.7143	-3.4965	5.4645
LEAD (14.7WT.%THALLIUM)				C	0.5121	0.4279	0.1733
10737	11.5950	15	77.0	S	8.1613	-3.7151	5.7703
LEAD (14.7WT.%THALLIUM)				C	0.4716	0.4060	0.1412
10738	11.3920	15	296.0	S	10.4223	-4.8216	7.0822
LEAD (17.4WT.%THALLIUM)				C	0.5188	0.4314	0.1804
10739	11.6450	15	4.2	S	7.8690	-3.5726	5.5432
LEAD (17.4WT.%THALLIUM)				C	0.5091	0.4271	0.1711
10740	11.6050	15	77.0	S	8.3746	-3.8205	5.8445

CUBIC

NAME CODE	DENSITY	REF	TEMP(K)		11	12	44
LEAD (17.4WT.%THALLIUM)				C	0.4680	0.4054	0.1395
10741	11.4020	15	296.0	S	10.9103	-5.0642	7.1685
LEAD(94.99)-THALLIUM(5.01)				C	0.5414	0.4414	0.1896
10742	11.3500	231	0.0	S	6.9007	-3.0993	5.2743
LEAD(94.99)-THALLIUM(5.01)				C	0.5230	0.4322	0.1756
10743	11.3500	231	100.0	S	7.5824	-3.4308	5.6948
LEAD(94.99)-THALLIUM(5.01)				C	0.5041	0.4229	0.1610
10744	11.3500	231	200.0	S	8.4571	-3.8582	6.2112
LEAD(94.99)-THALLIUM(5.01)				C	0.4854	0.4138	0.1464
10745	11.3500	231	300.0	S	9.5649	-4.4016	6.8306
LEAD(79.50)-THALLIUM(20.50)				C	0.5039	0.4195	0.1768
10746	11.4000	231	0.0	S	8.1471	-3.7012	5.6561
LEAD(79.50)-THALLIUM(20.50)				C	0.4879	0.4111	0.1645
10747	11.4000	231	100.0	S	8.9350	-4.0858	6.0790
LEAD(79.50)-THALLIUM(20.50)				C	0.4733	0.4043	0.1512
10748	11.4000	231	200.0	S	9.9219	-4.5709	6.6138
LEAD(79.50)-THALLIUM(20.50)				C	0.4576	0.3964	0.1380
10749	11.4000	231	300.0	S	11.1598	-5.1800	7.2464
LEAD(68.23)-THALLIUM(31.77)				C	0.4809	0.4087	0.1670
10750	11.4500	231	0.0	S	9.4904	-4.3601	5.9880
LEAD(68.23)-THALLIUM(31.77)				C	0.4668	0.4008	0.1550
10751	11.4500	231	100.0	S	10.3638	-4.7877	6.4516
LEAD(68.23)-THALLIUM(31.77)				C	0.4516	0.3924	0.1440
10752	11.4500	231	200.0	S	11.5309	-5.3610	6.9444
LEAD(68.23)-THALLIUM(31.77)				C	0.4363	0.3837	0.1330
10753	11.4500	231	300.0	S	12.9512	-6.0602	7.5188
LEAD(59.50)-THALLIUM(40.50)				C	0.4657	0.4023	0.1570
10754	11.4900	231	0.0	S	10.7777	-4.9952	6.3694
LEAD(59.50)-THALLIUM(40.50)				C	0.4545	0.3963	0.1472
10755	11.4900	231	100.0	S	11.7220	-5.4601	6.7935
LEAD(59.50)-THALLIUM(40.50)				C	0.4435	0.3905	0.1374
10756	11.4900	231	200.0	S	12.8508	-6.0171	7.2780
LEAD(59.50)-THALLIUM(40.50)				C	0.4329	0.3851	0.1275
10757	11.4900	231	300.0	S	14.2241	-6.6964	7.8431
LEAD(47.34)-THALLIUM(52.66)				C	0.4500	0.4010	0.1475
10758	11.5700	231	0.0	S	13.8717	-6.5365	6.7797
LEAD(47.34)-THALLIUM(52.66)				C	0.4388	0.3932	0.1394
10759	11.5700	231	100.0	S	14.8920	-7.0379	7.1736
LEAD(47.34)-THALLIUM(52.66)				C	0.4274	0.3854	0.1306
10760	11.5700	231	200.0	S	16.1512	-7.6583	7.6570

CUBIC

NAME CODE	DENSITY	REF	TEMP(K)		11	12	44
LEAD(47.34)-THALLIUM(52.66)				C	0.4158	0.3772	0.1220
10761	11.5700	231	300.0	S	17.5560	-8.3507	8.1967
LEAD(38.59)-THALLIUM(61.41)				C	0.4304	0.3926	0.1420
10762	11.6400	231	0.0	S	17.9109	-8.5441	7.0423
LEAD(38.59)-THALLIUM(61.41)				C	0.4265	0.3895	0.1350
10763	11.6400	231	100.0	S	18.2945	-8.7325	7.4074
LEAD(38.59)-THALLIUM(61.41)				C	0.4211	0.3849	0.1270
10764	11.6400	231	200.0	S	18.6961	-8.9282	7.8740
LEAD(38.59)-THALLIUM(61.41)				C	0.4157	0.3803	0.1195
10765	11.6400	231	300.0	S	19.1158	-9.1328	8.3682
LEAD(28.32)-THALLIUM(71.68)				C	0.4322	0.3928	0.1360
10766	11.7000	231	0.0	S	17.1942	-8.1865	7.3529
LEAD(28.32)-THALLIUM(71.68)				C	0.4266	0.3870	0.1300
10767	11.7000	231	100.0	S	17.1127	-8.1399	7.6923
LEAD(28.32)-THALLIUM(71.68)				C	0.4204	0.3806	0.1238
10768	11.7000	231	200.0	S	17.0325	-8.0931	8.0775
LEAD(28.32)-THALLIUM(71.68)				C	0.4145	0.3745	0.1175
10769	11.7000	231	300.0	S	16.9531	-8.0468	8.5106
LITHIUM				C	0.1480	0.1250	0.1080
10770	0.5454	190	78.0	S	29.8230	-13.6552	9.2593
LITHIUM				C	0.1401	0.1179	0.0997
10771	0.5421	190	155.0	S	30.9168	-14.1283	10.0301
LITHIUM				C	0.1342	0.1125	0.0960
10772	0.5398	190	195.0	S	31.6500	-14.4330	10.4167
LITHIUM				C	0.1350	0.1144	0.0878
10773	0.5326	255	298.0	S	33.2788	-15.2650	11.3895
LITHIUM BROMIDE				C	0.3940	0.1880	0.1910
10774	3.4700	4	ROOM	S	3.6691	-1.1852	5.2356
LITHIUM CHLORIDE				C	0.4940	0.2260	0.2490
10775	2.0750	4	ROOM	S	2.8399	-0.8914	4.0161
LITHIUM CHLORIDE				C	0.6074	0.2270	0.2692
10776	2.1110	152	4.2	S	2.0666	-0.5622	3.7147
LITHIUM CHLORIDE				C	0.4927	0.2310	0.2495
10777	2.0680	152	295.0	S	2.8966	-0.9246	4.0080
LITHIUM CHLORIDE				C	0.5860	0.2086	0.2671
10778	2.1321	176	4.2	S	2.0987	-0.5510	3.7439
LITHIUM CHLORIDE				C	0.5856	0.2084	0.2671
10779	2.1321	176	20.0	S	2.0989	-0.5508	3.7439
LITHIUM CHLORIDE				C	0.5824	0.2050	0.2670
10780	2.1317	176	40.0	S	2.1024	-0.5474	3.7453

CUBIC

CODE	NAME DENSITY	REF	TEMP(K)		11	12	44
LITHIUM CHLORIDE				C	0.5757	0.1989	0.2665
10781	2.1302	176	60.0	S	2.1117	-0.5422	3.7523
LITHIUM CHLORIDE				C	0.5676	0.1920	0.2656
10782	2.1277	176	80.0	S	2.1252	-0.5372	3.7651
LITHIUM CHLORIDE				C	0.5595	0.1859	0.2643
10783	2.1244	176	100.0	S	2.1424	-0.5343	3.7836
LITHIUM CHLORIDE				C	0.5516	0.1812	0.2627
10784	2.1205	176	120.0	S	2.1646	-0.5352	3.8066
LITHIUM CHLORIDE				C	0.5438	0.1776	0.2609
10785	2.1162	176	140.0	S	2.1913	-0.5395	3.8329
LITHIUM CHLORIDE				C	0.5361	0.1755	0.2592
10786	2.1115	176	160.0	S	2.2245	-0.5486	3.8580
LITHIUM CHLORIDE				C	0.5284	0.1754	0.2574
10787	2.1065	176	180.0	S	2.2677	-0.5652	3.8850
LITHIUM CHLORIDE				C	0.5209	0.1777	0.2556
10788	2.1013	176	200.0	S	2.3229	-0.5909	3.9124
LITHIUM CHLORIDE				C	0.5131	0.1807	0.2539
10789	2.0960	176	220.0	S	2.3868	-0.6216	3.9386
LITHIUM CHLORIDE				C	0.5054	0.1834	0.2521
10790	2.0906	176	240.0	S	2.4526	-0.6530	3.9667
LITHIUM CHLORIDE				C	0.4978	0.1858	0.2504
10791	2.0852	176	260.0	S	2.5202	-0.6850	3.9936
LITHIUM CHLORIDE				C	0.4903	0.1877	0.2486
10792	2.0796	176	280.0	S	2.5882	-0.7165	4.0225
LITHIUM FLUORIDE				C	1.1777	0.4336	0.6281
10793	2.6010	128	ROOM	S	1.0590	-0.2850	1.5920
LITHIUM FLUORIDE				C	0.9740	0.4040	0.5540
10794	2.6010	129	ROOM	S	1.3566	-0.3977	1.8051
LITHIUM FLUORIDE				C	1.1900	0.4580	0.5420
10795	2.6010	129	ROOM	S	1.0690	-0.2971	1.8450
LITHIUM FLUORIDE				C	1.1900	0.5380	0.5340
10796	2.6010	129	ROOM	S	1.1696	-0.3641	1.8727
LITHIUM FLUORIDE				C	0.9900	0.4300	0.5400
10797	2.6010	129	ROOM	S	1.3707	-0.4151	1.8519
LITHIUM FLUORIDE				C	1.1030	0.4480	0.6230
10798	2.6300	4	ROOM	S	1.1846	-0.3422	1.6051
LITHIUM FLUORIDE				C	1.1120	0.4200	0.6280
10799	2.6010	4	ROOM	S	1.1342	-0.3109	1.5924
LITHIUM FLUORIDE				C	1.2445	0.4264	0.6471
10800	2.6400	45	4.2	S	0.9738	-0.2485	1.5454

CODE / NAME	DENSITY	REF	TEMP(K)		11	12	44
LITHIUM FLUORIDE				C	1.2445	0.4264	0.6471
10801	2.6400	45	20.0	S	0.9738	-0.2485	1.5454
LITHIUM FLUORIDE				C	1.2445	0.4201	0.6471
10802	2.6400	45	40.0	S	0.9686	-0.2444	1.5454
LITHIUM FLUORIDE				C	1.2406	0.4217	0.6471
10803	2.6400	45	60.0	S	0.9741	-0.2471	1.5454
LITHIUM FLUORIDE				C	1.2366	0.4202	0.6471
10804	2.6400	45	80.0	S	0.9771	-0.2478	1.5454
LITHIUM FLUORIDE				C	1.2295	0.4213	0.6466
10805	2.6380	45	100.0	S	0.9857	-0.2516	1.5466
LITHIUM FLUORIDE				C	1.2203	0.4213	0.6461
10806	2.6360	45	120.0	S	0.9960	-0.2556	1.5477
LITHIUM FLUORIDE				C	1.2101	0.4222	0.6454
10807	2.6340	45	140.0	S	1.0084	-0.2608	1.5494
LITHIUM FLUORIDE				C	1.1983	0.4229	0.6444
10808	2.6300	45	160.0	S	1.0228	-0.2668	1.5518
LITHIUM FLUORIDE				C	1.1866	0.4233	0.6432
10809	2.6260	45	180.0	S	1.0373	-0.2728	1.5547
LITHIUM FLUORIDE				C	1.1742	0.4257	0.6417
10810	2.6220	45	200.0	S	1.0552	-0.2808	1.5584
LITHIUM FLUORIDE				C	1.1616	0.4263	0.6397
10811	2.6180	45	220.0	S	1.0722	-0.2878	1.5632
LITHIUM FLUORIDE				C	1.1490	0.4251	0.6374
10812	2.6140	45	240.0	S	1.0877	-0.2937	1.5689
LITHIUM FLUORIDE				C	1.1361	0.4250	0.6346
10813	2.6090	45	260.0	S	1.1053	-0.3009	1.5758
LITHIUM FLUORIDE				C	1.1234	0.4223	0.6321
10814	2.6050	45	280.0	S	1.1203	-0.3061	1.5820
LITHIUM FLUORIDE				C	1.1111	0.4208	0.6296
10815	2.6010	45	300.0	S	1.1365	-0.3122	1.5883
LITHIUM FLUORIDE				C	1.1300	0.4800	0.6250
10816	2.6000	181	293.0	S	1.1851	-0.3533	1.6000
LITHIUM FLUORIDE				C	1.0730	0.3964	0.6357
10817	2.6400	246	293.0	S	1.1640	-0.3140	1.5730
LITHIUM FLUORIDE				C	1.0009	0.3814	0.6188
10818	2.6330	246	373.0	S	1.2650	-0.3490	1.6160
LITHIUM FLUORIDE				C	0.9132	0.3604	0.5981
10819	2.6230	246	473.0	S	1.4100	-0.3990	1.6720
LITHIUM FLUORIDE				C	0.8177	0.3254	0.5767
10820	2.6120	246	573.0	S	1.5810	-0.4500	1.7340

CUBIC

CODE / NAME	DENSITY	REF	TEMP(K)		11	12	44
LITHIUM FLUORIDE				C	0.7218	0.2859	0.5565
10821	2.6000	246	673.0	S	1.7870	-0.5070	[illegible].7970
LITHIUM FLUORIDE				C	0.6233	0.2384	0.5359
10822	2.5870	246	773.0	S	2.0350	-0.5630	1.8660
LITHIUM FLUORIDE				C	0.5268	0.1887	0.5165
10823	2.5720	246	873.0	S	2.3400	-0.6170	1.9360
LITHIUM FLUORIDE				C	0.4380	0.1425	0.4968
10824	2.5570	246	973.0	S	2.7170	-0.6670	2.0130
LITHIUM FLUORIDE				C	0.3973	0.1215	0.4871
10825	2.5480	246	1023.0	S	2.9380	-0.6880	2.0530
LITHIUM FLUORIDE				C	1.1302	0.4766	0.6369
10826	2.6000	57	292.0	S	1.1800	-0.3500	1.5700
LITHIUM FLUORIDE				C	1.0281	0.4293	0.6173
10827	2.5900	57	373.0	S	1.2900	-0.3800	1.6200
LITHIUM FLUORIDE				C	0.9679	0.3932	0.6098
10828	2.5900	57	423.0	S	1.3500	-0.3900	1.6400
LITHIUM FLUORIDE				C	0.9140	0.3763	0.5952
10829	2.5800	57	473.0	S	1.4400	-0.4200	1.6800
LITHIUM FLUORIDE				C	0.8710	0.3710	0.5814
10830	2.5800	57	523.0	S	1.5400	-0.4600	1.7200
LITHIUM FLUORIDE				C	0.8180	0.3486	0.5650
10831	2.5700	57	573.0	S	1.6400	-0.4900	1.7700
LITHIUM FLUORIDE				C	0.7632	0.3226	0.5495
10832	2.5700	57	623.0	S	1.7500	-0.5200	1.8200
LITHIUM FLUORIDE				C	0.7050	0.2968	0.5348
10833	2.5600	57	673.0	S	1.8900	-0.5600	1.8700
LITHIUM FLUORIDE				C	0.6552	0.2676	0.5155
10834	2.5500	57	723.0	S	2.0000	-0.5800	1.9400
LITHIUM FLUORIDE				C	0.5960	0.2283	0.5000
10835	2.5500	57	773.0	S	2.1300	-0.5900	2.0000
LITHIUM FLUORIDE				C	1.1373	0.4759	0.6368
10836	2.6390	185	300.0	S	1.1675	-0.3444	1.5704
LITHIUM FLUORIDE				C	1.1397	0.4767	0.6364
10837	2.6380	79	ROOM	S	1.1648	-0.3435	1.5713
LITHIUM IODIDE				C	0.2850	0.1400	0.1350
10838	4.0610	4	ROOM	S	5.1877	-1.7089	7.4074
LITHIUM(98.91)-MAGNESIUM(1.09)				C	0.1374	0.1166	0.0891
10839	0.5485	255	298.0	S	32.9507	-15.1262	11.2233
LITHIUM(97.74)-MAGNESIUM(2.26)				C	0.1394	0.1184	0.0903
10840	0.5650	255	298.0	S	32.6320	-14.9869	11.0742

CUBIC

CODE	NAME / DENSITY	REF	TEMP(K)		11	12	44
LITHIUM(96.99)-MAGNESIUM(3.01)				C	0.1396	0.1185	0.0912
10841	0.5756	255	298.0	S	32.4807	-14.9127	10.9649
LITHIUM(95.72)-MAGNESIUM(4.28)				C	0.1429	0.1217	0.0924
10842	0.5936	255	298.0	S	32.3094	-14.8604	10.8225
MAGNESIUM OXIDE				C	2.8600	0.8700	1.4800
10843	3.5800	129	ROOM	S	0.4075	-0.0950	0.6757
MAGNESIUM OXIDE				C	2.8760	0.8740	1.5140
10844	3.5800	4	ROOM	S	0.4051	-0.0944	0.6605
MAGNESIUM OXIDE				C	2.9870	0.8560	1.5673
10845	3.5911	80	80.0	S	0.3838	-0.0855	0.6380
MAGNESIUM OXIDE				C	2.9870	0.8580	1.5671
10846	3.5911	80	90.0	S	0.3840	-0.0857	0.6381
MAGNESIUM OXIDE				C	2.9720	0.8640	1.5643
10847	3.5895	80	150.0	S	0.3872	-0.0872	0.6393
MAGNESIUM OXIDE				C	2.9670	0.8650	1.5635
10848	3.5889	80	160.0	S	0.3881	-0.0876	0.6396
MAGNESIUM OXIDE				C	2.9630	0.8660	1.5627
10849	3.5881	80	170.0	S	0.3889	-0.0880	0.6399
MAGNESIUM OXIDE				C	2.9580	0.8670	1.5618
10850	3.5873	80	180.0	S	0.3899	-0.0884	0.6403
MAGNESIUM OXIDE				C	2.9530	0.8680	1.5609
10851	3.5865	80	190.0	S	0.3908	-0.0888	0.6407
MAGNESIUM OXIDE				C	2.9480	0.8680	1.5599
10852	3.5857	80	200.0	S	0.3917	-0.0891	0.6411
MAGNESIUM OXIDE				C	2.9420	0.8690	1.5589
10853	3.5848	80	210.0	S	0.3928	-0.0896	0.6415
MAGNESIUM OXIDE				C	2.9370	0.8700	1.5578
10854	3.5838	80	220.0	S	0.3938	-0.0900	0.6419
MAGNESIUM OXIDE				C	2.9320	0.8710	1.5566
10855	3.5830	80	230.0	S	0.3948	-0.0904	0.6424
MAGNESIUM OXIDE				C	2.9270	0.8720	1.5554
10856	3.5824	80	240.0	S	0.3958	-0.0908	0.6429
MAGNESIUM OXIDE				C	2.9210	0.8730	1.5542
10857	3.5812	80	250.0	S	0.3969	-0.0913	0.6434
MAGNESIUM OXIDE				C	2.9160	0.8740	1.5529
10858	3.5803	80	260.0	S	0.3979	-0.0918	0.6440
MAGNESIUM OXIDE				C	2.9100	0.8740	1.5516
10859	3.5791	80	270.0	S	0.3990	-0.0922	0.6445
MAGNESIUM OXIDE				C	2.9050	0.8750	1.5503
10860	3.5782	80	280.0	S	0.4000	-0.0926	0.6450

CUBIC

CODE	NAME DENSITY	REF	TEMP(K)		11	12	44
MAGNESIUM OXIDE				C	2.8990	0.8760	1.5490
10861	3.5770	80	290.0	S	0.4012	-0.0931	0.6456
MAGNESIUM OXIDE				C	2.8930	0.8770	1.5477
10862	3.5761	80	300.0	S	0.4024	-0.0936	0.6461
MAGNESIUM OXIDE				C	2.8880	0.8780	1.5464
10863	3.5746	80	310.0	S	0.4035	-0.0941	0.6467
MAGNESIUM OXIDE				C	2.8820	0.8790	1.5451
10864	3.5737	80	320.0	S	0.4047	-0.0946	0.6472
MAGNESIUM OXIDE				C	2.8760	0.8800	1.5438
10865	3.5725	80	330.0	S	0.4059	-0.0951	0.6478
MAGNESIUM OXIDE				C	2.8700	0.8800	1.5424
10866	3.5714	80	340.0	S	0.4070	-0.0955	0.6483
MAGNESIUM OXIDE				C	2.8640	0.8810	1.5410
10867	3.5704	80	350.0	S	0.4082	-0.0960	0.6489
MAGNESIUM OXIDE				C	2.8580	0.8820	1.5396
10868	3.5691	80	360.0	S	0.4095	-0.0966	0.6495
MAGNESIUM OXIDE				C	2.8520	0.8830	1.5382
10869	3.5678	80	370.0	S	0.4108	-0.0971	0.6501
MAGNESIUM OXIDE				C	2.8460	0.8840	1.5368
10870	3.5664	80	380.0	S	0.4120	-0.0977	0.6507
MAGNESIUM OXIDE				C	2.8400	0.8850	1.5354
10871	3.5650	80	390.0	S	0.4133	-0.0982	0.6513
MAGNESIUM OXIDE				C	2.8330	0.8850	1.5339
10872	3.5640	80	400.0	S	0.4146	-0.0987	0.6519
MAGNESIUM OXIDE				C	2.8270	0.8860	1.5324
10873	3.5629	80	410.0	S	0.4159	-0.0993	0.6526
MAGNESIUM OXIDE				C	2.8210	0.8870	1.5309
10874	3.5612	80	420.0	S	0.4173	-0.0998	0.6532
MAGNESIUM OXIDE				C	2.8140	0.8880	1.5294
10875	3.5596	80	430.0	S	0.4188	-0.1004	0.6539
MAGNESIUM OXIDE				C	2.8080	0.8880	1.5279
10876	3.5587	80	440.0	S	0.4199	-0.1009	0.6545
MAGNESIUM OXIDE				C	2.8010	0.8890	1.5264
10877	3.5575	80	450.0	S	0.4215	-0.1015	0.6551
MAGNESIUM OXIDE				C	2.7950	0.8900	1.5249
10878	3.5560	80	460.0	S	0.4228	-0.1021	0.6558
MAGNESIUM OXIDE				C	2.7880	0.8910	1.5234
10879	3.5544	80	470.0	S	0.4244	-0.1028	0.6564
MAGNESIUM OXIDE				C	2.7810	0.8910	1.5219
10880	3.5533	80	480.0	S	0.4258	-0.1033	0.6571

CODE	NAME DENSITY	REF	TEMP(K)		11	12	44
MAGNESIUM	OXIDE			C	2.7740	0.8920	1.5204
10881	3.5522	80	490.0	S	0.4274	-0.1040	0.6577
MAGNESIUM	OXIDE			C	2.7670	0.8930	1.5188
10882	3.5511	80	500.0	S	0.4290	-0.1047	0.6584
MAGNESIUM	OXIDE			C	2.7600	0.8930	1.5172
10883	3.5500	80	510.0	S	0.4304	-0.1052	0.6591
MAGNESIUM	OXIDE			C	2.7540	0.8940	1.5156
10884	3.5490	80	520.0	S	0.4318	-0.1058	0.6598
MAGNESIUM	OXIDE			C	2.7470	0.8940	1.5140
10885	3.5480	80	530.0	S	0.4333	-0.1064	0.6605
MAGNESIUM	OXIDE			C	2.7400	0.8950	1.5124
10886	3.5462	80	540.0	S	0.4349	-0.1071	0.6612
MAGNESIUM	OXIDE			C	2.7320	0.8960	1.5108
10887	3.5444	80	550.0	S	0.4368	-0.1079	0.6619
MAGNESIUM	OXIDE			C	2.7250	0.8960	1.5092
10888	3.5426	80	560.0	S	0.4383	-0.1085	0.6626
MAGNESIUM	OXIDE			C	2.9980	0.9910	1.5750
10889	3.5900	246	293.0	S	0.3991	-0.0992	0.6349
MAGNESIUM	OXIDE			C	2.9240	0.9680	1.5640
10890	3.5800	246	373.0	S	0.4094	-0.1018	0.6394
MAGNESIUM	OXIDE			C	2.8460	0.9540	1.5490
10891	3.5670	246	473.0	S	0.4225	-0.1061	0.6456
MAGNESIUM	OXIDE			C	2.7620	0.9340	1.5340
10892	3.5540	246	573.0	S	0.4367	-0.1104	0.6519
MAGNESIUM	OXIDE			C	2.6780	0.9120	1.5180
10893	3.5390	246	673.0	S	0.4515	-0.1147	0.6588
MAGNESIUM	OXIDE			C	2.5980	0.8940	1.5020
10894	3.5240	246	773.0	S	0.4672	-0.1196	0.6658
MAGNESIUM	OXIDE			C	2.5300	0.8870	1.4860
10895	3.5080	246	873.0	S	0.4832	-0.1254	0.6729
MAGNESIUM	OXIDE			C	2.4440	0.8590	1.4690
10896	3.4920	246	973.0	S	0.5007	-0.1302	0.6807
MAGNESIUM	OXIDE			C	2.3680	0.8410	1.4530
10897	3.4770	246	1073.0	S	0.5189	-0.1360	0.6882
MAGNESIUM	OXIDE			C	2.2890	0.8180	1.4370
10898	3.4620	246	1173.0	S	0.5381	-0.1417	0.6959
MAGNESIUM	OXIDE			C	2.2170	0.8020	1.4200
10899	3.4460	246	1273.0	S	0.5584	-0.1483	0.7042
MAGNESIUM	OXIDE			C	2.1520	0.7920	1.4030
10900	3.4280	246	1373.0	S	0.5794	-0.1559	0.7128

NAME CODE	DENSITY	REF	TEMP(K)		11	12	44
MAGNESIUM OXIDE				C	2.0910	0.7870	1.3860
10901	3.4120	246	1473.0	S	0.6022	-0.1647	0.7215
MAGNESIUM OXIDE				C	2.9708	0.9536	1.5613
10902	3.5790	36	298.0	S	0.3988	-0.0969	0.6405
MAGNESIUM OXIDE				C	3.0631	0.9378	1.5763
10903	3.5995	20	73.0	S	0.3812	-0.0893	0.6344
MAGNESIUM OXIDE				C	3.0619	0.9382	1.5759
10904	3.5994	20	83.0	S	0.3814	-0.0895	0.6346
MAGNESIUM OXIDE				C	3.0601	0.9385	1.5756
10905	3.5992	20	93.0	S	0.3817	-0.0896	0.6347
MAGNESIUM OXIDE				C	3.0575	0.9388	1.5753
10906	3.5989	20	103.0	S	0.3822	-0.0898	0.6348
MAGNESIUM OXIDE				C	3.0550	0.9391	1.5749
10907	3.5986	20	113.0	S	0.3826	-0.0900	0.6350
MAGNESIUM OXIDE				C	3.0519	0.9395	1.5745
10908	3.5983	20	123.0	S	0.3832	-0.0902	0.6351
MAGNESIUM OXIDE				C	3.0487	0.9398	1.5741
10909	3.5977	20	133.0	S	0.3838	-0.0904	0.6353
MAGNESIUM OXIDE				C	3.0449	0.9403	1.5734
10910	3.5972	20	143.0	S	0.3844	-0.0907	0.6356
MAGNESIUM OXIDE				C	3.0412	0.9408	1.5727
10911	3.5966	20	153.0	S	0.3851	-0.0910	0.6358
MAGNESIUM OXIDE				C	3.0368	0.9413	1.5721
10912	3.5960	20	163.0	S	0.3859	-0.0913	0.6361
MAGNESIUM OXIDE				C	3.0325	0.9419	1.5713
10913	3.5953	20	173.0	S	0.3867	-0.0916	0.6364
MAGNESIUM OXIDE				C	3.0280	0.9426	1.5702
10914	3.5946	20	183.0	S	0.3875	-0.0920	0.6369
MAGNESIUM OXIDE				C	3.0280	0.9433	1.5695
10915	3.5938	20	193.0	S	0.3876	-0.0921	0.6371
MAGNESIUM OXIDE				C	3.1079	0.9441	1.5684
10916	3.5930	20	203.0	S	0.3748	-0.0873	0.6376
MAGNESIUM OXIDE				C	3.0124	0.9447	1.5677
10917	3.5921	20	213.0	S	0.3904	-0.0932	0.6379
MAGNESIUM OXIDE				C	3.0073	0.9457	1.5666
10918	3.5912	20	223.0	S	0.3914	-0.0936	0.6383
MAGNESIUM OXIDE				C	3.0022	0.9466	1.5658
10919	3.5902	20	233.0	S	0.3924	-0.0941	0.6387
MAGNESIUM OXIDE				C	2.9967	0.9475	1.5648
10920	3.5892	20	243.0	S	0.3935	-0.0945	0.6391

NAME / CODE	DENSITY	REF	TEMP(K)		11	12	44
MAGNESIUM OXIDE				C	2.9910	0.9485	1.5637
10921	3.5882	20	253.0	S	0.3946	-0.0950	0.6395
MAGNESIUM OXIDE				C	2.9847	0.9487	1.5626
10922	3.5872	20	263.0	S	0.3957	-0.0954	0.6400
MAGNESIUM OXIDE				C	2.9791	0.9496	1.5615
10923	3.5861	20	273.0	S	0.3968	-0.0959	0.6404
MAGNESIUM OXIDE				C	2.9728	0.9501	1.5604
10924	3.5850	20	283.0	S	0.3980	-0.0964	0.6409
MAGNESIUM OXIDE				C	2.9672	0.9505	1.5596
10925	3.5839	20	293.0	S	0.3990	-0.0968	0.6412
MAGNESIUM OXIDE				C	2.8917	0.8796	1.5461
10926	3.5760	57A	298.0	S	0.4030	-0.0940	0.6468
MAGNESIUM OXIDE				C	2.9664	0.9508	1.5581
10927	3.5790	54	298.0	S	0.3992	-0.0969	0.6418
MAGNESIUM OXIDE				C	2.9740	0.9557	1.5620
10928	3.5833	243	300.0	S	0.3985	-0.0969	0.6402
MAGNESIUM OXIDE				C	2.6630	0.9732	1.4870
10929	3.5206	243	800.0	S	0.4668	-0.1249	0.6725
MAGNESIUM OXIDE (P=8KB)				C	3.0420	0.9653	1.5710
10930	3.6010	243	300.0	S	0.3881	-0.0935	0.6365
MAGNESIUM OXIDE (P=8KB)				C	2.7320	0.9838	1.4960
10931	3.5396	243	800.0	S	0.4523	-0.1197	0.6684
MAGNESIUM(33.33)-COPPER(66.67)				C	1.2500	0.7170	0.4230
10932	5.8317	56	80.0	S	1.3750	-0.5012	2.3641
MAGNESIUM(33.33)-COPPER(66.67)				C	1.2490	0.7170	0.4220
10933	5.8252	56	100.0	S	1.3774	-0.5023	2.3697
MAGNESIUM(33.33)-COPPER(66.67)				C	1.2430	0.7140	0.4200
10934	5.8087	56	150.0	S	1.3850	-0.5053	2.3810
MAGNESIUM(33.33)-COPPER(66.67)				C	1.2390	0.7120	0.4180
10935	5.7924	56	200.0	S	1.3902	-0.5073	2.3923
MAGNESIUM(33.33)-COPPER(66.67)				C	1.2320	0.7090	0.4150
10936	5.7761	56	250.0	S	1.4005	-0.5116	2.4096
MAGNESIUM(33.33)-COPPER(66.67)				C	1.2280	0.7060	0.4120
10937	5.7600	56	300.0	S	1.4034	-0.5123	2.4272
MAGNESIUM(33.33)-COPPER(66.67)				C	1.2140	0.6970	0.4070
10938	5.7439	56	350.0	S	1.4173	-0.5169	2.4570
MAGNESIUM(33.33)-COPPER(66.67)				C	1.2000	0.6900	0.4020
10939	5.7279	56	400.0	S	1.4364	-0.5244	2.4876
MAGNESIUM(33.33)-COPPER(66.67)				C	1.1850	0.6810	0.3960
10940	5.7120	56	450.0	S	1.4536	-0.5305	2.5253

NAME / CODE	DENSITY	REF	TEMP(K)	CUBIC	11	12	44
MAGNESIUM(33.33)-COPPER(66.67)				C	1.1710	0.6750	0.3910
10941	5.6962	56	500.0	S	1.4763	-0.5398	2.5575
MAGNESIUM(66.67)-SILICON(33.33)				C	1.2100	0.2200	0.4640
10942	2.0000	277	300.0	S	0.8754	-0.1347	2.1552
MAGNESIUM(50.00)-SILVER(50.00)				C	0.8650	0.5790	0.5200
10943	6.1340	56	80.0	S	2.4958	-1.0007	1.9231
MAGNESIUM(50.00)-SILVER(50.00)				C	0.8630	0.5780	0.5170
10944	6.1253	56	100.0	S	2.5043	-1.0045	1.9342
MAGNESIUM(50.00)-SILVER(50.00)				C	0.8590	0.5750	0.5110
10945	6.1048	56	150.0	S	2.5133	-1.0078	1.9569
MAGNESIUM(50.00)-SILVER(50.00)				C	0.8550	0.5720	0.5040
10946	6.0833	56	200.0	S	2.5225	-1.0111	1.9841
MAGNESIUM(50.00)-SILVER(50.00)				C	0.8510	0.5700	0.4960
10947	6.0626	56	250.0	S	2.5399	-1.0188	2.0161
MAGNESIUM(50.00)-SILVER(50.00)				C	0.8460	0.5670	0.4850
10948	6.0420	56	300.0	S	2.5578	-1.0264	2.0619
MAGNESIUM(50.00)-SILVER(50.00)				C	0.8390	0.5600	0.4740
10949	6.0215	56	350.0	S	2.5596	-1.0246	2.1097
MAGNESIUM(50.00)-SILVER(50.00)				C	0.8310	0.5530	0.4640
10950	6.0011	56	400.0	S	2.5702	-1.0270	2.1552
MAGNESIUM(50.00)-SILVER(50.00)				C	0.8160	0.5400	0.4520
10951	5.9804	56	450.0	S	2.5913	-1.0319	2.2124
MAGNESIUM(50.00)-SILVER(50.00)				C	0.7980	0.5240	0.4380
10952	5.9609	56	500.0	S	2.6137	-1.0360	2.2831
MAGNESIUM(66.67)-TIN(33.33)				C	0.8240	0.2080	0.3660
10953	3.5920	70	300.0	S	1.3511	-0.2723	2.7322
MAGNETITE				C	2.7500	1.0400	0.9550
10954	5.1800	129	ROOM	S	0.4589	-0.1259	1.0471
MAGNETITE				C	2.7000	1.0800	0.9870
10955	5.1800	129	ROOM	S	0.4801	-0.1372	1.0132
MAGNETITE				C	2.7300	1.0600	0.9710
10956	5.1800	4	ROOM	S	0.4679	-0.1309	1.0299
MANGANESE OXIDE				C	2.2300	1.2000	0.7900
10957	5.3900	198	298.0	S	0.7192	-0.2516	1.2658
MERCURY TELLURIDE				C	0.5920	0.4140	0.2190
10958	8.0980	18	4.2	S	3.9801	-1.6379	4.5662
MERCURY TELLURIDE				C	0.5870	0.4100	0.2170
10959	8.0960	18	77.0	S	4.0034	-1.6463	4.6083
MERCURY TELLURIDE				C	0.5650	0.3940	0.2100
10960	8.0890	18	196.0	S	4.1450	-1.7030	4.7619

CUBIC

CODE	NAME DENSITY	REF	TEMP(K)		11	12	44
MERCURY TELLURIDE				C	0.5480	0.3810	0.2040
10961	8.0790	18	290.0	S	4.2465	-1.7416	4.9020
MOLYBDENUM				C	4.7900	1.6580	1.0850
10962	10.2480	39	77.0	S	0.2540	-0.0653	0.9217
MOLYBDENUM				C	4.7890	1.6570	1.0850
10963	10.2480	39	80.0	S	0.2540	-0.0653	0.9217
MOLYBDENUM				C	4.7860	1.6600	1.0840
10964	10.2470	39	100.0	S	0.2544	-0.0655	0.9225
MOLYBDENUM				C	4.7810	1.6630	1.0830
10965	10.2450	39	120.0	S	0.2549	-0.0658	0.9234
MOLYBDENUM				C	4.7750	1.6650	1.0820
10966	10.2440	39	140.0	S	0.2555	-0.0661	0.9242
MOLYBDENUM				C	4.7680	1.6700	1.0800
10967	10.2410	39	160.0	S	0.2563	-0.0665	0.9259
MOLYBDENUM				C	4.7580	1.6700	1.0780
10968	10.2380	39	180.0	S	0.2571	-0.0668	0.9276
MOLYBDENUM				C	4.7480	1.6700	1.0760
10969	10.2350	39	200.0	S	0.2578	-0.0671	0.9294
MOLYBDENUM				C	4.7380	1.6720	1.0750
10970	10.2320	39	220.0	S	0.2587	-0.0675	0.9302
MOLYBDENUM				C	4.7280	1.6720	1.0730
10971	10.2290	39	240.0	S	0.2594	-0.0678	0.9320
MOLYBDENUM				C	4.7180	1.6740	1.0710
10972	10.2260	39	260.0	S	0.2603	-0.0682	0.9337
MOLYBDENUM				C	4.7070	1.6750	1.0700
10973	10.2230	39	280.0	S	0.2612	-0.0686	0.9346
MOLYBDENUM				C	4.6960	1.6760	1.0680
10974	10.2190	39	300.0	S	0.2622	-0.0690	0.9363
MOLYBDENUM				C	4.6730	1.6770	1.0640
10975	10.2130	39	340.0	S	0.2640	-0.0697	0.9398
MOLYBDENUM				C	4.6500	1.6780	1.0590
10976	10.2070	39	380.0	S	0.2660	-0.0705	0.9443
MOLYBDENUM				C	4.6280	1.6800	1.0550
10977	10.2010	39	420.0	S	0.2679	-0.0713	0.9479
MOLYBDENUM				C	4.6040	1.6820	1.0510
10978	10.1950	39	460.0	S	0.2700	-0.0722	0.9515
MOLYBDENUM				C	4.5800	1.6820	1.0470
10979	10.1890	39	500.0	S	0.2720	-0.0731	0.9551
MOLYBDENUM				C	4.5002	1.7292	1.2503
10980	10.2622	89	0.0	S	0.2825	-0.0784	0.7998

CUBIC

CODE	NAME DENSITY	REF	TEMP(K)		11	12	44
MOLYBDENUM				C	4.4997	1.7287	1.2497
10981	10.2622	89	20.0	S	0.2825	-0.0784	0.8002
MOLYBDENUM				C	4.4962	1.7262	1.2488
10982	10.2622	89	40.0	S	0.2826	-0.0784	0.8008
MOLYBDENUM				C	4.4918	1.7238	1.2479
10983	10.2591	89	60.0	S	0.2828	-0.0784	0.8013
MOLYBDENUM				C	4.4900	1.7260	1.2468
10984	10.2560	89	80.0	S	0.2832	-0.0786	0.8021
MOLYBDENUM				C	4.4846	1.7232	1.2456
10985	10.2529	89	100.0	S	0.2835	-0.0787	0.8028
MOLYBDENUM				C	4.4713	1.7173	1.2446
10986	10.2499	89	120.0	S	0.2842	-0.0789	0.8035
MOLYBDENUM				C	4.4582	1.7144	1.2429
10987	10.2468	89	140.0	S	0.2852	-0.0792	0.8046
MOLYBDENUM				C	4.4482	1.7138	1.2402
10988	10.2437	89	160.0	S	0.2861	-0.0796	0.8063
MOLYBDENUM				C	4.4454	1.7130	1.2368
10989	10.2437	89	180.0	S	0.2863	-0.0796	0.8085
MOLYBDENUM				C	4.4451	1.7175	1.2333
10990	10.2407	89	200.0	S	0.2867	-0.0799	0.8108
MOLYBDENUM				C	4.4426	1.7236	1.2296
10991	10.2376	89	220.0	S	0.2874	-0.0803	0.8133
MOLYBDENUM				C	4.4344	1.7264	1.2257
10992	10.2345	89	240.0	S	0.2884	-0.0808	0.8159
MOLYBDENUM				C	4.4255	1.7281	1.2220
10993	10.2315	89	260.0	S	0.2894	-0.0813	0.8183
MOLYBDENUM				C	4.4161	1.7273	1.2188
10994	10.2284	89	280.0	S	0.2903	-0.0816	0.8205
MOLYBDENUM				C	4.4077	1.7243	1.2165
10995	10.2253	89	300.0	S	0.2909	-0.0818	0.8220
MOLYBDENUM				C	4.7300	1.5620	1.1090
10996	10.2620	75	73.0	S	0.2529	-0.0628	0.9017
MOLYBDENUM				C	4.6880	1.5690	1.1010
10997	10.2437	75	173.0	S	0.2563	-0.0643	0.9083
MOLYBDENUM				C	4.6370	1.5780	1.0920
10998	10.2284	75	273.0	S	0.2607	-0.0662	0.9158
MOLYBDENUM				C	4.5780	1.5810	1.0820
10999	10.2128	75	373.0	S	0.2655	-0.0682	0.9242
MOLYBDENUM				C	4.5220	1.5870	1.0720
11000	10.1946	75	473.0	S	0.2705	-0.0703	0.9328

NAME CODE	 DENSITY	 REF	 TEMP(K)		11	12	44
MOLYBDENUM				C	4.4620	1.5920	1.0620
11001	10.1793	75	573.0	S	0.2759	-0.0725	0.9416
MOLYBDENUM				C	4.4030	1.5980	1.0510
11002	10.1640	75	673.0	S	0.2815	-0.0750	0.9515
MOLYBDENUM				C	4.3450	1.6090	1.0400
11003	10.1459	75	773.0	S	0.2877	-0.0778	0.9615
MOLYBDENUM				C	4.2800	1.6110	1.0290
11004	10.1306	75	873.0	S	0.2942	-0.0805	0.9718
MOLYBDENUM				C	4.2150	1.6130	1.0180
11005	10.1155	75	973.0	S	0.3010	-0.0833	0.9823
MOLYBDENUM(93.0)-RHENIUM(7.0)				C	4.7780	1.7120	1.1670
11006	10.9722	69	83.0	S	0.2581	-0.0681	0.8569
MOLYBDENUM(93.0)-RHENIUM(7.0)				C	4.7780	1.7120	1.1670
11007	10.9722	69	93.0	S	0.2581	-0.0681	0.8569
MOLYBDENUM(93.0)-RHENIUM(7.0)				C	4.7740	1.7120	1.1660
11008	10.9722	69	103.0	S	0.2584	-0.0682	0.8576
MOLYBDENUM(93.0)-RHENIUM(7.0)				C	4.7700	1.7120	1.1650
11009	10.9689	69	113.0	S	0.2587	-0.0683	0.8584
MOLYBDENUM(93.0)-RHENIUM(7.0)				C	4.7650	1.7120	1.1650
11010	10.9689	69	123.0	S	0.2591	-0.0685	0.8584
MOLYBDENUM(93.0)-RHENIUM(7.0)				C	4.7600	1.7130	1.1640
11011	10.9689	69	133.0	S	0.2595	-0.0687	0.8591
MOLYBDENUM(93.0)-RHENIUM(7.0)				C	4.7480	1.7140	1.1620
11012	10.9656	69	153.0	S	0.2605	-0.0691	0.8606
MOLYBDENUM(93.0)-RHENIUM(7.0)				C	4.7370	1.7150	1.1600
11013	10.9623	69	173.0	S	0.2614	-0.0695	0.8621
MOLYBDENUM(93.0)-RHENIUM(7.0)				C	4.7270	1.7170	1.1580
11014	10.9590	69	193.0	S	0.2623	-0.0699	0.8636
MOLYBDENUM(93.0)-RHENIUM(7.0)				C	4.7140	1.7200	1.1560
11015	10.9557	69	213.0	S	0.2635	-0.0705	0.8651
MOLYBDENUM(93.0)-RHENIUM(7.0)				C	4.7030	1.7220	1.1540
11016	10.9524	69	233.0	S	0.2646	-0.0709	0.8666
MOLYBDENUM(93.0)-RHENIUM(7.0)				C	4.6930	1.7230	1.1520
11017	10.9458	69	253.0	S	0.2654	-0.0713	0.8681
MOLYBDENUM(93.0)-RHENIUM(7.0)				C	4.6810	1.7250	1.1500
11018	10.9428	69	273.0	S	0.2665	-0.0718	0.8696
MOLYBDENUM(93.0)-RHENIUM(7.0)				C	4.6680	1.7300	1.1480
11019	10.9428	69	293.0	S	0.2679	-0.0724	0.8711
MOLYBDENUM(93.0)-RHENIUM(7.0)				C	4.6650	1.7290	1.1480
11020	10.9362	69	298.0	S	0.2681	-0.0725	0.8711

CUBIC

NAME CODE	DENSITY	REF	TEMP(K)		11	12	44
MOLYBDENUM(93.0)-RHENIUM(7.0)				C	4.6560	1.7320	1.1460
11021	10.9329	69	313.0	S	0.2690	-0.0729	0.8726
MOLYBDENUM(93.0)-RHENIUM(7.0)				C	4.6450	1.7330	1.1440
11022	10.9296	69	333.0	S	0.2700	-0.0734	0.8741
MOLYBDENUM(93.0)-RHENIUM(7.0)				C	4.6310	1.7330	1.1420
11023	10.9263	69	353.0	S	0.2712	-0.0739	0.8757
MOLYBDENUM(93.0)-RHENIUM(7.0)				C	4.6190	1.7330	1.1390
11024	10.9230	69	373.0	S	0.2722	-0.0743	0.8780
MOLYBDENUM(83.4)-RHENIUM(16.6)				C	4.7580	1.8520	1.2570
11025	11.9670	69	83.0	S	0.2688	-0.0753	0.7955
MOLYBDENUM(83.4)-RHENIUM(16.6)				C	4.7550	1.8530	1.2570
11026	11.9670	69	93.0	S	0.2691	-0.0755	0.7955
MOLYBDENUM(83.4)-RHENIUM(16.6)				C	4.7510	1.8540	1.2560
11027	11.9670	69	103.0	S	0.2695	-0.0757	0.7962
MOLYBDENUM(83.4)-RHENIUM(16.6)				C	4.7470	1.8550	1.2560
11028	11.9634	69	113.0	S	0.2699	-0.0758	0.7962
MOLYBDENUM(83.4)-RHENIUM(16.6)				C	4.7420	1.8550	1.2550
11029	11.9634	69	123.0	S	0.2704	-0.0760	0.7968
MOLYBDENUM(83.4)-RHENIUM(16.6)				C	4.7390	1.8540	1.2540
11030	11.9634	69	133.0	S	0.2705	-0.0761	0.7974
MOLYBDENUM(83.4)-RHENIUM(16.6)				C	4.7300	1.8520	1.2530
11031	11.9598	69	153.0	S	0.2712	-0.0763	0.7981
MOLYBDENUM(83.4)-RHENIUM(16.6)				C	4.7210	1.8510	1.2510
11032	11.9562	69	173.0	S	0.2719	-0.0766	0.7994
MOLYBDENUM(83.4)-RHENIUM(16.6)				C	4.7110	1.8520	1.2490
11033	11.9526	69	193.0	S	0.2728	-0.0770	0.8006
MOLYBDENUM(83.4)-RHENIUM(16.6)				C	4.6990	1.8540	1.2470
11034	11.9490	69	213.0	S	0.2740	-0.0775	0.8019
MOLYBDENUM(83.4)-RHENIUM(16.6)				C	4.6870	1.8540	1.2440
11035	11.9454	69	233.0	S	0.2750	-0.0780	0.8039
MOLYBDENUM(83.4)-RHENIUM(16.6)				C	4.6760	1.8540	1.2420
11036	11.9382	69	253.0	S	0.2760	-0.0784	0.8052
MOLYBDENUM(83.4)-RHENIUM(16.6)				C	4.6640	1.8560	1.2400
11037	11.9349	69	273.0	S	0.2772	-0.0789	0.8065
MOLYBDENUM(83.4)-RHENIUM(16.6)				C	4.6540	1.8570	1.2370
11038	11.9313	69	293.0	S	0.2782	-0.0793	0.8084
MOLYBDENUM(83.4)-RHENIUM(16.6)				C	4.6500	1.8580	1.2370
11039	11.9277	69	298.0	S	0.2786	-0.0795	0.8084
MOLYBDENUM(83.4)-RHENIUM(16.6)				C	4.6260	1.8580	1.2350
11040	11.9241	69	313.0	S	0.2808	-0.0805	0.8097

CODE	NAME / DENSITY	REF	TEMP(K)		11	12	44
	MOLYBDENUM(83.4)-RHENIUM(16.6)			C	4.6140	1.8530	1.2330
11041	11.9205	69	333.0	S	0.2815	-0.0807	0.8110
	MOLYBDENUM(83.4)-RHENIUM(16.6)			C	4.6050	1.8540	1.2310
11042	11.9169	69	353.0	S	0.2824	-0.0811	0.8123
	MOLYBDENUM(83.4)-RHENIUM(16.6)			C	4.6000	1.8560	1.2280
11043	11.9133	69	373.0	S	0.2831	-0.0814	0.8143
	MOLYBDENUM(73.1)-RHENIUM(26.9)			C	4.7210	2.0210	1.3500
11044	13.0465	69	83.0	S	0.2850	-0.0854	0.7407
	MOLYBDENUM(73.1)-RHENIUM(26.9)			C	4.7180	2.0190	1.3490
11045	13.0465	69	93.0	S	0.2851	-0.0854	0.7413
	MOLYBDENUM(73.1)-RHENIUM(26.9)			C	4.7170	2.0160	1.3490
11046	13.0465	69	103.0	S	0.2849	-0.0853	0.7413
	MOLYBDENUM(73.1)-RHENIUM(26.9)			C	4.7120	2.0130	1.3470
11047	13.0426	69	113.0	S	0.2852	-0.0854	0.7424
	MOLYBDENUM(73.1)-RHENIUM(26.9)			C	4.7070	2.0100	1.3460
11048	13.0426	69	123.0	S	0.2854	-0.0854	0.7429
	MOLYBDENUM(73.1)-RHENIUM(26.9)			C	4.7010	2.0070	1.3450
11049	13.0426	69	133.0	S	0.2857	-0.0855	0.7435
	MOLYBDENUM(73.1)-RHENIUM(26.9)			C	4.6900	2.0010	1.3430
11050	13.0387	69	153.0	S	0.2863	-0.0856	0.7446
	MOLYBDENUM(73.1)-RHENIUM(26.9)			C	4.6780	1.9940	1.3410
11051	13.0348	69	173.0	S	0.2869	-0.0857	0.7457
	MOLYBDENUM(73.1)-RHENIUM(26.9)			C	4.6660	1.9890	1.3380
11052	13.0309	69	193.0	S	0.2876	-0.0860	0.7474
	MOLYBDENUM(73.1)-RHENIUM(26.9)			C	4.6540	1.9820	1.3360
11053	13.0270	69	213.0	S	0.2882	-0.0861	0.7485
	MOLYBDENUM(73.1)-RHENIUM(26.9)			C	4.6430	1.9770	1.3330
11054	13.0231	69	233.0	S	0.2888	-0.0863	0.7502
	MOLYBDENUM(73.1)-RHENIUM(26.9)			C	4.6320	1.9720	1.3300
11055	13.0153	69	253.0	S	0.2895	-0.0864	0.7519
	MOLYBDENUM(73.1)-RHENIUM(26.9)			C	4.6200	1.9650	1.3270
11056	13.0114	69	273.0	S	0.2901	-0.0866	0.7536
	MOLYBDENUM(73.1)-RHENIUM(26.9)			C	4.6100	1.9610	1.3240
11057	13.0072	69	293.0	S	0.2907	-0.0868	0.7553
	MOLYBDENUM(73.1)-RHENIUM(26.9)			C	4.6070	1.9590	1.3230
11058	13.0033	69	298.0	S	0.2909	-0.0868	0.7559
	MOLYBDENUM(73.1)-RHENIUM(26.9)			C	4.5980	1.9560	1.3210
11059	12.9995	69	313.0	S	0.2915	-0.0870	0.7570
	MOLYBDENUM(73.1)-RHENIUM(26.9)			C	4.5840	1.9490	1.3170
11060	12.9956	69	333.0	S	0.2923	-0.0872	0.7593

CUBIC

CODE	NAME DENSITY	REF	TEMP(K)		11	12	44
MOLYBDENUM(73.1)-RHENIUM(26.9)				C	4.5700	1.9440	1.3150
11061	12.9918	69	353.0	S	0.2933	-0.0875	0.7605
MOLYBDENUM(73.1)-RHENIUM(26.9)				C	4.5570	1.9370	1.3120
11062	12.9879	69	373.0	S	0.2940	-0.0877	0.7622
NICKEL				C	2.4400	1.5800	1.0200
11063	8.9000	129	ROOM	S	0.8347	-0.3281	0.9804
NICKEL				C	2.5200	1.5100	1.0400
11064	8.9000	129	ROOM	S	0.7202	-0.2699	0.9615
NICKEL				C	2.5120	1.5710	1.2130
11065	8.9000	129	ROOM	S	0.7674	-0.2953	0.8244
NICKEL				C	2.4360	1.4940	1.1960
11066	8.9000	129	ROOM	S	0.7692	-0.2924	0.8361
NICKEL				C	2.5260	1.5610	1.2260
11067	8.9000	129	ROOM	S	0.7499	-0.2864	0.8157
NICKEL				C	2.5280	1.5200	1.2380
11068	8.9000	129	ROOM	S	0.7212	-0.2708	0.8078
NICKEL				C	2.4650	1.4730	1.2470
11069	8.9000	129	ROOM	S	0.7336	-0.2744	0.8019
NICKEL				C	2.5000	1.6000	1.1850
11070	8.9000	177	ROOM	S	0.7992	-0.3119	0.8439
NICKEL				C	2.6120	1.5080	1.3170
11071	8.9600	17	0.0	S	0.6631	-0.2427	0.7593
NICKEL				C	2.6120	1.5080	1.3170
11072	8.9600	17	20.0	S	0.6631	-0.2427	0.7593
NICKEL				C	2.6100	1.5080	1.3160
11073	8.9600	17	40.0	S	0.6642	-0.2432	0.7599
NICKEL				C	2.6050	1.5070	1.3140
11074	8.9610	17	60.0	S	0.6665	-0.2443	0.7610
NICKEL				C	2.6010	1.5070	1.3090
11075	8.9600	17	80.0	S	0.6687	-0.2453	0.7639
NICKEL				C	2.5950	1.5070	1.3040
11076	8.9580	17	100.0	S	0.6722	-0.2469	0.7669
NICKEL				C	2.5870	1.5050	1.2980
11077	8.9550	17	120.0	S	0.6757	-0.2485	0.7704
NICKEL				C	2.5780	1.5040	1.2910
11078	8.9510	17	140.0	S	0.6804	-0.2507	0.7746
NICKEL				C	2.5690	1.5030	1.2840
11079	8.9470	17	160.0	S	0.6852	-0.2529	0.7788
NICKEL				C	2.5620	1.5020	1.2770
11080	8.9420	17	180.0	S	0.6888	-0.2546	0.7831

CUBIC

CODE / NAME	DENSITY	REF	TEMP(K)		11	12	44
NICKEL				C	2.5530	1.5010	1.2700
11081	8.9360	17	200.0	S	0.6937	-0.2569	0.7874
NICKEL				C	2.5450	1.5010	1.2630
11082	8.9310	17	220.0	S	0.6987	-0.2592	0.7918
NICKEL				C	2.5350	1.5010	1.2560
11083	8.9260	17	240.0	S	0.7049	-0.2622	0.7962
NICKEL				C	2.5260	1.5000	1.2490
11084	8.9200	17	260.0	S	0.7101	-0.2646	0.8006
NICKEL				C	2.5160	1.5000	1.2420
11085	8.9120	17	280.0	S	0.7166	-0.2677	0.8052
NICKEL				C	2.5080	1.5000	1.2350
11086	8.9070	17	300.0	S	0.7219	-0.2702	0.8097
NICKEL				C	2.4970	1.4990	1.2280
11087	8.8990	17	320.0	S	0.7287	-0.2733	0.8143
NICKEL				C	2.4880	1.5000	1.2200
11088	8.8910	17	340.0	S	0.7355	-0.2766	0.8197
NICKEL				C	2.4770	1.4990	1.2130
11089	8.8830	17	360.0	S	0.7425	-0.2799	0.8244
NICKEL				C	2.4660	1.4980	1.2050
11090	8.8750	17	380.0	S	0.7497	-0.2833	0.8299
NICKEL				C	2.4540	1.4960	1.1980
11091	8.8670	17	400.0	S	0.7571	-0.2867	0.8347
NICKEL				C	2.4430	1.4950	1.1900
11092	8.8590	17	420.0	S	0.7646	-0.2903	0.8403
NICKEL				C	2.4310	1.4930	1.1830
11093	8.8510	17	440.0	S	0.7723	-0.2938	0.8453
NICKEL				C	2.4190	1.4930	1.1750
11094	8.8430	17	460.0	S	0.7816	-0.2983	0.8511
NICKEL				C	2.4070	1.4910	1.1670
11095	8.8350	17	480.0	S	0.7897	-0.3020	0.8569
NICKEL				C	2.3960	1.4900	1.1590
11096	8.8270	17	500.0	S	0.7978	-0.3059	0.8628
NICKEL				C	2.3810	1.4870	1.1520
11097	8.8190	17	520.0	S	0.8080	-0.3106	0.8681
NICKEL				C	2.3690	1.4870	1.1430
11098	8.8110	17	540.0	S	0.8182	-0.3155	0.8749
NICKEL				C	2.3520	1.4840	1.1350
11099	8.8030	17	560.0	S	0.8307	-0.3214	0.8811
NICKEL				C	2.3370	1.4810	1.1260
11100	8.7950	17	580.0	S	0.8417	-0.3265	0.8881

CUBIC

CODE	NAME DENSITY	REF	TEMP(K)		11	12	44
NICKEL				C	2.3220	1.4780	1.1180
11101	8.7870	17	600.0	S	0.8530	-0.3318	0.8945
NICKEL				C	2.3090	1.4790	1.1090
11102	8.7790	17	620.0	S	0.8665	-0.3383	0.9017
NICKEL				C	2.2950	1.4790	1.1000
11103	8.7710	17	640.0	S	0.8804	-0.3450	0.9091
NICKEL				C	2.2820	1.4780	1.0930
11104	8.7640	17	660.0	S	0.8928	-0.3510	0.9149
NICKEL				C	2.2700	1.4760	1.0860
11105	8.7560	17	680.0	S	0.9035	-0.3560	0.9208
NICKEL				C	2.2610	1.4730	1.0790
11106	8.7480	17	700.0	S	0.9100	-0.3590	0.9268
NICKEL				C	2.2500	1.4700	1.0720
11107	8.7410	17	720.0	S	0.9189	-0.3631	0.9328
NICKEL				C	2.2410	1.4670	1.0650
11108	8.7330	17	740.0	S	0.9257	-0.3663	0.9390
NICKEL				C	2.2320	1.4640	1.0580
11109	8.7220	17	760.0	S	0.9327	-0.3694	0.9452
NICKEL				C	2.4810	1.5490	1.2420
11110	8.9100	85	298.0	S	0.7751	-0.2979	0.8052
NICKEL				C	2.4700	1.4400	1.2400
11111	8.9100	264	298.0	S	0.7096	-0.2613	0.8065
NICKEL(90.4)-ALUMINUM(9.6)				C	2.5200	1.5300	1.3200
11112	8.3800	264	298.0	S	0.7331	-0.2770	0.7576
NICKEL(49.4)-ALUMINUM(50.6)				C	2.1155	1.4323	1.1211
11113	5.9050	273	298.0	S	1.0428	-0.4210	0.8920
NICKEL(92.7)-COPPER(7.3)				C	2.4130	1.5110	1.2010
11114	8.9183	85	298.0	S	0.8004	-0.3082	0.8326
NICKEL(82.2)-COPPER(17.8)				C	2.3150	1.4750	1.1460
11115	8.9301	85	298.0	S	0.8570	-0.3335	0.8726
NICKEL(80.0)-COPPER(20.0)				C	2.2980	1.4940	1.1370
11116	8.9330	85	298.0	S	0.8922	-0.3515	0.8795
NICKEL(77.2)-COPPER(22.8)				C	2.2810	1.4730	1.1130
11117	8.9360	85	298.0	S	0.8889	-0.3488	0.8985
NICKEL(65.5)-COPPER(34.5)				C	2.1670	1.4620	1.0620
11118	8.9495	85	298.0	S	1.0111	-0.4073	0.9416
NICKEL(53.8)-COPPER(46.2)				C	2.0860	1.4280	1.0090
11119	8.9620	85	298.0	S	1.0806	-0.4391	0.9911
NICKEL(31.1)-COPPER(68.9)				C	1.8910	1.3190	0.8960
11120	8.9586	85	298.0	S	1.2391	-0.5092	1.1161

CODE	NAME DENSITY	REF	TEMP(K)		11	12	44
NICKEL(73.8)-IRON(26.2)				C	2.3040	1.4440	1.1920
11121	8.5390	82	300.0	S	0.8394	-0.3234	0.8389
NIOBIUM				C	2.4650	1.3450	0.2873
11122	8.5780	38	300.0	S	0.6599	-0.2330	3.4807
NIOBIUM (130 PPM TA)				C	2.3483	1.2068	0.2821
11123	8.5780	272	298.0	S	0.6540	-0.2220	3.5450
NIOBIUM (1175 PPM TA)				C	2.3542	1.2152	0.2825
11124	8.5780	272	298.0	S	0.6550	-0.2230	3.5400
NIOBIUM				C	2.5270	1.3320	0.3097
11125	8.5938	47	4.2	S	0.6221	-0.2147	3.2289
NIOBIUM				C	2.5290	1.3420	0.3096
11126	8.5938	47	20.0	S	0.6256	-0.2169	3.2300
NIOBIUM				C	2.5290	1.3450	0.3089
11127	8.5912	47	40.0	S	0.6269	-0.2177	3.2373
NIOBIUM				C	2.5210	1.3440	0.3075
11128	8.5912	47	60.0	S	0.6304	-0.2192	3.2520
NIOBIUM				C	2.5100	1.3520	0.3055
11129	8.5886	47	80.0	S	0.6396	-0.2239	3.2733
NIOBIUM				C	2.5030	1.3590	0.3033
11130	8.5861	47	100.0	S	0.6466	-0.2275	3.2971
NIOBIUM				C	2.4970	1.3530	0.3008
11131	8.5835	47	120.0	S	0.6468	-0.2273	3.3245
NIOBIUM				C	2.4920	1.3460	0.2989
11132	8.5809	47	140.0	S	0.6460	-0.2266	3.3456
NIOBIUM				C	2.4870	1.3380	0.2976
11133	8.5784	47	160.0	S	0.6448	-0.2255	3.3602
NIOBIUM				C	2.4810	1.3300	0.2964
11134	8.5732	47	180.0	S	0.6440	-0.2248	3.3738
NIOBIUM				C	2.4770	1.3250	0.2952
11135	8.5707	47	200.0	S	0.6437	-0.2243	3.3875
NIOBIUM				C	2.4720	1.3270	0.2944
11136	8.5681	47	220.0	S	0.6473	-0.2261	3.3967
NIOBIUM				C	2.4670	1.3370	0.2938
11137	8.5655	47	240.0	S	0.6548	-0.2301	3.4037
NIOBIUM				C	2.4620	1.3510	0.2934
11138	8.5630	47	260.0	S	0.6646	-0.2355	3.4083
NIOBIUM				C	2.4590	1.3690	0.2931
11139	8.5604	47	280.0	S	0.6758	-0.2417	3.4118
NIOBIUM				C	2.4560	1.3870	0.2930
11140	8.5553	47	300.0	S	0.6874	-0.2481	3.4130

CUBIC

CODE	NAME DENSITY	REF	TEMP(K)		11	12	44
NIOBIUM				C	2.4032	1.2092	0.2958
11141	8.5821	22	103.0	S	0.6275	-0.2100	3.3807
NIOBIUM				C	2.4063	1.2160	0.2946
11142	8.5815	22	113.0	S	0.6290	-0.2111	3.3944
NIOBIUM				C	2.4101	1.2230	0.2933
11143	8.5809	22	123.0	S	0.6302	-0.2122	3.4095
NIOBIUM				C	2.4156	1.2316	0.2922
11144	8.5802	22	133.0	S	0.6314	-0.2132	3.4223
NIOBIUM				C	2.4083	1.2274	0.2912
11145	8.5796	22	143.0	S	0.6331	-0.2137	3.4341
NIOBIUM				C	2.4005	1.2202	0.2901
11146	8.5790	22	153.0	S	0.6337	-0.2136	3.4471
NIOBIUM				C	2.3966	1.2182	0.2891
11147	8.5783	22	163.0	S	0.6347	-0.2139	3.4590
NIOBIUM				C	2.3926	1.2165	0.2881
11148	8.5777	22	173.0	S	0.6359	-0.2143	3.4710
NIOBIUM				C	2.3891	1.2156	0.2874
11149	8.5770	22	183.0	S	0.6373	-0.2149	3.4795
NIOBIUM				C	2.3880	1.2194	0.2860
11150	8.5758	22	203.0	S	0.6395	-0.2162	3.4965
NIOBIUM				C	2.3915	1.2253	0.2854
11151	8.5751	22	213.0	S	0.6405	-0.2170	3.5039
NIOBIUM				C	2.3897	1.2259	0.2848
11152	8.5745	22	223.0	S	0.6417	-0.2176	3.5112
NIOBIUM				C	2.3801	1.2178	0.2843
11153	8.5738	22	233.0	S	0.6428	-0.2176	3.5174
NIOBIUM				C	2.3852	1.2255	0.2838
11154	8.5732	22	243.0	S	0.6438	-0.2185	3.5236
NIOBIUM				C	2.3809	1.2254	0.2830
11155	8.5719	22	263.0	S	0.6459	-0.2195	3.5336
NIOBIUM				C	2.3818	1.2284	0.2827
11156	8.5713	22	273.0	S	0.6469	-0.2201	3.5373
NIOBIUM				C	2.3824	1.2315	0.2825
11157	8.5706	22	283.0	S	0.6481	-0.2208	3.5398
NIOBIUM				C	2.4019	1.2558	0.2822
11158	8.5700	22	293.0	S	0.6495	-0.2230	3.5436
NIOBIUM				C	2.3813	1.2347	0.2822
11159	8.5694	22	303.0	S	0.6501	-0.2220	3.5436
NIOBIUM				C	2.3783	1.2337	0.2822
11160	8.5687	22	313.0	S	0.6512	-0.2224	3.5436

CUBIC

CODE	NAME DENSITY	REF	TEMP(K)		11	12	44
NIOBIUM				C	2.3743	1.2335	0.2822
11161	8.5674	22	333.0	S	0.6532	-0.2233	3.5436
NIOBIUM				C	2.3683	1.2292	0.2823
11162	8.5668	22	343.0	S	0.6543	-0.2236	3.5423
NIOBIUM				C	2.3622	1.2250	0.2824
11163	8.5662	22	353.0	S	0.6555	-0.2238	3.5411
NIOBIUM				C	2.3553	1.2199	0.2826
11164	8.5655	22	363.0	S	0.6567	-0.2241	3.5386
NIOBIUM				C	2.3474	1.2139	0.2828
11165	8.5649	22	373.0	S	0.6580	-0.2243	3.5361
NIOBIUM				C	2.3365	1.2069	0.2832
11166	8.5636	22	393.0	S	0.6604	-0.2249	3.5311
NIOBIUM				C	2.3315	1.2037	0.2835
11167	8.5630	22	403.0	S	0.6615	-0.2252	3.5273
NIOBIUM				C	2.3232	1.1972	0.2838
11168	8.5623	22	413.0	S	0.6627	-0.2254	3.5236
NIOBIUM				C	2.3193	1.1906	0.2837
11169	8.5617	22	423.0	S	0.6616	-0.2244	3.5248
NIOBIUM				C	2.3142	1.1919	0.2844
11170	8.5610	22	433.0	S	0.6650	-0.2261	3.5162
NIOBIUM				C	2.3115	1.1911	0.2847
11171	8.5604	22	443.0	S	0.6660	-0.2265	3.5125
NIOBIUM				C	2.3099	1.1914	0.2851
11172	8.5598	22	453.0	S	0.6671	-0.2270	3.5075
NIOBIUM				C	2.3081	1.1916	0.2855
11173	8.5591	22	463.0	S	0.6682	-0.2275	3.5026
NIOBIUM				C	2.3063	1.1915	0.2860
11174	8.5585	22	473.0	S	0.6691	-0.2279	3.4965
NIOBIUM				C	2.3058	1.1929	0.2864
11175	8.5579	22	483.0	S	0.6701	-0.2285	3.4916
NIOBIUM				C	2.3053	1.1962	0.2874
11176	8.5566	22	503.0	S	0.6720	-0.2296	3.4795
NIOBIUM				C	2.3053	1.1980	0.2879
11177	8.5559	22	513.0	S	0.6730	-0.2301	3.4734
NIOBIUM				C	2.3027	1.1973	0.2884
11178	8.5553	22	523.0	S	0.6741	-0.2306	3.4674
NIOBIUM				C	2.3049	1.2015	0.2889
11179	8.5547	22	533.0	S	0.6750	-0.2313	3.4614
NIOBIUM				C	2.3056	1.2041	0.2895
11180	8.5540	22	543.0	S	0.6759	-0.2319	3.4542

CODE	NAME DENSITY	REF	TEMP(K)	CUBIC	11	12	44
NIOBIUM				C	2.3072	1.2078	0.2901
11181	8.5534	22	553.0	S	0.6770	-0.2326	3.4471
NIOBIUM				C	2.3083	1.2110	0.2901
11182	8.5527	22	563.0	S	0.6780	-0.2333	3.4471
NIOBIUM				C	2.3088	1.2133	0.2913
11183	8.5521	22	573.0	S	0.6789	-0.2339	3.4329
NIOBIUM				C	2.3099	1.2165	0.2919
11184	8.5515	22	583.0	S	0.6800	-0.2346	3.4258
NIOBIUM				C	2.3099	1.2185	0.2925
11185	8.5508	22	593.0	S	0.6811	-0.2352	3.4188
NIOBIUM				C	2.3098	1.2202	0.2931
11186	8.5502	22	603.0	S	0.6820	-0.2358	3.4118
NIOBIUM				C	2.3087	1.2209	0.2938
11187	8.5495	22	613.0	S	0.6830	-0.2363	3.4037
NIOBIUM				C	2.3099	1.2242	0.2944
11188	8.5489	22	623.0	S	0.6841	-0.2370	3.3967
NIOBIUM				C	2.3102	1.2264	0.2950
11189	8.5483	22	633.0	S	0.6851	-0.2376	3.3898
NIOBIUM				C	2.3091	1.2270	0.2957
11190	8.5476	22	643.0	S	0.6861	-0.2381	3.3818
NIOBIUM				C	2.3104	1.2303	0.2964
11191	8.5470	22	653.0	S	0.6871	-0.2387	3.3738
NIOBIUM				C	2.3105	1.2320	0.2970
11192	8.5463	22	663.0	S	0.6880	-0.2393	3.3670
NIOBIUM				C	2.3124	1.2359	0.2977
11193	8.5457	22	673.0	S	0.6890	-0.2400	3.3591
NIOBIUM				C	2.3101	1.2352	0.2984
11194	8.5451	22	683.0	S	0.6899	-0.2404	3.3512
NIOBIUM				C	2.3108	1.2377	0.2991
11195	8.5444	22	693.0	S	0.6909	-0.2410	3.3434
NIOBIUM				C	2.3102	1.2390	0.2998
11196	8.5438	22	703.0	S	0.6920	-0.2416	3.3356
NIOBIUM				C	2.3129	1.2436	0.3004
11197	8.5431	22	713.0	S	0.6929	-0.2423	3.3289
NIOBIUM				C	2.3086	1.2408	0.3012
11198	8.5425	22	723.0	S	0.6939	-0.2426	3.3201
NIOBIUM				C	2.3090	1.2428	0.3019
11199	8.5419	22	733.0	S	0.6948	-0.2431	3.3124
NIOBIUM				C	2.3071	1.2426	0.3026
11200	8.5412	22	743.0	S	0.6958	-0.2436	3.3047

CUBIC

CODE	NAME DENSITY	REF	TEMP(K)		11	12	44
NIOBIUM				C	2.3095	1.2467	0.3033
11201	8.5406	22	753.0	S	0.6967	-0.2442	3.2971
NIOBIUM				C	2.3110	1.2499	0.3040
11202	8.5400	22	763.0	S	0.6976	-0.2449	3.2895
NIOBIUM				C	2.3093	1.2498	0.3048
11203	8.5393	22	773.0	S	0.6985	-0.2453	3.2808
NIOBIUM(75.5)-TIN(24.5)				C	2.5380	1.1240	0.3957
11204	8.8340	141	300.0	S	0.5411	-0.1661	2.5272
NIOBIUM(75.5)-TIN(24.5)				C	1.6790	1.6370	0.2698
11205	8.8600	142	35.0	S	15.9404	-7.8692	3.7064
NIOBIUM(75.5)-TIN(24.5)				C	2.5380	1.1240	0.3960
11206	8.8600	142	300.0	S	0.5411	-0.1661	2.5253
PALLADIUM				C	2.3412	1.7614	0.7117
11207	12.0690	209	0.0	S	1.2067	-0.5181	1.4051
PALLADIUM				C	2.3409	1.7627	0.7112
11208	12.0690	209	20.0	S	1.2098	-0.5197	1.4061
PALLADIUM				C	2.3386	1.7666	0.7084
11209	12.0680	209	40.0	S	1.2223	-0.5260	1.4116
PALLADIUM				C	2.3345	1.7695	0.7050
11210	12.0670	209	60.0	S	1.2367	-0.5332	1.4184
PALLADIUM				C	2.3279	1.7695	0.7023
11211	12.0660	209	80.0	S	1.2507	-0.5401	1.4239
PALLADIUM				C	2.3192	1.7672	0.7008
11212	12.0650	209	100.0	S	1.2647	-0.5469	1.4269
PALLADIUM				C	2.3075	1.7615	0.7005
11213	12.0623	209	120.0	S	1.2782	-0.5533	1.4276
PALLADIUM				C	2.2960	1.7558	0.7011
11214	12.0596	209	140.0	S	1.2915	-0.5597	1.4263
PALLADIUM				C	2.2855	1.7501	0.7022
11215	12.0569	209	160.0	S	1.3028	-0.5650	1.4241
PALLADIUM				C	2.2777	1.7467	0.7038
11216	12.0542	209	180.0	S	1.3133	-0.5700	1.4209
PALLADIUM				C	2.2699	1.7429	0.7056
11217	12.0515	209	200.0	S	1.3229	-0.5746	1.4172
PALLADIUM				C	2.2651	1.7415	0.7077
11218	12.0488	209	220.0	S	1.3312	-0.5786	1.4130
PALLADIUM				C	2.2621	1.7419	0.7100
11219	12.0461	209	240.0	S	1.3396	-0.5828	1.4085
PALLADIUM				C	2.2620	1.7450	0.7125
11220	12.0434	209	260.0	S	1.3474	-0.5868	1.4035

CODE	NAME DENSITY	REF	TEMP(K)		11	12	44
PALLADIUM				C	2.2661	1.7523	0.7148
11221	12.0407	209	280.0	S	1.3553	-0.5910	1.3990
PALLADIUM				C	2.2710	1.7604	0.7173
11222	12.0380	209	300.0	S	1.3632	-0.5953	1.3941
PLATINUM				C	3.5800	2.5360	0.7740
11223	21.6200	158	0.0	S	0.6771	-0.2808	1.2920
PLATINUM				C	3.4670	2.5070	0.7650
11224	21.5000	158	300.0	S	0.7337	-0.3079	1.3072
POTASSIUM				C	0.0414	0.0331	0.0263
11225	0.8300	128	ROOM	S	83.3000	-37.0000	38.0000
POTASSIUM				C	0.0457	0.0374	0.0263
11226	0.8600	138	83.0	S	83.0876	-37.3944	38.0228
POTASSIUM				C	0.0416	0.0341	0.0286
11227	0.9030	172	4.2	S	91.9247	-41.4086	34.9650
POTASSIUM				C	0.0405	0.0336	0.0265
11228	0.8960	172	78.0	S	99.7134	-45.2142	37.7358
POTASSIUM				C	0.0415	0.0346	0.0258
11229	0.8960	172	78.0	S	99.6296	-45.2981	38.7597
POTASSIUM				C	0.0406	0.0340	0.0244
11230	0.8910	172	120.0	S	104.0795	-47.4357	40.9836
POTASSIUM				C	0.0398	0.0334	0.0229
11231	0.8830	172	160.0	S	107.2936	-48.9564	43.6681
POTASSIUM				C	0.0389	0.0326	0.0217
11232	0.8760	172	195.0	S	109.0222	-49.7081	46.0829
POTASSIUM				C	0.0370	0.0314	0.0188
11233	0.8510	239	295.0	S	122.3876	-56.1838	53.1915
POTASSIUM BROMIDE				C	0.3450	0.0540	0.0508
11234	2.7500	129	ROOM	S	3.0268	-0.4096	19.6850
POTASSIUM BROMIDE				C	0.3460	0.0580	0.0505
11235	2.7500	129	ROOM	S	3.0363	-0.4359	19.8020
POTASSIUM BROMIDE				C	0.3800	0.0600	0.0640
11236	2.7500	129	ROOM	S	2.7500	-0.3750	15.6250
POTASSIUM BROMIDE				C	0.3490	0.0580	0.0510
11237	2.7505	4	ROOM	S	3.0078	-0.4286	19.6078
POTASSIUM BROMIDE				C	0.3420	0.0560	0.0500
11238	2.7400	4	ROOM	S	3.0652	-0.4313	20.0000
POTASSIUM BROMIDE				C	0.3421	0.0436	0.0513
11239	2.7500	194	293.0	S	3.0100	-0.3400	19.5000
POTASSIUM BROMIDE				C	0.3224	0.0543	0.0503
11240	2.7340	194	373.0	S	3.2600	-0.4700	19.9000

CODE	NAME DENSITY	REF	TEMP(K)		11	12	44
POTASSIUM	BROMIDE			C	0.3109	0.0815	0.0485
11241	2.6900	194	473.0	S	3.6100	-0.7500	20.6000
POTASSIUM	BROMIDE			C	0.2941	0.0945	0.0469
11242	2.6510	194	573.0	S	4.0300	-0.9800	21.3000
POTASSIUM	BROMIDE			C	0.2602	0.0845	0.0457
11243	2.6210	194	673.0	S	4.5700	-1.1200	21.9000
POTASSIUM	BROMIDE			C	0.2286	0.0802	0.0442
11244	2.5820	194	773.0	S	5.3500	-1.3900	22.6000
POTASSIUM	BROMIDE			C	0.1929	0.0686	0.0427
11245	2.5430	194	873.0	S	6.3700	-1.6700	23.4000
POTASSIUM	BROMIDE			C	0.1535	0.0452	0.0415
11246	2.5030	194	973.0	S	7.5200	-1.7100	24.1000
POTASSIUM	BROMIDE			C	0.1511	0.0536	0.0410
11247	2.4650	194	1003.0	S	8.1300	-2.1300	24.4000
POTASSIUM	BROMIDE			C	0.3500	0.0620	0.0506
11248	2.7600	181	293.0	S	3.0180	-0.4542	19.7628
POTASSIUM	BROMIDE			C	0.3468	0.0580	0.0507
11249	2.7440	234	298.0	S	3.0287	-0.4339	19.7239
POTASSIUM	BROMIDE			C	0.3419	0.0520	0.0508
11250	2.7533	213	300.0	S	3.0472	-0.4023	19.6850
POTASSIUM	CHLORIDE			C	0.3669	0.0193	0.0641
11251	1.9840	128	ROOM	S	2.7400	-0.1370	15.6000
POTASSIUM	CHLORIDE			C	0.3694	0.0812	0.0787
11252	1.9840	128	ROOM	S	2.9400	-0.5300	12.7000
POTASSIUM	CHLORIDE			C	0.3980	0.0620	0.0625
11253	1.9880	129	ROOM	S	2.6227	-0.3535	16.0000
POTASSIUM	CHLORIDE			C	0.4300	0.0750	0.0790
11254	1.9840	129	ROOM	S	2.4526	-0.3643	12.6582
POTASSIUM	CHLORIDE			C	0.3990	0.0680	0.0630
11255	1.9900	4	ROOM	S	2.6372	-0.3840	15.8730
POTASSIUM	CHLORIDE			C	0.4807	0.0580	0.0664
11256	2.0253	80	80.0	S	2.1360	-0.2300	15.0700
POTASSIUM	CHLORIDE			C	0.4785	0.0601	0.0663
11257	2.0240	80	90.0	S	2.1500	-0.2400	15.0900
POTASSIUM	CHLORIDE			C	0.4653	0.0618	0.0659
11258	2.0181	80	130.0	S	2.2180	-0.2600	15.1800
POTASSIUM	CHLORIDE			C	0.4623	0.0635	0.0657
11259	2.0166	80	140.0	S	2.2370	-0.2700	15.2100
POTASSIUM	CHLORIDE			C	0.4580	0.0622	0.0656
11260	2.0150	80	150.0	S	2.2570	-0.2700	15.2400

CUBIC

CODE	NAME DENSITY	REF	TEMP(K)		11	12	44
POTASSIUM CHLORIDE				C	0.4534	0.0610	0.0655
11261	2.0133	80	160.0	S	2.2780	-0.2700	15.2700
POTASSIUM CHLORIDE				C	0.4502	0.0624	0.0653
11262	2.0116	80	170.0	S	2.2990	-0.2800	15.3100
POTASSIUM CHLORIDE				C	0.4458	0.0612	0.0651
11263	2.0099	80	180.0	S	2.3200	-0.2800	15.3500
POTASSIUM CHLORIDE				C	0.4413	0.0599	0.0650
11264	2.0082	80	190.0	S	2.3420	-0.2800	15.3900
POTASSIUM CHLORIDE				C	0.4380	0.0612	0.0648
11265	2.0063	80	200.0	S	2.3640	-0.2900	15.4300
POTASSIUM CHLORIDE				C	0.4337	0.0600	0.0646
11266	2.0045	80	210.0	S	2.3860	-0.2900	15.4700
POTASSIUM CHLORIDE				C	0.4304	0.0612	0.0645
11267	2.0026	80	220.0	S	2.4090	-0.3000	15.5100
POTASSIUM CHLORIDE				C	0.4262	0.0600	0.0643
11268	2.0007	80	230.0	S	2.4310	-0.3000	15.5500
POTASSIUM CHLORIDE				C	0.4229	0.0612	0.0641
11269	1.9987	80	240.0	S	2.4540	-0.3100	15.6000
POTASSIUM CHLORIDE				C	0.4198	0.0623	0.0639
11270	1.9967	80	250.0	S	2.4770	-0.3200	15.6400
POTASSIUM CHLORIDE				C	0.4156	0.0610	0.0637
11271	1.9947	80	260.0	S	2.5000	-0.3200	15.6900
POTASSIUM CHLORIDE				C	0.4126	0.0621	0.0636
11272	1.9926	80	270.0	S	2.5230	-0.3300	15.7300
POTASSIUM CHLORIDE				C	0.4085	0.0608	0.0634
11273	1.9905	80	280.0	S	2.5460	-0.3300	15.7800
POTASSIUM CHLORIDE (P= 0KB)				C	0.4095	0.0706	0.0630
11274	1.9860	149	298.0	S	2.5724	-0.3783	15.8730
POTASSIUM CHLORIDE (P= 2KB)				C	0.4335	0.0744	0.0621
11275	2.0082	149	298.0	S	2.4289	-0.3558	16.1031
POTASSIUM CHLORIDE (P= 4KB)				C	0.4584	0.0760	0.0610
11276	2.0298	149	298.0	S	2.2895	-0.3256	16.3934
POTASSIUM CHLORIDE (P= 6KB)				C	0.4813	0.0783	0.0603
11277	2.0505	149	298.0	S	2.1768	-0.3046	16.5837
POTASSIUM CHLORIDE (P= 8KB)				C	0.5000	0.0757	0.0594
11278	2.0705	149	298.0	S	2.0829	-0.2739	16.8350
POTASSIUM CHLORIDE (P=10KB)				C	0.5316	0.0815	0.0583
11279	2.0896	149	298.0	S	1.9610	-0.2607	17.1526
POTASSIUM CHLORIDE				C	0.4832	0.0540	0.0663
11280	2.0290	196	4.2	S	2.1171	-0.2128	15.0830

CUBIC

NAME CODE	DENSITY	REF	TEMP(K)		11	12	44
POTASSIUM CHLORIDE				C	0.4811	0.0540	0.0661
11281	2.0290	196	20.0	S	2.1267	-0.2146	15.1286
POTASSIUM CHLORIDE				C	0.4780	0.0550	0.0660
11282	2.0290	196	40.0	S	2.1429	-0.2211	15.1515
POTASSIUM CHLORIDE				C	0.4743	0.0550	0.0658
11283	2.0275	196	60.0	S	2.1604	-0.2245	15.1976
POTASSIUM CHLORIDE				C	0.4697	0.0550	0.0657
11284	2.0253	196	80.0	S	2.1826	-0.2288	15.2207
POTASSIUM CHLORIDE				C	0.4644	0.0570	0.0655
11285	2.0226	196	100.0	S	2.2127	-0.2419	15.2672
POTASSIUM CHLORIDE				C	0.4582	0.0580	0.0652
11286	2.0196	196	120.0	S	2.2464	-0.2524	15.3374
POTASSIUM CHLORIDE				C	0.4510	0.0590	0.0650
11287	2.0166	196	140.0	S	2.2865	-0.2645	15.3846
POTASSIUM CHLORIDE				C	0.4453	0.0590	0.0648
11288	2.0133	196	160.0	S	2.3175	-0.2711	15.4321
POTASSIUM CHLORIDE				C	0.4394	0.0600	0.0646
11289	2.0099	196	180.0	S	2.3530	-0.2827	15.4799
POTASSIUM CHLORIDE				C	0.4331	0.0610	0.0643
11290	2.0063	196	200.0	S	2.3921	-0.2953	15.5521
POTASSIUM CHLORIDE				C	0.4274	0.0620	0.0640
11291	2.0026	196	220.0	S	2.4290	-0.3077	15.6250
POTASSIUM CHLORIDE				C	0.4218	0.0620	0.0638
11292	1.9987	196	240.0	S	2.4636	-0.3157	15.6740
POTASSIUM CHLORIDE				C	0.4158	0.0630	0.0635
11293	1.9947	196	260.0	S	2.5049	-0.3296	15.7480
POTASSIUM CHLORIDE				C	0.4095	0.0650	0.0632
11294	1.9905	196	280.0	S	2.5530	-0.3497	15.8228
POTASSIUM CHLORIDE				C	0.4032	0.0660	0.0629
11295	1.9860	196	300.0	S	2.5999	-0.3657	15.8982
POTASSIUM CHLORIDE				C	0.4035	0.0651	0.0633
11296	1.9870	87	298.0	S	2.5946	-0.3605	15.7978
POTASSIUM CHLORIDE				C	0.3953	0.0487	0.0631
11297	1.9860	84	298.0	S	2.6000	-0.2850	15.8600
POTASSIUM CHLORIDE				C	0.3896	0.0511	0.0626
11298	1.9806	84	323.0	S	2.6470	-0.3070	15.9700
POTASSIUM CHLORIDE				C	0.3834	0.0534	0.0622
11299	1.9753	84	348.0	S	2.7000	-0.3300	16.0900
POTASSIUM CHLORIDE				C	0.3763	0.0555	0.0617
11300	1.9694	84	373.0	S	2.7620	-0.3550	16.2000

CUBIC

CODE	NAME / DENSITY	REF	TEMP(K)		11	12	44
POTASSIUM CHLORIDE				C	0.3687	0.0572	0.0613
11301	1.9641	84	398.0	S	2.8300	-0.3800	16.3100
POTASSIUM CHLORIDE				C	0.3615	0.0592	0.0609
11302	1.9583	84	423.0	S	2.9000	-0.4080	16.4300
POTASSIUM CHLORIDE				C	0.3543	0.0606	0.0605
11303	1.9526	84	448.0	S	2.9710	-0.4340	16.5400
POTASSIUM CHLORIDE				C	0.3465	0.0621	0.0601
11304	1.9460	84	473.0	S	3.0520	-0.4640	16.6500
POTASSIUM CHLORIDE				C	0.3390	0.0634	0.0597
11305	1.9402	84	498.0	S	3.1350	-0.4940	16.7600
POTASSIUM CHLORIDE				C	0.3316	0.0647	0.0593
11306	1.9337	84	523.0	S	3.2210	-0.5260	16.8700
POTASSIUM CHLORIDE				C	0.3243	0.0658	0.0589
11307	1.9281	84	548.0	S	3.3100	-0.5580	16.9900
POTASSIUM CHLORIDE				C	0.3170	0.0667	0.0584
11308	1.9217	84	573.0	S	3.4040	-0.5920	17.1100
POTASSIUM CHLORIDE				C	0.3099	0.0680	0.0580
11309	1.9149	84	598.0	S	3.5030	-0.6300	17.2270
POTASSIUM CHLORIDE				C	0.3016	0.0690	0.0576
11310	1.9081	84	623.0	S	3.6250	-0.6750	17.3500
POTASSIUM CHLORIDE				C	0.2944	0.0702	0.0573
11311	1.9021	84	648.0	S	3.7400	-0.7200	17.4650
POTASSIUM CHLORIDE				C	0.2873	0.0711	0.0568
11312	1.8952	84	673.0	S	3.8600	-0.7660	17.5950
POTASSIUM CHLORIDE				C	0.2797	0.0715	0.0564
11313	1.8884	84	698.0	S	3.9900	-0.8120	17.7280
POTASSIUM CHLORIDE				C	0.2726	0.0721	0.0560
11314	1.8818	84	723.0	S	4.1250	-0.8630	17.8620
POTASSIUM CHLORIDE				C	0.2648	0.0718	0.0556
11315	1.8752	84	748.0	S	4.2700	-0.9110	18.0000
POTASSIUM CHLORIDE				C	0.2569	0.0715	0.0551
11316	1.8681	84	773.0	S	4.4300	-0.9650	18.1360
POTASSIUM CHLORIDE				C	0.2496	0.0717	0.0547
11317	1.8609	84	798.0	S	4.5950	-1.0250	18.2800
POTASSIUM CHLORIDE				C	0.2428	0.0718	0.0543
11318	1.8533	84	823.0	S	4.7600	-1.0860	18.4260
POTASSIUM CHLORIDE				C	0.2363	0.0720	0.0538
11319	1.8462	84	848.0	S	4.9350	-1.1530	18.5710
POTASSIUM CHLORIDE				C	0.2298	0.0722	0.0534
11320	1.8386	84	873.0	S	5.1200	-1.2240	18.7220

CODE	NAME DENSITY	REF	TEMP(K)		11	12	44
POTASSIUM CHLORIDE				C	0.2234	0.0720	0.0530
11321	1.8317	84	898.0	S	5.3100	-1.2940	18.8740
POTASSIUM CHLORIDE				C	0.2167	0.0712	0.0526
11322	1.8242	84	923.0	S	5.5100	-1.3630	19.0000
POTASSIUM CHLORIDE				C	0.2107	0.0708	0.0521
11323	1.8168	84	948.0	S	5.7100	-1.4360	19.1900
POTASSIUM CHLORIDE				C	0.2047	0.0701	0.0517
11324	1.8089	84	973.0	S	5.9200	-1.5100	19.3600
POTASSIUM CHLORIDE				C	0.1986	0.0692	0.0512
11325	1.8015	84	998.0	S	6.1400	-1.5860	19.5300
POTASSIUM CHLORIDE				C	0.1919	0.0682	0.0507
11326	1.7932	84	1023.0	S	6.4050	-1.6800	19.7240
POTASSIUM CHLORIDE				C	0.1899	0.0680	0.0506
11327	1.7916	84	1028.0	S	6.4900	-1.7110	19.7720
POTASSIUM CHLORIDE				C	0.1896	0.0697	0.0505
11328	1.7900	84	1033.0	S	6.5700	-1.7650	19.8100
POTASSIUM CHLORIDE				C	0.4378	0.0658	0.0642
11329	2.0089	23	195.0	S	2.3775	-0.3106	15.5763
POTASSIUM CHLORIDE				C	0.4050	0.0698	0.0630
11330	1.9874	23	295.0	S	2.6009	-0.3824	15.8730
POTASSIUM CHLORIDE				C	0.4069	0.0711	0.0631
11331	1.9840	234	298.0	S	2.5924	-0.3856	15.8479
POTASSIUM CHLORIDE				C	0.4090	0.0704	0.0627
11332	1.9865	79	ROOM	S	2.5752	-0.3782	15.9490
POTASSIUM CHLORIDE(83.2)-BROMIDE(16.8)				C	0.3925	0.0688	0.0607
11333	2.1290	234	298.0	S	2.6883	-0.4009	16.4745
POTASSIUM CHLORIDE(82.9)-BROMIDE(17.1)				C	0.3922	0.0691	0.0608
11334	2.1260	234	298.0	S	2.6918	-0.4032	16.4474
POTASSIUM CHLORIDE(61.8)-BROMIDE(38.2)				C	0.3764	0.0658	0.0581
11335	2.3020	234	298.0	S	2.8026	-0.4170	17.2117
POTASSIUM CHLORIDE(61.3)-BROMIDE(38.7)				C	0.3762	0.0660	0.0580
11336	2.3000	234	298.0	S	2.8051	-0.4187	17.2414
POTASSIUM CHLORIDE(42.2)-BROMIDE(57.8)				C	0.3655	0.0632	0.0557
11337	2.4530	234	298.0	S	2.8830	-0.4250	17.9533
POTASSIUM CHLORIDE(40.2)-BROMIDE(59.8)				C	0.3630	0.0630	0.0552
11338	2.4730	234	298.0	S	2.9039	-0.4294	18.1159
POTASSIUM CHLORIDE(20.5)-BROMIDE(79.5)				C	0.3544	0.0615	0.0530
11339	2.6130	234	298.0	S	2.9743	-0.4398	18.8679
POTASSIUM CHLORIDE(20.0)-BROMIDE(80.0)				C	0.3545	0.0605	0.0531
11340	2.6130	234	298.0	S	2.9686	-0.4328	18.8324

CUBIC

CODE	NAME / DENSITY	REF	TEMP(K)		11	12	44
POTASSIUM CYANIDE				C	0.1940	0.1180	0.0150
11341	1.5530	4	ROOM	S	9.5471	-3.6108	66.6667
POTASSIUM FLUORIDE				C	0.6580	0.1490	0.1280
11342	2.5257	4	ROOM	S	1.6584	-0.3062	7.8125
POTASSIUM FLUORIDE				C	0.6485	0.1427	0.1281
11343	2.4800	145	298.0	S	1.6750	-0.3021	7.8064
POTASSIUM FLUORIDE				C	0.7570	0.1350	0.1336
11344	2.5300	152	4.2	S	1.3964	-0.2113	7.4850
POTASSIUM FLUORIDE				C	0.6490	0.1520	0.1232
11345	2.4800	152	295.0	S	1.6912	-0.3209	8.1169
POTASSIUM FLUORIDE				C	0.7585	0.1473	0.1293
11346	2.5858	175	4.2	S	1.4073	-0.2288	7.7340
POTASSIUM FLUORIDE				C	0.7571	0.1538	0.1293
11347	2.5852	175	20.0	S	1.4181	-0.2394	7.7340
POTASSIUM FLUORIDE				C	0.7105	0.1580	0.1284
11348	2.5618	175	140.0	S	1.5314	-0.2786	7.7882
POTASSIUM FLUORIDE				C	0.7024	0.1576	0.1281
11349	2.5570	175	160.0	S	1.5513	-0.2843	7.8064
POTASSIUM FLUORIDE				C	0.6943	0.1567	0.1277
11350	2.5524	175	180.0	S	1.5709	-0.2893	7.8309
POTASSIUM FLUORIDE				C	0.6863	0.1575	0.1273
11351	2.5476	175	200.0	S	1.5936	-0.2975	7.8555
POTASSIUM FLUORIDE				C	0.6776	0.1590	0.1269
11352	2.5427	175	220.0	S	1.6203	-0.3079	7.8802
POTASSIUM FLUORIDE				C	0.6706	0.1578	0.1265
11353	2.5377	175	240.0	S	1.6380	-0.3120	7.9051
POTASSIUM FLUORIDE				C	0.6622	0.1628	0.1260
11354	2.5329	175	260.0	S	1.6724	-0.3300	7.9365
POTASSIUM FLUORIDE				C	0.6547	0.1591	0.1256
11355	2.5280	175	280.0	S	1.6878	-0.3300	7.9618
POTASSIUM FLUORIDE				C	0.6480	0.1600	0.1252
11356	2.5256	175	300.0	S	1.7105	-0.3387	7.9872
POTASSIUM IODIDE				C	0.2668	0.0426	0.0420
11357	3.1300	128	ROOM	S	3.9200	-0.5400	23.8000
POTASSIUM IODIDE				C	0.2755	0.0470	0.0380
11358	3.1279	4	ROOM	S	3.8197	-0.5567	26.3158
POTASSIUM IODIDE				C	0.2700	0.0440	0.0370
11359	3.1300	4	ROOM	S	3.8810	-0.5438	27.0270
POTASSIUM IODIDE				C	0.2710	0.0450	0.0364
11360	3.1300	4	ROOM	S	3.8732	-0.5516	27.4725

CUBIC

CODE	NAME DENSITY	REF	TEMP(K)		11	12	44
POTASSIUM IODIDE				C	0.3380	0.0220	0.0368
11361	3.1690	196	4.2	S	2.9823	-0.1823	27.1739
POTASSIUM IODIDE				C	0.3360	0.0240	0.0368
11362	3.1690	196	20.0	S	3.0048	-0.2003	27.1739
POTASSIUM IODIDE				C	0.3320	0.0260	0.0368
11363	3.1680	196	40.0	S	3.0467	-0.2213	27.1739
POTASSIUM IODIDE				C	0.3280	0.0280	0.0369
11364	3.1660	196	60.0	S	3.0903	-0.2431	27.1003
POTASSIUM IODIDE				C	0.3230	0.0300	0.0369
11365	3.1640	196	80.0	S	3.1456	-0.2673	27.1003
POTASSIUM IODIDE				C	0.3190	0.0320	0.0369
11366	3.1620	196	100.0	S	3.1932	-0.2911	27.1003
POTASSIUM IODIDE				C	0.3140	0.0330	0.0369
11367	3.1590	196	120.0	S	3.2497	-0.3090	27.1003
POTASSIUM IODIDE				C	0.3100	0.0350	0.0369
11368	3.1560	196	140.0	S	3.3014	-0.3349	27.1003
POTASSIUM IODIDE				C	0.3050	0.0370	0.0368
11369	3.1530	196	160.0	S	3.3671	-0.3643	27.1739
POTASSIUM IODIDE				C	0.3010	0.0380	0.0368
11370	3.1500	196	180.0	S	3.4190	-0.3833	27.1739
POTASSIUM IODIDE				C	0.2960	0.0390	0.0368
11371	3.1470	196	200.0	S	3.4853	-0.4058	27.1739
POTASSIUM IODIDE				C	0.2920	0.0390	0.0367
11372	3.1440	196	220.0	S	3.5359	-0.4166	27.2480
POTASSIUM IODIDE				C	0.2870	0.0410	0.0367
11373	3.1400	196	240.0	S	3.6134	-0.4517	27.2480
POTASSIUM IODIDE				C	0.2820	0.0420	0.0366
11374	3.1360	196	260.0	S	3.6885	-0.4781	27.3224
POTASSIUM IODIDE				C	0.2760	0.0440	0.0365
11375	3.1320	196	280.0	S	3.7893	-0.5210	27.3972
POTASSIUM IODIDE				C	0.2710	0.0450	0.0364
11376	3.1280	196	300.0	S	3.8732	-0.5516	27.4725
POTASSIUM IODIDE				C	0.2836	0.0455	0.0417
11377	3.1260	212	303.0	S	3.6900	-0.5100	24.0000
POTASSIUM IODIDE				C	0.2677	0.0405	0.0369
11378	3.1257	213	300.0	S	3.8902	-0.5112	27.1003
POTASSIUM MAGNESIUM FLUORIDE				C	1.3200	0.3960	0.4850
11379	3.1500	219	290.0	S	0.8793	-0.2029	2.0619
POTASSIUM MANGANESE FLUORIDE				C	1.0789	0.4340	0.2665
11380	3.4397	9	200.0	S	1.2050	-0.3457	3.7523

CUBIC

NAME CODE	DENSITY	REF	TEMP(K)		11	12	44
POTASSIUM MANGANESE FLUORIDE				C	1.1412	0.4028	0.2730
11381	3.4377	9	210.0	S	1.0741	-0.2802	3.6630
POTASSIUM MANGANESE FLUORIDE				C	1.1692	0.3888	0.2746
11382	3.4357	9	220.0	S	1.0255	-0.2559	3.6417
POTASSIUM MANGANESE FLUORIDE				C	1.1823	0.3822	0.2751
11383	3.4337	9	230.0	S	1.0045	-0.2454	3.6350
POTASSIUM MANGANESE FLUORIDE				C	1.1890	0.3788	0.2753
11384	3.4318	9	240.0	S	0.9941	-0.2402	3.6324
POTASSIUM MANGANESE FLUORIDE				C	1.1930	0.3769	0.2755
11385	3.4298	9	250.0	S	0.9881	-0.2372	3.6298
POTASSIUM MANGANESE FLUORIDE				C	1.1957	0.3755	0.2756
11386	3.4278	9	260.0	S	0.9840	-0.2352	3.6284
POTASSIUM MANGANESE FLUORIDE				C	1.1979	0.3744	0.2757
11387	3.4259	9	270.0	S	0.9808	-0.2335	3.6271
POTASSIUM MANGANESE FLUORIDE				C	1.1999	0.3734	0.2758
11388	3.4239	9	280.0	S	0.9778	-0.2321	3.6258
POTASSIUM MANGANESE FLUORIDE				C	1.2018	0.3724	0.2760
11389	3.4220	9	290.0	S	0.9750	-0.2307	3.6232
POTASSIUM MANGANESE FLUORIDE				C	1.2037	0.3715	0.2761
11390	3.4200	9	300.0	S	0.9723	-0.2293	3.6219
PYRITE				C	3.6188	-0.4796	1.0549
11391	5.0180	128	ROOM	S	0.2880	0.0440	0.9480
PYRITE				C	3.6148	-0.4338	1.0406
11392	5.0180	128	ROOM	S	0.2860	0.0390	0.9610
PYRITE				C	3.6200	-0.4640	1.0520
11393	5.0100	129	ROOM	S	0.2871	0.0422	0.9506
PYRITE				C	3.6160	-0.4440	1.0840
11394	5.0100	129	ROOM	S	0.2864	0.0401	0.9225
PYRITE				C	3.5850	-0.5290	1.0350
11395	5.0100	129	ROOM	S	0.2940	0.0509	0.9662
PYRITE				C	3.5780	-0.4390	1.0850
11396	5.0100	129	ROOM	S	0.2894	0.0405	0.9217
PYRITE				C	3.7510	-0.4850	1.0140
11397	5.0100	129	ROOM	S	0.2772	0.0412	0.9862
PYRITE				C	3.7700	0.3200	1.0900
11398	5.0130	4	ROOM	S	0.2688	-0.0210	0.9174
PYRITE				C	3.4600	0.3100	1.0810
11399	5.0130	8	ROOM	S	0.2933	-0.0241	0.9251
PYRITE				C	3.6300	0.3400	1.0520
11400	5.0130	8	ROOM	S	0.2800	-0.0240	0.9506

CODE	NAME DENSITY	REF	TEMP(K)		11	12	44
PYRITE				C	3.8180	0.3100	1.0940
11401	5.0160	233	ROOM	S	0.2652	-0.0199	0.9141
PYRITE				C	3.7940	0.3080	1.0880
11402	4.9290	233	ROOM	S	0.2668	-0.0200	0.9191
PYRITE				C	3.6100	0.3360	1.0520
11403	4.9290	233	ROOM	S	0.2815	-0.0240	0.9506
RUBIDIUM				C	0.0296	0.0244	0.0160
11404	1.6070	217	80.0	S	132.4570	-59.8509	62.5000
RUBIDIUM				C	0.0342	0.0288	0.0221
11405	1.6200	122	4.2	S	127.0879	-58.0973	45.2489
RUBIDIUM				C	0.0325	0.0273	0.0198
11406	1.6100	122	78.0	S	132.0322	-60.2756	50.5051
RUBIDIUM				C	0.0314	0.0264	0.0189
11407	1.6000	122	110.0	S	137.2923	-62.7079	52.9100
RUBIDIUM				C	0.0306	0.0257	0.0180
11408	1.5900	122	140.0	S	140.1195	-63.9622	55.5556
RUBIDIUM				C	0.0296	0.0250	0.0171
11409	1.5800	122	170.0	S	149.1152	-68.2762	58.4795
RUBIDIUM BROMIDE				C	0.3185	0.0480	0.0385
11410	3.3510	4	ROOM	S	3.2688	-0.4281	25.9740
RUBIDIUM BROMIDE				C	0.3170	0.0420	0.0388
11411	3.3490	41	300.0	S	3.2555	-0.3809	25.7732
RUBIDIUM BROMIDE				C	0.3155	0.0493	0.0380
11412	3.3590	213	300.0	S	3.3093	-0.4472	26.3158
RUBIDIUM BROMIDE				C	0.3863	0.0474	0.0408
11413	3.4340	152	4.0	S	2.6600	-0.2907	24.4798
RUBIDIUM BROMIDE				C	0.3860	0.0474	0.0408
11414	3.4340	152	10.0	S	2.6622	-0.2912	24.4798
RUBIDIUM BROMIDE				C	0.3853	0.0474	0.0408
11415	3.4340	152	20.0	S	2.6673	-0.2922	24.4918
RUBIDIUM BROMIDE				C	0.3839	0.0474	0.0408
11416	3.4330	152	30.0	S	2.6775	-0.2943	24.5098
RUBIDIUM BROMIDE				C	0.3820	0.0474	0.0408
11417	3.4320	152	40.0	S	2.6915	-0.2971	24.5338
RUBIDIUM BROMIDE				C	0.3799	0.0474	0.0407
11418	3.4300	152	50.0	S	2.7072	-0.3003	24.5519
RUBIDIUM BROMIDE				C	0.3775	0.0474	0.0407
11419	3.4280	152	60.0	S	2.7254	-0.3040	24.5821
RUBIDIUM BROMIDE				C	0.3751	0.0474	0.0406
11420	3.4260	152	70.0	S	2.7438	-0.3078	24.6305

CUBIC

CODE	NAME DENSITY	REF	TEMP(K)		11	12	44
RUBIDIUM BROMIDE				C	0.3725	0.0474	0.0405
11421	3.4240	152	80.0	S	2.7640	-0.3120	24.6913
RUBIDIUM BROMIDE				C	0.3135	0.0512	0.0377
11422	3.3520	152	290.0	S	3.3431	-0.4693	26.4971
RUBIDIUM BROMIDE				C	0.3107	0.0515	0.0376
11423	3.3490	152	300.0	S	3.3778	-0.4803	26.5957
RUBIDIUM BROMIDE				C	0.3597	0.0444	0.0393
11424	3.4040	110	140.0	S	2.8576	-0.3140	25.4453
RUBIDIUM BROMIDE				C	0.3542	0.0451	0.0391
11425	3.3980	110	160.0	S	2.9069	-0.3283	25.5754
RUBIDIUM BROMIDE				C	0.3488	0.0459	0.0390
11426	3.3920	110	180.0	S	2.9575	-0.3439	25.6410
RUBIDIUM BROMIDE				C	0.3430	0.0465	0.0388
11427	3.3850	110	200.0	S	3.0130	-0.3597	25.7732
RUBIDIUM BROMIDE				C	0.3373	0.0473	0.0387
11428	3.3780	110	220.0	S	3.0706	-0.3776	25.8398
RUBIDIUM BROMIDE				C	0.3317	0.0480	0.0385
11429	3.3710	110	240.0	S	3.1293	-0.3956	25.9740
RUBIDIUM BROMIDE				C	0.3262	0.0486	0.0383
11430	3.3640	110	260.0	S	3.1888	-0.4135	26.1096
RUBIDIUM BROMIDE				C	0.3204	0.0494	0.0382
11431	3.3570	110	280.0	S	3.2552	-0.4348	26.1780
RUBIDIUM BROMIDE				C	0.3152	0.0500	0.0380
11432	3.3500	110	300.0	S	3.3167	-0.4541	26.3158
RUBIDIUM BROMIDE (P=0.001KB)				C	0.3488	0.0459	0.0390
11433	3.3920	110	180.0	S	2.9575	-0.3439	25.6410
RUBIDIUM BROMIDE (P=0.200KB)				C	0.3513	0.0462	0.0389
11434	3.3971	110	180.0	S	2.9363	-0.3413	25.7069
RUBIDIUM BROMIDE (P=0.400KB)				C	0.3540	0.0463	0.0387
11435	3.4013	110	180.0	S	2.9130	-0.3369	25.8398
RUBIDIUM BROMIDE (P=0.600KB)				C	0.3565	0.0464	0.0386
11436	3.4063	110	180.0	S	2.8917	-0.3330	25.9067
RUBIDIUM BROMIDE (P=0.800KB)				C	0.3590	0.0466	0.0385
11437	3.4103	110	180.0	S	2.8712	-0.3299	25.9740
RUBIDIUM BROMIDE (P=1.000KB)				C	0.3615	0.0466	0.0384
11438	3.4154	110	180.0	S	2.8502	-0.3255	26.0417
RUBIDIUM BROMIDE (P=1.200KB)				C	0.3640	0.0467	0.0383
11439	3.4196	110	180.0	S	2.8298	-0.3218	26.1096
RUBIDIUM BROMIDE (P=1.400KB)				C	0.3667	0.0470	0.0381
11440	3.4247	110	180.0	S	2.8088	-0.3191	26.2467

NAME CODE	DENSITY	REF	TEMP(K)		11	12	44
RUBIDIUM BROMIDE (P=1.600KB)				C	0.3693	0.0470	0.0380
11441	3.4288	110	180.0	S	2.7879	-0.3148	26.3158
RUBIDIUM BROMIDE(P=1.800KB)				C	0.3720	0.0475	0.0379
11442	3.4339	110	180.0	S	2.7682	-0.3134	26.3852
RUBIDIUM BROMIDE (P=2.000KB)				C	0.3748	0.0480	0.0378
11443	3.4380	110	180.0	S	2.7480	-0.3120	26.4550
RUBIDIUM BROMIDE (P=2.200KB)				C	0.3772	0.0478	0.0377
11444	3.4431	110	180.0	S	2.7289	-0.3069	26.5252
RUBIDIUM BROMIDE (P=2.400KB)				C	0.3802	0.0484	0.0376
11445	3.4472	110	180.0	S	2.7081	-0.3058	26.5957
RUBIDIUM BROMIDE (P=0.001KB)				C	0.3369	0.0471	0.0386
11446	3.3780	110	220.0	S	3.0737	-0.3770	25.9067
RUBIDIUM BROMIDE (P=0.200KB)				C	0.3397	0.0474	0.0385
11447	3.3831	110	220.0	S	3.0479	-0.3732	25.9740
RUBIDIUM BROMIDE (P=0.400KB)				C	0.3424	0.0478	0.0384
11448	3.3882	110	220.0	S	3.0240	-0.3704	26.0417
RUBIDIUM BROMIDE (P=0.600KB)				C	0.3453	0.0480	0.0383
11449	3.3921	110	220.0	S	2.9977	-0.3659	26.1096
RUBIDIUM BROMIDE (P=0.800KB)				C	0.3481	0.0483	0.0382
11450	3.3972	110	220.0	S	2.9733	-0.3623	26.1780
RUBIDIUM BROMIDE (P=1.000KB)				C	0.3511	0.0486	0.0380
11451	3.4023	110	220.0	S	2.9474	-0.3584	26.3158
RUBIDIUM BROMIDE (P=1.200KB)				C	0.3538	0.0490	0.0379
11452	3.4065	110	220.0	S	2.9250	-0.3558	26.3852
RUBIDIUM BROMIDE (P=1.400KB)				C	0.3568	0.0492	0.0378
11453	3.4115	110	220.0	S	2.8996	-0.3514	26.4550
RUBIDIUM BROMIDE (P=1.600KB)				C	0.3594	0.0495	0.0377
11454	3.4156	110	220.0	S	2.8784	-0.3484	26.5252
RUBIDIUM BROMIDE (P=1.800KB)				C	0.3622	0.0499	0.0376
11455	3.4207	110	220.0	S	2.8562	-0.3458	26.5957
RUBIDIUM BROMIDE (P=2.000KB)				C	0.3648	0.0502	0.0374
11456	3.4248	110	220.0	S	2.8356	-0.3430	26.7380
RUBIDIUM BROMIDE (P=2.200KB)				C	0.3676	0.0504	0.0373
11457	3.4299	110	220.0	S	2.8134	-0.3392	26.8096
RUBIDIUM BROMIDE (P=2.400KB)				C	0.3704	0.0506	0.0372
11458	3.4341	110	220.0	S	2.7915	-0.3355	26.8817
RUBIDIUM BROMIDE (P=2.615KB)				C	0.3732	0.0509	0.0371
11459	3.4392	110	220.0	S	2.7702	-0.3325	26.9542
RUBIDIUM BROMIDE (P=2.800KB)				C	0.3758	0.0508	0.0370
11460	3.4433	110	220.0	S	2.7495	-0.3274	27.0270

CODE	NAME DENSITY	REF	TEMP(K)	CUBIC	11	12	44
RUBIDIUM BROMIDE (P=3.000KB)				C	0.3785	0.0509	0.0368
11461	3.4474	110	220.0	S	2.7290	-0.3235	27.1739
RUBIDIUM BROMIDE (P=3.200KB)				C	0.3812	0.0510	0.0367
11462	3.4515	110	220.0	S	2.7088	-0.3196	27.2480
RUBIDIUM BROMIDE (P=3.415KB)				C	0.3837	0.0514	0.0366
11463	3.4566	110	220.0	S	2.6914	-0.3179	27.3224
RUBIDIUM BROMIDE (P=3.610KB)				C	0.3864	0.0515	0.0365
11464	3.4608	110	220.0	S	2.6718	-0.3142	27.3972
RUBIDIUM BROMIDE (P=3.810KB)				C	0.3892	0.0516	0.0363
11465	3.4649	110	220.0	S	2.6517	-0.3104	27.5482
RUBIDIUM BROMIDE (P=0.001KB)				C	0.3257	0.0482	0.0383
11466	3.3640	110	260.0	S	3.1921	-0.4115	26.1096
RUBIDIUM BROMIDE (P=0.210KB)				C	0.3286	0.0486	0.0382
11467	3.3691	110	260.0	S	3.1638	-0.4076	26.1780
RUBIDIUM BROMIDE (P=0.400KB)				C	0.3310	0.0488	0.0381
11468	3.3742	110	260.0	S	3.1401	-0.4035	26.2467
RUBIDIUM BROMIDE (P=0.600KB)				C	0.3338	0.0494	0.0380
11469	3.3792	110	260.0	S	3.1147	-0.4015	26.3158
RUBIDIUM BROMIDE (P=0.800KB)				C	0.3365	0.0496	0.0379
11470	3.3832	110	260.0	S	3.0887	-0.3968	26.3852
RUBIDIUM BROMIDE (P=1.000KB)				C	0.3395	0.0499	0.0377
11471	3.3883	110	260.0	S	3.0608	-0.3922	26.5252
RUBIDIUM BROMIDE (P=1.200KB)				C	0.3421	0.0502	0.0376
11472	3.3934	110	260.0	S	3.0372	-0.3886	26.5957
RUBIDIUM BROMIDE (P=1.400KB)				C	0.3450	0.0505	0.0375
11473	3.3984	110	260.0	S	3.0111	-0.3845	26.6667
RUBIDIUM BROMIDE (P=1.600KB)				C	0.3477	0.0508	0.0374
11474	3.4025	110	260.0	S	2.9873	-0.3808	26.7380
RUBIDIUM BROMIDE (P=1.800KB)				C	0.3503	0.0511	0.0373
11475	3.4076	110	260.0	S	2.9648	-0.3774	26.8096
RUBIDIUM BROMIDE (P=2.000KB)				C	0.3531	0.0515	0.0371
11476	3.4126	110	260.0	S	2.9413	-0.3744	26.9542
RUBIDIUM BROMIDE (P=2.200KB)				C	0.3558	0.0518	0.0370
11477	3.4168	110	260.0	S	2.9186	-0.3709	27.0270
RUBIDIUM BROMIDE (P=2.400KB)				C	0.3584	0.0521	0.0369
11478	3.4207	110	260.0	S	2.8971	-0.3677	27.1003
RUBIDIUM BROMIDE (P=2.600KB)				C	0.3612	0.0524	0.0368
11479	3.4259	110	260.0	S	2.8742	-0.3641	27.1739
RUBIDIUM BROMIDE (P=2.800KB)				C	0.3639	0.0527	0.0367
11480	3.4310	110	260.0	S	2.8525	-0.3608	27.2480

CODE	NAME / DENSITY	REF	TEMP(K)		11	12	44
RUBIDIUM	BROMIDE	(P=3.000KB)		C	0.3665	0.0530	0.0365
11481	3.4351	110	260.0	S	2.8320	-0.3578	27.3972
RUBIDIUM	BROMIDE	(P=3.200KB)		C	0.3687	0.0534	0.0364
11482	3.4393	110	260.0	S	2.8154	-0.3562	27.4725
RUBIDIUM	BROMIDE	(P=3.400KB)		C	0.3714	0.0537	0.0363
11483	3.4444	110	260.0	S	2.7946	-0.3530	27.5482
RUBIDIUM	BROMIDE	(P=3.600KB)		C	0.3741	0.0540	0.0362
11484	3.4485	110	260.0	S	2.7741	-0.3499	27.6243
RUBIDIUM	BROMIDE	(P=3.800KB)		C	0.3768	0.0542	0.0360
11485	3.4526	110	260.0	S	2.7535	-0.3463	27.7778
RUBIDIUM	BROMIDE	(P=0.001KB)		C	0.3151	0.0500	0.0380
11486	3.3500	110	300.0	S	3.3178	-0.4544	26.3158
RUBIDIUM	BROMIDE	(P=0.600KB)		C	0.3234	0.0508	0.0377
11487	3.3650	110	300.0	S	3.2299	-0.4385	26.5252
RUBIDIUM	BROMIDE	(P=1.200KB)		C	0.3318	0.0516	0.0373
11488	3.3803	110	300.0	S	3.1455	-0.4233	26.8096
RUBIDIUM	BROMIDE	(P=1.800KB)		C	0.3402	0.0526	0.0370
11489	3.3944	110	300.0	S	3.0664	-0.4106	27.0270
RUBIDIUM	BROMIDE	(P=2.400KB)		C	0.3484	0.0533	0.0366
11490	3.4086	110	300.0	S	2.9917	-0.3970	27.3224
RUBIDIUM	BROMIDE	(P=2.600KB)		C	0.3512	0.0536	0.0365
11491	3.4127	110	300.0	S	2.9673	-0.3929	27.3972
RUBIDIUM	BROMIDE	(P=2.800KB)		C	0.3539	0.0538	0.0364
11492	3.4177	110	300.0	S	2.9438	-0.3885	27.4725
RUBIDIUM	BROMIDE	(P=3.015KB)		C	0.3566	0.0542	0.0362
11493	3.4229	110	300.0	S	2.9214	-0.3854	27.6243
RUBIDIUM	BROMIDE	(P=3.220KB)		C	0.3590	0.0546	0.0361
11494	3.4270	110	300.0	S	2.9020	-0.3831	27.7008
RUBIDIUM	BROMIDE	(P=3.400KB)		C	0.3617	0.0550	0.0360
11495	3.4321	110	300.0	S	2.8803	-0.3802	27.7778
RUBIDIUM	BROMIDE	(P=3.610KB)		C	0.3646	0.0551	0.0359
11496	3.4362	110	300.0	S	2.8561	-0.3750	27.8551
RUBIDIUM	BROMIDE	(P=3.800KB)		C	0.3670	0.0555	0.0358
11497	3.4402	110	300.0	S	2.8375	-0.3727	27.9330
RUBIDIUM	BROMIDE	(P=4.010KB)		C	0.3697	0.0558	0.0356
11498	3.4454	110	300.0	S	2.8164	-0.3693	28.0899
RUBIDIUM	CHLORIDE			C	0.3645	0.0610	0.0475
11499	2.7990	4	ROOM	S	2.8818	-0.4131	21.0526
RUBIDIUM	CHLORIDE			C	0.3560	0.0600	0.0460
11500	2.7600	4	ROOM	S	2.9525	-0.4258	21.7391

CUBIC

CODE	NAME DENSITY	REF	TEMP(K)		11	12	44
RUBIDIUM CHLORIDE				C	0.4499	0.0676	0.0497
11501	2.8641	176	4.2	S	2.3135	-0.3022	20.1207
RUBIDIUM CHLORIDE				C	0.4475	0.0677	0.0496
11502	2.8639	176	30.0	S	2.3272	-0.3058	20.1613
RUBIDIUM CHLORIDE				C	0.4390	0.0677	0.0494
11503	2.8618	176	60.0	S	2.3758	-0.3174	20.2429
RUBIDIUM CHLORIDE				C	0.4298	0.0676	0.0492
11504	2.8572	176	90.0	S	2.4306	-0.3303	20.3252
RUBIDIUM CHLORIDE				C	0.4206	0.0676	0.0490
11505	2.8507	176	120.0	S	2.4883	-0.3446	20.4082
RUBIDIUM CHLORIDE				C	0.4116	0.0676	0.0488
11506	2.8430	176	150.0	S	2.5476	-0.3594	20.4918
RUBIDIUM CHLORIDE				C	0.4025	0.0676	0.0486
11507	2.8345	176	180.0	S	2.6106	-0.3754	20.5761
RUBIDIUM CHLORIDE				C	0.3933	0.0673	0.0484
11508	2.8256	176	210.0	S	2.6764	-0.3911	20.6611
RUBIDIUM CHLORIDE				C	0.3840	0.0666	0.0482
11509	2.8163	176	240.0	S	2.7449	-0.4057	20.7469
RUBIDIUM CHLORIDE				C	0.3746	0.0654	0.0480
11510	2.8068	176	270.0	S	2.8156	-0.4185	20.8333
RUBIDIUM CHLORIDE				C	0.3653	0.0645	0.0478
11511	2.7972	176	300.0	S	2.8907	-0.4338	20.9205
RUBIDIUM CHLORIDE				C	0.4297	0.0649	0.0493
11512	2.8200	152	4.2	S	2.4233	-0.3180	20.2840
RUBIDIUM CHLORIDE				C	0.3646	0.0647	0.0468
11513	2.7600	152	295.0	S	2.8977	-0.4367	21.3675
RUBIDIUM CHLORIDE				C	4.2394	0.5484	0.4858
11514	2.8538	110	100.0	S	0.2431	-0.0278	2.0585
RUBIDIUM CHLORIDE				C	4.1792	0.5516	0.4848
11515	2.8504	110	120.0	S	0.2469	-0.0288	2.0627
RUBIDIUM CHLORIDE				C	4.1194	0.5569	0.4831
11516	2.8458	110	140.0	S	0.2508	-0.0299	2.0700
RUBIDIUM CHLORIDE				C	4.0681	0.5630	0.4813
11517	2.8404	110	160.0	S	0.2544	-0.0309	2.0777
RUBIDIUM CHLORIDE				C	4.0071	0.5745	0.4794
11518	2.8346	110	180.0	S	0.2589	-0.0325	2.0859
RUBIDIUM CHLORIDE				C	3.9448	0.5812	0.4775
11519	2.8285	110	200.0	S	0.2635	-0.0338	2.0942
RUBIDIUM CHLORIDE				C	3.8831	0.5899	0.4754
11520	2.8223	110	220.0	S	0.2683	-0.0354	2.1035

CUBIC

CODE	NAME DENSITY	REF	TEMP(K)		11	12	44
RUBIDIUM CHLORIDE				C	3.8189	0.5959	0.4735
11521	2.8160	110	240.0	S	0.2734	-0.0369	2.1119
RUBIDIUM CHLORIDE				C	3.7519	0.6034	0.4714
11522	2.8097	110	260.0	S	0.2790	-0.0386	2.1213
RUBIDIUM CHLORIDE				C	3.6894	0.6053	0.4695
11523	2.8033	110	280.0	S	0.2842	-0.0401	2.1299
RUBIDIUM CHLORIDE				C	3.6242	0.6124	0.4678
11524	2.7969	110	300.0	S	0.2901	-0.0419	2.1377
RUBIDIUM CHLORIDE (P=0.001KB)				C	0.4001	0.0575	0.0479
11525	2.8346	110	181.0	S	2.5930	-0.3258	20.8768
RUBIDIUM CHLORIDE (P=0.350KB)				C	0.4048	0.0579	0.0477
11526	2.8406	110	181.0	S	2.5621	-0.3206	20.9644
RUBIDIUM CHLORIDE (P=0.695KB)				C	0.4096	0.0585	0.0475
11527	2.8466	110	181.0	S	2.5318	-0.3164	21.0526
RUBIDIUM CHLORIDE (P=0.982KB)				C	0.4137	0.0589	0.0473
11528	2.8508	110	181.0	S	2.5061	-0.3123	21.1416
RUBIDIUM CHLORIDE (P=1.326KB)				C	0.4186	0.0592	0.0471
11529	2.8568	110	181.0	S	2.4757	-0.3067	21.2314
RUBIDIUM CHLORIDE (P=1.730KB)				C	0.4240	0.0598	0.0468
11530	2.8636	110	181.0	S	2.4437	-0.3021	21.3675
RUBIDIUM CHLORIDE (P=2.020KB)				C	0.4280	0.0602	0.0466
11531	2.8679	110	181.0	S	2.4204	-0.2985	21.4592
RUBIDIUM CHLORIDE (P=2.490KB)				C	0.4346	0.0607	0.0464
11532	2.8756	110	181.0	S	2.3825	-0.2920	21.5517
RUBIDIUM CHLORIDE (P=0.001KB)				C	0.3869	0.0594	0.0475
11533	2.8222	110	222.0	S	2.6948	-0.3587	21.0526
RUBIDIUM CHLORIDE (P=0.325KB)				C	0.3912	0.0597	0.0473
11534	2.8282	110	222.0	S	2.6639	-0.3527	21.1416
RUBIDIUM CHLORIDE (P=0.690KB)				C	0.3963	0.0602	0.0471
11535	2.8342	110	222.0	S	2.6287	-0.3466	21.2314
RUBIDIUM CHLORIDE (P=1.010KB)				C	0.4010	0.0606	0.0469
11536	2.8392	110	222.0	S	2.5968	-0.3409	21.3220
RUBIDIUM CHLORIDE (P=1.490KB)				C	0.4077	0.0613	0.0466
11537	2.8477	110	222.0	S	2.5531	-0.3337	21.4592
RUBIDIUM CHLORIDE (P=2.015KB)				C	0.4146	0.0620	0.0462
11538	2.8562	110	222.0	S	2.5096	-0.3265	21.6450
RUBIDIUM CHLORIDE (P=2.515KB)				C	0.4216	0.0626	0.0459
11539	2.8648	110	222.0	S	2.4666	-0.3189	21.7865
RUBIDIUM CHLORIDE (P=3.035KB)				C	0.4284	0.0634	0.0456
11540	2.8732	110	222.0	S	2.4269	-0.3129	21.9298

CUBIC

CODE	NAME DENSITY	REF	TEMP(K)		11	12	44
RUBIDIUM CHLORIDE (P=3.530KB)				C	0.4350	0.0639	0.0453
11541	2.8810	110	222.0	S	2.3887	-0.3060	22.0750
RUBIDIUM CHLORIDE (P=0.001KB)				C	0.3749	0.0609	0.0471
11542	2.8097	110	259.0	S	2.7942	-0.3905	21.2314
RUBIDIUM CHLORIDE(P=0.320KB)				C	0.3792	0.0612	0.0469
11543	2.8157	110	259.0	S	2.7610	-0.3837	21.3220
RUBIDIUM CHLORIDE (P=0.602KB)				C	0.3830	0.0616	0.0468
11544	2.8208	110	259.0	S	2.7328	-0.3786	21.3675
RUBIDIUM CHLORIDE (P=0.998KB)				C	0.3887	0.0623	0.0465
11545	2.8274	110	259.0	S	2.6919	-0.3718	21.5054
RUBIDIUM CHLORIDE (P=1.515KB)				C	0.3958	0.0628	0.0462
11546	2.8359	110	259.0	S	2.6413	-0.3617	21.6450
RUBIDIUM CHLORIDE (P=1.995KB)				C	0.4023	0.0637	0.0459
11547	2.8444	110	259.0	S	2.5982	-0.3552	21.7865
RUBIDIUM CHLORIDE (P=2.485KB)				C	0.4089	0.0643	0.0456
11548	2.8520	110	259.0	S	2.5548	-0.3472	21.9298
RUBIDIUM CHLORIDE (P=3.015KB)				C	0.4158	0.0648	0.0452
11549	2.8613	110	259.0	S	2.5105	-0.3385	22.1239
RUBIDIUM CHLORIDE (P=3.540KB)				C	0.4229	0.0659	0.0450
11550	2.8699	110	259.0	S	2.4683	-0.3328	22.2222
RUBIDIUM CHLORIDE (P=4.035KB)				C	0.4291	0.0671	0.0446
11551	2.8777	110	259.0	S	2.4334	-0.3291	22.4215
RUBIDIUM CHLORIDE (P=0.001KB)				C	0.3609	0.0620	0.0466
11552	2.7966	110	304.0	S	2.9178	-0.4278	21.4592
RUBIDIUM CHLORIDE (P=0.295KB)				C	0.3644	0.0625	0.0465
11553	2.8017	110	304.0	S	2.8893	-0.4230	21.5054
RUBIDIUM CHLORIDE (P=0.575KB)				C	0.3683	0.0629	0.0463
11554	2.8068	110	304.0	S	2.8576	-0.4168	21.5983
RUBIDIUM CHLORIDE (P=0.975KB)				C	0.3741	0.0633	0.0461
11555	2.8143	110	304.0	S	2.8107	-0.4068	21.6920
RUBIDIUM CHLORIDE (P=1.500KB)				C	0.3817	0.0643	0.0457
11556	2.8235	110	304.0	S	2.7536	-0.3970	21.8818
RUBIDIUM CHLORIDE (P=2.030KB)				C	0.3887	0.0655	0.0454
11557	2.8328	110	304.0	S	2.7041	-0.3900	22.0264
RUBIDIUM CHLORIDE (P=2.490KB)				C	0.3953	0.0661	0.0451
11558	2.8404	110	304.0	S	2.6570	-0.3806	22.1729
RUBIDIUM CHLORIDE (P=2.995KB)				C	0.4018	0.0646	0.0448
11559	2.8489	110	304.0	S	2.6048	-0.3608	22.3214
RUBIDIUM CHLORIDE (P=3.525KB)				C	0.4088	0.0680	0.0445
11560	2.8574	110	304.0	S	2.5680	-0.3662	22.4719

CUBIC

CODE	NAME DENSITY	REF	TEMP(K)		11	12	44
RUBIDIUM FLUORIDE				C	0.5700	0.1250	0.0910
11561	2.8800	4	ROOM	S	1.9046	-0.3426	10.9890
RUBIDIUM IODIDE				C	0.2585	0.0375	0.0281
11562	3.5540	4	ROOM	S	4.0161	-0.5088	35.5872
RUBIDIUM IODIDE				C	0.2560	0.0310	0.0287
11563	3.5500	41	300.0	S	4.0112	-0.4333	34.8432
RUBIDIUM IODIDE				C	0.3210	0.0360	0.0292
11564	3.6680	152	4.0	S	3.1874	-0.3214	34.2466
RUBIDIUM IODIDE				C	0.2575	0.0340	0.0278
11565	3.5530	152	290.0	S	4.0069	-0.4674	35.9971
RUBIDIUM IODIDE				C	0.2550	0.0340	0.0277
11566	3.5490	152	300.0	S	4.0486	-0.4763	36.0620
RUBIDIUM IODIDE				C	0.2988	0.0314	0.0287
11567	3.6250	110	120.0	S	3.4150	-0.3247	34.8432
RUBIDIUM IODIDE				C	0.2948	0.0318	0.0286
11568	3.6180	110	140.0	S	3.4649	-0.3374	34.9650
RUBIDIUM IODIDE				C	0.2891	0.0329	0.0285
11569	3.6100	110	160.0	S	3.5414	-0.3618	35.0877
RUBIDIUM IODIDE				C	0.2850	0.0335	0.0284
11570	3.6020	110	180.0	S	3.5977	-0.3784	35.2113
RUBIDIUM IODIDE				C	0.2798	0.0342	0.0283
11571	3.5940	110	200.0	S	3.6717	-0.3999	35.3357
RUBIDIUM IODIDE				C	0.2747	0.0351	0.0282
11572	3.5850	110	220.0	S	3.7489	-0.4247	35.4610
RUBIDIUM IODIDE				C	0.2699	0.0358	0.0281
11573	3.5770	110	240.0	S	3.8239	-0.4478	35.5872
RUBIDIUM IODIDE				C	0.2652	0.0364	0.0280
11574	3.5690	110	260.0	S	3.8999	-0.4707	35.7143
RUBIDIUM IODIDE				C	0.2556	0.0382	0.0278
11575	3.5510	110	300.0	S	4.0706	-0.5293	35.9712
RUBIDIUM IODIDE (P=0.001KB)				C	0.2850	0.0336	0.0284
11576	3.6020	110	178.0	S	3.5982	-0.3795	35.2113
RUBIDIUM IODIDE (P=0.200KB)				C	0.2879	0.0338	0.0283
11577	3.6086	110	178.0	S	3.5613	-0.3742	35.3357
RUBIDIUM IODIDE (P=0.400KB)				C	0.2907	0.0342	0.0282
11578	3.6149	110	178.0	S	3.5273	-0.3713	35.4610
RUBIDIUM IODIDE (P=0.600KB)				C	0.2935	0.0344	0.0281
11579	3.6203	110	178.0	S	3.4931	-0.3665	35.5872
RUBIDIUM IODIDE (P=0.800KB)				C	0.2962	0.0347	0.0280
11580	3.6269	110	178.0	S	3.4611	-0.3630	35.7143

CUBIC

NAME CODE	DENSITY	REF	TEMP(K)		11	12	44
RUBIDIUM IODIDE (P=1.000KB)				C	0.2991	0.0350	0.0279
11581	3.6334	110	178.0	S	3.4274	-0.3591	35.8423
RUBIDIUM IODIDE (P=1.250KB)				C	0.3026	0.0354	0.0277
11582	3.6411	110	178.0	S	3.3877	-0.3548	36.1011
RUBIDIUM IODIDE (P=1.500KB)				C	0.3060	0.0357	0.0276
11583	3.6476	110	178.0	S	3.3496	-0.3500	36.2319
RUBIDIUM IODIDE (P=1.750KB)				C	0.3093	0.0361	0.0275
11584	3.6551	110	178.0	S	3.3140	-0.3464	36.3636
RUBIDIUM IODIDE (P=2.000KB)				C	0.3122	0.0359	0.0274
11585	3.6629	110	178.0	S	3.2809	-0.3384	36.4963
RUBIDIUM IODIDE (P=2.250KB)				C	0.3155	0.0363	0.0272
11586	3.6705	110	178.0	S	3.2467	-0.3350	36.7647
RUBIDIUM IODIDE (P=2.500KB)				C	0.3190	0.0366	0.0271
11587	3.6771	110	178.0	S	3.2106	-0.3305	36.9004
RUBIDIUM IODIDE (P=2.750KB)				C	0.3222	0.0368	0.0270
11588	3.6848	110	178.0	S	3.1781	-0.3258	37.0370
RUBIDIUM IODIDE (P=3.000KB)				C	0.3255	0.0371	0.0268
11589	3.6913	110	178.0	S	3.1456	-0.3218	37.3134
RUBIDIUM IODIDE (P=0.001KB)				C	0.2749	0.0351	0.0282
11590	3.5850	110	220.0	S	3.7460	-0.4241	35.4610
RUBIDIUM IODIDE (P=0.175KB)				C	0.2772	0.0352	0.0281
11591	3.5904	110	220.0	S	3.7138	-0.4185	35.5872
RUBIDIUM IODIDE (P=0.410KB)				C	0.2805	0.0355	0.0280
11592	3.5979	110	220.0	S	3.6694	-0.4122	35.7143
RUBIDIUM IODIDE (P=0.600KB)				C	0.2831	0.0357	0.0279
11593	3.6045	110	220.0	S	3.6350	-0.4071	35.8423
RUBIDIUM IODIDE (P=0.800KB)				C	0.2858	0.0360	0.0278
11594	3.6108	110	220.0	S	3.6004	-0.4028	35.9712
RUBIDIUM IODIDE (P=1.000KB)				C	0.2885	0.0362	0.0276
11595	3.6174	110	220.0	S	3.5660	-0.3976	36.2319
RUBIDIUM IODIDE (P=1.200KB)				C	0.2911	0.0364	0.0276
11596	3.6227	110	220.0	S	3.5335	-0.3927	36.2319
RUBIDIUM IODIDE (P=1.400KB)				C	0.2940	0.0367	0.0274
11597	3.6293	110	220.0	S	3.4983	-0.3882	36.4963
RUBIDIUM IODIDE (P=1.600KB)				C	0.2965	0.0368	0.0273
11598	3.6347	110	220.0	S	3.4677	-0.3829	36.6300
RUBIDIUM IODIDE (P=1.800KB)				C	0.2993	0.0372	0.0272
11599	3.6411	110	220.0	S	3.4355	-0.3798	36.7647
RUBIDIUM IODIDE (P=2.000KB)				C	0.3018	0.0374	0.0271
11600	3.6477	110	220.0	S	3.4065	-0.3756	36.9004

CUBIC

CODE	NAME DENSITY	REF	TEMP(K)		11	12	44
RUBIDIUM IODIDE (P=2.200KB)				C	0.3043	0.0375	0.0270
11601	3.6532	110	220.0	S	3.3776	-0.3706	37.0370
RUBIDIUM IODIDE (P=2.400KB)				C	0.3072	0.0379	0.0269
11602	3.6597	110	220.0	S	3.3459	-0.3675	37.1747
RUBIDIUM IODIDE (P=2.600KB)				C	0.3097	0.0380	0.0268
11603	3.6651	110	220.0	S	3.3179	-0.3626	37.3134
RUBIDIUM IODIDE (P=2.800KB)				C	0.3125	0.0385	0.0267
11604	3.6706	110	220.0	S	3.2889	-0.3607	37.4532
RUBIDIUM IODIDE (P=3.000KB)				C	0.3149	0.0384	0.0266
11605	3.6772	110	220.0	S	3.2621	-0.3546	37.5940
RUBIDIUM IODIDE (P=0.001KB)				C	0.2650	0.0368	0.0279
11606	3.5690	110	260.0	S	3.9059	-0.4763	35.8423
RUBIDIUM IODIDE (P=0.200KB)				C	0.2677	0.0369	0.0278
11607	3.5753	110	260.0	S	3.8646	-0.4682	35.9712
RUBIDIUM IODIDE (P=0.400KB)				C	0.2705	0.0371	0.0277
11608	3.5819	110	260.0	S	3.8234	-0.4611	36.1011
RUBIDIUM IODIDE (P=0.600KB)				C	0.2734	0.0373	0.0276
11609	3.5882	110	260.0	S	3.7815	-0.4540	36.2319
RUBIDIUM IODIDE (P=0.800KB)				C	0.2762	0.0376	0.0275
11610	3.5948	110	260.0	S	3.7427	-0.4485	36.3636
RUBIDIUM IODIDE (P=1.000KB)				C	0.2791	0.0378	0.0274
11611	3.6012	110	260.0	S	3.7026	-0.4416	36.4963
RUBIDIUM IODIDE (P=1.250KB)				C	0.2824	0.0381	0.0273
11612	3.6099	110	260.0	S	3.6584	-0.4349	36.6300
RUBIDIUM IODIDE (P=1.500KB)				C	0.2861	0.0384	0.0271
11613	3.6174	110	260.0	S	3.6100	-0.4272	36.9004
RUBIDIUM IODIDE (P=1.750KB)				C	0.2894	0.0386	0.0270
11614	3.6251	110	260.0	S	3.5674	-0.4198	37.0370
RUBIDIUM IODIDE (P=2.000KB)				C	0.2929	0.0390	0.0269
11615	3.6326	110	260.0	S	3.5244	-0.4141	37.1747
RUBIDIUM IODIDE (P=2.250KB)				C	0.2963	0.0392	0.0268
11616	3.6402	110	260.0	S	3.4826	-0.4069	37.3134
RUBIDIUM IODIDE (P=2.500KB)				C	0.2996	0.0395	0.0266
11617	3.6477	110	260.0	S	3.4436	-0.4011	37.5940
RUBIDIUM IODIDE (P=2.750KB)				C	0.3031	0.0397	0.0265
11618	3.6554	110	260.0	S	3.4025	-0.3940	37.7358
RUBIDIUM IODIDE (P=2.990KB)				C	0.3062	0.0401	0.0264
11619	3.6619	110	260.0	S	3.3680	-0.3900	37.8788
RUBIDIUM IODIDE (P=3.250KB)				C	0.3096	0.0404	0.0262
11620	3.6706	110	260.0	S	3.3303	-0.3844	38.1679

CUBIC

CODE	NAME DENSITY	REF	TEMP(K)		11	12	44
RUBIDIUM IODIDE (P=3.500KB)				C	0.3129	0.0407	0.0261
11621	3.6771	110	260.0	S	3.2946	-0.3792	38.3142
RUBIDIUM IODIDE (P=3.750KB)				C	0.3162	0.0409	0.0259
11622	3.6848	110	260.0	S	3.2591	-0.3733	38.6100
RUBIDIUM IODIDE (P=0.001KB)				C	0.2557	0.0377	0.0277
11623	3.5510	110	300.0	S	4.0649	-0.5223	36.1011
RUBIDIUM IODIDE (P=0.200KB)				C	0.2584	0.0379	0.0276
11624	3.5573	110	300.0	S	4.0208	-0.5143	36.2319
RUBIDIUM IODIDE (P=0.400KB)				C	0.2612	0.0382	0.0275
11625	3.5648	110	300.0	S	3.9769	-0.5074	36.3636
RUBIDIUM IODIDE (P=0.600KB)				C	0.2640	0.0384	0.0274
11626	3.5713	110	300.0	S	3.9332	-0.4995	36.4963
RUBIDIUM IODIDE (P=0.800KB)				C	0.2667	0.0387	0.0273
11627	3.5777	110	300.0	S	3.8927	-0.4933	36.6300
RUBIDIUM IODIDE (P=1.000KB)				C	0.2697	0.0389	0.0272
11628	3.5841	110	300.0	S	3.8477	-0.4850	36.7647
RUBIDIUM IODIDE (P=1.200KB)				C	0.2722	0.0391	0.0271
11629	3.5905	110	300.0	S	3.8113	-0.4787	36.9004
RUBIDIUM IODIDE (P=1.400KB)				C	0.2749	0.0395	0.0270
11630	3.5970	110	300.0	S	3.7739	-0.4741	37.0370
RUBIDIUM IODIDE (P=1.600KB)				C	0.2776	0.0397	0.0269
11631	3.6035	110	300.0	S	3.7360	-0.4674	37.1747
RUBIDIUM IODIDE (P=1.800KB)				C	0.2804	0.0400	0.0268
11632	3.6098	110	300.0	S	3.6981	-0.4617	37.3134
RUBIDIUM IODIDE (P=2.000KB)				C	0.2831	0.0403	0.0267
11633	3.6164	110	300.0	S	3.6623	-0.4564	37.4532
RUBIDIUM IODIDE (P=2.200KB)				C	0.2857	0.0405	0.0266
11634	3.6229	110	300.0	S	3.6279	-0.4504	37.5940
RUBIDIUM IODIDE (P=2.400KB)				C	0.2883	0.0408	0.0265
11635	3.6283	110	300.0	S	3.5947	-0.4457	37.7358
RUBIDIUM IODIDE (P=2.600KB)				C	0.2910	0.0410	0.0264
11636	3.6347	110	300.0	S	3.5603	-0.4397	37.8788
RUBIDIUM IODIDE (P=2.800KB)				C	0.2936	0.0413	0.0263
11637	3.6412	110	300.0	S	3.5284	-0.4351	38.0228
RUBIDIUM IODIDE (P=3.000KB)				C	0.2963	0.0416	0.0262
11638	3.6467	110	300.0	S	3.4958	-0.4304	38.1679
RUBIDIUM IODIDE (P=3.200KB)				C	0.2988	0.0418	0.0261
11639	3.6532	110	300.0	S	3.4657	-0.4253	38.3142
RUBIDIUM IODIDE (P=3.400KB)				C	0.3013	0.0421	0.0260
11640	3.6586	110	300.0	S	3.4367	-0.4213	38.4615

NAME / CODE	DENSITY	REF	TEMP(K)		11	12	44
RUBIDIUM IODIDE (P=3.600KB)				C	0.3040	0.0424	0.0258
11641	3.6650	110	300.0	S	3.4058	-0.4169	38.7597
RUBIDIUM IODIDE (P=3.800KB)				C	0.3066	0.0426	0.0258
11642	3.6706	110	300.0	S	3.3760	-0.4119	38.7597
SILICON				C	1.6570	0.6390	0.7960
11643	2.3400	129	293.0	S	0.7685	-0.2139	1.2563
SILICON				C	1.6800	0.6600	0.8400
11644	2.3400	129	ROOM	S	0.7647	-0.2157	1.1905
SILICON				C	1.6578	0.6394	0.7962
11645	2.3310	164	298.0	S	0.7681	-0.2138	1.2559
SILICON (P=30.000PSI)				C	1.6667	0.6480	0.7979
11646	2.3360	164	298.0	S	0.7669	-0.2147	1.2533
SILICON				C	1.6772	0.6498	0.8036
11647	2.3320	164	77.2	S	0.7609	-0.2125	1.2444
SILICON (P=30.000PSI)				C	1.6860	0.6583	0.8051
11648	2.3370	164	77.2	S	0.7597	-0.2133	1.2421
SILVER				C	1.1992	0.8974	0.4367
11649	10.5000	128	ROOM	S	2.3200	-0.9930	2.2900
SILVER				C	1.2400	0.9340	0.4610
11650	10.5300	129	ROOM	S	2.2859	-0.9821	2.1692
SILVER				C	1.2100	0.8990	0.4260
11651	10.5000	4	ROOM	S	2.2544	-0.9610	2.3474
SILVER				C	1.3149	0.9733	0.5109
11652	10.6330	191	0.0	S	2.0538	-0.8736	1.9573
SILVER				C	1.3144	0.9730	0.5108
11653	10.6330	191	10.0	S	2.0550	-0.8741	1.9577
SILVER				C	1.3141	0.9733	0.5098
11654	10.6330	191	25.0	S	2.0584	-0.8759	1.9616
SILVER				C	1.3112	0.9726	0.5072
11655	10.6300	191	50.0	S	2.0713	-0.8821	1.9716
SILVER				C	1.3054	0.9704	0.5028
11656	10.6210	191	75.0	S	2.0927	-0.8923	1.9889
SILVER				C	1.2980	0.9666	0.4982
11657	10.6110	191	100.0	S	2.1148	-0.9027	2.0072
SILVER				C	1.2908	0.9628	0.4936
11658	10.5980	191	125.0	S	2.1362	-0.9126	2.0259
SILVER				C	1.2835	0.9591	0.4890
11659	10.5850	191	150.0	S	2.1592	-0.9234	2.0450
SILVER				C	1.2762	0.9554	0.4844
11660	10.5730	191	175.0	S	2.1827	-0.9345	2.0644

				CUBIC			
	NAME				11	12	44
CODE	DENSITY	REF	TEMP(K)				
SILVER				C	1.2691	0.9517	0.4797
11661	10.5600	191	200.0	S	2.2055	-0.9451	2.0846
SILVER				C	1.2619	0.9481	0.4750
11662	10.5440	191	225.0	S	2.2300	-0.9567	2.1053
SILVER				C	1.2546	0.9444	0.4704
11663	10.5280	191	250.0	S	2.2552	-0.9685	2.1258
SILVER				C	1.2473	0.9405	0.4658
11664	10.5160	191	275.0	S	2.2795	-0.9799	2.1468
SILVER				C	1.2399	0.9367	0.4612
11665	10.5000	191	300.0	S	2.3058	-0.9923	2.1683
SILVER				C	1.2400	0.9400	0.4650
11666	10.4970	51	300.0	S	2.3291	-1.0043	2.1505
SILVER				C	1.2250	0.9300	0.4550
11667	10.4680	51	350.0	S	2.3679	-1.0219	2.1978
SILVER				C	1.2050	0.9200	0.4460
11668	10.4370	51	400.0	S	2.4487	-1.0601	2.2422
SILVER				C	1.1900	0.9100	0.4360
11669	10.4060	51	450.0	S	2.4917	-1.0797	2.2936
SILVER				C	1.1700	0.9000	0.4260
11670	10.3750	51	500.0	S	2.5814	-1.1223	2.3474
SILVER				C	1.1550	0.8900	0.4170
11671	10.3450	51	550.0	S	2.6293	-1.1443	2.3981
SILVER				C	1.1350	0.8800	0.4070
11672	10.3140	51	600.0	S	2.7295	-1.1920	2.4570
SILVER				C	1.1200	0.8700	0.3980
11673	10.2790	51	650.0	S	2.7832	-1.2168	2.5126
SILVER				C	1.1000	0.8600	0.3880
11674	10.2430	51	700.0	S	2.8960	-1.2707	2.5773
SILVER				C	1.0850	0.8550	0.3790
11675	10.2060	51	750.0	S	3.0178	-1.3300	2.6385
SILVER				C	1.0650	0.8450	0.3690
11676	10.1730	51	800.0	S	3.1513	-1.3942	2.7100
SILVER				C	1.2227	0.9177	0.4606
11677	10.4900	207	294.0	S	2.2950	-0.9840	2.1710
SILVER				C	1.2220	0.9070	0.4540
11678	10.5000	133	ROOM	S	2.2262	-0.9484	2.2026
SILVER BROMIDE				C	0.5620	0.3280	0.0728
11679	6.4730	129	ROOM	S	3.1227	-1.1508	13.7363
SILVER BROMIDE				C	0.5630	0.3300	0.0720
11680	6.4730	138	ROOM	S	3.1338	-1.1581	13.8889

CUBIC

NAME CODE	DENSITY	REF	TEMP(K)		11	12	44
SILVER CHLORIDE				C	0.6010	0.3620	0.0625
11681	5.5890	129	ROOM	S	3.0410	-1.1431	16.0000
SILVER CHLORIDE				C	0.7391	0.3907	0.0694
11682	5.6900	257	0.0	S	2.1327	-0.7375	14.4092
SILVER CHLORIDE				C	0.7382	0.3904	0.0693
11683	5.6900	257	20.0	S	2.1363	-0.7390	14.4300
SILVER CHLORIDE				C	0.7340	0.3888	0.0691
11684	5.6890	257	40.0	S	2.1518	-0.7451	14.4718
SILVER CHLORIDE				C	0.7268	0.3872	0.0687
11685	5.6870	257	60.0	S	2.1851	-0.7595	14.5560
SILVER CHLORIDE				C	0.7168	0.3850	0.0681
11686	5.6810	257	80.0	S	2.2334	-0.7804	14.6843
SILVER CHLORIDE				C	0.7053	0.3835	0.0676
11687	5.6730	257	100.0	S	2.2981	-0.8094	14.7929
SILVER CHLORIDE				C	0.6923	0.3809	0.0670
11688	5.6640	257	120.0	S	2.3701	-0.8412	14.9254
SILVER CHLORIDE				C	0.6802	0.3790	0.0665
11689	5.6560	257	140.0	S	2.4451	-0.8749	15.0376
SILVER CHLORIDE				C	0.6682	0.3766	0.0660
11690	5.6470	257	160.0	S	2.5207	-0.9086	15.1515
SILVER CHLORIDE				C	0.6561	0.3743	0.0654
11691	5.6360	257	180.0	S	2.6030	-0.9456	15.2905
SILVER CHLORIDE				C	0.6446	0.3726	0.0649
11692	5.6260	257	200.0	S	2.6908	-0.9856	15.4083
SILVER CHLORIDE				C	0.6322	0.3694	0.0644
11693	5.6130	257	220.0	S	2.7799	-1.0253	15.5279
SILVER CHLORIDE				C	0.6200	0.3662	0.0638
11694	5.6040	257	240.0	S	2.8732	-1.0669	15.6740
SILVER CHLORIDE				C	0.6082	0.3629	0.0633
11695	5.5930	257	260.0	S	2.9676	-1.1090	15.7978
SILVER CHLORIDE				C	0.5969	0.3605	0.0627
11696	5.5820	257	280.0	S	3.0730	-1.1571	15.9490
SILVER CHLORIDE				C	0.5860	0.3582	0.0622
11697	5.5690	257	295.0	S	3.1825	-1.2073	16.0772
SILVER CHLORIDE				C	0.7590	0.3908	0.0689
11698	5.6990	132	0.0	S	2.0270	-0.6889	14.5054
SILVER CHLORIDE				C	0.7585	0.3908	0.0689
11699	5.6990	132	10.0	S	2.0295	-0.6901	14.5096
SILVER CHLORIDE				C	0.7565	0.3907	0.0688
11700	5.6980	132	20.0	S	2.0392	-0.6945	14.5264

NAME CODE	DENSITY	REF	TEMP(K)		11	12	44
SILVER CHLORIDE				C	0.7540	0.3907	0.0687
11701	5.6970	132	30.0	S	2.0521	-0.7004	14.5560
SILVER CHLORIDE				C	0.7510	0.3905	0.0685
11702	5.6950	132	40.0	S	2.0669	-0.7071	14.5900
SILVER CHLORIDE				C	0.7465	0.3903	0.0683
11703	5.6930	132	50.0	S	2.0899	-0.7175	14.6413
SILVER CHLORIDE				C	0.7415	0.3900	0.0681
11704	5.6900	132	60.0	S	2.1157	-0.7292	14.6886
SILVER CHLORIDE				C	0.7355	0.3896	0.0678
11705	5.6860	132	70.0	S	2.1474	-0.7436	14.7449
SILVER CHLORIDE				C	0.7280	0.3894	0.0676
11706	5.6830	132	80.0	S	2.1901	-0.7632	14.7973
SILVER CHLORIDE				C	0.6185	0.3865	0.0659
11707	5.6560	132	150.0	S	3.1131	-1.1972	15.1791
SILVER CHLORIDE				C	0.6755	0.3852	0.0656
11708	5.6520	132	160.0	S	2.5270	-0.9177	15.2346
SILVER CHLORIDE				C	0.6700	0.3840	0.0654
11709	5.6480	132	170.0	S	2.5628	-0.9337	15.2999
SILVER CHLORIDE				C	0.6355	0.3765	0.0635
11710	5.6200	132	230.0	S	2.8141	-1.0469	15.7480
SILVER CHLORIDE				C	0.6295	0.3748	0.0632
11711	5.6160	132	240.0	S	2.8592	-1.0670	15.8228
SILVER CHLORIDE				C	0.6230	0.3730	0.0629
11712	5.6110	132	250.0	S	2.9102	-1.0898	15.8982
SILVER CHLORIDE				C	0.6170	0.3710	0.0627
11713	5.6060	132	260.0	S	2.9553	-1.1097	15.9490
SILVER CHLORIDE				C	0.6110	0.3690	0.0624
11714	5.6010	132	270.0	S	3.0019	-1.1303	16.0256
SILVER CHLORIDE				C	0.6045	0.3674	0.0622
11715	5.5960	132	280.0	S	3.0606	-1.1570	16.0772
SILVER CHLORIDE				C	0.5985	0.3655	0.0619
11716	5.5910	132	290.0	S	3.1120	-1.1799	16.1551
SILVER CHLORIDE				C	0.5965	0.3646	0.0618
11717	5.5890	132	293.0	S	3.1262	-1.1860	16.1812
SILVER CHLORIDE				C	0.5920	0.3640	0.0616
11718	5.5850	132	300.0	S	3.1765	-1.2095	16.2338
SILVER(98.4)-ALUMINUM(1.6)				C	1.2807	0.9830	0.4643
11719	10.3700	207	294.0	S	2.3420	-1.0170	2.1540
SILVER(96.1)-ALUMINUM(3.9)				C	1.2405	0.9472	0.4623
11720	10.2100	207	294.0	S	2.3790	-1.0300	2.1630

NAME / CODE	DENSITY	REF	TEMP(K)		11	12	44
SILVER(94.8)-ALUMINUM(5.2)				C	1.2288	0.9432	0.4713
11721	10.1200	207	294.0	S	2.4410	-1.0600	2.1220
SILVER(98.66)-CADMIUM(1.34)				C	1.2280	0.9250	0.4610
11722	10.4890	129	ROOM	S	2.3085	-0.9918	2.1692
SILVER(98.08)-CADMIUM(1.92)				C	1.2160	0.9130	0.4590
11723	10.4830	129	ROOM	S	2.3098	-0.9905	2.1786
SILVER(98.0)-GOLD(2.0)				C	1.2420	0.9318	0.4682
11724	10.6600	207	294.0	S	2.2560	-0.9670	2.1360
SILVER(96.0)-GOLD(4.0)				C	1.2427	0.9266	0.4728
11725	10.8500	207	294.0	S	2.2170	-0.9470	2.1150
SILVER(91.64)-INDIUM((8.36)				C	1.1660	0.8900	0.4500
11726	10.3540	129	ROOM	S	2.5286	-1.0946	2.2222
SILVER(96.93)-MAGNESIUM(3.07)				C	1.1980	0.8980	0.4600
11727	10.2360	129	ROOM	S	2.3336	-0.9998	2.1739
SILVER(92.67)-MAGNESIUM(7.33)				C	1.1590	0.8660	0.4520
11728	9.8620	129	ROOM	S	2.3906	-1.0224	2.2124
SILVER(50.0)-MAGNESIUM(50.0) (BETA')				C	0.8720	0.5930	0.5170
11729	6.1110	52	77.0	S	2.5515	-1.0328	1.9342
SILVER(50.0)-MAGNESIUM(50.0) (BETA')				C	0.8685	0.5900	0.5125
11730	6.1040	52	100.0	S	2.5565	-1.0342	1.9512
SILVER(50.0)-MAGNESIUM(50.0) (BETA')				C	0.8610	0.5835	0.5035
11731	6.0890	52	150.0	S	2.5668	-1.0368	1.9861
SILVER(50.0)-MAGNESIUM(50.0) (BETA')				C	0.8535	0.5765	0.4940
11732	6.0700	52	200.0	S	2.5729	-1.0372	2.0243
SILVER(50.0)-MAGNESIUM(50.0) (BETA')				C	0.8455	0.5700	0.4845
11733	6.0500	52	250.0	S	2.5877	-1.0420	2.0640
SILVER(50.0)-MAGNESIUM(50.0) (BETA')				C	0.8380	0.5635	0.4755
11734	6.0330	52	300.0	S	2.5983	-1.0447	2.1030
SILVER(50.0)-MAGNESIUM(50.0) (BETA')				C	0.8304	0.5565	0.4660
11735	6.0130	52	350.0	S	2.6055	-1.0455	2.1459
SILVER(50.0)-MAGNESIUM(50.0) (BETA')				C	0.8230	0.5500	0.4565
11736	5.9930	52	400.0	S	2.6153	-1.0477	2.1906
SILVER(50.0)-MAGNESIUM(50.0) (BETA')				C	0.8150	0.5435	0.4475
11737	5.9750	52	450.0	S	2.6307	-1.0525	2.2346
SILVER(50.0)-MAGNESIUM(50.0) (BETA')				C	0.8075	0.5365	0.4380
11738	5.9550	52	500.0	S	2.6373	-1.0528	2.2831
SILVER(50.0)-MAGNESIUM(50.0) (BETA')				C	0.8000	0.5300	0.4285
11739	5.9360	52	550.0	S	2.6483	-1.0554	2.3337
SILVER(93.78)-PALLADIUM(6.22)				C	1.2770	0.9580	0.4810
11740	10.6060	129	ROOM	S	2.1943	-0.9405	2.0790

CUBIC

CODE	NAME DENSITY	REF	TEMP(K)		11	12	44
SILVER(96.83)-TIN(3.17)				C	1.2100	0.9220	0.4580
11741	10.4340	129	ROOM	S	2.4240	-1.0483	2.1834
SILVER(97.60)-ZINC(2.40)				C	1.2090	0.9160	0.4580
11742	10.4390	129	ROOM	S	2.3849	-1.0280	2.1834
SILVER(96.47)-ZINC(3.53)				C	1.2300	0.9330	0.4610
11743	10.4080	129	ROOM	S	2.3523	-1.0147	2.1692
SODIUM				C	0.0526	0.0404	0.0426
11744	0.9712	128	ROOM	S	57.1000	-24.8000	23.5000
SODIUM				C	0.0945	0.0779	0.0618
11745	1.0080	4	90.0	S	41.4924	-18.7486	16.1812
SODIUM				C	0.0555	0.0425	0.0491
11746	0.9870	4	210.0	S	53.6545	-23.2686	20.3666
SODIUM				C	0.0615	0.0496	0.0592
11747	0.9710	4	ROOM	S	58.0967	-25.9370	16.8919
SODIUM				C	0.0739	0.0622	0.0419
11748	0.9710	67	299.0	S	58.6611	-26.8091	23.8663
SODIUM				C	0.0821	0.0683	0.0577
11749	1.0100	76	78.0	S	49.8335	-22.6305	17.3310
SODIUM				C	0.0803	0.0670	0.0553
11750	1.0060	76	115.0	S	51.6808	-23.5072	18.0832
SODIUM				C	0.0783	0.0653	0.0527
11751	0.9990	76	155.0	S	52.8777	-24.0453	18.9753
SODIUM				C	0.0766	0.0640	0.0500
11752	0.9910	76	195.0	S	54.5392	-24.8258	20.0000
SODIUM BROMATE				C	0.5732	0.1764	0.1523
11753	3.3390	4	ROOM	S	2.0400	-0.4800	6.5650
SODIUM BROMATE				C	0.5450	0.1910	0.1500
11754	3.3390	4	ROOM	S	2.2428	-0.5820	6.6667
SODIUM BROMIDE				C	0.3870	0.0970	0.0970
11755	3.2030	4	ROOM	S	2.8726	-0.5757	10.3093
SODIUM BROMIDE				C	0.4020	0.1150	0.0990
11756	3.2000	4	ROOM	S	2.8503	-0.6340	10.1010
SODIUM BROMIDE				C	0.3017	0.0875	0.0726
11757	3.2030	4	ROOM	S	3.8116	-0.8569	13.7741
SODIUM BROMIDE				C	0.3255	0.1314	0.1326
11758	3.2030	44	ROOM	S	4.0000	-1.1500	7.5400
SODIUM BROMIDE				C	0.4037	0.1013	0.1015
11759	3.2030	145	298.0	S	2.7544	-0.5525	9.8522
SODIUM BROMIDE				C	0.4800	0.0986	0.1070
11760	3.2990	152	4.0	S	2.2402	-0.3818	9.3458

CODE	NAME DENSITY	REF	TEMP(K)		11	12	44
SODIUM BROMIDE				C	0.4645	0.0988	0.1060
11761	3.2860	152	80.0	S	2.3264	-0.4080	9.4340
SODIUM BROMIDE				C	0.4153	0.0995	0.1017
11762	3.2250	152	240.0	S	2.6537	-0.5129	9.8328
SODIUM BROMIDE				C	0.4122	0.0996	0.1014
11763	3.2210	152	250.0	S	2.6778	-0.5211	9.8619
SODIUM BROMIDE				C	0.4092	0.0997	0.1011
11764	3.2180	152	260.0	S	2.7017	-0.5293	9.8912
SODIUM BROMIDE				C	0.4062	0.0998	0.1008
11765	3.2140	152	270.0	S	2.7260	-0.5377	9.9206
SODIUM BROMIDE				C	0.4031	0.0999	0.1005
11766	3.2100	152	280.0	S	2.7517	-0.5465	9.9502
SODIUM BROMIDE				C	0.4000	0.1000	0.1002
11767	3.2060	152	290.0	S	2.7778	-0.5556	9.9800
SODIUM BROMIDE				C	0.3970	0.1001	0.0998
11768	3.2020	152	300.0	S	2.8036	-0.5646	10.0200
SODIUM CHLORATE				C	0.6182	-0.2083	0.1196
11769	2.4900	128	ROOM	S	2.4600	1.2500	8.3600
SODIUM CHLORATE				C	0.5090	0.1550	0.1180
11770	2.4900	4	ROOM	S	2.2902	-0.5346	8.4746
SODIUM CHLORATE				C	0.4990	0.1410	0.1170
11771	2.4900	4	ROOM	S	2.2890	-0.5043	8.5470
SODIUM CHLORATE				C	0.4890	0.1390	0.1170
11772	2.4900	4	ROOM	S	2.3394	-0.5178	8.5470
SODIUM CHLORATE				C	0.4920	0.1420	0.1160
11773	2.4850	265	ROOM	S	2.3343	-0.5228	8.6207
SODIUM CHLORIDE				C	0.4641	0.1266	0.1278
11774	2.1650	128	ROOM	S	2.4400	-0.5230	7.8220
SODIUM CHLORIDE				C	0.4857	0.1194	0.1280
11775	2.1650	128	ROOM	S	2.2800	-0.4500	7.8100
SODIUM CHLORIDE				C	0.4870	0.1240	0.1260
11776	2.1650	129	ROOM	S	2.2901	-0.4648	7.9365
SODIUM CHLORIDE				C	0.4970	0.1270	0.1270
11777	2.1650	129	ROOM	S	2.2457	-0.4570	7.8740
SODIUM CHLORIDE				C	0.4850	0.1230	0.1265
11778	2.1650	129	ROOM	S	2.2976	-0.4648	7.9051
SODIUM CHLORIDE				C	0.4864	0.1258	0.1275
11779	2.1650	129	290.0	S	2.3004	-0.4727	7.8431
SODIUM CHLORIDE				C	0.4960	0.1240	0.1290
11780	2.1650	4	ROOM	S	2.2401	-0.4480	7.7519

CUBIC

CODE	NAME DENSITY	REF	TEMP(K)		11	12	44
SODIUM	CHLORIDE			C	0.4945	0.1374	0.1282
11781	2.1650	44	ROOM	S	2.3000	-0.5000	7.8000
SODIUM	CHLORIDE			C	0.5760	0.1170	0.1332
11782	2.2108	218	80.0	S	1.8640	-0.3147	7.5075
SODIUM	CHLORIDE			C	0.5730	0.1180	0.1330
11783	2.2088	218	90.0	S	1.8772	-0.3206	7.5188
SODIUM	CHLORIDE			C	0.5560	0.1220	0.1318
11784	2.1984	218	140.0	S	1.9528	-0.3514	7.5873
SODIUM	CHLORIDE			C	0.5520	0.1230	0.1315
11785	2.1962	218	150.0	S	1.9717	-0.3593	7.6046
SODIUM	CHLORIDE			C	0.5480	0.1240	0.1313
11786	2.1941	218	160.0	S	1.9911	-0.3674	7.6161
SODIUM	CHLORIDE			C	0.5440	0.1240	0.1310
11787	2.1919	218	170.0	S	2.0082	-0.3728	7.6336
SODIUM	CHLORIDE			C	0.5400	0.1250	0.1307
11788	2.1896	218	180.0	S	2.0284	-0.3813	7.6511
SODIUM	CHLORIDE			C	0.5370	0.1260	0.1304
11789	2.1874	218	190.0	S	2.0445	-0.3886	7.6687
SODIUM	CHLORIDE			C	0.5330	0.1260	0.1301
11790	2.1851	218	200.0	S	2.0626	-0.3944	7.6864
SODIUM	CHLORIDE			C	0.5290	0.1270	0.1298
11791	2.1829	218	210.0	S	2.0841	-0.4035	7.7042
SODIUM	CHLORIDE			C	0.5250	0.1270	0.1294
11792	2.1806	218	220.0	S	2.1029	-0.4096	7.7280
SODIUM	CHLORIDE			C	0.5220	0.1280	0.1291
11793	2.1782	218	230.0	S	2.1205	-0.4176	7.7459
SODIUM	CHLORIDE			C	0.5180	0.1280	0.1288
11794	2.1759	218	240.0	S	2.1401	-0.4240	7.7640
SODIUM	CHLORIDE			C	0.5140	0.1290	0.1285
11795	2.1735	218	250.0	S	2.1634	-0.4340	7.7821
SODIUM	CHLORIDE			C	0.5100	0.1290	0.1282
11796	2.1711	218	260.0	S	2.1838	-0.4409	7.8003
SODIUM	CHLORIDE			C	0.5060	0.1300	0.1278
11797	2.1687	80	270.0	S	2.2082	-0.4514	7.8247
SODIUM	CHLORIDE			C	0.5030	0.1300	0.1274
11798	2.1664	80	280.0	S	2.2242	-0.4568	7.8493
SODIUM	CHLORIDE			C	0.4990	0.1310	0.1271
11799	2.1639	80	290.0	S	2.2496	-0.4678	7.8678
SODIUM	CHLORIDE			C	0.4960	0.1310	0.1268
11800	2.1613	80	300.0	S	2.2662	-0.4735	7.8864

NAME / CODE	DENSITY	REF	TEMP(K)		11	12	44
SODIUM CHLORIDE				C	0.4920	0.1320	0.1265
11801	2.1586	80	310.0	S	2.2928	-0.4850	7.9051
SODIUM CHLORIDE				C	0.4880	0.1320	0.1261
11802	2.1558	80	320.0	S	2.3159	-0.4931	7.9302
SODIUM CHLORIDE				C	0.4840	0.1320	0.1258
11803	2.1528	80	330.0	S	2.3396	-0.5013	7.9491
SODIUM CHLORIDE				C	0.4810	0.1330	0.1255
11804	2.1497	80	340.0	S	2.3619	-0.5116	7.9681
SODIUM CHLORIDE				C	0.4770	0.1330	0.1251
11805	2.1464	80	350.0	S	2.3866	-0.5204	7.9936
SODIUM CHLORIDE				C	0.4730	0.1340	0.1248
11806	2.1431	80	360.0	S	2.4164	-0.5334	8.0128
SODIUM CHLORIDE				C	0.4690	0.1340	0.1244
11807	2.1396	80	370.0	S	2.4423	-0.5427	8.0386
SODIUM CHLORIDE				C	0.4660	0.1350	0.1241
11808	2.1360	80	380.0	S	2.4670	-0.5542	8.0580
SODIUM CHLORIDE				C	0.4620	0.1350	0.1238
11809	2.1322	80	390.0	S	2.4941	-0.5640	8.0775
SODIUM CHLORIDE				C	0.4580	0.1360	0.1234
11810	2.1283	80	400.0	S	2.5270	-0.5786	8.1037
SODIUM CHLORIDE				C	0.4550	0.1360	0.1231
11811	2.1243	80	410.0	S	2.5484	-0.5864	8.1235
SODIUM CHLORIDE				C	0.4510	0.1360	0.1227
11812	2.1202	80	420.0	S	2.5774	-0.5972	8.1500
SODIUM CHLORIDE				C	0.4470	0.1370	0.1223
11813	2.1159	80	430.0	S	2.6129	-0.6129	8.1766
SODIUM CHLORIDE				C	0.4430	0.1370	0.1219
11814	2.1115	80	440.0	S	2.6435	-0.6244	8.2034
SODIUM CHLORIDE				C	0.4390	0.1380	0.1215
11815	2.1069	80	450.0	S	2.6810	-0.6412	8.2305
SODIUM CHLORIDE				C	0.4360	0.1380	0.1211
11816	2.1023	80	460.0	S	2.7053	-0.6504	8.2576
SODIUM CHLORIDE				C	0.4320	0.1390	0.1207
11817	2.0975	80	470.0	S	2.7448	-0.6682	8.2850
SODIUM CHLORIDE				C	0.4280	0.1390	0.1203
11818	2.0926	80	480.0	S	2.7789	-0.6813	8.3126
SODIUM CHLORIDE (P= 0KB)				C	0.4911	0.1224	0.1281
11819	2.1620	149	298.0	S	2.2611	-0.4511	7.8064
SODIUM CHLORIDE (P= 2KB)				C	0.5182	0.1291	0.1287
11820	2.1804	149	298.0	S	2.1427	-0.4273	7.7700

CUBIC

NAME CODE	DENSITY	REF	TEMP(K)		11	12	44
SODIUM CHLORIDE (P= 4KB)				C	0.5379	0.1319	0.1294
11821	2.1982	149	298.0	S	2.0578	-0.4052	7.7280
SODIUM CHLORIDE (P= 6KB)				C	0.5630	0.1374	0.1300
11822	2.2153	149	298.0	S	1.9643	-0.3853	7.6923
SODIUM CHLORIDE (P= 8KB)				C	0.5897	0.1450	0.1304
11823	2.2319	149	298.0	S	1.8781	-0.3707	7.6687
SODIUM CHLORIDE (P=10KB)				C	0.6148	0.1522	0.1310
11824	2.2477	149	298.0	S	1.8038	-0.3579	7.6336
SODIUM CHLORIDE				C	0.5282	0.1228	0.1301
11825	2.1891	23	195.0	S	2.0752	-0.3915	7.6864
SODIUM CHLORIDE				C	0.4899	0.1257	0.1272
11826	2.1644	23	295.0	S	2.2802	-0.4656	7.8616
SODIUM CHLORIDE				C	0.4958	0.1306	0.1279
11827	2.1640	251	300.0	S	2.2658	-0.4724	7.8186
SODIUM CHLORIDE				C	0.5834	0.1192	0.1337
11828	2.2159	97	4.2	S	1.8418	-0.3125	7.4794
SODIUM CHLORIDE				C	0.5744	0.1200	0.1335
11829	2.2123	97	77.3	S	1.8764	-0.3243	7.4906
SODIUM CHLORIDE				C	0.4947	0.1288	0.1287
11830	2.1630	234	298.0	S	2.2651	-0.4679	7.7700
SODIUM CHLORIDE				C	0.4980	0.1300	0.1280
11831	2.1644	113	295.0	S	2.2513	-0.4660	7.8125
SODIUM CHLORIDE				C	0.5733	0.1123	0.1331
11832	2.2170	152	4.0	S	1.8639	-0.3053	7.5131
SODIUM CHLORIDE				C	0.5648	0.1142	0.1330
11833	2.2120	152	80.0	S	1.8997	-0.3195	7.5188
SODIUM CHLORIDE				C	0.5078	0.1294	0.1299
11834	2.1780	152	230.0	S	2.1966	-0.4461	7.6982
SODIUM CHLORIDE				C	0.5045	0.1299	0.1295
11835	2.1760	152	240.0	S	2.2158	-0.4537	7.7220
SODIUM CHLORIDE				C	0.5014	0.1302	0.1290
11836	2.1740	152	250.0	S	2.2335	-0.4604	7.7519
SODIUM CHLORIDE				C	0.4985	0.1305	0.1285
11837	2.1710	152	260.0	S	2.2505	-0.4669	7.7821
SODIUM CHLORIDE				C	0.4957	0.1306	0.1280
11838	2.1690	152	270.0	S	2.2664	-0.4726	7.8125
SODIUM CHLORIDE				C	0.4927	0.1308	0.1275
11839	2.1660	152	280.0	S	2.2840	-0.4792	7.8431
SODIUM CHLORIDE				C	0.4899	0.1309	0.1270
11840	2.1640	152	290.0	S	2.3004	-0.4851	7.8740

CODE	NAME DENSITY	REF	TEMP(K)		11	12	44
SODIUM CHLORIDE				C	0.4870	0.1311	0.1266
11841	2.1620	152	300.0	S	2.3181	-0.4917	7.8989
SODIUM CHLORIDE				C	0.4958	0.1306	0.1279
11842	2.1640	250	ROOM	S	2.2658	-0.4724	7.8186
SODIUM CHLORIDE				C	0.5838	0.1194	0.1327
11843	2.1669	201	4.2	S	1.8408	-0.3126	7.5358
SODIUM CHLORIDE				C	0.4942	0.1269	0.1281
11844	2.1640	79	ROOM	S	2.2607	-0.4619	7.8064
SODIUM FLUORIDE				C	0.9700	0.2560	0.2800
11845	2.7900	4	ROOM	S	1.1586	-0.2419	3.5714
SODIUM FLUORIDE				C	0.9710	0.2430	0.2800
11846	2.8050	4	ROOM	S	1.1445	-0.2291	3.5714
SODIUM FLUORIDE				C	0.9700	0.2380	0.2822
11847	2.8040	185	300.0	S	1.1413	-0.2249	3.5436
SODIUM FLUORIDE				C	1.1039	0.2242	0.2947
11848	2.8560	260	0.0	S	0.9726	-0.1642	3.3933
SODIUM FLUORIDE				C	0.9695	0.2363	0.2820
11849	2.8277	260	ROOM	S	1.1404	-0.2235	3.5461
SODIUM FLUORIDE				C	1.0850	0.2290	0.2899
11850	2.8510	152	4.0	S	0.9948	-0.1734	3.4495
SODIUM FLUORIDE				C	1.0710	0.2312	0.2897
11851	2.8470	152	80.0	S	1.0112	-0.1795	3.4518
SODIUM FLUORIDE				C	0.9862	0.2428	0.2819
11852	2.8080	152	260.0	S	1.1233	-0.2219	3.5474
SODIUM FLUORIDE				C	0.9807	0.2436	0.2813
11853	2.8050	152	270.0	S	1.1315	-0.2251	3.5549
SODIUM FLUORIDE				C	0.9749	0.2444	0.2807
11854	2.8020	152	280.0	S	1.1404	-0.2286	3.5625
SODIUM FLUORIDE				C	0.9690	0.2452	0.2801
11855	2.7990	152	290.0	S	1.1495	-0.2321	3.5702
SODIUM FLUORIDE				C	0.9630	0.2459	0.2794
11856	2.7960	152	300.0	S	1.1588	-0.2357	3.5791
SODIUM IODIDE				C	0.3035	0.0900	0.0720
11857	3.6714	4	ROOM	S	3.8120	-0.8719	13.8889
SODIUM IODIDE				C	0.2924	0.0778	0.0736
11858	3.6710	87	298.0	S	3.8506	-0.8092	13.5870
SODIUM IODIDE				C	0.3362	0.0832	0.0751
11859	3.6714	63	180.0	S	3.2983	-0.6543	13.3156
SODIUM IODIDE				C	0.3304	0.0865	0.0747
11860	3.6714	63	200.0	S	3.3955	-0.7045	13.3869

CODE	NAME DENSITY	REF	TEMP(K)		11	12	44
SODIUM IODIDE				C	0.3246	0.0887	0.0743
11861	3.6714	63	220.0	S	3.4901	-0.7490	13.4590
SODIUM IODIDE				C	0.3189	0.0898	0.0738
11862	3.6714	63	240.0	S	3.5786	-0.7863	13.5501
SODIUM IODIDE				C	0.3131	0.0899	0.0734
11863	3.6714	63	260.0	S	3.6631	-0.8172	13.6240
SODIUM IODIDE				C	0.3074	0.0890	0.0730
11864	3.6714	63	280.0	S	3.7392	-0.8395	13.6986
SODIUM IODIDE				C	0.3017	0.0875	0.0726
11865	3.6714	63	300.0	S	3.8116	-0.8569	13.7741
SODIUM IODIDE				C	0.3760	0.0798	0.0781
11866	3.7620	59	4.2	S	2.8731	-0.5030	12.8041
SODIUM IODIDE				C	0.3758	0.0798	0.0781
11867	3.7620	59	10.0	S	2.8748	-0.5035	12.8041
SODIUM IODIDE				C	0.3749	0.0799	0.0781
11868	3.7619	59	20.0	S	2.8833	-0.5065	12.8041
SODIUM IODIDE				C	0.3738	0.0799	0.0781
11869	3.7613	59	30.0	S	2.8930	-0.5095	12.8041
SODIUM IODIDE				C	0.3721	0.0805	0.0781
11870	3.7606	59	40.0	S	2.9115	-0.5178	12.8041
SODIUM IODIDE				C	0.3703	0.0809	0.0780
11871	3.7595	59	50.0	S	2.9301	-0.5254	12.8205
SODIUM IODIDE				C	0.3078	0.0900	0.0742
11872	3.6884	59	260.0	S	3.7443	-0.8471	13.4771
SODIUM IODIDE				C	0.3034	0.0905	0.0739
11873	3.6789	59	280.0	S	3.8195	-0.8775	13.5318
SODIUM IODIDE				C	0.3007	0.0912	0.0733
11874	3.6689	59	300.0	S	3.8722	-0.9011	13.6426
SPINEL (MAGNESIA-3.5ALUMINA)				C	3.0050	1.5370	1.5860
11875	3.6300	262	ROOM	S	0.5090	-0.1722	0.6305
SPINEL (MAGNESIUM ALUMINATE)				C	2.7900	1.5300	1.5300
11876	3.5810	153	ROOM	S	0.5861	-0.2076	0.6536
SPINEL (MAGNESIA-2.6ALUMINA) (P=1BAR)				C	2.9857	1.5372	1.5758
11877	3.6193	227	298.0	S	0.5153	-0.1751	0.6346
SPINEL (MAGNESIA-2.6ALUMINA) (P=2KB)				C	2.9955	1.5450	1.5775
11878	3.6229	227	298.0	S	0.5144	-0.1750	0.6339
STRONTIUM FLUORIDE				C	1.2870	0.4748	0.3308
11879	4.3210	105	4.2	S	0.9699	-0.2614	3.0230
STRONTIUM FLUORIDE				C	1.2880	0.4747	0.3308
11880	4.3210	105	20.0	S	0.9687	-0.2609	3.0230

CUBIC

CODE	NAME DENSITY	REF	TEMP(K)		11	12	44
STRONTIUM FLUORIDE				C	1.2860	0.4736	0.3304
11881	4.3200	105	40.0	S	0.9699	-0.2610	3.0266
STRONTIUM FLUORIDE				C	1.2850	0.4716	0.3297
11882	4.3200	105	60.0	S	0.9692	-0.2602	3.0331
STRONTIUM FLUORIDE				C	1.2820	0.4695	0.3291
11883	4.3190	105	80.0	S	0.9706	-0.2602	3.0386
STRONTIUM FLUORIDE				C	1.2790	0.4664	0.3280
11884	4.3160	105	100.0	S	0.9711	-0.2595	3.0488
STRONTIUM FLUORIDE				C	1.2750	0.4626	0.3264
11885	4.3140	105	120.0	S	0.9721	-0.2588	3.0637
STRONTIUM FLUORIDE				C	1.2710	0.4592	0.3250
11886	4.3110	105	140.0	S	0.9735	-0.2584	3.0769
STRONTIUM FLUORIDE				C	1.2660	0.4554	0.3235
11887	4.3080	105	160.0	S	0.9756	-0.2581	3.0912
STRONTIUM FLUORIDE				C	1.2620	0.4521	0.3219
11888	4.3040	105	180.0	S	0.9770	-0.2577	3.1066
STRONTIUM FLUORIDE				C	1.2580	0.4485	0.3205
11889	4.3000	105	200.0	S	0.9782	-0.2571	3.1201
STRONTIUM FLUORIDE				C	1.2530	0.4446	0.3191
11890	4.2950	105	220.0	S	0.9803	-0.2567	3.1338
STRONTIUM FLUORIDE				C	1.2480	0.4510	0.3176
11891	4.2900	105	240.0	S	0.9915	-0.2632	3.1486
STRONTIUM FLUORIDE				C	1.2440	0.4376	0.3161
11892	4.2860	105	260.0	S	0.9840	-0.2561	3.1636
STRONTIUM FLUORIDE				C	1.2390	0.4342	0.3144
11893	4.2810	105	280.0	S	0.9865	-0.2560	3.1807
STRONTIUM FLUORIDE				C	1.2350	0.4305	0.3128
11894	4.2770	105	300.0	S	0.9877	-0.2553	3.1969
STRONTIUM NITRATE				C	0.4730	0.2180	0.1460
11895	2.9860	129	ROOM	S	2.9811	-0.9405	6.8493
STRONTIUM NITRATE				C	0.4255	0.2921	0.1590
11896	2.9890	124	293.0	S	5.3276	-2.1686	6.2893
STRONTIUM OXIDE				C	1.8100	0.4100	0.5980
11897	4.9600	139	80.0	S	0.6029	-0.1114	1.6722
STRONTIUM OXIDE				C	1.8010	0.4110	0.5980
11898	4.9630	139	100.0	S	0.6067	-0.1127	1.6722
STRONTIUM OXIDE				C	1.7900	0.4120	0.5980
11899	4.9650	139	120.0	S	0.6113	-0.1144	1.6722
STRONTIUM OXIDE				C	1.7770	0.4140	0.5970
11900	4.9680	139	140.0	S	0.6171	-0.1166	1.6750

CUBIC

CODE	NAME DENSITY	REF	TEMP(K)		11	12	44
STRONTIUM OXIDE				C	1.7620	0.4160	0.5960
11901	4.9710	139	160.0	S	0.6238	-0.1191	1.6779
STRONTIUM OXIDE				C	1.7450	0.4180	0.5960
11902	4.9740	139	180.0	S	0.6315	-0.1220	1.6779
STRONTIUM OXIDE				C	1.7260	0.4200	0.5950
11903	4.9760	139	200.0	S	0.6404	-0.1253	1.6807
STRONTIUM OXIDE				C	1.7050	0.4230	0.5940
11904	4.9790	139	220.0	S	0.6507	-0.1293	1.6835
STRONTIUM OXIDE				C	1.6820	0.4250	0.5930
11905	4.9820	139	240.0	S	0.6620	-0.1335	1.6863
STRONTIUM OXIDE				C	1.6570	0.4280	0.5920
11906	4.9840	139	260.0	S	0.6751	-0.1386	1.6892
STRONTIUM OXIDE				C	1.6300	0.4320	0.5910
11907	4.9870	139	280.0	S	0.6901	-0.1446	1.6920
STRONTIUM OXIDE				C	1.6010	0.4350	0.5900
11908	4.9900	139	300.0	S	0.7067	-0.1510	1.6949
STRONTIUM TITANATE				C	3.4817	1.0064	4.5455
11909	5.1230	202	ROOM	S	0.3300	-0.0740	0.2200
TANTALUM				C	2.6670	1.6080	0.8249
11910	16.6780	38	300.0	S	0.6862	-0.2581	1.2123
TANTALUM				C	2.6680	1.6140	0.8249
11911	16.6780	38	300.0	S	0.6890	-0.2597	1.2123
TANTALUM				C	2.6632	1.5816	0.8736
11912	16.7129	89	0.0	S	0.6736	-0.2510	1.1447
TANTALUM				C	2.6633	1.5823	0.8728
11913	16.7129	89	20.0	S	0.6739	-0.2512	1.1457
TANTALUM				C	2.6641	1.5844	0.8709
11914	16.7079	89	40.0	S	0.6746	-0.2516	1.1482
TANTALUM				C	2.6649	1.5870	0.8682
11915	16.7029	89	60.0	S	0.6756	-0.2522	1.1518
TANTALUM				C	2.6652	1.5896	0.8647
11916	16.6979	89	80.0	S	0.6768	-0.2529	1.1565
TANTALUM				C	2.6638	1.5912	0.8609
11917	16.6928	89	100.0	S	0.6786	-0.2538	1.1616
TANTALUM				C	2.6603	1.5909	0.8564
11918	16.6879	89	120.0	S	0.6805	-0.2546	1.1677
TANTALUM				C	2.6562	1.5900	0.8522
11919	16.6829	89	140.0	S	0.6824	-0.2555	1.1734
TANTALUM				C	2.6518	1.5888	0.8480
11920	16.6778	89	160.0	S	0.6843	-0.2564	1.1792

CUBIC

CODE	NAME DENSITY	REF	TEMP(K)		11	12	44
TANTALUM				C	2.6467	1.5875	0.8436
11921	16.6728	89	180.0	S	0.6867	-0.2574	1.1854
TANTALUM				C	2.6412	1.5860	0.8393
11922	16.6679	89	200.0	S	0.6891	-0.2586	1.1915
TANTALUM				C	2.6350	1.5840	0.8352
11923	16.6577	89	220.0	S	0.6918	-0.2597	1.1973
TANTALUM				C	2.6286	1.5818	0.8309
11924	16.6529	89	240.0	S	0.6944	-0.2609	1.2035
TANTALUM				C	2.6221	1.5794	0.8266
11925	16.6478	89	260.0	S	0.6970	-0.2620	1.2098
TANTALUM				C	2.6155	1.5768	0.8224
11926	16.6413	89	280.0	S	0.6996	-0.2631	1.2160
TANTALUM				C	2.6091	1.5743	0.8182
11927	16.6330	89	300.0	S	0.7021	-0.2642	1.2222
TANTALUM				C	2.6023	1.5446	0.8255
11928	16.6260	241	298.0	S	0.6889	-0.2566	1.2114
TANTALUM				C	2.5988	1.5431	0.8226
11929	16.6209	241	313.0	S	0.6901	-0.2571	1.2156
TANTALUM				C	2.5948	1.5419	0.8195
11930	16.6158	241	333.0	S	0.6919	-0.2579	1.2203
TANTALUM				C	2.5896	1.5397	0.8162
11931	16.6110	241	353.0	S	0.6938	-0.2587	1.2252
TANTALUM				C	2.5864	1.5396	0.8131
11932	16.6008	241	373.0	S	0.6957	-0.2596	1.2298
TANTALUM				C	2.5822	1.5386	0.8100
11933	16.5960	241	393.0	S	0.6977	-0.2605	1.2346
TANTALUM				C	2.5780	1.5375	0.8067
11934	16.5909	241	413.0	S	0.6997	-0.2614	1.2396
TANTALUM				C	2.5737	1.5364	0.8041
11935	16.5861	241	433.0	S	0.7017	-0.2623	1.2437
TANTALUM				C	2.5695	1.5353	0.8012
11936	16.5761	241	453.0	S	0.7037	-0.2632	1.2482
TANTALUM				C	2.5654	1.5342	0.7983
11937	16.5711	241	473.0	S	0.7057	-0.2641	1.2527
TANTALUM				C	2.5601	1.5321	0.7955
11938	16.5661	241	493.0	S	0.7078	-0.2650	1.2570
TANTALUM				C	2.5548	1.5300	0.7928
11939	16.5562	241	513.0	S	0.7099	-0.2659	1.2613
TANTALUM				C	2.5516	1.5300	0.7902
11940	16.5512	241	533.0	S	0.7120	-0.2669	1.2655

CUBIC

CODE	NAME DENSITY	REF	TEMP(K)		11	12	44
TANTALUM				C	2.5473	1.5291	0.7875
11941	16.5463	241	553.0	S	0.7142	-0.2679	1.2698
TANTALUM				C	2.5430	1.5281	0.7849
11942	16.5413	241	573.0	S	0.7164	-0.2689	1.2740
TANTALUM				C	2.5376	1.5261	0.7824
11943	16.5314	241	593.0	S	0.7187	-0.2699	1.2781
TANTALUM				C	2.5342	1.5262	0.7798
11944	16.5265	241	613.0	S	0.7210	-0.2710	1.2823
TANTALUM				C	2.5289	1.5242	0.7774
11945	16.5216	241	633.0	S	0.7233	-0.2720	1.2864
TANTALUM				C	2.5245	1.5233	0.7750
11946	16.5117	241	653.0	S	0.7257	-0.2731	1.2903
TANTALUM				C	2.5201	1.5224	0.7724
11947	16.5066	241	673.0	S	0.7281	-0.2742	1.2946
TANTALUM				C	2.5146	1.5205	0.7702
11948	16.5018	241	693.0	S	0.7306	-0.2753	1.2984
TANTALUM				C	2.5092	1.5186	0.7680
11949	16.4968	241	713.0	S	0.7331	-0.2764	1.3021
TANTALUM				C	2.5046	1.5178	0.7660
11950	16.4919	241	733.0	S	0.7357	-0.2776	1.3055
TANTALUM				C	2.4991	1.5160	0.7639
11951	16.4821	241	753.0	S	0.7384	-0.2788	1.3090
TANTALUM				C	2.4955	1.5162	0.7620
11952	16.4722	241	773.0	S	0.7411	-0.2801	1.3123
TANTALUM MONOCARBIDE				C	5.0500	0.7300	0.7900
11953	14.6500	24	ROOM	S	0.2055	-0.0260	1.2658
THALLIUM BROMIDE				C	0.3780	0.1480	0.0756
11954	7.5600	125	293.0	S	3.3931	-0.9547	13.2275
THALLIUM BROMIDE				C	0.4469	0.1881	0.1007
11955	7.7490	259	4.2	S	2.9810	-0.8830	9.9305
THALLIUM BROMIDE				C	0.4447	0.1863	0.1003
11956	7.7390	259	30.0	S	2.9878	-0.8821	9.9701
THALLIUM BROMIDE				C	0.4389	0.1817	0.0989
11957	7.7350	259	60.0	S	3.0075	-0.8805	10.1112
THALLIUM BROMIDE				C	0.4314	0.1766	0.0963
11958	7.7200	259	90.0	S	3.0413	-0.8834	10.3842
THALLIUM BROMIDE				C	0.4227	0.1723	0.0932
11959	7.6870	259	120.0	S	3.0968	-0.8968	10.7296
THALLIUM BROMIDE				C	0.4139	0.1685	0.0898
11960	7.6440	259	150.0	S	3.1606	-0.9144	11.1359

CUBIC

NAME CODE	 DENSITY	 REF	 TEMP(K)		11	12	44
THALLIUM BROMIDE				C	0.4056	0.1646	0.0867
11961	7.6150	259	180.0	S	3.2199	-0.9295	11.5340
THALLIUM BROMIDE				C	0.3977	0.1611	0.0836
11962	7.5830	259	210.0	S	3.2807	-0.9458	11.9617
THALLIUM BROMIDE				C	0.3889	0.1571	0.0805
11963	7.5400	259	240.0	S	3.3501	-0.9639	12.4224
THALLIUM BROMIDE				C	0.3807	0.1533	0.0774
11964	7.5000	259	270.0	S	3.4167	-0.9809	12.9199
THALLIUM BROMIDE				C	0.3734	0.1498	0.0748
11965	7.4620	259	300.0	S	3.4768	-0.9955	13.3690
THALLIUM BROMIDE				C	0.4399	0.1660	0.1079
11966	7.7653	187	0.0	S	2.8658	-0.7852	9.2678
THALLIUM BROMIDE				C	0.4280	0.1618	0.0995
11967	7.6623	187	100.0	S	2.9479	-0.8087	10.0503
THALLIUM BROMIDE				C	0.4130	0.1580	0.0930
11968	7.6115	187	150.0	S	3.0716	-0.8499	10.7527
THALLIUM BROMIDE				C	0.4000	0.1544	0.0870
11969	7.5588	187	200.0	S	3.1847	-0.8869	11.4943
THALLIUM BROMIDE				C	0.3870	0.1496	0.0810
11970	7.5043	187	250.0	S	3.2940	-0.9183	12.3457
THALLIUM BROMIDE				C	0.3760	0.1458	0.0757
11971	7.4529	187	298.0	S	3.3953	-0.9487	13.2100
THALLIUM BROMIDE				C	0.3638	0.1400	0.0700
11972	7.3948	187	350.0	S	3.4966	-0.9717	14.2857
THALLIUM BROMIDE				C	0.3520	0.1338	0.0660
11973	7.3377	187	400.0	S	3.5933	-0.9897	15.1515
THALLIUM BROMIDE				C	0.3395	0.1283	0.0610
11974	7.2811	187	450.0	S	3.7158	-1.0191	16.3934
THALLIUM BROMIDE				C	0.3263	0.1220	0.0565
11975	7.2143	187	500.0	S	3.8477	-1.0471	17.6991
THALLIUM BROMIDE				C	0.3130	0.1153	0.0520
11976	7.1484	187	550.0	S	3.9853	-1.0729	19.2308
THALLIUM BROMIDE				C	0.2980	0.1098	0.0465
11977	7.0831	187	600.0	S	4.1863	-1.1272	21.5054
THALLIUM BROMIDE				C	0.2835	0.1026	0.0430
11978	7.0188	187	650.0	S	4.3674	-1.1606	23.2558
THALLIUM BROMIDE				C	0.2690	0.0958	0.0375
11979	6.9552	187	700.0	S	4.5728	-1.2009	26.6667
THALLIUM BROMIDE		(P=OKB)		C	0.3760	0.1458	0.0757
11980	7.4529	187	ROOM	S	3.3953	-0.9487	13.2100

CUBIC

CODE	NAME / DENSITY	REF	TEMP(K)		11	12	44
THALLIUM BROMIDE (P=0.981KB)				C	0.3870	0.1504	0.0790
11981	7.4888	187	ROOM	S	3.3023	-0.9242	12.6582
THALLIUM BROMIDE (P=1.962KB)				C	0.3945	0.1579	0.0820
11982	7.5236	187	ROOM	S	3.2870	-0.9396	12.1951
THALLIUM BROMIDE (P=2.943KB)				C	0.4020	0.1631	0.0855
11983	7.5579	187	ROOM	S	3.2483	-0.9375	11.6959
THALLIUM BROMIDE (P=3.924KB)				C	0.4095	0.1687	0.0880
11984	7.5918	187	ROOM	S	3.2148	-0.9380	11.3636
THALLIUM BROMIDE (P=4.905KB)				C	0.4165	0.1755	0.0905
11985	7.6244	187	ROOM	S	3.2006	-0.9488	11.0497
THALLIUM BROMIDE-CHLORIDE				C	0.3850	0.1490	0.0737
11986	7.2920	129	ROOM	S	3.3129	-0.9244	13.5685
THALLIUM BROMIDE-IODIDE				C	0.3310	0.1320	0.0579
11987	7.4110	129	ROOM	S	3.9103	-1.1148	17.2711
THALLIUM BROMIDE(0.6)-IODIDE(0.4)				C	0.3600	0.1500	0.0555
11988	7.4380	129	ROOM	S	3.6797	-1.0823	18.0180
THALLIUM CHLORIDE				C	0.4010	0.1530	0.0760
11989	7.0000	125	293.0	S	3.1596	-0.8726	13.1579
THORIUM				C	0.7770	0.4820	0.5110
11990	11.7790	21	80.0	S	2.4513	-0.9385	1.9569
THORIUM				C	0.7755	0.4820	0.5095
11991	11.7720	21	100.0	S	2.4631	-0.9441	1.9627
THORIUM				C	0.7740	0.4825	0.5075
11992	11.7650	21	120.0	S	2.4787	-0.9518	1.9704
THORIUM				C	0.7720	0.4830	0.5050
11993	11.7580	21	140.0	S	2.4986	-0.9616	1.9802
THORIUM				C	0.7700	0.4835	0.5020
11994	11.7500	21	160.0	S	2.5188	-0.9716	1.9920
THORIUM				C	0.7675	0.4840	0.4990
11995	11.7400	21	180.0	S	2.5436	-0.9837	2.0040
THORIUM				C	0.7655	0.4845	0.4960
11996	11.7330	21	200.0	S	2.5647	-0.9941	2.0161
THORIUM				C	0.7620	0.4855	0.4930
11997	11.7260	21	220.0	S	2.6034	-1.0132	2.0284
THORIUM				C	0.7605	0.4860	0.4890
11998	11.7190	21	240.0	S	2.6211	-1.0219	2.0450
THORIUM				C	0.7580	0.4870	0.4855
11999	11.7120	21	260.0	S	2.6525	-1.0376	2.0597
THORIUM				C	0.7560	0.4880	0.4820
12000	11.7030	21	280.0	S	2.6800	-1.0513	2.0747

CUBIC

CODE	NAME DENSITY	REF	TEMP(K)		11	12	44
THORIUM				C	0.7530	0.4890	0.4780
12001	11.6940	21	300.0	S	2.7178	-1.0701	2.0920
THORIUM				C	0.7510	0.4905	0.4745
12002	11.6870	21	320.0	S	2.7516	-1.0871	2.1075
THORIUM				C	0.7480	0.4915	0.4705
12003	11.6800	21	340.0	S	2.7917	-1.1070	2.1254
THORIUM DIOXIDE				C	3.6700	1.0600	0.7970
12004	9.9910	157	298.0	S	0.3130	-0.0701	1.2547
TIN TELLURIDE				C	1.2800	0.0540	0.1267
12005	6.4760	28	0.0	S	0.7839	-0.0317	7.8927
TIN TELLURIDE				C	1.2790	0.0540	0.1267
12006	6.4760	28	10.0	S	0.7845	-0.0318	7.8927
TIN TELLURIDE				C	1.2770	0.0550	0.1265
12007	6.4760	28	20.0	S	0.7859	-0.0325	7.9051
TIN TELLURIDE				C	1.2710	0.0560	0.1260
12008	6.4750	28	40.0	S	0.7897	-0.0333	7.9365
TIN TELLURIDE				C	1.2640	0.0580	0.1251
12009	6.4740	28	60.0	S	0.7943	-0.0348	7.9936
TIN TELLURIDE				C	1.2530	0.0590	0.1241
12010	6.4720	28	80.0	S	0.8015	-0.0360	8.0580
TIN TELLURIDE				C	1.2420	0.0610	0.1232
12011	6.4700	28	100.0	S	0.8089	-0.0379	8.1169
TIN TELLURIDE				C	1.2310	0.0620	0.1223
12012	6.4670	28	125.0	S	0.8163	-0.0391	8.1766
TIN TELLURIDE				C	1.2130	0.0640	0.1212
12013	6.4640	28	150.0	S	0.8288	-0.0415	8.2508
TIN TELLURIDE				C	1.1840	0.0680	0.1199
12014	6.4580	28	200.0	S	0.8499	-0.0462	8.3403
TIN TELLURIDE				C	1.1550	0.0720	0.1185
12015	6.4510	28	250.0	S	0.8722	-0.0512	8.4388
TIN TELLURIDE				C	1.1250	0.0750	0.1172
12016	6.4450	28	300.0	S	0.8964	-0.0560	8.5324
TITANIUM CARBIDE				C	5.0000	1.1300	1.7500
12017	4.9400	112	ROOM	S	0.2182	-0.0402	0.5714
TITANIUM CARBIDE				C	3.8910	0.4330	2.0320
12018	4.9220	73	ROOM	S	0.2629	-0.0263	0.4921
TITANIUM-0.91CARBIDE				C	5.1450	1.0600	1.7880
12019	4.9050	49	298.0	S	0.2091	-0.0357	0.5593
TUNGSTEN				C	5.1257	2.0582	1.5267
12020	19.3000	128	ROOM	S	0.2534	-0.0726	0.6550

CUBIC

NAME CODE	DENSITY	REF	TEMP(K)		11	12	44
TUNGSTEN				C	5.0085	1.9800	1.5267
12021	19.3000	128	ROOM	S	0.2573	-0.0729	0.6550
TUNGSTEN				C	5.3060	2.0400	1.6270
12022	19.3160	39	77.0	S	0.2396	-0.0665	0.6146
TUNGSTEN				C	5.3050	2.0390	1.6260
12023	19.3150	39	80.0	S	0.2396	-0.0665	0.6150
TUNGSTEN				C	5.2970	2.0330	1.6240
12024	19.3110	39	100.0	S	0.2398	-0.0665	0.6158
TUNGSTEN				C	5.2900	2.0280	1.6220
12025	19.3070	39	120.0	S	0.2400	-0.0665	0.6165
TUNGSTEN				C	5.2830	2.0250	1.6200
12026	19.3040	39	140.0	S	0.2403	-0.0666	0.6173
TUNGSTEN				C	5.2750	2.0210	1.6180
12027	19.3000	39	160.0	S	0.2407	-0.0667	0.6180
TUNGSTEN				C	5.2670	2.0190	1.6160
12028	19.2950	39	180.0	S	0.2411	-0.0668	0.6188
TUNGSTEN				C	5.2590	2.0170	1.6130
12029	19.2910	39	200.0	S	0.2415	-0.0669	0.6200
TUNGSTEN				C	5.2500	2.0160	1.6110
12030	19.2860	39	220.0	S	0.2421	-0.0672	0.6207
TUNGSTEN				C	5.2410	2.0170	1.6090
12031	19.2800	39	240.0	S	0.2427	-0.0675	0.6215
TUNGSTEN				C	5.2330	2.0170	1.6070
12032	19.2760	39	260.0	S	0.2433	-0.0677	0.6223
TUNGSTEN				C	5.2240	2.0180	1.6050
12033	19.2710	39	280.0	S	0.2439	-0.0680	0.6231
TUNGSTEN				C	5.2140	2.0180	1.6040
12034	19.2650	39	300.0	S	0.2446	-0.0683	0.6234
TUNGSTEN				C	5.1940	2.0180	1.5990
12035	19.2540	39	340.0	S	0.2460	-0.0688	0.6254
TUNGSTEN				C	5.1750	2.0190	1.5950
12036	19.2460	39	380.0	S	0.2474	-0.0694	0.6270
TUNGSTEN				C	5.1540	2.0200	1.5910
12037	19.2340	39	420.0	S	0.2490	-0.0701	0.6285
TUNGSTEN				C	5.1340	2.0210	1.5860
12038	19.2250	39	460.0	S	0.2505	-0.0708	0.6305
TUNGSTEN				C	5.1130	2.0230	1.5820
12039	19.2160	39	500.0	S	0.2521	-0.0715	0.6321
TUNGSTEN				C	5.3255	2.0495	1.6313
12040	19.3204	89	0.0	S	0.2389	-0.0664	0.6130

CUBIC

CODE	NAME DENSITY	REF	TEMP(K)		11	12	44
TUNGSTEN				C	5.3255	2.0495	1.6313
12041	19.3204	89	20.0	S	0.2389	-0.0664	0.6130
TUNGSTEN				C	5.3254	2.0494	1.6313
12042	19.3204	89	40.0	S	0.2389	-0.0664	0.6130
TUNGSTEN				C	5.3254	2.0496	1.6312
12043	19.3146	89	60.0	S	0.2389	-0.0664	0.6130
TUNGSTEN				C	5.3252	2.0496	1.6312
12044	19.3088	89	80.0	S	0.2389	-0.0664	0.6130
TUNGSTEN				C	5.3239	2.0503	1.6310
12045	19.3031	89	100.0	S	0.2390	-0.0665	0.6131
TUNGSTEN				C	5.3220	2.0516	1.6298
12046	19.2973	89	120.0	S	0.2392	-0.0666	0.6136
TUNGSTEN				C	5.3180	2.0550	1.6273
12047	19.2915	89	140.0	S	0.2397	-0.0668	0.6145
TUNGSTEN				C	5.3132	2.0592	1.6238
12048	19.2915	89	160.0	S	0.2402	-0.0671	0.6158
TUNGSTEN				C	5.3049	2.0609	1.6207
12049	19.2857	89	180.0	S	0.2409	-0.0674	0.6170
TUNGSTEN				C	5.2936	2.0592	1.6178
12050	19.2799	89	200.0	S	0.2415	-0.0676	0.6181
TUNGSTEN				C	5.2817	2.0569	1.6153
12051	19.2742	89	220.0	S	0.2422	-0.0679	0.6191
TUNGSTEN				C	5.2697	2.0543	1.6130
12052	19.2684	89	240.0	S	0.2429	-0.0681	0.6200
TUNGSTEN				C	5.2576	2.0514	1.6108
12053	19.2684	89	260.0	S	0.2435	-0.0684	0.6208
TUNGSTEN				C	5.2452	2.0484	1.6089
12054	19.2626	89	280.0	S	0.2442	-0.0686	0.6215
TUNGSTEN				C	5.2327	2.0453	1.6072
12055	19.2568	89	300.0	S	0.2449	-0.0688	0.6222
TUNGSTEN				C	5.2239	2.0437	1.6083
12056	19.2570	156	297.0	S	0.2454	-0.0690	0.6218
TUNGSTEN				C	5.1884	2.0414	1.5978
12057	19.2339	156	373.0	S	0.2478	-0.0700	0.6259
TUNGSTEN				C	5.1408	2.0390	1.5868
12058	19.2165	156	473.0	S	0.2511	-0.0713	0.6302
TUNGSTEN				C	5.0921	2.0372	1.5755
12059	19.1934	156	573.0	S	0.2546	-0.0728	0.6347
TUNGSTEN				C	5.0423	2.0361	1.5637
12060	19.1705	156	673.0	S	0.2583	-0.0743	0.6395

CUBIC

CODE	NAME DENSITY	REF	TEMP(K)		11	12	44
TUNGSTEN				C	4.9914	2.0356	1.5516
12061	19.1418	156	773.0	S	0.2623	-0.0760	0.6445
TUNGSTEN				C	4.9394	2.0358	1.5390
12062	19.1131	156	873.0	S	0.2666	-0.0778	0.6498
TUNGSTEN				C	4.8863	2.0366	1.5260
12063	19.0903	156	973.0	S	0.2711	-0.0798	0.6553
TUNGSTEN				C	4.8322	2.0381	1.5126
12064	19.0674	156	1073.0	S	0.2760	-0.0819	0.6611
TUNGSTEN				C	4.7769	2.0402	1.4988
12065	19.0332	156	1173.0	S	0.2812	-0.0842	0.6672
TUNGSTEN				C	4.7205	2.0430	1.4846
12066	19.0106	156	1273.0	S	0.2868	-0.0866	0.6736
TUNGSTEN				C	4.6631	2.0464	1.4699
12067	18.9822	156	1373.0	S	0.2928	-0.0893	0.6803
TUNGSTEN				C	4.6046	2.0505	1.4549
12068	18.9539	156	1473.0	S	0.2993	-0.0922	0.6873
TUNGSTEN				C	4.5449	2.0552	1.4394
12069	18.9144	156	1573.0	S	0.3063	-0.0954	0.6947
TUNGSTEN				C	4.4842	2.0606	1.4236
12070	18.8805	156	1673.0	S	0.3138	-0.0988	0.7024
TUNGSTEN				C	4.4224	2.0666	1.4073
12071	18.8469	156	1773.0	S	0.3220	-0.1025	0.7106
TUNGSTEN				C	4.3594	2.0733	1.3906
12072	18.8132	156	1873.0	S	0.3308	-0.1066	0.7191
TUNGSTEN				C	4.2955	2.0805	1.3735
12073	18.7796	156	1973.0	S	0.3404	-0.1111	0.7281
TUNGSTEN				C	4.2304	2.0885	1.3560
12074	18.7406	156	2073.0	S	0.3509	-0.1160	0.7375
URANIUM CARBIDE				C	3.2000	0.8500	0.6470
12075	13.6300	118	300.0	S	0.3517	-0.0738	1.5456
URANIUM DIOXIDE				C	3.9600	1.2100	0.6410
12076	10.9700	269	298.0	S	0.2947	-0.0690	1.5601
URANIUM-2.23OXIDE				C	3.1900	1.2100	0.4100
12077	10.9690	71	ROOM	S	0.3961	-0.1089	2.4390
VANADIUM				C	1.9600	1.3300	0.6700
12078	5.9600	4	ROOM	S	1.1304	-0.4570	1.4925
VANADIUM				C	2.3240	1.1936	0.4595
12079	6.0340	14	0.0	S	0.6605	-0.2241	2.1763
VANADIUM				C	2.3239	1.1940	0.4587
12080	6.0330	14	25.0	S	0.6608	-0.2243	2.1801

CUBIC

CODE	NAME DENSITY	REF	TEMP(K)		11	12	44
VANADIUM				C	2.3216	1.1930	0.4560
12081	6.0320	14	50.0	S	0.6615	-0.2245	2.1930
VANADIUM				C	2.3177	1.1920	0.4523
12082	6.0310	14	75.0	S	0.6631	-0.2252	2.2109
VANADIUM				C	2.3132	1.1906	0.4483
12083	6.0300	14	100.0	S	0.6649	-0.2259	2.2306
VANADIUM				C	2.3085	1.1914	0.4447
12084	6.0290	14	125.0	S	0.6678	-0.2273	2.2487
VANADIUM				C	2.3042	1.1908	0.4418
12085	6.0280	14	150.0	S	0.6699	-0.2282	2.2635
VANADIUM				C	2.2995	1.1916	0.4385
12086	6.0270	14	175.0	S	0.6729	-0.2297	2.2805
VANADIUM				C	2.2951	1.1910	0.4353
12087	6.0260	14	200.0	S	0.6751	-0.2306	2.2973
VANADIUM				C	2.2905	1.1910	0.4323
12088	6.0250	14	225.0	S	0.6777	-0.2318	2.3132
VANADIUM				C	2.2869	1.1898	0.4300
12089	6.0240	14	250.0	S	0.6791	-0.2324	2.3256
VANADIUM				C	2.2831	1.1884	0.4275
12090	6.0230	14	275.0	S	0.6805	-0.2330	2.3392
VANADIUM				C	2.2795	1.1870	0.4255
12091	6.0220	14	300.0	S	0.6819	-0.2335	2.3502
VANADIUM				C	2.2870	1.1900	0.4315
12092	6.0220	38	300.0	S	0.6791	-0.2324	2.3175
ZINC SELENIDE (CONSTANT E)				C	0.8096	0.4881	0.4405
12093	5.2620	32	298.0	S	2.2600	-0.8500	2.2700
ZINC SULFIDE				C	0.9441	0.5696	0.4367
12094	4.0830	128	ROOM	S	1.9400	-0.7300	2.2900
ZINC SULFIDE				C	1.0790	0.7220	0.4120
12095	4.0830	128	ROOM	S	1.9995	-0.8016	2.4272
ZINC SULFIDE				C	1.0000	0.6500	0.3400
12096	4.0870	204	ROOM	S	2.0497	-0.8075	2.9412
ZINC SULFIDE				C	0.9760	0.5900	0.4510
12097	4.0790	83	302.0	S	1.8817	-0.7090	2.2173
ZINC SULFIDE (CONSTANT E)				C	1.0711	0.6664	0.4655
12098	4.0990	32	77.0	S	1.7860	-0.6850	2.1480
ZINC SULFIDE (CONSTANT E)				C	1.0462	0.6534	0.4613
12099	4.0880	32	298.0	S	1.8390	-0.7070	2.1680
ZINC SULFIDE				C	1.0550	0.6540	0.4680
12100	4.0940	281	93.0	S	1.8036	-0.6902	2.1368

CODE	NAME DENSITY	REF	TEMP(K)		11	12	44
ZINC SULFIDE				C	1.0540	0.6535	0.4675
12101	4.0940	281	113.0	S	1.8058	-0.6911	2.1390
ZINC SULFIDE				C	1.0520	0.6530	0.4670
12102	4.0928	281	133.0	S	1.8122	-0.6941	2.1413
ZINC SULFIDE				C	1.0500	0.6520	0.4670
12103	4.0916	281	153.0	S	1.8166	-0.6959	2.1413
ZINC SULFIDE				C	1.0480	0.6515	0.4660
12104	4.0916	281	173.0	S	1.8232	-0.6989	2.1459
ZINC SULFIDE				C	1.0460	0.6510	0.4660
12105	4.0903	281	193.0	S	1.8297	-0.7019	2.1459
ZINC SULFIDE				C	1.0435	0.6505	0.4650
12106	4.0891	281	213.0	S	1.8385	-0.7060	2.1505
ZINC SULFIDE				C	1.0410	0.6500	0.4650
12107	4.0879	281	233.0	S	1.8474	-0.7101	2.1505
ZINC SULFIDE				C	1.0380	0.6490	0.4640
12108	4.0854	281	253.0	S	1.8565	-0.7142	2.1552
ZINC SULFIDE				C	1.0350	0.6485	0.4630
12109	4.0842	281	273.0	S	1.8678	-0.7195	2.1598
ZINC SULFIDE				C	1.0320	0.6475	0.4620
12110	4.0830	281	293.0	S	1.8771	-0.7237	2.1645
ZINC TELLURIDE (CONSTANT E)				C	0.7134	0.4078	0.3115
12111	5.6360	32	298.0	S	2.4000	-0.8730	3.2100
ZIRCONIUM CARBIDE				C	4.7200	0.9870	1.5930
12112	6.6060	49	298.0	S	0.2284	-0.0395	0.6277
ZIRCONIUM CARBIDE				C	4.2300	0.4080	1.4640
12113	6.5640	46	300.0	S	0.2405	-0.0212	0.6831
ZIRCONIUM-COBALT INTERMETALLIC				C	2.4137	1.1417	0.8823
12114	8.3240	229	0.0	S	0.5951	-0.1911	1.1334
ZIRCONIUM-COBALT INTERMETALLIC				C	2.4137	1.1417	0.8823
12115	8.3240	229	77.0	S	0.5951	-0.1911	1.1334
ZIRCONIUM-COBALT INTERMETALLIC				C	2.4126	1.1406	0.8823
12116	8.3220	229	100.0	S	0.5951	-0.1910	1.1334
ZIRCONIUM-COBALT INTERMETALLIC				C	2.4086	1.1382	0.8722
12117	8.3140	229	150.0	S	0.5959	-0.1912	1.1465
ZIRCONIUM-COBALT INTERMETALLIC				C	2.3867	1.1351	0.8591
12118	8.3020	229	200.0	S	0.6042	-0.1947	1.1640
ZIRCONIUM-COBALT INTERMETALLIC				C	2.3581	1.1301	0.8488
12119	8.2890	229	250.0	S	0.6151	-0.1993	1.1781
ZIRCONIUM-COBALT INTERMETALLIC				C	2.3311	1.1271	0.8366
12120	8.2770	229	300.0	S	0.6264	-0.2042	1.1953

CUBIC

CODE	NAME DENSITY	REF	TEMP(K)	PRESSURE DERIVATIVES	11	12	44
ALUMINUM				C'(P)	6.3500	3.4500	2.1000
12121	2.6970	254	298.0	S'(P)	-9.8560	3.6487	-26.1469
BARIUM FLUORIDE				C'(P)	4.6400	5.0300	0.7100
12122	4.9120	278	195.0	S'(P)	-0.5249	-2.0162	-10.1563
BARIUM FLUORIDE				C'(P)	4.8200	5.1700	0.7770
12123	4.8860	278	298.0	S'(P)	-0.7298	-2.1065	-11.7823
CALCIUM FLUORIDE				C'(P)	5.7300	5.3500	1.1700
12124	3.2085	134	77.4	S'(P)	-0.9473	-0.7002	-8.9878
CALCIUM FLUORIDE				C'(P)	6.3200	4.8200	1.2100
12125	3.1963	134	194.5	S'(P)	-1.4566	-0.4501	-9.8719
CALCIUM FLUORIDE				C'(P)	6.6500	5.6400	1.2600
12126	3.1835	134	273.1	S'(P)	-1.3758	-0.6846	-10.8549
CALCIUM FLUORIDE				C'(P)	6.6200	6.0800	1.3300
12127	3.1795	134	295.5	S'(P)	-1.2173	-0.8458	-11.6211
CALCIUM FLUORIDE				C'(P)	5.8500	4.2100	1.2900
12128	3.1980	278	195.0	S'(P)	-1.4461	-0.3504	-10.5910
CALCIUM FLUORIDE				C'(P)	6.0500	4.3500	1.3100
12129	3.1810	278	298.0	S'(P)	-1.5574	-0.3812	-11.5348
CESIUM BROMIDE				C'(P)	5.6810	4.3380	3.1860
12130	4.6124	213	77.0	S'(P)	-33.1050	-8.2392	-337.2197
CESIUM BROMIDE				C'(P)	5.8100	4.3700	3.2000
12131	4.4560	213	300.0	S'(P)	-40.2521	-11.8077	-573.4673
CESIUM BROMIDE				C'(P)	6.3000	4.9300	3.6800
12132	4.4560	55	286.0	S'(P)	-42.1077	-15.0460	-637.1194
CESIUM CHLORIDE				C'(P)	6.8200	5.0500	3.5600
12133	3.9880	55	286.0	S'(P)	-34.0156	-11.2770	-533.3416
CESIUM IODIDE				C'(P)	6.5700	4.9000	3.7200
12134	4.5250	55	286.0	S'(P)	-72.4374	-21.0655	-896.9553
COPPER				C'(P)	6.3600	5.2000	2.3500
12135	8.9300	68	298.0	S'(P)	-3.8797	1.4438	-4.1667
COPPER				C'(P)	5.9400	5.1900	2.6300
12136	8.9300	133	ROOM	S'(P)	-2.6727	0.8411	-4.6016
GALLIUM ANTIMONIDE				C'(P)	4.9600	4.6400	1.0100
12137	5.6137	170	298.0	S'(P)	-2.5846	-1.1992	-5.4220
GALLIUM ARSENIDE				C'(P)	4.6300	4.4200	1.1000
12138	5.3169	169	298.0	S'(P)	-1.2083	-0.7120	-3.1134
GARNET (ALMANDITE-PYROPE)				C'(P)	7.4800	4.4100	1.3100
12139	4.1602	242	298.0	S'(P)	-0.7379	0.0800	-1.5258
GOLD				C'(P)	7.0140	6.1380	1.7900
12140	19.3000	68	298.0	S'(P)	-6.9871	3.1338	-10.1377

CUBIC

CODE	NAME DENSITY	REF	TEMP(K)		11	12	44
					PRESSURE DERIVATIVES		
GOLD				C'(P)	5.7200	4.9600	1.5200
12141	19.3000	133	ROOM	S'(P)	-6.1757	2.7992	-8.8257
GOLD(100.0)				C'(P)	6.7340	5.8580	1.8410
12142	19.2830	114	296.5	S'(P)	-6.9578	3.1356	-10.4365
GOLD(97.05)-NICKEL(2.95)				C'(P)	6.9470	6.0670	1.8660
12143	19.0600	114	296.5	S'(P)	-6.7644	3.0264	-10.0919
GOLD(91.28)-NICKEL(9.72)				C'(P)	6.7950	5.9170	1.8650
12144	18.5330	114	296.5	S'(P)	-6.1867	2.7411	-9.2222
GOLD(75.80)-NICKEL(24.20)				C'(P)	6.6490	5.6610	1.9840
12145	17.3410	114	296.5	S'(P)	-4.9023	2.1085	-7.6338
GOLD(57.58)-NICKEL(42.42)				C'(P)	6.5780	5.6660	2.1060
12146	15.7120	114	296.5	S'(P)	-3.4413	1.3784	-5.5035
IRON				C'(P)	7.5200	5.2000	2.6600
12147	7.8720	220	300.0	S'(P)	-1.8922	0.5888	-1.9633
IRON				C'(P)	6.7200	4.5800	2.5900
12148	7.8672	121	300.0	S'(P)	-1.7766	0.5698	-1.9050
LEAD				C'(P)	5.8200	5.2600	1.9700
12149	11.4420	184	195.0	S'(P)	-0.5663	0.2408	-0.7159
LEAD				C'(P)	5.9310	5.3290	2.0600
12150	11.3440	184	296.0	S'(P)	-0.7732	0.3406	-0.9184
LITHIUM FLUORIDE				C'(P)	9.9700	2.7300	1.3800
12151	2.6390	185	300.0	S'(P)	-12.2122	4.3383	-3.4031
LITHIUM FLUORIDE				C'(P)	8.1800	2.6000	1.0000
12152	2.6380	79	ROOM	S'(P)	-9.4808	3.2134	-2.4691
MAGNESIUM OXIDE				C'(P)	9.9750	2.1660	1.4310
12153	3.5995	20	77.4	S'(P)	-1.3497	0.3814	-0.5763
MAGNESIUM OXIDE				C'(P)	9.4770	1.9920	1.1600
12154	3.5833	20	296.0	S'(P)	-1.4197	0.4256	-0.4773
MAGNESIUM OXIDE				C'(P)	9.1600	1.8200	1.1200
12155	3.5790	54	298.0	S'(P)	-1.3845	0.4222	-0.4613
MAGNESIUM OXIDE				C'(P)	8.7000	1.4200	1.0900
12156	3.5833	243	300.0	S'(P)	-1.3526	0.4345	-0.4467
MAGNESIUM OXIDE				C'(P)	8.5500	1.3400	1.1800
12157	3.5206	243	800.0	S'(P)	-1.8595	0.6655	-0.5337
MAGNESIUM OXIDE (P=8KB)				C'(P)	8.8000	1.4200	1.0900
12158	3.6010	243	300.0	S'(P)	-1.2977	0.4135	-0.4416
MAGNESIUM OXIDE (P=8KB)				C'(P)	8.6600	1.3200	1.1800
12159	3.5396	243	800.0	S'(P)	-1.7717	0.6300	-0.5273
POTASSIUM BROMIDE				C'(P)	12.9600	1.5910	-0.3280
12160	2.7533	213	300.0	S'(P)	-117.2471	18.0304	127.1001

CUBIC

CODE	NAME / DENSITY	REF	TEMP(K)	PRESSURE DERIVATIVES	11	12	44
	POTASSIUM CHLORIDE			C'(P)	12.8000	1.6200	-0.4200
12161	2.0089	23	195.0	S'(P)	-70.3508	10.4388	101.9012
	POTASSIUM CHLORIDE			C'(P)	12.8200	1.6000	-0.3900
12162	1.9874	23	95.0	S'(P)	-84.5769	15.2814	98.2614
	POTASSIUM CHLORIDE			C'(P)	13.0000	1.5600	-0.5600
12163	1.9865	79	ROOM	S'(P)	-84.2972	15.4846	142.4470
	POTASSIUM FLUORIDE			C'(P)	11.7400	1.6600	-0.4520
12164	2.4800	145	298.0	S'(P)	-32.0228	7.3778	27.5448
	POTASSIUM IODIDE			C'(P)	13.9900	2.4210	-0.2410
12165	3.1257	213	300.0	S'(P)	-201.0393	23.0801	176.9962
	RUBIDIUM BROMIDE			C'(P)	13.5200	3.0490	-0.3620
12166	3.3590	213	300.0	S'(P)	-136.6451	11.1201	250.6923
	RUBIDIUM BROMIDE			C'(P)	13.0200	0.9380	-0.5940
12167	3.3920	110	180.0	S'(P)	-113.3684	18.3178	390.5322
	RUBIDIUM BROMIDE			C'(P)	13.6900	1.1720	-0.6020
12168	3.3780	110	220.0	S'(P)	-127.8831	20.9636	401.9519
	RUBIDIUM BROMIDE			C'(P)	13.4600	1.5770	-0.5960
12169	3.3640	110	60.0	S'(P)	-133.6931	20.5079	406.3013
	RUBIDIUM BROMIDE			C'(P)	13.6200	1.4600	-0.5870
12170	3.3500	110	0.0	S'(P)	-147.2460	25.6507	406.5093
	RUBIDIUM CHLORIDE			C'(P)	13.9000	1.3130	-0.6160
12171	2.8346	110	80.0	S'(P)	-91.9301	14.8708	268.4783
	RUBIDIUM CHLORIDE			C'(P)	13.6600	1.3100	-0.6360
12172	2.8223	110	222.0	S'(P)	-97.0960	16.7933	281.8833
	RUBIDIUM CHLORIDE			C'(P)	13.5100	1.3950	-0.6320
12173	2.8097	110	260.0	S'(P)	-103.5523	18.6215	284.8884
	RUBIDIUM CHLORIDE			C'(P)	13.7400	1.3420	-0.6130
12174	2.7969	110	304.0	S'(P)	-114.4001	22.2600	279.8774
	RUBIDIUM IODIDE			C'(P)	13.4400	1.1340	-0.5270
12175	3.6020	110	180.0	S'(P)	-171.9613	22.5929	653.3914
	RUBIDIUM IODIDE			C'(P)	13.3700	1.1530	-0.5350
12176	3.5850	110	220.0	S'(P)	-185.7998	27.0099	672.7522
	RUBIDIUM IODIDE			C'(P)	13.7000	1.2610	-0.5290
12177	3.5690	110	260.0	S'(P)	-205.7414	31.8735	674.7444
	RUBIDIUM IODIDE			C'(P)	13.3900	1.3070	-0.5220
12178	3.5510	110	300.0	S'(P)	-218.8357	36.8202	675.4299
	SILICON			C'(P)	4.2900	4.2000	0.7500
12179	2.3320	164	77.2	S'(P)	-0.5397	-0.4530	-1.1781
	SILICON			C'(P)	4.3300	4.1900	0.8000
12180	2.3310	164	298.0	S'(P)	-0.5813	-0.4463	-1.2620

CUBIC

CODE	NAME DENSITY	REF	TEMP(K)		11	12	44
					PRESSURE DERIVATIVES		
SILVER				C'(P)	7.0320	5.7540	2.3100
12181	10.5000	68	298.0	S'(P)	-9.7627	3.9216	-10.8554
SILVER				C'(P)	5.1200	3.6100	3.0400
12182	10.5000	133	ROOM	S'(P)	-10.5915	4.6264	-14.7490
SODIUM BROMIDE				C'(P)	11.5000	1.6800	0.4230
12183	3.2030	145	298.0	S'(P)	-85.0655	22.3205	-41.0590
SODIUM CHLORIDE				C'(P)	11.4700	1.9500	0.3200
12184	2.1891	23	195.0	S'(P)	-47.1734	10.7520	-18.9058
SODIUM CHLORIDE				C'(P)	11.6600	2.0800	0.3700
12185	2.1644	23	295.0	S'(P)	-57.7458	14.4787	-22.8680
SODIUM CHLORIDE				C'(P)	11.8900	2.1300	0.3700
12186	2.1640	250	ROOM	S'(P)	-58.1803	14.9989	-22.6183
SODIUM CHLORIDE				C'(P)	11.6200	1.5800	0.1000
12187	2.1640	79	ROOM	S'(P)	-58.4189	16.0013	-6.0940
SODIUM FLUORIDE				C'(P)	11.5700	1.9900	0.2050
12188	2.8040	185	300.0	S'(P)	-14.3983	3.4807	-2.5742

CODE	NAME DENSITY	REF	TEMP(K)		11	12	44
					TEMPERATURE DERIVATIVES		
ALUMINUM				C'(T)	-0.3510	-0.0670	-0.1450
12189	2.6970	254	298.0	S'(T)	0.9129	-0.4096	1.8054
BARIUM FLUORIDE				C'(T)	-0.1960	-0.1300	-0.0700
12190	4.9120	278	195.0	S'(T)	0.2154	-0.0370	1.0013
BARIUM FLUORIDE				C'(T)	-0.2040	-0.1300	-0.0730
12191	4.8860	278	298.0	S'(T)	0.2445	-0.0466	1.1061
CALCIUM FLUORIDE				C'(T)	-0.2810	-0.0950	-0.1160
12192	3.1980	278	195.0	S'(T)	0.1065	-0.0178	0.9524
CALCIUM FLUORIDE				C'(T)	-0.3180	-0.1050	-0.1220
12193	3.1810	278	298.0	S'(T)	0.1259	-0.0214	1.0742
GARNET (ALMANDITE-PYROPE)				C'(T)	-0.3480	-0.1280	-0.1040
12194	4.1602	242	298.0	S'(T)	0.0462	-0.0124	0.1211
MAGNESIUM OXIDE				C'(T)	-0.6060	0.0740	-0.1030
12195	3.5833	243	300.0	S'(T)	0.1177	-0.0492	0.0422
MAGNESIUM OXIDE				C'(T)	-0.6210	0.0110	-0.1300
12196	3.5206	243	800.0	S'(T)	0.1570	-0.0644	0.0588
MAGNESIUM OXIDE (P=8KB)				C'(T)	-0.6020	0.0760	-0.1020
12197	3.6010	243	300.0	S'(T)	0.1109	-0.0463	0.0413
MAGNESIUM OXIDE (P=8KB)				C'(T)	-0.6220	0.0140	-0.1300
12198	3.5396	243	800.0	S'(T)	0.1477	-0.0604	0.0581
RUBIDIUM BROMIDE				C'(T)	-0.2818	0.0355	-0.0082
12199	3.3500	110	300.0	S'(T)	3.4153	-1.0962	5.6579
RUBIDIUM CHLORIDE				C'(T)	-0.3080	0.0340	-0.0093
12200	2.7969	110	300.0	S'(T)	2.8536	-0.9162	4.2644
RUBIDIUM IODIDE				C'(T)	-0.2416	0.0328	-0.0056
12201	3.5510	110	300.0	S'(T)	4.4028	-1.4030	7.1813
SODIUM CHLORIDE				C'(T)	0.3750	0.0280	-0.0330
12202	2.1640	250	ROOM	S'(T)	-1.9852	0.6165	2.0173

TETRAGONAL 1

CODE	NAME DENSITY	REF	TEMP(K)		11	12	13	33	44	66
AMMONIUM DIHYDROGEN ARSENATE				C	0.6747	-0.1066	0.1652	0.3022	0.0685	0.0639
22203	2.3110	1	298.0	S	1.9300	0.6500	-1.4100	4.8500	14.6000	15.6500
AMMONIUM DIHYDROGEN PHOSPHATE				C	0.6200	-0.0500	0.1400	0.3000	0.0910	0.0610
22204	1.8030	129	293.0	S	1.8844	0.3918	-1.0622	4.3247	10.9890	16.3934
AMMONIUM DIHYDROGEN PHOSPHATE				C	0.6890	0.0400	0.1890	0.3350	0.0856	0.0595
22205	1.8030	4	ROOM	S	1.7399	0.1991	-1.0939	4.2194	11.6822	16.8067
AMMONIUM DIHYDROGEN PHOSPHATE				C	0.6720	0.0620	0.1950	0.3420	0.0830	0.0604
22206	1.8030	4	ROOM	S	1.7969	0.1576	-1.1144	4.1948	12.0482	16.5563
AMMONIUM DIHYDROGEN PHOSPHATE				C	0.6610	0.0570	0.1940	0.3340	0.0858	0.0598
22207	1.8030	4	ROOM	S	1.8428	0.1871	-1.1790	4.3637	11.6550	16.7224
AMMONIUM DIHYDROGEN PHOSPHATE (D=0)				C	0.6763	0.0591	0.1995	0.3381	0.0867	0.0687
22208	1.8030	138	ROOM	S	1.8100	0.1900	-1.1800	4.3500	11.5300	14.5500
AMMONIUM DIHYDROGEN PHOSPHATE				C	0.6170	0.0720	0.1940	0.3280	0.0850	0.0590
22209	1.8030	138	ROOM	S	2.0055	0.1707	-1.2872	4.5714	11.7647	16.9491
AMMONIUM DIHYDROGEN PHOSPHATE				C	0.7580	-0.2430	0.1300	0.2960	0.0870	0.0614
22210	1.8030	138	ROOM	S	1.7470	0.7480	-1.0958	4.3409	11.4943	16.2866
AMMONIUM DIHYDROGEN PHOSPHATE				C	0.6763	0.0591	0.1995	0.3381	0.0862	0.0602
22211	1.8030	1	298.0	S	1.8100	0.1900	-1.1800	4.3500	11.6000	16.6000
BARIUM TITANATE (E=0)				C	2.7512	1.7897	1.5156	1.6486	0.5435	1.1312
22212	5.9988	31	298.0	S	0.8050	-0.2350	-0.5240	1.5700	1.8400	0.8840
BARIUM TITANATE (D=0)				C	2.8261	1.8645	1.4159	1.7807	0.8065	1.1312
22213	5.9988	31	298.0	S	0.7250	-0.3150	-0.3260	1.0800	1.2400	0.8840
INDIUM				C	0.4450	0.3950	0.4050	0.4440	0.0655	0.1220
22214	7.3000	138	ROOM	S	14.9432	-5.0568	-9.0180	18.7040	15.2672	8.1967
MAGNESIUM FLUORIDE				C	1.2370	0.7320	0.5360	1.7700	0.5520	0.9780
22215	3.1770	62	ROOM	S	1.2942	-0.6860	-0.1842	0.6765	1.8116	1.0225
NICKEL SULPHATE HEXAHYDRATE				C	0.3209	0.2315	0.0209	0.2931	0.1156	0.1779
22216	2.0700	177	ROOM	S	6.5000	-4.6800	-0.1300	3.4300	8.6500	5.6200
POTASSIUM DIHYDROGEN ARSENATE				C	0.5300	-0.0600	-0.0200	0.3700	0.1200	0.0700
22217	2.8670	129	ROOM	S	1.9162	0.2213	0.1155	2.7152	8.3333	14.2857
POTASSIUM DIHYDROGEN PHOSPHATE				C	0.7400	0.1800	0.2700	0.6800	0.1350	0.0630
22218	2.3880	129	280.0	S	1.6015	-0.1842	-0.5627	1.9175	7.4074	15.8730
POTASSIUM DIHYDROGEN PHOSPHATE				C	0.6910	-0.0600	0.2220	0.5560	0.1290	0.0600
22219	2.3880	4	ROOM	S	1.7678	0.4362	-0.8800	2.5013	7.7519	16.6667
POTASSIUM DIHYDROGEN PHOSPHATE				C	0.7080	-0.0383	0.1550	0.5840	0.1230	0.0633
22220	2.3880	4	ROOM	S	1.5211	0.1812	-0.4518	1.9522	8.1301	15.7978
POTASSIUM DIHYDROGEN PHOSPHATE				C	0.7040	-0.0419	0.1350	0.5670	0.1280	0.0620
22221	2.3880	4	ROOM	S	1.5067	0.1660	-0.3983	1.9533	7.8125	16.1290
POTASSIUM DIHYDROGEN PHOSPHATE				C	0.7140	-0.0490	0.1290	0.5620	0.1270	0.0628
22222	2.3880	138	ROOM	S	1.4807	0.1701	-0.3789	1.9533	7.8740	15.9236
POTASSIUM DIHYDROGEN PHOSPHATE				C	0.7850	0.3200	0.3870	0.7630	0.1230	0.0610
22223	2.3880	138	ROOM	S	1.7771	-0.3734	-0.7120	2.0328	8.1301	16.3934
RUBIDIUM DIHYDROGEN PHOSPHATE				C	0.6703	-0.0596	0.1419	0.5309	0.1028	0.0359
22224	2.8660	186	293.0	S	1.6200	0.2500	-0.5000	2.1510	9.7300	27.8200
RUBIDIUM DIHYDROGEN PHOSPHATE				C	0.5562	-0.0649	0.0279	0.4398	0.1142	0.0350
22225	2.8000	2	298.0	S	1.8300	0.2200	-0.1300	2.2900	8.7600	28.5800
RUTILE				C	2.4800	2.0000	1.4000	4.5200	1.2000	1.6000
22226	4.2600	263	ROOM	S	1.1801	-0.9033	-0.0857	0.2743	0.8333	0.6250
RUTILE				C	2.7300	1.7600	1.4900	4.8400	1.2500	1.9400
22227	4.2640	35	ROOM	S	0.6554	-0.3755	-0.0862	0.2597	0.8000	0.5155
RUTILE				C	2.6600	1.7330	1.3620	4.6990	1.2390	1.8860
22228	4.2500	267	ROOM	S	0.6781	-0.4006	-0.0804	0.2594	0.8071	0.5302
RUTILE				C	2.7143	1.7796	1.4957	4.8395	1.2443	1.9477
22229	4.2600	171	298.0	S	0.6750	-0.3949	-0.0866	0.2602	0.8037	0.5134
TELLURIUM OXIDE				C	0.5320	0.4860	0.2120	1.0850	0.2440	0.5520
22230	5.9900	256	ROOM	S	11.4042	-10.3349	-0.2089	1.0033	4.0984	1.8116

TETRAGONAL 1

CODE	NAME DENSITY	REF	TEMP(K)		11	12	13	33	44	66
TIN (WHITE)				C	0.8600	0.3500	0.3000	1.3300	0.4900	0.5300
22231	7.3100	129	ROOM	S	1.4457	-0.5151	-0.2099	0.8466	2.0408	1.8868
TIN (WHITE)				C	0.8391	0.4870	0.2810	0.9665	0.1754	0.0741
22232	7.2800	138	ROOM	S	1.8500	-0.9900	-0.2500	1.1800	5.7000	13.5000
TIN (WHITE)				C	0.7350	0.2340	0.2790	0.8700	0.2200	0.2265
22233	7.2800	138	ROOM	S	1.6309	-0.3651	-0.4059	1.4098	4.5455	4.4150
TIN (WHITE)				C	0.8274	0.5785	0.3421	1.0310	0.2695	0.2818
22234	7.3800	210	4.2	S	2.4330	-1.5847	-0.2815	1.1567	3.7106	3.5486
TIN (WHITE)				C	0.8152	0.5790	0.3642	1.0040	0.2620	0.2781
22235	7.3670	210	77.0	S	2.5593	-1.6744	-0.3210	1.2289	3.8168	3.5958
TIN (WHITE)				C	0.7230	0.5940	0.3578	0.8840	0.2203	0.2400
22236	7.2790	210	300.0	S	4.3627	-3.3893	-0.3940	1.4501	4.5393	4.1667
TIN (WHITE)				C	0.8021	0.5758	0.3638	0.9921	0.2551	0.2618
22237	7.2900	135	93.0	S	2.6600	-1.7600	-0.3300	1.2500	3.9200	3.8200
TIN (WHITE)				C	0.8075	0.5882	0.3766	0.9969	0.2525	0.2604
22238	7.2900	135	108.0	S	2.7300	-1.8300	-0.3400	1.2600	3.9600	3.8400
TIN (WHITE)				C	0.7784	0.5718	0.3739	0.9763	0.2475	0.2564
22239	7.2900	135	138.0	S	2.8900	-1.9500	-0.3600	1.3000	4.0400	3.9000
TIN (WHITE)				C	0.7616	0.5667	0.3795	0.9687	0.2421	0.2532
22240	7.2900	135	168.0	S	3.0500	-2.0800	-0.3800	1.3300	4.1300	3.9500
TIN (WHITE)				C	0.7620	0.5808	0.3949	0.9676	0.2364	0.2494
22241	7.2900	135	198.0	S	3.2500	-2.2700	-0.4000	1.3600	4.2300	4.0100
TIN (WHITE)				C	0.7610	0.5940	0.4065	0.9582	0.2309	0.2445
22242	7.2900	135	228.0	S	3.4900	-2.5000	-0.4200	1.4000	4.3300	4.0900
TIN (WHITE)				C	0.7267	0.5741	0.4065	0.9485	0.2247	0.2404
22243	7.2900	135	258.0	S	3.8000	-2.7500	-0.4500	1.4400	4.4500	4.1600
TIN (WHITE)				C	0.7529	0.6156	0.4409	0.9552	0.2193	0.2336
22244	7.2900	135	288.0	S	4.1600	-3.1200	-0.4800	1.4900	4.5600	4.2800
TIN (WHITE)				C	0.7177	0.5943	0.4345	0.9371	0.2132	0.2288
22245	7.2900	135	318.0	S	4.6000	-3.5000	-0.5100	1.5400	4.6900	4.3700
TIN (WHITE)				C	0.7347	0.6249	0.4788	0.9662	0.2062	0.2237
22246	7.2900	135	348.0	S	5.1200	-3.9900	-0.5600	1.5900	4.8500	4.4700
TIN (WHITE)				C	0.7336	0.6382	0.5072	0.9810	0.1980	0.2179
22247	7.2900	135	378.0	S	5.8300	-4.6500	-0.6100	1.6500	5.0500	4.5900
TIN (WHITE)				C	0.7865	0.7063	0.6059	1.0801	0.1887	0.2101
22248	7.2900	135	408.0	S	6.8500	-5.6200	-0.6900	1.7000	5.3000	4.7600
TIN (WHITE)				C	0.8655	0.8010	0.7250	1.1957	0.1792	0.2033
22249	7.2900	135	438.0	S	8.3800	-7.1100	-0.7700	1.7700	5.5800	4.9200
TIN (WHITE)				C	0.7619	0.7110	0.6768	1.1624	0.1704	0.1980
22250	7.2900	135	468.0	S	10.5600	-9.1000	-0.8500	1.8500	5.8700	5.0500
ZIRCON				C	0.7350	0.0900	-0.0540	0.4600	0.1380	0.1600
22251	4.5600	129	ROOM	S	1.3907	-0.1597	0.1445	2.2078	7.2464	6.2500
ZIRCON				C	2.5850	1.7910	1.5420	3.8050	0.7330	1.1130
22252	4.7000	223	ROOM	S	0.7897	-0.4698	-0.1296	0.3679	1.3643	0.8985

TETRAGONAL 1

CODE	NAME DENSITY	REF	TEMP(K)		11	12	13	33	44	66
					PRESSURE DERIVATIVES					
RUTILE				C'(P)	6.4700	9.1000	5.0200	8.3400	1.1000	6.4300
22253	4.2600	171	298.0	S'(P)	1.0751	-1.9352	0.1244	-0.3456	-0.7105	-1.6950

TETRAGONAL 1

CODE	NAME DENSITY	REF	TEMP(K)		11	12	13	33	44	66
					TEMPERATURE DERIVATIVES					
RUTILE				C'(T)	-0.5100	-0.5800	-0.3300	-0.9000	-0.2200	-0.7800
22254	4.2600	171	298.0	S'(T)	-0.0065	0.0736	-0.0177	0.0475	0.1421	0.2056

TETRAGONAL 2

CODE / NAME	DENSITY	REF	TEMP(K)		11	12	13	16	33	44	66
CALCIUM MOLYBDATE (+Z=-Z OF REF 266?)				C	1.4470	0.6640	0.4660	0.1340	1.2650	0.3690	0.4510
72255	4.2550	19	298.0	S	0.9937	-0.4280	-0.2084	-0.4224	0.9441	2.7100	2.4683
CALCIUM MOLYBDATE				C	1.4500	0.6180	0.4960	-0.1460	1.2820	0.3674	0.4558
72256	4.2550	266	298.0	S	0.9740	-0.3802	-0.2297	0.4338	0.9578	2.7218	2.4718

ORTHORHOMBIC

CODE	NAME / DENSITY	REF	TEMP(K)		11	12	13	22	23	33	44	55	66
ACENAPHTHENE				C	0.1380	0.0210	0.0410	0.1262	0.0460	0.1117	0.0265	0.0290	0.0185
32257	1.2200	183	293.0	S	8.1438	-0.3125	-2.8605	9.3354	-3.7298	11.5385	37.7358	34.4828	54.0540
AMMONIUM SULFATE				C	0.3607	0.1651	0.1580	0.2981	0.1456	0.3534	0.1025	0.0717	0.0974
32258	1.7740	126	293.0	S	4.0313	-1.6931	-1.1048	4.9108	-1.2663	3.8453	9.7561	13.9470	10.2669
ARAGONITE				C	1.5958	0.3663	0.0197	0.8697	0.1591	0.8503	0.4132	0.2564	0.4274
32259	2.9300	128	ROOM	S	0.6950	-0.3000	0.0400	1.3200	-0.2400	1.2200	2.4200	3.9000	2.3400
BARITE				C	0.8941	0.4614	0.2691	0.7842	0.2676	1.0548	0.1190	0.2874	0.2778
32260	4.4000	128	ROOM	S	1.6400	-0.9000	-0.1900	1.8900	-0.2500	1.0600	8.4000	3.4800	3.6000
BARITE				C	0.8620	0.5230	0.3410	0.9170	0.3560	1.0840	0.1200	0.2870	0.2740
32261	4.5000	129	ROOM	S	1.8396	-0.9450	-0.2683	1.7353	-0.2726	1.0965	8.3333	3.4843	3.6496
BARITE				C	0.8800	0.4770	0.2690	0.7810	0.2890	1.0400	0.1170	0.2790	0.2550
32262	4.5000	138	ROOM	S	1.7256	-0.9906	-0.1711	1.9959	-0.2984	1.0887	8.5470	3.5842	3.9216
BARITE				C	0.9261	0.4555	0.3880	0.7565	0.2772	1.1138	0.1218	0.2716	0.2704
32263	4.5170	214	308.0	S	1.6570	-0.8650	-0.3620	1.9060	-0.1730	1.0670	8.2120	3.6820	3.6980
BARITE				C	0.8923	0.4092	0.4736	0.7309	0.3691	1.2232	0.1240	0.2562	0.2794
32264	4.5170	214	338.0	S	1.6820	-0.7230	-0.4330	1.9250	-0.3010	1.0760	8.0630	3.9030	3.5790
BARITE				C	0.9289	0.5839	0.5563	1.0421	0.7862	1.5354	0.1207	0.2585	0.2695
32265	4.5170	214	378.0	S	1.7020	-0.7960	-0.2090	1.9360	-0.7030	1.0870	8.2880	3.8680	3.7100
BARITE				C	1.5870	1.8137	1.9506	3.2749	3.4423	4.5348	0.1189	0.2590	0.2631
32266	4.5170	214	413.0	S	1.7230	-0.8670	-0.0830	1.9470	-1.1050	1.0950	8.4070	3.8610	3.8010
BARITE				C	1.4605	1.5916	1.5763	2.8547	2.8197	3.6910	0.1189	0.2554	0.2570
32267	4.5170	214	438.0	S	1.7450	-0.9650	-0.0080	1.9610	-1.0860	1.1040	8.4090	3.9150	3.8910
BARITE				C	4.2611	6.6894	7.5237	12.1286	13.6170	16.1884	0.1188	0.2462	0.2567
32268	4.5170	214	458.0	S	1.7500	-0.9360	-0.0260	1.9830	-1.2330	1.1110	8.4190	4.0610	3.8950
BENZENE				C	0.0801	0.0385	0.0480	0.0926	0.0508	0.0788	0.0318	0.0553	0.0195
32269	1.0610	130	170.0	S	20.0371	-2.5296	-10.5746	17.0276	-9.4363	25.2150	31.4465	18.0832	51.2820
BENZENE				C	0.0781	0.0394	0.0466	0.0895	0.0507	0.0760	0.0304	0.0531	0.0189
32270	1.0610	130	180.0	S	20.7101	-3.0921	-10.6358	18.4221	-10.3936	26.6129	32.8947	18.8324	52.9100
BENZENE				C	0.0759	0.0389	0.0466	0.0869	0.0492	0.0733	0.0292	0.0508	0.0183
32271	1.0610	130	190.0	S	22.1257	-3.1299	-11.9654	19.0038	-10.7658	28.4756	34.2466	19.6850	54.6448
BENZENE				C	0.0735	0.0376	0.0461	0.0844	0.0476	0.0706	0.0278	0.0485	0.0178
32272	1.0610	130	200.0	S	23.4821	-2.9263	-13.3603	19.4825	-11.2247	30.4562	35.9712	20.6185	56.1798
BENZENE				C	0.0713	0.0370	0.0453	0.0812	0.0467	0.0682	0.0263	0.0464	0.0172
32273	1.0610	130	210.0	S	24.7011	-3.0013	-14.3519	20.6807	-12.1676	32.5274	38.0228	21.5517	58.1395
BENZENE				C	0.0690	0.0368	0.0443	0.0778	0.0444	0.0656	0.0249	0.0441	0.0168
32274	1.0610	130	220.0	S	26.2392	-3.7458	-15.1842	21.4777	-12.0072	33.6247	40.1606	22.6757	59.5238

ORTHORHOMBIC

CODE / NAME	DENSITY	REF	TEMP(K)		11	12	13	22	23	33	44	55	66
BENZENE				C	0.0666	0.0367	0.0441	0.0733	0.0434	0.0624	0.0232	0.0417	0.0163
32275	1.0610	130	230.0	S	28.9097	-4.0419	-17.6202	23.7590	-13.6682	37.9848	43.1034	23.9808	61.3497
BENZENE				C	0.0639	0.0357	0.0422	0.0699	0.0406	0.0601	0.0214	0.0399	0.0157
32276	1.0610	130	240.0	S	30.2564	-5.1235	-17.7838	24.4119	-12.8937	37.8363	46.7290	25.0627	63.6943
BENZENE				C	0.0614	0.0352	0.0401	0.0656	0.0390	0.0583	0.0197	0.0378	0.0153
32277	1.0610	130	250.0	S	31.2079	-6.6150	-17.0403	26.7117	-13.3189	37.7831	50.7614	26.4550	65.3595
BENZOPHENONE				C	0.1070	0.0550	0.0169	0.1000	0.0321	0.0710	0.0203	0.0155	0.0353
32278	1.2190	4	ROOM	S	13.0309	-7.2190	0.1621	15.6970	-5.3785	16.4776	49.2611	64.5161	28.3286
BRONZITE				C	1.8760	0.6860	0.6050	1.5780	0.5610	2.0850	0.7000	0.5920	0.5440
32279	3.3800	223	ROOM	S	0.6629	-0.2430	-0.1269	0.7898	-0.1420	0.5547	1.4286	1.6892	1.8382
BRONZITE (84.5% ENSTATITE)				C	2.2990	0.7010	0.5730	1.6540	0.4960	2.0570	0.8306	0.7637	0.7853
32280	3.3350	147	ROOM	S	0.5173	-0.1898	-0.0984	0.7213	-0.1211	0.5427	1.2039	1.3094	1.2734
CALCIUM SULFATE				C	0.9382	0.1650	0.1520	1.8545	0.3173	1.1180	0.3247	0.2653	0.0926
32281	2.9620	228	ROOM	S	1.1000	-0.0760	-0.1280	0.5720	-0.1520	0.9550	3.0800	3.7700	10.8000
CELESTITE				C	1.0440	0.7730	0.6050	1.0610	0.6190	1.2860	0.1350	0.2790	0.2660
32282	3.9600	129	ROOM	S	2.1970	-1.3872	-0.3659	2.1864	-0.3998	1.1422	7.4074	3.5842	3.7594
CESIUM SULFATE				C	0.4490	0.1958	0.1815	0.4283	0.1800	0.3785	0.1326	0.1319	0.1323
32283	4.2430	126	293.0	S	3.0582	-0.9770	-1.0018	3.2302	-1.0676	3.6301	7.5415	7.5815	7.5586
FORSTERITE				C	3.2840	0.6390	0.6880	1.9980	0.7380	2.3530	0.6515	0.8120	0.8088
32284	3.2240	148	298.0	S	0.3354	-0.0804	-0.0729	0.5853	-0.1601	0.4965	1.5349	1.2315	1.2364
FORSTERITE				C	3.2910	0.6630	0.6840	2.0050	0.7280	2.3630	0.6723	0.8144	0.8114
32285	3.2217	116	298.0	S	0.3358	-0.0853	-0.0709	0.5832	-0.1550	0.4915	1.4874	1.2279	1.2324
IODIC ACID				C	0.3030	0.1194	0.1169	0.5448	0.0548	0.4359	0.1835	0.2193	0.1736
32286	4.6300	177	ROOM	S	3.9800	-0.7750	-0.9700	2.0100	-0.0450	2.5600	5.4500	4.5600	5.7600
LITHIUM AMMONIUM TARTRATE				C	0.3864	0.1655	0.0875	0.5393	0.2007	0.3624	0.1190	0.0667	0.2326
32287	1.7100	129	ROOM	S	3.0000	-0.8200	-0.2700	2.5600	-1.2200	3.5000	8.4000	15.0000	4.3000
MAGNESIUM SULFATE HEPTAHYDRATE				C	0.6980	0.3900	0.2820	0.5290	0.2830	0.8220	0.1070	0.2330	0.2220
32288	1.6360	129	ROOM	S	2.4848	-1.6865	-0.2718	3.4618	-0.6133	1.5209	9.3458	4.2918	4.5045
MAGNESIUM SULFATE HEPTAHYDRATE				C	0.4030	0.2680	0.2360	0.3730	0.2420	0.3520	0.0770	0.1610	0.0830
32289	1.6800	4	ROOM	S	5.2452	-2.6845	-1.6711	6.2136	-2.4720	5.6608	12.9870	6.2112	12.0482
MAGNESIUM SULFATE HEPTAHYDRATE				C	0.3250	0.1740	0.1820	0.2880	0.1820	0.3150	0.0780	0.1560	0.0900
32290	1.6360	11	ROOM	S	5.1567	-1.9416	-1.8576	6.2002	-2.4605	5.6695	12.8205	6.4103	11.1111
NATROLITE				C	0.7160	0.2610	0.2970	0.6320	0.2970	1.3780	0.1960	0.2480	0.4230
32291	2.2500	223	ROOM	S	1.7133	-0.5942	-0.2412	1.9667	-0.2958	0.8414	5.1020	4.0323	2.3641
NICKEL SULFATE HEPTAHYDRATE				C	0.3530	0.1980	0.2010	0.3110	0.2010	0.3350	0.0910	0.1720	0.0990
32292	1.9480	11	ROOM	S	4.9767	-2.0231	-1.7722	6.0745	-2.4309	5.5069	10.9890	5.8140	10.1010

ORTHORHOMBIC

CODE	NAME / DENSITY	REF	TEMP(K)		11	12	13	22	23	33	44	55	66
	OLIVINE (91.7% FORSTERITE)			C	3.2400	0.5900	0.7900	1.9800	0.7800	2.4900	0.6670	0.8100	0.7930
32293	3.3240	262	ROOM	S	0.3423	-0.0675	-0.0874	0.5895	-0.1632	0.4805	1.4992	1.2346	1.2610
	OLIVINE (93.0% FORSTERITE)			C	3.2370	0.6640	0.7160	1.9760	0.7560	2.3510	0.6462	0.7805	0.7904
32294	3.3110	148	298.0	S	0.3436	-0.0860	-0.0770	0.5986	-0.1663	0.5023	1.5475	1.2812	1.2652
	POTASSIUM PENTABORATE			C	0.5820	0.2290	0.1740	0.3590	0.2310	0.2550	0.1640	0.0463	0.0570
32295	1.7400	138	ROOM	S	2.3297	-1.1105	-0.5837	7.2075	-5.7714	9.5481	6.0976	21.5983	17.5439
	POTASSIUM SULFATE			C	0.5357	0.1999	0.2095	0.5653	0.1990	0.5523	0.1950	0.1879	0.1424
32296	2.6650	126	293.0	S	2.3536	-0.5933	-0.6790	2.1755	-0.5588	2.2695	5.1282	5.3220	7.0225
	ROCHELLE SALT			C	1.1529	1.1358	1.5954	1.5802	1.8518	2.7350	0.1639	0.0328	0.1250
32297	1.7900	128	ROOM	S	4.7000	-0.8000	-2.2000	3.2000	-1.7000	2.8000	6.1000	30.5000	8.0000
	ROCHELLE SALT			C	0.5049	0.1839	0.4888	0.3349	0.2143	0.7484	0.1136	0.0278	0.0847
32298	1.7900	128	ROOM	S	5.6000	-0.9000	-3.4000	3.8000	-0.5000	3.7000	8.8000	36.0000	11.8000
	ROCHELLE SALT			C	0.6734	0.6710	0.6267	1.0153	0.7668	0.9503	0.1038	0.0296	0.0847
32299	1.7900	128	ROOM	S	5.2300	-2.1800	-1.6900	3.4300	-1.3300	3.2400	9.6300	33.8000	11.8000
	ROCHELLE SALT			C	0.4259	0.2928	0.3593	0.5165	0.3442	0.6326	0.1253	0.0305	0.0990
32300	1.7750	128	ROOM	S	5.1800	-1.5300	-2.1100	3.4900	-1.0300	3.3400	7.9800	32.8000	10.1000
	ROCHELLE SALT			C	0.2550	0.1410	0.1160	0.3810	0.1460	0.3710	0.1340	0.0321	0.0979
32301	1.7900	138	ROOM	S	5.2458	-1.5460	-1.0318	3.5464	-0.9122	3.3770	7.4627	31.1526	10.2145
	RUBIDIUM SULFATE			C	0.5029	0.1965	0.1999	0.5098	0.1925	0.4761	0.1626	0.1589	0.1407
32302	3.6210	126	293.0	S	2.5774	-0.6902	-0.8031	2.4998	-0.7210	2.7291	6.1501	6.2933	7.1073
	SODIUM AMMONIUM TARTRATE			C	0.3685	0.2725	0.3083	0.5092	0.3472	0.5541	0.1058	0.0303	0.0870
32303	1.5870	129	ROOM	S	5.7000	-1.5500	-2.2000	3.8500	-1.5500	4.0000	9.4500	33.0000	11.5000
	SODIUM TARTRATE			C	0.4610	0.2860	0.3200	0.5470	0.3520	0.6650	0.1240	0.0310	0.0980
32304	1.7940	129	ROOM	S	3.7132	-1.2006	-1.1513	3.1607	-1.0953	2.6376	8.0645	32.2581	10.2041
	STAUROLITE			C	3.4300	0.6700	0.6100	1.8500	0.1280	1.4700	0.4600	0.7000	0.9200
32305	3.3690	4	ROOM	S	0.3367	-0.1130	-0.1299	0.5817	-0.0038	0.7345	2.1739	1.4286	1.0870
	STRONTIUM FORMATE DIHYDRATE			C	0.4391	0.1037	-0.1491	0.3484	-0.0143	0.3746	0.1538	0.1075	0.1724
32306	2.2500	129	ROOM	S	2.8400	-0.8000	1.1000	3.1000	-0.2000	3.1000	6.5000	9.3000	5.8000
	SULPHUR			C	0.2400	0.1330	0.1710	0.2050	0.1590	0.4830	0.0430	0.0870	0.0760
32307	2.0700	129	ROOM	S	7.0965	-3.5658	-1.3386	8.3423	-1.4838	3.0327	23.2558	11.4943	13.1579
	TERPIN MONOHYDRATE			C	0.1250	0.0380	0.0620	0.0990	0.0410	0.1530	0.0243	0.0223	0.0346
32308	1.1100	4	ROOM	S	10.5619	-2.5664	-3.5923	11.9855	-2.1718	8.5736	41.1523	44.8430	28.9017
	THALLIUM SULFATE			C	0.4106	0.2573	0.2288	0.3885	0.2174	0.4268	0.1125	0.1068	0.0751
32309	6.7760	126	293.0	S	4.6164	-2.3394	-1.2832	4.7857	-1.1836	3.6338	8.8889	9.3633	13.3156
	TOPAZ			C	2.8136	1.2582	0.8464	3.4895	0.8815	2.9452	1.0811	1.3298	1.3089
32310	3.5200	128	ROOM	S	0.4430	-0.1380	-0.0860	0.3530	-0.0660	0.3840	0.9250	0.7520	0.7640

ORTHORHOMBIC

CODE	NAME	DENSITY	REF	TEMP(K)		11	12	13	22	23	33	44	55	66
	URANIUM (ALPHA)				C	1.9925	0.4032	0.2393	2.0928	1.0977	2.8755	1.3996	0.8964	0.8436
32311		19.2560	166	73.0	S	0.5223	-0.0973	-0.0063	0.6156	-0.2269	0.4349	0.7145	1.1156	1.1854
	URANIUM (ALPHA)				C	2.0254	0.4143	0.2343	2.0881	1.0979	2.8679	1.3939	0.8895	0.8395
32312		19.2472	166	83.0	S	0.5146	-0.1002	-0.0037	0.6191	-0.2288	0.4366	0.7174	1.1242	1.1912
	URANIUM (ALPHA)				C	2.0517	0.4212	0.2299	2.0834	1.0980	2.8600	1.3877	0.8826	0.8353
32313		19.2383	166	93.0	S	0.5085	-0.1019	-0.0018	0.6221	-0.2307	0.4383	0.7206	1.1330	1.1972
	URANIUM (ALPHA)				C	2.0728	0.4255	0.2264	2.0787	1.0980	2.8518	1.3810	0.8757	0.8309
32314		19.2294	166	103.0	S	0.5036	-0.1029	-0.0004	0.6249	-0.2324	0.4402	0.7241	1.1419	1.2035
	URANIUM (ALPHA)				C	2.0896	0.4283	0.2234	2.0741	1.0977	2.8435	1.3741	0.8687	0.8265
32315		19.2204	166	113.0	S	0.4997	-0.1036	0.0007	0.6274	-0.2341	0.4420	0.7277	1.1511	1.2099
	URANIUM (ALPHA)				C	2.1029	0.4304	0.2211	2.0695	1.0971	2.8350	1.3670	0.8617	0.8220
32316		19.2114	166	123.0	S	0.4967	-0.1041	0.0016	0.6298	-0.2356	0.4438	0.7315	1.1605	1.2165
	URANIUM (ALPHA)				C	2.1135	0.4323	0.2193	2.0649	1.0964	2.8264	1.3598	0.8546	0.8174
32317		19.2023	166	133.0	S	0.4943	-0.1047	0.0023	0.6321	-0.2371	0.4456	0.7354	1.1701	1.2234
	URANIUM (ALPHA)				C	2.1218	0.4343	0.2180	2.0603	1.0954	2.8176	1.3527	0.8475	0.8129
32318		19.1931	166	143.0	S	0.4926	-0.1053	0.0028	0.6344	-0.2385	0.4474	0.7393	1.1799	1.2302
	URANIUM (ALPHA)				C	2.1285	0.4363	0.2171	2.0557	1.0942	2.8088	1.3456	0.8403	0.8083
32319		19.1838	166	153.0	S	0.4912	-0.1060	0.0033	0.6366	-0.2398	0.4492	0.7432	1.1901	1.2372
	URANIUM (ALPHA)				C	2.1338	0.4386	0.2165	2.0510	1.0928	2.7999	1.3385	0.8330	0.8037
32320		19.1745	166	163.0	S	0.4902	-0.1069	0.0038	0.6389	-0.2411	0.4510	0.7471	1.2005	1.2442
	URANIUM (ALPHA)				C	2.1380	0.4408	0.2161	2.0463	1.0914	2.7909	1.3315	0.8257	0.7991
32321		19.1651	166	173.0	S	0.4895	-0.1077	0.0042	0.6412	-0.2424	0.4528	0.7510	1.2111	1.2514
	URANIUM (ALPHA)				C	2.1413	0.4431	0.2159	2.0416	1.0898	2.7818	1.3246	0.8184	0.7945
32322		19.1556	166	183.0	S	0.4890	-0.1086	0.0046	0.6434	-0.2436	0.4546	0.7549	1.2219	1.2587
	URANIUM (ALPHA)				C	2.1440	0.4453	0.2158	2.0369	1.0882	2.7726	1.3177	0.8111	0.7900
32323		19.1461	166	193.0	S	0.4887	-0.1095	0.0049	0.6457	-0.2449	0.4564	0.7589	1.2329	1.2658
	URANIUM (ALPHA)				C	2.1461	0.4473	0.2158	2.0321	1.0867	2.7634	1.3109	0.8038	0.7855
32324		19.1364	166	203.0	S	0.4884	-0.1103	0.0052	0.6481	-0.2462	0.4583	0.7628	1.2441	1.2731
	URANIUM (ALPHA)				C	2.1477	0.4491	0.2158	2.0274	1.0852	2.7540	1.3041	0.7964	0.7810
32325		19.1267	166	213.0	S	0.4883	-0.1111	0.0055	0.6504	-0.2476	0.4602	0.7668	1.2556	1.2804
	URANIUM (ALPHA)				C	2.1490	0.4507	0.2158	2.0226	1.0838	2.7446	1.2972	0.7892	0.7766
32326		19.1169	166	223.0	S	0.4882	-0.1119	0.0058	0.6528	-0.2490	0.4622	0.7709	1.2671	1.2877
	URANIUM (ALPHA)				C	2.1498	0.4521	0.2158	2.0177	1.0825	2.7351	1.2903	0.7819	0.7721
32327		19.1069	166	233.0	S	0.4882	-0.1127	0.0061	0.6552	-0.2504	0.4643	0.7750	1.2789	1.2952
	URANIUM (ALPHA)				C	2.1504	0.4535	0.2158	2.0129	1.0813	2.7255	1.2834	0.7747	0.7677
32328		19.0969	166	243.0	S	0.4883	-0.1134	0.0063	0.6577	-0.2519	0.4664	0.7792	1.2908	1.3026

ORTHORHOMBIC

CODE	NAME / DENSITY	REF	TEMP(K)		11	12	13	22	23	33	44	55	66
	URANIUM (ALPHA)			C	2.1506	0.4550	0.2158	2.0080	1.0803	2.7158	1.2763	0.7674	0.7633
32329	19.0868	166	253.0	S	0.4885	-0.1143	0.0066	0.6603	-0.2536	0.4686	0.7835	1.3031	1.3101
	URANIUM (ALPHA)			C	2.1506	0.4566	0.2159	2.0031	1.0793	2.7061	1.2692	0.7602	0.7589
32330	19.0766	166	263.0	S	0.4887	-0.1151	0.0069	0.6630	-0.2552	0.4708	0.7879	1.3154	1.3177
	URANIUM (ALPHA)			C	2.1502	0.4584	0.2162	1.9982	1.0784	2.6962	1.2621	0.7529	0.7544
32331	19.0663	166	273.0	S	0.4891	-0.1161	0.0072	0.6658	-0.2570	0.4731	0.7923	1.3282	1.3256
	URANIUM (ALPHA)			C	2.1495	0.4603	0.2167	1.9933	1.0775	2.6863	1.2550	0.7455	0.7499
32332	19.0558	166	283.0	S	0.4895	-0.1171	0.0075	0.6686	-0.2587	0.4754	0.7968	1.3414	1.3335
	URANIUM (ALPHA)			C	2.1486	0.4622	0.2176	1.9883	1.0764	2.6763	1.2479	0.7379	0.7454
32333	19.0453	166	293.0	S	0.4900	-0.1180	0.0076	0.6714	-0.2604	0.4778	0.8013	1.3552	1.3416
	URANIUM (ALPHA)			C	2.1474	0.4637	0.2189	1.9833	1.0751	2.6662	1.2409	0.7301	0.7407
32334	19.0347	166	303.0	S	0.4906	-0.1188	0.0076	0.6740	-0.2620	0.4801	0.8059	1.3697	1.3501
	URANIUM (ALPHA)			C	2.1459	0.4642	0.2210	1.9783	1.0734	2.6560	1.2343	0.7220	0.7360
32335	19.0239	166	313.0	S	0.4910	-0.1192	0.0073	0.6764	-0.2634	0.4824	0.8102	1.3850	1.3587
	URANIUM (ALPHA)			C	2.1441	0.4627	0.2240	1.9732	1.0711	2.6458	1.2280	0.7134	0.7311
32336	19.0130	166	323.0	S	0.4913	-0.1187	0.0065	0.6782	-0.2645	0.4845	0.8143	1.4017	1.3678
	URANIUM (ALPHA)			C	1.5000	0.2750	0.3450	2.0850	1.1230	2.8680	1.4070	0.8920	0.8490
32337	19.2539	91	44.0	S	0.6913	-0.0588	-0.0601	0.6128	-0.2329	0.4471	0.7107	1.1211	1.1779
	URANIUM (ALPHA)			C	1.6100	0.2940	0.3200	2.0900	1.1180	2.8750	1.4110	0.9010	0.8510
32338	19.2524	91	46.0	S	0.6423	-0.0658	-0.0459	0.6109	-0.2302	0.4425	0.7087	1.1099	1.1751
	URANIUM (ALPHA)			C	1.6850	0.3180	0.3050	2.0930	1.1150	2.8780	1.4120	0.9050	0.8510
32339	19.2510	91	48.0	S	0.6140	-0.0739	-0.0365	0.6109	-0.2289	0.4400	0.7082	1.1050	1.1751
	URANIUM (ALPHA)			C	1.7400	0.3390	0.2920	2.0940	1.1120	2.8790	1.4110	0.9060	0.8510
32340	19.2495	91	50.0	S	0.5954	-0.0809	-0.0291	0.6118	-0.2281	0.4384	0.7087	1.1038	1.1751
	URANIUM (ALPHA)			C	1.9000	0.3880	0.2560	2.0950	1.1060	2.8790	1.4070	0.9040	0.8480
32341	19.2424	91	60.0	S	0.5473	-0.0949	-0.0122	0.6152	-0.2279	0.4360	0.7107	1.1062	1.1792
	URANIUM (ALPHA)			C	1.9930	0.4030	0.2390	2.0930	1.0980	2.8750	1.4000	0.8960	0.8440
32342	19.2327	91	73.0	S	0.5222	-0.0973	-0.0063	0.6156	-0.2270	0.4351	0.7143	1.1161	1.1848
	URANIUM (ALPHA)			C	2.0630	0.4250	0.2260	2.0810	1.0980	2.8560	1.3840	0.8790	0.8330
32343	19.2142	91	98.0	S	0.5060	-0.1031	-0.0004	0.6238	-0.2317	0.4392	0.7225	1.1377	1.2005
	URANIUM (ALPHA)			C	2.1030	0.4300	0.2210	2.0700	1.0970	2.8350	1.3670	0.8620	0.8220
32344	19.1952	91	123.0	S	0.4966	-0.1040	0.0015	0.6295	-0.2355	0.4437	0.7315	1.1601	1.2165
	URANIUM (ALPHA)			C	2.1250	0.4350	0.2180	2.0580	1.0950	2.8130	1.3490	0.8440	0.8110
32345	19.1759	91	148.0	S	0.4919	-0.1055	0.0030	0.6355	-0.2392	0.4484	0.7413	1.1848	1.2330
	URANIUM (ALPHA)			C	2.1380	0.4410	0.2160	2.0460	1.0910	2.7910	1.3320	0.8260	0.7990
32346	19.1561	91	173.0	S	0.4895	-0.1078	0.0042	0.6412	-0.2423	0.4527	0.7508	1.2107	1.2516

ORTHORHOMBIC

CODE	NAME / DENSITY	REF	TEMP(K)		11	12	13	22	23	33	44	55	66
URANIUM (ALPHA)				C	2.1450	0.4460	0.2160	2.0350	1.0880	2.7680	1.3140	0.8070	0.7880
32347	19.1359	91	198.0	S	0.4885	-0.1097	0.0050	0.6468	-0.2457	0.4574	0.7610	1.2392	1.2690
URANIUM (ALPHA)				C	2.1490	0.4510	0.2160	2.0230	1.0840	2.7450	1.2970	0.7890	0.7770
32348	19.1153	91	223.0	S	0.4882	-0.1119	0.0058	0.6527	-0.2489	0.4621	0.7710	1.2674	1.2870
URANIUM (ALPHA)				C	2.1510	0.4540	0.2160	2.0110	1.0810	2.7210	1.2800	0.7710	0.7660
32349	19.0942	91	248.0	S	0.4882	-0.1137	0.0064	0.6588	-0.2527	0.4674	0.7813	1.2970	1.3055
URANIUM (ALPHA)				C	2.1510	0.4580	0.2160	1.9980	1.0780	2.6960	1.2620	0.7530	0.7540
32350	19.0727	91	273.0	S	0.4889	-0.1159	0.0072	0.6657	-0.2569	0.4731	0.7924	1.3280	1.3263
URANIUM (ALPHA)				C	2.1480	0.4650	0.2180	1.9860	1.0760	2.6710	1.2440	0.7340	0.7430
32351	19.0507	91	298.0	S	0.4905	-0.1192	0.0080	0.6731	-0.2614	0.4790	0.8039	1.3624	1.3459
URANIUM (ALPHA)				C	2.1440	0.4710	0.2200	1.9730	1.0740	2.6470	1.2280	0.7150	0.7310
32352	19.0283	91	323.0	S	0.4924	-0.1223	0.0087	0.6809	-0.2661	0.4850	0.8143	1.3986	1.3680
URANIUM (ALPHA)				C	2.1390	0.4760	0.2220	1.9610	1.0700	2.6230	1.2100	0.6950	0.7180
32353	19.0054	91	348.0	S	0.4944	-0.1250	0.0091	0.6875	-0.2699	0.4906	0.8264	1.4388	1.3928
URANIUM (ALPHA)				C	2.1320	0.4790	0.2230	1.9470	1.0690	2.5970	1.1930	0.6770	0.7080
32354	18.9821	91	373.0	S	0.4967	-0.1276	0.0099	0.6964	-0.2757	0.4977	0.8382	1.4771	1.4124
URANIUM (ALPHA)				C	2.1250	0.4830	0.2260	1.9340	1.0640	2.5710	1.1760	0.6590	0.6960
32355	18.9582	91	398.0	S	0.4991	-0.1301	0.0100	0.7034	-0.2797	0.5038	0.8503	1.5175	1.4368
URANIUM (ALPHA)				C	2.1170	0.4850	0.2280	1.9190	1.0580	2.5430	1.1580	0.6400	0.6860
32356	18.9338	91	423.0	S	0.5016	-0.1323	0.0101	0.7111	-0.2840	0.5105	0.8636	1.5625	1.4577
URANIUM (ALPHA)				C	2.1070	0.4880	0.2300	1.9040	1.0520	2.5130	1.1410	0.6200	0.6750
32357	18.9090	91	448.0	S	0.5048	-0.1351	0.0104	0.7194	-0.2888	0.5179	0.8764	1.6129	1.4815
URANIUM (ALPHA)				C	2.0970	0.4940	0.2350	1.8890	1.0440	2.4850	1.1220	0.6010	0.6610
32358	18.8836	91	473.0	S	0.5084	-0.1385	0.0101	0.7272	-0.2924	0.5243	0.8913	1.6639	1.5129
URANIUM (ALPHA)				C	2.0870	0.5010	0.2400	1.8740	1.0340	2.4560	1.1040	0.5790	0.6480
32359	18.8577	91	498.0	S	0.5122	-0.1424	0.0099	0.7347	-0.2954	0.5306	0.9058	1.7271	1.5432
URANIUM (ALPHA)				C	2.0760	0.5060	0.2450	1.8590	1.0270	2.4270	1.0840	0.5570	0.6320
32360	18.8313	91	523.0	S	0.5161	-0.1458	0.0096	0.7432	-0.2998	0.5379	0.9225	1.7953	1.5823
URANIUM (ALPHA)				C	2.0630	0.5110	0.2500	1.8430	1.0180	2.3980	1.0650	0.5350	0.6200
32361	18.8043	91	548.0	S	0.5206	-0.1494	0.0091	0.7517	-0.3035	0.5449	0.9390	1.8692	1.6129
URANIUM (ALPHA)				C	2.0490	0.5170	0.2570	1.8270	1.0090	2.3690	1.0450	0.5130	0.6070
32362	18.7768	91	573.0	S	0.5257	-0.1533	0.0083	0.7604	-0.3072	0.5521	0.9569	1.9493	1.6474
URANIUM (ALPHA)				C	2.0340	0.5230	0.2630	1.8110	1.0000	2.3380	1.0240	0.4910	0.5910
32363	18.7488	91	598.0	S	0.5312	-0.1576	0.0077	0.7697	-0.3115	0.5601	0.9766	2.0367	1.6920
URANIUM (ALPHA)				C	2.0180	0.5280	0.2700	1.7940	0.9940	2.3090	1.0030	0.4690	0.5760
32364	18.7202	91	623.0	S	0.5370	-0.1619	0.0069	0.7808	-0.3172	0.5688	0.9970	2.1322	1.7361

CODE	NAME / DENSITY	REF	TEMP(K)		11	12	13	22	23	33	44	55	66
	URANIUM (ALPHA)			C	2.0020	0.5330	0.2760	1.7770	0.9910	2.2780	0.9810	0.4480	0.5620
32365	18.6911	91	648.0	S	0.5429	-0.1666	0.0067	0.7941	-0.3253	0.5797	1.0194	2.2321	1.7794
	URANIUM (ALPHA)			C	1.9840	0.5370	0.2820	1.7600	0.9870	2.2490	0.9580	0.4260	0.5460
32366	18.6614	91	673.0	S	0.5495	-0.1711	0.0062	0.8070	-0.3327	0.5899	1.0438	2.3474	1.8315
	URANIUM (ALPHA)			C	1.9650	0.5400	0.2880	1.7430	0.9840	2.2190	0.9370	0.4040	0.5310
32367	18.6311	91	698.0	S	0.5563	-0.1755	0.0056	0.8207	-0.3412	0.6012	1.0672	2.4752	1.8832
	URANIUM (ALPHA)			C	1.9460	0.5430	0.2970	1.7240	0.9810	2.1870	0.9160	0.3830	0.5150
32368	18.6002	91	723.0	S	0.5634	-0.1798	0.0041	0.8362	-0.3507	0.6140	1.0917	2.6110	1.9417
	URANIUM (ALPHA)			C	1.9250	0.5520	0.3040	1.7030	0.9760	2.1530	0.8950	0.3610	0.4970
32369	18.5688	91	748.0	S	0.5727	-0.1882	0.0044	0.8551	-0.3611	0.6275	1.1173	2.7701	2.0121
	URANIUM (ALPHA)			C	1.9040	0.5660	0.3120	1.6810	0.9710	2.1180	0.8730	0.3400	0.4770
32370	18.5368	91	773.0	S	0.5837	-0.1998	0.0056	0.8775	-0.3729	0.6423	1.1455	2.9412	2.0964
	URANIUM (ALPHA)			C	1.8820	0.5780	0.3200	1.6580	0.9690	2.0820	0.8500	0.3180	0.4560
32371	18.5043	91	798.0	S	0.5951	-0.2116	0.0070	0.9037	-0.3881	0.6599	1.1765	3.1447	2.1930
	URANIUM (ALPHA)			C	1.8580	0.5920	0.3280	1.6350	0.9670	2.0490	0.8260	0.2950	0.4330
32372	18.4711	91	823.0	S	0.6085	-0.2257	0.0091	0.9322	-0.4038	0.6771	1.2107	3.3898	2.3095
	URANIUM (ALPHA)			C	1.8320	0.6030	0.3380	1.6100	0.9620	2.0130	0.8040	0.2730	0.4110
32373	18.4374	91	848.0	S	0.6227	-0.2390	0.0097	0.9611	-0.4192	0.6955	1.2438	3.6630	2.4331
	URANIUM (ALPHA)			C	1.8040	0.6120	0.3490	1.5840	0.9590	1.9770	0.7800	0.2530	0.3880
32374	18.4031	91	873.0	S	0.6381	-0.2525	0.0098	0.9937	-0.4375	0.7163	1.2821	3.9526	2.5773
	URANIUM (ALPHA)			C	1.7750	0.6240	0.3610	1.5620	0.9540	1.9420	0.7570	0.2320	0.3650
32375	18.3682	91	898.0	S	0.6556	-0.2678	0.0097	1.0240	-0.4533	0.7358	1.3210	4.3103	2.7397
	URANIUM (ALPHA)			C	1.7420	0.6300	0.3740	1.5350	0.9530	1.9070	0.7340	0.2110	0.3440
32376	18.3327	91	923.0	S	0.6742	-0.2822	0.0088	1.0626	-0.4757	0.7604	1.3624	4.7393	2.9070
	URANIUM (ALPHA)			C	1.1430	0.2860	0.3470	2.1110	1.1290	2.8600	1.3960	0.8200	0.8920
32377	19.3189	92	4.2	S	0.9195	-0.0823	-0.0791	0.6078	-0.2300	0.4500	0.7163	1.2195	1.1211
	ZINC SULFATE HEPTAHYDRATE			C	0.3320	0.1720	0.2000	0.2930	0.1980	0.3200	0.0780	0.1530	0.0830
32378	1.9700	11	ROOM	S	5.1721	-1.4638	-2.3269	6.2798	-2.9708	6.4175	12.8205	6.5359	12.0482

ORTHORHOMBIC

CODE	NAME / DENSITY	REF	TEMP(K)		11	12	13	22	23	33	44	55	66
FORSTERITE				C'(P)	8.4700	4.6700	4.8400	6.5600	4.1100	6.5700	2.1200	1.6600	2.3700
32379	3.2240	148	298.0	S'(P)	-0.5899	-0.1328	-0.1319	-1.3856	0.0633	-0.9381	-4.9947	-2.5176	-3.6230
FORSTERITE				C'(P)	8.3200	4.3000	4.2300	5.9300	3.5300	6.2100	2.1200	1.6500	2.3200
32380	3.2217	116	298.0	S'(P)	-0.6074	-0.1145	-0.0762	-1.2727	0.1140	-0.9462	-4.6904	-2.4878	-3.5238
OLIVINE (93.0% FORSTERITE)				C'(P)	7.9800	4.7400	4.4800	6.3700	3.7600	6.3800	2.1700	1.6400	2.3100
32381	3.3110	148	298.0	S'(P)	-0.5597	-0.1821	-0.0793	-1.4095	0.1665	-0.9798	-5.1967	-2.6921	-3.6976

ORTHORHOMBIC

CODE	NAME / DENSITY	REF	TEMP(K)		11	12	13	22	23	33	44	55	66
								TEMPERATURE DERIVATIVES					
AMMONIUM SULFATE				C'(T)	-0.1900	-0.1260	-0.1640	-0.1750	-0.1200	-0.2510	-0.0410	0.0516	-0.0350
32382	1.7779	126	273.0	S'(T)	1.1468	-0.4638	0.1693	2.2396	-0.7339	1.9659	3.8422	-10.3231	3.6369
CESIUM SULFATE				C'(T)	-0.2310	-0.0730	-0.1080	-0.2250	-0.1000	-0.3080	-0.0760	-0.0760	-0.0720
32383	4.2523	126	273.0	S'(T)	1.7389	-0.7256	-0.6779	1.9443	-0.8560	3.0357	4.2263	4.2707	4.0278
FORSTERITE				C'(T)	-0.3310	-0.1040	-0.0820	-0.2810	-0.0460	-0.2830	-0.1300	-0.1320	-0.1510
32384	3.2240	148	298.0	S'(T)	0.0315	-0.0030	-0.0069	0.0894	-0.0378	0.0679	0.3063	0.2002	0.2308
FORSTERITE				C'(T)	-0.3890	-0.1170	-0.0870	-0.3110	-0.0920	-0.2690	-0.1300	-0.1440	-0.1630
32385	3.2078	116	473.0	S'(T)	0.0377	-0.0065	-0.0080	0.0891	-0.0237	0.0569	0.2876	0.2171	0.2476
OLIVINE (93.0% FORSTERITE)				C'(T)	-0.3400	-0.1050	-0.0940	-0.2850	-0.0510	-0.2860	-0.1280	-0.1300	-0.1570
32386	3.3110	148	298.0	S'(T)	0.0334	-0.0051	-0.0061	0.0943	-0.0395	0.0689	0.3065	0.2134	0.2513
POTASSIUM SULFATE				C'(T)	-0.2000	-0.0460	-0.1360	-0.2910	-0.1020	-0.4600	-0.1040	-0.1030	-0.0200
32387	2.6708	126	273.0	S'(T)	0.9253	-0.4572	-0.2782	1.2823	-0.4868	1.8450	2.6771	2.8532	0.9808
RUBIDIUM SULFATE				C'(T)	-0.2340	-0.0550	-0.1240	-0.2630	-0.1090	-0.4020	-0.0900	-0.0790	-0.0600
32388	3.6289	126	273.0	S'(T)	1.3216	-0.5666	-0.4831	1.4683	-0.5607	2.2923	3.3300	3.0667	2.9798
THALLIUM SULFATE				C'(T)	-0.2580	-0.1350	-0.2190	-0.2400	-0.1700	-0.4130	-0.0880	-0.0880	-0.0570
32389	6.7909	126	273.0	S'(T)	2.9185	-2.1580	-0.2850	3.6308	-1.1021	3.0062	6.7476	7.4614	9.8167

CODE	NAME	DENSITY	REF	TEMP(K)		11	12	13	15	22	23	25	33	35	44	46	55	66
42390	AEGIRITE	3.5000	7	ROOM	C	1.8580	0.6850	0.7070	0.0980	1.8130	0.6260	0.0940	2.3440	0.2140	0.6920	0.0770	0.5100	0.4740
					S	0.6693	-0.2011	-0.1454	-0.0305	0.6690	-0.1147	-0.0365	0.5164	-0.1676	1.4717	-0.2391	2.0437	2.1485
42391	AEGIRITE-AUGITE	3.4200	10	ROOM	C	1.5560	0.8110	0.6600	0.2530	1.5180	0.6840	0.2600	2.1610	0.1920	0.4000	0.0410	0.4650	0.4920
					S	0.9516	-0.4046	-0.1419	-0.2330	0.9965	-0.1681	-0.2677	0.5648	-0.0620	2.5215	-0.2101	2.4526	2.0500
42392	ANTHRACENE	1.2580	3	ROOM	C	0.0852	0.0672	0.0590	-0.0192	0.1170	0.0375	-0.0170	0.1522	-0.0187	0.0272	0.0138	0.0282	0.0399
					S	28.4314	-13.1072	-6.9507	6.8469	15.8627	1.3620	1.5418	9.1980	2.1881	44.5890	-15.4217	42.5031	30.3965
42393	AUGITE	3.3200	10	ROOM	C	1.8160	0.7340	0.7240	0.1990	1.5070	0.3390	0.1660	2.1780	0.2460	0.6970	0.0430	0.5110	0.5580
					S	0.7683	-0.3193	-0.1942	-0.1020	0.8367	-0.0079	-0.1437	0.5458	-0.1846	1.4416	-0.1111	2.1322	1.8007
42394	COBALT SULFATE HEPTAHYDRATE	1.9480	11	ROOM	C	0.3350	0.2050	0.1580	0.0160	0.3780	0.1580	-0.0180	0.3710	-0.0470	0.0600	0.0160	0.0580	0.1010
					S	5.4654	-2.3984	-1.7735	-3.6892	4.2717	-0.6085	1.4942	4.1767	3.6850	17.4018	-2.7567	21.7089	10.3377
42395	DIALLAGE	3.3000	10	ROOM	C	1.5390	0.5690	0.3740	0.1460	1.4960	0.3050	0.1420	2.1080	0.1190	0.6390	-0.0860	0.6220	0.5230
					S	0.7817	-0.2683	-0.0940	-0.1043	0.7930	-0.0611	-0.1064	0.5033	-0.0603	1.6004	0.2632	1.6680	1.9553
42396	DIBENZYL	1.1040	253	ROOM	C	0.0945	0.0395	0.0415	-0.0240	0.0680	0.0335	-0.0080	0.0720	-0.0070	0.0310	0.0080	0.0255	0.0260
					S	20.2786	-6.4472	-7.2225	15.0804	21.4646	-6.3764	-1.0843	20.7161	-3.1113	35.0404	-10.7817	52.2148	41.7790
42397	DIOPSIDE (BAIKALITE)	3.3100	10	ROOM	C	2.0400	0.8440	0.8830	-0.1930	1.7500	0.4820	-0.1960	2.3800	-0.3360	0.6750	-0.1130	0.5880	0.7050
					S	0.6925	-0.2771	-0.1977	0.0220	0.7273	-0.0251	0.1371	0.5311	0.2302	1.5223	0.2440	1.8852	1.4575
42398	DIPOTASSIUM TARTRATE	1.9700	129	ROOM	C	0.3068	0.1731	0.1655	0.0317	0.3962	0.1624	0.0039	0.5581	0.0760	0.0884	0.0049	0.1105	0.0819
					S	4.7500	-1.7400	-0.8000	-0.7500	3.5300	-0.6200	0.8000	2.4000	-1.4000	11.3500	-0.6800	10.2000	12.2500
42399	DIPOTASSIUM TARTRATE	1.9700	129	ROOM	C	0.4294	0.1399	0.3129	-0.0105	0.3460	0.1173	0.0176	0.6816	0.0294	0.0961	-0.0044	0.1270	0.0841
					S	3.8900	-1.0600	-1.6400	0.8500	3.3700	-0.0700	-0.5400	2.2600	-0.6500	10.4300	0.5500	8.1700	11.9200
42400	EPIDOTE	3.4000	223	ROOM	C	2.1180	0.6630	0.4520	0.0000	2.3870	0.4560	-0.0820	2.0200	-0.1430	0.3910	-0.0340	0.4320	0.7750
					S	0.5335	-0.1322	-0.0935	-0.0560	0.4717	-0.0723	0.0656	0.5441	0.1664	2.5673	0.1126	2.3823	1.2953
42401	ETHYLENE DIAMINE TARTRATE	1.5380	129	ROOM	C	1.3310	0.3195	0.8037	0.1518	0.3829	0.2504	0.0412	0.6020	0.1079	0.0535	0.0006	0.0778	0.0575
					S	3.9000	0.2000	-5.2000	-0.5000	3.6000	-1.8000	0.2000	9.8000	-2.5000	18.7000	-0.2000	17.2000	17.4000
42402	ETHYLENE DIAMINE TARTRATE	1.5380	129	ROOM	C	0.5683	0.1069	0.2213	0.1190	0.3296	0.0897	-0.0064	0.2000	0.0659	0.0521	-0.0010	0.1185	0.0524
					S	3.3400	-0.3000	-3.0000	-1.7000	3.6500	-1.8000	1.5000	10.0000	-2.6500	19.2000	0.3800	11.7000	19.1000
42403	FELDSPAR (LABRADORITE)	2.6800	6	ROOM	C	1.0100	0.6170	0.4800	-0.0030	1.5820	0.2600	-0.0800	1.5100	0.0960	0.2140	-0.0560	0.3350	0.3700
					S	1.4888	-0.5175	-0.3841	-0.0002	0.8426	0.0071	0.1946	0.7978	-0.2304	4.8656	0.7364	3.0976	2.8142
42404	FELDSPAR (MICROCLINE)	2.5600	6	ROOM	C	0.6640	0.4380	0.2590	-0.0330	1.7100	0.1920	-0.1480	1.2150	-0.1310	0.1430	-0.0150	0.2380	0.3610
					S	1.9531	-0.4790	-0.3652	-0.2281	0.7396	0.0294	0.4097	0.9491	0.4900	7.0236	0.2918	4.6946	2.7822
42405	FELDSPAR (OLIGOCLASE)	2.6400	6	ROOM	C	0.8060	0.4170	0.5380	0.1610	1.6300	0.3740	0.1710	1.2420	-0.0740	0.1770	0.0100	0.2740	0.3620
					S	2.3538	-0.1990	-1.0516	-1.5429	0.7421	-0.1605	-0.3895	1.3739	1.0891	5.6585	-0.1563	5.0934	2.7667
42406	FELDSPAR (PLAGIOCLASE-9%AN)	2.6100	221	ROOM	C	0.7490	0.3630	0.3760	-0.0910	1.3750	0.3260	-0.1040	1.2890	-0.1910	0.1720	-0.0130	0.3030	0.3110
					S	1.7190	-0.3498	-0.3908	0.1499	0.8517	-0.0944	0.1277	0.9833	0.4701	5.8324	0.2438	3.6855	3.2256
42407	FELDSPAR (PLAGIOCLASE-24%AN)	2.6400	221	ROOM	C	0.8180	0.3930	0.4070	-0.0900	1.4490	0.3410	-0.0790	1.3280	-0.1850	0.1770	-0.0080	0.3120	0.3330
					S	1.5882	-0.3331	-0.3806	0.1481	0.8063	-0.0980	0.0500	0.9549	0.4316	5.6559	0.1359	3.5164	3.0063

CODE / NAME	DENSITY	REF	TEMP(K)		11	12	13	15	22	23	25	33	35	44	46	55	66
FELDSPAR (PLAGIOCLASE-29%AN)				C	0.8450	0.4170	0.4090	-0.0870	1.5050	0.3300	-0.0690	1.3250	-0.1850	0.1850	-0.0110	0.3140	0.3430
42408	2.6400	221	ROOM	S	1.5460	-0.3403	-0.3739	0.1333	0.7786	-0.0852	0.0266	0.9526	0.4390	5.4157	0.1737	3.4861	2.9210
FELDSPAR (PLAGIOCLASE-53%AN)				C	0.9700	0.5070	0.4420	-0.0960	1.6290	0.3700	-0.0510	1.4100	-0.1500	0.1960	-0.0160	0.3300	0.3700
42409	2.6800	221	ROOM	S	1.3795	-0.3506	-0.3189	0.2022	0.7421	-0.0877	-0.0272	0.8626	0.2858	5.1201	0.2214	3.2148	2.7123
FELDSPAR (PLAGIOCLASE-56%AN)				C	0.9890	0.5210	0.4410	-0.0810	1.7200	0.3660	-0.0510	1.4140	-0.1910	0.1990	-0.0190	0.3410	0.3760
42410	2.6900	221	ROOM	S	1.3365	-0.3345	-0.3182	0.0892	0.6990	-0.0792	-0.0193	0.8819	0.4065	5.0495	0.2552	3.1786	2.6725
FELDSPAR (53.5%OR-34.6%AB-9.15%AN)				C	0.6300	0.3590	0.4900	-0.1290	1.5220	0.3610	-0.0180	1.1790	-0.1810	0.1010	-0.0260	0.2680	0.3560
42411	2.5800	222	ROOM	S	2.6267	-0.4061	-0.8673	0.6513	0.7742	-0.1007	-0.2115	1.3102	0.4606	10.0907	0.7370	4.3417	2.8628
FELDSPAR (60.7%OR-35.6%AB-1.64%AN)				C	0.5960	0.3440	0.2800	-0.1700	1.5680	0.2160	-0.0590	1.1950	-0.1290	0.1360	-0.0180	0.2260	0.3420
42412	2.5700	222	ROOM	S	2.5612	-0.4538	-0.3441	1.6116	0.7370	-0.0458	-0.1751	0.9554	0.2745	7.4045	0.3897	5.7481	2.9445
FELDSPAR (66.6%OR-28.6%AB)				C	0.5840	0.3330	0.3400	-0.1070	1.4680	0.2160	-0.0430	0.9880	-0.1300	0.1240	-0.0300	0.1850	0.3430
42413	2.5400	222	ROOM	S	2.5080	-0.4475	-0.6480	0.8912	0.7843	-0.0304	-0.0978	1.3129	0.5408	8.2389	0.7206	6.2781	2.9785
FELDSPAR (74%OR-18.9%AB-1.95%AN)				C	0.6190	0.4340	0.3680	-0.1000	1.5830	0.2180	-0.0180	1.0020	-0.1210	0.1410	-0.0230	0.2030	0.3600
42414	2.5700	222	ROOM	S	2.5993	-0.6034	-0.7275	0.7933	0.7915	0.0237	-0.2129	1.3114	0.4254	7.1669	0.4579	5.5516	2.8070
FELDSPAR (75%OR-22%AB)				C	0.5720	0.3280	0.3330	-0.1240	1.4830	0.1930	-0.0610	1.0260	-0.1120	0.1370	-0.0250	0.1800	0.3230
42415	2.5400	222	ROOM	S	2.6558	-0.4518	-0.6373	1.2799	0.7727	-0.0044	-0.0521	1.2169	0.3167	7.4038	0.5731	6.6167	3.1403
FELDSPAR (78.5%OR-19.4%AB-2.1%AN)				C	0.6250	0.4280	0.3580	-0.1540	1.7200	0.2410	-0.1430	1.2440	-0.1150	0.1430	-0.0280	0.2230	0.3740
42416	2.5600	222	ROOM	S	2.4952	-0.4508	-0.5231	1.1643	0.7046	0.0065	0.1439	0.9662	0.1412	7.0970	0.5313	5.4534	2.7136
FELDSPAR (64.9%OR-26.6%AB-3.60%AN)				C	0.5960	0.3620	0.3600	-0.1180	1.5810	0.2850	-0.0570	1.0490	-0.1290	0.1390	-0.0260	0.2030	0.3700
42417	2.5700	222	ROOM	S	2.5006	-0.4257	-0.6275	0.9353	0.7387	-0.0646	-0.0811	1.2357	0.4023	7.2901	0.5123	5.7027	2.7387
FERROUS SULPHATE HEPTAHYDRATE				C	0.3490	0.2080	0.1740	-0.0200	0.3760	0.1720	-0.0190	0.3600	-0.0140	0.0640	0.0010	0.0560	0.0960
42418	1.8980	11	ROOM	S	4.7468	-1.9883	-1.3174	0.6913	4.2682	-1.0599	0.4731	3.9268	0.1516	15.6275	-0.1628	18.3024	10.4184
HORNBLENDE				C	1.1600	0.4490	0.6140	0.0430	1.5970	0.6550	-0.0250	1.9160	0.1000	0.5740	-0.0620	0.3180	0.3680
42419	3.1200	7	ROOM	S	1.0868	-0.1925	-0.2786	-0.0745	0.7684	-0.2089	0.1521	0.6929	-0.1966	1.7745	0.2990	3.2285	2.7678
HORNBLENDE				C	1.3010	0.6140	0.5920	0.0950	1.8770	0.6140	-0.0690	1.9840	-0.4060	0.6110	-0.0090	0.3870	0.4500
42420	3.1500	7	ROOM	S	1.1368	-0.2648	-0.4126	-0.7591	0.6582	-0.1112	0.0656	0.8637	0.9875	1.6371	0.0327	3.8180	2.2229
LITHIUM SULPHATE MONOHYDRATE				C	0.5250	0.1715	0.1730	-0.0196	0.5060	0.0368	0.0571	0.5400	-0.0254	0.1400	-0.0054	0.1565	0.2770
42421	2.2210	4	ROOM	S	2.4225	-0.8259	-0.6967	0.4916	2.3597	0.0589	-0.9548	2.0818	0.2291	7.1482	0.1394	6.8369	3.6128
LITHIUM SULPHATE MONOHYDRATE				C	0.5490	0.2630	0.1140	0.0650	0.7050	0.1710	0.1570	0.6180	-0.0516	0.1395	-0.0265	0.2420	0.2700
42422	2.2210	4	ROOM	S	2.2448	-0.7469	-0.2213	-0.1656	2.1202	-0.5569	-1.2937	1.8817	0.8219	7.3047	0.7169	5.1912	3.7741
L-RHAMNOSE MONOHYDRATE				C	0.3820	0.1600	0.1660	-0.0030	0.2190	0.0888	0.0122	0.1980	-0.0118	0.0537	0.0022	0.0502	0.0911
42423	1.4710	4	ROOM	S	4.9521	-2.3793	-3.0757	0.1512	6.9237	-1.2365	-2.1155	8.3070	2.0693	18.6404	-0.4502	20.9299	10.9878
NAPHTHALENE				C	0.0780	0.0445	0.0340	-0.0060	0.0990	0.0230	-0.0270	0.1190	0.0290	0.0330	-0.0050	0.0210	0.0415
42424	1.1270	13	ROOM	S	19.0368	-4.6907	-6.6142	8.5421	55.9402	-40.2020	126.1000	46.9053	-118.3521	30.8665	3.7189	375.6267	24.5444
POTASSIUM TARTRATE				C	0.3060	0.1740	0.1690	0.0318	0.3770	0.1620	0.0038	0.5650	0.0950	0.0880	0.0049	0.1105	0.0830
42425	1.9870	4	ROOM	S	4.8510	-1.8879	-0.8018	-0.6418	3.8141	-0.6993	1.0134	2.5337	-1.9235	11.4011	-0.6731	10.8533	12.0879

MONOCLINIC

CODE	NAME DENSITY	REF	TEMP(K)		11	12	13	15	22	23	25	33	35	44	46	55	66
POTASSIUM TARTRATE				C	0.4520	0.1880	0.3360	-0.0110	0.3700	0.1060	0.0190	0.7370	0.0310	0.0940	0.0041	0.1350	0.0838
42426	1.9870	4	ROOM	S	4.2253	-1.7037	-1.7225	0.9796	3.5185	0.3003	-0.7030	2.1272	-0.6711	10.6610	-0.5216	7.7403	11.9587
POTASSIUM TARTRATE				C	0.3110	0.1720	0.1690	0.0287	0.3900	0.1330	0.0182	0.5540	0.0710	0.0870	0.0072	0.1040	0.0826
42427	1.9870	4	ROOM	S	4.7451	-1.7406	-0.9872	-0.3309	3.4313	-0.3040	0.0874	2.3424	-1.2735	11.5778	-1.0092	10.5608	12.1945
SODIUM THIOSULPHATE				C	0.3323	0.1814	0.1875	0.0225	0.2953	0.1713	0.0983	0.4590	-0.0678	0.0569	-0.0268	0.1070	0.0598
42428	1.7499	129	ROOM	S	5.0200	-3.2300	-0.6210	1.5200	15.6000	-7.1900	-18.2000	6.7400	11.0000	22.3000	10.0000	32.7000	21.2000
STILBENE				C	0.0930	0.0570	0.0670	-0.0030	0.0920	0.0485	-0.0050	0.0790	-0.0050	0.0325	0.0050	0.0640	0.0245
42429	1.6000	253	ROOM	S	31.1682	-7.9741	-21.5920	-0.8488	18.1260	-4.3205	0.7048	33.7041	1.2835	31.7666	-6.4830	15.7405	42.1394
TARTARIC ACID				C	0.9304	0.2025	0.3667	-0.1200	0.1927	0.1390	-0.0397	0.4649	-0.0036	0.0812	0.0138	0.0820	0.1063
42430	1.7600	4	ROOM	S	2.1600	-0.6100	-1.5000	2.8000	7.7000	-1.8000	2.7600	3.8500	-2.9000	12.6000	-1.6400	17.5000	9.6200
TAURINE				C	0.3178	0.1631	0.1347	-0.0135	0.4692	0.1495	0.0390	0.1399	0.0090	0.0722	-0.0008	0.0960	0.1000
42431	1.7120	127	293.0	S	5.5337	-0.4786	-4.9087	1.4328	3.3704	-3.0669	-1.1490	15.2076	-0.8701	13.8517	0.1108	11.1665	10.0009
TOLANE				C	0.0785	0.0350	0.0115	0.0030	0.0855	0.0350	0.0250	0.0645	0.0090	0.0290	0.0010	0.0545	0.0185
42432	1.1280	253	ROOM	S	15.9028	-7.6217	0.9568	2.4628	20.6286	-8.7753	-7.5940	20.0016	0.6697	34.5471	-1.8674	21.5859	54.1550
TRIGLYCINE SULFATE				C	0.4550	0.1720	0.1980	-0.0300	0.3210	0.2080	-0.0036	0.2630	-0.0500	0.0950	-0.0026	0.1110	0.0620
42433	1.6800	4	ROOM	S	3.2869	-0.2904	-2.2724	-0.1447	6.9813	-5.7686	-2.4505	10.8523	4.0872	10.5384	0.4419	10.7315	16.1476

HEXAGONAL

CODE / NAME	DENSITY	REF	TEMP(K)		11	12	13	33	55
APATITE				C	1.6670	0.1310	0.6550	1.3960	0.6630
52434	3.2180	129	ROOM	S	0.7481	0.0970	-0.3965	1.0884	1.5083
BERYL				C	2.6915	0.9644	0.6690	2.3725	0.6536
52435	2.6600	128	ROOM	S	0.4420	-0.1370	-0.0860	0.4700	1.5300
BERYL				C	2.7810	1.0010	0.6770	2.4800	0.6610
52436	2.6600	129	ROOM	S	0.4274	-0.1344	-0.0800	0.4469	1.5129
BERYL				C	2.9710	1.0260	0.7390	2.6500	0.7540
52437	2.6400	129	ROOM	S	0.3965	-0.1176	-0.0778	0.4207	1.3263
BERYL				C	2.8000	0.9900	0.6700	2.4800	0.6580
52438	2.6800	129	ROOM	S	0.4221	-0.1304	-0.0788	0.4458	1.5198
BERYL				C	2.8730	0.9910	0.7290	2.4180	0.7020
52439	2.7200	129	ROOM	S	0.4117	-0.1197	-0.0880	0.4667	1.4245
BERYLLIUM				C	3.0800	-0.5800	0.8700	3.5700	1.1000
52440	1.8000	129	ROOM	S	0.3775	0.1042	-0.1174	0.3373	0.9091
BERYLLIUM				C	2.9940	0.2760	0.1100	3.4220	1.6620
52441	1.8477	237	0.0	S	0.3372	-0.0307	-0.0099	0.2929	0.6017
BERYLLIUM				C	2.9940	0.2760	0.1100	3.4220	1.6620
52442	1.8477	237	75.0	S	0.3372	-0.0307	-0.0099	0.2929	0.6017
BERYLLIUM				C	2.9920	0.2760	0.1100	3.4210	1.6620
52443	1.8477	237	100.0	S	0.3374	-0.0308	-0.0099	0.2929	0.6017
BERYLLIUM				C	2.9900	0.2760	0.1200	3.4200	1.6610
52444	1.8477	237	125.0	S	0.3377	-0.0307	-0.0108	0.2932	0.6020
BERYLLIUM				C	2.9870	0.2770	0.1200	3.4180	1.6580
52445	1.8477	237	150.0	S	0.3381	-0.0309	-0.0108	0.2933	0.6031
BERYLLIUM				C	2.9810	0.2750	0.1200	3.4140	1.6550
52446	1.8477	237	175.0	S	0.3387	-0.0308	-0.0108	0.2937	0.6042
BERYLLIUM				C	2.9750	0.2770	0.1300	3.4080	1.6500
52447	1.8477	237	200.0	S	0.3395	-0.0311	-0.0118	0.2943	0.6061
BERYLLIUM				C	2.9640	0.2740	0.1300	3.4040	1.6460
52448	1.8477	237	225.0	S	0.3408	-0.0310	-0.0118	0.2947	0.6075
BERYLLIUM				C	2.9530	0.2710	0.1400	3.3900	1.6400
52449	1.8477	237	250.0	S	0.3421	-0.0308	-0.0129	0.2960	0.6098
BERYLLIUM				C	2.9390	0.2690	0.1400	3.3780	1.6330
52450	1.8477	237	275.0	S	0.3437	-0.0308	-0.0130	0.2971	0.6124
BERYLLIUM				C	2.9230	0.2670	0.1400	3.3640	1.6250
52451	1.8477	237	300.0	S	0.3456	-0.0309	-0.0131	0.2984	0.6154
BERYLLIUM OXIDE				C	4.7000	1.6800	1.1900	4.9400	1.5300
52452	3.0100	30	ROOM	S	0.2517	-0.0795	-0.0415	0.2224	0.6536
BERYLLIUM OXIDE				C	4.6060	1.2650	0.8848	4.9160	1.4770
52453	3.0100	60	298.0	S	0.2397	-0.0596	-0.0324	0.2151	0.6770
BIOTITE				C	1.8600	0.3240	0.1160	0.5400	0.0580
52454	3.0500	5	ROOM	S	0.5598	-0.0912	-0.1007	1.8951	17.2414
CADMIUM				C	1.0921	0.3977	0.3755	0.4603	0.1563
52455	8.6420	4	ROOM	S	1.2900	-0.1500	-0.9300	3.6900	6.4000
CADMIUM				C	1.2100	0.4810	0.4420	0.5130	0.1850
52456	8.6420	4	ROOM	S	1.2239	-0.1479	-0.9271	3.5469	5.4054
CADMIUM				C	1.0800	0.3890	0.3750	0.4600	0.1560
52457	8.6420	4	ROOM	S	1.3066	-0.1406	-0.9506	3.7238	6.4103
CADMIUM				C	1.2120	0.4810	0.4420	0.4450	0.1850
52458	8.6420	177	ROOM	S	1.2975	-0.0705	-1.2188	4.6683	5.4054

HEXAGONAL

CODE	NAME DENSITY	REF	TEMP(K)		11	12	13	33	55
CADMIUM				C	1.2923	0.3999	0.4095	0.5668	0.2420
52459	8.8415	101	0.0	S	1.0146	-0.1059	-0.6565	2.7129	4.1322
CADMIUM				C	1.2913	0.3995	0.4093	0.5656	0.2414
52460	8.8415	101	20.0	S	1.0159	-0.1055	-0.6588	2.7216	4.1425
CADMIUM				C	1.2864	0.3984	0.4076	0.5622	0.2395
52461	8.8361	101	40.0	S	1.0202	-0.1059	-0.6629	2.7399	4.1754
CADMIUM				C	1.2814	0.3995	0.4077	0.5576	0.2369
52462	8.8254	101	60.0	S	1.0279	-0.1060	-0.6741	2.7791	4.2212
CADMIUM				C	1.2735	0.3992	0.4079	0.5533	0.2342
52463	8.8121	101	80.0	S	1.0386	-0.1052	-0.6882	2.8220	4.2699
CADMIUM				C	1.2646	0.3993	0.4077	0.5486	0.2314
52464	8.7988	101	100.0	S	1.0505	-0.1052	-0.7025	2.8669	4.3215
CADMIUM				C	1.2538	0.3982	0.4071	0.5443	0.2287
52465	8.7856	101	120.0	S	1.0638	-0.1050	-0.7171	2.9099	4.3725
CADMIUM				C	1.2429	0.3972	0.4072	0.5399	0.2261
52466	8.7697	101	140.0	S	1.0786	-0.1038	-0.7352	2.9612	4.4228
CADMIUM				C	1.2320	0.3968	0.4071	0.5355	0.2235
52467	8.7539	101	160.0	S	1.0938	-0.1035	-0.7528	3.0120	4.4743
CADMIUM				C	1.2212	0.3970	0.4066	0.5311	0.2205
52468	8.7381	101	180.0	S	1.1088	-0.1045	-0.7689	3.0603	4.5351
CADMIUM				C	1.2093	0.3966	0.4064	0.5272	0.2177
52469	8.7248	101	200.0	S	1.1258	-0.1047	-0.7871	3.1103	4.5935
CADMIUM				C	1.1965	0.3958	0.4059	0.5229	0.2146
52470	8.7065	101	220.0	S	1.1441	-0.1048	-0.8068	3.1650	4.6598
CADMIUM				C	1.1826	0.3946	0.4051	0.5190	0.2114
52471	8.6934	101	240.0	S	1.1637	-0.1053	-0.8261	3.2164	4.7304
CADMIUM				C	1.1678	0.3932	0.4034	0.5149	0.2079
52472	8.6752	101	260.0	S	1.1838	-0.1072	-0.8434	3.2637	4.8100
CADMIUM				C	1.1530	0.3924	0.4022	0.5107	0.2040
52473	8.6595	101	280.0	S	1.2057	-0.1091	-0.8636	3.3184	4.9020
CADMIUM				C	1.1382	0.3924	0.4004	0.5061	0.2001
52474	8.6440	101	300.0	S	1.2278	-0.1130	-0.8819	3.3714	4.9975
CADMIUM				C	1.1450	0.3950	0.3990	0.5085	0.1985
52475	8.6520	50	300.0	S	1.2138	-0.1195	-0.8586	3.3141	5.0378
CADMIUM				C	1.1290	0.3930	0.4000	0.5045	0.1928
52476	8.6330	50	320.0	S	1.2426	-0.1161	-0.8932	3.3985	5.1867
CADMIUM				C	1.1140	0.3930	0.4000	0.5005	0.1872
52477	8.6141	50	340.0	S	1.2698	-0.1172	-0.9211	3.4703	5.3419
CADMIUM				C	1.0970	0.3930	0.3990	0.4960	0.1818
52478	8.5953	50	360.0	S	1.2998	-0.1206	-0.9486	3.5422	5.5005
CADMIUM				C	1.0800	0.3930	0.3980	0.4915	0.1759
52479	8.5766	50	380.0	S	1.3314	-0.1242	-0.9775	3.6176	5.6850
CADMIUM				C	1.0610	0.3930	0.3980	0.4870	0.1703
52480	8.5579	50	400.0	S	1.3708	-0.1262	-1.0171	3.7159	5.8720
CADMIUM				C	1.0420	0.3940	0.3970	0.4820	0.1646
52481	8.5394	50	420.0	S	1.4110	-0.1322	-1.0532	3.8097	6.0753
CADMIUM				C	1.0220	0.3950	0.3960	0.4770	0.1590
52482	8.5188	50	440.0	S	1.4558	-0.1391	-1.0931	3.9114	6.2893
CADMIUM				C	1.0000	0.3950	0.3960	0.4720	0.1532
52483	8.4984	50	460.0	S	1.5109	-0.1420	-1.1485	4.0457	6.5274

CODE (NAME)	DENSITY	REF	TEMP(K)		11	12	13	33	55
CADMIUM				C	0.9780	0.3960	0.3950	0.4670	0.1474
52484	8.4780	50	480.0	S	1.5675	-0.1507	-1.1984	4.1686	6.7843
CADMIUM				C	0.9540	0.3970	0.3950	0.4615	0.1415
52485	8.4578	50	500.0	S	1.6386	-0.1567	-1.2683	4.3380	7.0671
CADMIUM				C	0.9290	0.3980	0.3950	0.4565	0.1353
52486	8.4377	50	520.0	S	1.7187	-0.1645	-1.3448	4.5178	7.3910
CADMIUM				C	0.9020	0.3980	0.3960	0.4510	0.1293
52487	8.4176	50	540.0	S	1.8191	-0.1650	-1.4523	4.7677	7.7340
CADMIUM				C	0.8730	0.3960	0.3960	0.4455	0.1232
52488	8.3977	50	560.0	S	1.9332	-0.1633	-1.5733	5.0416	8.1169
CADMIUM				C	0.8500	0.3940	0.3960	0.4410	0.1187
52489	8.3828	50	575.0	S	2.0349	-0.1581	-1.6853	5.2943	8.4246
CADMIUM SELENIDE (CONSTANT E)				C	0.7406	0.4516	0.3930	0.8355	0.1317
52490	5.6550	32	298.0	S	2.3380	-1.1220	-0.5720	1.7350	7.5950
CADMIUM SELENIDE (HEXAGONAL)				C	0.7490	0.4609	0.3926	0.8451	0.1315
52491	5.6840	60	298.0	S	2.3271	-1.1439	-0.5497	1.6940	7.6046
CADMIUM SULFIDE				C	0.8160	0.4950	0.4790	0.8080	0.1430
52492	4.8200	4	ROOM	S	2.2305	-0.8848	-0.7978	2.1835	6.9930
CADMIUM SULFIDE				C	0.8100	0.4900	0.4800	0.8000	0.1430
52493	4.8250	123	ROOM	S	2.2531	-0.8719	-0.8287	2.2445	6.9930
CADMIUM SULFIDE				C	0.8432	0.5212	0.4638	0.9397	0.1489
52494	4.8250	43	300.0	S	2.1043	-1.0013	-0.5444	1.6016	6.7159
CADMIUM SULFIDE (CONSTANT E)				C	0.9068	0.5809	0.5094	0.9380	0.1504
52495	4.8250	32	298.0	S	2.0690	-0.9990	-0.5810	1.6970	6.6490
CADMIUM SULFIDE				C	0.8565	0.5321	0.4614	0.9361	0.8565
52496	4.8200	61	296.0	S	2.0768	-1.0058	-0.5279	1.5886	1.1675
CADMIUM SULFIDE				C	0.8431	0.5208	0.4567	0.9183	0.1458
52497	4.8240	107	298.0	S	2.1010	-1.0017	-0.5467	1.6328	6.8587
CALCIUM(33.33)-MAGNESIUM(66.67)				C	0.6124	0.1760	0.1500	0.6552	0.1927
52498	1.8064	248	100.0	S	1.8405	-0.4510	-0.3181	1.6719	5.1894
CALCIUM(33.33)-MAGNESIUM(66.67)				C	0.6045	0.1770	0.1500	0.6482	0.1913
52499	1.8000	248	125.0	S	1.8718	-0.4674	-0.3250	1.6931	5.2274
CALCIUM(33.33)-MAGNESIUM(66.67)				C	0.5996	0.1690	0.1500	0.6452	0.1896
52500	1.7939	248	150.0	S	1.8766	-0.4457	-0.3327	1.7046	5.2743
CALCIUM(33.33)-MAGNESIUM(66.67)				C	0.5932	0.1650	0.1500	0.6406	0.1881
52501	1.7877	248	175.0	S	1.8945	-0.4409	-0.3404	1.7204	5.3163
CALCIUM(33.33)-MAGNESIUM(66.67)				C	0.5871	0.1630	0.1500	0.6357	0.1866
52502	1.7814	248	200.0	S	1.9150	-0.4429	-0.3474	1.7370	5.3591
CALCIUM(33.33)-MAGNESIUM(66.67)				C	0.5811	0.1610	0.1500	0.6310	0.1850
52503	1.7753	248	225.0	S	1.9356	-0.4448	-0.3544	1.7533	5.4054
CALCIUM(33.33)-MAGNESIUM(66.67)				C	0.5750	0.1590	0.1500	0.6260	0.1835
52504	1.7691	248	250.0	S	1.9571	-0.4468	-0.3619	1.7709	5.4496
CALCIUM(33.33)-MAGNESIUM(66.67)				C	0.5688	0.1590	0.1500	0.6212	0.1820
52505	1.7630	248	275.0	S	1.9830	-0.4572	-0.3685	1.7877	5.4945
CALCIUM(33.33)-MAGNESIUM(66.67)				C	0.5625	0.1590	0.1500	0.6163	0.1805
52506	1.7570	248	300.0	S	2.0102	-0.4681	-0.3753	1.8053	5.5402
CANCRINITE				C	0.5200	0.0860	0.1240	0.8260	0.2380
52507	2.4600	4	ROOM	S	2.0312	-0.2730	-0.2639	1.2899	4.2017
COBALT				C	2.6300	1.5200	1.3300	2.8300	0.5200
52508	8.9000	129	ROOM	S	0.6229	-0.2780	-0.1621	0.5057	1.9231

HEXAGONAL

CODE	NAME DENSITY	REF	TEMP(K)		11	12	13	33	55
COBALT				C	3.0710	1.6500	1.0270	3.5810	0.7550
52509	8.8360	160	298.0	S	0.4729	-0.2309	-0.0694	0.3191	1.3245
COBALT				C	2.7080	1.4850	1.2800	2.9120	0.6310
52510	8.8630	178	293.0	S	0.5718	-0.2458	-0.1433	0.4694	1.5848
COBALT				C	3.1950	1.6610	1.0210	3.7360	0.8240
52511	8.9332	94	4.0	S	0.4423	-0.2096	-0.0636	0.3024	1.2136
COBALT				C	3.1950	1.6610	1.0200	3.7350	0.8230
52512	8.9332	94	23.0	S	0.4423	-0.2096	-0.0635	0.3024	1.2151
COBALT				C	3.1860	1.6620	1.0190	3.7240	0.8170
52513	8.9208	94	73.0	S	0.4446	-0.2115	-0.0638	0.3034	1.2240
COBALT				C	3.1630	1.6570	1.0220	3.6970	0.8060
52514	8.9023	94	123.0	S	0.4495	-0.2145	-0.0650	0.3064	1.2407
COBALT				C	3.1380	1.6560	1.0210	3.6670	0.7920
52515	8.8837	94	173.0	S	0.4557	-0.2191	-0.0659	0.3094	1.2626
COBALT				C	3.1090	1.6550	1.0210	3.6310	0.7770
52516	8.8643	94	223.0	S	0.4632	-0.2245	-0.0671	0.3131	1.2870
COBALT				C	3.0790	1.6530	1.0200	3.5940	0.7610
52517	8.8466	94	273.0	S	0.4710	-0.2302	-0.0683	0.3170	1.3141
COBALT				C	3.0630	1.6510	1.0190	3.5740	0.7530
52518	8.8360	94	298.0	S	0.4751	-0.2331	-0.0690	0.3191	1.3280
COBALT				C	3.0470	1.6490	1.0180	3.5560	0.7450
52519	8.8254	94	323.0	S	0.4792	-0.2361	-0.0696	0.3211	1.3423
COBALT				C	3.0150	1.6430	1.0140	3.5210	0.7300
52520	8.8068	94	373.0	S	0.4872	-0.2417	-0.0707	0.3247	1.3699
COBALT				C	2.9820	1.6360	1.0110	3.4830	0.7130
52521	8.7909	94	423.0	S	0.4955	-0.2474	-0.0720	0.3289	1.4025
COBALT				C	2.9460	1.6280	1.0090	3.4400	0.6940
52522	8.7724	94	473.0	S	0.5049	-0.2538	-0.0737	0.3339	1.4409
COBALT				C	2.9100	1.6180	1.0030	3.3860	0.6750
52523	8.7565	94	523.0	S	0.5141	-0.2599	-0.0753	0.3399	1.4815
DYSPROSIUM				C	0.7466	0.2616	0.2233	0.7871	0.2427
52524	8.5600	94	298.0	S	1.5981	-0.4637	-0.3218	1.4531	4.1203
DYSPROSIUM				C	0.7423	0.2625	0.2228	0.7847	0.2411
52525	8.5275	94	323.0	S	1.6114	-0.4728	-0.3233	1.4579	4.1477
DYSPROSIUM				C	0.7337	0.2641	0.2218	0.7803	0.2378
52526	8.4649	94	373.0	S	1.6383	-0.4911	-0.3261	1.4669	4.2052
DYSPROSIUM				C	0.7252	0.2658	0.2209	0.7762	0.2338
52527	8.4018	94	423.0	S	1.6662	-0.5105	-0.3289	1.4755	4.2772
DYSPROSIUM				C	0.7161	0.2669	0.2202	0.7710	0.2298
52528	8.3394	94	473.0	S	1.6964	-0.5298	-0.3332	1.4873	4.3516
DYSPROSIUM				C	0.7073	0.2689	0.2194	0.7654	0.2260
52529	8.2779	94	523.0	S	1.7285	-0.5526	-0.3371	1.4997	4.4248
DYSPROSIUM				C	0.6984	0.2708	0.2184	0.7598	0.2222
52530	8.2173	94	573.0	S	1.7620	-0.5766	-0.3407	1.5120	4.5005
DYSPROSIUM				C	0.6896	0.2724	0.2177	0.7543	0.2185
52531	8.1576	94	623.0	S	1.7963	-0.6006	-0.3451	1.5249	4.5767
DYSPROSIUM				C	0.6810	0.2740	0.2168	0.7488	0.2149
52532	8.0986	94	673.0	S	1.8313	-0.6257	-0.3491	1.5376	4.6533
DYSPROSIUM				C	0.6768	0.2748	0.2165	0.7425	0.2131
52533	8.0694	94	698.0	S	1.8496	-0.6380	-0.3533	1.5528	4.6926

HEXAGONAL

CODE	NAME DENSITY	REF	TEMP(K)		11	12	13	33	55
ERBIUM				C	0.9087	0.3083	0.2260	0.8728	0.2969
52534	9.1093	94	81.0	S	1.2873	-0.3782	-0.2354	1.2676	3.3681
ERBIUM				C	0.9082	0.3080	0.2280	0.8742	0.2969
52535	9.1093	94	83.0	S	1.2887	-0.3774	-0.2377	1.2679	3.3681
ERBIUM				C	0.9056	0.3074	0.2280	0.8774	0.2960
52536	9.1066	94	98.0	S	1.2927	-0.3790	-0.2374	1.2631	3.3784
ERBIUM				C	0.9007	0.3071	0.2280	0.8770	0.2951
52537	9.1012	94	123.0	S	1.3013	-0.3833	-0.2387	1.2644	3.3887
ERBIUM				C	0.8955	0.3065	0.2280	0.8757	0.2933
52538	9.0948	94	148.0	S	1.3105	-0.3873	-0.2403	1.2671	3.4095
ERBIUM				C	0.8902	0.3060	0.2280	0.8730	0.2917
52539	9.0894	94	173.0	S	1.3201	-0.3917	-0.2425	1.2721	3.4282
ERBIUM				C	0.8793	0.3055	0.2270	0.8664	0.2875
52540	9.0794	94	223.0	S	1.3405	-0.4023	-0.2458	1.2830	3.4783
ERBIUM				C	0.8688	0.3054	0.2270	0.8588	0.2831
52541	9.0694	94	273.0	S	1.3618	-0.4132	-0.2507	1.2970	3.5323
ERBIUM				C	0.8634	0.3050	0.2270	0.8554	0.2809
52542	9.0640	94	298.0	S	1.3726	-0.4183	-0.2532	1.3034	3.5600
FLUORAPATITE				C	1.4340	0.4450	0.5750	1.8050	0.4150
52543	3.2147	280	ROOM	S	0.8361	-0.1750	-0.2106	0.6882	2.4096
GADOLINIUM				C	0.7400	0.2822	0.2034	0.7901	0.2382
52544	7.8880	93	4.0	S	1.6372	-0.5472	-0.2806	1.4101	4.1982
GADOLINIUM				C	0.7374	0.2822	0.2028	0.7862	0.2369
52545	7.8880	93	50.0	S	1.6449	-0.5520	-0.2819	1.4174	4.2212
GADOLINIUM				C	0.7287	0.2805	0.2008	0.7792	0.2337
52546	7.8880	93	100.0	S	1.6676	-0.5635	-0.2845	1.4300	4.2790
GADOLINIUM				C	0.7202	0.2790	0.2026	0.7723	0.2293
52547	7.8880	93	150.0	S	1.6932	-0.5733	-0.2938	1.4490	4.3611
GADOLINIUM				C	0.7105	0.2740	0.2035	0.7569	0.2235
52548	7.8880	93	200.0	S	1.7169	-0.5741	-0.3072	1.4864	4.4743
GADOLINIUM				C	0.7050	0.2703	0.2052	0.7481	0.2209
52549	7.8880	93	218.0	S	1.7298	-0.5707	-0.3179	1.5111	4.5269
GADOLINIUM				C	0.7025	0.2689	0.2020	0.7514	0.2197
52550	7.8880	93	225.0	S	1.7326	-0.5736	-0.3116	1.4984	4.5517
GADOLINIUM				C	0.6922	0.2632	0.2034	0.7387	0.2158
52551	7.8880	93	250.0	S	1.7583	-0.5727	-0.3265	1.5335	4.6339
GADOLINIUM				C	0.6690	0.2494	0.2018	0.7036	0.2088
52552	7.8880	93	286.0	S	1.8146	-0.5687	-0.3573	1.6262	4.7893
GADOLINIUM				C	0.6691	0.2504	0.2100	0.7150	0.2080
52553	7.8880	93	289.0	S	1.8222	-0.5661	-0.3689	1.6153	4.8077
GADOLINIUM				C	0.6667	0.2499	0.2132	0.7191	0.2069
52554	7.8880	93	298.0	S	1.8324	-0.5668	-0.3752	1.6131	4.8333
HAFNIUM				C	1.9010	0.7450	0.6550	2.0440	0.6000
52555	12.7730	95	4.2	S	0.6571	-0.2079	-0.1439	0.5815	1.6667
HAFNIUM				C	1.9000	0.7440	0.6550	2.0430	0.5990
52556	12.7690	95	23.0	S	0.6573	-0.2077	-0.1442	0.5819	1.6694
HAFNIUM				C	1.8910	0.7470	0.6580	2.0350	0.5950
52557	12.7610	95	73.0	S	0.6631	-0.2111	-0.1461	0.5859	1.6807
HAFNIUM				C	1.8750	0.7510	0.6590	2.0220	0.5880
52558	12.7540	95	123.0	S	0.6725	-0.2172	-0.1484	0.5913	1.7007

HEXAGONAL

CODE / NAME	DENSITY	REF	TEMP(K)		11	12	13	33	55
HAFNIUM				C	1.8590	0.7570	0.6600	2.0080	0.5790
52559	12.7460	95	173.0	S	0.6829	-0.2246	-0.1506	0.5970	1.7271
HAFNIUM				C	1.8420	0.7650	0.6610	1.9930	0.5710
52560	12.7380	95	223.0	S	0.6948	-0.2337	-0.1529	0.6032	1.7513
HAFNIUM				C	1.8220	0.7700	0.6610	1.9770	0.5620
52561	12.7310	95	273.0	S	0.7078	-0.2427	-0.1555	0.6098	1.7794
HAFNIUM				C	1.8110	0.7720	0.6610	1.9690	0.5570
52562	12.7270	95	298.0	S	0.7150	-0.2475	-0.1569	0.6132	1.7953
ICE				C	0.1385	0.0707	0.0581	0.1499	0.0319
52563	1.0640	129	257.0	S	10.4204	-4.3289	-2.3610	8.5013	31.3480
ICE				C	0.1333	0.0603	0.0508	0.1428	0.0326
52564	1.0000	4	257.0	S	10.0248	-3.6738	-2.2593	8.6103	30.6748
ICE				C	0.1330	0.0630	0.0460	0.1420	0.0306
52565	1.0000	4	257.0	S	10.1513	-4.1344	-1.9492	8.3051	32.6797
ICE				C	0.1338	0.0623	0.0454	0.1430	0.0311
52566	0.9188	244	248.0	S	9.9800	-4.0000	-1.9000	8.2000	32.2000
ICE				C	0.1337	0.0634	0.0454	0.1423	0.0308
52567	0.9183	244	253.0	S	10.0800	-4.1300	-1.9000	8.2400	32.5000
ICE				C	0.1334	0.0635	0.0451	0.1411	0.0305
52568	0.9178	244	258.0	S	10.1300	-4.1800	-1.9000	8.3000	32.8000
ICE				C	0.1315	0.0624	0.0440	0.1395	0.0303
52569	0.9173	244	263.0	S	10.2400	-4.2200	-1.9000	8.3700	33.0000
ICE				C	0.1299	0.0615	0.0431	0.1381	0.0301
52570	0.9168	244	268.0	S	10.3500	-4.2700	-1.9000	8.4300	33.2000
ICE				C	0.1337	0.0679	0.0778	0.1683	0.0336
52571	0.9020	37	258.0	S	11.4547	-3.7429	-3.5649	9.2377	29.7619
ICE				C	0.1208	0.0668	0.0713	0.1599	0.0309
52572	0.9017	37	263.0	S	13.2911	-5.2275	-3.5956	9.4605	32.3625
ICE				C	0.0982	0.0633	0.0516	0.1302	0.0283
52573	0.9017	37	273.0	S	18.4725	-10.1807	-3.2862	10.2852	35.3357
ICE				C	0.1679	0.0852	0.0692	0.1792	0.0363
52574	0.9333	205	50.0	S	8.5470	-3.5370	-1.9360	7.0760	27.5790
ICE				C	0.1634	0.0833	0.0673	0.1747	0.0356
52575	0.9333	205	100.0	S	8.8100	-3.6770	-1.9780	7.2490	28.1290
ICE				C	0.1574	0.0795	0.0655	0.1686	0.0341
52576	0.9322	205	150.0	S	9.1070	-3.7280	-2.0910	7.5590	29.3600
ICE				C	0.1499	0.0737	0.0639	0.1608	0.0318
52577	0.9261	205	200.0	S	9.4530	-3.6670	-2.2990	8.0450	31.4470
ICE				C	0.1410	0.0660	0.0624	0.1515	0.0288
52578	0.9200	205	250.0	S	9.8790	-3.4490	-2.6500	8.7870	34.7710
INDIUM				C	0.5392	0.3871	0.4513	0.5162	0.0797
52579	7.4713	48	4.2	S	6.9321	0.3575	-6.3732	13.0810	12.5471
INDIUM				C	0.5260	0.4056	0.4457	0.5080	0.0764
52580	7.4244	48	77.0	S	7.4969	-0.8088	-5.8679	12.2650	13.0890
INDIUM				C	0.4535	0.4006	0.4151	0.4515	0.0651
52581	7.2788	48	300.0	S	14.9565	-3.9471	-10.1219	20.8265	15.3610
MAGNESIUM				C	0.5649	0.2316	0.1810	0.5873	0.1681
52582	1.7400	128	ROOM	S	2.2300	-0.7700	-0.4500	1.9800	5.9500
MAGNESIUM				C	0.5858	0.2502	0.2079	0.6111	0.1658
52583	1.7400	128	ROOM	S	2.2100	-0.7700	-0.4900	1.9700	6.0300

CODE	NAME DENSITY	REF	TEMP(K)		11	12	13	33	55
MAGNESIUM				C	0.6348	0.2594	0.2170	0.6645	0.1842
52584	1.7752	236	0.0	S	1.9964	-0.6674	-0.4340	1.7883	5.4289
MAGNESIUM				C	0.6340	0.2594	0.2170	0.6640	0.1840
52585	1.7752	236	20.0	S	2.0000	-0.6695	-0.4348	1.7902	5.4348
MAGNESIUM				C	0.6330	0.2593	0.2169	0.6635	0.1835
52586	1.7746	236	40.0	S	2.0042	-0.6717	-0.4356	1.7919	5.4496
MAGNESIUM				C	0.6315	0.2592	0.2169	0.6616	0.1829
52587	1.7741	236	60.0	S	2.0110	-0.6750	-0.4380	1.7987	5.4675
MAGNESIUM				C	0.6300	0.2591	0.2168	0.6595	0.1820
52588	1.7723	236	80.0	S	2.0178	-0.6783	-0.4403	1.8058	5.4945
MAGNESIUM				C	0.6275	0.2590	0.2167	0.6566	0.1810
52589	1.7714	236	100.0	S	2.0294	-0.6843	-0.4439	1.8160	5.5249
MAGNESIUM				C	0.6249	0.2588	0.2166	0.6534	0.1798
52590	1.7690	236	120.0	S	2.0413	-0.6902	-0.4479	1.8274	5.5617
MAGNESIUM				C	0.6219	0.2586	0.2165	0.6495	0.1783
52591	1.7676	236	140.0	S	2.0555	-0.6971	-0.4528	1.8415	5.6085
MAGNESIUM				C	0.6189	0.2585	0.2163	0.6455	0.1768
52592	1.7655	236	160.0	S	2.0700	-0.7047	-0.4575	1.8558	5.6561
MAGNESIUM				C	0.6154	0.2582	0.2162	0.6411	0.1750
52593	1.7627	236	180.0	S	2.0868	-0.7128	-0.4634	1.8723	5.7143
MAGNESIUM				C	0.6118	0.2580	0.2160	0.6370	0.1732
52594	1.7606	236	200.0	S	2.1045	-0.7220	-0.4688	1.8878	5.7737
MAGNESIUM				C	0.6084	0.2576	0.2157	0.6326	0.1715
52595	1.7580	236	220.0	S	2.1208	-0.7298	-0.4743	1.9042	5.8309
MAGNESIUM				C	0.6049	0.2573	0.2155	0.6281	0.1697
52596	1.7552	236	240.0	S	2.1384	-0.7385	-0.4803	1.9217	5.8928
MAGNESIUM				C	0.6011	0.2569	0.2152	0.6243	0.1678
52597	1.7528	236	260.0	S	2.1572	-0.7481	-0.4857	1.9367	5.9595
MAGNESIUM				C	0.5974	0.2565	0.2148	0.6200	0.1659
52598	1.7501	236	280.0	S	2.1759	-0.7576	-0.4914	1.9534	6.0277
MAGNESIUM				C	0.5940	0.2561	0.2144	0.6160	0.1640
52599	1.7474	236	300.0	S	2.1931	-0.7663	-0.4966	1.9691	6.0976
MAGNESIUM				C	0.5912	0.2552	0.2131	0.6175	0.1648
52600	1.7370	225	294.0	S	2.2031	-0.7731	-0.4935	1.9600	6.0680
MAGNESIUM				C	0.5950	0.2612	0.2180	0.6155	0.1635
52601	1.7364	275	298.0	S	2.2104	-0.7854	-0.5047	1.9822	6.1162
MAGNESIUM (PURE)				C	0.6332	0.2631	0.2165	0.6561	0.1807
52602	1.7620	88	78.0	S	2.0146	-0.6873	-0.4380	1.8132	5.5340
MAGNESIUM (PURE)				C	0.5928	0.2590	0.2157	0.6135	0.1632
52603	1.7380	88	298.0	S	2.2121	-0.7837	-0.5022	1.9831	6.1274
MAGNESIUM(98.98)-INDIUM(1.02)				C	0.6344	0.2658	0.2229	0.6682	0.1796
52604	1.8272	88	78.0	S	2.0218	-0.6911	-0.4439	1.7927	5.5679
MAGNESIUM(98.98)-INDIUM(1.02)				C	0.5941	0.2620	0.2234	0.6276	0.1620
52605	1.8017	88	298.0	S	2.2229	-0.7883	-0.5107	1.9569	6.1728
MAGNESIUM(98.65)-INDIUM(1.35)				C	0.6391	0.2663	0.2201	0.6618	0.1821
52606	1.8498	88	4.0	S	2.0000	-0.6824	-0.4382	1.8025	5.4915
MAGNESIUM(98.65)-INDIUM(1.35)				C	0.6350	0.2652	0.2201	0.6562	0.1800
52607	1.8481	88	78.0	S	2.0165	-0.6877	-0.4457	1.8229	5.5556
MAGNESIUM(98.65)-INDIUM(1.35)				C	0.5955	0.2626	0.2201	0.6148	0.1625
52608	1.8223	88	298.0	S	2.2157	-0.7882	-0.5111	1.9925	6.1538

HEXAGONAL

NAME / CODE	DENSITY	REF	TEMP(K)		11	12	13	33	55
MAGNESIUM(98.04)-INDIUM(1.96)				C	0.6352	0.2675	0.2222	0.6598	0.1788
52609	1.8866	88	78.0	S	2.0238	-0.6958	-0.4472	1.8168	5.5928
MAGNESIUM(98.04)-INDIUM(1.96)				C	0.5944	0.2634	0.2225	0.6190	0.1611
52610	1.8603	88	298.0	S	2.2271	-0.7941	-0.5151	1.9858	6.2073
MAGNESIUM(98.49)-LITHIUM(1.51)				C	0.5924	0.2600	0.2176	0.6128	0.1625
52611	1.7204	275	298.0	S	2.2207	-0.7877	-0.5088	1.9932	6.1538
MAGNESIUM(96.98)-LITHIUM(3.02)				C	0.5892	0.2588	0.2165	0.6095	0.1614
52612	1.7032	275	298.0	S	2.2336	-0.7931	-0.5117	2.0042	6.1958
MAGNESIUM(94.90)-LITHIUM(5.10)				C	0.5853	0.2572	0.2158	0.6055	0.1601
52613	1.6815	275	298.0	S	2.2500	-0.7979	-0.5175	2.0204	6.2461
MAGNESIUM(93.00)-LITHIUM(7.00)				C	0.5795	0.2539	0.2126	0.5994	0.1587
52614	1.6693	275	298.0	S	2.2681	-0.8031	-0.5196	2.0369	6.3012
MAGNESIUM(90.00)-LITHIUM(10.00)				C	0.5724	0.2510	0.2103	0.5921	0.1568
52615	1.6497	275	298.0	S	2.2975	-0.8139	-0.5270	2.0632	6.3776
MAGNESIUM(87.95)-LITHIUM(12.05)				C	0.5672	0.2479	0.2076	0.5868	0.1555
52616	1.6167	275	298.0	S	2.3142	-0.8177	-0.5294	2.0788	6.4309
MAGNESIUM(85.85)-LITHIUM(14.15)				C	0.5627	0.2464	0.2067	0.5821	0.1539
52617	1.5830	275	298.0	S	2.3357	-0.8258	-0.5361	2.0987	6.4977
MAGNESIUM(84.06)-LITHIUM(15.94)				C	0.5549	0.2415	0.2022	0.5742	0.1525
52618	1.5542	275	298.0	S	2.3599	-0.8309	-0.5384	2.1208	6.5574
MAGNESIUM(99.93)-SILVER(0.07)				C	0.5950	0.2614	0.2210	0.6290	0.1640
52619	1.7425	154	298.0	S	2.2120	-0.7856	-0.5011	1.9420	6.0976
MAGNESIUM(99.74)-SILVER(0.26)				C	0.6020	0.2674	0.2210	0.6200	0.1664
52620	1.7547	154	298.0	S	2.1967	-0.7919	-0.5007	1.9699	6.0096
MAGNESIUM(99.63)-SILVER(0.37)				C	0.5969	0.2615	0.2170	0.6170	0.1640
52621	1.7618	154	298.0	S	2.1992	-0.7823	-0.4983	1.9713	6.0976
MAGNESIUM(99.74)-SILVER(0.26)				C	0.6374	0.2654	0.2181	0.6603	0.1825
52622	1.7764	88	78.0	S	2.0031	-0.6851	-0.4353	1.8021	5.4795
MAGNESIUM(99.74)-SILVER(0.26)				C	0.5968	0.2615	0.2172	0.6168	0.1649
52623	1.7547	88	298.0	S	2.2001	-0.7823	-0.4993	1.9729	6.0643
MAGNESIUM(99.63)-SILVER(0.37)				C	0.6389	0.2666	0.2191	0.6631	0.1829
52624	1.7867	88	78.0	S	2.0003	-0.6857	-0.4344	1.7951	5.4675
MAGNESIUM(99.63)-SILVER(0.37)				C	0.5975	0.2620	0.2179	0.6201	0.1652
52625	1.7618	88	298.0	S	2.1982	-0.7825	-0.4975	1.9623	6.0533
MAGNESIUM(99.79)-TIN(0.21)				C	0.5972	0.2636	0.2220	0.6240	0.1630
52626	1.7516	154	298.0	S	2.2102	-0.7874	-0.5062	1.9627	6.1350
MAGNESIUM(99.73)-TIN(0.27)				C	0.5948	0.2608	0.2200	0.6200	0.1638
52627	1.7554	154	298.0	S	2.2118	-0.7822	-0.5073	1.9729	6.1050
MAGNESIUM(99.57)-TIN(0.43)				C	0.5981	0.2635	0.2240	0.6240	0.1632
52628	1.7658	154	298.0	S	2.2078	-0.7808	-0.5122	1.9703	6.1274
MAGNESIUM(99.48)-TIN(0.52)				C	0.5945	0.2625	0.2200	0.6200	0.1621
52629	1.7716	154	298.0	S	2.2194	-0.7926	-0.5063	1.9722	6.1690
MAGNESIUM(99.28)-TIN(0.72)				C	0.5988	0.2666	0.2200	0.6130	0.1623
52630	1.7845	154	298.0	S	2.2118	-0.7984	-0.5073	1.9954	6.1614
MAGNESIUM(99.00)-TIN(1.00)				C	0.5982	0.2694	0.2220	0.6120	0.1610
52631	1.8025	154	298.0	S	2.2284	-0.8130	-0.5134	2.0065	6.2112
MAGNESIUM(99.79)-TIN(0.21)				C	0.6346	0.2647	0.2177	0.6588	0.1810
52632	1.7764	88	78.0	S	2.0136	-0.6898	-0.4374	1.8070	5.5249
MAGNESIUM(99.79)-TIN(0.21)				C	0.5945	0.2611	0.2174	0.6154	0.1634
52633	1.7516	88	298.0	S	2.2120	-0.7874	-0.5032	1.9805	6.1200

NAME CODE	DENSITY	REF	TEMP(K)		11	12	13	33	55
MAGNESIUM(99.54)-TIN(0.46)				C	0.6333	0.2652	0.2191	0.6574	0.1798
52634	1.7927	88	78.0	S	2.0228	-0.6938	-0.4429	1.8164	5.5617
MAGNESIUM(99.54)-TIN(0.46)				C	0.5932	0.2613	0.2180	0.6148	0.1624
52635	1.7677	88	298.0	S	2.2209	-0.7921	-0.5066	1.9858	6.1576
MAGNESIUM(99.33)-TIN(0.67)				C	0.6371	0.2674	0.2191	0.6560	0.1807
52636	1.8065	88	78.0	S	2.0120	-0.6929	-0.4405	1.8187	5.5340
MAGNESIUM(99.33)-TIN(0.67)				C	0.5964	0.2629	0.2189	0.6138	0.1631
52637	1.7813	88	298.0	S	2.2103	-0.7882	-0.5072	1.9909	6.1312
MAGNESIUM(99.06)-TIN(0.94)				C	0.6350	0.2673	0.2217	0.6606	0.1790
52638	1.8241	88	78.0	S	2.0234	-0.6962	-0.4454	1.8127	5.5866
MAGNESIUM(99.06)-TIN(0.94)				C	0.5951	0.2642	0.2211	0.6177	0.1615
52639	1.7987	88	298.0	S	2.2243	-0.7978	-0.5106	1.9844	6.1920
MUSCOVITE				C	1.7800	0.4240	0.1450	0.5490	0.1220
52640	2.7900	5	ROOM	S	0.6038	-0.1337	-0.1241	1.8871	8.1967
PHLOGOPITE				C	1.7900	0.3240	0.2580	0.5170	0.0560
52641	2.8000	5	ROOM	S	0.6104	-0.0717	-0.2688	2.2025	17.8571
PHLOGOPITE				C	1.7800	0.3020	0.1520	0.5100	0.0650
52642	2.8200	5	ROOM	S	0.5894	-0.0872	-0.1497	2.0500	15.3846
QUARTZ (BETA)				C	1.1660	0.1670	0.3280	1.1040	0.3606
52643	2.5330	129	873.0	S	0.9398	-0.0612	-0.2611	1.0609	2.7732
RHENIUM				C	6.3440	2.6600	2.0200	7.0160	1.6910
52644	21.0240	230	4.2	S	0.1995	-0.0720	-0.0367	0.1637	0.5914
RHENIUM				C	6.3420	2.6600	2.0000	7.0120	1.6880
52645	21.0240	230	20.0	S	0.1994	-0.0722	-0.0363	0.1633	0.5924
RHENIUM				C	6.3300	2.6600	1.9800	7.0080	1.6870
52646	21.0240	230	40.0	S	0.1998	-0.0727	-0.0359	0.1630	0.5928
RHENIUM				C	6.3170	2.6700	1.9600	7.0020	1.6850
52647	21.0240	230	60.0	S	0.2005	-0.0737	-0.0355	0.1627	0.5935
RHENIUM				C	6.3060	2.6700	1.9400	6.9970	1.6840
52648	21.0240	230	78.0	S	0.2008	-0.0742	-0.0351	0.1624	0.5938
RHENIUM				C	6.2840	2.6800	1.9400	6.9830	1.6780
52649	21.0240	230	100.0	S	0.2021	-0.0753	-0.0352	0.1628	0.5959
RHENIUM				C	6.2680	2.7000	1.9300	6.9720	1.6720
52650	21.0240	230	120.0	S	0.2034	-0.0768	-0.0350	0.1628	0.5981
RHENIUM				C	6.2500	2.7000	1.9300	6.9610	1.6660
52651	21.0240	230	140.0	S	0.2043	-0.0774	-0.0352	0.1632	0.6002
RHENIUM				C	6.2340	2.7100	1.9400	6.9470	1.6610
52652	21.0240	230	160.0	S	0.2055	-0.0783	-0.0355	0.1638	0.6020
RHENIUM				C	6.2190	2.7200	1.9400	6.9360	1.6550
52653	21.0240	230	180.0	S	0.2066	-0.0792	-0.0356	0.1641	0.6042
RHENIUM				C	6.2040	2.7200	1.9500	6.9200	1.6490
52654	21.0240	230	200.0	S	0.2074	-0.0796	-0.0360	0.1648	0.6064
RHENIUM				C	6.1860	2.7200	1.9700	6.9050	1.6440
52655	21.0240	230	220.0	S	0.2085	-0.0800	-0.0367	0.1657	0.6083
RHENIUM				C	6.1720	2.7200	1.9800	6.8880	1.6380
52656	21.0240	230	240.0	S	0.2093	-0.0804	-0.0371	0.1665	0.6105
RHENIUM				C	6.1570	2.7200	2.0000	6.8700	1.6330
52657	21.0240	230	260.0	S	0.2103	-0.0806	-0.0377	0.1675	0.6124
RHENIUM				C	6.1400	2.7100	2.0300	6.8490	1.6270
52658	21.0240	230	280.0	S	0.2112	-0.0804	-0.0388	0.1690	0.6146

HEXAGONAL

NAME CODE	DENSITY	REF	TEMP(K)		11	12	13	33	55
RHENIUM				C	6.1260	2.7000	2.0600	6.8270	1.6250
52659	21.0240	230	298.0	S	0.2119	-0.0800	-0.0398	0.1705	0.6154
RHENIUM				C	6.4460	2.7700	1.9590	7.1700	1.6850
52660	21.1207	94	4.0	S	0.1974	-0.0746	-0.0335	0.1578	0.5935
RHENIUM				C	6.4390	2.7670	1.9620	7.1700	1.6850
52661	21.1207	94	23.0	S	0.1976	-0.0747	-0.0336	0.1579	0.5935
RHENIUM				C	6.4100	2.7580	1.9740	7.1350	1.6790
52662	21.1207	94	73.0	S	0.1988	-0.0750	-0.0343	0.1591	0.5956
RHENIUM				C	6.3640	2.7510	2.0000	7.0720	1.6630
52663	21.1102	94	123.0	S	0.2010	-0.0758	-0.0354	0.1614	0.6013
RHENIUM				C	6.3110	2.7450	2.0260	7.0010	1.6480
52664	21.0850	94	173.0	S	0.2036	-0.0768	-0.0367	0.1641	0.6068
RHENIUM				C	6.2580	2.7480	2.0480	6.9330	1.6320
52665	21.0576	94	223.0	S	0.2066	-0.0783	-0.0379	0.1666	0.6127
RHENIUM				C	6.2060	2.7500	2.0680	6.8670	1.6150
52666	21.0345	94	273.0	S	0.2095	-0.0798	-0.0391	0.1691	0.6192
RHENIUM				C	6.1820	2.7530	2.0780	6.8350	1.6060
52667	21.0240	94	298.0	S	0.2110	-0.0806	-0.0396	0.1704	0.6227
RHENIUM				C	6.1540	2.7530	2.0840	6.8040	1.5980
52668	21.0111	94	323.0	S	0.2125	-0.0815	-0.0401	0.1716	0.6258
RHENIUM				C	6.1030	2.7520	2.0980	6.7480	1.5820
52669	20.9923	94	373.0	S	0.2154	-0.0830	-0.0412	0.1738	0.6321
RHENIUM				C	6.0530	2.7510	2.1100	6.6960	1.5660
52670	20.9735	94	423.0	S	0.2183	-0.0845	-0.0422	0.1759	0.6386
RHENIUM				C	6.0020	2.7500	2.1180	6.6410	1.5490
52671	20.9546	94	473.0	S	0.2213	-0.0862	-0.0431	0.1781	0.6456
RHENIUM				C	5.9540	2.7500	2.1280	6.5900	1.5320
52672	20.9357	94	523.0	S	0.2243	-0.0878	-0.0441	0.1802	0.6527
RHENIUM				C	5.9030	2.7450	2.1380	6.5410	1.5140
52673	20.9170	94	573.0	S	0.2273	-0.0894	-0.0451	0.1824	0.6605
RHENIUM				C	5.8610	2.7480	2.1460	6.4930	1.4960
52674	20.8983	94	623.0	S	0.2302	-0.0911	-0.0460	0.1844	0.6684
RHENIUM				C	5.8220	2.7560	2.1540	6.4500	1.4790
52675	20.8732	94	673.0	S	0.2331	-0.0930	-0.0468	0.1863	0.6761
RHENIUM				C	5.7810	2.7590	2.1600	6.4060	1.4610
52676	20.8482	94	723.0	S	0.2360	-0.0949	-0.0476	0.1882	0.6845
RHENIUM				C	5.7400	2.7600	2.1680	6.3650	1.4440
52677	20.8296	94	773.0	S	0.2390	-0.0966	-0.0485	0.1901	0.6925
RHENIUM				C	5.7000	2.7640	2.1740	6.3210	1.4250
52678	20.8046	94	823.0	S	0.2421	-0.0985	-0.0494	0.1922	0.7018
RHENIUM				C	5.6600	2.7650	2.1780	6.2790	1.4080
52679	20.7861	94	873.0	S	0.2450	-0.1004	-0.0502	0.1941	0.7102
RHENIUM				C	5.6190	2.7620	2.1840	6.2370	1.3910
52680	20.7612	94	923.0	S	0.2480	-0.1020	-0.0511	0.1961	0.7189
RUTHENIUM				C	5.7630	1.8720	1.6730	6.4050	1.8910
52681	12.3875	94	4.0	S	0.2025	-0.0545	-0.0386	0.1763	0.5288
RUTHENIUM				C	5.7620	1.8710	1.6730	6.4050	1.8900
52682	12.3875	94	23.0	S	0.2025	-0.0545	-0.0386	0.1763	0.5291
RUTHENIUM				C	5.7530	1.8770	1.6750	6.3940	1.8800
52683	12.3875	94	73.0	S	0.2030	-0.0550	-0.0388	0.1767	0.5319

HEXAGONAL

CODE / NAME	DENSITY	REF	TEMP(K)		11	12	13	33	55
RUTHENIUM				C	5.7280	1.8730	1.6770	6.3690	1.8680
52684	12.3862	94	123.0	S	0.2041	-0.0553	-0.0392	0.1777	0.5353
RUTHENIUM				C	5.6990	1.8730	1.6790	6.3370	1.8530
52685	12.3800	94	173.0	S	0.2055	-0.0559	-0.0396	0.1788	0.5397
RUTHENIUM				C	5.6690	1.8750	1.6810	6.2990	1.8340
52686	12.3714	94	223.0	S	0.2070	-0.0566	-0.0401	0.1802	0.5453
RUTHENIUM				C	5.6410	1.8780	1.6810	6.2600	1.8150
52687	12.3627	94	273.0	S	0.2084	-0.0573	-0.0406	0.1815	0.5510
RUTHENIUM				C	5.6260	1.8780	1.6820	6.2420	1.8060
52688	12.3615	94	298.0	S	0.2092	-0.0576	-0.0408	0.1822	0.5537
RUTHENIUM				C	5.6110	1.8790	1.6810	6.2210	1.7960
52689	12.3504	94	323.0	S	0.2099	-0.0580	-0.0411	0.1829	0.5568
RUTHENIUM				C	5.5800	1.8820	1.6810	6.1810	1.7780
52690	12.3392	94	373.0	S	0.2116	-0.0588	-0.0415	0.1844	0.5624
RUTHENIUM				C	5.5480	1.8820	1.6810	6.1380	1.7570
52691	12.3273	94	423.0	S	0.2132	-0.0596	-0.0421	0.1860	0.5692
RUTHENIUM				C	5.5160	1.8830	1.6830	6.0960	1.7350
52692	12.3176	94	473.0	S	0.2149	-0.0603	-0.0427	0.1876	0.5764
RUTHENIUM				C	5.4810	1.8830	1.6830	6.0500	1.7120
52693	12.3065	94	523.0	S	0.2168	-0.0612	-0.0433	0.1894	0.5841
RUTHENIUM				C	5.4480	1.8820	1.6850	6.0080	1.6910
52694	12.2954	94	573.0	S	0.2185	-0.0619	-0.0439	0.1911	0.5914
RUTHENIUM				C	5.4110	1.8830	1.6890	5.9630	1.6690
52695	12.2843	94	623.0	S	0.2206	-0.0628	-0.0447	0.1930	0.5992
RUTHENIUM				C	5.3780	1.8850	1.6890	5.9180	1.6470
52696	12.2725	94	673.0	S	0.2225	-0.0638	-0.0453	0.1948	0.6072
RUTHENIUM				C	5.3410	1.8810	1.6890	5.8720	1.6250
52697	12.2579	94	723.0	S	0.2245	-0.0645	-0.0460	0.1968	0.6154
RUTHENIUM				C	5.3060	1.8800	1.6890	5.8270	1.6030
52698	12.2434	94	773.0	S	0.2265	-0.0654	-0.0467	0.1987	0.6238
RUTHENIUM				C	5.2720	1.8800	1.6890	5.7810	1.5810
52699	12.2289	94	823.0	S	0.2285	-0.0663	-0.0474	0.2007	0.6325
RUTHENIUM				C	5.2390	1.8800	1.6910	5.7370	1.5570
52700	12.2145	94	873.0	S	0.2305	-0.0672	-0.0481	0.2027	0.6423
RUTHENIUM				C	5.2040	1.8800	1.6910	5.6910	1.5340
52701	12.1985	94	923.0	S	0.2327	-0.0682	-0.0489	0.2048	0.6519
SILVER(66.67)-ALUMINUM(33.33)				C	1.4850	0.8650	0.8040	1.7850	0.3715
52702	8.2596	53	77.0	S	1.1140	-0.4989	-0.2771	0.8098	2.6918
SILVER(66.67)-ALUMINUM(33.33)				C	1.4750	0.8600	0.8020	1.7750	0.3681
52703	8.2533	53	100.0	S	1.1235	-0.5025	-0.2806	0.8169	2.7167
SILVER(66.67)-ALUMINUM(33.33)				C	1.4610	0.8570	0.7930	1.7530	0.3612
52704	8.2362	53	150.0	S	1.1402	-0.5154	-0.2826	0.8262	2.7685
SILVER(66.67)-ALUMINUM(33.33)				C	1.4450	0.8530	0.7880	1.7300	0.3542
52705	8.2165	53	200.0	S	1.1610	-0.5282	-0.2883	0.8406	2.8233
SILVER(66.67)-ALUMINUM(33.33)				C	1.4300	0.8510	0.7680	1.7070	0.3472
52706	8.2165	53	250.0	S	1.1780	-0.5491	-0.2830	0.8405	2.8802
SILVER(66.67)-ALUMINUM(33.33)				C	1.4150	0.8470	0.7460	1.6850	0.3408
52707	8.1700	53	298.0	S	1.1925	-0.5681	-0.2765	0.8383	2.9343
SILVER(66.67)-ALUMINUM(33.33)				C	1.3960	0.8420	0.7300	1.6590	0.3333
52708	8.1391	53	350.0	S	1.2159	-0.5892	-0.2758	0.8455	3.0003

CODE	NAME DENSITY	REF	TEMP(K)		11	12	13	33	55
SILVER(66.67)-ALUMINUM(33.33)				C	1.3750	0.8370	0.7300	1.6340	0.3260
52709	8.1137	53	400.0	S	1.2499	-0.6088	-0.2864	0.8679	3.0675
SILVER(66.67)-ALUMINUM(33.33)				C	1.3460	0.8260	0.7400	1.6070	0.3184
52710	8.0911	53	450.0	S	1.2970	-0.6261	-0.3090	0.9068	3.1407
SILVER(66.67)-ALUMINUM(33.33)				C	1.3120	0.8100	0.7570	1.5800	0.3105
52711	8.0711	53	500.0	S	1.3540	-0.6380	-0.3431	0.9616	3.2206
SILVER(66.67)-ALUMINUM(33.33)				C	1.2750	0.7910	0.7760	1.5550	0.3022
52712	8.0526	53	550.0	S	1.4202	-0.6459	-0.3864	1.0287	3.3091
THALLIUM				C	0.4440	0.3760	0.3000	0.6020	0.0880
52713	11.8170	90	4.2	S	8.3126	-6.3932	-0.9565	2.6145	11.3636
THALLIUM				C	0.4395	0.3730	0.3000	0.5940	0.0859
52714	11.7940	90	50.0	S	8.5002	-6.5374	-0.9913	2.6848	11.6414
THALLIUM				C	0.4350	0.3700	0.3000	0.5885	0.0837
52715	11.7740	90	75.0	S	8.6940	-6.6906	-1.0213	2.7405	11.9474
THALLIUM				C	0.4300	0.3660	0.2900	0.5820	0.0810
52716	11.7510	90	100.0	S	8.7987	-6.8263	-0.9828	2.6976	12.3457
THALLIUM				C	0.4215	0.3600	0.2800	0.5700	0.0762
52717	11.7050	90	150.0	S	9.1174	-7.1427	-0.9700	2.7074	13.1234
THALLIUM				C	0.4160	0.3550	0.2800	0.5565	0.0743
52718	11.6560	90	200.0	S	9.2187	-7.1747	-1.0284	2.8318	13.4590
THALLIUM				C	0.4080	0.3540	0.2900	0.5280	0.0726
52719	11.5600	90	300.0	S	10.3868	-8.1317	-1.2386	3.2545	13.7741
TITANIUM				C	1.7610	0.8690	0.6830	1.9050	0.5080
52720	4.5225	95	4.0	S	0.7942	-0.3269	-0.1675	0.6451	1.9685
TITANIUM				C	1.7590	0.8670	0.6820	1.9050	0.5080
52721	4.5225	95	23.0	S	0.7944	-0.3266	-0.1675	0.6448	1.9685
TITANIUM				C	1.7490	0.8710	0.6800	1.8940	0.5050
52722	4.5198	95	73.0	S	0.8040	-0.3349	-0.1684	0.6489	1.9802
TITANIUM				C	1.7260	0.8770	0.6810	1.8760	0.4990
52723	4.5171	95	123.0	S	0.8261	-0.3518	-0.1722	0.6580	2.0040
TITANIUM				C	1.6990	0.8890	0.6840	1.8570	0.4900
52724	4.5144	95	173.0	S	0.8572	-0.3774	-0.1767	0.6687	2.0408
TITANIUM				C	1.6680	0.9010	0.6870	1.8370	0.4810
52725	4.5117	95	223.0	S	0.8952	-0.4086	-0.1820	0.6805	2.0790
TITANIUM				C	1.6390	0.9130	0.6890	1.8160	0.4720
52726	4.5090	95	273.0	S	0.9351	-0.4423	-0.1870	0.6925	2.1186
TITANIUM				C	1.6240	0.9200	0.6900	1.8070	0.4670
52727	4.5063	95	298.0	S	0.9581	-0.4623	-0.1893	0.6980	2.1413
TITANIUM				C	1.6090	0.9250	0.6910	1.7950	0.4620
52728	4.5022	95	323.0	S	0.9807	-0.4812	-0.1923	0.7051	2.1645
TITANIUM				C	1.5790	0.9340	0.6940	1.7740	0.4530
52729	4.4982	95	373.0	S	1.0290	-0.5214	-0.1986	0.7191	2.2075
TITANIUM				C	1.5510	0.9430	0.6950	1.7530	0.4440
52730	4.4928	95	423.0	S	1.0797	-0.5650	-0.2041	0.7323	2.2523
TITANIUM				C	1.5220	0.9520	0.6950	1.7340	0.4340
52731	4.4874	95	473.0	S	1.1380	-0.6164	-0.2091	0.7443	2.3041
TITANIUM				C	1.4950	0.9610	0.6920	1.7150	0.4240
52732	4.4834	95	523.0	S	1.1998	-0.6728	-0.2126	0.7547	2.3585
TITANIUM				C	1.4680	0.9670	0.6920	1.6960	0.4140
52733	4.4754	95	573.0	S	1.2653	-0.7307	-0.2182	0.7676	2.4155

HEXAGONAL

CODE / NAME	DENSITY	REF	TEMP(K)		11	12	13	33	55
TITANIUM				C	1.4420	0.9730	0.6910	1.6780	0.4030
52734	4.4700	95	623.0	S	1.3370	-0.7952	-0.2231	0.7797	2.4814
TITANIUM				C	1.4160	0.9780	0.6900	1.6610	0.3920
52735	4.4634	95	673.0	S	1.4162	-0.8669	-0.2282	0.7916	2.5510
TITANIUM				C	1.3920	0.9830	0.6920	1.6440	0.3810
52736	4.4568	95	723.0	S	1.5014	-0.9435	-0.2348	0.8060	2.6247
TITANIUM				C	1.3680	0.9850	0.6880	1.6270	0.3700
52737	4.4502	95	773.0	S	1.5878	-1.0232	-0.2388	0.8165	2.7027
TITANIUM				C	1.3450	0.9880	0.6880	1.6100	0.3590
52738	4.4434	95	823.0	S	1.6871	-1.1140	-0.2449	0.8304	2.7855
TITANIUM				C	1.3220	0.9910	0.6880	1.5930	0.3480
52739	4.4366	95	873.0	S	1.8015	-1.2197	-0.2513	0.8448	2.8736
TITANIUM				C	1.2990	0.9920	0.6880	1.5760	0.3370
52740	4.4300	95	923.0	S	1.9245	-1.3329	-0.2583	0.8600	2.9674
TITANIUM				C	1.2760	0.9930	0.6880	1.5600	0.3260
52741	4.4234	95	973.0	S	2.0676	-1.4660	-0.2653	0.8751	3.0675
TITANIUM				C	1.2530	0.9940	0.6880	1.5450	0.3160
52742	4.4169	95	1023.0	S	2.2364	-1.6245	-0.2725	0.8899	3.1646
TITANIUM				C	1.2310	0.9960	0.6880	1.5290	0.3070
52743	4.4089	95	1073.0	S	2.4386	-1.8167	-0.2799	0.9059	3.2573
TITANIUM DIBORIDE				C	6.9000	4.1000	3.2000	4.4000	2.5000
52744	4.9500	112	ROOM	S	0.2574	-0.0998	-0.1146	0.3940	0.4000
YTTRIUM				C	0.8340	0.2910	0.1900	0.8010	0.2690
52745	4.4990	238	4.2	S	1.4040	-0.4377	-0.2292	1.3572	3.7175
YTTRIUM				C	0.8340	0.2910	0.1900	0.8010	0.2684
52746	4.4990	238	20.0	S	1.4040	-0.4377	-0.2292	1.3572	3.7258
YTTRIUM				C	0.8330	0.2930	0.1900	0.8000	0.2670
52747	4.4980	238	50.0	S	1.4087	-0.4432	-0.2293	1.3589	3.7453
YTTRIUM				C	0.8300	0.2930	0.1800	0.7980	0.2656
52748	4.4970	238	75.0	S	1.4110	-0.4512	-0.2165	1.3508	3.7651
YTTRIUM				C	0.8270	0.2950	0.1800	0.7960	0.2635
52749	4.4950	238	100.0	S	1.4203	-0.4594	-0.2173	1.3546	3.7951
YTTRIUM				C	0.8220	0.2940	0.1700	0.7930	0.2613
52750	4.4930	238	125.0	S	1.4263	-0.4676	-0.2055	1.3491	3.8270
YTTRIUM				C	0.8160	0.2930	0.1700	0.7910	0.2586
52751	4.4900	238	150.0	S	1.4387	-0.4734	-0.2075	1.3534	3.8670
YTTRIUM				C	0.8100	0.2920	0.1800	0.7880	0.2557
52752	4.4870	238	175.0	S	1.4556	-0.4749	-0.2240	1.3714	3.9108
YTTRIUM				C	0.8030	0.2910	0.1800	0.7840	0.2530
52753	4.4840	238	200.0	S	1.4710	-0.4822	-0.2270	1.3798	3.9526
YTTRIUM				C	0.7970	0.2890	0.1900	0.7810	0.2511
52754	4.4810	238	225.0	S	1.4875	-0.4810	-0.2449	1.3995	3.9825
YTTRIUM				C	0.7900	0.2870	0.2000	0.7770	0.2480
52755	4.4780	238	250.0	S	1.5074	-0.4807	-0.2643	1.4230	4.0323
YTTRIUM				C	0.7840	0.2860	0.2000	0.7730	0.2456
52756	4.4750	238	275.0	S	1.5213	-0.4867	-0.2677	1.4322	4.0717
YTTRIUM				C	0.7790	0.2850	0.2100	0.7690	0.2431
52757	4.4720	238	300.0	S	1.5388	-0.4854	-0.2877	1.4575	4.1135
YTTRIUM				C	0.7750	0.2850	0.2100	0.7670	0.2405
52758	4.4690	238	325.0	S	1.5495	-0.4913	-0.2897	1.4624	4.1580

HEXAGONAL

CODE	NAME DENSITY	REF	TEMP(K)		11	12	13	33	55
YTTRIUM				C	0.7710	0.2850	0.2100	0.7660	0.2382
52759	4.4660	238	350.0	S	1.5602	-0.4974	-0.2914	1.4653	4.1982
YTTRIUM				C	0.7670	0.2840	0.2100	0.7650	0.2357
52760	4.4630	238	375.0	S	1.5696	-0.5008	-0.2934	1.4683	4.2427
YTTRIUM				C	0.7630	0.2840	0.2100	0.7640	0.2333
52761	4.4590	238	400.0	S	1.5806	-0.5071	-0.2951	1.4711	4.2863
ZINC				C	1.5825	0.3151	0.4744	0.6160	0.4000
52762	7.1400	128	ROOM	S	0.8230	0.0340	-0.6600	2.6400	2.5000
ZINC				C	1.6168	0.4404	0.4450	0.5471	0.4000
52763	7.1400	128	ROOM	S	0.8000	-0.0500	-0.6100	2.8200	2.5000
ZINC				C	1.6163	0.2465	0.4998	0.6166	0.3788
52764	7.1400	128	ROOM	S	0.8400	0.1100	-0.7700	2.8700	2.6400
ZINC				C	1.7850	0.2418	0.6088	0.7460	0.3984
52765	7.1400	128	ROOM	S	0.8080	0.1600	-0.7900	2.6300	2.5100
ZINC				C	1.6516	0.2723	0.4445	0.5669	0.4098
52766	7.1400	128	ROOM	S	0.7700	0.0450	-0.6390	2.7660	2.4400
ZINC				C	1.6110	0.3420	0.5030	0.6110	0.3830
52767	7.1400	129	ROOM	S	0.8385	0.0505	-0.7319	2.8417	2.6110
ZINC				C	1.5900	0.3230	0.4820	0.6210	0.4000
52768	7.1400	4	ROOM	S	0.8239	0.0346	-0.6664	2.6447	2.5000
ZINC				C	1.7909	0.3750	0.5540	0.6880	0.4595
52769	7.2810	16	4.2	S	0.7457	0.0394	-0.6322	2.4716	2.1763
ZINC				C	1.7677	0.3680	0.5520	0.6766	0.4479
52770	7.2620	16	77.0	S	0.7621	0.0476	-0.6606	2.5559	2.2326
ZINC				C	1.7336	0.3660	0.5480	0.6620	0.4296
52771	7.2220	16	150.0	S	0.7849	0.0537	-0.6943	2.6600	2.3277
ZINC				C	1.7047	0.3670	0.5450	0.6523	0.4158
52772	7.1920	16	200.0	S	0.8044	0.0569	-0.7196	2.7356	2.4050
ZINC				C	1.6730	0.3660	0.5370	0.6428	0.4018
52773	7.1620	16	250.0	S	0.8205	0.0554	-0.7317	2.7782	2.4888
ZINC				C	1.6368	0.3640	0.5300	0.6347	0.3879
52774	7.1340	16	295.0	S	0.8410	0.0553	-0.7485	2.8256	2.5780
ZINC				C	1.6002	0.3610	0.5210	0.6257	0.3724
52775	7.0970	16	350.0	S	0.8607	0.0537	-0.7614	2.8662	2.6853
ZINC				C	1.5590	0.3580	0.5150	0.6168	0.3573
52776	7.0640	16	400.0	S	0.8894	0.0567	-0.7899	2.9404	2.7988
ZINC				C	1.5138	0.3550	0.5110	0.6077	0.3417
52777	7.0320	16	450.0	S	0.9268	0.0639	-0.8330	3.0465	2.9265
ZINC				C	1.4648	0.3510	0.5080	0.5982	0.3261
52778	6.9980	16	500.0	S	0.9736	0.0757	-0.8911	3.1851	3.0665
ZINC				C	1.4219	0.3540	0.5060	0.5887	0.3098
52779	6.9660	16	550.0	S	1.0200	0.0836	-0.9486	3.3294	3.2279
ZINC				C	1.3843	0.3650	0.5040	0.5793	0.2933
52780	6.9310	16	600.0	S	1.0637	0.0826	-0.9974	3.4616	3.4095
ZINC				C	1.3395	0.4080	0.5020	0.5661	0.2667
52781	6.8780	16	670.0	S	1.1201	0.0465	-1.0345	3.6012	3.7495
ZINC				C	1.4300	0.1700	0.3300	0.5000	0.4000
52782	7.1000	189	ROOM	S	0.8262	0.0326	-0.5668	2.7482	2.5000
ZINC OXIDE (CONSTANT E)				C	2.0970	1.2110	1.0510	2.1090	0.4247
52783	5.6760	26	298.0	S	0.7855	-0.3431	-0.2205	0.6939	2.3546

CODE	NAME DENSITY	REF	TEMP(K)		11	12	13	33	55
ZINC SULFIDE (CONSTANT E)				C	1.1389	0.5012	0.2711	1.2703	0.2907
52784	4.0900	143	ROOM	S	1.1120	-0.4560	-0.1400	0.8470	3.4400
ZINC SULFIDE (CONSTANT D)				C	1.2400	0.6000	0.4500	1.4040	0.2850
52785	4.0900	282	ROOM	S	1.1035	-0.4590	-0.2066	0.8447	3.5088
ZINC SULFIDE (HEXAGONAL)				C	1.2420	0.6015	0.4554	1.4000	0.2864
52786	4.0890	60	298.0	S	1.1038	-0.4575	-0.2102	0.8511	3.4916
ZINC SULFIDE (ALPHA)				C	1.3120	0.6630	0.5090	1.4080	0.2860
52787	4.0890	74	ROOM	S	1.0816	-0.4593	-0.2250	0.8729	3.4965
ZIRCONIUM				C	1.5540	0.6720	0.6460	1.7250	0.3630
52788	6.5280	95	4.0	S	0.8539	-0.2799	-0.2150	0.7407	2.7548
ZIRCONIUM				C	1.5530	0.6730	0.6460	1.7250	0.3620
52789	6.5260	95	23.0	S	0.8552	-0.2812	-0.2150	0.7407	2.7624
ZIRCONIUM				C	1.5420	0.6780	0.6480	1.7160	0.3580
52790	6.5230	95	73.0	S	0.8676	-0.2898	-0.2182	0.7475	2.7933
ZIRCONIUM				C	1.5220	0.6850	0.6480	1.7020	0.3510
52791	6.5190	95	123.0	S	0.8892	-0.3056	-0.2222	0.7567	2.8490
ZIRCONIUM				C	1.4950	0.6920	0.6480	1.6870	0.3420
52792	6.5150	95	173.0	S	0.9187	-0.3267	-0.2274	0.7675	2.9240
ZIRCONIUM				C	1.4700	0.7050	0.6490	1.6720	0.3330
52793	6.5110	95	223.0	S	0.9528	-0.3544	-0.2323	0.7784	3.0030
ZIRCONIUM				C	1.4450	0.7190	0.6510	1.6550	0.3240
52794	6.5070	95	273.0	S	0.9914	-0.3860	-0.2381	0.7916	3.0864
ZIRCONIUM				C	1.4340	0.7280	0.6530	1.6480	0.3200
52795	6.5050	95	298.0	S	1.0123	-0.4042	-0.2409	0.7977	3.1250
ZIRCONIUM				C	1.4200	0.7340	0.6530	1.6390	0.3160
52796	6.5030	95	323.0	S	1.0349	-0.4228	-0.2439	0.8045	3.1646
ZIRCONIUM				C	1.3960	0.7460	0.6540	1.6230	0.3080
52797	6.4970	95	373.0	S	1.0788	-0.4596	-0.2495	0.8172	3.2468
ZIRCONIUM				C	1.3700	0.7560	0.6540	1.6070	0.3010
52798	6.4950	95	423.0	S	1.1281	-0.5006	-0.2554	0.8301	3.3223
ZIRCONIUM				C	1.3470	0.7670	0.6540	1.5910	0.2950
52799	6.4890	95	473.0	S	1.1793	-0.5449	-0.2608	0.8429	3.3898
ZIRCONIUM				C	1.3230	0.7780	0.6560	1.5740	0.2880
52800	6.4840	95	523.0	S	1.2391	-0.5957	-0.2682	0.8588	3.4722
ZIRCONIUM				C	1.3010	0.7860	0.6570	1.5590	0.2820
52801	6.4780	95	573.0	S	1.2970	-0.6448	-0.2749	0.8731	3.5461
ZIRCONIUM				C	1.2780	0.7930	0.6580	1.5430	0.2760
52802	6.4740	95	623.0	S	1.3621	-0.6998	-0.2824	0.8890	3.6232
ZIRCONIUM				C	1.2570	0.7990	0.6590	1.5260	0.2700
52803	6.4680	95	673.0	S	1.4280	-0.7554	-0.2905	0.9062	3.7037
ZIRCONIUM				C	1.2350	0.8060	0.6590	1.5100	0.2640
52804	6.4620	95	723.0	S	1.5066	-0.8244	-0.2977	0.9221	3.7879
ZIRCONIUM				C	1.2140	0.8120	0.6590	1.4930	0.2570
52805	6.4560	95	773.0	S	1.5900	-0.8976	-0.3056	0.9396	3.8910
ZIRCONIUM				C	1.1920	0.8160	0.6590	1.4760	0.2510
52806	6.4510	95	823.0	S	1.6820	-0.9776	-0.3145	0.9584	3.9841
ZIRCONIUM				C	1.1680	0.8190	0.6590	1.4600	0.2450
52807	6.4430	95	873.0	S	1.7918	-1.0735	-0.3242	0.9776	4.0816
ZIRCONIUM				C	1.4700	0.8240	0.6580	1.4430	0.2390
52808	6.4370	95	923.0	S	1.0692	-0.4788	-0.2692	0.9385	4.1841

CODE	NAME DENSITY	REF	TEMP(K)		11	12	13	14	33	44
ALUMINUM OXIDE				C	4.6600	1.2700	1.1700	0.9400	5.0600	2.3500
62809	3.9700	129	ROOM	S	0.2823	-0.0968	-0.0429	-0.1516	0.2175	0.5468
ALUMINUM OXIDE				C	4.6500	1.2400	1.1700	1.0100	5.6300	2.3300
62810	3.9700	129	ROOM	S	0.2898	-0.1048	-0.0385	-0.1710	0.1936	0.5775
ALUMINUM OXIDE				C	4.9600	1.3500	1.1700	-0.2300	5.0200	0.4100
62811	3.9800	180	300.0	S	0.2359	-0.0624	-0.0404	0.1674	0.2180	2.6268
ALUMINUM OXIDE				C	4.9020	1.6540	1.1300	-0.2320	4.9020	1.4540
62812	3.9860	33	300.0	S	0.2404	-0.0747	-0.0382	0.0503	0.2216	0.7038
ALUMINUM OXIDE				C	4.9200	1.6840	1.1640	-0.2340	4.9200	1.4680
62813	3.9860	33	300.0	S	0.2408	-0.0756	-0.0391	0.0504	0.2217	0.6973
ALUMINUM OXIDE				C	4.9672	1.6340	1.1073	-0.2349	4.9798	1.4740
62814	3.9860	268	298.0	S	0.2353	-0.0716	-0.0364	0.0489	0.2170	0.6940
ALUMINUM OXIDE				C	5.0007	1.6167	1.1137	-0.2326	5.0239	1.5102
62815	3.9929	252	0.0	S	0.2326	-0.0693	-0.0362	0.0465	0.2151	0.6765
ALUMINUM OXIDE				C	5.0056	1.6234	1.1182	-0.2331	5.0241	1.5079
62816	3.9929	252	100.0	S	0.2326	-0.0695	-0.0363	0.0467	0.2152	0.6776
ALUMINUM OXIDE				C	5.0011	1.6259	1.1199	-0.2337	5.0210	1.5025
62817	3.9929	252	150.0	S	0.2330	-0.0698	-0.0364	0.0471	0.2154	0.6802
ALUMINUM OXIDE				C	4.9960	1.6330	1.1212	0.2345	5.0132	1.4945
62818	3.9905	252	200.0	S	0.2336	-0.0704	-0.0365	-0.0477	C.2158	0.6841
ALUMINUM OXIDE				C	4.9853	1.6366	1.1205	-0.2348	5.0024	1.4845
62819	3.9881	252	250.0	S	0.2344	-0.0710	-0.0366	0.0483	0.2163	0.6889
ALUMINUM OXIDE				C	4.9737	1.6398	1.1226	-0.2359	4.9936	1.4743
62820	3.9860	252	298.0	S	0.2353	-0.0716	-0.0368	0.0491	0.2168	0.6940
ALUMINUM OXIDE				C	4.9735	1.6397	1.1220	-0.2358	4.9911	1.4739
62821	3.9860	252	300.0	S	0.2353	-0.0716	-0.0368	0.0491	0.2169	0.6942
ALUMINUM OXIDE				C	4.9565	1.6404	1.1187	-0.2369	4.9750	1.4624
62822	3.9824	252	350.0	S	0.2364	-0.0723	-0.0369	0.0500	0.2176	0.7000
ALUMINUM OXIDE				C	4.9413	1.6438	1.1161	-0.2377	4.9592	1.4505
62823	3.9788	252	400.0	S	0.2375	-0.0731	-0.0370	0.0509	0.2183	0.7061
ALUMINUM OXIDE				C	4.9264	1.6481	1.1102	-0.2388	4.9391	1.4381
62824	3.9740	252	450.0	S	0.2386	-0.0740	-0.0370	0.0519	0.2191	0.7126
ALUMINUM OXIDE				C	4.9061	1.6476	1.1052	-0.2401	4.9182	1.4256
62825	3.9692	252	500.0	S	0.2399	-0.0748	-0.0371	0.0530	0.2200	0.7193
ALUMINUM OXIDE				C	4.8876	1.6496	1.1009	-0.2412	4.8977	1.4128
62826	3.9656	252	550.0	S	0.2412	-0.0757	-0.0372	0.0541	0.2209	0.7263
ALUMINUM OXIDE				C	4.8678	1.6502	1.0932	-0.2422	4.8753	1.4000
62827	3.9608	252	600.0	S	0.2425	-0.0766	-0.0372	0.0552	0.2218	0.7334
ALUMINUM OXIDE				C	4.8497	1.6518	1.0884	-0.2435	4.8528	1.3872
62828	3.9560	252	650.0	S	0.2438	-0.0775	-0.0373	0.0564	0.2228	0.7407
ALUMINUM OXIDE				C	4.8252	1.6489	1.0795	-0.2445	4.8303	1.3742
62829	3.9502	252	700.0	S	0.2453	-0.0784	-0.0373	0.0576	0.2237	0.7482
ALUMINUM OXIDE				C	4.8095	1.6567	1.0763	-0.2448	4.8087	1.3610
62830	3.9454	252	750.0	S	0.2467	-0.0796	-0.0374	0.0587	0.2247	0.7559
ALUMINUM OXIDE				C	4.7902	1.6610	1.0685	-0.2458	4.7827	1.3479
62831	3.9408	252	800.0	S	0.2482	-0.0808	-0.0374	0.0600	0.2258	0.7638
ALUMINUM OXIDE				C	4.7712	1.6653	1.0614	-0.2467	4.7592	1.3347
62832	3.9360	252	850.0	S	0.2497	-0.0820	-0.0374	0.0613	0.2268	0.7719
ALUMINUM OXIDE				C	4.7537	1.6706	1.0571	-0.2473	4.7358	1.3212
62833	3.9302	252	900.0	S	0.2512	-0.0832	-0.0375	0.0626	0.2279	0.7803
ALUMINUM OXIDE				C	4.9760	1.6260	1.1718	-0.2290	5.0185	1.4724
62834	3.9860	111	298.0	S	0.2351	-0.0699	-0.0386	0.0474	0.2173	0.6939
ALUMINUM PHOSPHATE				C	1.0503	0.2934	0.6927	-0.1271	1.3353	0.2314
62835	2.5660	177	ROOM	S	1.6100	-0.0100	-0.8300	0.8900	1.6100	5.3000

CODE	NAME DENSITY	REF	TEMP(K)		11	12	13	14	33	44
ANTIMONY				C	0.7916	0.2474	0.2613	0.1062	0.4273	0.2853
62836	6.6910	128	ROOM	S	1.7700	-0.3800	-0.8500	-0.8000	3.3800	4.1000
ANTIMONY				C	0.9940	0.3090	0.2640	0.2160	0.4450	0.3950
62837	6.7000	86	ROOM	S	1.6193	-0.6090	-0.5994	-1.2185	2.9584	3.8643
ANTIMONY				C	1.0130	0.3450	0.2920	0.2090	0.4500	0.3930
62838	6.7000	72	295.0	S	1.6325	-0.6111	-0.6628	-1.1932	3.0824	3.8136
BISMUTH				C	0.6280	0.3503	0.2114	-0.0424	0.4397	0.1084
62839	9.7470	4	ROOM	S	2.6900	-1.4000	-0.6200	1.6000	2.8700	10.4800
BISMUTH				C	0.6350	0.2470	0.2450	0.0723	0.3810	0.1130
62840	9.8000	81	300.0	S	2.5741	-0.8102	-1.1343	-2.1654	4.0835	11.6205
BISMUTH				C	0.6370	0.2490	0.2470	0.0717	0.3820	0.1123
62841	9.8000	72	295.0	S	2.5691	-0.8042	-1.1412	-2.1538	4.0936	11.6549
CALCITE				C	1.3704	0.4521	0.4478	-0.2073	0.7915	0.3418
62842	2.7170	128	ROOM	S	1.1300	-0.3700	-0.4300	0.9100	1.7500	4.0300
CALCITE				C	1.3740	0.4400	0.4500	-0.2030	0.8010	0.3420
62843	2.7170	129	ROOM	S	1.1036	-0.3393	-0.4294	0.8565	1.7309	3.9408
CALCITE				C	1.3800	0.4550	0.4490	-0.2070	0.7920	0.3420
62844	2.7170	4	ROOM	S	1.1185	-0.3643	-0.4276	0.8975	1.7474	4.0104
CALCITE				C	1.4450	0.5710	0.5340	-0.2050	0.8310	0.3265
62845	2.7120	200	298.0	S	1.1870	-0.4349	-0.4833	1.0183	1.8245	4.3415
CALCITE (P=0KB)				C	1.4626	0.5970	0.5076	-0.2076	0.8531	0.3405
62846	2.7120	64	298.0	S	1.1599	-0.4729	-0.4088	0.9955	1.6587	4.1507
CALCITE (P=1KB)				C	1.4638	0.5974	0.5122	-0.2076	0.8529	0.3433
62847	2.7158	64	298.0	S	1.1584	-0.4668	-0.4153	0.9828	1.6713	4.1015
CALCITE (P=2KB)				C	1.4650	0.5980	0.5190	-0.2078	0.8526	0.3453
62848	2.7193	64	298.0	S	1.1599	-0.4611	-0.4253	0.9755	1.6907	4.0702
CALCITE (P=3KB)				C	1.4622	0.5944	0.5288	-0.2086	0.8524	0.3465
62849	2.7231	64	298.0	S	1.1679	-0.4538	-0.4430	0.9763	1.7228	4.0615
CALCITE (P=4KB)				C	1.4674	0.5990	0.5419	-0.2101	0.8522	0.3468
62850	2.7267	64	298.0	S	1.1776	-0.4515	-0.4617	0.9870	1.7606	4.0793
CALCITE (P=5KB)				C	1.4685	0.5993	0.5574	-0.2116	0.8520	0.3460
62851	2.7302	64	298.0	S	1.1927	-0.4456	-0.4888	1.0019	1.8133	4.1156
CALCITE (P=6KB)				C	1.4697	0.5999	0.5757	-0.2137	0.8518	0.3449
62852	2.7341	64	298.0	S	1.2136	-0.4393	-0.5234	1.0242	1.8814	4.1685
CALCITE				C	1.4797	0.5541	0.5446	-0.2085	0.8568	0.3274
62853	2.7120	65	298.0	S	1.1303	-0.3848	-0.4738	0.9648	1.7695	4.2833
CALCITE				C	1.4637	0.5447	0.5400	-0.2080	0.8532	0.3241
62854	2.7110	65	323.0	S	1.1442	-0.3894	-0.4777	0.9843	1.7768	4.3489
CALCITE				C	1.4480	0.5356	0.5332	-0.2058	0.8494	0.3210
62855	2.7100	65	348.0	S	1.1515	-0.3905	-0.4777	0.9886	1.7770	4.3829
CALCITE				C	1.4311	0.5259	0.5260	-0.2043	0.8451	0.3175
62856	2.7080	65	373.0	S	1.1624	-0.3945	-0.4780	1.0018	1.7783	4.4389
CALCITE				C	1.4158	0.5172	0.5198	-0.2016	0.8415	0.3138
62857	2.7070	65	398.0	S	1.1691	-0.3944	-0.4785	1.0045	1.7795	4.4774
CALCITE				C	1.4019	0.5095	0.5131	-0.1991	0.8374	0.3103
62858	2.7060	65	423.0	S	1.1749	-0.3952	-0.4777	1.0074	1.7796	4.5155
CALCITE				C	1.3880	0.5028	0.5055	-0.1973	0.8346	0.3064
62859	2.7040	65	448.0	S	1.1834	-0.4012	-0.4738	1.0203	1.7721	4.5777
CALCITE				C	1.3747	0.4961	0.4989	-0.1944	0.8312	0.3029
62860	2.7030	65	473.0	S	1.1879	-0.4017	-0.4719	1.0202	1.7696	4.6110
CALCITE				C	1.3610	0.4916	0.4944	-0.1931	0.8274	0.2984
62861	2.7010	65	498.0	S	1.2034	-0.4109	-0.4736	1.0446	1.7746	4.7032
CALCITE				C	1.3475	0.4857	0.4386	-0.1908	0.8237	0.2957
62862	2.7000	65	523.0	S	1.1783	-0.4462	-0.3898	1.0482	1.6291	4.7345

CODE / NAME	DENSITY	REF	TEMP(K)		11	12	13	14	33	44
CALCITE				C	1.5593	0.6013	0.5791	-0.2157	0.8722	0.3457
62863	2.7170	66	160.0	S	1.0852	-0.3666	-0.4771	0.9058	1.7801	4.0231
CALCITE				C	1.5507	0.5967	0.5748	-0.2145	0.8702	0.3427
62864	2.7170	66	180.0	S	1.0896	-0.3692	-0.4759	0.9131	1.7778	4.0610
CALCITE				C	1.5378	0.5886	0.5701	-0.2132	0.8683	0.3404
62865	2.7160	66	200.0	S	1.0959	-0.3701	-0.4765	0.9182	1.7774	4.0879
CALCITE				C	1.5275	0.5845	0.5657	-0.2114	0.8662	0.3381
62866	2.7150	66	220.0	S	1.1009	-0.3726	-0.4756	0.9213	1.7757	4.1099
CALCITE				C	1.5160	0.5780	0.5601	-0.2102	0.8636	0.3356
62867	2.7150	66	240.0	S	1.1067	-0.3754	-0.4743	0.9283	1.7731	4.1427
CALCITE				C	1.5041	0.5697	0.5566	-0.2085	0.8608	0.3315
62868	2.7140	66	260.0	S	1.1132	-0.3746	-0.4776	0.9358	1.7793	4.1937
CALCITE				C	1.4923	0.5659	0.5496	-0.2070	0.8578	0.3300
62869	2.7130	66	280.0	S	1.1192	-0.3807	-0.4732	0.9408	1.7722	4.2106
CALCITE				C	1.4806	0.5578	0.5464	-0.2058	0.8557	0.3269
62870	2.7120	66	300.0	S	1.1263	-0.3804	-0.4763	0.9486	1.7769	4.2534
DEXTROSE SODIUM BROMIDE				C	0.2063	0.0533	0.0794	0.0033	0.2398	0.0634
62871	1.6900	129	ROOM	S	5.6900	-0.8600	-1.6000	-0.3400	5.2300	15.8000
DEXTROSE SODIUM CHLORIDE				C	0.2202	0.1087	0.0750	-0.0031	0.1766	0.0771
62872	1.5640	129	ROOM	S	6.3800	-2.6100	-1.6000	0.3600	7.0200	13.0000
DEXTROSE SODIUM IODIDE				C	0.2579	0.1518	0.0492	-0.0031	0.2056	0.0771
62873	1.8640	129	ROOM	S	6.0200	-3.4300	-0.6200	0.3800	5.1600	13.0000
GUANIDINE AL SELENATE HEXAHYDRATE				C	0.4293	0.2688	0.1380	0.0316	0.1361	0.0815
62874	2.1239	4	ROOM	S	4.8721	-2.4811	-2.4243	-2.8511	12.2639	14.4808
GUANIDINE AL SULFATE HEXAHYDRATE				C	0.4475	0.2835	0.1345	0.0295	0.1438	0.0920
62875	1.8064	4	ROOM	S	4.4893	-2.4034	-1.9510	-2.2102	10.6038	12.2869
GUANIDINE GA SELENATE HEXAHYDRATE				C	0.4040	0.2540	0.1428	0.0304	0.1353	0.0810
62876	2.2798	4	ROOM	S	5.3337	-2.5291	-2.9600	-2.9510	13.6391	14.5607
GUANIDINE GA SULFATE HEXAHYDRATE				C	0.4378	0.2853	0.1413	0.0305	0.1446	0.0920
62877	1.9801	4	ROOM	S	4.8986	-2.6612	-2.1863	-2.5063	11.1885	12.5313
HEMATITE				C	2.4243	0.5464	0.1542	-0.1247	2.2734	0.8569
62878	5.2400	128	ROOM	S	0.4410	-0.1020	-0.0230	0.0790	0.4430	1.1900
LITHIUM NIOBATE (CONSTANT D)				C	2.1900	0.3700	0.7600	-0.1500	2.5200	0.9500
62879	4.7000	271	ROOM	S	0.5200	-0.0442	-0.1435	0.0891	0.4834	1.0808
LITHIUM NIOBATE (CONSTANT E)				C	2.0300	0.5300	0.7500	0.0900	2.4500	0.6000
62880	4.7000	271	ROOM	S	0.5774	-0.1014	-0.1457	-0.1018	0.4974	1.6972
LITHIUM TANTALATE (CONSTANT E)				C	2.2980	0.4410	0.8110	-0.1040	2.7810	0.9680
62881	7.4540	240	ROOM	S	0.4932	-0.0519	-0.1287	0.0586	0.4346	1.0456
LITHIUM TANTALATE (CONSTANT E)				C	2.3300	0.4700	0.8000	-0.1100	2.7500	0.9400
62882	7.4500	271	ROOM	S	0.4868	-0.0584	-0.1246	0.0638	0.4361	1.0788
LITHIUM TANTALATE (CONSTANT D)				C	2.3900	0.4100	0.8000	-0.2200	2.8400	1.1300
62883	7.4500	271	ROOM	S	0.4768	-0.0511	-0.1199	0.1028	0.4197	0.9250
MERCURY				C	0.3600	0.2890	0.3030	0.0470	0.5050	0.1290
62884	14.4690	129	83.0	S	15.3566	-11.8528	-2.1023	-9.9135	4.5030	14.9757
QUARTZ				C	0.8501	0.0696	0.1412	0.1678	1.0535	0.5722
62885	2.6500	128	ROOM	S	1.3000	-0.1660	-0.1520	-0.4300	0.9900	2.0000
QUARTZ				C	0.8605	0.0480	0.1045	0.1832	1.0701	0.5882
62886	2.6500	128	ROOM	S	1.2790	-0.1530	-0.1100	-0.4460	0.9560	1.9780
QUARTZ				C	0.8683	0.0709	0.1193	-0.1806	1.0594	0.5826
62887	2.6540	4	ROOM	S	1.2775	-0.1814	-0.1234	0.4523	0.9717	1.9968
QUARTZ				C	0.8750	0.0762	0.1510	0.1720	1.0770	0.5730
62888	2.6600	138	ROOM	S	1.2690	-0.1687	-0.1543	-0.4316	0.9718	2.0043
QUARTZ				C	0.8755	0.0607	0.1330	0.1725	1.0680	0.5719
62889	2.6600	138	ROOM	S	1.2572	-0.1498	-0.1379	-0.4244	0.9707	2.0046

CODE	NAME DENSITY	REF	TEMP(K)		11	12	13	14	33	44
QUARTZ				C	0.8683	0.0709	0.1193	-0.1806	1.0594	0.5826
62890	2.6487	144	293.0	S	1.2776	-0.1815	-0.1235	0.4524	0.9717	1.9971
QUARTZ (CONSTANT E)				C	0.8734	0.0610	0.1268	0.1809	1.0621	0.5789
62891	2.6540	179	293.0	S	1.2680	-0.1620	-0.1320	-0.4470	0.9730	2.0070
QUARTZ (CONSTANT E)				C	0.8674	0.0699	0.1191	-0.1791	1.0720	0.5794
62892	2.6540	29	293.0	S	1.2770	-0.1791	-0.1220	0.4501	0.9599	2.0042
QUARTZ				C	0.8726	0.0990	0.1302	-0.1773	1.0919	0.5972
62893	2.6638	168	77.2	S	1.2797	-0.2165	-0.1268	0.4442	0.9461	1.9382
QUARTZ				C	0.8680	0.0704	0.1191	-0.1804	1.0575	0.5820
62894	2.6485	168	298.0	S	1.2776	-0.1806	-0.1235	0.4520	0.9735	1.9984
SELENIUM				C	0.1870	0.0710	0.2620	0.0620	0.7410	0.1490
62895	4.8380	188	300.0	S	14.6388	-0.8885	-4.8618	-6.4610	4.7875	12.0884
SODIUM NITRATE				C	0.8670	0.1630	0.1600	0.0820	0.3740	0.2130
62896	2.2700	129	ROOM	S	1.3400	-0.2204	-0.4790	-0.6007	3.0836	5.1574
TOURMALINE				C	2.7066	0.6927	0.0872	-0.0774	1.6070	0.6682
62897	3.0500	128	ROOM	S	0.3980	-0.1030	-0.0160	0.0580	0.6240	1.5100
TOURMALINE				C	2.7200	0.4000	0.3500	-0.0680	1.6500	0.6500
62898	3.1000	129	ROOM	S	0.3851	-0.0486	-0.0714	0.0454	0.6363	1.5480
TOURMALINE				C	2.6300	0.6100	0.4900	-0.0900	1.5100	0.5950
62899	3.0500	129	ROOM	S	0.4220	-0.0798	-0.1111	0.0759	0.7343	1.7036
TOURMALINE				C	3.0400	0.8800	0.3500	-0.0400	1.7600	0.6500
62900	3.0000	129	ROOM	S	0.3643	-0.0998	-0.0526	0.0286	0.5891	1.5420

TRIGONAL 1

CODE	NAME DENSITY	REF	TEMP(K)		11	12	13	14	33	44
					PRESSURE DERIVATIVES					
ALUMINUM OXIDE				C'(P)	6.1740	3.2820	3.6530	0.1300	4.9980	2.2430
62901	3.9860	111	298.0	S'(P)	-0.2332	0.0534	-0.0398	-0.1438	-0.1416	-1.1102
QUARTZ				C'(P)	3.4000	7.4000	5.4100	2.0000	10.2600	1.7700
62902	2.6638	168	77.2	S'(P)	-3.3410	-6.2804	-2.9293	-5.4549	-6.9348	-11.9590
QUARTZ				C'(P)	3.2800	8.6600	5.9700	1.9300	10.8400	2.6600
62903	2.6485	168	298.0	S'(P)	-3.0992	-8.3640	-3.6355	-5.2696	-7.7646	-15.3983

Calculated Aggregate Constants

CUBIC

CODE	NAME / DENSITY	REF	TEMP(K)		E	G	SIGMA	K	BETA	VS	VP
				V	0.198	0.077	0.286	0.154	6.494	2.164	3.954
	ALUM (AMMONIUM)			H	0.197	0.077	0.286	0.154	6.494	2.163	3.953
10001	1.6400	129	ROOM	S	0.197	0.077	0.286	0.154	6.494	2.163	3.953
				R	0.197	0.077	0.287	0.154	6.493	2.161	3.952
				V	0.199	0.077	0.287	0.156	6.424	2.170	3.971
	ALUM (AMMONIUM)			H	0.198	0.077	0.288	0.156	6.424	2.168	3.970
10002	1.6400	129	ROOM	S	0.198	0.077	0.288	0.156	6.424	2.168	3.969
				R	0.198	0.077	0.288	0.156	6.424	2.166	3.968
				V	0.205	0.080	0.281	0.156	6.410	2.135	3.872
	ALUM (POTASSIUM)			H	0.204	0.080	0.282	0.156	6.410	2.133	3.871
10003	1.7500	129	ROOM	S	0.204	0.080	0.282	0.156	6.410	2.133	3.871
				R	0.204	0.079	0.282	0.156	6.410	2.131	3.869
				V	0.207	0.081	0.280	0.157	6.383	2.143	3.876
	ALUM (POTASSIUM)			H	0.206	0.081	0.280	0.157	6.383	2.141	3.874
10004	1.7600	129	ROOM	S	0.206	0.081	0.280	0.157	6.383	2.140	3.874
				R	0.206	0.080	0.281	0.157	6.383	2.138	3.873
				V	0.208	0.081	0.279	0.157	6.383	2.151	3.882
	ALUM (100.0K- 0.0CR)			H	0.208	0.081	0.279	0.157	6.383	2.148	3.880
10005	1.7600	129	ROOM	S	0.208	0.081	0.279	0.157	6.383	2.148	3.880
				R	0.207	0.081	0.279	0.157	6.383	2.145	3.878
				V	0.200	0.078	0.283	0.154	6.494	2.098	3.816
	ALUM (86.5K- 13.5CR)			H	0.200	0.078	0.284	0.154	6.494	2.097	3.815
10006	1.7720	129	ROOM	S	0.200	0.078	0.284	0.154	6.494	2.097	3.815
				R	0.200	0.078	0.284	0.154	6.494	2.096	3.814
				V	0.195	0.076	0.283	0.150	6.682	2.057	3.738
	ALUM (60.0K- 40.0CR)			H	0.195	0.076	0.283	0.150	6.682	2.057	3.738
10007	1.7960	129	ROOM	S	0.195	0.076	0.283	0.150	6.682	2.057	3.738
				R	0.195	0.076	0.283	0.150	6.682	2.056	3.738
				V	0.194	0.076	0.282	0.148	6.757	2.048	3.716
	ALUM (54.5K- 45.5CR)			H	0.194	0.076	0.282	0.148	6.757	2.048	3.715
10008	1.8020	129	ROOM	S	0.194	0.076	0.282	0.148	6.757	2.047	3.715
				R	0.194	0.075	0.282	0.148	6.757	2.047	3.715
				V	0.191	0.075	0.274	0.141	7.092	2.016	3.614
	ALUM (0.0K-100.0CR)			H	0.191	0.075	0.274	0.141	7.092	2.016	3.614
10009	1.8450	129	ROOM	S	0.191	0.075	0.274	0.141	7.092	2.016	3.614
				R	0.191	0.075	0.274	0.141	7.092	2.015	3.613
				V	0.219	0.084	0.301	0.183	5.454	1.905	3.570
	ALUM (THALLIUM)			H	0.219	0.084	0.301	0.183	5.455	1.904	3.569
10010	2.3200	129	ROOM	S	0.219	0.084	0.301	0.183	5.455	1.904	3.569
				R	0.219	0.084	0.301	0.183	5.455	1.903	3.568
				V	0.217	0.083	0.305	0.185	5.400	1.894	3.572
	ALUM (THALLIUM)			H	0.217	0.083	0.305	0.185	5.400	1.892	3.572
10011	2.3200	245	ROOM	S	0.217	0.083	0.305	0.185	5.400	1.892	3.571
				R	0.216	0.083	0.305	0.185	5.400	1.891	3.570
				V	0.164	0.060	0.372	0.215	4.662	1.940	4.303
	ALUM (CH3NH3-AL-S)			H	0.164	0.060	0.372	0.215	4.662	1.940	4.303
10012	1.5890	4	ROOM	S	0.164	0.060	0.372	0.215	4.662	1.940	4.303
				R	0.164	0.060	0.373	0.215	4.662	1.939	4.303
				V	0.217	0.082	0.325	0.206	4.844	2.024	3.973
	ALUM (CS-AL-S)			H	0.217	0.082	0.325	0.206	4.844	2.023	3.973
10013	1.9990	4	ROOM	S	0.217	0.082	0.325	0.206	4.844	2.023	3.973
				R	0.217	0.082	0.325	0.206	4.844	2.023	3.973
				V	0.203	0.080	0.276	0.151	6.608	2.134	3.837
	ALUM (K-AL-S)			H	0.203	0.079	0.277	0.151	6.608	2.130	3.834
10014	1.7500	4	ROOM	S	0.203	0.079	0.277	0.151	6.608	2.130	3.833
				R	0.202	0.079	0.278	0.151	6.608	2.125	3.830
				V	0.205	0.081	0.273	0.150	6.645	2.146	3.837
	ALUM (K-AL-S)			H	0.205	0.080	0.273	0.150	6.645	2.142	3.834
10015	1.7530	4	ROOM	S	0.205	0.080	0.273	0.150	6.645	2.141	3.834
				R	0.204	0.080	0.274	0.151	6.644	2.137	3.831

CUBIC

CODE	NAME DENSITY	REF	TEMP(K)		E	G	SIGMA	K	BETA	VS	VP
				V	0.199	0.072	0.374	0.264	3.789	2.023	4.513
ALUM (NA-AL-S)				H	0.199	0.072	0.375	0.264	3.789	2.020	4.512
10016	1.7700	4	ROOM	S	0.199	0.072	0.375	0.264	3.789	2.020	4.511
				R	0.198	0.072	0.375	0.264	3.789	2.017	4.509
				V	0.199	0.077	0.288	0.157	6.383	2.169	3.977
ALUM (NH4-AL-S)				H	0.199	0.077	0.289	0.157	6.383	2.167	3.975
10017	1.6420	4	ROOM	S	0.199	0.077	0.289	0.157	6.383	2.167	3.975
				R	0.198	0.077	0.289	0.157	6.383	2.165	3.974
				V	0.206	0.081	0.276	0.153	6.520	2.069	3.722
ALUM (RB-AL-S)				H	0.206	0.081	0.277	0.153	6.520	2.068	3.720
10018	1.8840	4	ROOM	S	0.206	0.081	0.277	0.153	6.520	2.068	3.720
				R	0.205	0.080	0.277	0.153	6.520	2.066	3.719
				V	0.199	0.077	0.293	0.160	6.250	1.821	3.364
ALUM (TL-AL-S)				H	0.199	0.077	0.293	0.160	6.250	1.819	3.362
10019	2.3220	4	ROOM	S	0.199	0.077	0.293	0.160	6.250	1.819	3.362
				R	0.198	0.077	0.293	0.160	6.250	1.817	3.360
				V	0.176	0.067	0.314	0.158	6.313	1.917	3.684
ALUM (CH3NH3-AL-SE)				H	0.173	0.066	0.318	0.158	6.313	1.896	3.669
10020	1.8270	4	ROOM	S	0.172	0.065	0.319	0.158	6.313	1.891	3.666
				R	0.169	0.064	0.323	0.158	6.313	1.868	3.650
				V	0.191	0.073	0.307	0.165	6.044	1.813	3.439
ALUM (CS-AL-SE)				H	0.191	0.073	0.307	0.165	6.044	1.813	3.438
10021	2.2240	4	ROOM	S	0.191	0.073	0.307	0.165	6.044	1.813	3.438
				R	0.191	0.073	0.307	0.165	6.043	1.813	3.438
				V	0.189	0.074	0.279	0.142	7.026	1.926	3.481
ALUM (K-AL-SE)				H	0.188	0.074	0.280	0.142	7.026	1.925	3.479
10022	1.9860	4	ROOM	S	0.188	0.074	0.280	0.142	7.026	1.924	3.479
				R	0.188	0.073	0.280	0.142	7.026	1.922	3.478
				V	0.186	0.072	0.292	0.149	6.720	1.953	3.601
ALUM (NH4-AL-SE)				H	0.186	0.072	0.292	0.149	6.720	1.951	3.600
10023	1.8880	4	ROOM	S	0.186	0.072	0.292	0.149	6.720	1.951	3.600
				R	0.186	0.072	0.292	0.149	6.720	1.950	3.599
				V	0.193	0.075	0.282	0.147	6.790	1.888	3.423
ALUM (RB-AL-SE)				H	0.193	0.075	0.282	0.147	6.790	1.887	3.423
10024	2.1130	4	ROOM	S	0.193	0.075	0.282	0.147	6.790	1.887	3.423
				R	0.193	0.075	0.282	0.147	6.790	1.886	3.422
				V	0.215	0.082	0.321	0.200	4.995	1.987	3.868
ALUM (CS-FE-S)				H	0.215	0.081	0.321	0.200	4.995	1.986	3.867
10025	2.0650	4	ROOM	S	0.215	0.081	0.321	0.200	4.995	1.986	3.867
				R	0.215	0.081	0.321	0.200	4.995	1.986	3.867
				V	0.159	0.058	0.373	0.209	4.785	1.837	4.083
ALUM (CH3NH3-GA-S)				H	0.159	0.058	0.373	0.209	4.785	1.837	4.083
10026	1.7170	4	ROOM	S	0.159	0.058	0.373	0.209	4.785	1.837	4.083
				R	0.159	0.058	0.373	0.209	4.785	1.836	4.083
				V	0.212	0.080	0.328	0.205	4.890	1.935	3.822
ALUM (CS-GA-S)				H	0.211	0.080	0.328	0.204	4.890	1.935	3.822
10027	2.1270	4	ROOM	S	0.211	0.080	0.328	0.204	4.890	1.935	3.822
				R	0.211	0.080	0.328	0.205	4.890	1.935	3.822
				V	0.199	0.078	0.271	0.145	6.906	2.030	3.622
ALUM (K-GA-S)				H	0.198	0.078	0.272	0.145	6.906	2.024	3.618
10028	1.8980	4	ROOM	S	0.198	0.078	0.272	0.145	6.906	2.024	3.618
				R	0.197	0.077	0.273	0.145	6.906	2.018	3.614
				V	0.194	0.076	0.282	0.148	6.737	2.059	3.738
ALUM (NH4-GA-S)				H	0.193	0.075	0.283	0.148	6.737	2.056	3.736
10029	1.7840	4	ROOM	S	0.193	0.075	0.283	0.148	6.737	2.056	3.736
				R	0.193	0.075	0.283	0.148	6.737	2.052	3.733
				V	0.204	0.080	0.270	0.148	6.754	1.991	3.549
ALUM (RB-GA-S)				H	0.203	0.080	0.271	0.148	6.754	1.988	3.547
10030	2.0250	4	ROOM	S	0.203	0.080	0.271	0.148	6.754	1.988	3.547
				R	0.203	0.080	0.272	0.148	6.754	1.985	3.545

CUBIC

CODE	NAME / DENSITY	REF	TEMP(K)		E	G	SIGMA	K	BETA	VS	VP
	ALUM (CS-GA-SE)			V	0.191	0.073	0.301	0.160	6.263	1.770	3.316
10031	2.3420	4	ROOM	H	0.191	0.073	0.301	0.160	6.263	1.769	3.315
				S	0.191	0.073	0.301	0.160	6.263	1.769	3.315
				R	0.191	0.073	0.301	0.160	6.263	1.769	3.315
	ALUM (CS-IN-S)			V	0.211	0.080	0.317	0.192	5.198	1.901	3.676
10032	2.2120	4	ROOM	H	0.211	0.080	0.318	0.192	5.198	1.901	3.676
				S	0.211	0.080	0.318	0.192	5.198	1.901	3.676
				R	0.211	0.080	0.318	0.192	5.198	1.901	3.676
	ALUM (RB-IN-S)			V	0.197	0.078	0.269	0.142	7.019	1.922	3.418
10033	2.1070	4	ROOM	H	0.197	0.078	0.270	0.142	7.019	1.919	3.416
				S	0.197	0.078	0.270	0.142	7.019	1.919	3.416
				R	0.196	0.077	0.270	0.142	7.019	1.916	3.414
	ALUMINUM			V	0.708	0.263	0.348	0.775	1.290	3.118	6.454
10034	2.7020	128	ROOM	H	0.705	0.261	0.348	0.775	1.290	3.110	6.449
				S	0.705	0.261	0.348	0.775	1.290	3.110	6.449
				R	0.701	0.260	0.349	0.775	1.290	3.101	6.443
	ALUMINUM			V	0.703	0.259	0.356	0.813	1.230	3.098	6.550
10035	2.7020	129	293.0	H	0.701	0.258	0.356	0.813	1.230	3.092	6.546
				S	0.701	0.258	0.356	0.813	1.230	3.092	6.546
				R	0.698	0.257	0.357	0.813	1.230	3.085	6.541
	ALUMINUM			V	0.707	0.263	0.346	0.764	1.309	3.119	6.424
10036	2.6990	4	ROOM	H	0.704	0.262	0.346	0.764	1.309	3.113	6.420
				S	0.704	0.261	0.346	0.764	1.309	3.112	6.419
				R	0.701	0.260	0.347	0.764	1.309	3.104	6.414
	ALUMINUM (P= 0KB)			V	0.710	0.265	0.341	0.745	1.342	3.131	6.379
10037	2.6990	149	298.0	H	0.707	0.264	0.342	0.745	1.342	3.125	6.375
				S	0.707	0.263	0.342	0.745	1.342	3.124	6.374
				R	0.704	0.262	0.343	0.745	1.342	3.116	6.369
	ALUMINUM (P= 2KB)			V	0.711	0.264	0.345	0.767	1.303	3.125	6.432
10038	2.7063	149	298.0	H	0.707	0.263	0.346	0.767	1.303	3.116	6.426
				S	0.707	0.263	0.346	0.767	1.303	3.115	6.425
				R	0.702	0.261	0.347	0.767	1.303	3.103	6.418
	ALUMINUM (P= 4KB)			V	0.729	0.272	0.341	0.766	1.305	3.165	6.449
10039	2.7134	149	298.0	H	0.726	0.271	0.342	0.766	1.305	3.158	6.444
				S	0.726	0.270	0.342	0.766	1.305	3.157	6.444
				R	0.723	0.269	0.343	0.766	1.305	3.149	6.439
	ALUMINUM (P= 6KB)			V	0.753	0.282	0.335	0.761	1.313	3.218	6.465
10040	2.7206	149	298.0	H	0.751	0.281	0.336	0.761	1.313	3.215	6.463
				S	0.751	0.281	0.336	0.761	1.313	3.215	6.462
				R	0.749	0.280	0.336	0.761	1.313	3.211	6.460
	ALUMINUM (P= 8KB)			V	0.768	0.288	0.333	0.768	1.302	3.248	6.498
10041	2.7276	149	298.0	H	0.766	0.287	0.334	0.768	1.302	3.245	6.496
				S	0.766	0.287	0.334	0.768	1.302	3.245	6.496
				R	0.765	0.287	0.334	0.768	1.302	3.241	6.494
	ALUMINUM (P=10KB)			V	0.794	0.299	0.329	0.775	1.291	3.305	6.549
10042	2.7346	149	298.0	H	0.793	0.298	0.329	0.775	1.291	3.303	6.548
				S	0.793	0.298	0.329	0.775	1.291	3.303	6.548
				R	0.793	0.298	0.329	0.775	1.291	3.302	6.547
	ALUMINUM			V	0.783	0.290	0.352	0.882	1.134	3.256	6.812
10043	2.7333	249	0.0	H	0.781	0.289	0.352	0.882	1.134	3.251	6.808
				S	0.781	0.289	0.352	0.882	1.134	3.251	6.808
				R	0.779	0.288	0.353	0.882	1.134	3.245	6.804
	ALUMINUM			V	0.780	0.288	0.352	0.880	1.137	3.249	6.803
10044	2.7311	249	63.2	H	0.778	0.287	0.353	0.880	1.137	3.244	6.800
				S	0.778	0.287	0.353	0.880	1.137	3.244	6.800
				R	0.775	0.286	0.353	0.880	1.137	3.238	6.796
	ALUMINUM			V	0.779	0.288	0.352	0.880	1.137	3.247	6.802
10045	2.7302	249	70.2	H	0.777	0.287	0.353	0.880	1.137	3.242	6.799
				S	0.776	0.287	0.353	0.880	1.137	3.242	6.799
				R	0.774	0.286	0.353	0.880	1.137	3.236	6.795

CUBIC

CODE	NAME DENSITY	REF	TEMP(K)		E	G	SIGMA	K	BETA	VS	VP
				V	0.777	0.287	0.352	0.877	1.140	3.244	6.795
ALUMINUM				H	0.775	0.286	0.353	0.877	1.140	3.239	6.791
10046	2.7300	249	77.2	S	0.775	0.286	0.353	0.877	1.140	3.239	6.791
				R	0.772	0.285	0.353	0.877	1.140	3.233	6.788
				V	0.761	0.281	0.354	0.868	1.152	3.212	6.753
ALUMINUM				H	0.759	0.280	0.354	0.868	1.152	3.207	6.750
10047	2.7249	249	133.2	S	0.759	0.280	0.354	0.868	1.152	3.206	6.750
				R	0.756	0.279	0.355	0.868	1.152	3.201	6.746
				V	0.758	0.280	0.354	0.868	1.152	3.206	6.751
ALUMINUM				H	0.756	0.279	0.355	0.868	1.152	3.201	6.748
10048	2.7233	249	143.2	S	0.756	0.279	0.355	0.868	1.152	3.200	6.748
				R	0.753	0.278	0.355	0.868	1.152	3.195	6.744
				V	0.755	0.279	0.354	0.861	1.161	3.200	6.731
ALUMINUM				H	0.753	0.278	0.354	0.861	1.161	3.195	6.728
10049	2.7216	249	153.2	S	0.752	0.278	0.354	0.861	1.161	3.195	6.727
				R	0.750	0.277	0.355	0.861	1.161	3.189	6.724
				V	0.748	0.276	0.354	0.857	1.167	3.187	6.713
ALUMINUM				H	0.746	0.275	0.355	0.857	1.167	3.182	6.710
10050	2.7187	249	173.2	S	0.746	0.275	0.355	0.857	1.167	3.182	6.710
				R	0.743	0.274	0.355	0.857	1.167	3.176	6.706
				V	0.741	0.274	0.355	0.850	1.176	3.174	6.689
ALUMINUM				H	0.739	0.273	0.355	0.850	1.176	3.169	6.686
10051	2.7157	249	193.2	S	0.739	0.273	0.355	0.850	1.176	3.168	6.686
				R	0.736	0.272	0.356	0.850	1.176	3.162	6.682
				V	0.734	0.271	0.355	0.844	1.185	3.160	6.665
ALUMINUM				H	0.732	0.270	0.355	0.844	1.185	3.155	6.662
10052	2.7125	249	213.2	S	0.732	0.270	0.355	0.844	1.185	3.155	6.662
				R	0.729	0.269	0.356	0.844	1.185	3.149	6.658
				V	0.727	0.268	0.356	0.840	1.191	3.146	6.648
ALUMINUM				H	0.725	0.267	0.356	0.840	1.191	3.141	6.644
10053	2.7092	249	233.2	S	0.725	0.267	0.356	0.840	1.191	3.141	6.644
				R	0.722	0.266	0.357	0.840	1.191	3.135	6.640
				V	0.720	0.265	0.356	0.833	1.200	3.132	6.625
ALUMINUM				H	0.718	0.265	0.356	0.833	1.200	3.127	6.621
10054	2.7055	249	253.2	S	0.718	0.265	0.356	0.833	1.200	3.127	6.621
				R	0.715	0.263	0.357	0.833	1.200	3.121	6.617
				V	0.712	0.263	0.356	0.825	1.212	3.118	6.595
ALUMINUM				H	0.710	0.262	0.357	0.825	1.212	3.112	6.592
10055	2.7022	249	273.2	S	0.710	0.262	0.357	0.825	1.212	3.112	6.591
				R	0.707	0.261	0.357	0.825	1.212	3.105	6.587
				V	0.705	0.260	0.357	0.821	1.218	3.103	6.578
ALUMINUM				H	0.703	0.259	0.357	0.821	1.218	3.098	6.575
10056	2.6976	249	293.2	S	0.703	0.259	0.357	0.821	1.218	3.098	6.575
				R	0.700	0.258	0.358	0.821	1.218	3.091	6.571
				V	0.698	0.257	0.357	0.815	1.227	3.088	6.555
ALUMINUM				H	0.695	0.256	0.358	0.815	1.227	3.083	6.552
10057	2.6938	249	313.2	S	0.695	0.256	0.358	0.815	1.227	3.083	6.552
				R	0.692	0.255	0.358	0.815	1.227	3.076	6.547
				V	0.690	0.254	0.358	0.807	1.239	3.074	6.528
ALUMINUM				H	0.687	0.253	0.358	0.807	1.239	3.068	6.524
10058	2.6890	249	333.2	S	0.687	0.253	0.358	0.807	1.239	3.068	6.524
				R	0.685	0.252	0.359	0.807	1.239	3.061	6.520
				V	0.682	0.251	0.357	0.797	1.254	3.058	6.493
ALUMINUM				H	0.680	0.250	0.358	0.797	1.254	3.053	6.490
10059	2.6857	249	353.2	S	0.680	0.250	0.358	0.797	1.254	3.052	6.490
				R	0.677	0.249	0.359	0.797	1.254	3.045	6.485
				V	0.674	0.248	0.357	0.786	1.272	3.043	6.454
ALUMINUM				H	0.672	0.247	0.358	0.786	1.272	3.037	6.450
10060	2.6823	249	373.2	S	0.672	0.247	0.358	0.786	1.272	3.037	6.450
				R	0.669	0.246	0.358	0.786	1.272	3.029	6.446

CUBIC

CODE	NAME DENSITY	REF	TEMP(K)		E	G	SIGMA	K	BETA	VS	VP
				V	0.658	0.243	0.357	0.768	1.302	3.012	6.388
ALUMINUM				H	0.656	0.242	0.358	0.768	1.302	3.005	6.384
10061	2.6744	249	413.2	S	0.656	0.241	0.358	0.768	1.302	3.005	6.384
				R	0.653	0.240	0.358	0.768	1.302	2.997	6.379
				V	0.642	0.237	0.357	0.751	1.332	2.979	6.323
ALUMINUM				H	0.640	0.236	0.358	0.751	1.332	2.972	6.319
10062	2.6666	249	453.2	S	0.640	0.236	0.358	0.751	1.332	2.972	6.319
				R	0.637	0.234	0.359	0.751	1.332	2.964	6.314
				V	0.626	0.231	0.357	0.728	1.374	2.947	6.242
ALUMINUM				H	0.623	0.230	0.357	0.728	1.374	2.940	6.238
10063	2.6577	249	493.2	S	0.623	0.230	0.357	0.728	1.374	2.939	6.237
				R	0.620	0.228	0.358	0.728	1.374	2.931	6.232
				V	0.610	0.225	0.357	0.711	1.407	2.912	6.175
ALUMINUM				H	0.607	0.224	0.358	0.711	1.407	2.905	6.170
10064	2.6504	249	533.2	S	0.607	0.224	0.358	0.711	1.407	2.904	6.169
				R	0.604	0.222	0.358	0.711	1.407	2.895	6.164
				V	0.593	0.219	0.356	0.689	1.452	2.879	6.094
ALUMINUM				H	0.591	0.218	0.357	0.689	1.452	2.871	6.089
10065	2.6403	249	573.2	S	0.590	0.217	0.357	0.689	1.452	2.870	6.088
				R	0.587	0.216	0.358	0.689	1.452	2.860	6.082
				V	0.577	0.213	0.356	0.668	1.497	2.843	6.013
ALUMINUM				H	0.574	0.211	0.357	0.668	1.497	2.834	6.008
10066	2.6317	249	613.2	S	0.573	0.211	0.357	0.668	1.497	2.833	6.007
				R	0.569	0.210	0.358	0.668	1.497	2.823	6.001
				V	0.560	0.206	0.357	0.651	1.536	2.805	5.943
ALUMINUM				H	0.557	0.205	0.358	0.651	1.536	2.796	5.937
10067	2.6223	249	653.2	S	0.556	0.205	0.358	0.651	1.536	2.795	5.936
				R	0.552	0.203	0.359	0.651	1.536	2.784	5.930
				V	0.542	0.200	0.357	0.631	1.584	2.766	5.862
ALUMINUM				H	0.539	0.198	0.358	0.631	1.584	2.756	5.856
10068	2.6131	249	693.2	S	0.539	0.198	0.358	0.631	1.584	2.755	5.855
				R	0.534	0.197	0.359	0.631	1.584	2.743	5.847
				V	0.524	0.193	0.357	0.613	1.632	2.724	5.782
ALUMINUM				H	0.521	0.192	0.358	0.613	1.632	2.714	5.776
10069	2.6029	249	733.2	S	0.520	0.192	0.358	0.613	1.632	2.713	5.775
				R	0.516	0.190	0.360	0.613	1.632	2.700	5.768
				V	0.506	0.186	0.358	0.595	1.680	2.680	5.703
ALUMINUM				H	0.503	0.185	0.359	0.595	1.680	2.670	5.697
10070	2.5937	249	773.2	S	0.502	0.185	0.359	0.595	1.680	2.668	5.696
				R	0.498	0.183	0.361	0.595	1.680	2.655	5.688
				V	0.786	0.294	0.335	0.794	1.260	3.282	6.588
ALUMINUM				H	0.784	0.293	0.335	0.794	1.260	3.276	6.584
10071	2.7334	140	0.0	S	0.783	0.293	0.336	0.794	1.260	3.276	6.584
				R	0.780	0.292	0.336	0.794	1.260	3.268	6.579
				V	0.786	0.294	0.335	0.794	1.260	3.282	6.587
ALUMINUM				H	0.783	0.293	0.335	0.794	1.260	3.275	6.583
10072	2.7334	140	20.0	S	0.783	0.293	0.336	0.794	1.260	3.275	6.583
				R	0.780	0.292	0.336	0.794	1.260	3.268	6.578
				V	0.785	0.294	0.335	0.794	1.260	3.278	6.586
ALUMINUM				H	0.782	0.293	0.336	0.794	1.260	3.272	6.582
10073	2.7331	140	40.0	S	0.782	0.293	0.336	0.794	1.260	3.272	6.581
				R	0.779	0.291	0.337	0.794	1.260	3.264	6.577
				V	0.782	0.293	0.336	0.793	1.261	3.273	6.581
ALUMINUM				H	0.779	0.292	0.336	0.793	1.261	3.267	6.577
10074	2.7325	140	60.0	S	0.779	0.291	0.336	0.793	1.261	3.266	6.576
				R	0.776	0.290	0.337	0.793	1.261	3.259	6.571
				V	0.778	0.291	0.336	0.792	1.263	3.266	6.573
ALUMINUM				H	0.776	0.290	0.337	0.792	1.263	3.260	6.569
10075	2.7314	140	80.0	S	0.776	0.290	0.337	0.792	1.263	3.259	6.569
				R	0.773	0.289	0.337	0.792	1.263	3.252	6.564

CUBIC

CODE	NAME DENSITY	REF	TEMP(K)		E	G	SIGMA	K	BETA	VS	VP
				V	0.773	0.289	0.337	0.790	1.266	3.255	6.563
ALUMINUM				H	0.771	0.288	0.337	0.790	1.266	3.249	6.559
10076	2.7296	140	100.0	S	0.770	0.288	0.337	0.790	1.266	3.248	6.559
				R	0.767	0.287	0.338	0.790	1.266	3.241	6.554
				V	0.767	0.287	0.338	0.788	1.269	3.243	6.551
ALUMINUM				H	0.765	0.286	0.338	0.788	1.269	3.237	6.547
10077	2.7274	140	120.0	S	0.765	0.286	0.338	0.788	1.269	3.236	6.547
				R	0.762	0.284	0.339	0.788	1.269	3.229	6.542
				V	0.761	0.284	0.339	0.786	1.273	3.230	6.538
ALUMINUM				H	0.758	0.283	0.339	0.786	1.273	3.224	6.534
10078	2.7249	140	140.0	S	0.758	0.283	0.339	0.786	1.273	3.223	6.533
				R	0.755	0.282	0.340	0.786	1.273	3.216	6.529
				V	0.754	0.282	0.339	0.783	1.277	3.216	6.524
ALUMINUM				H	0.752	0.281	0.340	0.783	1.277	3.210	6.520
10079	2.7220	140	160.0	S	0.752	0.280	0.340	0.783	1.277	3.210	6.519
				R	0.749	0.279	0.341	0.783	1.277	3.203	6.515
				V	0.748	0.279	0.340	0.780	1.282	3.203	6.508
ALUMINUM				H	0.745	0.278	0.341	0.780	1.282	3.197	6.504
10080	2.7190	140	180.0	S	0.745	0.278	0.341	0.780	1.282	3.196	6.504
				R	0.742	0.277	0.341	0.780	1.282	3.189	6.499
				V	0.740	0.276	0.341	0.777	1.288	3.188	6.492
ALUMINUM				H	0.738	0.275	0.342	0.777	1.288	3.182	6.488
10081	2.7158	140	200.0	S	0.738	0.275	0.342	0.777	1.288	3.182	6.488
				R	0.735	0.274	0.342	0.777	1.288	3.175	6.483
				V	0.733	0.273	0.342	0.774	1.292	3.174	6.477
ALUMINUM				H	0.731	0.272	0.343	0.774	1.292	3.167	6.473
10082	2.7125	140	220.0	S	0.731	0.272	0.343	0.774	1.292	3.167	6.473
				R	0.728	0.271	0.343	0.774	1.292	3.160	6.468
				V	0.726	0.270	0.343	0.771	1.298	3.159	6.462
ALUMINUM				H	0.723	0.269	0.344	0.771	1.298	3.153	6.458
10083	2.7090	140	240.0	S	0.723	0.269	0.344	0.771	1.298	3.152	6.457
				R	0.720	0.268	0.344	0.771	1.298	3.145	6.453
				V	0.719	0.267	0.344	0.767	1.303	3.143	6.445
ALUMINUM				H	0.716	0.266	0.344	0.767	1.303	3.137	6.441
10084	2.7055	140	260.0	S	0.716	0.266	0.345	0.767	1.303	3.136	6.440
				R	0.713	0.265	0.345	0.767	1.303	3.129	6.435
				V	0.711	0.264	0.345	0.764	1.309	3.128	6.428
ALUMINUM				H	0.708	0.263	0.346	0.764	1.309	3.121	6.424
10085	2.7019	140	280.0	S	0.708	0.263	0.346	0.764	1.309	3.121	6.424
				R	0.705	0.262	0.346	0.764	1.309	3.113	6.419
				V	0.703	0.261	0.346	0.761	1.314	3.112	6.412
ALUMINUM				H	0.701	0.260	0.347	0.761	1.314	3.105	6.408
10086	2.6982	140	300.0	S	0.700	0.260	0.347	0.761	1.314	3.105	6.407
				R	0.697	0.259	0.347	0.761	1.314	3.097	6.402
				V	0.774	0.288	0.343	0.820	1.220	3.245	6.631
ALUMINUM				H	0.772	0.287	0.343	0.820	1.220	3.239	6.627
10087	2.7390	261	4.0	S	0.772	0.287	0.343	0.820	1.220	3.239	6.627
				R	0.769	0.286	0.344	0.820	1.220	3.232	6.622
				V	0.766	0.286	0.341	0.801	1.248	3.233	6.575
ALUMINUM				H	0.764	0.285	0.341	0.801	1.248	3.227	6.571
10088	2.7350	261	77.0	S	0.764	0.285	0.341	0.801	1.248	3.227	6.571
				R	0.761	0.284	0.342	0.801	1.248	3.220	6.567
				V	0.707	0.263	0.346	0.763	1.311	3.119	6.419
ALUMINUM				H	0.705	0.262	0.346	0.763	1.311	3.112	6.415
10089	2.7020	261	300.0	S	0.704	0.262	0.346	0.763	1.311	3.112	6.415
				R	0.701	0.260	0.347	0.763	1.311	3.104	6.410
				V	0.704	0.262	0.345	0.758	1.320	3.116	6.406
ALUMINUM				H	0.702	0.261	0.346	0.758	1.320	3.109	6.402
10090	2.6970	109	293.0	S	0.701	0.261	0.346	0.758	1.320	3.109	6.401
				R	0.698	0.259	0.346	0.758	1.320	3.101	6.396

CUBIC

CODE	NAME DENSITY	REF	TEMP(K)		E	G	SIGMA	K	BETA	VS	VP
				V	0.703	0.261	0.345	0.757	1.321	3.115	6.405
ALUMINUM				H	0.701	0.260	0.346	0.757	1.321	3.108	6.401
10091	2.6950	109	300.0	S	0.700	0.260	0.346	0.757	1.321	3.107	6.400
				R	0.697	0.259	0.347	0.757	1.321	3.099	6.395
				V	0.697	0.259	0.346	0.753	1.328	3.102	6.389
ALUMINUM				H	0.694	0.258	0.346	0.753	1.328	3.095	6.385
10092	2.6900	109	325.0	S	0.694	0.258	0.346	0.753	1.328	3.094	6.384
				R	0.690	0.256	0.347	0.753	1.328	3.086	6.379
				V	0.689	0.256	0.347	0.749	1.335	3.085	6.370
ALUMINUM				H	0.686	0.255	0.347	0.749	1.335	3.078	6.366
10093	2.6860	109	350.0	S	0.686	0.254	0.347	0.749	1.335	3.078	6.365
				R	0.682	0.253	0.348	0.749	1.335	3.070	6.360
				V	0.679	0.252	0.348	0.745	1.343	3.065	6.347
ALUMINUM				H	0.677	0.251	0.349	0.745	1.343	3.058	6.343
10094	2.6820	109	375.0	S	0.676	0.251	0.349	0.745	1.343	3.058	6.342
				R	0.673	0.249	0.349	0.745	1.343	3.049	6.337
				V	0.670	0.248	0.349	0.740	1.352	3.045	6.324
ALUMINUM				H	0.667	0.247	0.350	0.740	1.352	3.038	6.320
10095	2.6770	109	400.0	S	0.667	0.247	0.350	0.740	1.352	3.037	6.319
				R	0.663	0.245	0.351	0.740	1.352	3.028	6.314
				V	0.660	0.244	0.350	0.735	1.360	3.024	6.302
ALUMINUM				H	0.657	0.243	0.351	0.735	1.360	3.017	6.297
10096	2.6710	109	425.0	S	0.656	0.243	0.351	0.735	1.360	3.016	6.297
				R	0.653	0.242	0.352	0.735	1.360	3.007	6.291
				V	0.650	0.240	0.352	0.730	1.370	3.002	6.276
ALUMINUM				H	0.647	0.239	0.352	0.730	1.370	2.994	6.271
10097	2.6670	109	450.0	S	0.646	0.239	0.352	0.730	1.370	2.993	6.270
				R	0.643	0.238	0.353	0.730	1.370	2.984	6.264
				V	0.640	0.236	0.353	0.725	1.379	2.980	6.251
ALUMINUM				H	0.637	0.235	0.354	0.725	1.379	2.972	6.246
10098	2.6620	109	475.0	S	0.636	0.235	0.354	0.725	1.379	2.971	6.246
				R	0.633	0.234	0.355	0.725	1.379	2.962	6.239
				V	0.629	0.232	0.354	0.720	1.389	2.957	6.227
ALUMINUM				H	0.626	0.231	0.355	0.720	1.389	2.949	6.222
10099	2.6560	109	500.0	S	0.626	0.231	0.355	0.720	1.389	2.948	6.221
				R	0.622	0.229	0.356	0.720	1.389	2.938	6.215
				V	0.619	0.228	0.356	0.715	1.399	2.934	6.201
ALUMINUM				H	0.616	0.227	0.356	0.715	1.399	2.926	6.196
10100	2.6510	109	525.0	S	0.615	0.227	0.357	0.715	1.399	2.925	6.195
				R	0.611	0.225	0.357	0.715	1.399	2.914	6.188
				V	0.609	0.224	0.357	0.709	1.410	2.911	6.173
ALUMINUM				H	0.605	0.223	0.358	0.709	1.410	2.903	6.168
10101	2.6460	109	550.0	S	0.605	0.223	0.358	0.709	1.410	2.902	6.167
				R	0.601	0.221	0.359	0.709	1.410	2.891	6.161
				V	0.599	0.220	0.358	0.704	1.420	2.889	6.148
ALUMINUM				H	0.595	0.219	0.359	0.704	1.420	2.880	6.143
10102	2.6400	109	575.0	S	0.595	0.219	0.359	0.704	1.420	2.879	6.142
				R	0.591	0.217	0.360	0.704	1.420	2.868	6.135
				V	0.588	0.216	0.360	0.699	1.430	2.865	6.122
ALUMINUM				H	0.585	0.215	0.361	0.699	1.430	2.856	6.116
10103	2.6350	109	600.0	S	0.584	0.215	0.361	0.699	1.430	2.855	6.115
				R	0.580	0.213	0.362	0.699	1.430	2.844	6.109
				V	0.578	0.212	0.361	0.693	1.443	2.842	6.094
ALUMINUM				H	0.575	0.211	0.362	0.693	1.443	2.833	6.089
10104	2.6290	109	625.0	S	0.574	0.211	0.362	0.693	1.443	2.832	6.088
				R	0.570	0.209	0.363	0.693	1.443	2.821	6.081
				V	0.567	0.208	0.363	0.688	1.454	2.818	6.069
ALUMINUM				H	0.564	0.207	0.363	0.688	1.454	2.809	6.063
10105	2.6220	109	650.0	S	0.564	0.207	0.363	0.688	1.454	2.807	6.062
				R	0.559	0.205	0.365	0.688	1.454	2.796	6.055

NAME / CODE	DENSITY	REF	TEMP(K)		E	G	SIGMA	K	BETA	VS	VP
				V	0.557	0.204	0.364	0.683	1.464	2.792	6.041
ALUMINUM				H	0.553	0.203	0.365	0.683	1.464	2.783	6.035
10106	2.6170	109	675.0	S	0.553	0.202	0.365	0.683	1.464	2.781	6.034
				R	0.548	0.201	0.366	0.683	1.464	2.769	6.027
				V	0.546	0.200	0.366	0.677	1.477	2.767	6.010
ALUMINUM				H	0.542	0.199	0.366	0.677	1.477	2.757	6.004
10107	2.6120	109	700.0	S	0.542	0.198	0.367	0.677	1.477	2.755	6.003
				R	0.538	0.197	0.368	0.677	1.477	2.743	5.996
				V	0.535	0.196	0.367	0.672	1.489	2.741	5.983
ALUMINUM				H	0.532	0.194	0.368	0.672	1.489	2.731	5.977
10108	2.6050	109	725.0	S	0.531	0.194	0.368	0.672	1.489	2.730	5.976
				R	0.527	0.192	0.369	0.672	1.489	2.717	5.969
				V	0.524	0.192	0.369	0.666	1.502	2.715	5.954
ALUMINUM				H	0.521	0.190	0.370	0.666	1.502	2.705	5.948
10109	2.5990	109	750.0	S	0.520	0.190	0.370	0.666	1.502	2.703	5.947
				R	0.516	0.188	0.371	0.666	1.502	2.691	5.939
				V	0.513	0.187	0.370	0.660	1.514	2.687	5.923
ALUMINUM				H	0.510	0.186	0.371	0.660	1.514	2.676	5.917
10110	2.5940	109	775.0	S	0.509	0.186	0.372	0.660	1.514	2.675	5.916
				R	0.504	0.184	0.373	0.660	1.514	2.661	5.908
				V	0.502	0.183	0.372	0.654	1.528	2.660	5.893
ALUMINUM				H	0.498	0.181	0.373	0.654	1.528	2.649	5.887
10111	2.5870	109	800.0	S	0.498	0.181	0.373	0.654	1.528	2.647	5.886
				R	0.493	0.179	0.374	0.654	1.528	2.634	5.878
				V	0.491	0.179	0.374	0.648	1.543	2.630	5.859
ALUMINUM				H	0.487	0.177	0.375	0.648	1.543	2.619	5.852
10112	2.5820	109	825.0	S	0.486	0.177	0.375	0.648	1.543	2.617	5.851
				R	0.482	0.175	0.376	0.648	1.543	2.603	5.843
				V	0.479	0.174	0.376	0.642	1.558	2.601	5.826
ALUMINUM				H	0.475	0.173	0.377	0.642	1.558	2.589	5.819
10113	2.5750	109	850.0	S	0.475	0.172	0.377	0.642	1.558	2.587	5.818
				R	0.470	0.170	0.378	0.642	1.558	2.572	5.809
				V	0.467	0.170	0.377	0.635	1.574	2.571	5.793
ALUMINUM				H	0.463	0.168	0.378	0.635	1.574	2.558	5.786
10114	2.5680	109	875.0	S	0.463	0.168	0.379	0.635	1.574	2.556	5.784
				R	0.458	0.166	0.380	0.636	1.574	2.541	5.775
				V	0.455	0.165	0.379	0.629	1.590	2.539	5.758
ALUMINUM				H	0.451	0.163	0.380	0.629	1.590	2.526	5.750
10115	2.5610	109	900.0	S	0.450	0.163	0.381	0.629	1.590	2.524	5.749
				R	0.445	0.161	0.382	0.629	1.590	2.508	5.739
				V	0.443	0.160	0.381	0.622	1.607	2.506	5.722
ALUMINUM				H	0.439	0.159	0.382	0.622	1.607	2.493	5.714
10116	2.5540	109	925.0	S	0.438	0.158	0.383	0.622	1.607	2.490	5.713
				R	0.432	0.156	0.384	0.622	1.607	2.473	5.702
				V	0.707	0.263	0.345	0.759	1.318	3.121	6.412
ALUMINUM				H	0.704	0.262	0.345	0.759	1.318	3.114	6.408
10117	2.6970	254	298.0	S	0.704	0.262	0.345	0.759	1.318	3.114	6.407
				R	0.700	0.260	0.346	0.759	1.318	3.106	6.402
				V	0.855	0.340	0.260	0.593	1.686	2.791	4.898
ALUMINUM ANTIMONIDE				H	0.830	0.328	0.267	0.593	1.686	2.741	4.860
10118	4.3600	42	300.0	S	0.822	0.324	0.269	0.593	1.686	2.726	4.849
				R	0.794	0.311	0.277	0.593	1.686	2.670	4.807
				V	0.199	0.079	0.259	0.138	7.246	1.806	3.167
AMMONIUM BROMIDE				H	0.189	0.074	0.272	0.138	7.246	1.748	3.124
10119	2.4290	129	ROOM	S	0.185	0.072	0.277	0.138	7.246	1.727	3.108
				R	0.175	0.068	0.288	0.138	7.246	1.674	3.069
				V	0.231	0.090	0.278	0.173	5.769	1.927	3.474
AMMONIUM BROMIDE				H	0.224	0.087	0.284	0.173	5.769	1.893	3.449
10120	2.4360	4	ROOM	S	0.222	0.086	0.286	0.173	5.769	1.884	3.442
				R	0.216	0.083	0.293	0.173	5.769	1.850	3.417

CUBIC

CODE	NAME DENSITY	REF	TEMP(K)		E	G	SIGMA	K	BETA	VS	VP
				V	0.247	0.101	0.226	0.150	6.652	2.027	3.408
AMMONIUM BROMIDE				H	0.241	0.098	0.232	0.150	6.652	1.999	3.386
10121	2.4497	102	237.0	S	0.240	0.097	0.234	0.150	6.652	1.993	3.382
				R	0.235	0.095	0.240	0.150	6.652	1.967	3.361
				V	0.248	0.101	0.229	0.152	6.570	2.028	3.421
AMMONIUM BROMIDE				H	0.242	0.098	0.235	0.152	6.570	2.000	3.398
10122	2.4489	102	238.0	S	0.241	0.097	0.236	0.152	6.570	1.994	3.394
				R	0.235	0.095	0.242	0.152	6.570	1.967	3.373
				V	0.248	0.101	0.233	0.155	6.447	2.028	3.438
AMMONIUM BROMIDE				H	0.243	0.098	0.239	0.155	6.447	2.000	3.416
10123	2.4482	102	240.0	S	0.241	0.097	0.241	0.155	6.447	1.993	3.411
				R	0.236	0.095	0.246	0.155	6.447	1.966	3.390
				V	0.249	0.101	0.240	0.160	6.261	2.027	3.465
AMMONIUM BROMIDE				H	0.243	0.098	0.246	0.160	6.261	1.998	3.442
10124	2.4475	102	245.0	S	0.242	0.097	0.247	0.160	6.261	1.991	3.437
				R	0.236	0.094	0.253	0.160	6.261	1.963	3.415
				V	0.249	0.100	0.244	0.162	6.170	2.025	3.477
AMMONIUM BROMIDE				H	0.243	0.097	0.250	0.162	6.170	1.994	3.454
10125	2.4460	102	250.0	S	0.242	0.097	0.251	0.162	6.170	1.987	3.449
				R	0.236	0.094	0.257	0.162	6.170	1.958	3.426
				V	0.249	0.100	0.249	0.165	6.066	2.018	3.490
AMMONIUM BROMIDE				H	0.242	0.096	0.255	0.165	6.066	1.987	3.466
10126	2.4431	102	260.0	S	0.241	0.096	0.257	0.165	6.066	1.979	3.460
				R	0.234	0.093	0.263	0.165	6.066	1.949	3.437
				V	0.247	0.099	0.252	0.166	6.021	2.011	3.493
AMMONIUM BROMIDE				H	0.240	0.095	0.259	0.166	6.021	1.978	3.468
10127	2.4402	102	270.0	S	0.239	0.095	0.260	0.166	6.021	1.970	3.461
				R	0.232	0.092	0.267	0.166	6.021	1.939	3.437
				V	0.245	0.098	0.254	0.167	6.006	2.003	3.490
AMMONIUM BROMIDE				H	0.238	0.094	0.261	0.167	6.006	1.969	3.464
10128	2.4373	102	280.0	S	0.237	0.094	0.263	0.167	6.006	1.960	3.458
				R	0.230	0.091	0.270	0.167	6.006	1.927	3.433
				V	0.243	0.097	0.256	0.166	6.010	1.995	3.485
AMMONIUM BROMIDE				H	0.236	0.093	0.263	0.166	6.010	1.959	3.457
10129	2.4343	102	290.0	S	0.234	0.093	0.265	0.166	6.010	1.950	3.450
				R	0.227	0.089	0.272	0.166	6.010	1.916	3.425
				V	0.241	0.096	0.258	0.166	6.025	1.987	3.477
AMMONIUM BROMIDE				H	0.234	0.092	0.265	0.166	6.025	1.949	3.449
10130	2.4314	102	300.0	S	0.232	0.091	0.267	0.166	6.025	1.939	3.441
				R	0.225	0.088	0.275	0.166	6.025	1.903	3.414
				V	0.239	0.095	0.259	0.165	6.049	1.978	3.467
AMMONIUM BROMIDE				H	0.231	0.091	0.267	0.165	6.049	1.939	3.438
10131	2.4285	102	310.0	S	0.229	0.090	0.269	0.165	6.049	1.928	3.430
				R	0.222	0.087	0.277	0.165	6.049	1.891	3.402
				V	0.237	0.094	0.260	0.164	6.082	1.969	3.456
AMMONIUM BROMIDE				H	0.229	0.090	0.268	0.164	6.082	1.928	3.425
10132	2.4256	102	320.0	S	0.226	0.089	0.271	0.164	6.082	1.916	3.417
				R	0.219	0.085	0.278	0.164	6.082	1.877	3.388
				V	0.249	0.100	0.247	0.164	6.106	2.021	3.485
AMMONIUM BROMIDE (P= 0KB)				H	0.243	0.097	0.253	0.164	6.106	1.991	3.462
10133	2.4445	102	255.0	S	0.241	0.096	0.254	0.164	6.106	1.983	3.456
				R	0.235	0.093	0.261	0.164	6.106	1.954	3.433
				V	0.266	0.106	0.259	0.184	5.425	2.067	3.627
AMMONIUM BROMIDE (P= 2KB)				H	0.261	0.103	0.264	0.184	5.425	2.044	3.609
10134	2.4725	102	255.0	S	0.260	0.103	0.265	0.184	5.425	2.039	3.605
				R	0.255	0.100	0.269	0.184	5.425	2.016	3.588
				V	0.281	0.111	0.266	0.200	4.989	2.108	3.735
AMMONIUM BROMIDE (P= 4KB)				H	0.277	0.109	0.270	0.200	4.989	2.090	3.722
10135	2.4982	102	255.0	S	0.277	0.109	0.270	0.200	4.989	2.087	3.719
				R	0.273	0.107	0.273	0.200	4.989	2.070	3.706

CUBIC

CODE	NAME / DENSITY	REF	TEMP(K)		E	G	SIGMA	K	BETA	VS	VP
	AMMONIUM BROMIDE (P= 6KB)			V	0.295	0.116	0.271	0.215	4.657	2.145	3.827
10136	2.5222	102	255.0	H	0.292	0.115	0.273	0.215	4.657	2.132	3.818
				S	0.292	0.114	0.274	0.215	4.657	2.130	3.816
				R	0.289	0.113	0.276	0.215	4.657	2.117	3.807
	AMMONIUM BROMIDE (P= 8KB)			V	0.308	0.121	0.274	0.228	4.391	2.180	3.909
10137	2.5450	102	255.0	H	0.306	0.120	0.276	0.228	4.391	2.171	3.903
				S	0.306	0.120	0.276	0.228	4.391	2.169	3.902
				R	0.303	0.119	0.278	0.228	4.391	2.160	3.895
	AMMONIUM BROMIDE (P=10KB)			V	0.321	0.126	0.277	0.240	4.171	2.212	3.983
10138	2.5667	102	255.0	H	0.319	0.125	0.278	0.240	4.171	2.205	3.978
				S	0.319	0.125	0.278	0.240	4.171	2.205	3.978
				R	0.317	0.124	0.279	0.240	4.171	2.198	3.973
	AMMONIUM BROMIDE (P= 0KB)			V	0.250	0.100	0.249	0.166	6.018	2.030	3.511
10139	2.4328	102	295.0	H	0.245	0.097	0.255	0.166	6.018	2.002	3.489
				S	0.243	0.097	0.256	0.166	6.018	1.996	3.484
				R	0.238	0.094	0.262	0.166	6.018	1.968	3.463
	AMMONIUM BROMIDE (P= 2KB)			V	0.258	0.102	0.264	0.182	5.481	2.036	3.597
10140	2.4607	102	295.0	H	0.252	0.099	0.270	0.182	5.481	2.008	3.576
				S	0.251	0.099	0.271	0.182	5.481	2.002	3.571
				R	0.245	0.096	0.276	0.182	5.481	1.974	3.551
	AMMONIUM BROMIDE (P= 4KB)			V	0.272	0.107	0.270	0.197	5.072	2.077	3.699
10141	2.4866	102	295.0	H	0.268	0.105	0.274	0.197	5.072	2.055	3.683
				S	0.267	0.105	0.275	0.197	5.072	2.051	3.680
				R	0.262	0.102	0.279	0.197	5.072	2.030	3.664
	AMMONIUM BROMIDE (P= 6KB)			V	0.286	0.112	0.274	0.211	4.746	2.115	3.789
10142	2.5109	102	295.0	H	0.282	0.111	0.277	0.211	4.746	2.099	3.777
				S	0.282	0.110	0.277	0.211	4.746	2.096	3.775
				R	0.278	0.109	0.280	0.211	4.746	2.080	3.763
	AMMONIUM BROMIDE (P= 8KB)			V	0.299	0.117	0.277	0.223	4.478	2.150	3.870
10143	2.5340	102	295.0	H	0.296	0.116	0.279	0.223	4.478	2.138	3.861
				S	0.296	0.116	0.279	0.223	4.478	2.137	3.860
				R	0.293	0.114	0.281	0.223	4.478	2.124	3.851
	AMMONIUM BROMIDE (P=10KB)			V	0.312	0.122	0.279	0.235	4.255	2.185	3.945
10144	2.5561	102	295.0	H	0.310	0.121	0.280	0.235	4.255	2.177	3.939
				S	0.310	0.121	0.280	0.235	4.255	2.176	3.938
				R	0.308	0.120	0.282	0.235	4.255	2.167	3.932
	AMMONIUM CHLORIDE			V	0.262	0.104	0.255	0.178	5.618	2.615	4.558
10145	1.5270	129	ROOM	H	0.247	0.097	0.269	0.178	5.618	2.523	4.488
				S	0.241	0.094	0.275	0.178	5.618	2.487	4.462
				R	0.227	0.088	0.287	0.178	5.618	2.403	4.400
	AMMONIUM CHLORIDE			V	0.269	0.106	0.265	0.191	5.236	2.638	4.669
10146	1.5260	4	ROOM	H	0.262	0.103	0.271	0.191	5.236	2.599	4.639
				S	0.261	0.102	0.273	0.191	5.236	2.590	4.632
				R	0.254	0.099	0.278	0.191	5.236	2.552	4.604
	AMMONIUM CHLORIDE			V	0.195	0.071	0.372	0.254	3.936	2.223	4.929
10147	1.4350	124	293.0	H	0.195	0.071	0.372	0.254	3.936	2.223	4.929
				S	0.195	0.071	0.372	0.254	3.936	2.223	4.929
				R	0.195	0.071	0.372	0.254	3.937	2.223	4.929
	AMMONIUM CHLORIDE			V	0.270	0.108	0.252	0.182	5.494	2.656	4.614
10148	1.5310	98	300.0	H	0.265	0.105	0.258	0.182	5.495	2.621	4.588
				S	0.263	0.105	0.259	0.182	5.495	2.614	4.582
				R	0.258	0.102	0.264	0.182	5.494	2.580	4.557
	AMMONIUM CHLORIDE			V	0.347	0.136	0.276	0.257	3.885	2.938	5.278
10149	1.5743	100	155.0	H	0.345	0.135	0.277	0.257	3.886	2.929	5.272
				S	0.345	0.135	0.277	0.257	3.886	2.929	5.271
				R	0.343	0.134	0.278	0.257	3.886	2.919	5.264
	AMMONIUM CHLORIDE			V	0.343	0.135	0.275	0.255	3.926	2.926	5.257
10150	1.5710	100	170.0	H	0.341	0.134	0.277	0.255	3.926	2.917	5.249
				S	0.341	0.134	0.277	0.255	3.926	2.916	5.249
				R	0.339	0.133	0.278	0.255	3.926	2.906	5.241

CUBIC

CODE	NAME DENSITY	REF	TEMP(K)		E	G	SIGMA	K	BETA	VS	VP
				V	0.338	0.133	0.274	0.249	4.012	2.910	5.215
AMMONIUM CHLORIDE				H	0.336	0.132	0.275	0.249	4.012	2.899	5.207
10151	1.5666	100	190.0	S	0.336	0.132	0.276	0.249	4.012	2.897	5.206
				R	0.333	0.130	0.277	0.249	4.012	2.886	5.197
				V	0.331	0.130	0.271	0.241	4.155	2.889	5.151
AMMONIUM CHLORIDE				H	0.329	0.129	0.272	0.241	4.155	2.876	5.141
10152	1.5621	100	210.0	S	0.328	0.129	0.273	0.241	4.155	2.874	5.140
				R	0.326	0.128	0.274	0.241	4.155	2.860	5.130
				V	0.320	0.127	0.257	0.219	4.569	2.858	4.996
AMMONIUM CHLORIDE				H	0.316	0.126	0.259	0.219	4.569	2.841	4.983
10153	1.5564	100	230.0	S	0.316	0.125	0.259	0.219	4.569	2.839	4.981
				R	0.313	0.124	0.262	0.219	4.569	2.822	4.968
				V	0.312	0.126	0.243	0.203	4.936	2.845	4.882
AMMONIUM CHLORIDE				H	0.309	0.124	0.246	0.203	4.936	2.826	4.868
10154	1.5532	100	236.0	S	0.309	0.124	0.246	0.203	4.936	2.824	4.866
				R	0.305	0.122	0.249	0.203	4.936	2.805	4.851
				V	0.311	0.125	0.239	0.198	5.050	2.842	4.850
AMMONIUM CHLORIDE				H	0.307	0.124	0.241	0.198	5.050	2.823	4.835
10155	1.5524	100	237.0	S	0.307	0.123	0.242	0.198	5.050	2.820	4.833
				R	0.303	0.122	0.245	0.198	5.050	2.802	4.819
				V	0.309	0.125	0.233	0.193	5.185	2.839	4.814
AMMONIUM CHLORIDE				H	0.305	0.123	0.236	0.193	5.185	2.820	4.799
10156	1.5517	100	238.0	S	0.305	0.123	0.237	0.193	5.185	2.817	4.797
				R	0.301	0.121	0.240	0.193	5.185	2.798	4.782
				V	0.306	0.125	0.226	0.186	5.370	2.835	4.767
AMMONIUM CHLORIDE				H	0.302	0.123	0.229	0.186	5.370	2.815	4.751
10157	1.5510	100	239.0	S	0.302	0.123	0.230	0.186	5.370	2.812	4.749
				R	0.298	0.121	0.233	0.186	5.370	2.793	4.734
				V	0.303	0.124	0.218	0.179	5.586	2.831	4.715
AMMONIUM CHLORIDE				H	0.299	0.122	0.222	0.179	5.586	2.811	4.699
10158	1.5501	100	240.0	S	0.299	0.122	0.222	0.179	5.586	2.807	4.697
				R	0.295	0.120	0.225	0.179	5.586	2.788	4.681
				V	0.299	0.124	0.209	0.171	5.835	2.826	4.659
AMMONIUM CHLORIDE				H	0.296	0.122	0.213	0.171	5.835	2.805	4.642
10159	1.5491	100	241.0	S	0.295	0.122	0.213	0.171	5.835	2.801	4.640
				R	0.292	0.120	0.216	0.171	5.835	2.781	4.623
				V	0.282	0.119	0.189	0.152	6.599	2.777	4.487
AMMONIUM CHLORIDE				H	0.278	0.116	0.194	0.152	6.599	2.750	4.464
10160	1.5391	100	242.5	S	0.277	0.116	0.195	0.152	6.599	2.745	4.461
				R	0.273	0.114	0.200	0.152	6.599	2.720	4.440
				V	0.283	0.119	0.195	0.155	6.466	2.777	4.509
AMMONIUM CHLORIDE				H	0.279	0.116	0.199	0.155	6.466	2.749	4.487
10161	1.5389	100	243.0	S	0.278	0.116	0.200	0.155	6.466	2.745	4.483
				R	0.274	0.114	0.205	0.155	6.466	2.719	4.462
				V	0.285	0.118	0.202	0.159	6.284	2.775	4.540
AMMONIUM CHLORIDE				H	0.280	0.116	0.206	0.159	6.284	2.748	4.518
10162	1.5386	100	244.0	S	0.279	0.116	0.207	0.159	6.284	2.743	4.514
				R	0.275	0.114	0.212	0.159	6.284	2.717	4.493
				V	0.285	0.118	0.206	0.162	6.177	2.774	4.559
AMMONIUM CHLORIDE				H	0.281	0.116	0.211	0.162	6.177	2.746	4.536
10163	1.5384	100	245.0	S	0.280	0.116	0.212	0.162	6.177	2.741	4.532
				R	0.276	0.113	0.216	0.162	6.177	2.715	4.511
				V	0.287	0.118	0.219	0.170	5.875	2.766	4.613
AMMONIUM CHLORIDE				H	0.282	0.115	0.224	0.170	5.875	2.738	4.590
10164	1.5371	100	250.0	S	0.281	0.115	0.225	0.170	5.875	2.732	4.586
				R	0.277	0.112	0.229	0.170	5.875	2.705	4.564
				V	0.869	0.619	-0.298	0.181	5.515	6.355	8.104
AMMONIUM CHLORIDE				H	0.774	0.491	-0.211	0.181	5.515	5.659	7.384
10165	1.5326	100	270.0	S	0.667	0.376	-0.113	0.181	5.515	4.952	6.673
				R	0.579	0.299	-0.032	0.181	5.515	4.418	6.153

CUBIC

CODE	NAME / DENSITY	REF	TEMP(K)		E	G	SIGMA	K	BETA	VS	VP
				V	0.282	0.113	0.246	0.185	5.409	2.721	4.688
AMMONIUM CHLORIDE				H	0.276	0.110	0.251	0.185	5.409	2.686	4.660
10166	1.5279	100	290.0	S	0.275	0.110	0.252	0.185	5.409	2.679	4.655
				R	0.269	0.107	0.258	0.185	5.409	2.645	4.629
				V	0.278	0.111	0.251	0.186	5.384	2.700	4.681
AMMONIUM CHLORIDE				H	0.271	0.108	0.257	0.186	5.384	2.661	4.652
10167	1.5228	100	310.0	S	0.270	0.107	0.258	0.186	5.384	2.652	4.645
				R	0.263	0.104	0.264	0.186	5.384	2.615	4.617
				V	0.321	0.128	0.254	0.218	4.591	2.866	4.991
AMMONIUM CHLORIDE (P= 2KB)				H	0.319	0.127	0.256	0.218	4.591	2.851	4.979
10168	1.5606	100	250.7	S	0.318	0.127	0.256	0.218	4.591	2.849	4.978
				R	0.316	0.125	0.259	0.218	4.591	2.834	4.966
				V	0.340	0.134	0.274	0.251	3.991	2.911	5.214
AMMONIUM CHLORIDE (P= 4KB)				H	0.339	0.133	0.275	0.251	3.991	2.902	5.207
10169	1.5769	100	250.7	S	0.338	0.133	0.275	0.251	3.991	2.901	5.207
				R	0.336	0.132	0.276	0.251	3.991	2.891	5.200
				V	0.354	0.138	0.281	0.269	3.717	2.949	5.341
AMMONIUM CHLORIDE (P= 6KB)				H	0.353	0.138	0.281	0.269	3.717	2.943	5.336
10170	1.5892	100	250.7	S	0.353	0.138	0.281	0.269	3.717	2.943	5.336
				R	0.351	0.137	0.282	0.269	3.717	2.937	5.332
				V	0.366	0.143	0.285	0.284	3.524	2.983	5.438
AMMONIUM CHLORIDE (P= 8KB)				H	0.366	0.142	0.285	0.284	3.524	2.979	5.436
10171	1.6019	100	250.7	S	0.365	0.142	0.285	0.284	3.524	2.979	5.436
				R	0.365	0.142	0.286	0.284	3.524	2.976	5.433
				V	0.378	0.147	0.288	0.297	3.367	3.015	5.525
AMMONIUM CHLORIDE (P=10KB)				H	0.378	0.147	0.288	0.297	3.367	3.014	5.524
10172	1.6133	100	250.7	S	0.378	0.147	0.288	0.297	3.367	3.014	5.524
				R	0.377	0.146	0.288	0.297	3.367	3.012	5.523
				V	0.389	0.151	0.291	0.309	3.232	3.045	5.604
AMMONIUM CHLORIDE (P=12KB)				H	0.389	0.151	0.291	0.309	3.232	3.044	5.603
10173	1.6248	100	250.7	S	0.389	0.151	0.291	0.309	3.232	3.044	5.603
				R	0.389	0.151	0.291	0.309	3.232	3.044	5.603
				V	0.305	0.122	0.247	0.201	4.988	2.807	4.842
AMMONIUM CHLORIDE (P= 2KB)				H	0.301	0.120	0.250	0.201	4.988	2.786	4.826
10174	1.5495	100	265.0	S	0.300	0.120	0.251	0.201	4.988	2.782	4.823
				R	0.296	0.118	0.254	0.201	4.988	2.761	4.807
				V	0.329	0.131	0.260	0.229	4.376	2.886	5.066
AMMONIUM CHLORIDE (P= 4KB)				H	0.327	0.130	0.261	0.229	4.376	2.874	5.057
10175	1.5695	100	265.0	S	0.327	0.130	0.262	0.229	4.376	2.873	5.056
				R	0.324	0.128	0.263	0.229	4.376	2.860	5.047
				V	0.347	0.136	0.276	0.258	3.878	2.931	5.266
AMMONIUM CHLORIDE (P= 6KB)				H	0.346	0.135	0.277	0.258	3.878	2.924	5.261
10176	1.5840	100	265.0	S	0.346	0.135	0.277	0.258	3.878	2.923	5.260
				R	0.344	0.135	0.278	0.258	3.878	2.916	5.255
				V	0.361	0.141	0.283	0.276	3.617	2.968	5.391
AMMONIUM CHLORIDE (P= 8KB)				H	0.360	0.140	0.283	0.276	3.617	2.964	5.388
10177	1.5967	100	265.0	S	0.360	0.140	0.283	0.276	3.617	2.963	5.387
				R	0.359	0.140	0.284	0.276	3.617	2.959	5.384
				V	0.373	0.145	0.286	0.291	3.436	3.002	5.488
AMMONIUM CHLORIDE (P=10KB)				H	0.373	0.145	0.287	0.291	3.436	3.000	5.487
10178	1.6081	100	265.0	S	0.372	0.145	0.287	0.291	3.436	3.000	5.486
				R	0.372	0.145	0.287	0.291	3.436	2.998	5.485
				V	0.385	0.149	0.289	0.304	3.287	3.033	5.571
AMMONIUM CHLORIDE (P=12KB)				H	0.384	0.149	0.289	0.304	3.287	3.032	5.570
10179	1.6210	100	265.0	S	0.384	0.149	0.289	0.304	3.287	3.032	5.570
				R	0.384	0.149	0.290	0.304	3.287	3.031	5.570
				V	0.302	0.120	0.255	0.206	4.862	2.788	4.865
AMMONIUM CHLORIDE (P= 2KB)				H	0.297	0.118	0.259	0.206	4.862	2.764	4.846
10180	1.5460	100	280.1	S	0.297	0.118	0.260	0.206	4.862	2.759	4.843
				R	0.292	0.116	0.263	0.206	4.862	2.735	4.825

CUBIC

CODE / NAME	DENSITY	REF	TEMP(K)		E	G	SIGMA	K	BETA	VS	VP
				V	0.317	0.126	0.261	0.221	4.526	2.838	4.990
AMMONIUM CHLORIDE (P= 4KB)				H	0.314	0.124	0.263	0.221	4.526	2.821	4.977
10181	1.5604	100	280.1	S	0.313	0.124	0.264	0.221	4.526	2.819	4.976
				R	0.310	0.122	0.266	0.221	4.526	2.802	4.962
				V	0.336	0.133	0.263	0.236	4.243	2.902	5.116
AMMONIUM CHLORIDE (P= 6KB)				H	0.334	0.132	0.264	0.236	4.243	2.893	5.108
10182	1.5776	100	280.1	S	0.333	0.132	0.264	0.236	4.243	2.892	5.107
				R	0.332	0.131	0.266	0.236	4.243	2.882	5.100
				V	0.353	0.138	0.277	0.264	3.784	2.947	5.309
AMMONIUM CHLORIDE (P= 8KB)				H	0.352	0.138	0.278	0.264	3.784	2.942	5.305
10183	1.5917	100	280.1	S	0.352	0.138	0.278	0.264	3.784	2.941	5.305
				R	0.351	0.137	0.279	0.264	3.784	2.936	5.300
				V	0.367	0.143	0.284	0.283	3.534	2.986	5.435
AMMONIUM CHLORIDE (P=10KB)				H	0.366	0.143	0.284	0.283	3.534	2.983	5.433
10184	1.6026	100	280.1	S	0.366	0.143	0.284	0.283	3.534	2.983	5.433
				R	0.366	0.142	0.285	0.283	3.534	2.980	5.431
				V	0.380	0.147	0.288	0.298	3.357	3.021	5.532
AMMONIUM CHLORIDE (P=12KB)				H	0.379	0.147	0.288	0.298	3.357	3.020	5.532
10185	1.6154	100	280.1	S	0.379	0.147	0.288	0.298	3.357	3.020	5.532
				R	0.379	0.147	0.288	0.298	3.357	3.018	5.531
				V	0.298	0.118	0.258	0.205	4.869	2.770	4.852
AMMONIUM CHLORIDE (P= 2KB)				H	0.293	0.116	0.262	0.205	4.869	2.743	4.832
10186	1.5423	100	295.0	S	0.292	0.116	0.263	0.205	4.869	2.738	4.828
				R	0.287	0.113	0.267	0.205	4.869	2.711	4.808
				V	0.313	0.124	0.266	0.223	4.479	2.818	4.993
AMMONIUM CHLORIDE (P= 4KB)				H	0.310	0.122	0.269	0.223	4.479	2.799	4.979
10187	1.5567	100	295.0	S	0.309	0.122	0.269	0.223	4.479	2.796	4.977
				R	0.305	0.120	0.272	0.223	4.479	2.776	4.962
				V	0.327	0.129	0.270	0.238	4.208	2.865	5.106
AMMONIUM CHLORIDE (P= 6KB)				H	0.325	0.128	0.272	0.238	4.208	2.851	5.096
10188	1.5706	100	295.0	S	0.325	0.128	0.272	0.238	4.208	2.849	5.095
				R	0.322	0.126	0.274	0.238	4.208	2.836	5.085
				V	0.341	0.135	0.264	0.241	4.148	2.916	5.151
AMMONIUM CHLORIDE (P= 8KB)				H	0.339	0.134	0.265	0.241	4.148	2.908	5.145
10189	1.5861	100	295.0	S	0.339	0.134	0.266	0.241	4.148	2.907	5.145
				R	0.338	0.133	0.267	0.241	4.148	2.898	5.138
				V	0.360	0.141	0.279	0.271	3.689	2.966	5.357
AMMONIUM CHLORIDE (P=10KB)				H	0.359	0.140	0.280	0.271	3.689	2.962	5.354
10190	1.5978	100	295.0	S	0.359	0.140	0.280	0.271	3.689	2.961	5.354
				R	0.358	0.140	0.280	0.271	3.689	2.957	5.350
				V	0.374	0.146	0.284	0.289	3.457	3.008	5.480
AMMONIUM CHLORIDE (P=12KB)				H	0.374	0.145	0.285	0.289	3.457	3.006	5.478
10191	1.6102	100	295.0	S	0.374	0.145	0.285	0.289	3.457	3.006	5.478
				R	0.373	0.145	0.285	0.289	3.457	3.004	5.477
				V	0.295	0.117	0.260	0.205	4.880	2.756	4.842
AMMONIUM CHLORIDE (P= 2KB)				H	0.289	0.114	0.265	0.205	4.880	2.727	4.820
10192	1.5389	100	308.0	S	0.288	0.114	0.265	0.205	4.880	2.722	4.816
				R	0.283	0.112	0.270	0.205	4.880	2.693	4.794
				V	0.310	0.122	0.267	0.222	4.507	2.805	4.977
AMMONIUM CHLORIDE (P= 4KB)				H	0.306	0.120	0.270	0.222	4.507	2.784	4.962
10193	1.5532	100	308.0	S	0.305	0.120	0.271	0.222	4.507	2.780	4.959
				R	0.301	0.118	0.274	0.222	4.507	2.759	4.943
				V	0.324	0.127	0.272	0.237	4.219	2.850	5.095
AMMONIUM CHLORIDE (P= 6KB)				H	0.321	0.126	0.274	0.237	4.219	2.835	5.084
10194	1.5671	100	308.0	S	0.321	0.126	0.275	0.237	4.219	2.833	5.082
				R	0.318	0.124	0.277	0.237	4.219	2.818	5.071
				V	0.337	0.132	0.276	0.251	3.983	2.893	5.201
AMMONIUM CHLORIDE (P= 8KB)				H	0.335	0.131	0.277	0.251	3.983	2.883	5.194
10195	1.5797	100	308.0	S	0.335	0.131	0.277	0.251	3.983	2.882	5.193
				R	0.333	0.130	0.279	0.251	3.983	2.871	5.185

CUBIC

CODE / NAME	DENSITY	REF	TEMP(K)		E	G	SIGMA	K	BETA	VS	VP
				V	0.350	0.137	0.275	0.259	3.862	2.935	5.268
AMMONIUM CHLORIDE (P=10KB)				H	0.348	0.137	0.276	0.259	3.862	2.929	5.263
10196	1.5919	100	308.0	S	0.348	0.136	0.276	0.259	3.862	2.928	5.263
				R	0.347	0.136	0.277	0.259	3.863	2.921	5.258
				V	0.366	0.143	0.280	0.278	3.601	2.983	5.399
AMMONIUM CHLORIDE (P=12KB)				H	0.365	0.143	0.281	0.278	3.601	2.980	5.397
10197	1.6066	100	308.0	S	0.365	0.143	0.281	0.278	3.601	2.980	5.397
				R	0.365	0.142	0.281	0.278	3.601	2.977	5.395
				V	0.217	0.088	0.233	0.136	7.371	1.915	3.247
AMMONIUM CHLOROSTANNATE				H	0.212	0.085	0.240	0.136	7.371	1.886	3.224
10198	2.4000	197	292.0	S	0.210	0.085	0.242	0.136	7.371	1.878	3.218
				R	0.204	0.082	0.249	0.136	7.371	1.847	3.194
				V	0.043	0.017	0.232	0.027	37.547	0.992	1.678
ARGON				H	0.043	0.017	0.233	0.027	37.547	0.989	1.676
10199	1.7710	120	4.2	S	0.043	0.017	0.233	0.027	37.547	0.989	1.676
				R	0.043	0.017	0.234	0.027	37.547	0.986	1.674
				V	0.026	0.010	0.259	0.018	55.866	0.790	1.386
ARGON				H	0.026	0.010	0.261	0.018	55.866	0.787	1.384
10200	1.6451	120	76.8	S	0.026	0.010	0.261	0.018	55.866	0.787	1.384
				R	0.026	0.010	0.262	0.018	55.866	0.785	1.382
				V	0.652	0.249	0.309	0.569	1.757	2.271	4.319
BARIUM FLUORIDE				H	0.652	0.249	0.309	0.569	1.757	2.271	4.319
10201	4.8300	4	ROOM	S	0.652	0.249	0.309	0.569	1.757	2.271	4.319
				R	0.652	0.249	0.309	0.569	1.757	2.271	4.319
				V	0.662	0.253	0.310	0.581	1.722	2.274	4.334
BARIUM FLUORIDE				H	0.662	0.253	0.310	0.581	1.722	2.274	4.334
10202	4.8840	124	293.0	S	0.662	0.253	0.310	0.581	1.722	2.274	4.334
				R	0.662	0.253	0.310	0.581	1.722	2.274	4.334
				V	0.654	0.250	0.307	0.564	1.773	2.264	4.287
BARIUM FLUORIDE				H	0.654	0.250	0.307	0.564	1.773	2.264	4.286
10203	4.8860	103	ROOM	S	0.654	0.250	0.307	0.564	1.773	2.264	4.286
				R	0.654	0.250	0.307	0.564	1.773	2.263	4.286
				V	0.683	0.261	0.310	0.599	1.670	2.305	4.390
BARIUM FLUORIDE				H	0.683	0.261	0.310	0.599	1.670	2.305	4.390
10204	4.9120	278	195.0	S	0.683	0.261	0.310	0.599	1.670	2.305	4.390
				R	0.683	0.261	0.310	0.599	1.670	2.304	4.390
				V	0.668	0.255	0.309	0.584	1.713	2.284	4.348
BARIUM FLUORIDE				H	0.668	0.255	0.309	0.584	1.713	2.284	4.348
10205	4.8860	278	298.0	S	0.668	0.255	0.309	0.584	1.713	2.284	4.348
				R	0.668	0.255	0.309	0.584	1.713	2.284	4.348
				V	0.656	0.252	0.303	0.555	1.801	2.269	4.269
BARIUM FLUORIDE				H	0.656	0.252	0.303	0.555	1.801	2.269	4.269
10206	4.8860	108	295.0	S	0.656	0.252	0.303	0.555	1.801	2.269	4.269
				R	0.656	0.252	0.303	0.555	1.801	2.269	4.269
				V	0.403	0.156	0.293	0.325	3.080	2.192	4.051
BARIUM NITRATE				H	0.392	0.151	0.299	0.325	3.080	2.158	4.027
10207	3.2440	129	ROOM	S	0.390	0.150	0.300	0.325	3.080	2.150	4.021
				R	0.379	0.145	0.305	0.325	3.080	2.116	3.998
				V	0.405	0.157	0.292	0.325	3.074	2.199	4.059
BARIUM NITRATE				H	0.395	0.152	0.298	0.325	3.074	2.166	4.035
10208	3.2440	129	ROOM	S	0.392	0.151	0.299	0.325	3.074	2.158	4.029
				R	0.382	0.146	0.304	0.325	3.074	2.124	4.006
				V	0.397	0.153	0.295	0.324	3.090	2.175	4.035
BARIUM NITRATE				H	0.388	0.149	0.300	0.324	3.090	2.145	4.014
10209	3.2440	129	ROOM	S	0.386	0.148	0.301	0.324	3.090	2.138	4.009
				R	0.376	0.144	0.306	0.324	3.090	2.108	3.988
				V	0.248	0.094	0.324	0.235	4.252	1.697	3.326
BARIUM NITRATE				H	0.229	0.086	0.337	0.235	4.252	1.623	3.276
10210	3.2560	124	293.0	S	0.217	0.081	0.346	0.235	4.252	1.575	3.245
				R	0.195	0.071	0.362	0.235	4.252	1.481	3.186

CUBIC

CODE	NAME / DENSITY	REF	TEMP(K)		E	G	SIGMA	K	BETA	VS	VP
	BARIUM TITANATE			V	2.000	0.831	0.203	1.122	0.891	3.711	6.079
10211	6.0350	31	423.0	H	1.893	0.776	0.219	1.122	0.891	3.587	5.979
				S	1.847	0.753	0.226	1.122	0.891	3.533	5.936
				R	1.732	0.697	0.243	1.122	0.891	3.398	5.830
	CADMIUM TELLURIDE			V	0.410	0.153	0.339	0.424	2.360	1.617	3.275
10212	5.8544	167	298.0	H	0.388	0.144	0.348	0.424	2.360	1.567	3.242
				S	0.377	0.139	0.352	0.424	2.360	1.543	3.227
				R	0.349	0.128	0.363	0.424	2.360	1.480	3.187
	CADMIUM TELLURIDE (CONSTANT E)			V	0.420	0.154	0.359	0.498	2.010	1.623	3.464
10213	5.8600	32	77.0	H	0.402	0.147	0.365	0.498	2.010	1.585	3.441
				S	0.395	0.144	0.368	0.498	2.010	1.569	3.431
				R	0.372	0.135	0.375	0.498	2.010	1.519	3.401
	CALCIUM FLUORIDE			V	1.121	0.439	0.275	0.831	1.203	3.717	6.676
10214	3.1800	4	ROOM	H	1.088	0.424	0.282	0.831	1.203	3.652	6.628
				S	1.079	0.420	0.284	0.831	1.203	3.635	6.615
				R	1.046	0.406	0.290	0.831	1.203	3.571	6.569
	CALCIUM FLUORIDE			V	1.179	0.462	0.275	0.873	1.145	3.812	6.844
10215	3.1800	4	ROOM	H	1.153	0.451	0.280	0.873	1.145	3.764	6.808
				S	1.148	0.448	0.281	0.873	1.145	3.754	6.801
				R	1.123	0.437	0.286	0.873	1.145	3.705	6.765
	CALCIUM FLUORIDE			V	1.127	0.441	0.278	0.847	1.181	3.723	6.716
10216	3.1800	4	ROOM	H	1.095	0.426	0.284	0.847	1.181	3.661	6.671
				S	1.087	0.423	0.286	0.847	1.181	3.646	6.659
				R	1.056	0.409	0.292	0.847	1.181	3.585	6.615
	CALCIUM FLUORIDE			V	1.152	0.449	0.281	0.878	1.139	3.759	6.816
10217	3.1800	4	ROOM	H	1.127	0.438	0.286	0.878	1.139	3.711	6.780
				S	1.121	0.435	0.287	0.878	1.139	3.700	6.773
				R	1.096	0.424	0.292	0.878	1.139	3.652	6.737
	CALCIUM FLUORIDE			V	1.124	0.437	0.288	0.883	1.133	3.705	6.787
10218	3.1800	4	ROOM	H	1.099	0.425	0.292	0.883	1.133	3.657	6.752
				S	1.094	0.423	0.293	0.883	1.133	3.646	6.744
				R	1.069	0.412	0.298	0.883	1.133	3.598	6.709
	CALCIUM FLUORIDE			V	1.170	0.452	0.295	0.953	1.049	3.750	6.960
10219	3.2107	137	4.2	H	1.144	0.440	0.300	0.953	1.049	3.702	6.926
				S	1.138	0.438	0.301	0.953	1.049	3.691	6.918
				R	1.112	0.426	0.306	0.953	1.049	3.642	6.883
	CALCIUM FLUORIDE			V	1.177	0.456	0.291	0.940	1.064	3.767	6.942
10220	3.2107	137	20.0	H	1.149	0.443	0.296	0.940	1.064	3.715	6.905
				S	1.143	0.440	0.297	0.940	1.064	3.703	6.897
				R	1.115	0.428	0.302	0.940	1.064	3.651	6.859
	CALCIUM FLUORIDE			V	1.167	0.451	0.294	0.943	1.060	3.749	6.937
10221	3.2106	137	40.0	H	1.141	0.440	0.298	0.943	1.060	3.700	6.902
				S	1.136	0.437	0.299	0.943	1.060	3.689	6.894
				R	1.109	0.425	0.304	0.943	1.060	3.640	6.859
	CALCIUM FLUORIDE			V	1.173	0.455	0.290	0.930	1.075	3.763	6.917
10222	3.2102	137	60.0	H	1.145	0.442	0.295	0.930	1.075	3.711	6.880
				S	1.138	0.439	0.296	0.930	1.075	3.698	6.871
				R	1.110	0.427	0.301	0.930	1.075	3.645	6.833
	CALCIUM FLUORIDE			V	1.166	0.452	0.290	0.927	1.079	3.752	6.902
10223	3.2094	137	80.0	H	1.138	0.439	0.295	0.927	1.079	3.700	6.865
				S	1.132	0.437	0.296	0.927	1.079	3.688	6.856
				R	1.104	0.424	0.301	0.927	1.079	3.636	6.819
	CALCIUM FLUORIDE			V	1.156	0.447	0.293	0.930	1.075	3.732	6.897
10224	3.2081	137	100.0	H	1.130	0.435	0.298	0.930	1.075	3.684	6.862
				S	1.124	0.433	0.299	0.930	1.075	3.673	6.854
				R	1.098	0.421	0.303	0.930	1.075	3.623	6.818
	CALCIUM FLUORIDE			V	1.152	0.446	0.291	0.920	1.087	3.730	6.873
10225	3.2063	137	120.0	H	1.126	0.434	0.296	0.920	1.087	3.680	6.837
				S	1.120	0.432	0.297	0.920	1.087	3.669	6.829
				R	1.093	0.420	0.302	0.920	1.087	3.618	6.793

CUBIC

CODE	NAME / DENSITY	REF	TEMP(K)		E	G	SIGMA	K	BETA	VS	VP
				V	1.138	0.439	0.296	0.930	1.075	3.702	6.877
CALCIUM FLUORIDE				H	1.114	0.428	0.300	0.930	1.075	3.657	6.845
10226	3.2042	137	140.0	S	1.109	0.426	0.301	0.930	1.075	3.647	6.838
				R	1.085	0.415	0.306	0.930	1.075	3.601	6.805
				V	1.135	0.438	0.294	0.920	1.087	3.700	6.855
CALCIUM FLUORIDE				H	1.111	0.428	0.299	0.920	1.087	3.655	6.822
10227	3.2017	137	160.0	S	1.105	0.425	0.300	0.920	1.087	3.644	6.815
				R	1.081	0.414	0.304	0.920	1.087	3.598	6.782
				V	1.124	0.433	0.297	0.923	1.083	3.680	6.850
CALCIUM FLUORIDE				H	1.101	0.423	0.301	0.923	1.083	3.637	6.819
10228	3.1991	137	180.0	S	1.096	0.421	0.302	0.923	1.083	3.627	6.812
				R	1.073	0.411	0.306	0.923	1.083	3.583	6.781
				V	1.109	0.426	0.302	0.933	1.071	3.650	6.853
CALCIUM FLUORIDE				H	1.088	0.417	0.306	0.933	1.071	3.611	6.826
10229	3.1962	137	200.0	S	1.084	0.415	0.306	0.933	1.071	3.603	6.820
				R	1.063	0.406	0.310	0.933	1.071	3.562	6.791
				V	1.105	0.425	0.300	0.923	1.083	3.648	6.831
CALCIUM FLUORIDE				H	1.084	0.416	0.304	0.923	1.083	3.608	6.802
10230	3.1931	137	220.0	S	1.080	0.414	0.305	0.923	1.083	3.600	6.797
				R	1.058	0.404	0.309	0.923	1.083	3.558	6.767
				V	1.103	0.424	0.301	0.923	1.083	3.645	6.831
CALCIUM FLUORIDE				H	1.081	0.414	0.305	0.923	1.083	3.604	6.802
10231	3.1898	137	240.0	S	1.077	0.412	0.306	0.923	1.083	3.595	6.796
				R	1.055	0.403	0.310	0.923	1.083	3.553	6.766
				V	1.091	0.419	0.304	0.927	1.079	3.624	6.826
CALCIUM FLUORIDE				H	1.071	0.410	0.307	0.927	1.079	3.586	6.799
10232	3.1864	137	260.0	S	1.067	0.408	0.308	0.927	1.079	3.578	6.793
				R	1.046	0.399	0.312	0.927	1.079	3.538	6.765
				V	1.099	0.423	0.298	0.907	1.103	3.647	6.799
CALCIUM FLUORIDE				H	1.076	0.413	0.302	0.907	1.103	3.603	6.767
10233	3.1830	137	280.0	S	1.071	0.411	0.303	0.907	1.103	3.593	6.760
				R	1.047	0.400	0.308	0.907	1.103	3.547	6.728
				V	1.100	0.424	0.296	0.900	1.111	3.653	6.789
CALCIUM FLUORIDE				H	1.075	0.413	0.301	0.900	1.111	3.605	6.756
10234	3.1795	137	300.0	S	1.070	0.411	0.302	0.900	1.111	3.595	6.748
				R	1.045	0.400	0.307	0.900	1.111	3.546	6.714
				V	1.129	0.443	0.276	0.839	1.192	3.731	6.703
CALCIUM FLUORIDE				H	1.097	0.428	0.282	0.839	1.192	3.669	6.657
10235	3.1804	124	293.0	S	1.089	0.424	0.284	0.839	1.192	3.653	6.645
				R	1.058	0.410	0.290	0.839	1.192	3.591	6.600
				V	1.199	0.469	0.277	0.895	1.118	3.825	6.884
CALCIUM FLUORIDE (P=0.001KB)				H	1.165	0.454	0.283	0.895	1.118	3.762	6.838
10236	3.2085	134	77.4	S	1.157	0.450	0.284	0.895	1.118	3.747	6.827
				R	1.124	0.436	0.291	0.895	1.118	3.685	6.782
				V	1.199	0.470	0.277	0.896	1.117	3.825	6.886
CALCIUM FLUORIDE (P=0.158KB)				H	1.166	0.454	0.283	0.895	1.117	3.763	6.840
10237	3.2091	134	77.4	S	1.158	0.451	0.285	0.895	1.117	3.747	6.828
				R	1.125	0.436	0.291	0.896	1.117	3.685	6.783
				V	1.199	0.470	0.277	0.896	1.116	3.825	6.886
CALCIUM FLUORIDE (P=0.272KB)				H	1.166	0.454	0.283	0.896	1.116	3.763	6.840
10238	3.2095	134	77.4	S	1.158	0.451	0.285	0.896	1.116	3.747	6.829
				R	1.125	0.436	0.291	0.896	1.116	3.685	6.783
				V	1.202	0.471	0.277	0.900	1.111	3.827	6.895
CALCIUM FLUORIDE (P=1.072KB)				H	1.169	0.455	0.283	0.900	1.111	3.765	6.850
10239	3.2124	134	77.4	S	1.161	0.452	0.285	0.900	1.111	3.750	6.838
				R	1.128	0.437	0.291	0.900	1.111	3.688	6.793
				V	1.204	0.471	0.278	0.902	1.108	3.829	6.900
CALCIUM FLUORIDE (P=1.538KB)				H	1.171	0.456	0.284	0.902	1.108	3.767	6.855
10240	3.2141	134	77.4	S	1.163	0.452	0.285	0.902	1.108	3.751	6.843
				R	1.130	0.438	0.291	0.902	1.108	3.690	6.799

CUBIC

NAME / CODE	DENSITY	REF	TEMP(K)		E	G	SIGMA	K	BETA	VS	VP
				V	1.205	0.471	0.278	0.904	1.106	3.829	6.904
CALCIUM FLUORIDE (P=1.848KB)				H	1.172	0.456	0.284	0.904	1.106	3.767	6.858
10241	3.2152	134	77.4	S	1.164	0.453	0.285	0.904	1.106	3.752	6.847
				R	1.131	0.438	0.291	0.904	1.106	3.690	6.802
				V	1.206	0.472	0.278	0.906	1.103	3.830	6.909
CALCIUM FLUORIDE (P=2.381KB)				H	1.174	0.457	0.284	0.906	1.103	3.769	6.864
10242	3.2171	134	77.4	S	1.165	0.453	0.286	0.906	1.103	3.753	6.853
				R	1.133	0.438	0.292	0.907	1.103	3.692	6.808
				V	1.209	0.473	0.278	0.909	1.100	3.832	6.915
CALCIUM FLUORIDE (P=2.936KB)				H	1.176	0.458	0.284	0.909	1.100	3.770	6.870
10243	3.2191	134	77.4	S	1.167	0.454	0.286	0.909	1.100	3.755	6.859
				R	1.135	0.439	0.292	0.909	1.100	3.694	6.814
				V	1.210	0.473	0.279	0.912	1.097	3.833	6.920
CALCIUM FLUORIDE (P=3.409KB)				H	1.177	0.458	0.285	0.912	1.097	3.772	6.875
10244	3.2209	134	77.4	S	1.169	0.454	0.286	0.912	1.097	3.756	6.864
				R	1.137	0.440	0.292	0.912	1.097	3.695	6.819
				V	1.163	0.456	0.276	0.865	1.156	3.776	6.787
CALCIUM FLUORIDE (P=0.001KB)				H	1.130	0.441	0.282	0.865	1.156	3.714	6.742
10245	3.1963	134	194.5	S	1.122	0.437	0.284	0.865	1.156	3.698	6.730
				R	1.091	0.423	0.290	0.865	1.156	3.637	6.686
				V	1.164	0.456	0.276	0.866	1.154	3.776	6.790
CALCIUM FLUORIDE (P=0.241KB)				H	1.131	0.441	0.282	0.866	1.154	3.715	6.745
10246	3.1972	134	194.5	S	1.123	0.437	0.284	0.866	1.154	3.699	6.733
				R	1.092	0.423	0.290	0.866	1.154	3.638	6.689
				V	1.165	0.456	0.276	0.868	1.152	3.778	6.794
CALCIUM FLUORIDE (P=0.574KB)				H	1.133	0.442	0.282	0.868	1.152	3.716	6.748
10247	3.1984	134	194.5	S	1.125	0.438	0.284	0.868	1.152	3.700	6.737
				R	1.093	0.424	0.290	0.868	1.152	3.639	6.692
				V	1.166	0.457	0.276	0.869	1.150	3.778	6.797
CALCIUM FLUORIDE (P=0.924KB)				H	1.134	0.442	0.283	0.869	1.150	3.717	6.752
10248	3.1998	134	194.5	S	1.126	0.438	0.284	0.869	1.150	3.701	6.741
				R	1.094	0.424	0.290	0.869	1.150	3.640	6.696
				V	1.167	0.457	0.277	0.871	1.148	3.779	6.801
CALCIUM FLUORIDE (P=1.273KB)				H	1.135	0.442	0.283	0.871	1.148	3.718	6.756
10249	3.2011	134	194.5	S	1.127	0.439	0.284	0.871	1.148	3.702	6.744
				R	1.095	0.424	0.290	0.871	1.148	3.641	6.700
				V	1.169	0.458	0.277	0.873	1.146	3.780	6.805
CALCIUM FLUORIDE (P=1.615KB)				H	1.136	0.443	0.283	0.873	1.146	3.719	6.759
10250	3.2023	134	194.5	S	1.128	0.439	0.285	0.873	1.146	3.703	6.748
				R	1.097	0.425	0.291	0.873	1.146	3.642	6.704
				V	1.171	0.458	0.277	0.875	1.143	3.782	6.810
CALCIUM FLUORIDE (P=2.129KB)				H	1.138	0.444	0.283	0.875	1.143	3.721	6.765
10251	3.2042	134	194.5	S	1.130	0.440	0.285	0.875	1.143	3.705	6.754
				R	1.099	0.426	0.291	0.875	1.143	3.644	6.709
				V	1.172	0.459	0.277	0.878	1.139	3.783	6.816
CALCIUM FLUORIDE (P=2.677KB)				H	1.140	0.444	0.283	0.878	1.139	3.722	6.771
10252	3.2063	134	194.5	S	1.132	0.440	0.285	0.878	1.139	3.706	6.759
				R	1.100	0.426	0.291	0.878	1.139	3.646	6.715
				V	1.174	0.459	0.278	0.880	1.136	3.784	6.821
CALCIUM FLUORIDE (P=3.189KB)				H	1.142	0.445	0.284	0.880	1.136	3.723	6.776
10253	3.2082	134	194.5	S	1.134	0.441	0.285	0.880	1.136	3.708	6.765
				R	1.102	0.427	0.291	0.880	1.136	3.647	6.721
				V	1.176	0.460	0.278	0.883	1.133	3.785	6.826
CALCIUM FLUORIDE (P=3.732KB)				H	1.144	0.445	0.284	0.882	1.133	3.725	6.782
10254	3.2102	134	194.5	S	1.136	0.442	0.286	0.882	1.133	3.710	6.770
				R	1.104	0.428	0.291	0.882	1.133	3.649	6.727
				V	1.177	0.460	0.278	0.884	1.131	3.787	6.830
CALCIUM FLUORIDE (P=4.164KB)				H	1.145	0.446	0.284	0.884	1.131	3.726	6.786
10255	3.2118	134	194.5	S	1.137	0.442	0.286	0.884	1.131	3.711	6.775
				R	1.106	0.428	0.292	0.884	1.131	3.651	6.731

CUBIC

CODE	NAME / DENSITY	REF	TEMP(K)		E	G	SIGMA	K	BETA	VS	VP
				V	1.140	0.446	0.277	0.852	1.173	3.744	6.743
CALCIUM FLUORIDE (P=0.001KB)				H	1.107	0.431	0.284	0.852	1.173	3.680	6.696
10256	3.1835	134	273.1	S	1.099	0.427	0.285	0.852	1.173	3.664	6.684
				R	1.066	0.413	0.291	0.852	1.173	3.601	6.638
				V	1.142	0.447	0.277	0.853	1.173	3.748	6.746
CALCIUM FLUORIDE (P=0.265KB)				H	1.109	0.432	0.283	0.853	1.173	3.683	6.698
10257	3.1845	134	273.1	S	1.100	0.428	0.285	0.853	1.173	3.667	6.686
				R	1.068	0.413	0.291	0.853	1.173	3.603	6.640
				V	1.142	0.447	0.278	0.855	1.169	3.745	6.749
CALCIUM FLUORIDE (P=0.578KB)				H	1.109	0.432	0.284	0.855	1.169	3.682	6.703
10258	3.1857	134	273.1	S	1.101	0.428	0.286	0.855	1.169	3.666	6.691
				R	1.068	0.414	0.292	0.855	1.169	3.603	6.645
				V	1.142	0.447	0.278	0.858	1.166	3.744	6.753
CALCIUM FLUORIDE (P=0.911KB)				H	1.109	0.432	0.284	0.858	1.166	3.681	6.707
10259	3.1870	134	273.1	S	1.101	0.428	0.286	0.858	1.166	3.665	6.695
				R	1.069	0.414	0.292	0.858	1.166	3.603	6.650
				V	1.143	0.447	0.278	0.860	1.163	3.744	6.758
CALCIUM FLUORIDE (P=1.281KB)				H	1.111	0.432	0.285	0.860	1.163	3.682	6.712
10260	3.1884	134	273.1	S	1.102	0.428	0.286	0.860	1.163	3.666	6.700
				R	1.070	0.414	0.293	0.860	1.163	3.604	6.655
				V	1.145	0.448	0.279	0.863	1.159	3.746	6.764
CALCIUM FLUORIDE (P=1.741KB)				H	1.113	0.433	0.285	0.863	1.159	3.684	6.718
10261	3.1901	134	273.1	S	1.104	0.429	0.287	0.863	1.159	3.668	6.706
				R	1.072	0.415	0.293	0.863	1.159	3.606	6.661
				V	1.146	0.448	0.279	0.864	1.157	3.747	6.767
CALCIUM FLUORIDE (P=2.026KB)				H	1.114	0.433	0.285	0.864	1.157	3.684	6.722
10262	3.1912	134	273.1	S	1.105	0.429	0.287	0.864	1.157	3.668	6.710
				R	1.073	0.415	0.293	0.864	1.157	3.607	6.665
				V	1.148	0.449	0.279	0.867	1.153	3.748	6.774
CALCIUM FLUORIDE (P=2.520KB)				H	1.115	0.434	0.286	0.867	1.153	3.686	6.728
10263	3.1931	134	273.1	S	1.107	0.430	0.287	0.867	1.153	3.670	6.717
				R	1.075	0.416	0.293	0.867	1.153	3.608	6.672
				V	1.150	0.449	0.280	0.870	1.149	3.750	6.781
CALCIUM FLUORIDE (P=3.030KB)				H	1.118	0.435	0.286	0.870	1.149	3.688	6.735
10264	3.1950	134	273.1	S	1.109	0.431	0.287	0.870	1.149	3.672	6.724
				R	1.078	0.416	0.294	0.870	1.149	3.610	6.679
				V	1.151	0.450	0.280	0.873	1.146	3.751	6.787
CALCIUM FLUORIDE (P=3.505KB)				H	1.119	0.435	0.286	0.873	1.146	3.689	6.742
10265	3.1968	134	273.1	S	1.111	0.431	0.288	0.873	1.146	3.673	6.730
				R	1.079	0.417	0.294	0.873	1.146	3.612	6.686
				V	1.153	0.450	0.281	0.876	1.142	3.752	6.794
CALCIUM FLUORIDE (P=4.028KB)				H	1.121	0.436	0.287	0.876	1.142	3.690	6.749
10266	3.1988	134	273.1	S	1.113	0.432	0.288	0.876	1.142	3.675	6.737
				R	1.081	0.418	0.294	0.876	1.142	3.614	6.693
				V	1.133	0.443	0.277	0.847	1.180	3.735	6.726
CALCIUM FLUORIDE (P=0.001KB)				H	1.100	0.428	0.284	0.847	1.180	3.670	6.679
10267	3.1795	134	295.5	S	1.091	0.424	0.285	0.847	1.180	3.654	6.667
				R	1.059	0.410	0.292	0.847	1.180	3.590	6.621
				V	1.137	0.445	0.277	0.848	1.179	3.741	6.733
CALCIUM FLUORIDE (P=0.264KB)				H	1.103	0.430	0.283	0.848	1.179	3.676	6.685
10268	3.1805	134	295.5	S	1.094	0.426	0.285	0.848	1.179	3.659	6.673
				R	1.062	0.411	0.291	0.848	1.179	3.595	6.626
				V	1.136	0.445	0.277	0.851	1.176	3.739	6.736
CALCIUM FLUORIDE (P=0.582KB)				H	1.103	0.430	0.284	0.851	1.176	3.675	6.689
10269	3.1817	134	295.5	S	1.095	0.426	0.285	0.851	1.176	3.658	6.677
				R	1.062	0.411	0.292	0.851	1.176	3.595	6.630
				V	1.137	0.445	0.278	0.853	1.172	3.738	6.740
CALCIUM FLUORIDE (P=0.950KB)				H	1.104	0.430	0.284	0.853	1.172	3.675	6.693
10270	3.1832	134	295.5	S	1.096	0.426	0.286	0.853	1.172	3.658	6.681
				R	1.063	0.411	0.292	0.853	1.172	3.595	6.636

CUBIC

CODE	NAME / DENSITY	REF	TEMP(K)		E	G	SIGMA	K	BETA	VS	VP
				V	1.138	0.445	0.278	0.855	1.169	3.739	6.745
CALCIUM FLUORIDE (P=1.269KB)				H	1.105	0.430	0.285	0.855	1.169	3.675	6.698
10271	3.1844	134	295.5	S	1.097	0.426	0.286	0.855	1.169	3.659	6.686
				R	1.065	0.412	0.293	0.855	1.169	3.596	6.641
				V	1.139	0.445	0.279	0.857	1.166	3.739	6.749
CALCIUM FLUORIDE (P=1.669KB)				H	1.106	0.430	0.285	0.857	1.166	3.676	6.703
10272	3.1859	134	295.5	S	1.098	0.427	0.287	0.857	1.166	3.660	6.691
				R	1.066	0.412	0.293	0.857	1.166	3.597	6.646
				V	1.141	0.446	0.279	0.860	1.162	3.740	6.756
CALCIUM FLUORIDE (P=2.126KB)				H	1.108	0.431	0.285	0.860	1.162	3.677	6.710
10273	3.1876	134	295.5	S	1.100	0.427	0.287	0.860	1.162	3.661	6.698
				R	1.068	0.413	0.293	0.860	1.162	3.599	6.653
				V	1.142	0.446	0.279	0.863	1.158	3.741	6.763
CALCIUM FLUORIDE (P=2.629KB)				H	1.110	0.432	0.286	0.863	1.158	3.679	6.717
10274	3.1895	134	295.5	S	1.102	0.428	0.287	0.863	1.158	3.663	6.705
				R	1.070	0.413	0.294	0.863	1.158	3.600	6.660
				V	1.145	0.447	0.280	0.867	1.154	3.743	6.770
CALCIUM FLUORIDE (P=3.139KB)				H	1.112	0.432	0.286	0.867	1.154	3.681	6.724
10275	3.1915	134	295.5	S	1.104	0.429	0.288	0.867	1.154	3.665	6.713
				R	1.072	0.414	0.294	0.867	1.154	3.602	6.668
				V	1.725	0.745	0.158	0.841	1.190	4.839	7.592
CALCIUM FLUORIDE				H	1.708	0.736	0.161	0.841	1.190	4.809	7.567
10276	3.1810	279	298.0	S	1.706	0.734	0.162	0.841	1.190	4.804	7.563
				R	1.689	0.725	0.165	0.841	1.190	4.774	7.538
				V	1.144	0.443	0.291	0.913	1.095	3.732	6.876
CALCIUM FLUORIDE				H	1.113	0.429	0.297	0.913	1.095	3.673	6.833
10277	3.1810	193	293.0	S	1.105	0.426	0.298	0.913	1.095	3.658	6.823
				R	1.074	0.412	0.304	0.913	1.095	3.597	6.780
				V	1.136	0.439	0.293	0.913	1.095	3.721	6.872
CALCIUM FLUORIDE				H	1.104	0.425	0.298	0.913	1.095	3.661	6.829
10278	3.1740	193	323.0	S	1.096	0.422	0.300	0.913	1.095	3.645	6.818
				R	1.065	0.408	0.306	0.913	1.095	3.585	6.775
				V	1.122	0.434	0.294	0.906	1.104	3.700	6.845
CALCIUM FLUORIDE				H	1.090	0.419	0.299	0.906	1.104	3.639	6.802
10279	3.1666	193	373.0	S	1.082	0.416	0.301	0.906	1.104	3.623	6.790
				R	1.050	0.402	0.307	0.906	1.104	3.562	6.747
				V	1.106	0.428	0.293	0.891	1.122	3.682	6.805
CALCIUM FLUORIDE				H	1.075	0.414	0.299	0.891	1.122	3.620	6.761
10280	3.1562	193	423.0	S	1.066	0.410	0.301	0.891	1.122	3.604	6.749
				R	1.034	0.396	0.307	0.891	1.122	3.541	6.705
				V	1.092	0.422	0.295	0.887	1.128	3.662	6.787
CALCIUM FLUORIDE				H	1.060	0.407	0.301	0.887	1.128	3.599	6.742
10281	3.1457	193	473.0	S	1.051	0.404	0.302	0.887	1.128	3.582	6.730
				R	1.019	0.389	0.308	0.887	1.128	3.518	6.685
				V	1.078	0.416	0.295	0.875	1.143	3.644	6.753
CALCIUM FLUORIDE				H	1.045	0.401	0.301	0.875	1.143	3.578	6.707
10282	3.1352	193	523.0	S	1.036	0.397	0.303	0.875	1.143	3.561	6.694
				R	1.003	0.383	0.309	0.875	1.143	3.495	6.648
				V	1.061	0.410	0.293	0.855	1.170	3.624	6.698
CALCIUM FLUORIDE				H	1.028	0.395	0.300	0.855	1.170	3.557	6.650
10283	3.1251	193	573.0	S	1.019	0.391	0.301	0.855	1.170	3.539	6.637
				R	0.985	0.377	0.308	0.855	1.170	3.472	6.590
				V	1.045	0.404	0.293	0.842	1.188	3.603	6.659
CALCIUM FLUORIDE				H	1.011	0.389	0.300	0.842	1.188	3.535	6.611
10284	3.1129	193	623.0	S	1.002	0.385	0.302	0.842	1.188	3.517	6.598
				R	0.969	0.370	0.308	0.842	1.188	3.449	6.550
				V	1.026	0.397	0.294	0.829	1.206	3.577	6.619
CALCIUM FLUORIDE				H	0.993	0.382	0.300	0.829	1.206	3.509	6.570
10285	3.1001	193	673.0	S	0.984	0.378	0.302	0.829	1.206	3.490	6.557
				R	0.951	0.363	0.309	0.829	1.206	3.422	6.509

CUBIC

CODE	NAME DENSITY	REF	TEMP(K)		E	G	SIGMA	K	BETA	VS	VP
				V	1.021	0.393	0.298	0.844	1.185	3.564	6.648
CALCIUM FLUORIDE				H	0.988	0.378	0.305	0.844	1.185	3.497	6.600
10286	3.0955	193	693.0	S	0.978	0.374	0.307	0.844	1.185	3.478	6.587
				R	0.945	0.360	0.313	0.844	1.185	3.410	6.539
				V	1.158	0.454	0.275	0.859	1.164	3.768	6.768
CALCIUM FLUORIDE				H	1.126	0.439	0.282	0.859	1.164	3.707	6.722
10287	3.1980	278	195.0	S	1.118	0.436	0.283	0.859	1.164	3.691	6.711
				R	1.087	0.421	0.289	0.859	1.164	3.630	6.666
				V	1.130	0.443	0.276	0.841	1.190	3.730	6.706
CALCIUM FLUORIDE				H	1.096	0.427	0.283	0.841	1.190	3.666	6.659
10288	3.1810	278	298.0	S	1.088	0.424	0.284	0.841	1.190	3.649	6.647
				R	1.055	0.409	0.291	0.841	1.190	3.585	6.600
				V	0.027	0.010	0.304	0.023	43.290	0.722	1.361
CESIUM				H	0.023	0.009	0.336	0.023	43.290	0.652	1.312
10289	2.0000	146	4.2	S	0.017	0.006	0.380	0.023	43.290	0.550	1.248
				R	0.012	0.004	0.411	0.023	43.289	0.468	1.203
				V	0.026	0.010	0.304	0.022	45.455	0.707	1.333
CESIUM				H	0.022	0.008	0.336	0.022	45.455	0.640	1.287
10290	1.9800	146	63.0	S	0.016	0.006	0.377	0.022	45.455	0.546	1.228
				R	0.012	0.004	0.407	0.022	45.454	0.468	1.185
				V	0.025	0.010	0.307	0.022	45.627	0.700	1.327
CESIUM				H	0.021	0.008	0.338	0.022	45.627	0.634	1.281
10291	1.9800	146	78.0	S	0.016	0.006	0.379	0.022	45.627	0.540	1.223
				R	0.012	0.004	0.409	0.022	45.627	0.463	1.180
				V	0.228	0.090	0.262	0.159	6.276	1.424	2.507
CESIUM BROMIDE				H	0.224	0.089	0.266	0.159	6.276	1.411	2.497
10292	4.4500	125	293.0	S	0.224	0.088	0.266	0.159	6.276	1.408	2.495
				R	0.220	0.087	0.270	0.159	6.276	1.395	2.485
				V	0.225	0.090	0.253	0.152	6.579	1.423	2.476
CESIUM BROMIDE				H	0.222	0.088	0.257	0.152	6.579	1.411	2.467
10293	4.4330	41	300.0	S	0.221	0.088	0.257	0.152	6.579	1.410	2.465
				R	0.218	0.087	0.260	0.152	6.579	1.398	2.457
				V	0.267	0.107	0.253	0.180	5.556	1.523	2.647
CESIUM BROMIDE				H	0.267	0.106	0.253	0.180	5.556	1.521	2.645
10294	4.5992	173	4.2	S	0.267	0.106	0.253	0.180	5.556	1.521	2.645
				R	0.266	0.106	0.254	0.180	5.556	1.519	2.644
				V	0.267	0.107	0.252	0.180	5.565	1.522	2.645
CESIUM BROMIDE				H	0.266	0.106	0.253	0.180	5.565	1.520	2.644
10295	4.5992	173	10.0	S	0.266	0.106	0.253	0.180	5.565	1.520	2.643
				R	0.266	0.106	0.254	0.180	5.565	1.518	2.642
				V	0.266	0.106	0.253	0.179	5.572	1.518	2.641
CESIUM BROMIDE				H	0.265	0.106	0.254	0.179	5.572	1.516	2.640
10296	4.5992	173	20.0	S	0.265	0.106	0.254	0.179	5.572	1.516	2.640
				R	0.265	0.105	0.254	0.179	5.572	1.514	2.638
				V	0.265	0.105	0.254	0.179	5.586	1.515	2.637
CESIUM BROMIDE				H	0.264	0.105	0.254	0.179	5.586	1.513	2.636
10297	4.5962	173	30.0	S	0.264	0.105	0.254	0.179	5.586	1.513	2.636
				R	0.263	0.105	0.255	0.179	5.586	1.511	2.634
				V	0.263	0.105	0.254	0.178	5.612	1.513	2.633
CESIUM BROMIDE				H	0.263	0.105	0.254	0.178	5.612	1.510	2.631
10298	4.5922	173	40.0	S	0.263	0.105	0.254	0.178	5.612	1.510	2.631
				R	0.262	0.104	0.255	0.178	5.612	1.508	2.629
				V	0.262	0.104	0.254	0.178	5.619	1.509	2.630
CESIUM BROMIDE				H	0.262	0.104	0.255	0.178	5.619	1.507	2.628
10299	4.5872	173	50.0	S	0.261	0.104	0.255	0.178	5.619	1.507	2.628
				R	0.261	0.104	0.256	0.178	5.619	1.505	2.626
				V	0.236	0.094	0.257	0.162	6.161	1.448	2.533
CESIUM BROMIDE				H	0.234	0.093	0.260	0.162	6.161	1.439	2.526
10300	4.4820	173	220.0	S	0.233	0.093	0.260	0.162	6.161	1.437	2.525
				R	0.231	0.091	0.263	0.162	6.161	1.428	2.518

CUBIC

CODE	NAME / DENSITY	REF	TEMP(K)		E	G	SIGMA	K	BETA	VS	VP
				V	0.233	0.093	0.258	0.160	6.231	1.440	2.521
CESIUM BROMIDE				H	0.230	0.091	0.261	0.160	6.231	1.430	2.514
10301	4.4696	173	240.0	S	0.230	0.091	0.261	0.160	6.231	1.429	2.512
				R	0.227	0.090	0.264	0.161	6.231	1.419	2.505
				V	0.230	0.091	0.259	0.159	6.275	1.432	2.512
CESIUM BROMIDE				H	0.227	0.090	0.262	0.159	6.275	1.421	2.504
10302	4.4571	173	260.0	S	0.227	0.090	0.263	0.159	6.275	1.420	2.503
				R	0.224	0.088	0.266	0.159	6.275	1.409	2.494
				V	0.228	0.091	0.258	0.157	6.373	1.428	2.500
CESIUM BROMIDE				H	0.225	0.089	0.261	0.157	6.373	1.416	2.490
10303	4.4446	173	280.0	S	0.224	0.089	0.262	0.157	6.373	1.414	2.489
				R	0.221	0.087	0.265	0.157	6.374	1.401	2.480
				V	0.226	0.090	0.254	0.154	6.510	1.426	2.486
CESIUM BROMIDE				H	0.223	0.088	0.259	0.154	6.510	1.412	2.475
10304	4.4322	173	300.0	S	0.222	0.088	0.259	0.154	6.510	1.410	2.473
				R	0.218	0.086	0.263	0.154	6.510	1.396	2.462
				V	0.263	0.105	0.255	0.178	5.605	1.514	2.638
CESIUM BROMIDE				H	0.262	0.104	0.256	0.178	5.605	1.510	2.635
10305	4.5692	216	80.0	S	0.262	0.104	0.256	0.178	5.605	1.510	2.635
				R	0.261	0.104	0.257	0.178	5.605	1.507	2.633
				V	0.259	0.103	0.256	0.177	5.647	1.505	2.628
CESIUM BROMIDE				H	0.258	0.103	0.257	0.177	5.647	1.501	2.625
10306	4.5568	216	100.0	S	0.258	0.103	0.257	0.177	5.647	1.501	2.625
				R	0.257	0.102	0.258	0.177	5.647	1.497	2.622
				V	0.256	0.102	0.257	0.176	5.689	1.496	2.618
CESIUM BROMIDE				H	0.255	0.101	0.259	0.176	5.689	1.492	2.614
10307	4.5443	216	120.0	S	0.254	0.101	0.259	0.176	5.689	1.491	2.614
				R	0.253	0.100	0.260	0.176	5.689	1.487	2.611
				V	0.252	0.100	0.259	0.174	5.734	1.487	2.607
CESIUM BROMIDE				H	0.251	0.100	0.260	0.174	5.734	1.482	2.603
10308	4.5318	216	140.0	S	0.251	0.099	0.260	0.174	5.734	1.482	2.603
				R	0.249	0.099	0.262	0.174	5.734	1.476	2.599
				V	0.249	0.099	0.260	0.173	5.780	1.478	2.596
CESIUM BROMIDE				H	0.247	0.098	0.262	0.173	5.780	1.472	2.592
10309	4.5194	216	160.0	S	0.247	0.098	0.262	0.173	5.780	1.472	2.592
				R	0.245	0.097	0.264	0.173	5.780	1.466	2.587
				V	0.245	0.097	0.262	0.172	5.826	1.469	2.586
CESIUM BROMIDE				H	0.244	0.096	0.263	0.172	5.826	1.462	2.581
10310	4.5069	216	180.0	S	0.243	0.096	0.264	0.172	5.826	1.462	2.580
				R	0.242	0.095	0.265	0.172	5.826	1.455	2.575
				V	0.242	0.096	0.263	0.170	5.873	1.459	2.575
CESIUM BROMIDE				H	0.240	0.095	0.265	0.170	5.873	1.452	2.569
10311	4.4945	216	200.0	S	0.240	0.095	0.265	0.170	5.873	1.451	2.569
				R	0.238	0.094	0.267	0.170	5.873	1.444	2.563
				V	0.238	0.094	0.265	0.169	5.921	1.450	2.564
CESIUM BROMIDE				H	0.236	0.093	0.267	0.169	5.921	1.442	2.558
10312	4.4820	216	220.0	S	0.236	0.093	0.267	0.169	5.921	1.441	2.557
				R	0.234	0.092	0.269	0.169	5.921	1.433	2.551
				V	0.235	0.093	0.266	0.168	5.968	1.440	2.552
CESIUM BROMIDE				H	0.232	0.092	0.269	0.168	5.968	1.432	2.546
10313	4.4696	216	240.0	S	0.232	0.091	0.269	0.168	5.968	1.430	2.545
				R	0.230	0.090	0.272	0.168	5.968	1.422	2.538
				V	0.231	0.091	0.268	0.166	6.018	1.431	2.541
CESIUM BROMIDE				H	0.229	0.090	0.271	0.166	6.018	1.421	2.534
10314	4.4571	216	260.0	S	0.228	0.090	0.271	0.166	6.018	1.420	2.533
				R	0.226	0.089	0.274	0.166	6.018	1.410	2.526
				V	0.228	0.090	0.270	0.165	6.063	1.421	2.530
CESIUM BROMIDE				H	0.225	0.088	0.273	0.165	6.063	1.410	2.522
10315	4.4446	216	280.0	S	0.225	0.088	0.273	0.165	6.063	1.409	2.521
				R	0.222	0.087	0.276	0.165	6.063	1.398	2.513

CUBIC

CODE	NAME / DENSITY	REF	TEMP(K)		E	G	SIGMA	K	BETA	VS	VP
				V	0.226	0.089	0.270	0.163	6.119	1.416	2.521
CESIUM BROMIDE				H	0.223	0.087	0.273	0.163	6.119	1.404	2.513
10316	4.4353	216	295.0	S	0.222	0.087	0.273	0.163	6.119	1.403	2.512
				R	0.219	0.086	0.276	0.163	6.119	1.391	2.503
				V	0.225	0.088	0.275	0.167	6.000	1.408	2.529
CESIUM BROMIDE				H	0.222	0.087	0.278	0.167	6.000	1.399	2.523
10317	4.4400	195	293.0	S	0.222	0.087	0.278	0.167	6.000	1.398	2.522
				R	0.219	0.086	0.281	0.167	6.000	1.389	2.515
				V	0.221	0.086	0.279	0.167	6.000	1.398	2.526
CESIUM BROMIDE				H	0.218	0.085	0.282	0.167	6.000	1.387	2.518
10318	4.4170	195	323.0	S	0.218	0.085	0.282	0.167	6.000	1.386	2.517
				R	0.215	0.083	0.285	0.167	6.000	1.375	2.509
				V	0.212	0.083	0.278	0.159	6.300	1.374	2.477
CESIUM BROMIDE				H	0.208	0.081	0.281	0.159	6.300	1.361	2.467
10319	4.3870	195	373.0	S	0.208	0.081	0.282	0.159	6.300	1.358	2.465
				R	0.204	0.079	0.286	0.159	6.300	1.345	2.456
				V	0.202	0.079	0.278	0.152	6.600	1.347	2.428
CESIUM BROMIDE				H	0.198	0.077	0.282	0.152	6.600	1.331	2.417
10320	4.3560	195	423.0	S	0.197	0.077	0.283	0.152	6.600	1.328	2.414
				R	0.193	0.075	0.288	0.152	6.600	1.312	2.403
				V	0.196	0.075	0.304	0.167	6.000	1.324	2.493
CESIUM BROMIDE				H	0.191	0.073	0.309	0.167	6.000	1.305	2.480
10321	4.2960	195	473.0	S	0.190	0.073	0.310	0.167	6.000	1.300	2.477
				R	0.185	0.070	0.315	0.167	6.000	1.280	2.463
				V	0.187	0.072	0.304	0.159	6.300	1.295	2.439
CESIUM BROMIDE				H	0.181	0.069	0.310	0.159	6.300	1.272	2.423
10322	4.2730	195	523.0	S	0.179	0.068	0.312	0.159	6.300	1.265	2.419
				R	0.174	0.066	0.318	0.159	6.300	1.242	2.402
				V	0.177	0.069	0.287	0.139	7.200	1.273	2.332
CESIUM BROMIDE				H	0.170	0.066	0.296	0.139	7.200	1.243	2.310
10323	4.2410	195	573.0	S	0.168	0.065	0.299	0.139	7.200	1.234	2.303
				R	0.161	0.062	0.307	0.139	7.200	1.204	2.282
				V	0.165	0.064	0.285	0.128	7.800	1.235	2.251
CESIUM BROMIDE				H	0.158	0.061	0.295	0.128	7.800	1.200	2.226
10324	4.2240	195	623.0	S	0.155	0.060	0.299	0.128	7.800	1.188	2.218
				R	0.147	0.056	0.308	0.128	7.800	1.155	2.194
				V	0.157	0.061	0.288	0.123	8.100	1.207	2.213
CESIUM BROMIDE				H	0.148	0.057	0.300	0.123	8.100	1.166	2.183
10325	4.1800	195	673.0	S	0.144	0.055	0.305	0.123	8.100	1.150	2.172
				R	0.136	0.052	0.317	0.123	8.100	1.111	2.145
				V	0.146	0.057	0.288	0.115	8.700	1.173	2.150
CESIUM BROMIDE				H	0.136	0.052	0.303	0.115	8.700	1.125	2.115
10326	4.1230	195	723.0	S	0.131	0.050	0.309	0.115	8.700	1.103	2.100
				R	0.122	0.046	0.323	0.115	8.700	1.059	2.069
				V	0.271	0.108	0.254	0.184	5.448	1.523	2.653
CESIUM BROMIDE				H	0.270	0.108	0.255	0.184	5.448	1.520	2.650
10327	4.6550	258	4.2	S	0.270	0.108	0.255	0.184	5.448	1.520	2.650
				R	0.269	0.107	0.256	0.184	5.448	1.517	2.648
				V	0.262	0.105	0.252	0.176	5.685	1.504	2.610
CESIUM BROMIDE				H	0.261	0.104	0.253	0.176	5.685	1.499	2.607
10328	4.6290	258	77.0	S	0.261	0.104	0.253	0.176	5.685	1.499	2.607
				R	0.259	0.103	0.254	0.176	5.685	1.495	2.603
				V	0.224	0.089	0.260	0.156	6.421	1.410	2.477
CESIUM BROMIDE				H	0.220	0.087	0.264	0.156	6.421	1.395	2.465
10329	4.4720	258	300.0	S	0.219	0.087	0.265	0.156	6.421	1.392	2.463
				R	0.215	0.085	0.270	0.156	6.421	1.377	2.451
				V	0.227	0.090	0.258	0.156	6.414	1.422	2.489
CESIUM BROMIDE				H	0.223	0.088	0.261	0.156	6.414	1.409	2.479
10330	4.4560	234	298.0	S	0.223	0.088	0.262	0.156	6.414	1.407	2.478
				R	0.219	0.087	0.266	0.156	6.414	1.394	2.468

CUBIC

NAME / CODE	DENSITY	REF	TEMP(K)		E	G	SIGMA	K	BETA	VS	VP
				V	0.263	0.105	0.255	0.179	5.575	1.507	2.630
CESIUM BROMIDE				H	0.262	0.104	0.256	0.179	5.575	1.505	2.628
10331	4.6124	213	77.0	S	0.262	0.104	0.256	0.179	5.575	1.504	2.628
				R	0.261	0.104	0.257	0.179	5.575	1.502	2.626
				V	0.227	0.090	0.262	0.159	6.285	1.420	2.502
CESIUM BROMIDE				H	0.223	0.088	0.266	0.159	6.285	1.407	2.492
10332	4.4560	213	300.0	S	0.223	0.088	0.267	0.159	6.285	1.405	2.490
				R	0.219	0.086	0.270	0.159	6.285	1.392	2.480
				V	0.228	0.091	0.259	0.158	6.341	1.426	2.500
CESIUM BROMIDE				H	0.225	0.089	0.262	0.158	6.341	1.414	2.491
10333	4.4560	55	286.0	S	0.224	0.089	0.263	0.158	6.341	1.412	2.489
				R	0.221	0.087	0.266	0.158	6.341	1.400	2.480
				V	0.259	0.102	0.264	0.183	5.474	1.602	2.828
CESIUM CHLORIDE				H	0.252	0.099	0.270	0.183	5.474	1.578	2.811
10334	3.9900	125	293.0	S	0.251	0.099	0.271	0.183	5.474	1.573	2.806
				R	0.245	0.096	0.277	0.183	5.474	1.549	2.789
				V	0.262	0.104	0.259	0.181	5.527	1.614	2.830
CESIUM CHLORIDE				H	0.255	0.101	0.265	0.181	5.527	1.588	2.811
10335	3.9880	234	298.0	S	0.253	0.100	0.267	0.181	5.527	1.582	2.806
				R	0.246	0.097	0.273	0.181	5.527	1.557	2.788
				V	0.264	0.105	0.259	0.182	5.485	1.621	2.842
CESIUM CHLORIDE				H	0.257	0.102	0.265	0.182	5.485	1.597	2.823
10336	3.9880	55	286.0	S	0.256	0.101	0.266	0.182	5.485	1.591	2.819
				R	0.249	0.098	0.272	0.182	5.485	1.567	2.801
				V	0.182	0.072	0.265	0.129	7.752	1.262	2.231
CESIUM IODIDE				H	0.180	0.071	0.267	0.129	7.752	1.254	2.225
10337	4.5200	125	293.0	S	0.180	0.071	0.268	0.129	7.752	1.253	2.224
				R	0.178	0.070	0.270	0.129	7.752	1.245	2.218
				V	0.184	0.073	0.258	0.127	7.895	1.274	2.230
CESIUM IODIDE				H	0.182	0.072	0.261	0.127	7.895	1.265	2.224
10338	4.5090	41	300.0	S	0.182	0.072	0.261	0.127	7.895	1.264	2.222
				R	0.179	0.071	0.264	0.127	7.895	1.255	2.216
				V	0.215	0.087	0.241	0.139	7.209	1.365	2.338
CESIUM IODIDE				H	0.215	0.086	0.242	0.139	7.209	1.363	2.336
10339	4.6525	216	80.0	S	0.215	0.086	0.242	0.139	7.209	1.363	2.336
				R	0.214	0.086	0.243	0.139	7.209	1.360	2.334
				V	0.212	0.086	0.242	0.137	7.287	1.358	2.327
CESIUM IODIDE				H	0.212	0.085	0.243	0.137	7.287	1.355	2.325
10340	4.6393	216	100.0	S	0.212	0.085	0.243	0.137	7.287	1.355	2.325
				R	0.211	0.085	0.244	0.137	7.287	1.352	2.323
				V	0.210	0.084	0.243	0.136	7.360	1.350	2.317
CESIUM IODIDE				H	0.209	0.084	0.244	0.136	7.360	1.347	2.314
10341	4.6261	216	120.0	S	0.209	0.084	0.244	0.136	7.360	1.347	2.314
				R	0.208	0.084	0.245	0.136	7.360	1.344	2.312
				V	0.204	0.081	0.259	0.141	7.086	1.326	2.325
CESIUM IODIDE				H	0.204	0.081	0.259	0.141	7.086	1.324	2.323
10342	4.6129	216	140.0	S	0.204	0.081	0.259	0.141	7.086	1.324	2.323
				R	0.203	0.081	0.260	0.141	7.086	1.322	2.322
				V	0.204	0.082	0.244	0.133	7.520	1.335	2.295
CESIUM IODIDE				H	0.203	0.081	0.246	0.133	7.520	1.330	2.292
10343	4.5997	216	160.0	S	0.203	0.081	0.246	0.133	7.520	1.330	2.291
				R	0.202	0.081	0.247	0.133	7.520	1.326	2.288
				V	0.201	0.081	0.245	0.132	7.599	1.327	2.284
CESIUM IODIDE				H	0.200	0.080	0.247	0.132	7.599	1.322	2.280
10344	4.5864	216	180.0	S	0.200	0.080	0.247	0.132	7.599	1.322	2.280
				R	0.199	0.080	0.248	0.132	7.599	1.317	2.276
				V	0.198	0.080	0.246	0.130	7.683	1.319	2.273
CESIUM IODIDE				H	0.197	0.079	0.248	0.130	7.683	1.314	2.269
10345	4.5732	216	200.0	S	0.197	0.079	0.248	0.130	7.683	1.313	2.268
				R	0.195	0.078	0.250	0.130	7.683	1.308	2.264

CUBIC

CODE	NAME DENSITY	REF	TEMP(K)		E	G	SIGMA	K	BETA	VS	VP
				V	0.195	0.078	0.247	0.129	7.768	1.311	2.261
CESIUM IODIDE				H	0.194	0.078	0.249	0.129	7.768	1.305	2.257
10346	4.5600	216	220.0	S	0.194	0.078	0.249	0.129	7.768	1.304	2.256
				R	0.192	0.077	0.251	0.129	7.768	1.298	2.252
				V	0.193	0.077	0.248	0.127	7.848	1.302	2.250
CESIUM IODIDE				H	0.191	0.076	0.250	0.127	7.848	1.296	2.245
10347	4.5468	216	240.0	S	0.191	0.076	0.251	0.127	7.848	1.295	2.245
				R	0.189	0.075	0.253	0.127	7.848	1.288	2.240
				V	0.190	0.076	0.249	0.126	7.941	1.294	2.238
CESIUM IODIDE				H	0.188	0.075	0.251	0.126	7.941	1.287	2.233
10348	4.5336	216	260.0	S	0.188	0.075	0.252	0.126	7.941	1.286	2.232
				R	0.186	0.074	0.254	0.126	7.941	1.278	2.226
				V	0.187	0.075	0.250	0.125	8.028	1.286	2.227
CESIUM IODIDE				H	0.185	0.074	0.253	0.125	8.028	1.278	2.221
10349	4.5204	216	280.0	S	0.185	0.074	0.253	0.125	8.028	1.276	2.220
				R	0.183	0.073	0.256	0.125	8.028	1.268	2.214
				V	0.185	0.074	0.251	0.124	8.095	1.279	2.217
CESIUM IODIDE				H	0.183	0.073	0.254	0.124	8.095	1.270	2.211
10350	4.5155	216	295.0	S	0.182	0.073	0.254	0.124	8.095	1.269	2.210
				R	0.180	0.072	0.257	0.124	8.095	1.260	2.203
				V	0.220	0.088	0.245	0.144	6.940	1.370	2.358
CESIUM IODIDE				H	0.220	0.088	0.246	0.144	6.940	1.367	2.356
10351	4.7120	258	4.2	S	0.220	0.088	0.246	0.144	6.940	1.367	2.356
				R	0.219	0.088	0.247	0.144	6.940	1.365	2.354
				V	0.213	0.085	0.247	0.141	7.109	1.350	2.329
CESIUM IODIDE				H	0.213	0.085	0.248	0.141	7.109	1.347	2.327
10352	4.6930	258	77.0	S	0.212	0.085	0.248	0.141	7.109	1.347	2.327
				R	0.212	0.085	0.249	0.141	7.109	1.344	2.325
				V	0.185	0.074	0.253	0.125	7.998	1.277	2.220
CESIUM IODIDE				H	0.183	0.073	0.256	0.125	7.998	1.267	2.213
10353	4.5370	258	300.0	S	0.183	0.073	0.256	0.125	7.998	1.266	2.212
				R	0.180	0.072	0.259	0.125	7.998	1.257	2.205
				V	0.184	0.073	0.255	0.126	7.962	1.274	2.223
CESIUM IODIDE				H	0.182	0.072	0.258	0.126	7.962	1.265	2.216
10354	4.5250	234	298.0	S	0.182	0.072	0.259	0.126	7.962	1.264	2.215
				R	0.180	0.071	0.261	0.126	7.962	1.255	2.208
				V	0.187	0.075	0.252	0.126	7.937	1.285	2.233
CESIUM IODIDE				H	0.185	0.074	0.255	0.126	7.937	1.277	2.227
10355	4.5250	55	286.0	S	0.185	0.074	0.255	0.126	7.937	1.276	2.226
				R	0.183	0.073	0.258	0.126	7.937	1.268	2.220
				V	0.226	0.092	0.235	0.142	7.044	1.394	2.367
CESIUM IODIDE				H	0.226	0.091	0.235	0.142	7.044	1.393	2.366
10356	4.7120	174	4.2	S	0.226	0.091	0.235	0.142	7.044	1.393	2.366
				R	0.225	0.091	0.235	0.142	7.044	1.392	2.365
				V	2.704	1.058	0.278	2.033	0.492	4.876	8.797
CHROMITE				H	2.687	1.050	0.280	2.033	0.492	4.857	8.783
10357	4.4500	129	ROOM	S	2.685	1.049	0.280	2.033	0.492	4.855	8.781
				R	2.665	1.040	0.281	2.033	0.492	4.834	8.766
				V	3.019	1.222	0.235	1.901	0.526	4.113	6.991
CHROMIUM				H	2.980	1.203	0.239	1.901	0.526	4.081	6.965
10358	7.2230	40	77.0	S	2.973	1.200	0.239	1.901	0.526	4.075	6.961
				R	2.935	1.181	0.243	1.901	0.526	4.043	6.936
				V	2.827	1.169	0.209	1.619	0.618	4.030	6.643
CHROMIUM				H	2.798	1.154	0.212	1.619	0.618	4.004	6.623
10359	7.2000	40	298.0	S	2.794	1.152	0.212	1.619	0.618	4.000	6.620
				R	2.766	1.138	0.215	1.619	0.618	3.976	6.600
				V	2.767	1.132	0.222	1.661	0.602	3.975	6.653
CHROMIUM				H	2.742	1.119	0.225	1.661	0.602	3.953	6.635
10360	7.1630	40	500.0	S	2.739	1.118	0.225	1.661	0.602	3.950	6.633
				R	2.715	1.106	0.228	1.661	0.602	3.929	6.616

CUBIC

CODE	NAME / DENSITY	REF	TEMP(K)		E	G	SIGMA	K	BETA	VS	VP
				V	2.769	1.156	0.197	1.523	0.656	4.008	6.525
CHROMIUM				H	2.738	1.140	0.200	1.523	0.656	3.980	6.502
10361	7.2000	247	298.0	S	2.733	1.138	0.201	1.523	0.656	3.976	6.499
				R	2.704	1.123	0.204	1.523	0.656	3.949	6.477
				V	1.497	0.552	0.357	1.739	0.575	2.927	6.199
COBALT OXIDE				H	1.299	0.472	0.376	1.739	0.575	2.708	6.065
10362	6.4400	12	253.0	S	1.081	0.387	0.396	1.739	0.575	2.452	5.918
				R	0.860	0.303	0.418	1.739	0.575	2.170	5.769
				V	1.480	0.544	0.360	1.758	0.569	2.907	6.210
COBALT OXIDE				H	1.275	0.462	0.379	1.758	0.569	2.680	6.072
10363	6.4400	12	263.0	S	1.035	0.369	0.402	1.758	0.569	2.395	5.911
				R	0.810	0.284	0.423	1.758	0.569	2.102	5.761
				V	1.478	0.543	0.361	1.771	0.565	2.904	6.224
COBALT OXIDE				H	1.277	0.463	0.380	1.771	0.565	2.681	6.089
10364	6.4400	12	273.0	S	1.046	0.373	0.402	1.771	0.565	2.407	5.935
				R	0.822	0.289	0.423	1.771	0.565	2.118	5.786
				V	1.495	0.550	0.359	1.772	0.564	2.922	6.237
COBALT OXIDE				H	1.309	0.475	0.377	1.772	0.564	2.716	6.111
10365	6.4400	12	277.0	S	1.114	0.399	0.395	1.772	0.564	2.490	5.981
				R	0.899	0.318	0.415	1.772	0.564	2.221	5.839
				V	1.503	0.553	0.358	1.767	0.566	2.931	6.237
COBALT OXIDE				H	1.323	0.481	0.375	1.767	0.566	2.733	6.116
10366	6.4400	12	278.0	S	1.143	0.411	0.392	1.767	0.566	2.525	5.996
				R	0.934	0.331	0.412	1.767	0.566	2.266	5.856
				V	1.533	0.566	0.356	1.771	0.565	2.963	6.262
COBALT OXIDE				H	1.375	0.502	0.371	1.771	0.565	2.791	6.155
10367	6.4400	12	281.0	S	1.240	0.448	0.383	1.771	0.565	2.638	6.065
				R	1.050	0.375	0.401	1.771	0.565	2.412	5.938
				V	1.573	0.582	0.352	1.772	0.564	3.005	6.289
COBALT OXIDE				H	1.435	0.526	0.365	1.772	0.564	2.857	6.196
10368	6.4400	12	283.0	S	1.334	0.485	0.374	1.772	0.564	2.745	6.129
				R	1.166	0.419	0.390	1.772	0.564	2.552	6.016
				V	1.635	0.607	0.346	1.775	0.563	3.071	6.336
COBALT OXIDE				H	1.522	0.561	0.357	1.775	0.563	2.950	6.259
10369	6.4400	12	285.0	S	1.454	0.533	0.363	1.775	0.563	2.878	6.214
				R	1.315	0.477	0.377	1.775	0.563	2.723	6.120
				V	1.731	0.646	0.340	1.801	0.555	3.167	6.430
COBALT OXIDE				H	1.653	0.613	0.347	1.801	0.555	3.086	6.377
10370	6.4400	12	287.0	S	1.620	0.600	0.350	1.801	0.555	3.052	6.355
				R	1.524	0.561	0.359	1.801	0.555	2.951	6.291
				V	1.867	0.704	0.327	1.797	0.557	3.305	6.517
COBALT OXIDE				H	1.832	0.689	0.330	1.797	0.557	3.270	6.493
10371	6.4400	12	289.0	S	1.824	0.685	0.331	1.797	0.557	3.262	6.488
				R	1.782	0.668	0.335	1.797	0.556	3.220	6.460
				V	1.863	0.699	0.333	1.855	0.539	3.294	6.578
COBALT OXIDE				H	1.825	0.683	0.336	1.855	0.539	3.256	6.553
10372	6.4400	12	290.0	S	1.815	0.679	0.337	1.855	0.539	3.247	6.547
				R	1.769	0.660	0.341	1.855	0.539	3.200	6.516
				V	1.886	0.709	0.330	1.854	0.539	3.318	6.593
COBALT OXIDE				H	1.854	0.695	0.333	1.854	0.539	3.286	6.572
10373	6.4400	12	291.0	S	1.847	0.692	0.334	1.854	0.539	3.279	6.567
				R	1.808	0.676	0.337	1.854	0.539	3.240	6.541
				V	1.908	0.718	0.329	1.855	0.539	3.339	6.608
COBALT OXIDE				H	1.880	0.706	0.331	1.855	0.539	3.312	6.590
10374	6.4400	12	293.0	S	1.875	0.704	0.332	1.855	0.539	3.306	6.587
				R	1.842	0.690	0.334	1.855	0.539	3.274	6.565
				V	1.927	0.726	0.326	1.851	0.540	3.359	6.617
COBALT OXIDE				H	1.904	0.716	0.329	1.851	0.540	3.335	6.601
10375	6.4400	12	298.0	S	1.899	0.715	0.329	1.851	0.540	3.331	6.598
				R	1.871	0.703	0.331	1.851	0.540	3.303	6.579

CUBIC

CODE	NAME DENSITY	REF	TEMP(K)		E	G	SIGMA	K	BETA	VS	VP
				V	1.940	0.733	0.324	1.839	0.544	3.373	6.612
COBALT OXIDE				H	1.918	0.723	0.326	1.839	0.544	3.351	6.598
10376	6.4400	12	303.0	S	1.915	0.722	0.326	1.839	0.544	3.348	6.595
				R	1.889	0.711	0.329	1.839	0.544	3.322	6.578
				V	1.857	0.694	0.338	1.907	0.524	3.575	7.222
COBALT ZINC FERRITE				H	1.839	0.687	0.339	1.907	0.524	3.556	7.209
10377	5.4300	129	303.0	S	1.837	0.686	0.339	1.907	0.524	3.553	7.208
				R	1.816	0.677	0.341	1.907	0.524	3.531	7.193
				V	1.104	0.397	0.392	1.711	0.585	2.150	5.110
COLUMBIUM				H	1.060	0.379	0.397	1.711	0.585	2.103	5.083
10378	8.5780	119	298.0	S	1.044	0.373	0.398	1.711	0.585	2.086	5.074
				R	0.994	0.354	0.403	1.711	0.585	2.033	5.045
				V	1.106	0.397	0.392	1.710	0.585	2.151	5.109
COLUMBIUM				H	1.061	0.380	0.397	1.710	0.585	2.105	5.083
10379	8.5780	119	298.0	S	1.045	0.374	0.398	1.710	0.585	2.088	5.074
				R	0.996	0.355	0.403	1.710	0.585	2.034	5.045
				V	1.277	0.476	0.340	1.333	0.750	2.311	4.698
COPPER				H	1.218	0.452	0.348	1.333	0.750	2.250	4.658
10380	8.9200	128	ROOM	S	1.192	0.441	0.351	1.333	0.750	2.224	4.641
				R	1.119	0.411	0.360	1.333	0.750	2.147	4.593
				V	1.454	0.545	0.333	1.449	0.690	2.473	4.940
COPPER				H	1.332	0.494	0.347	1.449	0.690	2.354	4.862
10381	8.9200	128	ROOM	S	1.248	0.460	0.356	1.449	0.690	2.271	4.809
				R	1.101	0.401	0.373	1.449	0.690	2.120	4.716
				V	1.289	0.479	0.346	1.393	0.718	2.316	4.770
COPPER				H	1.198	0.442	0.357	1.393	0.718	2.224	4.711
10382	8.9300	129	ROOM	S	1.143	0.419	0.363	1.393	0.718	2.167	4.676
				R	1.031	0.375	0.377	1.393	0.718	2.048	4.604
				V	1.447	0.546	0.324	1.371	0.730	2.469	4.840
COPPER				H	1.324	0.494	0.339	1.371	0.730	2.349	4.760
10383	8.9600	129	ROOM	S	1.240	0.460	0.349	1.371	0.730	2.265	4.705
				R	1.094	0.400	0.367	1.371	0.730	2.114	4.610
				V	1.448	0.546	0.326	1.383	0.723	2.475	4.865
COPPER				H	1.326	0.495	0.340	1.383	0.723	2.355	4.786
10384	8.9200	4	ROOM	S	1.243	0.460	0.350	1.383	0.723	2.272	4.732
				R	1.098	0.401	0.368	1.383	0.723	2.121	4.638
				V	1.451	0.547	0.327	1.397	0.716	2.473	4.876
COPPER (P= 0KB)				H	1.328	0.495	0.342	1.397	0.716	2.352	4.796
10385	8.9410	149	298.0	S	1.243	0.460	0.352	1.397	0.716	2.268	4.741
				R	1.096	0.400	0.369	1.397	0.716	2.116	4.646
				V	1.456	0.548	0.327	1.406	0.711	2.475	4.885
COPPER (P= 2KB)				H	1.332	0.496	0.342	1.406	0.711	2.355	4.805
10386	8.9542	149	298.0	S	1.248	0.461	0.352	1.406	0.711	2.270	4.751
				R	1.101	0.402	0.369	1.406	0.711	2.118	4.656
				V	1.461	0.550	0.328	1.414	0.707	2.477	4.894
COPPER (P= 4KB)				H	1.337	0.498	0.342	1.414	0.707	2.357	4.815
10387	8.9673	149	298.0	S	1.253	0.463	0.352	1.414	0.707	2.273	4.760
				R	1.105	0.404	0.370	1.414	0.707	2.121	4.666
				V	1.464	0.551	0.329	1.423	0.703	2.477	4.902
COPPER (P= 6KB)				H	1.341	0.499	0.343	1.423	0.703	2.358	4.823
10388	8.9803	149	298.0	S	1.257	0.465	0.353	1.423	0.703	2.274	4.769
				R	1.110	0.405	0.370	1.423	0.703	2.124	4.676
				V	1.468	0.553	0.328	1.425	0.702	2.479	4.903
COPPER (P= 8KB)				H	1.345	0.501	0.343	1.425	0.702	2.360	4.824
10389	8.9933	149	298.0	S	1.262	0.467	0.352	1.425	0.702	2.278	4.771
				R	1.115	0.407	0.370	1.425	0.702	2.128	4.678
				V	1.473	0.554	0.329	1.434	0.697	2.481	4.912
COPPER (P=10KB)				H	1.350	0.503	0.343	1.434	0.697	2.362	4.833
10390	9.0062	149	298.0	S	1.267	0.468	0.353	1.434	0.697	2.281	4.781
				R	1.121	0.409	0.370	1.434	0.697	2.131	4.688

CUBIC

CODE	NAME DENSITY	REF	TEMP(K)		E	G	SIGMA	K	BETA	VS	VP
				V	1.562	0.593	0.317	1.420	0.704	2.563	4.948
COPPER				H	1.430	0.537	0.332	1.420	0.704	2.438	4.863
10391	9.0305	199	0.0	S	1.342	0.500	0.343	1.420	0.704	2.352	4.807
				R	1.186	0.436	0.361	1.420	0.704	2.197	4.708
				V	1.562	0.593	0.317	1.420	0.704	2.563	4.948
COPPER				H	1.430	0.537	0.332	1.420	0.704	2.438	4.863
10392	9.0305	199	10.0	S	1.341	0.500	0.343	1.420	0.704	2.352	4.807
				R	1.186	0.436	0.361	1.420	0.704	2.197	4.708
				V	1.561	0.593	0.317	1.420	0.704	2.562	4.948
COPPER				H	1.430	0.537	0.332	1.420	0.704	2.438	4.863
10393	9.0305	199	20.0	S	1.341	0.499	0.343	1.420	0.704	2.352	4.806
				R	1.186	0.436	0.361	1.420	0.704	2.197	4.707
				V	1.561	0.593	0.317	1.420	0.704	2.562	4.947
COPPER				H	1.429	0.536	0.332	1.420	0.704	2.437	4.863
10394	9.0278	199	30.0	S	1.341	0.499	0.343	1.420	0.704	2.352	4.806
				R	1.186	0.436	0.361	1.420	0.704	2.197	4.707
				V	1.559	0.592	0.317	1.419	0.705	2.561	4.946
COPPER				H	1.428	0.536	0.332	1.419	0.705	2.436	4.861
10395	9.0278	199	40.0	S	1.340	0.499	0.343	1.419	0.705	2.351	4.805
				R	1.185	0.435	0.361	1.419	0.705	2.196	4.706
				V	1.557	0.591	0.317	1.418	0.705	2.560	4.945
COPPER				H	1.426	0.535	0.332	1.418	0.705	2.435	4.860
10396	9.0251	199	50.0	S	1.338	0.498	0.343	1.418	0.705	2.350	4.804
				R	1.184	0.435	0.361	1.418	0.705	2.195	4.705
				V	1.555	0.590	0.317	1.417	0.706	2.558	4.943
COPPER				H	1.424	0.534	0.332	1.417	0.706	2.434	4.858
10397	9.0224	199	60.0	S	1.336	0.498	0.343	1.417	0.706	2.348	4.802
				R	1.182	0.434	0.361	1.417	0.706	2.194	4.704
				V	1.552	0.589	0.317	1.416	0.706	2.556	4.940
COPPER				H	1.422	0.533	0.333	1.416	0.706	2.432	4.856
10398	9.0197	199	70.0	S	1.334	0.497	0.343	1.416	0.706	2.347	4.800
				R	1.180	0.434	0.361	1.416	0.706	2.192	4.702
				V	1.549	0.588	0.317	1.414	0.707	2.553	4.938
COPPER				H	1.419	0.532	0.333	1.414	0.707	2.430	4.854
10399	9.0171	199	80.0	S	1.331	0.496	0.343	1.414	0.707	2.344	4.797
				R	1.178	0.433	0.361	1.414	0.707	2.190	4.699
				V	1.545	0.586	0.318	1.413	0.708	2.551	4.934
COPPER				H	1.416	0.531	0.333	1.413	0.708	2.427	4.851
10400	9.0144	199	90.0	S	1.328	0.494	0.343	1.413	0.708	2.342	4.795
				R	1.175	0.432	0.361	1.413	0.708	2.188	4.697
				V	1.542	0.585	0.318	1.411	0.709	2.548	4.931
COPPER				H	1.412	0.530	0.333	1.411	0.708	2.424	4.847
10401	9.0117	199	100.0	S	1.325	0.493	0.344	1.411	0.708	2.339	4.792
				R	1.172	0.430	0.362	1.411	0.708	2.186	4.694
				V	1.538	0.583	0.318	1.410	0.709	2.544	4.928
COPPER				H	1.408	0.528	0.333	1.410	0.709	2.421	4.844
10402	9.0090	199	110.0	S	1.322	0.492	0.344	1.410	0.709	2.336	4.788
				R	1.169	0.429	0.362	1.410	0.709	2.183	4.691
				V	1.534	0.582	0.318	1.408	0.710	2.541	4.924
COPPER				H	1.405	0.527	0.334	1.408	0.710	2.418	4.841
10403	9.0063	199	120.0	S	1.318	0.490	0.344	1.408	0.710	2.333	4.785
				R	1.166	0.428	0.362	1.408	0.710	2.180	4.687
				V	1.529	0.580	0.319	1.406	0.711	2.538	4.921
COPPER				H	1.401	0.525	0.334	1.406	0.711	2.415	4.838
10404	9.0010	199	130.0	S	1.314	0.489	0.344	1.406	0.711	2.331	4.782
				R	1.163	0.427	0.362	1.406	0.711	2.177	4.685
				V	1.525	0.578	0.319	1.405	0.712	2.535	4.917
COPPER				H	1.397	0.523	0.334	1.405	0.712	2.412	4.834
10405	8.9983	199	140.0	S	1.310	0.487	0.344	1.405	0.712	2.327	4.778
				R	1.159	0.425	0.362	1.405	0.712	2.174	4.681

CUBIC

CODE	NAME DENSITY	REF	TEMP(K)		E	G	SIGMA	K	BETA	VS	VP
COPPER 10406	8.9956	199	150.0	V	1.521	0.576	0.319	1.403	0.713	2.531	4.913
				H	1.393	0.522	0.335	1.403	0.713	2.408	4.830
				S	1.307	0.486	0.345	1.403	0.713	2.324	4.774
				R	1.155	0.424	0.363	1.403	0.713	2.171	4.677
COPPER 10407	8.9902	199	160.0	V	1.516	0.574	0.320	1.401	0.714	2.528	4.909
				H	1.389	0.520	0.335	1.401	0.714	2.405	4.826
				S	1.303	0.484	0.345	1.401	0.714	2.321	4.771
				R	1.152	0.422	0.363	1.401	0.714	2.168	4.674
COPPER 10408	8.9875	199	170.0	V	1.512	0.573	0.320	1.399	0.715	2.524	4.905
				H	1.384	0.518	0.335	1.399	0.715	2.402	4.822
				S	1.298	0.483	0.345	1.399	0.715	2.317	4.767
				R	1.148	0.421	0.363	1.399	0.715	2.164	4.670
COPPER 10409	8.9849	199	180.0	V	1.507	0.571	0.320	1.397	0.716	2.520	4.901
				H	1.380	0.517	0.335	1.397	0.716	2.398	4.818
				S	1.294	0.481	0.346	1.397	0.716	2.314	4.763
				R	1.144	0.420	0.363	1.397	0.716	2.161	4.666
COPPER 10410	8.9795	199	190.0	V	1.502	0.569	0.321	1.395	0.717	2.517	4.897
				H	1.376	0.515	0.336	1.395	0.717	2.395	4.814
				S	1.290	0.479	0.346	1.395	0.717	2.310	4.759
				R	1.140	0.418	0.364	1.395	0.717	2.157	4.663
COPPER 10411	8.9768	199	200.0	V	1.498	0.567	0.321	1.393	0.718	2.513	4.892
				H	1.371	0.513	0.336	1.393	0.718	2.391	4.810
				S	1.286	0.478	0.346	1.393	0.718	2.306	4.755
				R	1.136	0.416	0.364	1.393	0.718	2.154	4.658
COPPER 10412	8.9714	199	210.0	V	1.493	0.565	0.321	1.391	0.719	2.509	4.888
				H	1.367	0.511	0.336	1.391	0.719	2.388	4.806
				S	1.281	0.476	0.346	1.391	0.719	2.303	4.751
				R	1.132	0.415	0.364	1.391	0.719	2.151	4.655
COPPER 10413	8.9661	199	220.0	V	1.488	0.563	0.321	1.389	0.720	2.506	4.884
				H	1.362	0.510	0.337	1.388	0.720	2.384	4.802
				S	1.277	0.474	0.347	1.388	0.720	2.299	4.747
				R	1.128	0.413	0.365	1.388	0.720	2.147	4.651
COPPER 10414	8.9634	199	230.0	V	1.483	0.561	0.322	1.386	0.721	2.502	4.880
				H	1.357	0.508	0.337	1.386	0.721	2.380	4.798
				S	1.272	0.472	0.347	1.386	0.721	2.295	4.742
				R	1.124	0.412	0.365	1.386	0.721	2.143	4.646
COPPER 10415	8.9580	199	240.0	V	1.478	0.559	0.322	1.384	0.722	2.498	4.875
				H	1.353	0.506	0.337	1.384	0.722	2.376	4.794
				S	1.268	0.470	0.347	1.384	0.722	2.292	4.739
				R	1.119	0.410	0.365	1.384	0.722	2.139	4.643
COPPER 10416	8.9553	199	250.0	V	1.473	0.557	0.322	1.382	0.724	2.494	4.870
				H	1.348	0.504	0.337	1.382	0.724	2.372	4.789
				S	1.263	0.469	0.348	1.382	0.724	2.288	4.734
				R	1.115	0.408	0.365	1.382	0.724	2.136	4.638
COPPER 10417	8.9500	199	260.0	V	1.468	0.555	0.323	1.380	0.725	2.490	4.866
				H	1.343	0.502	0.338	1.380	0.725	2.368	4.785
				S	1.259	0.467	0.348	1.380	0.725	2.284	4.730
				R	1.111	0.407	0.366	1.380	0.725	2.132	4.634
COPPER 10418	8.9446	199	270.0	V	1.463	0.553	0.323	1.377	0.726	2.486	4.862
				H	1.339	0.500	0.338	1.377	0.726	2.365	4.781
				S	1.254	0.465	0.348	1.377	0.726	2.280	4.726
				R	1.107	0.405	0.366	1.378	0.726	2.128	4.630
COPPER 10419	8.9419	199	280.0	V	1.457	0.551	0.323	1.375	0.727	2.481	4.857
				H	1.334	0.498	0.338	1.375	0.727	2.360	4.776
				S	1.249	0.463	0.349	1.375	0.727	2.276	4.721
				R	1.103	0.403	0.366	1.375	0.727	2.124	4.626
COPPER 10420	8.9366	199	290.0	V	1.452	0.548	0.324	1.373	0.728	2.477	4.852
				H	1.329	0.496	0.339	1.373	0.728	2.357	4.772
				S	1.245	0.461	0.349	1.373	0.728	2.272	4.717
				R	1.098	0.402	0.367	1.373	0.728	2.120	4.622

CUBIC

CODE	NAME DENSITY	REF	TEMP(K)		E	G	SIGMA	K	BETA	VS	VP
				V	1.447	0.546	0.324	1.371	0.730	2.473	4.848
COPPER				H	1.324	0.494	0.339	1.371	0.730	2.353	4.767
10421	8.9312	199	300.0	S	1.240	0.459	0.349	1.371	0.730	2.268	4.712
				R	1.094	0.400	0.367	1.371	0.730	2.117	4.618
				V	1.441	0.544	0.325	1.370	0.730	2.460	4.828
COPPER				H	1.318	0.492	0.340	1.370	0.730	2.340	4.748
10422	8.9883	226	300.0	S	1.234	0.457	0.350	1.370	0.730	2.255	4.693
				R	1.088	0.398	0.368	1.370	0.730	2.104	4.598
				V	1.452	0.549	0.323	1.368	0.731	2.478	4.849
COPPER				H	1.329	0.496	0.338	1.368	0.731	2.357	4.768
10423	8.9320	85	298.0	S	1.245	0.462	0.348	1.368	0.731	2.273	4.713
				R	1.098	0.402	0.366	1.368	0.731	2.121	4.617
				V	1.448	0.546	0.324	1.374	0.728	2.474	4.852
COPPER				H	1.325	0.495	0.339	1.374	0.728	2.353	4.772
10424	8.9320	251	300.0	S	1.242	0.460	0.349	1.374	0.728	2.269	4.717
				R	1.096	0.401	0.367	1.374	0.728	2.119	4.622
				V	1.456	0.550	0.325	1.383	0.723	2.481	4.867
COPPER				H	1.334	0.498	0.339	1.383	0.723	2.361	4.787
10425	8.9340	51	300.0	S	1.250	0.463	0.349	1.383	0.723	2.277	4.733
				R	1.104	0.404	0.367	1.383	0.723	2.126	4.638
				V	1.429	0.539	0.326	1.368	0.731	2.459	4.839
COPPER				H	1.307	0.487	0.341	1.368	0.731	2.338	4.758
10426	8.9130	51	350.0	S	1.222	0.452	0.351	1.368	0.731	2.252	4.703
				R	1.076	0.393	0.369	1.368	0.731	2.100	4.608
				V	1.403	0.529	0.327	1.355	0.738	2.438	4.813
COPPER				H	1.282	0.478	0.342	1.355	0.738	2.318	4.733
10427	8.8920	51	400.0	S	1.198	0.443	0.353	1.355	0.738	2.232	4.678
				R	1.055	0.385	0.370	1.355	0.738	2.080	4.584
				V	1.379	0.519	0.329	1.342	0.745	2.419	4.788
COPPER				H	1.259	0.469	0.344	1.342	0.745	2.299	4.709
10428	8.8680	51	450.0	S	1.176	0.434	0.354	1.342	0.745	2.213	4.654
				R	1.034	0.377	0.372	1.342	0.745	2.061	4.560
				V	1.353	0.508	0.330	1.328	0.753	2.398	4.763
COPPER				H	1.235	0.459	0.345	1.328	0.753	2.278	4.684
10429	8.8440	51	500.0	S	1.152	0.425	0.355	1.328	0.753	2.193	4.629
				R	1.012	0.369	0.373	1.328	0.753	2.042	4.536
				V	1.328	0.499	0.332	1.315	0.760	2.378	4.738
COPPER				H	1.212	0.450	0.346	1.315	0.760	2.259	4.660
10430	8.8200	51	550.0	S	1.130	0.416	0.357	1.315	0.760	2.173	4.605
				R	0.991	0.361	0.374	1.315	0.760	2.022	4.512
				V	1.302	0.488	0.333	1.302	0.768	2.355	4.710
COPPER				H	1.188	0.441	0.348	1.302	0.768	2.237	4.633
10431	8.8000	51	600.0	S	1.106	0.407	0.358	1.302	0.768	2.151	4.578
				R	0.970	0.352	0.376	1.302	0.768	2.001	4.487
				V	1.276	0.478	0.335	1.288	0.776	2.333	4.684
COPPER				H	1.163	0.431	0.350	1.288	0.776	2.216	4.607
10432	8.7760	51	650.0	S	1.083	0.398	0.360	1.288	0.776	2.130	4.553
				R	0.948	0.344	0.377	1.288	0.776	1.981	4.462
				V	1.251	0.468	0.336	1.275	0.784	2.313	4.659
COPPER				H	1.140	0.422	0.351	1.275	0.784	2.196	4.583
10433	8.7500	51	700.0	S	1.060	0.389	0.361	1.275	0.784	2.109	4.528
				R	0.927	0.336	0.379	1.275	0.784	1.960	4.438
				V	1.225	0.458	0.338	1.262	0.793	2.290	4.632
COPPER				H	1.116	0.412	0.353	1.262	0.793	2.174	4.557
10434	8.7240	51	750.0	S	1.036	0.380	0.363	1.262	0.793	2.088	4.503
				R	0.906	0.328	0.380	1.262	0.793	1.939	4.413
				V	1.200	0.448	0.340	1.248	0.801	2.269	4.605
COPPER				H	1.092	0.403	0.354	1.248	0.801	2.153	4.531
10435	8.7010	51	800.0	S	1.014	0.371	0.365	1.248	0.801	2.066	4.476
				R	0.884	0.320	0.382	1.248	0.801	1.918	4.388

CUBIC

CODE	NAME / DENSITY	REF	TEMP(K)		E	G	SIGMA	K	BETA	VS	VP
				V	1.444	0.546	0.322	1.353	0.739	2.473	4.827
COPPER				H	1.318	0.493	0.338	1.353	0.739	2.349	4.744
10436	8.9300	133	ROOM	S	1.231	0.456	0.348	1.353	0.739	2.261	4.687
				R	1.082	0.396	0.367	1.353	0.739	2.106	4.590
				V	0.738	0.253	0.456	2.778	0.360	2.944	10.323
COPPER(5.0)-ALUMINUM(95.0)				H	0.736	0.253	0.456	2.778	0.360	2.940	10.321
10437	2.9240	128	ROOM	S	0.736	0.253	0.456	2.778	0.360	2.940	10.321
				R	0.733	0.252	0.456	2.778	0.360	2.934	10.319
				V	1.426	0.538	0.326	1.363	0.733	2.494	4.905
COPPER(95.19)-ALUMINUM(4.81)				H	1.297	0.483	0.341	1.363	0.733	2.364	4.818
10438	8.6479	192	ROOM	S	1.203	0.445	0.353	1.363	0.733	2.268	4.756
				R	1.051	0.383	0.372	1.363	0.733	2.104	4.655
				V	1.433	0.543	0.319	1.316	0.760	2.560	4.961
COPPER(90.02)-ALUMINUM(9.98)				H	1.291	0.483	0.337	1.316	0.760	2.414	4.862
10439	8.2916	192	ROOM	S	1.181	0.437	0.350	1.316	0.760	2.297	4.786
				R	1.017	0.371	0.371	1.316	0.760	2.115	4.673
				V	1.423	0.537	0.324	1.345	0.744	2.447	4.793
COPPER(98.42)-GALLIUM(1.58)				H	1.300	0.486	0.339	1.345	0.744	2.326	4.712
10440	8.9730	192	ROOM	S	1.216	0.450	0.349	1.345	0.744	2.241	4.656
				R	1.071	0.392	0.367	1.345	0.744	2.089	4.561
				V	1.414	0.533	0.326	1.357	0.737	2.442	4.809
COPPER(95.85)-GALLIUM(4.15)				H	1.288	0.480	0.342	1.357	0.737	2.316	4.726
10441	8.9403	192	ROOM	S	1.197	0.442	0.353	1.357	0.737	2.224	4.667
				R	1.047	0.382	0.371	1.357	0.737	2.066	4.569
				V	1.434	0.541	0.325	1.362	0.734	2.455	4.817
COPPER(98.97)-GERMANIUM(1.03)				H	1.308	0.488	0.340	1.362	0.734	2.332	4.735
10442	8.9781	192	ROOM	S	1.220	0.452	0.351	1.362	0.734	2.243	4.678
				R	1.072	0.391	0.369	1.362	0.734	2.088	4.581
				V	1.427	0.540	0.321	1.332	0.751	2.453	4.782
COPPER(98.29)-GERMANIUM(1.71)				H	1.300	0.486	0.337	1.332	0.751	2.327	4.698
10443	8.9700	192	ROOM	S	1.209	0.448	0.349	1.332	0.751	2.236	4.638
				R	1.059	0.387	0.367	1.332	0.751	2.078	4.539
				V	1.361	0.503	0.354	1.558	0.642	2.029	4.271
COPPER(75.08)-GOLD(24.92)				H	1.280	0.469	0.363	1.558	0.642	1.961	4.228
10444	12.2132	232	293.0	S	1.236	0.452	0.368	1.558	0.642	1.923	4.206
				R	1.134	0.411	0.379	1.558	0.642	1.835	4.152
				V	1.331	0.490	0.358	1.558	0.642	2.007	4.263
COPPER(75.08)-GOLD(24.92)				H	1.251	0.458	0.366	1.558	0.642	1.940	4.221
10445	12.1656	232	373.0	S	1.207	0.440	0.371	1.558	0.642	1.903	4.199
				R	1.106	0.400	0.382	1.558	0.642	1.814	4.146
				V	1.291	0.474	0.361	1.543	0.648	1.979	4.239
COPPER(75.08)-GOLD(24.92)				H	1.212	0.442	0.369	1.543	0.648	1.912	4.198
10446	12.1069	232	473.0	S	1.168	0.425	0.374	1.543	0.648	1.874	4.175
				R	1.068	0.386	0.385	1.543	0.648	1.785	4.122
				V	1.269	0.466	0.362	1.529	0.654	1.965	4.220
COPPER(75.08)-GOLD(24.92)				H	1.191	0.434	0.370	1.529	0.654	1.897	4.179
10447	12.0740	232	523.0	S	1.147	0.417	0.375	1.529	0.654	1.859	4.156
				R	1.048	0.378	0.386	1.529	0.654	1.769	4.103
				V	1.240	0.455	0.364	1.515	0.660	1.943	4.198
COPPER(75.08)-GOLD(24.92)				H	1.162	0.424	0.372	1.515	0.660	1.876	4.156
10448	12.0410	232	573.0	S	1.119	0.406	0.377	1.515	0.660	1.837	4.133
				R	1.020	0.368	0.388	1.515	0.660	1.747	4.081
				V	1.223	0.448	0.365	1.515	0.660	1.930	4.192
COPPER(75.08)-GOLD(24.92)				H	1.145	0.417	0.374	1.515	0.660	1.862	4.150
10449	12.0239	232	598.0	S	1.100	0.399	0.379	1.515	0.660	1.822	4.126
				R	1.001	0.360	0.390	1.515	0.660	1.731	4.074
				V	1.205	0.441	0.366	1.502	0.666	1.917	4.173
COPPER(75.08)-GOLD(24.92)				H	1.126	0.410	0.375	1.502	0.666	1.847	4.130
10450	12.0019	232	623.0	S	1.080	0.391	0.380	1.502	0.666	1.806	4.106
				R	0.980	0.352	0.391	1.502	0.666	1.713	4.053

CUBIC

CODE	NAME / DENSITY	REF	TEMP(K)		E	G	SIGMA	K	BETA	VS	VP
				V	1.196	0.438	0.365	1.481	0.675	1.911	4.150
COPPER(75.C8)-GCLD(24.92)				H	1.116	0.406	0.374	1.481	0.675	1.840	4.107
10451	11.9909	232	633.C	S	1.069	0.388	0.380	1.481	0.675	1.798	4.082
				R	0.969	0.348	0.391	1.481	0.675	1.704	4.028
				V	1.187	0.434	0.366	1.475	0.678	1.904	4.141
CCPPER(75.C8)-GCLC(24.92)				H	1.107	0.402	0.375	1.475	0.678	1.833	4.098
10452	11.S763	232	643.C	S	1.059	0.384	0.380	1.475	0.678	1.790	4.072
				R	0.958	0.344	0.392	1.475	0.678	1.695	4.018
				V	1.176	0.431	0.365	1.449	0.690	1.898	4.114
CCPPER(75.C8)-GCLD(24.92)				H	1.095	0.398	0.374	1.449	0.690	1.826	4.070
10453	11.9555	232	653.C	S	1.046	0.379	0.380	1.449	C.690	1.781	4.044
				R	0.945	0.340	0.391	1.449	0.690	1.685	3.989
				V	1.167	0.428	0.364	1.431	0.699	1.893	4.094
COPPER(75.C8)-GOLD(24.92)				H	1.086	0.395	0.374	1.431	0.699	1.820	4.050
10454	11.9372	232	658.0	S	1.036	0.375	0.379	1.431	0.699	1.774	4.022
				R	0.933	0.335	0.391	1.431	0.699	1.676	3.966
				V	1.156	0.424	0.362	1.401	0.714	1.886	4.059
COPPER(75.C8)-GOLD(24.92)				H	1.075	0.392	0.372	1.401	0.714	1.812	4.014
10455	11.9335	232	660.5	S	1.025	0.372	0.378	1.401	0.714	1.766	3.987
				R	0.923	0.332	0.390	1.401	0.714	1.668	3.930
				V	1.111	0.407	0.364	1.366	0.732	1.847	4.000
COPPER(75.C8)-GOLD(24.92)				H	1.020	0.371	0.376	1.366	0.732	1.763	3.949
10456	11.9335	232	66C.5	S	0.957	0.346	0.383	1.366	0.732	1.702	3.913
				R	0.844	0.302	0.397	1.366	0.732	1.591	3.850
				V	1.108	0.406	0.364	1.355	0.738	1.845	3.987
COPPER(75.08)-GOLD(24.92)				H	1.017	0.370	0.375	1.355	0.738	1.761	3.936
10457	11.9299	232	663.C	S	0.953	0.345	0.383	1.355	0.738	1.700	3.900
				R	0.840	0.301	0.397	1.355	0.738	1.588	3.837
				V	1.104	0.404	0.365	1.366	0.732	1.842	3.997
COPPER(75.08)-GOLD(24.92)				H	1.014	0.368	0.376	1.366	0.732	1.757	3.946
10458	11.9262	232	668.0	S	0.949	0.343	0.384	1.366	0.732	1.696	3.910
				R	0.836	0.299	0.398	1.366	0.732	1.583	3.847
				V	1.101	0.403	0.365	1.361	0.735	1.839	3.990
COPPER(75.C8)-GOLD(24.92)				H	1.010	0.367	0.376	1.361	0.735	1.754	3.939
10459	11.9225	232	673.0	S	0.945	0.342	0.384	1.361	0.735	1.692	3.903
				R	0.832	0.298	0.398	1.361	0.735	1.580	3.839
				V	1.094	0.400	0.366	1.361	0.735	1.833	3.987
COPPER(75.C8)-GOLD(24.92)				H	1.003	0.364	0.377	1.361	0.735	1.748	3.936
10460	11.9152	232	683.0	S	0.938	0.339	0.385	1.361	0.735	1.686	3.900
				R	0.825	0.295	0.399	1.361	0.735	1.573	3.836
				V	1.088	0.398	0.367	1.361	0.735	1.828	3.985
COPPER(75.C8)-GCLD(24.92)				H	0.997	0.362	0.378	1.361	0.735	1.743	3.934
10461	11.9079	232	6S3.C	S	0.932	0.336	0.386	1.361	0.735	1.680	3.897
				R	0.818	0.292	0.400	1.361	0.735	1.567	3.834
				V	1.070	0.391	0.367	1.339	0.747	1.814	3.956
COPPER(75.08)-GOLD(24.92)				H	0.979	0.355	0.378	1.339	0.747	1.729	3.905
10462	11.E871	232	723.C	S	0.914	0.329	0.386	1.339	0.747	1.665	3.867
				R	0.801	0.286	0.400	1.339	0.747	1.551	3.804
				V	1.486	0.556	0.336	1.506	0.664	2.124	4.269
COPPER(75.C)-GOLD(25.0) (DISORDERED)				H	1.385	0.514	0.347	1.506	0.664	2.042	4.216
10463	12.3310	96	4.2	S	1.327	0.490	0.353	1.506	0.664	1.994	4.185
				R	1.204	0.440	0.367	1.506	0.664	1.890	4.120
				V	1.379	0.511	0.349	1.518	0.659	2.046	4.244
COPPER(75.C)-GOLD(25.0) (DISORDERED)				H	1.284	0.472	0.359	1.518	0.659	1.966	4.194
10464	12.2140	96	300.C	S	1.227	0.449	0.365	1.518	0.659	1.918	4.164
				R	1.109	0.402	0.378	1.518	0.659	1.815	4.101
				V	1.487	0.556	0.336	1.510	0.662	2.124	4.274
COPPER(75.C)-GOLD(25.0) (ORDEREC)				H	1.395	0.518	0.346	1.510	0.662	2.050	4.225
10465	12.3310	96	4.2	S	1.347	0.498	0.351	1.510	0.662	2.010	4.199
				R	1.235	0.453	0.364	1.510	0.662	1.916	4.140

CUBIC

NAME / CODE	DENSITY	REF	TEMP(K)		E	G	SIGMA	K	BETA	VS	VP
				V	1.379	0.511	0.349	1.523	0.657	2.046	4.249
COPPER(75.0)-GOLD(25.0) (ORDERED)				H	1.293	0.476	0.359	1.523	0.657	1.974	4.203
10466	12.2140	96	300.0	S	1.245	0.457	0.364	1.523	0.657	1.933	4.178
				R	1.138	0.414	0.375	1.523	0.657	1.840	4.121
				V	1.463	0.553	0.323	1.377	0.726	2.480	4.850
COPPER(97.66)-NICKEL(2.34)				H	1.338	0.500	0.338	1.377	0.726	2.358	4.768
10467	8.9886	226	300.0	S	1.253	0.465	0.348	1.377	0.726	2.273	4.713
				R	1.105	0.404	0.366	1.377	0.726	2.121	4.617
				V	1.469	0.555	0.322	1.377	0.726	2.486	4.854
COPPER(96.98)-NICKEL(3.02)				H	1.343	0.502	0.337	1.377	0.726	2.363	4.771
10468	8.9892	226	300.0	S	1.257	0.466	0.348	1.377	0.726	2.277	4.715
				R	1.108	0.406	0.366	1.377	0.726	2.124	4.619
				V	1.480	0.560	0.321	1.380	0.725	2.495	4.861
COPPER(95.51)-NICKEL(4.49)				H	1.355	0.507	0.336	1.380	0.725	2.373	4.779
10469	9.0000	226	300.0	S	1.269	0.471	0.347	1.380	0.725	2.288	4.724
				R	1.120	0.410	0.365	1.380	0.725	2.135	4.627
				V	1.495	0.566	0.320	1.386	0.722	2.509	4.880
COPPER(93.96)-NICKEL(6.04)				H	1.368	0.512	0.336	1.386	0.722	2.387	4.797
10470	8.9894	226	300.0	S	1.281	0.476	0.346	1.386	0.722	2.301	4.741
				R	1.131	0.415	0.364	1.386	0.722	2.148	4.644
				V	1.514	0.574	0.319	1.392	0.718	2.526	4.898
COPPER(90.27)-NICKEL(9.73)				H	1.387	0.520	0.334	1.392	0.718	2.404	4.815
10471	8.9949	226	300.0	S	1.302	0.484	0.344	1.392	0.718	2.320	4.760
				R	1.152	0.423	0.362	1.392	0.718	2.168	4.663
				V	1.425	0.536	0.329	1.387	0.721	2.474	4.899
COPPER(95.83)-SILICON(4.17)				H	1.294	0.481	0.345	1.387	0.721	2.344	4.813
10472	8.7600	192	ROOM	S	1.198	0.442	0.356	1.387	0.721	2.245	4.749
				R	1.043	0.379	0.375	1.387	0.721	2.081	4.649
				V	1.411	0.533	0.323	1.328	0.753	2.475	4.840
COPPER(94.84)-SILICON(5.16)				H	1.276	0.476	0.340	1.328	0.753	2.339	4.749
10473	8.7050	192	ROOM	S	1.173	0.434	0.353	1.328	0.753	2.232	4.679
				R	1.015	0.370	0.373	1.328	0.753	2.060	4.573
				V	1.396	0.523	0.333	1.395	0.717	2.472	4.943
COPPER(92.31)-SILICON(7.69)				H	1.255	0.465	0.350	1.395	0.717	2.329	4.850
10474	8.5662	192	ROOM	S	1.140	0.418	0.364	1.395	0.717	2.209	4.774
				R	0.974	0.352	0.384	1.395	0.717	2.027	4.666
				V	1.695	1.075	-0.212	0.397	2.520	3.588	4.682
COPPER(52.8)-ZINC(47.2)				H	1.393	0.761	-0.085	0.397	2.520	3.019	4.112
10475	8.3500	115	297.0	S	0.772	0.328	0.176	0.397	2.520	1.983	3.162
				R	0.545	0.214	0.271	0.397	2.520	1.602	2.859
				V	1.764	1.074	-0.179	0.433	2.310	3.608	4.755
COPPER(52.8)-ZINC(47.2)				H	1.446	0.767	-0.057	0.433	2.310	3.048	4.200
10476	8.2500	115	468.0	S	0.823	0.348	0.183	0.433	2.310	2.054	3.297
				R	0.583	0.229	0.275	0.433	2.310	1.665	2.991
				V	1.519	1.058	-0.282	0.324	3.090	3.594	4.601
COPPER(52.8)-ZINC(47.2)				H	1.267	0.748	-0.153	0.324	3.090	3.021	4.014
10477	8.1940	115	566.0	S	0.738	0.330	0.120	0.324	3.090	2.006	3.052
				R	0.534	0.218	0.225	0.324	3.090	1.632	2.738
				V	1.433	0.997	-0.281	0.306	3.270	3.500	4.482
COPPER(52.8)-ZINC(47.2)				H	1.195	0.704	-0.151	0.306	3.270	2.941	3.910
10478	8.1380	115	662.0	S	0.691	0.308	0.123	0.306	3.270	1.945	2.967
				R	0.499	0.203	0.228	0.306	3.270	1.581	2.663
				V	1.400	0.907	-0.228	0.321	3.120	3.350	4.351
COPPER(52.8)-ZINC(47.2)				H	1.154	0.641	-0.100	0.321	3.120	2.817	3.814
10479	8.0800	115	721.0	S	0.642	0.275	0.166	0.321	3.120	1.846	2.917
				R	0.454	0.180	0.264	0.321	3.120	1.492	2.633
				V	1.384	0.856	-0.192	0.333	3.000	3.257	4.275
COPPER(52.8)-ZINC(47.2)				H	1.132	0.606	-0.066	0.333	3.000	2.740	3.761
10480	8.0710	115	739.0	S	0.614	0.258	0.193	0.333	3.000	1.786	2.896
				R	0.429	0.167	0.285	0.333	3.000	1.439	2.625

CUBIC

CODE	NAME / DENSITY	REF	TEMP(K)		E	G	SIGMA	K	BETA	VS	VP
				V	1.351	0.854	-0.209	0.317	3.150	3.254	4.249
COPPER(52.8)-ZINC(47.2)				H	1.109	0.604	-0.082	0.317	3.150	2.736	3.730
10481	8.0680	115	741.0	S	0.607	0.257	0.181	0.317	3.150	1.785	2.861
				R	0.427	0.167	0.276	0.317	3.150	1.440	2.588
				V	1.322	0.840	-0.214	0.309	3.240	3.232	4.215
COPPER(52.8)-ZINC(47.2)				H	1.085	0.594	-0.086	0.309	3.240	2.717	3.699
10482	8.0440	115	779.0	S	0.596	0.253	0.178	0.309	3.240	1.772	2.833
				R	0.419	0.164	0.274	0.309	3.240	1.430	2.562
				V	1.388	0.534	0.301	1.161	0.862	2.535	4.748
COPPER(51.06WT%)-ZINC(48.94WT%) (P= 0KB)				H	1.148	0.430	0.335	1.161	0.862	2.275	4.569
10483	8.3040	149	298.0	S	0.805	0.291	0.384	1.161	0.862	1.872	4.318
				R	0.586	0.207	0.416	1.161	0.862	1.578	4.159
				V	1.401	0.539	0.300	1.170	0.855	2.545	4.764
COPPER(51.06WT%)-ZINC(48.94WT%) (P= 2KB)				H	1.158	0.434	0.335	1.170	0.855	2.283	4.584
10484	8.3190	149	298.0	S	0.810	0.293	0.385	1.170	0.855	1.875	4.330
				R	0.589	0.208	0.416	1.170	0.855	1.580	4.170
				V	1.410	0.542	0.300	1.177	0.850	2.551	4.775
COPPER(51.06WT%)-ZINC(48.94WT%) (P= 4KB)				H	1.165	0.436	0.335	1.177	0.850	2.288	4.594
10485	8.3339	149	298.0	S	0.816	0.295	0.384	1.177	0.850	1.880	4.340
				R	0.593	0.209	0.416	1.177	0.850	1.585	4.180
				V	1.421	0.546	0.301	1.187	0.842	2.558	4.790
COPPER(51.06WT%)-ZINC(48.94WT%) (P= 6KB)				H	1.174	0.440	0.335	1.187	0.842	2.295	4.609
10486	8.3487	149	298.0	S	0.821	0.296	0.385	1.187	0.842	1.884	4.354
				R	0.596	0.210	0.416	1.187	0.842	1.588	4.193
				V	1.429	0.549	0.301	1.196	0.836	2.563	4.802
COPPER(51.06WT%)-ZINC(48.94WT%) (P= 8KB)				H	1.181	0.442	0.335	1.196	0.836	2.299	4.620
10487	8.3634	149	298.0	S	0.825	0.298	0.385	1.196	0.836	1.887	4.364
				R	0.599	0.211	0.417	1.196	0.836	1.590	4.203
				V	1.438	0.552	0.302	1.210	0.826	2.567	4.820
COPPER(51.06WT%)-ZINC(48.94WT%) (P=10KB)				H	1.188	0.444	0.336	1.210	0.826	2.303	4.638
10488	8.3780	149	298.0	S	0.829	0.299	0.386	1.210	0.826	1.889	4.382
				R	0.601	0.212	0.417	1.210	0.826	1.591	4.222
				V	1.413	0.534	0.324	1.339	0.747	2.444	4.791
COPPER(95.41)-ZINC(4.59)				H	1.287	0.480	0.340	1.339	0.747	2.318	4.707
10489	8.9346	192	ROOM	S	1.196	0.443	0.351	1.339	0.747	2.225	4.647
				R	1.046	0.382	0.370	1.339	0.747	2.067	4.549
				V	1.448	0.547	0.324	1.371	0.730	2.467	4.833
COPPER(100.0)				H	1.325	0.495	0.339	1.371	0.730	2.346	4.753
10490	8.9900	208	300.0	S	1.241	0.460	0.349	1.371	0.730	2.262	4.698
				R	1.095	0.401	0.367	1.371	0.730	2.111	4.603
				V	1.421	0.538	0.322	1.328	0.753	2.452	4.782
COPPER(95.9)-ZINC(4.1)				H	1.298	0.485	0.337	1.328	0.753	2.330	4.700
10491	8.9410	208	300.0	S	1.213	0.450	0.348	1.328	0.753	2.243	4.643
				R	1.067	0.390	0.366	1.328	0.753	2.090	4.547
				V	1.376	0.521	0.321	1.282	0.780	2.422	4.718
COPPER(90.9)-ZINC(9.1)				H	1.253	0.469	0.337	1.282	0.780	2.298	4.635
10492	8.8760	208	300.0	S	1.167	0.433	0.348	1.282	0.780	2.208	4.576
				R	1.023	0.374	0.367	1.282	0.780	2.053	4.479
				V	1.343	0.509	0.318	1.231	0.812	2.411	4.669
COPPER(82.9)-ZINC(17.1)				H	1.215	0.455	0.336	1.231	0.812	2.278	4.579
10493	8.7640	208	300.0	S	1.118	0.414	0.349	1.231	0.812	2.174	4.511
				R	0.968	0.353	0.369	1.231	0.812	2.008	4.407
				V	1.324	0.503	0.316	1.196	0.836	2.406	4.636
COPPER(77.3)-ZINC(22.7)				H	1.188	0.445	0.334	1.196	0.836	2.264	4.539
10494	8.6860	208	300.0	S	1.080	0.400	0.350	1.196	0.836	2.146	4.463
				R	0.924	0.337	0.371	1.196	0.836	1.969	4.352
				V	8.840	3.542	0.248	5.848	0.171	10.044	17.351
DIAMOND				H	8.676	3.463	0.253	5.848	0.171	9.931	17.265
10495	3.5110	128	ROOM	S	8.640	3.446	0.254	5.848	0.171	9.907	17.246
				R	8.460	3.360	0.259	5.848	0.171	9.783	17.151

CUBIC

CODE	NAME DENSITY	REF	TEMP(K)		E	G	SIGMA	K	BETA	VS	VP
				V	9.144	3.700	0.236	5.767	0.173	10.266	17.457
DIAMOND				H	9.001	3.630	0.240	5.767	0.173	10.168	17.381
10496	3.5110	129	ROOM	S	8.973	3.616	0.241	5.767	0.173	10.149	17.366
				R	8.818	3.541	0.245	5.767	0.173	10.043	17.284
				V	10.133	4.180	0.212	5.867	0.170	10.924	18.071
DIAMOND				H	10.117	4.171	0.213	5.867	0.170	10.912	18.062
10497	3.5030	129	300.0	S	10.116	4.171	0.213	5.867	0.170	10.912	18.062
				R	10.098	4.162	0.213	5.867	0.170	10.900	18.053
				V	10.284	4.722	0.089	4.170	0.240	11.597	17.265
DIAMOND				H	10.223	4.683	0.091	4.170	0.240	11.549	17.223
10498	3.5110	4	ROOM	S	10.216	4.679	0.092	4.170	0.240	11.544	17.218
				R	10.157	4.642	0.094	4.170	0.240	11.499	17.177
				V	11.446	5.357	0.068	4.420	0.226	12.350	18.145
DIAMOND				H	11.411	5.333	0.070	4.420	0.226	12.323	18.120
10499	3.5120	165	300.0	S	11.408	5.332	0.070	4.420	0.226	12.321	18.118
				R	11.375	5.310	0.071	4.420	0.226	12.296	18.096
				V	0.721	0.278	0.297	0.593	1.685	1.917	3.570
GALENA				H	0.717	0.276	0.299	0.593	1.685	1.911	3.565
10500	7.5640	129	ROOM	S	0.717	0.276	0.299	0.593	1.685	1.910	3.565
				R	0.712	0.274	0.300	0.593	1.685	1.903	3.560
				V	0.870	0.343	0.267	0.622	1.608	2.130	3.778
GALENA				H	0.836	0.328	0.276	0.622	1.608	2.081	3.741
10501	7.5640	129	ROOM	S	0.826	0.323	0.279	0.622	1.608	2.066	3.730
				R	0.794	0.308	0.287	0.622	1.608	2.019	3.696
				V	0.882	0.356	0.239	0.564	1.772	2.517	4.300
GALLIUM ANTIMONIDE				H	0.858	0.344	0.247	0.564	1.772	2.475	4.267
10502	5.6190	27	298.0	S	0.851	0.341	0.249	0.564	1.772	2.463	4.258
				R	0.824	0.328	0.257	0.564	1.772	2.416	4.222
				V	0.880	0.354	0.244	0.574	1.743	2.509	4.313
GALLIUM ANTIMONIDE				H	0.857	0.343	0.251	0.574	1.743	2.470	4.283
10503	5.6190	83	302.0	S	0.851	0.340	0.253	0.574	1.743	2.459	4.275
				R	0.826	0.328	0.260	0.574	1.743	2.415	4.241
				V	0.880	0.355	0.240	0.564	1.775	2.515	4.298
GALLIUM ANTIMONIDE				H	0.856	0.343	0.247	0.563	1.775	2.473	4.265
10504	5.6137	170	298.0	S	0.849	0.340	0.249	0.563	1.775	2.461	4.256
				R	0.823	0.327	0.257	0.563	1.775	2.415	4.221
				V	1.201	0.486	0.235	0.755	1.325	3.027	5.142
GALLIUM ARSENIDE				H	1.166	0.469	0.242	0.755	1.325	2.974	5.100
10505	5.3070	27	298.0	S	1.156	0.465	0.245	0.755	1.325	2.959	5.088
				R	1.118	0.446	0.253	0.755	1.325	2.900	5.043
				V	1.204	0.488	0.235	0.756	1.323	3.028	5.142
GALLIUM ARSENIDE				H	1.169	0.471	0.242	0.756	1.323	2.975	5.101
10506	5.3160	78	ROOM	S	1.159	0.466	0.244	0.756	1.323	2.960	5.089
				R	1.121	0.447	0.253	0.756	1.323	2.901	5.044
				V	1.202	0.487	0.234	0.754	1.326	3.026	5.137
GALLIUM ARSENIDE				H	1.167	0.470	0.242	0.754	1.326	2.972	5.095
10507	5.3169	169	298.0	S	1.157	0.465	0.244	0.754	1.326	2.957	5.083
				R	1.119	0.447	0.253	0.754	1.326	2.898	5.038
				V	1.217	0.490	0.241	0.784	1.275	3.040	5.206
GALLIUM ARSENIDE				H	1.182	0.473	0.249	0.784	1.275	2.986	5.164
10508	5.3070	99	77.4	S	1.172	0.468	0.251	0.784	1.275	2.971	5.152
				R	1.133	0.450	0.259	0.784	1.275	2.911	5.107
				V	1.199	0.486	0.233	0.748	1.336	3.027	5.130
GALLIUM ARSENIDE				H	1.164	0.469	0.241	0.748	1.336	2.973	5.088
10509	5.3070	99	300.0	S	1.154	0.464	0.243	0.748	1.336	2.958	5.076
				R	1.116	0.446	0.251	0.748	1.336	2.899	5.030
				V	1.429	0.580	0.232	0.888	1.127	3.748	6.342
GALLIUM PHOSPHIDE				H	1.390	0.561	0.239	0.888	1.127	3.686	6.293
10510	4.1297	276	300.0	S	1.379	0.556	0.241	0.888	1.127	3.668	6.280
				R	1.337	0.535	0.249	0.888	1.127	3.600	6.227

CUBIC

CODE	NAME DENSITY	REF	TEMP(K)		E	G	SIGMA	K	BETA	VS	VP
				V	1.454	0.556	0.307	1.257	0.796	3.846	7.291
GARNET	(21.8%FERRCUS OXIDE)			H	1.453	0.556	0.307	1.257	0.796	3.845	7.290
10511	3.7590	129	ROOM	S	1.453	0.556	0.307	1.257	0.796	3.845	7.290
				R	1.452	0.555	0.307	1.257	0.796	3.844	7.289
				V	1.423	0.540	0.318	1.300	0.769	3.834	7.416
GARNET	(22.7%FERROUS OXIDE)			H	1.416	0.537	0.319	1.300	0.769	3.823	7.408
10512	3.6730	129	ROOM	S	1.415	0.536	0.319	1.300	0.769	3.822	7.407
				R	1.406	0.533	0.320	1.300	0.769	3.808	7.398
				V	1.610	0.616	0.307	1.387	0.721	4.119	7.799
GARNET	(23.6%FERROUS OXIDE)			H	1.602	0.613	0.307	1.387	0.721	4.108	7.791
10513	3.6300	129	ROOM	S	1.601	0.612	0.308	1.387	0.721	4.107	7.791
				R	1.593	0.609	0.309	1.387	0.721	4.095	7.782
				V	1.708	0.656	0.301	1.433	0.698	4.228	7.930
GARNET	(23.0%FERROUS OXIDE)			H	1.703	0.654	0.302	1.433	0.698	4.221	7.925
10514	3.6700	129	ROOM	S	1.702	0.654	0.302	1.433	0.698	4.221	7.925
				R	1.697	0.651	0.303	1.433	0.698	4.213	7.920
				V	1.533	0.572	0.340	1.593	0.628	3.906	7.926
GARNET	(26.2%FERROUS OXIDE)			H	1.526	0.569	0.340	1.593	0.628	3.896	7.920
10515	3.7500	129	ROOM	S	1.525	0.569	0.340	1.593	0.628	3.895	7.920
				R	1.517	0.566	0.341	1.593	0.628	3.884	7.912
				V	1.731	0.640	0.353	1.957	0.511	3.937	8.249
GARNET	(28.7%FERRCUS OXIDE)			H	1.727	0.638	0.353	1.957	0.511	3.931	8.245
10516	4.1300	129	ROOM	S	1.727	0.638	0.353	1.957	0.511	3.931	8.245
				R	1.722	0.636	0.353	1.957	0.511	3.925	8.241
				V	2.424	0.940	0.289	1.917	0.522	4.665	8.566
GARNET	(33.5%FERRCUS OXIDE)			H	2.420	0.938	0.290	1.917	0.522	4.660	8.563
10517	4.3200	129	ROOM	S	2.419	0.938	0.290	1.917	0.522	4.660	8.563
				R	2.415	0.936	0.290	1.917	0.522	4.655	8.559
				V	2.182	0.865	0.261	1.522	0.657	4.863	8.552
GARNET	(ALMANDINE-PYROPE)			H	2.182	0.865	0.261	1.522	0.657	4.863	8.552
10518	3.6580	224	ROOM	S	2.182	0.865	0.261	1.522	0.657	4.863	8.552
				R	2.182	0.865	0.261	1.522	0.657	4.863	8.552
				V	2.225	0.889	0.251	1.487	0.672	4.711	8.168
GARNET	(ALMANDINE-PYROPE)			H	2.225	0.889	0.251	1.487	0.672	4.711	8.168
10519	4.0070	224	ROOM	S	2.225	0.889	0.251	1.487	0.672	4.711	8.168
				R	2.225	0.889	0.251	1.487	0.672	4.711	8.168
				V	2.397	0.956	0.253	1.621	0.617	4.853	8.445
GARNET	(ALMANDINE-PYROPE)			H	2.397	0.956	0.253	1.621	0.617	4.853	8.445
10520	4.0600	224	ROOM	S	2.397	0.956	0.253	1.621	0.617	4.853	8.445
				R	2.397	0.956	0.254	1.621	0.617	4.853	8.445
				V	2.285	0.906	0.261	1.595	0.627	4.670	8.216
GARNET	(ALMANDINE-PYROPE)			H	2.284	0.905	0.261	1.595	0.627	4.669	8.215
10521	4.1520	224	ROOM	S	2.284	0.905	0.261	1.595	0.627	4.669	8.215
				R	2.283	0.905	0.262	1.595	0.627	4.668	8.214
				V	2.122	0.839	0.265	1.508	0.663	4.545	8.042
GARNET	(ALMANDINE-PYROPE)			H	2.122	0.839	0.265	1.508	0.663	4.545	8.042
10522	4.0600	224	ROOM	S	2.122	0.839	0.265	1.508	0.663	4.545	8.042
				R	2.122	0.838	0.265	1.508	0.663	4.544	8.042
				V	2.419	0.951	0.271	1.765	0.567	4.769	8.515
GARNET	(ALMANDITE)			H	2.419	0.951	0.272	1.765	0.567	4.769	8.515
10523	4.1830	262	ROOM	S	2.419	0.951	0.272	1.765	0.567	4.769	8.515
				R	2.419	0.951	0.272	1.765	0.567	4.769	8.515
				V	2.403	0.943	0.274	1.770	0.565	4.762	8.532
GARNET	(ALMANDITE-PYROPE)			H	2.403	0.943	0.274	1.770	0.565	4.762	8.531
10524	4.1602	242	298.0	S	2.403	0.943	0.274	1.770	0.565	4.762	8.531
				R	2.402	0.943	0.274	1.770	0.565	4.761	8.531
				V	1.927	0.745	0.293	1.550	0.645	3.445	6.364
GARNET	(EUROPIUM IRON)			H	1.926	0.745	0.293	1.550	0.645	3.444	6.364
10525	6.2800	25	298.0	S	1.926	0.745	0.293	1.550	0.645	3.444	6.364
				R	1.926	0.745	0.293	1.550	0.645	3.443	6.363

CUBIC

CODE	NAME / DENSITY	REF	TEMP(K)		E	G	SIGMA	K	BETA	VS	VP
				V	2.263	0.883	0.281	1.720	0.581	3.530	6.394
GARNET (GADOLINIUM GALLIUM)				H	2.262	0.883	0.281	1.720	0.581	3.529	6.393
10526	7.0900	117	298.0	S	2.262	0.883	0.281	1.720	0.581	3.529	6.393
				R	2.261	0.883	0.281	1.720	0.581	3.529	6.393
				V	2.305	0.926	0.245	1.504	0.665	5.073	8.724
GARNET (GROSSULAR)				H	2.303	0.925	0.245	1.504	0.665	5.071	8.723
10527	3.5980	224	ROOM	S	2.303	0.925	0.245	1.504	0.665	5.071	8.723
				R	2.302	0.924	0.245	1.504	0.665	5.069	8.721
				V	2.450	0.966	0.267	1.756	0.570	4.770	8.466
GARNET (SPESSARTITE-ALMANDITE)				H	2.449	0.966	0.267	1.756	0.570	4.770	8.466
10528	4.2470	262	ROOM	S	2.449	0.966	0.267	1.756	0.570	4.770	8.466
				R	2.449	0.966	0.268	1.756	0.570	4.769	8.466
				V	2.006	0.774	0.295	1.631	0.613	3.870	7.177
GARNET (YTTRIUM IRON)				H	2.006	0.774	0.295	1.631	0.613	3.870	7.177
10529	5.1700	25	298.0	S	2.006	0.774	0.295	1.631	0.613	3.870	7.177
				R	2.005	0.774	0.295	1.631	0.613	3.870	7.177
				V	2.018	0.781	0.292	1.615	0.619	3.887	7.168
GARNET (YTTRIUM IRON)				H	2.017	0.781	0.292	1.615	0.619	3.886	7.167
10530	5.1700	58	296.0	S	2.017	0.781	0.292	1.615	0.619	3.886	7.167
				R	2.016	0.780	0.292	1.615	0.619	3.885	7.167
				V	1.360	0.563	0.209	0.778	1.285	3.243	5.345
GERMANIUM				H	1.332	0.548	0.215	0.778	1.285	3.201	5.311
10531	5.3500	129	298.0	S	1.325	0.545	0.216	0.778	1.285	3.191	5.303
				R	1.295	0.530	0.223	0.778	1.285	3.146	5.267
				V	1.354	0.565	0.199	0.750	1.333	3.249	5.300
GERMANIUM				H	1.326	0.550	0.205	0.750	1.333	3.207	5.266
10532	5.3500	162	ROOM	S	1.320	0.547	0.207	0.750	1.333	3.197	5.258
				R	1.291	0.532	0.213	0.750	1.333	3.154	5.223
				V	1.379	0.574	0.201	0.768	1.302	3.280	5.360
GERMANIUM				H	1.350	0.559	0.207	0.768	1.302	3.236	5.325
10533	5.3390	161	73.0	S	1.343	0.556	0.209	0.768	1.302	3.226	5.316
				R	1.312	0.540	0.215	0.768	1.302	3.180	5.279
				V	1.376	0.573	0.200	0.765	1.307	3.277	5.354
GERMANIUM				H	1.346	0.558	0.207	0.765	1.307	3.233	5.318
10534	5.3360	161	123.0	S	1.340	0.554	0.208	0.765	1.307	3.223	5.310
				R	1.309	0.539	0.215	0.765	1.307	3.177	5.273
				V	1.369	0.570	0.200	0.762	1.313	3.271	5.343
GERMANIUM				H	1.340	0.555	0.207	0.762	1.313	3.227	5.308
10535	5.3310	161	173.0	S	1.333	0.552	0.208	0.762	1.313	3.217	5.300
				R	1.303	0.536	0.215	0.762	1.313	3.171	5.263
				V	1.363	0.568	0.200	0.758	1.320	3.264	5.332
GERMANIUM				H	1.334	0.553	0.207	0.758	1.320	3.221	5.297
10536	5.3280	161	223.0	S	1.327	0.549	0.208	0.758	1.320	3.211	5.289
				R	1.297	0.534	0.215	0.758	1.320	3.165	5.252
				V	1.356	0.565	0.200	0.754	1.327	3.258	5.320
GERMANIUM				H	1.327	0.550	0.206	0.754	1.327	3.214	5.285
10537	5.3240	161	273.0	S	1.321	0.547	0.208	0.754	1.327	3.204	5.277
				R	1.290	0.531	0.215	0.754	1.327	3.158	5.240
				V	1.353	0.564	0.200	0.752	1.330	3.254	5.314
GERMANIUM				H	1.324	0.549	0.206	0.752	1.330	3.211	5.279
10538	5.3230	161	298.0	S	1.318	0.545	0.208	0.752	1.330	3.201	5.271
				R	1.287	0.530	0.215	0.752	1.330	3.155	5.234
				V	1.349	0.562	0.200	0.750	1.334	3.250	5.308
GERMANIUM				H	1.320	0.547	0.206	0.750	1.334	3.207	5.272
10539	5.3210	161	323.0	S	1.314	0.544	0.208	0.750	1.334	3.197	5.264
				R	1.283	0.528	0.215	0.750	1.334	3.151	5.227
				V	1.342	0.559	0.200	0.746	1.341	3.243	5.297
GERMANIUM				H	1.313	0.544	0.207	0.746	1.341	3.200	5.261
10540	5.3150	161	373.0	S	1.306	0.541	0.208	0.746	1.341	3.190	5.253
				R	1.276	0.525	0.215	0.746	1.341	3.144	5.216

CUBIC

CODE	NAME / DENSITY	REF	TEMP(K)		E	G	SIGMA	K	BETA	VS	VP
				V	1.335	0.556	0.200	0.742	1.347	3.235	5.285
GERMANIUM				H	1.306	0.541	0.207	0.742	1.347	3.192	5.250
10541	5.3120	161	423.0	S	1.300	0.538	0.208	0.742	1.347	3.182	5.242
				R	1.270	0.523	0.215	0.742	1.347	3.137	5.205
				V	1.327	0.553	0.201	0.739	1.354	3.227	5.273
GERMANIUM				H	1.299	0.538	0.207	0.739	1.354	3.184	5.238
10542	5.3070	161	473.0	S	1.292	0.535	0.208	0.739	1.354	3.174	5.230
				R	1.262	0.519	0.215	0.739	1.354	3.128	5.193
				V	1.319	0.549	0.201	0.735	1.360	3.218	5.261
GERMANIUM				H	1.291	0.535	0.207	0.735	1.360	3.175	5.225
10543	5.3020	161	523.0	S	1.284	0.531	0.209	0.735	1.360	3.165	5.217
				R	1.254	0.516	0.216	0.735	1.360	3.119	5.181
				V	1.312	0.546	0.201	0.732	1.367	3.212	5.250
GERMANIUM				H	1.284	0.532	0.208	0.732	1.367	3.168	5.215
10544	5.2970	161	573.0	S	1.277	0.528	0.209	0.732	1.367	3.158	5.207
				R	1.248	0.513	0.216	0.732	1.367	3.112	5.170
				V	1.345	0.560	0.201	0.749	1.334	3.247	5.307
GERMANIUM				H	1.317	0.545	0.207	0.749	1.334	3.204	5.272
10545	5.3130	36	298.0	S	1.310	0.542	0.209	0.749	1.334	3.194	5.264
				R	1.280	0.527	0.215	0.749	1.334	3.149	5.227
				V	1.357	0.566	0.200	0.754	1.327	3.259	5.321
GERMANIUM				H	1.329	0.551	0.206	0.754	1.327	3.217	5.287
10546	5.3250	251	300.0	S	1.323	0.548	0.208	0.754	1.327	3.207	5.279
				R	1.293	0.532	0.214	0.754	1.327	3.162	5.243
				V	1.375	0.573	0.200	0.765	1.307	3.276	5.352
GERMANIUM (P=0)				H	1.346	0.558	0.207	0.765	1.307	3.232	5.317
10547	5.3380	163	77.2	S	1.339	0.554	0.208	0.765	1.307	3.222	5.308
				R	1.309	0.538	0.215	0.765	1.307	3.176	5.271
				V	1.382	0.575	0.203	0.775	1.291	3.277	5.366
GERMANIUM (P=30,000PSI)				H	1.353	0.559	0.209	0.775	1.291	3.233	5.330
10548	5.3524	163	77.2	S	1.346	0.556	0.210	0.775	1.291	3.223	5.322
				R	1.315	0.540	0.217	0.775	1.291	3.176	5.285
				V	1.348	0.561	0.201	0.750	1.333	3.247	5.306
GERMANIUM (P=0)				H	1.319	0.547	0.207	0.750	1.333	3.204	5.271
10549	5.3230	163	298.0	S	1.313	0.543	0.208	0.750	1.333	3.194	5.263
				R	1.282	0.528	0.215	0.750	1.333	3.149	5.226
				V	1.355	0.563	0.203	0.760	1.317	3.249	5.320
GERMANIUM (P=30,000PSI)				H	1.326	0.548	0.209	0.760	1.317	3.205	5.285
10550	5.3377	163	298.0	S	1.319	0.545	0.211	0.760	1.317	3.195	5.277
				R	1.289	0.529	0.217	0.760	1.317	3.149	5.240
				V	0.906	0.323	0.405	1.587	0.630	1.293	3.233
GOLD				H	0.841	0.298	0.412	1.587	0.630	1.243	3.207
10551	19.3000	128	ROOM	S	0.797	0.281	0.416	1.587	0.630	1.207	3.189
				R	0.710	0.249	0.425	1.587	0.630	1.136	3.154
				V	0.881	0.311	0.416	1.754	0.570	1.269	3.352
GOLD				H	0.819	0.288	0.422	1.754	0.570	1.221	3.328
10552	19.3000	128	ROOM	S	0.776	0.272	0.426	1.754	0.570	1.187	3.312
				R	0.692	0.241	0.434	1.754	0.570	1.118	3.280
				V	0.890	0.316	0.407	1.587	0.630	1.281	3.227
GOLD				H	0.827	0.293	0.413	1.587	0.630	1.232	3.201
10553	19.3000	128	ROOM	S	0.785	0.277	0.418	1.587	0.630	1.198	3.184
				R	0.701	0.246	0.426	1.587	0.630	1.129	3.150
				V	0.841	0.296	0.420	1.753	0.570	1.238	3.336
GOLD				H	0.781	0.274	0.426	1.753	0.570	1.192	3.313
10554	19.3000	4	ROOM	S	0.740	0.259	0.430	1.753	0.570	1.158	3.298
				R	0.660	0.230	0.437	1.753	0.570	1.090	3.267
				V	0.906	0.321	0.409	1.657	0.604	1.290	3.287
GOLD				H	0.847	0.299	0.415	1.657	0.604	1.245	3.264
10555	19.3000	4	ROOM	S	0.810	0.286	0.418	1.657	0.604	1.217	3.249
				R	0.732	0.257	0.426	1.657	0.604	1.153	3.218

CUBIC

CODE	NAME DENSITY	REF	TEMP(K)		E	G	SIGMA	K	BETA	VS	VP
GOLD 10556	19.4920	191	0.0	V	0.951	0.337	0.412	1.803	0.555	1.314	3.399
				H	0.884	0.312	0.418	1.803	0.555	1.264	3.374
				S	0.838	0.295	0.423	1.803	0.555	1.230	3.357
				R	0.748	0.262	0.431	1.803	0.555	1.158	3.323
GOLD 10557	19.4920	191	10.0	V	0.950	0.336	0.412	1.803	0.555	1.314	3.399
				H	0.883	0.311	0.418	1.803	0.555	1.264	3.374
				S	0.838	0.295	0.423	1.803	0.555	1.229	3.356
				R	0.748	0.261	0.431	1.803	0.555	1.158	3.322
GOLD 10558	19.4890	191	25.0	V	0.948	0.336	0.412	1.802	0.555	1.313	3.398
				H	0.882	0.311	0.418	1.802	0.555	1.263	3.373
				S	0.837	0.294	0.423	1.802	0.555	1.228	3.356
				R	0.747	0.261	0.431	1.802	0.555	1.157	3.322
GOLD 10559	19.4860	191	50.0	V	0.944	0.334	0.413	1.799	0.556	1.309	3.394
				H	0.877	0.309	0.419	1.799	0.556	1.260	3.369
				S	0.832	0.292	0.423	1.799	0.556	1.225	3.352
				R	0.743	0.259	0.431	1.799	0.556	1.154	3.318
GOLD 10560	19.4690	191	75.0	V	0.937	0.331	0.413	1.793	0.558	1.305	3.388
				H	0.871	0.307	0.419	1.793	0.558	1.255	3.363
				S	0.826	0.290	0.423	1.793	0.558	1.221	3.346
				R	0.737	0.257	0.432	1.793	0.558	1.150	3.313
GOLD 10561	19.4570	191	100.0	V	0.930	0.329	0.413	1.786	0.560	1.301	3.382
				H	0.865	0.305	0.419	1.786	0.560	1.251	3.357
				S	0.820	0.288	0.424	1.786	0.560	1.217	3.340
				R	0.731	0.255	0.432	1.786	0.560	1.146	3.306
GOLD 10562	19.4390	191	125.0	V	0.924	0.327	0.413	1.779	0.562	1.297	3.375
				H	0.859	0.302	0.420	1.779	0.562	1.247	3.351
				S	0.814	0.286	0.424	1.779	0.562	1.213	3.334
				R	0.726	0.254	0.432	1.779	0.562	1.142	3.300
GOLD 10563	19.4160	191	150.0	V	0.917	0.324	0.414	1.772	0.564	1.293	3.369
				H	0.852	0.300	0.420	1.772	0.564	1.243	3.345
				S	0.808	0.284	0.424	1.772	0.564	1.209	3.328
				R	0.721	0.252	0.432	1.772	0.564	1.138	3.294
GOLD 10564	19.3990	191	175.0	V	0.911	0.322	0.414	1.765	0.567	1.288	3.363
				H	0.846	0.298	0.420	1.765	0.567	1.239	3.338
				S	0.802	0.282	0.424	1.765	0.567	1.205	3.321
				R	0.715	0.250	0.432	1.765	0.567	1.134	3.288
GOLD 10565	19.3810	191	200.0	V	0.904	0.320	0.414	1.757	0.569	1.284	3.357
				H	0.840	0.296	0.420	1.757	0.569	1.235	3.332
				S	0.796	0.279	0.425	1.757	0.569	1.201	3.315
				R	0.710	0.248	0.433	1.757	0.569	1.130	3.282
GOLD 10566	19.3580	191	225.0	V	0.898	0.317	0.415	1.750	0.571	1.280	3.351
				H	0.834	0.294	0.421	1.750	0.571	1.231	3.326
				S	0.790	0.277	0.425	1.750	0.571	1.197	3.309
				R	0.704	0.246	0.433	1.750	0.571	1.127	3.276
GOLD 10567	19.3400	191	250.0	V	0.891	0.315	0.415	1.743	0.574	1.276	3.344
				H	0.828	0.291	0.421	1.743	0.574	1.227	3.320
				S	0.784	0.275	0.425	1.743	0.574	1.193	3.303
				R	0.699	0.244	0.433	1.743	0.574	1.123	3.270
GOLD 10568	19.3170	191	275.0	V	0.884	0.312	0.415	1.736	0.576	1.272	3.338
				H	0.821	0.289	0.421	1.736	0.576	1.223	3.314
				S	0.778	0.273	0.425	1.736	0.576	1.189	3.297
				R	0.693	0.242	0.433	1.736	0.576	1.119	3.264
GOLD 10569	19.3000	191	300.0	V	0.878	0.310	0.415	1.729	0.578	1.268	3.332
				H	0.815	0.287	0.421	1.729	0.578	1.219	3.307
				S	0.772	0.271	0.426	1.729	0.578	1.185	3.291
				R	0.688	0.240	0.434	1.729	0.578	1.115	3.258
GOLD 10570	19.3000	51	300.0	V	0.887	0.313	0.415	1.728	0.579	1.274	3.335
				H	0.823	0.290	0.421	1.728	0.579	1.225	3.310
				S	0.780	0.274	0.425	1.728	0.579	1.191	3.293
				R	0.694	0.242	0.433	1.728	0.579	1.120	3.260

CUBIC

NAME / CODE	DENSITY	REF	TEMP(K)		E	G	SIGMA	K	BETA	VS	VP
GOLD				V	0.875	0.309	0.415	1.713	0.584	1.267	3.322
10571	19.2590	51	350.0	H	0.814	0.286	0.421	1.713	0.584	1.220	3.298
				S	0.773	0.271	0.425	1.713	0.584	1.187	3.282
				R	0.690	0.241	0.433	1.713	0.584	1.118	3.250
GOLD				V	0.858	0.303	0.416	1.700	0.588	1.256	3.309
10572	19.2190	51	400.0	H	0.797	0.280	0.422	1.700	0.588	1.207	3.285
				S	0.754	0.265	0.426	1.700	0.588	1.173	3.268
				R	0.672	0.234	0.434	1.700	0.588	1.104	3.236
GOLD				V	0.846	0.299	0.416	1.685	0.593	1.248	3.296
10573	19.1790	51	450.0	H	0.787	0.277	0.422	1.685	0.593	1.201	3.273
				S	0.747	0.262	0.426	1.685	0.593	1.169	3.257
				R	0.668	0.233	0.434	1.685	0.593	1.102	3.226
GOLD				V	0.828	0.292	0.417	1.670	0.599	1.236	3.281
10574	19.1340	51	500.0	H	0.767	0.270	0.423	1.670	0.599	1.187	3.257
				S	0.724	0.254	0.428	1.670	0.599	1.151	3.240
				R	0.642	0.223	0.436	1.670	0.599	1.081	3.207
GOLD				V	0.819	0.289	0.418	1.657	0.604	1.230	3.271
10575	19.0880	51	550.0	H	0.762	0.268	0.423	1.657	0.604	1.184	3.248
				S	0.723	0.253	0.427	1.657	0.604	1.152	3.232
				R	0.645	0.225	0.435	1.657	0.604	1.085	3.201
GOLD				V	0.802	0.283	0.419	1.643	0.609	1.219	3.257
10576	19.0430	51	600.0	H	0.745	0.261	0.424	1.643	0.609	1.172	3.234
				S	0.705	0.247	0.429	1.643	0.609	1.138	3.218
				R	0.627	0.218	0.436	1.643	0.609	1.070	3.187
GOLD				V	0.791	0.279	0.419	1.628	0.614	1.211	3.245
10577	18.9960	51	650.0	H	0.736	0.258	0.425	1.628	0.614	1.166	3.222
				S	0.698	0.244	0.429	1.628	0.614	1.134	3.207
				R	0.622	0.217	0.436	1.628	0.614	1.068	3.177
GOLD				V	0.774	0.272	0.420	1.615	0.619	1.199	3.230
10578	18.9560	51	700.0	H	0.718	0.252	0.426	1.615	0.619	1.153	3.208
				S	0.679	0.237	0.430	1.615	0.619	1.119	3.192
				R	0.604	0.210	0.438	1.615	0.619	1.052	3.162
GOLD				V	0.759	0.267	0.421	1.603	0.624	1.189	3.219
10579	18.9120	51	750.0	H	0.705	0.247	0.427	1.603	0.624	1.143	3.197
				S	0.666	0.233	0.431	1.603	0.624	1.110	3.181
				R	0.592	0.206	0.438	1.603	0.624	1.043	3.151
GOLD				V	0.748	0.263	0.422	1.588	0.630	1.181	3.206
10580	18.8660	51	800.0	H	0.696	0.244	0.427	1.588	0.630	1.137	3.185
				S	0.660	0.230	0.431	1.588	0.630	1.105	3.170
				R	0.588	0.204	0.438	1.588	0.630	1.041	3.141
GOLD				V	0.870	0.307	0.416	1.735	0.576	1.262	3.333
10581	19.3000	133	ROOM	H	0.809	0.284	0.422	1.735	0.576	1.214	3.310
				S	0.767	0.269	0.426	1.735	0.576	1.180	3.293
				R	0.684	0.238	0.434	1.735	0.576	1.111	3.261
GOLD(100.0)				V	0.880	0.311	0.415	1.728	0.579	1.270	3.333
10582	19.2830	114	296.5	H	0.818	0.288	0.421	1.728	0.579	1.222	3.309
				S	0.776	0.272	0.425	1.728	0.579	1.188	3.293
				R	0.692	0.241	0.433	1.728	0.579	1.119	3.260
GOLD(97.05)-NICKEL(2.95)				V	0.899	0.318	0.413	1.726	0.579	1.292	3.359
10583	19.0600	114	296.5	H	0.835	0.294	0.419	1.726	0.579	1.242	3.334
				S	0.791	0.278	0.424	1.726	0.579	1.207	3.316
				R	0.705	0.246	0.432	1.726	0.579	1.136	3.283
GOLD(90.28)-NICKEL(9.72)				V	0.937	0.333	0.409	1.714	0.583	1.340	3.412
10584	18.5330	114	296.5	H	0.870	0.307	0.415	1.714	0.583	1.288	3.385
				S	0.825	0.290	0.420	1.714	0.583	1.252	3.367
				R	0.735	0.257	0.428	1.714	0.583	1.178	3.332
GOLD(75.80)-NICKEL(24.20)				V	1.064	0.381	0.396	1.707	0.586	1.482	3.574
10585	17.3410	114	296.5	H	0.993	0.354	0.403	1.707	0.586	1.429	3.545
				S	0.949	0.337	0.407	1.707	0.586	1.395	3.527
				R	0.856	0.302	0.416	1.707	0.586	1.320	3.488

CUBIC

CODE	NAME DENSITY	REF	TEMP(K)		E	G	SIGMA	K	BETA	VS	VP
				V	1.261	0.458	0.377	1.705	0.587	1.708	3.839
GOLD(57.58)-NICKEL(42.42)				H	1.171	0.423	0.386	1.705	0.587	1.640	3.800
10586	15.7120	114	296.5	S	1.114	0.400	0.391	1.705	0.587	1.596	3.775
				R	0.998	0.356	0.402	1.705	0.587	1.505	3.725
				V	0.994	0.359	0.385	1.437	0.696	1.443	3.333
GOLD(75.0)-SILVER(25.0)				H	0.922	0.331	0.393	1.437	0.696	1.385	3.300
10587	17.2440	128	ROOM	S	0.875	0.313	0.399	1.437	0.696	1.347	3.279
				R	0.782	0.278	0.409	1.437	0.696	1.269	3.237
				V	1.024	0.375	0.364	1.253	0.798	1.580	3.416
GOLD(50.0)-SILVER(50.0)				H	0.950	0.346	0.374	1.253	0.798	1.517	3.377
10588	15.0300	128	ROOM	S	0.903	0.327	0.380	1.253	0.798	1.475	3.353
				R	0.810	0.291	0.392	1.253	0.798	1.391	3.304
				V	0.979	0.360	0.359	1.157	0.864	1.678	3.577
GOLD(25.0)-SILVER(75.0)				H	0.907	0.331	0.369	1.157	0.864	1.609	3.535
10589	12.7990	128	ROOM	S	0.862	0.313	0.376	1.157	0.864	1.564	3.508
				R	0.772	0.278	0.389	1.157	0.864	1.473	3.455
				V	2.147	0.828	0.297	1.760	0.568	2.668	4.963
HAFNIUM-COBALT INTERMETALLIC				H	2.131	0.821	0.298	1.760	0.568	2.657	4.955
10590	11.6280	229	0.0	S	2.129	0.820	0.298	1.760	0.568	2.655	4.954
				R	2.111	0.812	0.300	1.760	0.568	2.642	4.945
				V	2.147	0.828	0.297	1.760	0.568	2.668	4.963
HAFNIUM-COBALT INTERMETALLIC				H	2.131	0.821	0.298	1.760	0.568	2.657	4.955
10591	11.6280	229	77.0	S	2.129	0.820	0.298	1.760	0.568	2.655	4.954
				R	2.111	0.812	0.300	1.760	0.568	2.642	4.945
				V	2.147	0.828	0.297	1.759	0.569	2.668	4.961
HAFNIUM-COBALT INTERMETALLIC				H	2.131	0.821	0.298	1.759	0.569	2.657	4.953
10592	11.6280	229	100.0	S	2.129	0.820	0.298	1.759	0.569	2.655	4.952
				R	2.111	0.812	0.300	1.759	0.569	2.642	4.943
				V	2.142	0.826	0.297	1.758	0.569	2.665	4.958
HAFNIUM-COBALT INTERMETALLIC				H	2.125	0.818	0.298	1.758	0.569	2.653	4.950
10593	11.6280	229	150.0	S	2.123	0.817	0.299	1.758	0.569	2.651	4.948
				R	2.104	0.809	0.300	1.758	0.569	2.638	4.939
				V	2.124	0.820	0.295	1.731	0.578	2.655	4.928
HAFNIUM-COBALT INTERMETALLIC				H	2.107	0.812	0.297	1.731	0.578	2.643	4.919
10594	11.6280	229	200.0	S	2.104	0.811	0.297	1.731	0.578	2.641	4.918
				R	2.085	0.802	0.299	1.731	0.578	2.627	4.908
				V	2.103	0.813	0.294	1.703	0.587	2.644	4.895
HAFNIUM-COBALT INTERMETALLIC				H	2.086	0.805	0.296	1.703	0.587	2.631	4.886
10595	11.6280	229	250.0	S	2.083	0.804	0.296	1.703	0.587	2.629	4.885
				R	2.064	0.795	0.298	1.703	0.587	2.615	4.874
				V	2.084	0.806	0.292	1.674	0.597	2.633	4.862
HAFNIUM-COBALT INTERMETALLIC				H	2.066	0.798	0.294	1.674	0.597	2.620	4.852
10596	11.6280	229	300.0	S	2.063	0.797	0.295	1.674	0.597	2.618	4.851
				R	2.043	0.788	0.297	1.674	0.597	2.603	4.840
				V	0.151	0.066	0.141	0.070	14.286	2.220	3.435
HEXAMETHYLENETETRAMINE				H	0.150	0.066	0.142	0.070	14.286	2.217	3.432
10597	1.3390	129	ROOM	S	0.150	0.066	0.142	0.070	14.286	2.217	3.432
				R	0.150	0.066	0.143	0.070	14.286	2.214	3.430
				V	0.136	0.055	0.230	0.084	11.957	2.029	3.425
HEXAMETHYLENETETRAMINE				H	0.135	0.055	0.231	0.084	11.957	2.026	3.423
10598	1.3390	4	ROOM	S	0.135	0.055	0.231	0.084	11.957	2.025	3.423
				R	0.135	0.055	0.231	0.084	11.957	2.022	3.420
				V	0.642	0.255	0.260	0.445	2.246	2.097	3.681
INDIUM ANTIMONIDE				H	0.621	0.245	0.268	0.445	2.246	2.056	3.651
10599	5.7907	203	0.0	S	0.614	0.242	0.270	0.445	2.246	2.044	3.641
				R	0.591	0.231	0.279	0.445	2.246	1.998	3.607
				V	0.641	0.254	0.260	0.445	2.247	2.095	3.679
INDIUM ANTIMONIDE				H	0.620	0.244	0.268	0.445	2.247	2.054	3.649
10600	5.7907	203	100.0	S	0.613	0.241	0.270	0.445	2.247	2.041	3.639
				R	0.590	0.231	0.279	0.445	2.247	1.995	3.605

CUBIC

CODE	NAME / DENSITY	REF	TEMP(K)		E	G	SIGMA	K	BETA	VS	VP
				V	0.635	0.252	0.261	0.443	2.256	2.086	3.670
INDIUM ANTIMONIDE				H	0.614	0.242	0.269	0.443	2.256	2.045	3.639
10601	5.7835	203	200.0	S	0.608	0.239	0.272	0.443	2.256	2.032	3.629
				R	0.584	0.228	0.280	0.443	2.256	1.986	3.595
				V	0.626	0.248	0.259	0.433	2.307	2.074	3.639
INDIUM ANTIMONIDE				H	0.605	0.239	0.267	0.433	2.307	2.033	3.608
10602	5.7751	203	300.0	S	0.598	0.236	0.270	0.433	2.307	2.020	3.598
				R	0.575	0.225	0.279	0.433	2.307	1.973	3.563
				V	0.616	0.245	0.257	0.421	2.373	2.061	3.602
INDIUM ANTIMONIDE				H	0.595	0.235	0.265	0.421	2.373	2.019	3.570
10603	5.7664	203	400.0	S	0.588	0.232	0.267	0.421	2.373	2.006	3.560
				R	0.565	0.221	0.277	0.421	2.373	1.959	3.525
				V	0.605	0.242	0.252	0.407	2.457	2.049	3.559
INDIUM ANTIMONIDE				H	0.585	0.232	0.261	0.407	2.457	2.007	3.527
10604	5.7577	203	500.0	S	0.578	0.229	0.263	0.407	2.457	1.993	3.516
				R	0.555	0.218	0.273	0.407	2.457	1.946	3.481
				V	0.593	0.238	0.246	0.389	2.568	2.034	3.506
INDIUM ANTIMONIDE				H	0.572	0.228	0.255	0.389	2.568	1.991	3.472
10605	5.7490	203	600.0	S	0.565	0.225	0.258	0.389	2.568	1.977	3.462
				R	0.542	0.214	0.268	0.389	2.568	1.928	3.425
				V	0.641	0.251	0.279	0.483	2.070	2.080	3.757
INDIUM ANTIMONIDE				H	0.617	0.240	0.287	0.483	2.070	2.034	3.723
10606	5.7907	235	0.0	S	0.609	0.236	0.290	0.483	2.070	2.018	3.711
				R	0.581	0.224	0.299	0.483	2.070	1.965	3.673
				V	0.639	0.250	0.279	0.482	2.076	2.077	3.751
INDIUM ANTIMONIDE				H	0.615	0.239	0.287	0.482	2.076	2.032	3.718
10607	5.7907	235	50.0	S	0.607	0.235	0.290	0.482	2.076	2.016	3.706
				R	0.580	0.223	0.299	0.482	2.076	1.963	3.668
				V	0.638	0.249	0.278	0.479	2.087	2.075	3.744
INDIUM ANTIMONIDE				H	0.614	0.239	0.287	0.479	2.087	2.030	3.711
10608	5.7907	235	77.6	S	0.606	0.235	0.289	0.479	2.087	2.014	3.699
				R	0.579	0.223	0.299	0.479	2.087	1.961	3.661
				V	0.636	0.249	0.278	0.477	2.095	2.073	3.738
INDIUM ANTIMONIDE				H	0.612	0.238	0.286	0.477	2.095	2.028	3.705
10609	5.7907	235	100.0	S	0.604	0.234	0.289	0.477	2.095	2.012	3.693
				R	0.577	0.222	0.298	0.477	2.095	1.959	3.655
				V	0.633	0.248	0.277	0.473	2.114	2.069	3.726
INDIUM ANTIMONIDE				H	0.609	0.237	0.285	0.473	2.114	2.023	3.692
10610	5.7871	235	150.0	S	0.601	0.233	0.288	0.473	2.114	2.008	3.681
				R	0.574	0.221	0.298	0.473	2.114	1.955	3.643
				V	0.628	0.246	0.277	0.469	2.130	2.063	3.713
INDIUM ANTIMONIDE				H	0.605	0.235	0.285	0.469	2.130	2.017	3.680
10611	5.7835	235	200.0	S	0.597	0.232	0.288	0.469	2.130	2.001	3.669
				R	0.570	0.220	0.298	0.469	2.130	1.949	3.631
				V	0.624	0.244	0.277	0.467	2.142	2.055	3.703
INDIUM ANTIMONIDE				H	0.600	0.233	0.286	0.467	2.142	2.010	3.670
10612	5.7784	235	250.0	S	0.592	0.230	0.289	0.467	2.142	1.994	3.658
				R	0.566	0.218	0.298	0.467	2.142	1.942	3.620
				V	0.618	0.242	0.279	0.465	2.149	2.046	3.693
INDIUM ANTIMONIDE				H	0.595	0.231	0.287	0.465	2.149	2.001	3.660
10613	5.7751	235	300.0	S	0.587	0.228	0.290	0.465	2.149	1.985	3.649
				R	0.561	0.216	0.299	0.465	2.149	1.933	3.611
				V	0.620	0.242	0.280	0.469	2.134	2.045	3.698
INDIUM ANTIMONIDE				H	0.597	0.232	0.288	0.469	2.134	2.002	3.666
10614	5.7890	27	298.0	S	0.590	0.228	0.290	0.469	2.134	1.987	3.655
				R	0.564	0.217	0.299	0.469	2.134	1.936	3.618
				V	0.619	0.242	0.279	0.467	2.143	2.048	3.697
INDIUM ANTIMONIDE				H	0.597	0.232	0.287	0.467	2.143	2.004	3.665
10615	5.7747	77	ROOM	S	0.589	0.228	0.290	0.467	2.143	1.989	3.654
				R	0.563	0.217	0.299	0.467	2.143	1.938	3.618

CUBIC

CODE	NAME / DENSITY	REF	TEMP(K)		E	G	SIGMA	K	BETA	VS	VP
				V	0.797	0.314	0.271	0.579	1.726	2.351	4.194
INDIUM ARSENIDE				H	0.764	0.299	0.280	0.579	1.726	2.294	4.151
10616	5.6720	104	ROOM	S	0.752	0.293	0.284	0.579	1.726	2.273	4.136
				R	0.715	0.276	0.294	0.579	1.726	2.207	4.088
				V	0.810	0.315	0.287	0.634	1.576	2.355	4.310
INDIUM ARSENIDE				H	0.780	0.301	0.295	0.634	1.576	2.305	4.274
10617	5.6720	106	4.2	S	0.770	0.297	0.298	0.634	1.576	2.287	4.261
				R	0.736	0.282	0.307	0.634	1.576	2.228	4.219
				V	0.791	0.307	0.286	0.616	1.625	2.328	4.252
INDIUM ARSENIDE				H	0.751	0.289	0.297	0.616	1.625	2.259	4.202
10618	5.6720	106	77.4	S	0.734	0.282	0.301	0.616	1.625	2.229	4.181
				R	0.688	0.262	0.314	0.616	1.625	2.148	4.124
				V	0.797	0.314	0.271	0.579	1.726	2.351	4.194
INDIUM ARSENIDE				H	0.764	0.299	0.280	0.579	1.726	2.294	4.151
10619	5.6720	106	293.0	S	0.752	0.293	0.284	0.579	1.726	2.273	4.136
				R	0.715	0.276	0.294	0.579	1.726	2.207	4.088
				V	0.825	0.321	0.284	0.638	1.569	2.380	4.336
INDIUM ARSENIDE				H	0.792	0.306	0.293	0.638	1.569	2.325	4.296
10620	5.6670	215	4.2	S	0.780	0.301	0.296	0.638	1.569	2.304	4.281
				R	0.742	0.284	0.306	0.638	1.569	2.239	4.235
				V	0.823	0.320	0.284	0.633	1.579	2.378	4.326
INDIUM ARSENIDE				H	0.790	0.306	0.292	0.633	1.579	2.323	4.286
10621	5.6670	215	77.0	S	0.778	0.300	0.295	0.633	1.579	2.302	4.272
				R	0.741	0.284	0.305	0.633	1.579	2.238	4.226
				V	0.814	0.317	0.282	0.623	1.605	2.367	4.297
INDIUM ARSENIDE				H	0.782	0.303	0.291	0.623	1.605	2.312	4.256
10622	5.6670	215	195.0	S	0.770	0.297	0.294	0.623	1.605	2.291	4.242
				R	0.733	0.281	0.304	0.623	1.605	2.227	4.195
				V	0.804	0.314	0.281	0.613	1.631	2.353	4.266
INDIUM ARSENIDE				H	0.772	0.299	0.290	0.613	1.631	2.297	4.225
10623	5.6670	215	300.0	S	0.760	0.294	0.294	0.613	1.631	2.276	4.210
				R	0.722	0.277	0.304	0.613	1.631	2.211	4.163
				V	0.938	0.365	0.284	0.725	1.380	2.764	5.035
INDIUM PHOSPHIDE				H	0.900	0.348	0.293	0.725	1.380	2.699	4.987
10624	4.7800	131	ROOM	S	0.887	0.342	0.296	0.725	1.380	2.675	4.970
				R	0.843	0.323	0.306	0.725	1.380	2.599	4.916
				V	5.717	2.300	0.243	3.704	0.270	3.196	5.483
IRIDIUM				H	5.620	2.253	0.247	3.704	0.270	3.163	5.458
10625	22.5200	206	294.0	S	5.600	2.244	0.248	3.704	0.270	3.156	5.453
				R	5.495	2.193	0.253	3.704	0.270	3.121	5.425
				V	5.494	2.212	0.242	3.547	0.282	3.134	5.371
IRIDIUM				H	5.413	2.173	0.246	3.547	0.282	3.106	5.349
10626	22.5200	159	300.0	S	5.397	2.165	0.246	3.547	0.282	3.101	5.345
				R	5.309	2.123	0.251	3.547	0.282	3.070	5.321
				V	2.186	0.854	0.279	1.650	0.606	3.297	5.957
IRON				H	2.071	0.802	0.291	1.650	0.606	3.195	5.883
10627	7.8600	128	ROOM	S	2.020	0.779	0.296	1.650	0.606	3.149	5.849
				R	1.888	0.721	0.309	1.650	0.606	3.029	5.764
				V	2.155	0.860	0.253	1.456	0.687	3.307	5.753
IRON				H	2.043	0.807	0.266	1.456	0.687	3.204	5.675
10628	7.8600	128	ROOM	S	1.995	0.785	0.272	1.456	0.687	3.159	5.642
				R	1.871	0.727	0.286	1.456	0.687	3.042	5.555
				V	2.275	0.889	0.280	1.727	0.579	3.362	6.087
IRON				H	2.150	0.831	0.293	1.727	0.579	3.253	6.007
10629	7.8600	128	ROOM	S	2.091	0.805	0.298	1.727	0.579	3.201	5.970
				R	1.946	0.742	0.312	1.727	0.579	3.072	5.878
				V	2.229	0.863	0.292	1.783	0.561	3.314	6.110
IRON				H	2.112	0.811	0.303	1.783	0.561	3.212	6.037
10630	7.8600	4	ROOM	S	2.059	0.787	0.308	1.783	0.561	3.165	6.004
				R	1.923	0.728	0.320	1.783	0.561	3.044	5.919

CUBIC

CODE	NAME DENSITY	REF	TEMP(K)		E	G	SIGMA	K	BETA	VS	VP
				V	2.275	0.888	0.281	1.730	0.578	3.361	6.089
IRON				H	2.148	0.831	0.293	1.730	0.578	3.251	6.009
10631	7.8600	4	ROOM	S	2.090	0.805	0.299	1.730	0.578	3.199	5.971
				R	1.944	0.740	0.313	1.730	0.578	3.069	5.880
				V	2.391	0.941	0.270	1.731	0.578	3.447	6.139
IRON				H	2.268	0.885	0.282	1.731	0.578	3.341	6.061
10632	7.9240	211	4.2	S	2.214	0.860	0.287	1.731	0.578	3.295	6.027
				R	2.074	0.797	0.300	1.731	0.578	3.172	5.938
				V	2.391	0.941	0.270	1.731	0.578	3.447	6.139
IRON				H	2.268	0.885	0.282	1.731	0.578	3.341	6.061
10633	7.9238	211	20.0	S	2.214	0.860	0.287	1.731	0.578	3.295	6.027
				R	2.074	0.797	0.300	1.731	0.578	3.172	5.938
				V	2.389	0.941	0.270	1.729	0.578	3.445	6.136
IRON				H	2.265	0.884	0.282	1.729	0.578	3.340	6.058
10634	7.9238	211	40.0	S	2.212	0.859	0.287	1.729	0.578	3.293	6.024
				R	2.071	0.796	0.300	1.729	0.578	3.170	5.935
				V	2.385	0.939	0.270	1.727	0.579	3.443	6.132
IRON				H	2.262	0.882	0.282	1.727	0.579	3.337	6.054
10635	7.9226	211	60.0	S	2.208	0.858	0.287	1.727	0.579	3.291	6.019
				R	2.068	0.795	0.300	1.727	0.579	3.168	5.931
				V	2.380	0.937	0.270	1.721	0.581	3.440	6.124
IRON				H	2.256	0.880	0.282	1.721	0.581	3.334	6.045
10636	7.9214	211	80.0	S	2.202	0.856	0.287	1.721	0.581	3.287	6.011
				R	2.061	0.793	0.300	1.721	0.581	3.163	5.922
				V	2.374	0.935	0.269	1.716	0.583	3.436	6.117
IRON				H	2.250	0.878	0.281	1.716	0.583	3.330	6.038
10637	7.9190	211	100.0	S	2.196	0.853	0.287	1.716	0.583	3.283	6.003
				R	2.055	0.790	0.300	1.716	0.583	3.159	5.914
				V	2.368	0.933	0.269	1.711	0.585	3.433	6.109
IRON				H	2.244	0.876	0.281	1.711	0.585	3.326	6.030
10638	7.9166	211	120.0	S	2.190	0.851	0.287	1.711	0.585	3.278	5.995
				R	2.049	0.788	0.300	1.711	0.585	3.154	5.905
				V	2.361	0.930	0.269	1.707	0.586	3.428	6.102
IRON				H	2.236	0.873	0.282	1.707	0.586	3.320	6.022
10639	7.9143	211	140.0	S	2.182	0.848	0.287	1.707	0.586	3.272	5.987
				R	2.040	0.784	0.301	1.707	0.586	3.148	5.897
				V	2.353	0.927	0.270	1.703	0.587	3.423	6.095
IRON				H	2.229	0.869	0.282	1.703	0.587	3.315	6.015
10640	7.9095	211	160.0	S	2.173	0.844	0.287	1.703	0.587	3.267	5.980
				R	2.031	0.781	0.301	1.703	0.587	3.141	5.889
				V	2.345	0.923	0.270	1.699	0.589	3.418	6.088
IRON				H	2.221	0.866	0.282	1.699	0.589	3.310	6.008
10641	7.9047	211	180.0	S	2.165	0.841	0.288	1.699	0.589	3.261	5.972
				R	2.022	0.777	0.302	1.699	0.589	3.135	5.882
				V	2.338	0.920	0.270	1.695	0.590	3.413	6.081
IRON				H	2.212	0.863	0.282	1.695	0.590	3.304	6.001
10642	7.8999	211	200.0	S	2.156	0.837	0.288	1.695	0.590	3.255	5.965
				R	2.013	0.773	0.302	1.695	0.590	3.128	5.874
				V	2.330	0.917	0.270	1.692	0.591	3.408	6.076
IRON				H	2.205	0.859	0.283	1.692	0.591	3.299	5.995
10643	7.8952	211	220.0	S	2.148	0.834	0.288	1.692	0.591	3.249	5.959
				R	2.005	0.770	0.302	1.692	0.591	3.122	5.867
				V	2.322	0.914	0.271	1.689	0.592	3.403	6.069
IRON				H	2.196	0.856	0.283	1.689	0.592	3.293	5.988
10644	7.8906	211	240.0	S	2.139	0.830	0.289	1.689	0.592	3.243	5.952
				R	1.996	0.766	0.303	1.689	0.592	3.116	5.860
				V	2.313	0.910	0.271	1.686	0.593	3.397	6.063
IRON				H	2.187	0.852	0.284	1.686	0.593	3.286	5.981
10645	7.8858	211	260.0	S	2.129	0.826	0.289	1.686	0.593	3.236	5.944
				R	1.985	0.761	0.304	1.686	0.593	3.107	5.852

CUBIC

CODE	NAME DENSITY	REF	TEMP(K)		E	G	SIGMA	K	BETA	VS	VP
				V	2.305	0.906	0.272	1.682	0.594	3.391	6.057
IRON				H	2.178	0.848	0.284	1.682	0.594	3.281	5.975
10646	7.8786	211	280.0	S	2.120	0.822	0.290	1.682	0.594	3.229	5.938
				R	1.975	0.757	0.304	1.682	0.594	3.100	5.845
				V	2.296	0.902	0.272	1.680	0.595	3.385	6.051
IRON				H	2.169	0.844	0.285	1.680	0.595	3.274	5.969
10647	7.8740	211	300.0	S	2.110	0.817	0.291	1.680	0.595	3.222	5.931
				R	1.965	0.753	0.305	1.680	0.595	3.092	5.838
				V	2.330	0.918	0.269	1.680	0.595	3.406	6.058
IRON				H	2.209	0.862	0.281	1.680	0.595	3.301	5.980
10648	7.9123	155	77.0	S	2.155	0.838	0.286	1.680	0.595	3.254	5.946
				R	2.018	0.776	0.300	1.680	0.595	3.132	5.858
				V	2.263	0.891	0.270	1.640	0.610	3.365	5.996
IRON				H	2.137	0.833	0.283	1.640	0.610	3.253	5.913
10649	7.8672	155	300.0	S	2.078	0.806	0.289	1.640	0.610	3.201	5.874
				R	1.933	0.742	0.304	1.640	0.610	3.070	5.781
				V	2.271	0.892	0.273	1.669	0.599	3.366	6.026
IRON (P= 0KB)				H	2.146	0.834	0.286	1.669	0.599	3.256	5.945
10650	7.8720	220	300.0	S	2.088	0.808	0.292	1.669	0.599	3.205	5.908
				R	1.945	0.745	0.306	1.669	0.599	3.076	5.815
				V	2.282	0.896	0.274	1.681	0.595	3.371	6.040
IRON (P= 2KB)				H	2.156	0.838	0.286	1.681	0.595	3.261	5.959
10651	7.8810	220	300.0	S	2.099	0.812	0.292	1.681	0.595	3.210	5.922
				R	1.955	0.748	0.306	1.681	0.595	3.081	5.830
				V	2.293	0.900	0.274	1.693	0.591	3.377	6.054
IRON (P= 4KB)				H	2.167	0.842	0.287	1.693	0.591	3.267	5.973
10652	7.8910	220	300.0	S	2.109	0.816	0.292	1.693	0.591	3.216	5.936
				R	1.965	0.752	0.307	1.693	0.591	3.087	5.844
				V	2.267	0.891	0.273	1.664	0.601	3.365	6.021
IRON (P= 0KB)				H	2.139	0.832	0.286	1.664	0.601	3.252	5.937
10653	7.8672	121	300.0	S	2.079	0.805	0.292	1.664	0.601	3.198	5.899
				R	1.932	0.739	0.307	1.664	0.601	3.066	5.804
				V	2.278	0.894	0.273	1.675	0.597	3.370	6.033
IRON (P= 2KB)				H	2.149	0.835	0.286	1.675	0.597	3.257	5.950
10654	7.8766	121	300.0	S	2.088	0.808	0.292	1.675	0.597	3.203	5.911
				R	1.941	0.743	0.307	1.675	0.597	3.071	5.817
				V	2.289	0.898	0.274	1.685	0.593	3.375	6.047
IRON (P= 4KB)				H	2.159	0.839	0.286	1.685	0.593	3.262	5.963
10655	7.8861	121	300.0	S	2.099	0.812	0.292	1.685	0.593	3.209	5.924
				R	1.951	0.746	0.307	1.685	0.593	3.076	5.830
				V	2.300	0.903	0.274	1.696	0.590	3.381	6.060
IRON (P= 6KB)				H	2.170	0.843	0.287	1.696	0.590	3.268	5.976
10656	7.8956	121	300.0	S	2.109	0.816	0.293	1.696	0.590	3.214	5.937
				R	1.960	0.750	0.307	1.696	0.590	3.081	5.842
				V	2.311	0.907	0.274	1.707	0.586	3.387	6.073
IRON (P= 8KB)				H	2.180	0.847	0.287	1.707	0.586	3.273	5.990
10657	7.9051	121	300.0	S	2.119	0.819	0.293	1.707	0.586	3.219	5.951
				R	1.969	0.753	0.308	1.707	0.586	3.086	5.856
				V	2.321	0.910	0.275	1.717	0.582	3.391	6.085
IRON (P=10KB)				H	2.190	0.850	0.287	1.717	0.582	3.278	6.002
10658	7.9154	121	300.0	S	2.128	0.823	0.293	1.717	0.582	3.224	5.963
				R	1.977	0.756	0.308	1.717	0.582	3.090	5.868
				V	2.223	0.868	0.280	1.687	0.593	3.322	6.012
IRON				H	2.074	0.801	0.295	1.687	0.593	3.190	5.917
10659	7.8672	151	298.0	S	1.994	0.765	0.303	1.687	0.593	3.119	5.866
				R	1.824	0.691	0.320	1.687	0.593	2.963	5.757
				V	2.201	0.864	0.274	1.620	0.617	3.318	5.944
IRON				H	2.068	0.803	0.287	1.620	0.617	3.200	5.856
10660	7.8458	151	373.0	S	2.003	0.774	0.294	1.620	0.617	3.141	5.814
				R	1.851	0.707	0.310	1.620	0.617	3.001	5.715

CUBIC

CODE	NAME	DENSITY	REF	TEMP(K)		E	G	SIGMA	K	BETA	VS	VP
10661	IRON	7.8318	151	423.0	V	2.176	0.854	0.274	1.603	0.624	3.302	5.917
					H	2.041	0.793	0.288	1.603	0.624	3.181	5.828
					S	1.974	0.762	0.295	1.603	0.624	3.120	5.784
					R	1.820	0.694	0.311	1.603	0.624	2.978	5.683
10662	IRON	7.8154	151	473.0	V	2.164	0.852	0.270	1.570	0.637	3.302	5.884
					H	2.038	0.794	0.284	1.570	0.637	3.187	5.799
					S	1.977	0.766	0.290	1.570	0.637	3.131	5.759
					R	1.833	0.702	0.305	1.570	0.637	2.997	5.663
10663	IRON	7.7990	151	523.0	V	2.138	0.838	0.275	1.587	0.630	3.278	5.888
					H	2.005	0.777	0.289	1.587	0.630	3.157	5.799
					S	1.937	0.747	0.297	1.587	0.630	3.095	5.755
					R	1.785	0.680	0.313	1.587	0.630	2.953	5.654
10664	IRON	7.7804	151	573.0	V	2.097	0.820	0.278	1.577	0.634	3.246	5.858
					H	1.952	0.755	0.294	1.577	0.634	3.114	5.761
					S	1.873	0.719	0.302	1.577	0.634	3.041	5.709
					R	1.708	0.647	0.319	1.577	0.634	2.884	5.599
10665	IRON	7.7618	151	623.0	V	2.083	0.818	0.274	1.533	0.652	3.246	5.814
					H	1.947	0.756	0.288	1.533	0.652	3.120	5.722
					S	1.876	0.724	0.296	1.533	0.652	3.054	5.673
					R	1.721	0.655	0.313	1.533	0.652	2.906	5.569
10666	IRON	7.7433	151	673.0	V	2.072	0.816	0.270	1.500	0.667	3.246	5.781
					H	1.944	0.757	0.284	1.500	0.667	3.127	5.693
					S	1.880	0.728	0.291	1.500	0.667	3.066	5.649
					R	1.734	0.663	0.307	1.500	0.667	2.926	5.549
10667	IRON	7.7249	151	723.0	V	2.047	0.806	0.270	1.483	0.674	3.230	5.754
					H	1.917	0.746	0.285	1.483	0.674	3.108	5.664
					S	1.851	0.716	0.292	1.483	0.674	3.045	5.619
					R	1.704	0.651	0.309	1.483	0.674	2.903	5.517
10668	IRON	7.7042	151	773.0	V	2.003	0.786	0.274	1.480	0.676	3.194	5.728
					H	1.870	0.725	0.289	1.480	0.676	3.068	5.635
					S	1.799	0.693	0.297	1.480	0.676	3.000	5.586
					R	1.646	0.626	0.315	1.480	0.676	2.851	5.481
10669	IRON(96.03)-ALUMINUM(3.97)	7.6617	150	300.0	V	2.274	0.898	0.266	1.619	0.618	3.424	6.063
					H	2.120	0.827	0.282	1.619	0.618	3.285	5.960
					S	2.037	0.789	0.290	1.619	0.618	3.210	5.905
					R	1.863	0.712	0.308	1.619	0.618	3.048	5.790
10670	IRON(90.35)-ALUMINUM(9.65)	7.3619	150	300.0	V	2.240	0.892	0.256	1.529	0.654	3.480	6.076
					H	2.054	0.805	0.276	1.529	0.654	3.306	5.946
					S	1.937	0.751	0.289	1.529	0.654	3.195	5.864
					R	1.730	0.660	0.311	1.529	0.654	2.993	5.720
10671	IRON(85.50)-ALUMINUM(14.50)	7.1097	150	300.0	V	2.210	0.884	0.251	1.479	0.676	3.525	6.113
					H	1.995	0.782	0.275	1.479	0.676	3.317	5.956
					S	1.840	0.712	0.293	1.479	0.676	3.164	5.843
					R	1.603	0.608	0.319	1.479	0.676	2.923	5.674
10672	IRON(82.15)-ALUMINUM(17.85)	6.9422	150	300.0	V	2.180	0.873	0.249	1.449	0.690	3.545	6.134
					H	1.938	0.759	0.277	1.449	0.690	3.306	5.954
					S	1.741	0.670	0.300	1.449	0.690	3.106	5.808
					R	1.480	0.556	0.330	1.449	0.690	2.831	5.618
10673	IRON(80.17)-ALUMINUM(19.83)	6.8459	150	300.0	V	2.154	0.862	0.249	1.430	0.699	3.549	6.139
					H	1.892	0.739	0.280	1.430	0.699	3.286	5.941
					S	1.656	0.634	0.307	1.430	0.699	3.042	5.765
					R	1.379	0.515	0.339	1.430	0.699	2.742	5.560
10674	IRON(77.55)-ALUMINUM(22.45)	6.7441	150	300.0	V	2.136	0.855	0.249	1.417	0.706	3.561	6.158
					H	1.852	0.722	0.282	1.417	0.706	3.272	5.941
					S	1.572	0.598	0.315	1.417	0.706	2.977	5.730
					R	1.278	0.473	0.350	1.417	0.706	2.649	5.511
10675	IRON(76.43)-ALUMINUM(23.57)	6.7020	150	300.0	V	2.152	0.859	0.252	1.449	0.690	3.581	6.222
					H	1.846	0.717	0.288	1.449	0.690	3.271	5.990
					S	1.520	0.573	0.325	1.449	0.690	2.925	5.747
					R	1.209	0.444	0.361	1.449	0.690	2.575	5.519

CUBIC

CODE	NAME / DENSITY	REF	TEMP(K)		E	G	SIGMA	K	BETA	VS	VP
				V	2.175	0.871	0.248	1.441	0.694	3.621	6.258
IRON(74.95)-ALUMINUM(25.05)				H	1.841	0.715	0.287	1.441	0.694	3.281	6.003
10676	6.6436	150	300.0	S	1.451	0.544	0.332	1.441	0.694	2.863	5.711
				R	1.124	0.410	0.370	1.441	0.694	2.485	5.470
				V	2.152	0.867	0.242	1.388	0.720	3.633	6.225
IRON(73.03)-ALUMINUM(26.97)				H	1.827	0.714	0.281	1.388	0.720	3.297	5.970
10677	6.5645	150	300.0	S	1.458	0.550	0.325	1.388	0.720	2.895	5.685
				R	1.140	0.418	0.363	1.388	0.720	2.523	5.444
				V	2.161	0.873	0.237	1.372	0.729	3.659	6.236
IRON(71.92)-ALUMINUM(28.08)				H	1.846	0.723	0.276	1.372	0.729	3.330	5.985
10678	6.5228	150	300.0	S	1.501	0.569	0.318	1.372	0.729	2.955	5.716
				R	1.187	0.438	0.356	1.372	0.729	2.590	5.475
				V	2.190	0.893	0.225	1.329	0.752	3.765	6.324
IRON(66.00)-ALUMINUM(34.00)				H	1.931	0.767	0.258	1.329	0.752	3.490	6.110
10679	6.3023	150	300.0	S	1.707	0.664	0.286	1.329	0.752	3.245	5.927
				R	1.435	0.544	0.320	1.329	0.752	2.937	5.709
				V	2.207	0.897	0.230	1.361	0.735	3.854	6.506
IRON(59.89)-ALUMINUM(40.11)				H	1.983	0.789	0.257	1.361	0.735	3.613	6.319
10680	6.0420	150	300.0	S	1.816	0.711	0.278	1.361	0.735	3.430	6.182
				R	1.575	0.602	0.307	1.361	0.735	3.158	5.986
				V	0.281	0.101	0.390	0.427	2.340	0.945	2.227
LEAD				H	0.252	0.090	0.402	0.427	2.340	0.891	2.197
10681	11.3400	128	ROOM	S	0.226	0.080	0.412	0.427	2.340	0.840	2.170
				R	0.190	0.067	0.426	0.427	2.340	0.766	2.134
				V	0.281	0.101	0.388	0.417	2.400	0.945	2.205
LEAD				H	0.252	0.090	0.399	0.417	2.400	0.891	2.175
10682	11.3437	128	ROOM	S	0.226	0.080	0.410	0.417	2.400	0.841	2.148
				R	0.190	0.067	0.424	0.417	2.400	0.767	2.111
				V	0.375	0.137	0.372	0.488	2.049	1.086	2.405
LEAD				H	0.337	0.122	0.385	0.488	2.049	1.025	2.368
10683	11.5900	270	0.0	S	0.304	0.109	0.396	0.488	2.049	0.969	2.337
				R	0.257	0.091	0.412	0.488	2.049	0.886	2.293
				V	0.373	0.136	0.372	0.488	2.050	1.083	2.403
LEAD				H	0.335	0.121	0.385	0.488	2.050	1.022	2.367
10684	11.5860	270	20.0	S	0.302	0.108	0.397	0.488	2.050	0.966	2.336
				R	0.256	0.091	0.413	0.488	2.050	0.884	2.292
				V	0.368	0.134	0.374	0.485	2.061	1.075	2.394
LEAD				H	0.331	0.119	0.386	0.485	2.061	1.015	2.359
10685	11.5760	270	40.0	S	0.298	0.107	0.398	0.485	2.061	0.960	2.328
				R	0.253	0.089	0.413	0.485	2.061	0.879	2.285
				V	0.361	0.131	0.375	0.483	2.072	1.066	2.386
LEAD				H	0.324	0.117	0.388	0.483	2.072	1.006	2.350
10686	11.5580	270	60.0	S	0.292	0.104	0.399	0.483	2.072	0.950	2.320
				R	0.247	0.087	0.415	0.483	2.072	0.869	2.276
				V	0.355	0.129	0.377	0.480	2.084	1.057	2.376
LEAD				H	0.319	0.115	0.389	0.480	2.084	0.997	2.341
10687	11.5440	270	80.0	S	0.286	0.102	0.401	0.480	2.084	0.941	2.310
				R	0.242	0.085	0.416	0.480	2.084	0.860	2.268
				V	0.349	0.127	0.378	0.477	2.097	1.049	2.367
LEAD				H	0.313	0.113	0.391	0.477	2.097	0.989	2.332
10688	11.5270	270	100.0	S	0.281	0.100	0.402	0.477	2.097	0.933	2.302
				R	0.237	0.084	0.417	0.477	2.097	0.851	2.259
				V	0.343	0.124	0.379	0.474	2.110	1.040	2.358
LEAD				H	0.308	0.111	0.392	0.474	2.110	0.980	2.323
10689	11.5100	270	120.0	S	0.276	0.098	0.403	0.474	2.110	0.924	2.293
				R	0.232	0.082	0.418	0.474	2.110	0.843	2.251
				V	0.337	0.122	0.381	0.471	2.124	1.031	2.349
LEAD				H	0.302	0.109	0.393	0.471	2.124	0.972	2.315
10690	11.4890	270	140.0	S	0.271	0.096	0.404	0.471	2.124	0.916	2.284
				R	0.228	0.080	0.419	0.471	2.124	0.835	2.242

CUBIC

CODE	NAME DENSITY	REF	TEMP(K)		E	G	SIGMA	K	BETA	VS	VP
				V	0.331	0.120	0.382	0.468	2.135	1.022	2.340
LEAD				H	0.297	0.106	0.394	0.468	2.135	0.963	2.306
10691	11.4720	270	160.0	S	0.265	0.094	0.406	0.468	2.135	0.907	2.276
				R	0.223	0.078	0.421	0.468	2.135	0.827	2.235
				V	0.325	0.118	0.383	0.465	2.151	1.013	2.330
LEAD				H	0.291	0.104	0.396	0.465	2.151	0.955	2.296
10692	11.4540	270	180.0	S	0.260	0.092	0.407	0.465	2.151	0.899	2.266
				R	0.218	0.077	0.422	0.465	2.151	0.818	2.225
				V	0.319	0.115	0.385	0.462	2.165	1.004	2.320
LEAD				H	0.286	0.102	0.397	0.462	2.165	0.946	2.287
10693	11.4340	270	200.0	S	0.255	0.091	0.408	0.462	2.165	0.890	2.257
				R	0.213	0.075	0.423	0.462	2.165	0.810	2.217
				V	0.313	0.113	0.386	0.459	2.179	0.995	2.311
LEAD				H	0.280	0.100	0.398	0.459	2.179	0.937	2.278
10694	11.4170	270	220.0	S	0.249	0.088	0.409	0.459	2.179	0.880	2.248
				R	0.208	0.073	0.424	0.459	2.179	0.801	2.208
				V	0.307	0.111	0.388	0.456	2.193	0.986	2.301
LEAD				H	0.275	0.098	0.400	0.456	2.193	0.928	2.269
10695	11.3960	270	240.0	S	0.244	0.087	0.411	0.456	2.193	0.872	2.239
				R	0.204	0.072	0.425	0.456	2.193	0.793	2.200
				V	0.301	0.108	0.389	0.453	2.208	0.976	2.291
LEAD				H	0.269	0.096	0.401	0.453	2.208	0.919	2.259
10696	11.3790	270	260.0	S	0.239	0.085	0.412	0.453	2.208	0.862	2.229
				R	0.199	0.070	0.427	0.453	2.208	0.783	2.190
				V	0.295	0.106	0.391	0.450	2.222	0.967	2.282
LEAD				H	0.263	0.094	0.402	0.450	2.222	0.909	2.251
10697	11.3580	270	280.0	S	0.233	0.082	0.414	0.450	2.222	0.852	2.221
				R	0.194	0.068	0.428	0.450	2.222	0.773	2.182
				V	0.289	0.104	0.392	0.447	2.237	0.957	2.273
LEAD				H	0.258	0.092	0.404	0.447	2.237	0.900	2.241
10698	11.3380	270	300.0	S	0.228	0.081	0.415	0.447	2.237	0.843	2.212
				R	0.190	0.066	0.429	0.447	2.237	0.765	2.173
				V	3.752	1.368	0.372	4.879	0.205	3.435	7.605
LEAD				H	3.370	1.217	0.385	4.879	0.205	3.240	7.490
10699	11.5880	15	4.2	S	3.037	1.088	0.396	4.879	0.205	3.063	7.391
				R	2.569	0.910	0.412	4.879	0.205	2.802	7.251
				V	3.562	1.294	0.376	4.803	0.208	3.347	7.519
LEAD				H	3.196	1.150	0.389	4.803	0.208	3.156	7.408
10700	11.5480	15	77.0	S	2.873	1.026	0.400	4.803	0.208	2.981	7.310
				R	2.424	0.856	0.416	4.803	0.208	2.723	7.174
				V	2.904	1.043	0.392	4.477	0.223	3.032	7.192
LEAD				H	2.591	0.923	0.404	4.477	0.223	2.853	7.093
10701	11.3450	15	296.0	S	2.293	0.810	0.415	4.477	0.223	2.672	6.999
				R	1.905	0.667	0.429	4.477	0.223	2.424	6.877
				V	3.217	1.162	0.384	4.629	0.216	3.187	7.348
LEAD				H	2.879	1.031	0.396	4.629	0.216	3.001	7.243
10702	11.4420	184	195.0	S	2.569	0.912	0.408	4.629	0.216	2.824	7.148
				R	2.152	0.756	0.423	4.629	0.216	2.571	7.019
				V	2.910	1.046	0.392	4.476	0.223	3.036	7.194
LEAD				H	2.599	0.926	0.403	4.476	0.223	2.857	7.095
10703	11.3440	184	296.0	S	2.305	0.815	0.414	4.476	0.223	2.680	7.002
				R	1.919	0.672	0.429	4.476	0.223	2.433	6.881
				V	0.614	0.230	0.332	0.611	1.638	1.720	3.433
LEAD FLUORIDE				H	0.612	0.230	0.333	0.611	1.638	1.717	3.431
10704	7.7900	274	300.0	S	0.612	0.230	0.333	0.611	1.638	1.717	3.431
				R	0.611	0.229	0.333	0.611	1.638	1.714	3.429
				V	0.303	0.112	0.359	0.358	2.793	1.570	3.345
LEAD NITRATE				H	0.294	0.108	0.363	0.358	2.793	1.543	3.328
10705	4.5300	34	ROOM	S	0.291	0.107	0.365	0.358	2.793	1.534	3.322
				R	0.279	0.102	0.370	0.358	2.793	1.499	3.301

CUBIC

CODE	NAME	DENSITY	REF	TEMP(K)		E	G	SIGMA	K	BETA	VS	VP
10706	LEAD NITRATE	4.5470	124	293.0	V	0.271	0.100	0.354	0.309	3.240	1.484	3.118
					H	0.252	0.092	0.364	0.309	3.240	1.425	3.082
					S	0.240	0.088	0.370	0.309	3.240	1.389	3.060
					R	0.217	0.078	0.383	0.309	3.240	1.313	3.014
10707	LEAD NITRATE	4.5300	182	ROOM	V	0.272	0.100	0.355	0.312	3.205	1.487	3.136
					H	0.252	0.092	0.365	0.312	3.205	1.428	3.100
					S	0.241	0.088	0.371	0.312	3.205	1.391	3.077
					R	0.217	0.078	0.384	0.312	3.205	1.314	3.032
10708	LEAD TELLURIDE	8.1600	83	302.0	V	0.678	0.277	0.223	0.409	2.448	1.843	3.088
					H	0.595	0.237	0.257	0.409	2.448	1.703	2.979
					S	0.544	0.213	0.278	0.409	2.448	1.614	2.912
					R	0.481	0.185	0.304	0.409	2.448	1.504	2.832
10709	LEAD TELLURIDE	8.3672	136	0.0	V	0.813	0.338	0.203	0.456	2.195	2.010	3.291
					H	0.707	0.285	0.241	0.456	2.195	1.845	3.160
					S	0.639	0.252	0.266	0.456	2.195	1.736	3.076
					R	0.562	0.217	0.295	0.456	2.195	1.610	2.984
10710	LEAD TELLURIDE	8.3672	136	10.0	V	0.811	0.337	0.203	0.455	2.196	2.007	3.289
					H	0.706	0.284	0.242	0.455	2.196	1.843	3.158
					S	0.637	0.252	0.267	0.455	2.196	1.734	3.074
					R	0.561	0.216	0.295	0.455	2.196	1.608	2.982
10711	LEAD TELLURIDE	8.3672	136	20.0	V	0.809	0.336	0.203	0.454	2.201	2.005	3.284
					H	0.704	0.283	0.242	0.454	2.201	1.841	3.154
					S	0.636	0.251	0.267	0.454	2.201	1.732	3.071
					R	0.560	0.216	0.295	0.454	2.201	1.607	2.979
10712	LEAD TELLURIDE	8.3646	136	30.0	V	0.806	0.335	0.203	0.453	2.207	2.001	3.279
					H	0.702	0.283	0.242	0.453	2.207	1.838	3.150
					S	0.634	0.250	0.267	0.453	2.207	1.730	3.067
					R	0.558	0.216	0.295	0.453	2.207	1.605	2.975
10713	LEAD TELLURIDE	8.3571	136	50.0	V	0.800	0.332	0.204	0.450	2.224	1.994	3.268
					H	0.696	0.280	0.242	0.450	2.224	1.832	3.139
					S	0.630	0.249	0.267	0.450	2.224	1.725	3.057
					R	0.554	0.214	0.295	0.450	2.224	1.601	2.966
10714	LEAD TELLURIDE	8.3345	136	100.0	V	0.781	0.324	0.205	0.441	2.266	1.971	3.237
					H	0.680	0.274	0.243	0.441	2.266	1.812	3.110
					S	0.616	0.243	0.267	0.441	2.266	1.707	3.030
					R	0.543	0.209	0.295	0.441	2.266	1.585	2.940
10715	LEAD TELLURIDE	8.3121	136	150.0	V	0.760	0.315	0.208	0.434	2.306	1.945	3.204
					H	0.663	0.266	0.245	0.434	2.306	1.790	3.081
					S	0.602	0.237	0.269	0.434	2.306	1.689	3.004
					R	0.531	0.205	0.296	0.434	2.306	1.570	2.916
10716	LEAD TELLURIDE	8.2872	136	200.0	V	0.736	0.304	0.212	0.426	2.345	1.914	3.167
					H	0.644	0.258	0.248	0.426	2.345	1.765	3.049
					S	0.587	0.231	0.271	0.426	2.345	1.669	2.976
					R	0.519	0.200	0.297	0.426	2.345	1.553	2.891
10717	LEAD TELLURIDE	8.2625	136	250.0	V	0.713	0.293	0.216	0.419	2.389	1.883	3.130
					H	0.626	0.250	0.251	0.419	2.389	1.740	3.017
					S	0.572	0.225	0.272	0.419	2.389	1.649	2.948
					R	0.507	0.195	0.298	0.419	2.389	1.537	2.866
10718	LEAD TELLURIDE	8.2379	136	303.2	V	0.687	0.281	0.221	0.411	2.434	1.848	3.088
					H	0.605	0.241	0.254	0.411	2.434	1.711	2.982
					S	0.556	0.218	0.275	0.411	2.434	1.627	2.918
					R	0.494	0.190	0.300	0.411	2.434	1.518	2.839
10719	LEAD (0.17WT.%THALLIUM)	11.5880	15	4.2	V	0.375	0.136	0.373	0.490	2.039	1.085	2.409
					H	0.337	0.122	0.386	0.490	2.039	1.024	2.373
					S	0.304	0.109	0.397	0.490	2.039	0.969	2.342
					R	0.258	0.091	0.412	0.490	2.039	0.887	2.298
10720	LEAD (0.17WT.%THALLIUM)	11.5480	15	77.0	V	0.353	0.128	0.377	0.479	2.086	1.054	2.373
					H	0.317	0.114	0.390	0.479	2.086	0.995	2.339
					S	0.286	0.102	0.401	0.479	2.086	0.940	2.308
					R	0.241	0.085	0.416	0.479	2.086	0.859	2.266

CUBIC

CODE	NAME DENSITY	REF	TEMP(K)		E	G	SIGMA	K	BETA	VS	VP
				V	0.369	0.134	0.375	0.492	2.034	1.076	2.405
LEAD (1.04WT.%THALLIUM)				H	0.333	0.120	0.387	0.492	2.034	1.017	2.371
10721	11.5910	15	4.2	S	0.302	0.108	0.398	0.492	2.034	0.965	2.341
				R	0.256	0.091	0.413	0.492	2.034	0.885	2.299
				V	0.352	0.128	0.377	0.477	2.095	1.052	2.368
LEAD (1.04WT.%THALLIUM)				H	0.316	0.114	0.390	0.477	2.095	0.992	2.334
10722	11.5510	15	77.0	S	0.285	0.102	0.401	0.477	2.095	0.938	2.304
				R	0.241	0.085	0.416	0.477	2.094	0.858	2.262
				V	0.286	0.103	0.393	0.446	2.244	0.952	2.266
LEAD (1.04WT.%THALLIUM)				H	0.256	0.091	0.404	0.446	2.244	0.896	2.236
10723	11.3480	15	296.0	S	0.227	0.080	0.415	0.446	2.244	0.841	2.207
				R	0.189	0.066	0.429	0.446	2.244	0.764	2.169
				V	0.367	0.134	0.374	0.486	2.057	1.074	2.394
LEAD (1.75WT.%THALLIUM)				H	0.330	0.119	0.387	0.486	2.057	1.013	2.358
10724	11.5930	15	4.2	S	0.298	0.106	0.398	0.486	2.057	0.958	2.328
				R	0.252	0.089	0.414	0.486	2.057	0.876	2.284
				V	0.350	0.127	0.378	0.479	2.089	1.048	2.368
LEAD (1.75WT.%THALLIUM)				H	0.314	0.113	0.391	0.479	2.089	0.989	2.334
10725	11.5530	15	77.0	S	0.283	0.101	0.402	0.479	2.089	0.934	2.304
				R	0.239	0.084	0.417	0.479	2.089	0.854	2.262
				V	0.286	0.103	0.393	0.445	2.246	0.951	2.265
LEAD (1.75WT.%THALLIUM)				H	0.255	0.091	0.404	0.445	2.246	0.895	2.234
10726	11.3500	15	296.0	S	0.226	0.080	0.415	0.445	2.246	0.839	2.205
				R	0.188	0.066	0.430	0.445	2.246	0.762	2.167
				V	0.367	0.133	0.374	0.486	2.058	1.073	2.393
LEAD (2.32WT.%THALLIUM)				H	0.330	0.119	0.387	0.486	2.058	1.013	2.358
10727	11.5950	15	4.2	S	0.298	0.107	0.398	0.486	2.058	0.959	2.327
				R	0.253	0.089	0.413	0.486	2.058	0.878	2.284
				V	0.350	0.127	0.378	0.477	2.095	1.048	2.365
LEAD (2.32WT.%THALLIUM)				H	0.314	0.113	0.390	0.477	2.095	0.988	2.331
10728	11.5560	15	77.0	S	0.282	0.101	0.401	0.477	2.095	0.934	2.300
				R	0.238	0.084	0.417	0.477	2.095	0.853	2.258
				V	0.287	0.103	0.393	0.447	2.237	0.952	2.269
LEAD (2.32WT.%THALLIUM)				H	0.256	0.091	0.404	0.447	2.237	0.897	2.238
10729	11.3520	15	296.0	S	0.228	0.081	0.415	0.447	2.237	0.843	2.210
				R	0.190	0.067	0.429	0.447	2.237	0.766	2.173
				V	0.362	0.132	0.375	0.482	2.075	1.066	2.381
LEAD (3.45WT.%THALLIUM)				H	0.326	0.117	0.387	0.482	2.075	1.006	2.346
10730	11.5990	15	4.2	S	0.294	0.105	0.398	0.482	2.075	0.952	2.316
				R	0.249	0.088	0.414	0.482	2.075	0.872	2.274
				V	0.345	0.125	0.379	0.474	2.109	1.040	2.355
LEAD (3.45WT.%THALLIUM)				H	0.310	0.111	0.391	0.474	2.109	0.982	2.321
10731	11.5590	15	77.0	S	0.279	0.100	0.402	0.474	2.109	0.929	2.292
				R	0.236	0.083	0.417	0.474	2.109	0.850	2.251
				V	0.282	0.101	0.394	0.443	2.258	0.945	2.256
LEAD (3.45WT.%THALLIUM)				H	0.252	0.090	0.405	0.443	2.258	0.889	2.226
10732	11.3560	15	296.0	S	0.224	0.079	0.416	0.443	2.258	0.835	2.198
				R	0.187	0.065	0.430	0.443	2.258	0.758	2.160
				V	0.359	0.131	0.376	0.481	2.079	1.061	2.376
LEAD (6.02WT.%THALLIUM)				H	0.323	0.116	0.388	0.481	2.079	1.001	2.341
10733	11.6070	15	4.2	S	0.291	0.104	0.399	0.481	2.079	0.946	2.310
				R	0.246	0.087	0.415	0.481	2.079	0.865	2.268
				V	0.341	0.124	0.380	0.472	2.117	1.034	2.347
LEAD (6.02WT.%THALLIUM)				H	0.306	0.110	0.392	0.472	2.117	0.975	2.313
10734	11.5670	15	77.0	S	0.275	0.098	0.403	0.472	2.117	0.921	2.284
				R	0.232	0.082	0.418	0.472	2.117	0.841	2.242
				V	0.280	0.101	0.394	0.439	2.280	0.941	2.245
LEAD (6.02WT.%THALLIUM)				H	0.250	0.089	0.405	0.439	2.280	0.885	2.215
10735	11.3640	15	296.0	S	0.222	0.078	0.416	0.439	2.280	0.830	2.186
				R	0.185	0.065	0.430	0.439	2.280	0.754	2.149

CUBIC

CODE	NAME DENSITY	REF	TEMP(K)		E	G	SIGMA	K	BETA	VS	VP
				V	0.351	0.128	0.374	0.462	2.164	1.047	2.331
LEAD (14.7WT.%THALLIUM)				H	0.313	0.113	0.387	0.462	2.164	0.984	2.294
10736	11.6350	15	4.2	S	0.277	0.099	0.400	0.462	2.164	0.923	2.260
				R	0.231	0.082	0.417	0.462	2.164	0.838	2.215
				V	0.333	0.121	0.378	0.456	2.193	1.021	2.307
LEAD (14.7WT.%THALLIUM)				H	0.297	0.107	0.391	0.456	2.193	0.959	2.271
10737	11.5950	15	77.0	S	0.263	0.094	0.404	0.456	2.193	0.899	2.238
				R	0.219	0.077	0.420	0.456	2.193	0.816	2.195
				V	0.273	0.098	0.394	0.428	2.337	0.927	2.214
LEAD (14.7WT.%THALLIUM)				H	0.242	0.086	0.406	0.428	2.337	0.870	2.183
10738	11.3920	15	296.0	S	0.212	0.075	0.418	0.428	2.337	0.810	2.152
				R	0.174	0.061	0.432	0.428	2.337	0.731	2.114
				V	0.346	0.126	0.375	0.461	2.171	1.039	2.323
LEAD (17.4WT.%THALLIUM)				H	0.308	0.111	0.389	0.461	2.171	0.976	2.286
10739	11.6450	15	4.2	S	0.273	0.097	0.401	0.461	2.171	0.914	2.252
				R	0.227	0.080	0.418	0.461	2.171	0.830	2.207
				V	0.328	0.119	0.380	0.454	2.201	1.013	2.299
LEAD (17.4WT.%THALLIUM)				H	0.292	0.105	0.393	0.454	2.201	0.951	2.263
10740	11.6050	15	77.0	S	0.258	0.092	0.405	0.454	2.201	0.890	2.230
				R	0.214	0.075	0.421	0.454	2.201	0.806	2.187
				V	0.268	0.096	0.395	0.426	2.346	0.919	2.205
LEAD (17.4WT.%THALLIUM)				H	0.238	0.084	0.407	0.426	2.346	0.860	2.174
10741	11.4020	15	296.0	S	0.206	0.072	0.420	0.426	2.346	0.797	2.141
				R	0.168	0.059	0.434	0.426	2.346	0.717	2.103
				V	0.367	0.134	0.371	0.475	2.106	1.086	2.399
LEAD(94.99)-THALLIUM(5.01)				H	0.330	0.119	0.384	0.475	2.106	1.025	2.363
10742	11.3500	231	0.0	S	0.298	0.107	0.395	0.475	2.106	0.970	2.332
				R	0.253	0.090	0.411	0.475	2.106	0.888	2.288
				V	0.340	0.124	0.377	0.462	2.162	1.043	2.351
LEAD(94.99)-THALLIUM(5.01)				H	0.305	0.110	0.390	0.462	2.162	0.984	2.316
10743	11.3500	231	100.0	S	0.275	0.098	0.401	0.462	2.162	0.929	2.286
				R	0.232	0.082	0.417	0.462	2.162	0.849	2.244
				V	0.312	0.113	0.384	0.450	2.222	0.997	2.300
LEAD(94.99)-THALLIUM(5.01)				H	0.280	0.100	0.396	0.450	2.222	0.939	2.267
10744	11.3500	231	200.0	S	0.250	0.089	0.407	0.450	2.222	0.884	2.238
				R	0.210	0.074	0.422	0.450	2.222	0.806	2.198
				V	0.284	0.102	0.392	0.438	2.285	0.949	2.249
LEAD(94.99)-THALLIUM(5.01)				H	0.254	0.090	0.403	0.438	2.285	0.893	2.218
10745	11.3500	231	300.0	S	0.225	0.079	0.414	0.438	2.285	0.837	2.189
				R	0.187	0.065	0.429	0.438	2.285	0.760	2.151
				V	0.338	0.123	0.374	0.448	2.234	1.039	2.316
LEAD(79.50)-THALLIUM(20.50)				H	0.301	0.108	0.388	0.448	2.234	0.975	2.279
10746	11.4000	231	0.0	S	0.265	0.095	0.401	0.448	2.234	0.911	2.244
				R	0.220	0.078	0.418	0.448	2.234	0.826	2.199
				V	0.315	0.114	0.380	0.437	2.290	1.000	2.273
LEAD(79.50)-THALLIUM(20.50)				H	0.279	0.100	0.393	0.437	2.290	0.938	2.237
10747	11.4000	231	100.0	S	0.245	0.087	0.406	0.437	2.290	0.874	2.202
				R	0.202	0.071	0.423	0.437	2.290	0.790	2.159
				V	0.290	0.105	0.387	0.427	2.340	0.958	2.230
LEAD(79.50)-THALLIUM(20.50)				H	0.257	0.092	0.400	0.427	2.340	0.897	2.196
10748	11.4000	231	200.0	S	0.224	0.079	0.413	0.427	2.340	0.833	2.162
				R	0.184	0.064	0.428	0.427	2.340	0.751	2.121
				V	0.265	0.095	0.394	0.417	2.399	0.913	2.184
LEAD(79.50)-THALLIUM(20.50)				H	0.234	0.083	0.406	0.417	2.399	0.855	2.152
10749	11.4000	231	300.0	S	0.202	0.071	0.419	0.417	2.399	0.790	2.119
				R	0.165	0.057	0.434	0.417	2.399	0.710	2.080
				V	0.316	0.115	0.378	0.433	2.311	1.001	2.262
LEAD(68.23)-THALLIUM(31.77)				H	0.278	0.100	0.393	0.433	2.311	0.934	2.223
10750	11.4500	231	0.0	S	0.239	0.085	0.408	0.433	2.311	0.861	2.184
				R	0.194	0.068	0.425	0.433	2.311	0.771	2.139

CUBIC

NAME / CODE	DENSITY	REF	TEMP(K)		E	G	SIGMA	K	BETA	VS	VP
				V	0.294	0.106	0.384	0.423	2.365	0.963	2.220
LEAD(68.23)-THALLIUM(31.77)				H	0.258	0.092	0.398	0.423	2.365	0.899	2.184
10751	11.4500	231	100.0	S	0.221	0.078	0.413	0.423	2.365	0.826	2.146
				R	0.179	0.063	0.430	0.423	2.365	0.739	2.103
				V	0.273	0.098	0.390	0.412	2.426	0.926	2.178
LEAD(68.23)-THALLIUM(31.77)				H	0.239	0.085	0.403	0.412	2.426	0.863	2.143
10752	11.4500	231	200.0	S	0.202	0.071	0.418	0.412	2.426	0.789	2.105
				R	0.162	0.057	0.434	0.412	2.426	0.703	2.064
				V	0.252	0.090	0.395	0.401	2.492	0.888	2.134
LEAD(68.23)-THALLIUM(31.77)				H	0.220	0.078	0.409	0.401	2.492	0.826	2.101
10753	11.4500	231	300.0	S	0.184	0.065	0.424	0.401	2.492	0.751	2.063
				R	0.146	0.051	0.439	0.401	2.492	0.665	2.024
				V	0.296	0.107	0.384	0.423	2.362	0.964	2.219
LEAD(59.50)-THALLIUM(40.50)				H	0.259	0.092	0.398	0.423	2.362	0.897	2.181
10754	11.4900	231	0.0	S	0.218	0.077	0.414	0.423	2.362	0.818	2.140
				R	0.174	0.061	0.431	0.423	2.362	0.728	2.096
				V	0.278	0.100	0.389	0.416	2.406	0.933	2.186
LEAD(59.50)-THALLIUM(40.50)				H	0.242	0.086	0.403	0.416	2.406	0.867	2.149
10755	11.4900	231	100.0	S	0.202	0.071	0.419	0.416	2.406	0.788	2.109
				R	0.161	0.056	0.435	0.416	2.406	0.699	2.066
				V	0.259	0.093	0.394	0.408	2.450	0.900	2.152
LEAD(59.50)-THALLIUM(40.50)				H	0.226	0.080	0.408	0.408	2.450	0.836	2.117
10756	11.4900	231	200.0	S	0.187	0.066	0.424	0.408	2.450	0.756	2.077
				R	0.148	0.051	0.440	0.408	2.450	0.669	2.037
				V	0.241	0.086	0.400	0.401	2.494	0.865	2.119
LEAD(59.50)-THALLIUM(40.50)				H	0.209	0.074	0.413	0.401	2.494	0.803	2.086
10757	11.4900	231	300.0	S	0.172	0.060	0.429	0.401	2.494	0.723	2.046
				R	0.135	0.047	0.444	0.401	2.494	0.637	2.008
				V	0.273	0.098	0.391	0.417	2.396	0.922	2.177
LEAD(47.34)-THALLIUM(52.66)				H	0.235	0.083	0.406	0.417	2.396	0.849	2.137
10758	11.5700	231	0.0	S	0.185	0.065	0.426	0.417	2.396	0.748	2.087
				R	0.142	0.049	0.443	0.417	2.396	0.651	2.043
				V	0.259	0.093	0.394	0.408	2.449	0.895	2.144
LEAD(47.34)-THALLIUM(52.66)				H	0.222	0.079	0.410	0.408	2.449	0.825	2.106
10759	11.5700	231	100.0	S	0.173	0.061	0.429	0.408	2.449	0.724	2.056
				R	0.132	0.046	0.446	0.408	2.449	0.629	2.014
				V	0.243	0.087	0.399	0.399	2.504	0.866	2.110
LEAD(47.34)-THALLIUM(52.66)				H	0.208	0.074	0.413	0.399	2.504	0.797	2.073
10760	11.5700	231	200.0	S	0.161	0.056	0.433	0.399	2.504	0.698	2.025
				R	0.123	0.042	0.449	0.399	2.504	0.605	1.985
				V	0.227	0.081	0.403	0.390	2.564	0.836	2.075
LEAD(47.34)-THALLIUM(52.66)				H	0.194	0.069	0.417	0.390	2.564	0.770	2.040
10761	11.5700	231	300.0	S	0.150	0.052	0.436	0.390	2.564	0.671	1.993
				R	0.113	0.039	0.452	0.390	2.564	0.581	1.955
				V	0.259	0.093	0.394	0.405	2.468	0.893	2.132
LEAD(38.59)-THALLIUM(61.41)				H	0.218	0.077	0.411	0.405	2.468	0.814	2.089
10762	11.6400	231	0.0	S	0.157	0.055	0.436	0.405	2.468	0.684	2.026
				R	0.114	0.039	0.453	0.405	2.468	0.582	1.983
				V	0.247	0.088	0.398	0.402	2.489	0.871	2.113
LEAD(38.59)-THALLIUM(61.41)				H	0.209	0.074	0.414	0.402	2.489	0.796	2.073
10763	11.6400	231	100.0	S	0.152	0.053	0.437	0.402	2.489	0.674	2.014
				R	0.112	0.038	0.454	0.402	2.489	0.574	1.973
				V	0.234	0.083	0.402	0.397	2.519	0.847	2.090
LEAD(38.59)-THALLIUM(61.41)				H	0.198	0.070	0.417	0.397	2.519	0.775	2.052
10764	11.6400	231	200.0	S	0.147	0.051	0.438	0.397	2.519	0.661	1.998
				R	0.108	0.037	0.454	0.397	2.519	0.566	1.959
				V	0.222	0.079	0.406	0.392	2.550	0.823	2.067
LEAD(38.59)-THALLIUM(61.41)				H	0.188	0.066	0.420	0.392	2.550	0.755	2.032
10765	11.6400	231	300.0	S	0.141	0.049	0.440	0.392	2.550	0.649	1.983
				R	0.105	0.036	0.455	0.392	2.550	0.558	1.945

CUBIC

CODE	NAME / DENSITY	REF	TEMP(K)		E	G	SIGMA	K	BETA	VS	VP
				V	0.250	0.089	0.397	0.406	2.463	0.875	2.119
LEAD(28.32)-THALLIUM(71.68)				H	0.212	0.075	0.413	0.406	2.463	0.801	2.080
10766	11.7000	231	0.0	S	0.158	0.055	0.435	0.406	2.463	0.686	2.024
				R	0.117	0.040	0.452	0.406	2.463	0.588	1.983
				V	0.241	0.086	0.400	0.400	2.499	0.857	2.098
LEAD(28.32)-THALLIUM(71.68)				H	0.205	0.072	0.415	0.400	2.499	0.787	2.061
10767	11.7000	231	100.0	S	0.156	0.054	0.435	0.400	2.499	0.681	2.010
				R	0.117	0.040	0.451	0.400	2.499	0.587	1.970
				V	0.231	0.082	0.402	0.394	2.539	0.838	2.075
LEAD(28.32)-THALLIUM(71.68)				H	0.198	0.070	0.416	0.394	2.539	0.772	2.040
10768	11.7000	231	200.0	S	0.153	0.053	0.435	0.394	2.539	0.675	1.994
				R	0.116	0.040	0.451	0.394	2.539	0.585	1.955
				V	0.221	0.078	0.405	0.388	2.578	0.819	2.052
LEAD(28.32)-THALLIUM(71.68)				H	0.190	0.067	0.418	0.388	2.578	0.757	2.019
10769	11.7000	231	300.0	S	0.150	0.052	0.435	0.388	2.578	0.669	1.978
				R	0.116	0.040	0.450	0.388	2.578	0.583	1.941
				V	0.177	0.069	0.277	0.133	7.538	3.567	6.426
LITHIUM				H	0.145	0.055	0.318	0.133	7.538	3.179	6.148
10770	0.5454	190	78.0	S	0.098	0.036	0.377	0.133	7.538	2.551	5.745
				R	0.070	0.025	0.412	0.133	7.538	2.132	5.512
				V	0.165	0.064	0.281	0.125	7.981	3.443	6.239
LITHIUM				H	0.135	0.051	0.320	0.125	7.981	3.075	5.977
10771	0.5421	190	155.0	S	0.093	0.034	0.376	0.125	7.981	2.496	5.605
				R	0.067	0.024	0.411	0.125	7.981	2.094	5.382
				V	0.158	0.062	0.279	0.120	8.352	3.387	6.122
LITHIUM				H	0.130	0.049	0.318	0.120	8.352	3.028	5.865
10772	0.5398	190	195.0	S	0.090	0.033	0.374	0.120	8.352	2.467	5.504
				R	0.065	0.023	0.409	0.120	8.352	2.073	5.283
				V	0.147	0.057	0.297	0.121	8.246	3.266	6.082
LITHIUM				H	0.122	0.046	0.333	0.121	8.246	2.929	5.849
10773	0.5326	255	298.0	S	0.085	0.031	0.383	0.121	8.246	2.406	5.521
				R	0.062	0.022	0.415	0.121	8.246	2.028	5.315
				V	0.389	0.156	0.248	0.257	3.896	2.119	3.658
LITHIUM BROMIDE				H	0.377	0.150	0.255	0.257	3.896	2.080	3.629
10774	3.4700	4	ROOM	S	0.373	0.148	0.257	0.257	3.896	2.069	3.620
				R	0.360	0.142	0.266	0.257	3.896	2.025	3.587
				V	0.501	0.203	0.235	0.315	3.171	3.128	5.314
LITHIUM CHLORIDE				H	0.486	0.196	0.243	0.315	3.171	3.070	5.269
10775	2.0750	4	ROOM	S	0.482	0.193	0.245	0.315	3.171	3.053	5.256
				R	0.465	0.185	0.254	0.315	3.171	2.989	5.207
				V	0.582	0.238	0.226	0.354	2.826	3.355	5.636
LITHIUM CHLORIDE				H	0.576	0.235	0.229	0.354	2.826	3.333	5.619
10776	2.1110	152	4.2	S	0.575	0.234	0.229	0.354	2.826	3.330	5.616
				R	0.569	0.231	0.232	0.354	2.826	3.307	5.598
				V	0.500	0.202	0.238	0.318	3.142	3.126	5.331
LITHIUM CHLORIDE				H	0.484	0.194	0.247	0.318	3.142	3.064	5.282
10777	2.0680	152	295.0	S	0.479	0.192	0.249	0.318	3.142	3.044	5.267
				R	0.461	0.183	0.259	0.318	3.142	2.975	5.215
				V	0.573	0.236	0.215	0.334	2.990	3.325	5.516
LITHIUM CHLORIDE				H	0.567	0.233	0.218	0.334	2.990	3.304	5.499
10778	2.1321	176	4.2	S	0.566	0.232	0.218	0.334	2.990	3.300	5.496
				R	0.559	0.229	0.221	0.334	2.990	3.278	5.478
				V	0.573	0.236	0.214	0.334	2.992	3.325	5.515
LITHIUM CHLORIDE				H	0.567	0.233	0.217	0.334	2.992	3.304	5.498
10779	2.1321	176	20.0	S	0.566	0.232	0.218	0.334	2.992	3.300	5.495
				R	0.559	0.229	0.221	0.334	2.992	3.278	5.477
				V	0.571	0.236	0.212	0.331	3.023	3.325	5.501
LITHIUM CHLORIDE				H	0.565	0.233	0.215	0.331	3.023	3.303	5.484
10780	2.1317	176	40.0	S	0.564	0.232	0.216	0.331	3.023	3.300	5.481
				R	0.558	0.229	0.219	0.331	3.023	3.278	5.463

CUBIC

CODE / NAME	DENSITY	REF	TEMP(K)		E	G	SIGMA	K	BETA	VS	VP
				V	0.568	0.235	0.208	0.324	3.082	3.323	5.473
LITHIUM CHLORIDE				H	0.562	0.232	0.211	0.324	3.082	3.302	5.456
10781	2.1302	176	60.0	S	0.562	0.232	0.212	0.324	3.082	3.298	5.453
				R	0.555	0.229	0.215	0.324	3.082	3.276	5.435
				V	0.564	0.234	0.203	0.317	3.153	3.320	5.441
LITHIUM CHLORIDE				H	0.558	0.231	0.207	0.317	3.153	3.298	5.423
10782	2.1277	176	80.0	S	0.558	0.231	0.207	0.317	3.153	3.295	5.421
				R	0.551	0.228	0.210	0.317	3.153	3.272	5.402
				V	0.560	0.233	0.200	0.310	3.221	3.314	5.409
LITHIUM CHLORIDE				H	0.554	0.230	0.203	0.310	3.221	3.292	5.391
10783	2.1244	176	100.0	S	0.553	0.230	0.203	0.310	3.221	3.289	5.389
				R	0.547	0.227	0.206	0.310	3.221	3.267	5.370
				V	0.555	0.232	0.197	0.305	3.282	3.306	5.379
LITHIUM CHLORIDE				H	0.549	0.229	0.200	0.305	3.282	3.284	5.361
10784	2.1205	176	120.0	S	0.548	0.228	0.200	0.305	3.282	3.280	5.359
				R	0.542	0.225	0.204	0.305	3.282	3.258	5.340
				V	0.549	0.230	0.195	0.300	3.337	3.295	5.351
LITHIUM CHLORIDE				H	0.543	0.227	0.198	0.300	3.337	3.273	5.333
10785	2.1162	176	140.0	S	0.542	0.226	0.198	0.300	3.337	3.269	5.330
				R	0.536	0.223	0.202	0.300	3.337	3.246	5.311
				V	0.543	0.228	0.194	0.296	3.382	3.283	5.327
LITHIUM CHLORIDE				H	0.537	0.224	0.197	0.296	3.382	3.260	5.308
10786	2.1115	176	160.0	S	0.536	0.224	0.198	0.296	3.382	3.256	5.305
				R	0.530	0.221	0.201	0.296	3.382	3.232	5.285
				V	0.538	0.225	0.194	0.293	3.412	3.269	5.306
LITHIUM CHLORIDE				H	0.531	0.222	0.198	0.293	3.412	3.244	5.286
10787	2.1065	176	180.0	S	0.530	0.221	0.199	0.293	3.412	3.239	5.282
				R	0.523	0.218	0.202	0.293	3.412	3.213	5.261
				V	0.531	0.222	0.197	0.292	3.423	3.250	5.290
LITHIUM CHLORIDE				H	0.524	0.218	0.201	0.292	3.423	3.223	5.268
10788	2.1013	176	200.0	S	0.523	0.218	0.202	0.292	3.423	3.218	5.264
				R	0.516	0.214	0.206	0.292	3.423	3.189	5.241
				V	0.525	0.219	0.200	0.292	3.431	3.231	5.275
LITHIUM CHLORIDE				H	0.517	0.215	0.204	0.291	3.431	3.200	5.250
10789	2.0960	176	220.0	S	0.516	0.214	0.205	0.291	3.431	3.195	5.246
				R	0.507	0.210	0.210	0.291	3.431	3.163	5.220
				V	0.519	0.216	0.203	0.291	3.440	3.212	5.259
LITHIUM CHLORIDE				H	0.510	0.211	0.208	0.291	3.440	3.178	5.232
10790	2.0906	176	240.0	S	0.508	0.210	0.209	0.291	3.440	3.171	5.226
				R	0.499	0.206	0.214	0.291	3.440	3.136	5.198
				V	0.513	0.213	0.205	0.290	3.451	3.193	5.244
LITHIUM CHLORIDE				H	0.503	0.208	0.211	0.290	3.451	3.156	5.214
10791	2.0852	176	260.0	S	0.501	0.207	0.212	0.290	3.451	3.148	5.207
				R	0.491	0.202	0.218	0.290	3.451	3.109	5.176
				V	0.506	0.210	0.208	0.289	3.465	3.175	5.227
LITHIUM CHLORIDE				H	0.496	0.204	0.213	0.289	3.465	3.135	5.194
10792	2.0796	176	280.0	S	0.494	0.203	0.215	0.289	3.465	3.126	5.187
				R	0.483	0.198	0.221	0.289	3.465	3.084	5.153
				V	1.255	0.526	0.193	0.682	1.467	4.496	7.291
LITHIUM FLUORIDE				H	1.227	0.511	0.200	0.682	1.467	4.433	7.239
10793	2.6010	128	ROOM	S	1.220	0.508	0.202	0.682	1.467	4.418	7.227
				R	1.191	0.493	0.209	0.682	1.467	4.351	7.173
				V	1.071	0.446	0.200	0.594	1.684	4.143	6.762
LITHIUM FLUORIDE				H	1.035	0.428	0.210	0.594	1.684	4.055	6.690
10794	2.6010	129	ROOM	S	1.023	0.422	0.213	0.594	1.684	4.027	6.668
				R	0.984	0.402	0.224	0.594	1.683	3.932	6.592
				V	1.156	0.472	0.226	0.702	1.425	4.258	7.153
LITHIUM FLUORIDE				H	1.141	0.464	0.229	0.702	1.425	4.224	7.126
10795	2.6010	129	ROOM	S	1.138	0.463	0.230	0.702	1.425	4.217	7.121
				R	1.122	0.455	0.234	0.702	1.424	4.180	7.092

CUBIC

CODE	NAME DENSITY	REF	TEMP(K)		E	G	SIGMA	K	BETA	VS	VP
				V	1.128	0.451	0.251	0.755	1.324	4.163	7.221
LITHIUM FLUORIDE				H	1.105	0.440	0.256	0.755	1.324	4.112	7.183
10796	2.6010	129	ROOM	S	1.100	0.437	0.257	0.755	1.324	4.101	7.174
				R	1.075	0.425	0.263	0.755	1.324	4.044	7.131
				V	1.059	0.436	0.214	0.617	1.622	4.094	6.787
LITHIUM FLUORIDE				H	1.023	0.418	0.223	0.617	1.622	4.009	6.719
10797	2.6010	129	ROOM	S	1.012	0.413	0.226	0.617	1.622	3.983	6.698
				R	0.974	0.394	0.237	0.617	1.622	3.891	6.625
				V	1.209	0.505	0.198	0.666	1.501	4.381	7.136
LITHIUM FLUORIDE				H	1.170	0.485	0.207	0.666	1.501	4.293	7.065
10798	2.6300	4	ROOM	S	1.159	0.479	0.210	0.666	1.501	4.266	7.043
				R	1.117	0.458	0.220	0.666	1.501	4.172	6.967
				V	1.223	0.515	0.187	0.651	1.537	4.451	7.171
LITHIUM FLUORIDE				H	1.189	0.497	0.196	0.651	1.537	4.372	7.106
10799	2.6010	4	ROOM	S	1.179	0.492	0.198	0.651	1.537	4.350	7.088
				R	1.143	0.474	0.207	0.651	1.537	4.267	7.021
				V	1.311	0.552	0.188	0.699	1.430	4.572	7.372
LITHIUM FLUORIDE				H	1.288	0.540	0.193	0.699	1.430	4.522	7.331
10800	2.6400	45	4.2	S	1.283	0.537	0.194	0.699	1.430	4.511	7.322
				R	1.260	0.525	0.200	0.699	1.430	4.459	7.280
				V	1.311	0.552	0.188	0.699	1.430	4.572	7.372
LITHIUM FLUORIDE				H	1.288	0.540	0.193	0.699	1.430	4.522	7.331
10801	2.6400	45	20.0	S	1.283	0.537	0.194	0.699	1.430	4.511	7.322
				R	1.260	0.525	0.200	0.699	1.430	4.459	7.280
				V	1.311	0.553	0.185	0.695	1.439	4.577	7.366
LITHIUM FLUORIDE				H	1.289	0.541	0.191	0.695	1.439	4.528	7.326
10802	2.6400	45	40.0	S	1.285	0.539	0.192	0.695	1.439	4.519	7.317
				R	1.262	0.527	0.197	0.695	1.439	4.468	7.276
				V	1.309	0.552	0.186	0.695	1.440	4.573	7.362
LITHIUM FLUORIDE				H	1.287	0.540	0.191	0.695	1.440	4.523	7.320
10803	2.6400	45	60.0	S	1.282	0.538	0.192	0.695	1.440	4.512	7.312
				R	1.258	0.525	0.198	0.695	1.440	4.460	7.269
				V	1.307	0.552	0.185	0.692	1.444	4.571	7.354
LITHIUM FLUORIDE				H	1.284	0.539	0.191	0.692	1.444	4.520	7.312
10804	2.6400	45	80.0	S	1.280	0.537	0.192	0.692	1.444	4.509	7.303
				R	1.256	0.524	0.198	0.692	1.444	4.457	7.260
				V	1.303	0.550	0.186	0.691	1.448	4.564	7.346
LITHIUM FLUORIDE				H	1.279	0.537	0.191	0.691	1.448	4.512	7.302
10805	2.6380	45	100.0	S	1.274	0.534	0.192	0.691	1.448	4.501	7.293
				R	1.250	0.521	0.198	0.691	1.448	4.446	7.248
				V	1.298	0.547	0.185	0.688	1.454	4.557	7.333
LITHIUM FLUORIDE				H	1.273	0.534	0.191	0.688	1.454	4.502	7.288
10806	2.6360	45	120.0	S	1.268	0.532	0.193	0.688	1.454	4.491	7.278
				R	1.242	0.518	0.199	0.688	1.454	4.434	7.232
				V	1.292	0.545	0.186	0.685	1.460	4.548	7.320
LITHIUM FLUORIDE				H	1.266	0.531	0.192	0.685	1.460	4.491	7.272
10807	2.6340	45	140.0	S	1.260	0.528	0.193	0.685	1.460	4.478	7.262
				R	1.234	0.514	0.200	0.685	1.460	4.418	7.213
				V	1.285	0.542	0.186	0.681	1.468	4.538	7.306
LITHIUM FLUORIDE				H	1.258	0.527	0.192	0.681	1.468	4.478	7.256
10808	2.6300	45	160.0	S	1.252	0.524	0.194	0.681	1.468	4.464	7.244
				R	1.223	0.509	0.201	0.681	1.468	4.401	7.193
				V	1.277	0.539	0.186	0.678	1.476	4.529	7.291
LITHIUM FLUORIDE				H	1.249	0.524	0.193	0.678	1.476	4.466	7.239
10809	2.6260	45	180.0	S	1.243	0.520	0.194	0.678	1.476	4.451	7.226
				R	1.213	0.505	0.202	0.678	1.476	4.384	7.172
				V	1.269	0.535	0.187	0.675	1.481	4.516	7.276
LITHIUM FLUORIDE				H	1.239	0.519	0.194	0.675	1.481	4.449	7.221
10810	2.6220	45	200.0	S	1.232	0.515	0.196	0.675	1.481	4.433	7.208
				R	1.201	0.499	0.203	0.675	1.481	4.363	7.150

CUBIC

CODE	NAME DENSITY	REF	TEMP(K)		E	G	SIGMA	K	BETA	VS	VP
				V	1.260	0.531	0.187	0.671	1.489	4.503	7.258
LITHIUM FLUORIDE				H	1.230	0.515	0.195	0.671	1.489	4.433	7.201
10811	2.6180	45	220.0	S	1.222	0.510	0.197	0.671	1.489	4.415	7.186
				R	1.189	0.494	0.205	0.671	1.489	4.342	7.126
				V	1.252	0.527	0.187	0.666	1.501	4.491	7.238
LITHIUM FLUORIDE				H	1.220	0.510	0.195	0.666	1.501	4.419	7.178
10812	2.6140	45	240.0	S	1.211	0.506	0.197	0.666	1.501	4.400	7.163
				R	1.178	0.489	0.205	0.666	1.501	4.324	7.101
				V	1.242	0.523	0.187	0.662	1.510	4.477	7.218
LITHIUM FLUORIDE				H	1.209	0.506	0.196	0.662	1.510	4.402	7.156
10813	2.6090	45	260.0	S	1.200	0.501	0.198	0.662	1.510	4.382	7.140
				R	1.166	0.483	0.207	0.662	1.510	4.303	7.075
				V	1.233	0.519	0.187	0.656	1.524	4.466	7.195
LITHIUM FLUORIDE				H	1.199	0.502	0.195	0.656	1.524	4.388	7.131
10814	2.6050	45	280.0	S	1.190	0.497	0.198	0.656	1.524	4.367	7.114
				R	1.155	0.478	0.207	0.656	1.524	4.285	7.048
				V	1.224	0.516	0.187	0.651	1.536	4.453	7.174
LITHIUM FLUORIDE				H	1.189	0.497	0.195	0.651	1.536	4.373	7.108
10815	2.6010	45	300.0	S	1.180	0.492	0.198	0.651	1.536	4.351	7.090
				R	1.143	0.474	0.207	0.651	1.536	4.267	7.021
				V	1.220	0.505	0.208	0.697	1.435	4.407	7.259
LITHIUM FLUORIDE				H	1.180	0.484	0.218	0.697	1.435	4.316	7.186
10816	2.6000	181	293.0	S	1.167	0.478	0.221	0.697	1.435	4.288	7.163
				R	1.124	0.456	0.231	0.697	1.435	4.190	7.085
				V	1.214	0.517	0.175	0.622	1.608	4.424	7.047
LITHIUM FLUORIDE				H	1.177	0.497	0.185	0.622	1.608	4.337	6.974
10817	2.6400	246	293.0	S	1.166	0.491	0.188	0.622	1.608	4.312	6.953
				R	1.127	0.470	0.198	0.622	1.608	4.221	6.878
				V	1.160	0.495	0.171	0.588	1.701	4.337	6.885
LITHIUM FLUORIDE				H	1.118	0.473	0.183	0.588	1.701	4.236	6.801
10818	2.6330	246	373.0	S	1.104	0.465	0.187	0.588	1.701	4.204	6.774
				R	1.061	0.442	0.199	0.588	1.701	4.099	6.688
				V	1.094	0.469	0.165	0.545	1.836	4.230	6.680
LITHIUM FLUORIDE				H	1.047	0.444	0.180	0.545	1.836	4.112	6.581
10819	2.6230	246	473.0	S	1.029	0.434	0.185	0.545	1.836	4.068	6.545
				R	0.980	0.408	0.200	0.545	1.836	3.944	6.443
				V	1.024	0.444	0.151	0.489	2.043	4.125	6.437
LITHIUM FLUORIDE				H	0.972	0.416	0.169	0.489	2.043	3.989	6.321
10820	2.6120	246	573.0	S	0.950	0.404	0.176	0.489	2.043	3.932	6.273
				R	0.897	0.375	0.195	0.489	2.043	3.790	6.156
				V	0.953	0.421	0.132	0.431	2.319	4.024	6.179
LITHIUM FLUORIDE				H	0.897	0.389	0.153	0.431	2.319	3.867	6.044
10821	2.6000	246	673.0	S	0.871	0.374	0.163	0.431	2.319	3.795	5.982
				R	0.814	0.343	0.185	0.431	2.319	3.633	5.847
				V	0.878	0.399	0.101	0.367	2.727	3.925	5.892
LITHIUM FLUORIDE				H	0.819	0.363	0.128	0.367	2.727	3.747	5.735
10822	2.5870	246	773.0	S	0.789	0.346	0.141	0.367	2.727	3.656	5.656
				R	0.730	0.313	0.168	0.367	2.727	3.477	5.504
				V	0.799	0.378	0.058	0.301	3.318	3.831	5.594
LITHIUM FLUORIDE				H	0.740	0.339	0.091	0.301	3.318	3.630	5.412
10823	2.5720	246	873.0	S	0.706	0.318	0.110	0.301	3.318	3.517	5.311
				R	0.648	0.284	0.142	0.301	3.318	3.320	5.140
				V	0.717	0.357	0.004	0.241	4.149	3.737	5.296
LITHIUM FLUORIDE				H	0.659	0.315	0.044	0.241	4.149	3.512	5.086
10824	2.5570	246	973.0	S	0.622	0.291	0.070	0.241	4.149	3.373	4.959
				R	0.566	0.255	0.108	0.241	4.149	3.161	4.769
				V	0.676	0.347	-0.028	0.213	4.686	3.693	5.153
LITHIUM FLUORIDE				H	0.618	0.304	0.017	0.213	4.686	3.455	4.928
10825	2.5480	246	1023.0	S	0.581	0.278	0.046	0.213	4.686	3.301	4.786
				R	0.527	0.242	0.089	0.213	4.686	3.082	4.587

CUBIC

CODE / NAME	DENSITY	REF	TEMP(K)		E	G	SIGMA	K	BETA	VS	VP
				V	1.235	0.513	0.204	0.694	1.440	4.441	7.281
LITHIUM FLUORIDE				H	1.192	0.491	0.214	0.694	1.440	4.346	7.204
10826	2.6000	57	292.0	S	1.179	0.484	0.217	0.694	1.440	4.316	7.180
				R	1.134	0.462	0.228	0.694	1.440	4.214	7.098
				V	1.167	0.490	0.191	0.629	1.590	4.350	7.037
LITHIUM FLUORIDE				H	1.121	0.466	0.203	0.629	1.590	4.242	6.948
10827	2.5900	57	373.0	S	1.106	0.458	0.207	0.629	1.590	4.205	6.918
				R	1.057	0.433	0.220	0.629	1.590	4.090	6.826
				V	1.132	0.481	0.177	0.585	1.710	4.309	6.880
LITHIUM FLUORIDE				H	1.085	0.456	0.191	0.585	1.710	4.194	6.785
10828	2.5900	57	423.0	S	1.068	0.447	0.196	0.585	1.710	4.152	6.750
				R	1.018	0.421	0.210	0.585	1.710	4.031	6.652
				V	1.090	0.465	0.173	0.556	1.800	4.244	6.749
LITHIUM FLUORIDE				H	1.040	0.438	0.188	0.556	1.800	4.120	6.645
10829	2.5800	57	473.0	S	1.021	0.428	0.194	0.556	1.800	4.072	6.606
				R	0.969	0.401	0.209	0.556	1.800	3.941	6.499
				V	1.053	0.449	0.173	0.538	1.860	4.171	6.636
LITHIUM FLUORIDE				H	1.000	0.420	0.190	0.538	1.860	4.036	6.524
10830	2.5800	57	523.0	S	0.978	0.409	0.197	0.538	1.860	3.980	6.478
				R	0.923	0.380	0.214	0.538	1.860	3.837	6.362
				V	1.010	0.433	0.167	0.505	1.980	4.104	6.489
LITHIUM FLUORIDE				H	0.956	0.403	0.185	0.505	1.980	3.962	6.370
10831	2.5700	57	573.0	S	0.932	0.391	0.192	0.505	1.980	3.900	6.320
				R	0.876	0.362	0.211	0.505	1.980	3.751	6.197
				V	0.967	0.418	0.157	0.469	2.130	4.032	6.320
LITHIUM FLUORIDE				H	0.911	0.387	0.176	0.469	2.130	3.882	6.194
10832	2.5700	57	623.0	S	0.886	0.374	0.185	0.469	2.130	3.815	6.137
				R	0.829	0.344	0.206	0.469	2.130	3.658	6.009
				V	0.922	0.402	0.145	0.433	2.310	3.965	6.154
LITHIUM FLUORIDE				H	0.865	0.371	0.167	0.433	2.310	3.804	6.017
10833	2.5600	57	673.0	S	0.838	0.356	0.178	0.433	2.310	3.727	5.952
				R	0.779	0.324	0.200	0.433	2.310	3.560	5.815
				V	0.876	0.387	0.132	0.397	2.520	3.895	5.982
LITHIUM FLUORIDE				H	0.821	0.355	0.155	0.397	2.520	3.732	5.842
10834	2.5500	57	723.0	S	0.794	0.340	0.167	0.397	2.520	3.653	5.775
				R	0.737	0.310	0.190	0.397	2.520	3.485	5.636
				V	0.827	0.374	0.107	0.351	2.850	3.827	5.770
LITHIUM FLUORIDE				H	0.774	0.342	0.132	0.351	2.850	3.660	5.623
10835	2.5500	57	773.0	S	0.747	0.326	0.145	0.351	2.850	3.577	5.552
				R	0.693	0.296	0.171	0.351	2.850	3.408	5.408
				V	1.238	0.514	0.204	0.696	1.436	4.415	7.237
LITHIUM FLUORIDE				H	1.197	0.493	0.213	0.696	1.436	4.323	7.163
10836	2.6390	185	300.0	S	1.184	0.487	0.217	0.696	1.436	4.295	7.140
				R	1.140	0.465	0.227	0.696	1.436	4.196	7.062
				V	1.239	0.514	0.204	0.698	1.433	4.416	7.242
LITHIUM FLUORIDE				H	1.198	0.494	0.214	0.698	1.433	4.325	7.169
10837	2.6380	79	ROOM	S	1.185	0.487	0.217	0.698	1.433	4.297	7.146
				R	1.142	0.465	0.227	0.698	1.433	4.200	7.068
				V	0.276	0.110	0.256	0.188	5.310	1.646	2.872
LITHIUM IODIDE				H	0.268	0.106	0.263	0.188	5.310	1.616	2.849
10838	4.0610	4	ROOM	S	0.265	0.105	0.265	0.188	5.310	1.606	2.842
				R	0.256	0.100	0.274	0.188	5.310	1.572	2.817
				V	0.150	0.058	0.298	0.124	8.095	3.241	6.044
LITHIUM(98.91)-MAGNESIUM(1.09)				H	0.124	0.046	0.333	0.124	8.095	2.906	5.812
10839	0.5485	255	298.0	S	0.086	0.031	0.384	0.124	8.095	2.384	5.487
				R	0.063	0.022	0.415	0.124	8.095	2.008	5.282
				V	0.152	0.058	0.298	0.125	7.974	3.214	5.998
LITHIUM(97.74)-MAGNESIUM(2.26)				H	0.125	0.047	0.334	0.125	7.974	2.882	5.768
10840	0.5650	255	298.0	S	0.087	0.032	0.384	0.125	7.974	2.362	5.444
				R	0.063	0.022	0.416	0.125	7.974	1.989	5.241

CUBIC

CODE	NAME / DENSITY	REF	TEMP(K)		E	G	SIGMA	K	BETA	VS	VP
				V	0.153	0.059	0.297	0.126	7.966	3.200	5.955
LITHIUM(96.99)-MAGNESIUM(3.01)				H	0.126	0.047	0.332	0.126	7.966	2.868	5.725
10841	0.5756	255	298.0	S	0.088	0.032	0.383	0.126	7.966	2.348	5.400
				R	0.064	0.022	0.416	0.126	7.966	1.976	5.198
				V	0.155	0.060	0.299	0.129	7.766	3.171	5.924
LITHIUM(95.72)-MAGNESIUM(4.28)				H	0.128	0.048	0.334	0.129	7.766	2.841	5.697
10842	0.5936	255	298.0	S	0.089	0.032	0.385	0.129	7.766	2.321	5.374
				R	0.064	0.023	0.417	0.129	7.766	1.952	5.174
				V	3.015	1.286	0.172	1.533	0.652	5.993	9.525
MAGNESIUM OXIDE				H	2.975	1.264	0.177	1.533	0.652	5.943	9.482
10843	3.5800	129	ROOM	S	2.968	1.260	0.177	1.533	0.652	5.934	9.475
				R	2.927	1.239	0.182	1.533	0.652	5.882	9.432
				V	3.060	1.309	0.169	1.541	0.649	6.046	9.581
MAGNESIUM OXIDE				H	3.016	1.285	0.174	1.541	0.649	5.991	9.535
10844	3.5800	4	ROOM	S	3.009	1.281	0.175	1.541	0.649	5.981	9.526
				R	2.964	1.256	0.180	1.541	0.649	5.924	9.479
				V	3.176	1.367	0.162	1.566	0.638	6.169	9.714
MAGNESIUM OXIDE				H	3.136	1.345	0.166	1.566	0.638	6.119	9.672
10845	3.5911	80	80.0	S	3.130	1.341	0.167	1.566	0.638	6.111	9.665
				R	3.089	1.319	0.171	1.566	0.638	6.060	9.622
				V	3.176	1.366	0.162	1.568	0.638	6.168	9.715
MAGNESIUM OXIDE				H	3.136	1.344	0.167	1.568	0.638	6.118	9.672
10846	3.5911	80	90.0	S	3.129	1.340	0.167	1.568	0.638	6.109	9.665
				R	3.089	1.318	0.172	1.568	0.638	6.059	9.623
				V	3.165	1.360	0.163	1.567	0.638	6.156	9.704
MAGNESIUM OXIDE				H	3.123	1.337	0.168	1.567	0.638	6.104	9.660
10847	3.5895	80	150.0	S	3.116	1.333	0.168	1.567	0.638	6.095	9.653
				R	3.074	1.311	0.173	1.567	0.638	6.042	9.609
				V	3.161	1.358	0.163	1.566	0.639	6.152	9.700
MAGNESIUM OXIDE				H	3.119	1.335	0.168	1.566	0.639	6.100	9.656
10848	3.5889	80	160.0	S	3.112	1.331	0.169	1.566	0.639	6.091	9.648
				R	3.070	1.308	0.173	1.566	0.639	6.038	9.604
				V	3.158	1.357	0.164	1.565	0.639	6.150	9.698
MAGNESIUM OXIDE				H	3.116	1.334	0.168	1.565	0.639	6.097	9.653
10849	3.5881	80	170.0	S	3.109	1.330	0.169	1.565	0.639	6.088	9.645
				R	3.066	1.306	0.173	1.565	0.639	6.034	9.600
				V	3.155	1.355	0.164	1.564	0.639	6.147	9.694
MAGNESIUM OXIDE				H	3.112	1.332	0.168	1.564	0.639	6.093	9.649
10850	3.5873	80	180.0	S	3.105	1.328	0.169	1.564	0.639	6.084	9.641
				R	3.062	1.304	0.174	1.564	0.639	6.030	9.595
				V	3.151	1.354	0.164	1.563	0.640	6.143	9.690
MAGNESIUM OXIDE				H	3.108	1.330	0.169	1.563	0.640	6.089	9.645
10851	3.5865	80	190.0	S	3.101	1.326	0.169	1.563	0.640	6.080	9.637
				R	3.057	1.302	0.174	1.563	0.640	6.025	9.591
				V	3.147	1.352	0.164	1.561	0.640	6.140	9.686
MAGNESIUM OXIDE				H	3.104	1.328	0.169	1.561	0.640	6.086	9.640
10852	3.5857	80	200.0	S	3.096	1.324	0.169	1.561	0.640	6.076	9.632
				R	3.053	1.300	0.174	1.561	0.640	6.021	9.586
				V	3.143	1.350	0.164	1.560	0.641	6.137	9.681
MAGNESIUM OXIDE				H	3.099	1.326	0.169	1.560	0.641	6.082	9.635
10853	3.5848	80	210.0	S	3.092	1.322	0.170	1.560	0.641	6.072	9.627
				R	3.047	1.297	0.174	1.560	0.641	6.016	9.580
				V	3.139	1.348	0.164	1.559	0.641	6.133	9.678
MAGNESIUM OXIDE				H	3.095	1.324	0.169	1.559	0.641	6.078	9.631
10854	3.5838	80	220.0	S	3.087	1.319	0.170	1.559	0.641	6.068	9.622
				R	3.043	1.295	0.175	1.559	0.641	6.011	9.575
				V	3.135	1.346	0.165	1.558	0.642	6.129	9.674
MAGNESIUM OXIDE				H	3.091	1.322	0.169	1.558	0.642	6.073	9.626
10855	3.5830	80	230.0	S	3.083	1.317	0.170	1.558	0.642	6.063	9.618
				R	3.038	1.293	0.175	1.558	0.642	6.006	9.570

CUBIC

CODE	NAME DENSITY	REF	TEMP(K)		E	G	SIGMA	K	BETA	VS	VP
				V	3.132	1.344	0.165	1.557	0.642	6.126	9.669
MAGNESIUM OXIDE				H	3.087	1.320	0.170	1.557	0.642	6.069	9.622
10856	3.5824	80	240.0	S	3.078	1.315	0.170	1.557	0.642	6.059	9.613
				R	3.033	1.290	0.175	1.557	0.642	6.001	9.565
				V	3.127	1.342	0.165	1.556	0.643	6.122	9.665
MAGNESIUM OXIDE				H	3.082	1.317	0.170	1.556	0.643	6.065	9.617
10857	3.5812	80	250.0	S	3.073	1.313	0.171	1.556	0.643	6.054	9.608
				R	3.027	1.288	0.176	1.556	0.643	5.996	9.559
				V	3.123	1.340	0.165	1.555	0.643	6.118	9.661
MAGNESIUM OXIDE				H	3.077	1.315	0.170	1.555	0.643	6.060	9.612
10858	3.5803	80	260.0	S	3.069	1.310	0.171	1.555	0.643	6.050	9.603
				R	3.023	1.285	0.176	1.555	0.643	5.991	9.554
				V	3.119	1.338	0.165	1.553	0.644	6.115	9.656
MAGNESIUM OXIDE				H	3.073	1.313	0.170	1.553	0.644	6.056	9.607
10859	3.5791	80	270.0	S	3.064	1.308	0.171	1.553	0.644	6.046	9.598
				R	3.017	1.283	0.176	1.553	0.644	5.986	9.548
				V	3.115	1.336	0.165	1.552	0.644	6.111	9.652
MAGNESIUM OXIDE				H	3.068	1.311	0.170	1.552	0.644	6.052	9.602
10860	3.5782	80	280.0	S	3.059	1.306	0.171	1.552	0.644	6.041	9.593
				R	3.012	1.280	0.176	1.552	0.644	5.982	9.543
				V	3.110	1.334	0.166	1.550	0.645	6.107	9.647
MAGNESIUM OXIDE				H	3.063	1.308	0.171	1.550	0.645	6.048	9.597
10861	3.5770	80	290.0	S	3.054	1.303	0.172	1.550	0.645	6.036	9.588
				R	3.007	1.277	0.177	1.550	0.645	5.976	9.537
				V	3.105	1.332	0.166	1.549	0.646	6.103	9.642
MAGNESIUM OXIDE				H	3.058	1.306	0.171	1.549	0.646	6.043	9.592
10862	3.5761	80	300.0	S	3.049	1.301	0.172	1.549	0.646	6.031	9.582
				R	3.001	1.275	0.177	1.549	0.646	5.970	9.531
				V	3.101	1.330	0.166	1.548	0.646	6.099	9.639
MAGNESIUM OXIDE				H	3.054	1.304	0.171	1.548	0.646	6.039	9.588
10863	3.5746	80	310.0	S	3.045	1.299	0.172	1.548	0.646	6.027	9.578
				R	2.996	1.272	0.177	1.548	0.646	5.966	9.527
				V	3.097	1.328	0.166	1.547	0.647	6.095	9.634
MAGNESIUM OXIDE				H	3.049	1.301	0.171	1.547	0.647	6.034	9.583
10864	3.5737	80	320.0	S	3.039	1.296	0.172	1.547	0.647	6.022	9.573
				R	2.990	1.269	0.178	1.547	0.647	5.960	9.521
				V	3.092	1.325	0.166	1.545	0.647	6.091	9.629
MAGNESIUM OXIDE				H	3.044	1.299	0.172	1.545	0.647	6.029	9.577
10865	3.5725	80	330.0	S	3.034	1.294	0.173	1.545	0.647	6.018	9.568
				R	2.985	1.267	0.178	1.545	0.647	5.955	9.515
				V	3.088	1.323	0.167	1.543	0.648	6.087	9.624
MAGNESIUM OXIDE				H	3.039	1.297	0.172	1.543	0.648	6.025	9.572
10866	3.5714	80	340.0	S	3.029	1.291	0.173	1.543	0.648	6.013	9.562
				R	2.979	1.264	0.178	1.543	0.648	5.950	9.508
				V	3.083	1.321	0.167	1.542	0.649	6.083	9.619
MAGNESIUM OXIDE				H	3.034	1.294	0.172	1.542	0.649	6.020	9.566
10867	3.5704	80	350.0	S	3.024	1.289	0.173	1.542	0.649	6.008	9.556
				R	2.973	1.261	0.179	1.542	0.649	5.944	9.502
				V	3.078	1.319	0.167	1.541	0.649	6.079	9.615
MAGNESIUM OXIDE				H	3.028	1.292	0.172	1.541	0.649	6.016	9.561
10868	3.5691	80	360.0	S	3.019	1.286	0.173	1.541	0.649	6.003	9.551
				R	2.968	1.259	0.179	1.541	0.649	5.938	9.496
				V	3.074	1.317	0.167	1.539	0.650	6.075	9.610
MAGNESIUM OXIDE				H	3.023	1.289	0.173	1.539	0.650	6.011	9.556
10869	3.5678	80	370.0	S	3.013	1.284	0.174	1.539	0.650	5.998	9.545
				R	2.962	1.256	0.179	1.539	0.650	5.933	9.491
				V	3.069	1.314	0.167	1.538	0.650	6.071	9.606
MAGNESIUM OXIDE				H	3.018	1.287	0.173	1.538	0.650	6.006	9.551
10870	3.5664	80	380.0	S	3.008	1.281	0.174	1.538	0.650	5.993	9.540
				R	2.956	1.253	0.180	1.538	0.650	5.927	9.485

CODE	NAME / DENSITY	REF	TEMP(K)		E	G	SIGMA	K	BETA	VS	VP
				V	3.064	1.312	0.168	1.537	0.651	6.067	9.601
MAGNESIUM	OXIDE			H	3.013	1.284	0.173	1.537	0.651	6.002	9.546
10871	3.5650	80	390.0	S	3.003	1.278	0.174	1.537	0.651	5.988	9.535
				R	2.950	1.250	0.180	1.537	0.651	5.921	9.479
				V	3.059	1.310	0.168	1.534	0.652	6.063	9.595
MAGNESIUM	OXIDE			H	3.007	1.282	0.173	1.534	0.652	5.996	9.539
10872	3.5640	80	400.0	S	2.997	1.276	0.174	1.534	0.652	5.983	9.528
				R	2.944	1.247	0.180	1.534	0.652	5.915	9.471
				V	3.054	1.308	0.168	1.533	0.652	6.058	9.590
MAGNESIUM	OXIDE			H	3.002	1.279	0.174	1.533	0.652	5.991	9.534
10873	3.5629	80	410.0	S	2.991	1.273	0.175	1.533	0.652	5.978	9.522
				R	2.938	1.244	0.181	1.533	0.652	5.910	9.465
				V	3.050	1.305	0.168	1.532	0.653	6.054	9.586
MAGNESIUM	OXIDE			H	2.997	1.276	0.174	1.532	0.653	5.987	9.529
10874	3.5612	80	420.0	S	2.986	1.271	0.175	1.532	0.653	5.973	9.517
				R	2.932	1.241	0.181	1.532	0.653	5.904	9.460
				V	3.044	1.303	0.168	1.530	0.654	6.050	9.580
MAGNESIUM	OXIDE			H	2.991	1.274	0.174	1.530	0.654	5.982	9.523
10875	3.5596	80	430.0	S	2.980	1.268	0.175	1.530	0.654	5.968	9.511
				R	2.925	1.238	0.181	1.530	0.654	5.898	9.453
				V	3.040	1.301	0.168	1.528	0.654	6.046	9.575
MAGNESIUM	OXIDE			H	2.986	1.271	0.174	1.528	0.654	5.977	9.517
10876	3.5587	80	440.0	S	2.975	1.265	0.176	1.528	0.654	5.963	9.505
				R	2.920	1.236	0.182	1.528	0.654	5.892	9.446
				V	3.034	1.298	0.169	1.526	0.655	6.041	9.569
MAGNESIUM	OXIDE			H	2.980	1.269	0.175	1.526	0.655	5.971	9.510
10877	3.5575	80	450.0	S	2.969	1.262	0.176	1.526	0.655	5.957	9.498
				R	2.913	1.232	0.182	1.526	0.655	5.886	9.439
				V	3.030	1.296	0.169	1.525	0.656	6.037	9.564
MAGNESIUM	OXIDE			H	2.975	1.266	0.175	1.525	0.656	5.967	9.505
10878	3.5560	80	460.0	S	2.963	1.260	0.176	1.525	0.656	5.952	9.493
				R	2.907	1.229	0.182	1.525	0.656	5.880	9.433
				V	3.024	1.293	0.169	1.523	0.656	6.032	9.559
MAGNESIUM	OXIDE			H	2.969	1.263	0.175	1.523	0.656	5.961	9.499
10879	3.5544	80	470.0	S	2.957	1.257	0.176	1.523	0.656	5.946	9.487
				R	2.900	1.226	0.183	1.523	0.656	5.873	9.426
				V	3.019	1.291	0.169	1.521	0.657	6.028	9.553
MAGNESIUM	OXIDE			H	2.963	1.261	0.175	1.521	0.657	5.956	9.492
10880	3.5533	80	480.0	S	2.951	1.254	0.177	1.521	0.657	5.941	9.480
				R	2.894	1.223	0.183	1.521	0.657	5.867	9.418
				V	3.014	1.289	0.169	1.519	0.658	6.023	9.547
MAGNESIUM	OXIDE			H	2.957	1.258	0.176	1.519	0.658	5.950	9.486
10881	3.5522	80	490.0	S	2.945	1.251	0.177	1.519	0.658	5.935	9.473
				R	2.887	1.220	0.183	1.519	0.658	5.860	9.411
				V	3.008	1.286	0.170	1.518	0.659	6.018	9.541
MAGNESIUM	OXIDE			H	2.951	1.255	0.176	1.518	0.659	5.945	9.479
10882	3.5511	80	500.0	S	2.939	1.248	0.177	1.518	0.659	5.929	9.466
				R	2.880	1.217	0.184	1.518	0.659	5.853	9.403
				V	3.003	1.284	0.170	1.515	0.660	6.013	9.534
MAGNESIUM	OXIDE			H	2.945	1.252	0.176	1.515	0.660	5.939	9.472
10883	3.5500	80	510.0	S	2.933	1.245	0.177	1.515	0.660	5.923	9.459
				R	2.874	1.214	0.184	1.515	0.660	5.847	9.395
				V	2.998	1.281	0.170	1.514	0.661	6.009	9.529
MAGNESIUM	OXIDE			H	2.940	1.250	0.176	1.514	0.661	5.934	9.466
10884	3.5490	80	520.0	S	2.927	1.243	0.178	1.514	0.661	5.918	9.452
				R	2.868	1.211	0.184	1.514	0.660	5.841	9.389
				V	2.993	1.279	0.170	1.512	0.662	6.004	9.522
MAGNESIUM	OXIDE			H	2.934	1.247	0.176	1.512	0.662	5.929	9.459
10885	3.5480	80	530.0	S	2.921	1.240	0.178	1.512	0.662	5.912	9.445
				R	2.861	1.208	0.185	1.512	0.662	5.834	9.380

CUBIC

CODE	NAME DENSITY	REF	TEMP(K)		E	G	SIGMA	K	BETA	VS	VP
				V	2.988	1.276	0.170	1.510	0.662	6.000	9.517
MAGNESIUM OXIDE				H	2.928	1.244	0.177	1.510	0.662	5.923	9.453
10886	3.5462	80	540.0	S	2.915	1.237	0.178	1.510	0.662	5.906	9.439
				R	2.854	1.204	0.185	1.510	0.662	5.828	9.374
				V	2.982	1.274	0.170	1.508	0.663	5.995	9.511
MAGNESIUM OXIDE				H	2.922	1.241	0.177	1.508	0.663	5.917	9.446
10887	3.5444	80	550.0	S	2.908	1.234	0.179	1.508	0.663	5.900	9.432
				R	2.847	1.201	0.185	1.508	0.663	5.820	9.366
				V	2.976	1.271	0.171	1.506	0.664	5.991	9.505
MAGNESIUM OXIDE				H	2.916	1.238	0.177	1.506	0.664	5.912	9.440
10888	3.5426	80	560.0	S	2.902	1.231	0.179	1.506	0.664	5.895	9.425
				R	2.840	1.198	0.186	1.506	0.664	5.814	9.358
				V	3.180	1.346	0.181	1.660	0.602	6.124	9.810
MAGNESIUM OXIDE				H	3.126	1.318	0.186	1.660	0.602	6.059	9.756
10889	3.5900	246	293.0	S	3.115	1.312	0.187	1.660	0.602	6.045	9.745
				R	3.060	1.283	0.193	1.660	0.602	5.978	9.689
				V	3.132	1.330	0.178	1.620	0.617	6.094	9.735
MAGNESIUM OXIDE				H	3.075	1.299	0.184	1.620	0.617	6.024	9.676
10890	3.5800	246	373.0	S	3.063	1.293	0.185	1.620	0.617	6.009	9.664
				R	3.005	1.262	0.191	1.620	0.617	5.936	9.604
				V	3.077	1.308	0.176	1.585	0.631	6.055	9.660
MAGNESIUM OXIDE				H	3.016	1.275	0.183	1.585	0.631	5.979	9.596
10891	3.5670	246	473.0	S	3.002	1.268	0.184	1.585	0.631	5.962	9.582
				R	2.940	1.234	0.191	1.585	0.631	5.882	9.516
				V	3.019	1.286	0.174	1.543	0.648	6.015	9.575
MAGNESIUM OXIDE				H	2.954	1.251	0.181	1.543	0.648	5.932	9.505
10892	3.5540	246	573.0	S	2.939	1.242	0.183	1.543	0.648	5.913	9.489
				R	2.871	1.207	0.190	1.543	0.648	5.827	9.418
				V	2.961	1.264	0.171	1.501	0.666	5.976	9.488
MAGNESIUM OXIDE				H	2.891	1.226	0.179	1.501	0.666	5.887	9.413
10893	3.5390	246	673.0	S	2.874	1.217	0.181	1.501	0.666	5.864	9.395
				R	2.803	1.179	0.189	1.501	0.666	5.772	9.318
				V	2.904	1.242	0.169	1.462	0.684	5.937	9.406
MAGNESIUM OXIDE				H	2.830	1.202	0.177	1.462	0.684	5.840	9.325
10894	3.5240	246	773.0	S	2.811	1.191	0.180	1.462	0.684	5.815	9.304
				R	2.735	1.151	0.188	1.462	0.684	5.715	9.221
				V	2.852	1.220	0.169	1.435	0.697	5.898	9.342
MAGNESIUM OXIDE				H	2.774	1.177	0.178	1.435	0.697	5.794	9.255
10895	3.5080	246	873.0	S	2.752	1.166	0.180	1.435	0.697	5.765	9.231
				R	2.671	1.123	0.190	1.435	0.697	5.657	9.142
				V	2.791	1.198	0.165	1.387	0.721	5.858	9.246
MAGNESIUM OXIDE				H	2.709	1.153	0.175	1.387	0.721	5.747	9.152
10896	3.4920	246	973.0	S	2.686	1.141	0.177	1.387	0.721	5.715	9.126
				R	2.601	1.095	0.188	1.387	0.721	5.600	9.030
				V	2.736	1.177	0.162	1.350	0.741	5.819	9.163
MAGNESIUM OXIDE				H	2.650	1.130	0.173	1.350	0.741	5.700	9.063
10897	3.4770	246	1073.0	S	2.624	1.115	0.176	1.350	0.741	5.664	9.033
				R	2.534	1.067	0.187	1.350	0.741	5.541	8.931
				V	2.680	1.156	0.159	1.308	0.764	5.780	9.073
MAGNESIUM OXIDE				H	2.589	1.106	0.170	1.308	0.764	5.652	8.966
10898	3.4620	246	1173.0	S	2.560	1.091	0.174	1.308	0.764	5.613	8.933
				R	2.467	1.040	0.186	1.308	0.764	5.481	8.823
				V	2.625	1.135	0.156	1.274	0.785	5.739	8.993
MAGNESIUM OXIDE				H	2.530	1.082	0.169	1.274	0.785	5.604	8.879
10899	3.4460	246	1273.0	S	2.499	1.065	0.173	1.274	0.785	5.560	8.842
				R	2.401	1.012	0.186	1.274	0.785	5.420	8.725
				V	2.574	1.114	0.156	1.245	0.803	5.700	8.925
MAGNESIUM OXIDE				H	2.474	1.058	0.169	1.245	0.803	5.557	8.803
10900	3.4280	246	1373.0	S	2.440	1.040	0.173	1.245	0.803	5.507	8.762
				R	2.337	0.984	0.187	1.245	0.803	5.359	8.638

CUBIC

CODE	NAME / DENSITY	REF	TEMP(K)		E	G	SIGMA	K	BETA	VS	VP
				V	2.525	1.092	0.156	1.222	0.819	5.658	8.860
MAGNESIUM OXIDE				H	2.420	1.034	0.170	1.222	0.819	5.506	8.730
10901	3.4120	246	1473.0	S	2.382	1.014	0.175	1.222	0.819	5.451	8.684
				R	2.274	0.956	0.190	1.222	0.819	5.292	8.553
				V	3.154	1.340	0.177	1.626	0.615	6.119	9.765
MAGNESIUM OXIDE				H	3.104	1.313	0.182	1.626	0.615	6.057	9.714
10902	3.5790	36	298.0	S	3.094	1.308	0.183	1.626	0.615	6.045	9.704
				R	3.043	1.281	0.188	1.626	0.615	5.982	9.651
				V	3.219	1.371	0.174	1.646	0.607	6.171	9.824
MAGNESIUM OXIDE				H	3.177	1.348	0.178	1.646	0.607	6.120	9.781
10903	3.5995	20	73.0	S	3.170	1.344	0.179	1.646	0.607	6.111	9.774
				R	3.127	1.321	0.183	1.646	0.607	6.058	9.730
				V	3.218	1.370	0.174	1.646	0.607	6.170	9.823
MAGNESIUM OXIDE				H	3.176	1.347	0.178	1.646	0.607	6.118	9.780
10904	3.5994	20	83.0	S	3.168	1.343	0.179	1.646	0.607	6.109	9.772
				R	3.125	1.320	0.184	1.646	0.607	6.056	9.728
				V	3.217	1.370	0.174	1.646	0.608	6.169	9.822
MAGNESIUM OXIDE				H	3.174	1.347	0.179	1.646	0.608	6.117	9.778
10905	3.5992	20	93.0	S	3.167	1.343	0.179	1.646	0.608	6.108	9.771
				R	3.124	1.319	0.184	1.646	0.608	6.055	9.726
				V	3.215	1.369	0.174	1.645	0.608	6.167	9.820
MAGNESIUM OXIDE				H	3.172	1.346	0.179	1.645	0.608	6.115	9.776
10906	3.5989	20	103.0	S	3.165	1.342	0.179	1.645	0.608	6.106	9.768
				R	3.121	1.318	0.184	1.645	0.608	6.053	9.724
				V	3.213	1.368	0.174	1.644	0.608	6.166	9.818
MAGNESIUM OXIDE				H	3.170	1.345	0.179	1.644	0.608	6.113	9.774
10907	3.5986	20	113.0	S	3.163	1.341	0.179	1.644	0.608	6.104	9.766
				R	3.119	1.317	0.184	1.644	0.608	6.051	9.722
				V	3.211	1.367	0.174	1.644	0.608	6.164	9.815
MAGNESIUM OXIDE				H	3.168	1.344	0.179	1.644	0.608	6.111	9.771
10908	3.5983	20	123.0	S	3.161	1.340	0.180	1.644	0.608	6.102	9.763
				R	3.117	1.316	0.184	1.644	0.608	6.048	9.718
				V	3.209	1.366	0.174	1.643	0.609	6.162	9.813
MAGNESIUM OXIDE				H	3.166	1.343	0.179	1.643	0.609	6.109	9.769
10909	3.5977	20	133.0	S	3.158	1.339	0.180	1.643	0.609	6.100	9.761
				R	3.114	1.315	0.184	1.643	0.609	6.046	9.716
				V	3.206	1.365	0.175	1.642	0.609	6.160	9.810
MAGNESIUM OXIDE				H	3.163	1.341	0.179	1.642	0.609	6.106	9.765
10910	3.5972	20	143.0	S	3.155	1.337	0.180	1.642	0.609	6.097	9.757
				R	3.110	1.313	0.184	1.642	0.609	6.042	9.712
				V	3.204	1.364	0.175	1.641	0.609	6.158	9.807
MAGNESIUM OXIDE				H	3.160	1.340	0.179	1.641	0.609	6.104	9.762
10911	3.5966	20	153.0	S	3.152	1.336	0.180	1.641	0.609	6.094	9.754
				R	3.107	1.312	0.184	1.641	0.609	6.039	9.708
				V	3.201	1.362	0.175	1.640	0.610	6.155	9.804
MAGNESIUM OXIDE				H	3.156	1.338	0.179	1.640	0.610	6.101	9.758
10912	3.5960	20	163.0	S	3.149	1.334	0.180	1.640	0.610	6.091	9.750
				R	3.103	1.310	0.185	1.640	0.610	6.035	9.704
				V	3.198	1.361	0.175	1.639	0.610	6.152	9.801
MAGNESIUM OXIDE				H	3.153	1.337	0.179	1.639	0.610	6.098	9.755
10913	3.5953	20	173.0	S	3.145	1.332	0.180	1.639	0.610	6.088	9.747
				R	3.099	1.308	0.185	1.639	0.610	6.032	9.700
				V	3.194	1.359	0.175	1.638	0.611	6.149	9.797
MAGNESIUM OXIDE				H	3.149	1.335	0.180	1.638	0.611	6.094	9.751
10914	3.5946	20	183.0	S	3.141	1.331	0.180	1.638	0.611	6.084	9.742
				R	3.095	1.306	0.185	1.638	0.611	6.027	9.695
				V	3.193	1.359	0.175	1.638	0.610	6.149	9.797
MAGNESIUM OXIDE				H	3.148	1.334	0.180	1.638	0.610	6.093	9.751
10915	3.5938	20	193.0	S	3.140	1.330	0.181	1.638	0.610	6.083	9.743
				R	3.094	1.305	0.185	1.638	0.610	6.027	9.696

CUBIC

CODE	NAME DENSITY	REF	TEMP(K)		E	G	SIGMA	K	BETA	VS	VP
				V	3.233	1.374	0.176	1.665	0.600	6.183	9.866
MAGNESIUM OXIDE				H	3.195	1.353	0.180	1.665	0.600	6.137	9.827
10916	3.5930	20	203.0	S	3.189	1.350	0.181	1.665	0.600	6.130	9.821
				R	3.150	1.329	0.185	1.665	0.600	6.083	9.782
				V	3.183	1.354	0.175	1.634	0.612	6.140	9.785
MAGNESIUM OXIDE				H	3.137	1.329	0.180	1.634	0.612	6.083	9.738
10917	3.5921	20	213.0	S	3.128	1.325	0.181	1.634	0.612	6.073	9.729
				R	3.081	1.299	0.186	1.634	0.612	6.014	9.681
				V	3.179	1.352	0.176	1.633	0.612	6.136	9.781
MAGNESIUM OXIDE				H	3.133	1.327	0.180	1.633	0.612	6.079	9.734
10918	3.5912	20	223.0	S	3.124	1.322	0.181	1.633	0.612	6.068	9.725
				R	3.076	1.297	0.186	1.633	0.612	6.010	9.676
				V	3.176	1.351	0.176	1.632	0.613	6.133	9.778
MAGNESIUM OXIDE				H	3.129	1.325	0.180	1.632	0.613	6.075	9.730
10919	3.5902	20	233.0	S	3.120	1.320	0.181	1.632	0.613	6.065	9.721
				R	3.072	1.295	0.186	1.632	0.613	6.005	9.671
				V	3.172	1.349	0.176	1.631	0.613	6.130	9.774
MAGNESIUM OXIDE				H	3.124	1.323	0.181	1.631	0.613	6.071	9.725
10920	3.5892	20	243.0	S	3.115	1.318	0.182	1.631	0.613	6.060	9.716
				R	3.067	1.292	0.187	1.631	0.613	6.000	9.666
				V	3.167	1.347	0.176	1.629	0.614	6.126	9.770
MAGNESIUM OXIDE				H	3.120	1.321	0.181	1.629	0.614	6.067	9.721
10921	3.5882	20	253.0	S	3.110	1.316	0.182	1.629	0.614	6.056	9.711
				R	3.061	1.290	0.187	1.629	0.614	5.995	9.661
				V	3.163	1.345	0.176	1.627	0.614	6.123	9.765
MAGNESIUM OXIDE				H	3.115	1.319	0.181	1.627	0.614	6.063	9.715
10922	3.5872	20	263.0	S	3.105	1.314	0.182	1.627	0.614	6.052	9.705
				R	3.056	1.287	0.187	1.627	0.614	5.990	9.654
				V	3.159	1.343	0.176	1.626	0.615	6.119	9.761
MAGNESIUM OXIDE				H	3.110	1.316	0.181	1.626	0.615	6.059	9.710
10923	3.5861	20	273.0	S	3.101	1.311	0.182	1.626	0.615	6.047	9.701
				R	3.051	1.285	0.187	1.626	0.615	5.985	9.649
				V	3.154	1.341	0.176	1.624	0.616	6.116	9.756
MAGNESIUM OXIDE				H	3.105	1.314	0.181	1.624	0.616	6.054	9.705
10924	3.5850	20	283.0	S	3.096	1.309	0.182	1.624	0.616	6.043	9.695
				R	3.045	1.282	0.188	1.624	0.616	5.980	9.643
				V	3.151	1.339	0.176	1.623	0.616	6.113	9.752
MAGNESIUM OXIDE				H	3.101	1.312	0.182	1.623	0.616	6.051	9.700
10925	3.5839	20	293.0	S	3.091	1.307	0.183	1.623	0.616	6.039	9.690
				R	3.040	1.280	0.188	1.623	0.616	5.976	9.638
				V	3.103	1.330	0.166	1.550	0.645	6.099	9.641
MAGNESIUM OXIDE				H	3.055	1.304	0.172	1.550	0.645	6.039	9.590
10926	3.5760	57A	298.0	S	3.046	1.299	0.172	1.550	0.645	6.027	9.581
				R	2.998	1.273	0.178	1.550	0.645	5.966	9.529
				V	3.149	1.338	0.177	1.623	0.616	6.114	9.756
MAGNESIUM OXIDE				H	3.099	1.311	0.182	1.623	0.616	6.053	9.705
10927	3.5790	54	298.0	S	3.089	1.306	0.183	1.623	0.616	6.041	9.695
				R	3.038	1.279	0.188	1.623	0.616	5.977	9.643
				V	3.156	1.341	0.177	1.628	0.614	6.117	9.764
MAGNESIUM OXIDE				H	3.106	1.314	0.182	1.628	0.614	6.055	9.713
10928	3.5833	243	300.0	S	3.097	1.309	0.183	1.628	0.614	6.043	9.703
				R	3.045	1.281	0.188	1.628	0.614	5.980	9.650
				V	2.913	1.230	0.184	1.536	0.651	5.911	9.499
MAGNESIUM OXIDE				H	2.839	1.191	0.192	1.536	0.651	5.816	9.420
10929	3.5206	243	800.0	S	2.820	1.181	0.194	1.536	0.651	5.791	9.400
				R	2.743	1.140	0.203	1.536	0.651	5.691	9.318
				V	3.200	1.358	0.178	1.658	0.603	6.141	9.814
MAGNESIUM OXIDE (P=8KB)				H	3.154	1.333	0.183	1.658	0.603	6.085	9.767
10930	3.6010	243	300.0	S	3.146	1.329	0.184	1.658	0.603	6.074	9.758
				R	3.098	1.304	0.188	1.658	0.603	6.017	9.711

CUBIC

CODE / NAME	DENSITY	REF	TEMP(K)		E	G	SIGMA	K	BETA	VS	VP
				V	2.957	1.247	0.185	1.567	0.638	5.936	9.552
MAGNESIUM OXIDE (P=8KB)				H	2.888	1.211	0.193	1.567	0.638	5.849	9.480
10931	3.5396	243	800.0	S	2.871	1.202	0.194	1.567	0.638	5.827	9.462
				R	2.800	1.165	0.202	1.567	0.638	5.736	9.388
				V	0.953	0.360	0.322	0.895	1.118	2.486	4.856
MAGNESIUM(33.33)-COPPER(66.67)				H	0.936	0.353	0.326	0.895	1.118	2.460	4.839
10932	5.8317	56	80.0	S	0.932	0.351	0.326	0.895	1.118	2.454	4.835
				R	0.911	0.343	0.330	0.895	1.118	2.424	4.814
				V	0.951	0.360	0.323	0.894	1.118	2.485	4.856
MAGNESIUM(33.33)-COPPER(66.67)				H	0.934	0.352	0.326	0.894	1.118	2.459	4.839
10933	5.8252	56	100.0	S	0.930	0.351	0.327	0.894	1.118	2.453	4.835
				R	0.910	0.342	0.330	0.894	1.118	2.422	4.814
				V	0.947	0.358	0.323	0.890	1.123	2.482	4.852
MAGNESIUM(33.33)-COPPER(66.67)				H	0.929	0.350	0.326	0.890	1.123	2.456	4.834
10934	5.8087	56	150.0	S	0.925	0.349	0.327	0.890	1.123	2.450	4.830
				R	0.905	0.340	0.331	0.890	1.123	2.419	4.810
				V	0.943	0.356	0.323	0.888	1.127	2.480	4.850
MAGNESIUM(33.33)-COPPER(66.67)				H	0.925	0.349	0.326	0.888	1.127	2.454	4.833
10935	5.7924	56	200.0	S	0.922	0.347	0.327	0.888	1.127	2.448	4.829
				R	0.901	0.339	0.331	0.888	1.127	2.418	4.808
				V	0.936	0.354	0.323	0.883	1.132	2.474	4.843
MAGNESIUM(33.33)-COPPER(66.67)				H	0.919	0.346	0.327	0.883	1.132	2.449	4.826
10936	5.7761	56	250.0	S	0.915	0.345	0.327	0.883	1.132	2.443	4.822
				R	0.895	0.336	0.331	0.883	1.132	2.412	4.801
				V	0.931	0.352	0.324	0.880	1.136	2.471	4.839
MAGNESIUM(33.33)-COPPER(66.67)				H	0.914	0.345	0.327	0.880	1.136	2.446	4.822
10937	5.7600	56	300.0	S	0.911	0.343	0.328	0.880	1.136	2.440	4.818
				R	0.891	0.335	0.331	0.880	1.136	2.410	4.798
				V	0.920	0.348	0.324	0.869	1.150	2.460	4.817
MAGNESIUM(33.33)-COPPER(66.67)				H	0.904	0.341	0.327	0.869	1.150	2.435	4.800
10938	5.7439	56	350.0	S	0.900	0.339	0.327	0.869	1.150	2.430	4.797
				R	0.881	0.331	0.331	0.869	1.150	2.400	4.777
				V	0.909	0.343	0.324	0.860	1.163	2.448	4.796
MAGNESIUM(33.33)-COPPER(66.67)				H	0.893	0.336	0.327	0.860	1.163	2.423	4.779
10939	5.7279	56	400.0	S	0.889	0.335	0.328	0.860	1.163	2.418	4.776
				R	0.870	0.327	0.331	0.860	1.163	2.388	4.756
				V	0.896	0.338	0.324	0.849	1.178	2.434	4.771
MAGNESIUM(33.33)-COPPER(66.67)				H	0.880	0.332	0.327	0.849	1.178	2.410	4.755
10940	5.7120	56	450.0	S	0.877	0.330	0.328	0.849	1.178	2.404	4.751
				R	0.858	0.322	0.331	0.849	1.178	2.375	4.732
				V	0.884	0.334	0.325	0.840	1.190	2.421	4.750
MAGNESIUM(33.33)-COPPER(66.67)				H	0.869	0.327	0.328	0.840	1.190	2.396	4.734
10941	5.6962	56	500.0	S	0.865	0.326	0.328	0.840	1.190	2.391	4.730
				R	0.846	0.318	0.332	0.840	1.190	2.362	4.711
				V	1.109	0.476	0.164	0.550	1.818	4.881	7.698
MAGNESIUM(66.67)-SILICON(33.33)				H	1.109	0.476	0.164	0.550	1.818	4.879	7.697
10942	2.0000	277	300.0	S	1.109	0.476	0.164	0.550	1.818	4.879	7.697
				R	1.108	0.476	0.164	0.550	1.818	4.878	7.696
				V	0.937	0.369	0.268	0.674	1.483	2.453	4.361
MAGNESIUM(50.00)-SILVER(50.00)				H	0.845	0.327	0.291	0.674	1.483	2.309	4.255
10943	6.1340	56	80.0	S	0.777	0.297	0.308	0.674	1.483	2.201	4.177
				R	0.675	0.253	0.333	0.674	1.483	2.031	4.061
				V	0.932	0.367	0.269	0.673	1.486	2.448	4.357
MAGNESIUM(50.00)-SILVER(50.00)				H	0.841	0.325	0.292	0.673	1.486	2.305	4.251
10944	6.1253	56	100.0	S	0.774	0.296	0.308	0.673	1.486	2.197	4.174
				R	0.672	0.252	0.334	0.673	1.486	2.028	4.059
				V	0.923	0.363	0.270	0.670	1.493	2.440	4.348
MAGNESIUM(50.00)-SILVER(50.00)				H	0.834	0.322	0.293	0.670	1.493	2.298	4.244
10945	6.1048	56	150.0	S	0.768	0.293	0.309	0.670	1.493	2.193	4.169
				R	0.668	0.251	0.334	0.670	1.493	2.026	4.055

CUBIC

NAME / CODE	DENSITY	REF	TEMP(K)		E	G	SIGMA	K	BETA	VS	VP
MAGNESIUM(50.00)-SILVER(50.00) 10946	6.0833	56	200.0	V	0.913	0.359	0.272	0.666	1.501	2.429	4.338
				H	0.825	0.319	0.294	0.666	1.501	2.290	4.236
				S	0.762	0.291	0.309	0.666	1.501	2.187	4.163
				R	0.664	0.249	0.334	0.666	1.501	2.023	4.051
MAGNESIUM(50.00)-SILVER(50.00) 10947	6.0626	56	250.0	V	0.901	0.354	0.274	0.664	1.507	2.416	4.328
				H	0.816	0.315	0.295	0.664	1.507	2.279	4.227
				S	0.754	0.288	0.311	0.664	1.507	2.178	4.156
				R	0.658	0.247	0.335	0.664	1.507	2.016	4.046
MAGNESIUM(50.00)-SILVER(50.00) 10948	6.0420	56	300.0	V	0.885	0.347	0.276	0.660	1.515	2.396	4.310
				H	0.803	0.309	0.297	0.660	1.515	2.263	4.213
				S	0.744	0.283	0.312	0.660	1.515	2.166	4.145
				R	0.651	0.244	0.336	0.660	1.515	2.008	4.037
MAGNESIUM(50.00)-SILVER(50.00) 10949	6.0215	56	350.0	V	0.870	0.340	0.278	0.653	1.531	2.377	4.287
				H	0.790	0.304	0.298	0.653	1.531	2.248	4.193
				S	0.735	0.280	0.312	0.653	1.531	2.157	4.129
				R	0.646	0.242	0.335	0.653	1.531	2.004	4.025
MAGNESIUM(50.00)-SILVER(50.00) 10950	6.0011	56	400.0	V	0.855	0.334	0.279	0.646	1.549	2.359	4.264
				H	0.778	0.300	0.299	0.646	1.549	2.234	4.173
				S	0.726	0.277	0.312	0.646	1.549	2.147	4.112
				R	0.640	0.240	0.335	0.646	1.549	1.999	4.011
MAGNESIUM(50.00)-SILVER(50.00) 10951	5.9804	56	450.0	V	0.835	0.326	0.280	0.632	1.582	2.336	4.224
				H	0.763	0.294	0.299	0.632	1.582	2.215	4.137
				S	0.714	0.272	0.312	0.632	1.582	2.133	4.078
				R	0.631	0.237	0.334	0.632	1.582	1.989	3.980
MAGNESIUM(50.00)-SILVER(50.00) 10952	5.9609	56	500.0	V	0.813	0.318	0.280	0.615	1.625	2.308	4.175
				H	0.744	0.287	0.298	0.615	1.625	2.192	4.090
				S	0.699	0.267	0.311	0.615	1.625	2.115	4.036
				R	0.621	0.233	0.332	0.615	1.625	1.978	3.942
MAGNESIUM(66.67)-TIN(33.33) 10953	3.5920	70	300.0	V	0.806	0.343	0.175	0.413	2.419	3.089	4.923
				H	0.803	0.342	0.176	0.413	2.419	3.084	4.918
				S	0.803	0.342	0.176	0.413	2.419	3.084	4.918
				R	0.801	0.340	0.177	0.413	2.419	3.078	4.913
MAGNETITE 10954	5.1800	129	ROOM	V	2.308	0.915	0.261	1.610	0.621	4.203	7.391
				H	2.305	0.914	0.261	1.610	0.621	4.200	7.389
				S	2.305	0.914	0.261	1.610	0.621	4.200	7.389
				R	2.302	0.912	0.262	1.610	0.621	4.197	7.387
MAGNETITE 10955	5.1800	129	ROOM	V	2.313	0.916	0.262	1.620	0.617	4.206	7.407
				H	2.304	0.912	0.263	1.620	0.617	4.197	7.400
				S	2.304	0.912	0.263	1.620	0.617	4.196	7.399
				R	2.294	0.908	0.264	1.620	0.617	4.186	7.392
MAGNETITE 10956	5.1800	4	ROOM	V	2.313	0.917	0.262	1.617	0.619	4.207	7.403
				H	2.308	0.914	0.262	1.617	0.619	4.201	7.399
				S	2.307	0.914	0.262	1.617	0.619	4.201	7.399
				R	2.302	0.912	0.263	1.617	0.619	4.195	7.394
MANGANESE OXIDE 10957	5.3900	198	298.0	V	1.779	0.680	0.308	1.543	0.648	3.552	6.742
				H	1.751	0.668	0.311	1.543	0.648	3.520	6.719
				S	1.745	0.665	0.312	1.543	0.648	3.513	6.715
				R	1.712	0.651	0.315	1.543	0.648	3.475	6.689
MERCURY TELLURIDE 10958	8.0980	18	4.2	V	0.448	0.167	0.342	0.473	2.113	1.436	2.932
				H	0.423	0.156	0.351	0.473	2.113	1.390	2.902
				S	0.410	0.151	0.356	0.473	2.113	1.366	2.887
				R	0.378	0.138	0.367	0.473	2.113	1.307	2.850
MERCURY TELLURIDE 10959	8.0960	18	77.0	V	0.444	0.166	0.342	0.469	2.132	1.430	2.919
				H	0.419	0.155	0.351	0.469	2.132	1.384	2.889
				S	0.406	0.150	0.356	0.469	2.132	1.361	2.874
				R	0.375	0.137	0.367	0.469	2.132	1.302	2.838
MERCURY TELLURIDE 10960	8.0890	18	196.0	V	0.430	0.160	0.341	0.451	2.217	1.407	2.866
				H	0.405	0.150	0.350	0.451	2.217	1.362	2.837
				S	0.393	0.145	0.355	0.451	2.217	1.339	2.822
				R	0.363	0.133	0.366	0.451	2.217	1.281	2.786

CUBIC

CODE	NAME / DENSITY	REF	TEMP(K)		E	G	SIGMA	K	BETA	VS	VP
				V	0.418	0.156	0.341	0.437	2.290	1.389	2.824
MERCURY TELLURIDE				H	0.394	0.146	0.350	0.437	2.290	1.344	2.795
10961	8.0790	18	290.0	S	0.382	0.141	0.354	0.437	2.290	1.322	2.781
				R	0.353	0.129	0.365	0.437	2.290	1.265	2.746
				V	3.311	1.277	0.296	2.702	0.370	3.531	6.556
MOLYBDENUM				H	3.269	1.259	0.298	2.702	0.370	3.505	6.538
10962	10.2480	39	77.0	S	3.263	1.256	0.299	2.702	0.370	3.501	6.535
				R	3.220	1.237	0.301	2.702	0.370	3.474	6.516
				V	3.310	1.277	0.296	2.701	0.370	3.531	6.556
MOLYBDENUM				H	3.269	1.259	0.298	2.701	0.370	3.505	6.537
10963	10.2480	39	80.0	S	3.262	1.256	0.299	2.701	0.370	3.501	6.534
				R	3.219	1.237	0.301	2.701	0.370	3.474	6.515
				V	3.306	1.276	0.296	2.702	0.370	3.528	6.555
MOLYBDENUM				H	3.266	1.257	0.299	2.702	0.370	3.503	6.537
10964	10.2470	39	100.0	S	3.259	1.254	0.299	2.702	0.370	3.499	6.534
				R	3.216	1.235	0.302	2.702	0.370	3.472	6.515
				V	3.302	1.273	0.296	2.702	0.370	3.526	6.554
MOLYBDENUM				H	3.261	1.255	0.299	2.702	0.370	3.500	6.536
10965	10.2450	39	120.0	S	3.254	1.252	0.299	2.702	0.370	3.496	6.533
				R	3.212	1.234	0.302	2.702	0.370	3.470	6.514
				V	3.297	1.271	0.297	2.702	0.370	3.523	6.551
MOLYBDENUM				H	3.256	1.253	0.299	2.702	0.370	3.498	6.533
10966	10.2440	39	140.0	S	3.250	1.250	0.300	2.702	0.370	3.494	6.531
				R	3.208	1.232	0.302	2.702	0.370	3.468	6.512
				V	3.289	1.268	0.297	2.703	0.370	3.518	6.549
MOLYBDENUM				H	3.249	1.250	0.300	2.703	0.370	3.494	6.532
10967	10.2410	39	160.0	S	3.243	1.247	0.300	2.703	0.370	3.490	6.529
				R	3.201	1.229	0.303	2.703	0.370	3.464	6.511
				V	3.281	1.264	0.297	2.699	0.370	3.514	6.545
MOLYBDENUM				H	3.242	1.247	0.300	2.699	0.370	3.490	6.527
10968	10.2380	39	180.0	S	3.235	1.244	0.300	2.699	0.370	3.486	6.524
				R	3.194	1.226	0.303	2.699	0.370	3.461	6.506
				V	3.273	1.261	0.298	2.696	0.371	3.510	6.540
MOLYBDENUM				H	3.234	1.244	0.300	2.696	0.371	3.486	6.523
10969	10.2350	39	200.0	S	3.228	1.241	0.300	2.696	0.371	3.482	6.520
				R	3.188	1.223	0.303	2.696	0.371	3.457	6.502
				V	3.266	1.258	0.298	2.694	0.371	3.507	6.536
MOLYBDENUM				H	3.228	1.241	0.300	2.694	0.371	3.483	6.519
10970	10.2320	39	220.0	S	3.222	1.239	0.301	2.694	0.371	3.479	6.517
				R	3.182	1.221	0.303	2.694	0.371	3.454	6.499
				V	3.258	1.255	0.298	2.691	0.372	3.503	6.532
MOLYBDENUM				H	3.221	1.238	0.301	2.691	0.372	3.479	6.515
10971	10.2290	39	240.0	S	3.215	1.236	0.301	2.691	0.372	3.476	6.512
				R	3.175	1.218	0.303	2.691	0.372	3.451	6.495
				V	3.250	1.251	0.299	2.689	0.372	3.498	6.528
MOLYBDENUM				H	3.213	1.235	0.301	2.689	0.372	3.475	6.511
10972	10.2260	39	260.0	S	3.207	1.232	0.301	2.689	0.372	3.471	6.508
				R	3.168	1.215	0.304	2.689	0.372	3.447	6.491
				V	3.243	1.248	0.299	2.686	0.372	3.495	6.523
MOLYBDENUM				H	3.206	1.232	0.301	2.686	0.372	3.472	6.507
10973	10.2230	39	280.0	S	3.201	1.230	0.301	2.686	0.372	3.468	6.505
				R	3.162	1.213	0.304	2.686	0.372	3.444	6.487
				V	3.234	1.245	0.299	2.683	0.373	3.490	6.519
MOLYBDENUM				H	3.198	1.229	0.301	2.683	0.373	3.468	6.503
10974	10.2190	39	300.0	S	3.193	1.226	0.302	2.683	0.373	3.464	6.500
				R	3.155	1.210	0.304	2.683	0.373	3.441	6.483
				V	3.217	1.238	0.300	2.676	0.374	3.481	6.508
MOLYBDENUM				H	3.182	1.222	0.302	2.676	0.374	3.459	6.493
10975	10.2130	39	340.0	S	3.176	1.220	0.302	2.676	0.374	3.456	6.490
				R	3.140	1.203	0.304	2.676	0.374	3.433	6.474

CUBIC

CODE	NAME / DENSITY	REF	TEMP(K)		E	G	SIGMA	K	BETA	VS	VP
				V	3.198	1.230	0.300	2.669	0.375	3.471	6.497
MOLYBDENUM				H	3.164	1.215	0.302	2.669	0.375	3.450	6.482
10976	10.2070	39	380.0	S	3.159	1.212	0.303	2.669	0.375	3.446	6.479
				R	3.123	1.197	0.305	2.669	0.375	3.424	6.463
				V	3.181	1.223	0.301	2.663	0.376	3.462	6.487
MOLYBDENUM				H	3.148	1.208	0.303	2.663	0.376	3.441	6.472
10977	10.2010	39	420.0	S	3.143	1.206	0.303	2.663	0.376	3.438	6.470
				R	3.108	1.190	0.305	2.663	0.376	3.416	6.455
				V	3.163	1.215	0.302	2.656	0.377	3.452	6.476
MOLYBDENUM				H	3.131	1.201	0.304	2.656	0.377	3.432	6.462
10978	10.1950	39	460.0	S	3.126	1.199	0.304	2.656	0.377	3.429	6.460
				R	3.092	1.184	0.306	2.656	0.377	3.408	6.445
				V	3.145	1.208	0.302	2.648	0.378	3.443	6.465
MOLYBDENUM				H	3.114	1.194	0.304	2.648	0.378	3.423	6.451
10979	10.1890	39	500.0	S	3.110	1.192	0.304	2.648	0.378	3.421	6.449
				R	3.077	1.178	0.306	2.648	0.378	3.400	6.434
				V	3.362	1.304	0.289	2.653	0.377	3.565	6.542
MOLYBDENUM				H	3.359	1.303	0.289	2.653	0.377	3.563	6.540
10980	10.2622	89	0.0	S	3.359	1.303	0.289	2.653	0.377	3.563	6.540
				R	3.355	1.301	0.289	2.653	0.377	3.561	6.539
				V	3.361	1.304	0.289	2.652	0.377	3.565	6.541
MOLYBDENUM				H	3.358	1.302	0.289	2.652	0.377	3.563	6.540
10981	10.2622	89	20.0	S	3.358	1.302	0.289	2.652	0.377	3.562	6.540
				R	3.354	1.301	0.289	2.652	0.377	3.560	6.538
				V	3.359	1.303	0.289	2.650	0.377	3.564	6.538
MOLYBDENUM				H	3.356	1.302	0.289	2.650	0.377	3.562	6.537
10982	10.2622	89	40.0	S	3.355	1.302	0.289	2.650	0.377	3.561	6.537
				R	3.352	1.300	0.289	2.650	0.377	3.559	6.535
				V	3.356	1.302	0.289	2.646	0.378	3.563	6.536
MOLYBDENUM				H	3.353	1.301	0.289	2.646	0.378	3.561	6.535
10983	10.2591	89	60.0	S	3.353	1.301	0.289	2.646	0.378	3.561	6.535
				R	3.349	1.299	0.289	2.646	0.378	3.558	6.533
				V	3.353	1.301	0.289	2.647	0.378	3.561	6.536
MOLYBDENUM				H	3.350	1.299	0.289	2.647	0.378	3.559	6.535
10984	10.2560	89	80.0	S	3.350	1.299	0.289	2.647	0.378	3.559	6.535
				R	3.346	1.298	0.289	2.647	0.378	3.557	6.533
				V	3.350	1.300	0.289	2.644	0.378	3.560	6.533
MOLYBDENUM				H	3.347	1.298	0.289	2.644	0.378	3.558	6.532
10985	10.2529	89	100.0	S	3.346	1.298	0.289	2.644	0.378	3.558	6.532
				R	3.343	1.296	0.289	2.644	0.378	3.556	6.530
				V	3.344	1.298	0.289	2.635	0.379	3.558	6.526
MOLYBDENUM				H	3.341	1.296	0.289	2.635	0.379	3.556	6.525
10986	10.2499	89	120.0	S	3.340	1.296	0.289	2.635	0.379	3.556	6.524
				R	3.337	1.294	0.289	2.635	0.379	3.554	6.523
				V	3.336	1.294	0.289	2.629	0.380	3.554	6.519
MOLYBDENUM				H	3.333	1.293	0.289	2.629	0.380	3.552	6.518
10987	10.2468	89	140.0	S	3.333	1.293	0.289	2.629	0.380	3.552	6.518
				R	3.329	1.291	0.289	2.629	0.380	3.550	6.516
				V	3.328	1.291	0.289	2.625	0.381	3.550	6.514
MOLYBDENUM				H	3.324	1.290	0.289	2.625	0.381	3.548	6.513
10988	10.2437	89	160.0	S	3.324	1.290	0.289	2.625	0.381	3.548	6.513
				R	3.321	1.288	0.289	2.625	0.381	3.546	6.511
				V	3.322	1.289	0.289	2.624	0.381	3.547	6.510
MOLYBDENUM				H	3.319	1.287	0.289	2.624	0.381	3.545	6.509
10989	10.2437	89	180.0	S	3.319	1.287	0.289	2.624	0.381	3.545	6.509
				R	3.315	1.286	0.289	2.624	0.381	3.542	6.507
				V	3.316	1.285	0.290	2.627	0.381	3.543	6.511
MOLYBDENUM				H	3.312	1.284	0.290	2.627	0.381	3.541	6.509
10990	10.2407	89	200.0	S	3.312	1.284	0.290	2.627	0.381	3.541	6.509
				R	3.309	1.282	0.290	2.627	0.381	3.539	6.507

CUBIC

CODE	NAME / DENSITY	REF	TEMP(K)		E	G	SIGMA	K	BETA	VS	VP
				V	3.307	1.282	0.290	2.630	0.380	3.538	6.510
MOLYBDENUM				H	3.304	1.280	0.291	2.630	0.380	3.536	6.509
10991	10.2376	89	220.0	S	3.304	1.280	0.291	2.630	0.380	3.536	6.508
				R	3.301	1.278	0.291	2.630	0.380	3.534	6.507
				V	3.297	1.277	0.291	2.629	0.380	3.532	6.506
MOLYBDENUM				H	3.294	1.276	0.291	2.629	0.380	3.530	6.504
10992	10.2345	89	240.0	S	3.294	1.276	0.291	2.629	0.380	3.530	6.504
				R	3.290	1.274	0.291	2.629	0.380	3.528	6.503
				V	3.287	1.273	0.291	2.627	0.381	3.527	6.501
MOLYBDENUM				H	3.284	1.271	0.292	2.627	0.381	3.525	6.500
10993	10.2315	89	260.0	S	3.284	1.271	0.292	2.627	0.381	3.525	6.500
				R	3.281	1.270	0.292	2.627	0.381	3.523	6.498
				V	3.279	1.269	0.292	2.624	0.381	3.522	6.496
MOLYBDENUM				H	3.275	1.268	0.292	2.624	0.381	3.520	6.494
10994	10.2284	89	280.0	S	3.275	1.268	0.292	2.624	0.381	3.520	6.494
				R	3.272	1.266	0.292	2.624	0.381	3.518	6.493
				V	3.272	1.267	0.292	2.619	0.382	3.519	6.490
MOLYBDENUM				H	3.269	1.265	0.292	2.619	0.382	3.518	6.489
10995	10.2253	89	300.0	S	3.269	1.265	0.292	2.619	0.382	3.517	6.489
				R	3.266	1.264	0.292	2.619	0.382	3.515	6.488
				V	3.344	1.299	0.287	2.618	0.382	3.558	6.511
MOLYBDENUM				H	3.305	1.281	0.290	2.618	0.382	3.533	6.493
10996	10.2620	75	73.0	S	3.298	1.278	0.290	2.618	0.382	3.530	6.490
				R	3.258	1.260	0.293	2.618	0.382	3.504	6.472
				V	3.310	1.284	0.289	2.609	0.383	3.541	6.495
MOLYBDENUM				H	3.273	1.268	0.291	2.609	0.383	3.518	6.478
10997	10.2437	75	173.0	S	3.267	1.265	0.291	2.609	0.383	3.514	6.476
				R	3.228	1.248	0.294	2.609	0.383	3.490	6.458
				V	3.269	1.267	0.290	2.598	0.385	3.520	6.474
MOLYBDENUM				H	3.235	1.251	0.292	2.598	0.385	3.498	6.458
10998	10.2284	75	273.0	S	3.230	1.249	0.293	2.598	0.385	3.495	6.456
				R	3.194	1.233	0.295	2.598	0.385	3.472	6.440
				V	3.225	1.249	0.292	2.580	0.388	3.497	6.447
MOLYBDENUM				H	3.194	1.234	0.294	2.580	0.388	3.476	6.432
10999	10.2128	75	373.0	S	3.189	1.232	0.294	2.580	0.388	3.474	6.430
				R	3.156	1.217	0.296	2.580	0.388	3.452	6.415
				V	3.182	1.230	0.293	2.565	0.390	3.474	6.423
MOLYBDENUM				H	3.153	1.217	0.295	2.565	0.390	3.455	6.409
11000	10.1946	75	473.0	S	3.149	1.215	0.295	2.565	0.390	3.453	6.408
				R	3.118	1.202	0.297	2.565	0.390	3.433	6.394
				V	3.137	1.211	0.295	2.549	0.392	3.449	6.396
MOLYBDENUM				H	3.110	1.199	0.297	2.549	0.392	3.432	6.383
11001	10.1793	75	573.0	S	3.107	1.198	0.297	2.549	0.392	3.430	6.382
				R	3.078	1.185	0.299	2.549	0.392	3.412	6.369
				V	3.090	1.192	0.297	2.533	0.395	3.424	6.368
MOLYBDENUM				H	3.066	1.181	0.298	2.533	0.395	3.408	6.357
11002	10.1640	75	673.0	S	3.063	1.179	0.298	2.533	0.395	3.406	6.356
				R	3.037	1.168	0.300	2.533	0.395	3.390	6.344
				V	3.042	1.171	0.299	2.521	0.397	3.398	6.343
MOLYBDENUM				H	3.021	1.162	0.300	2.521	0.397	3.384	6.333
11003	10.1459	75	773.0	S	3.018	1.160	0.300	2.521	0.397	3.382	6.332
				R	2.995	1.150	0.302	2.521	0.397	3.367	6.322
				V	2.994	1.151	0.300	2.501	0.400	3.371	6.312
MOLYBDENUM				H	2.975	1.143	0.302	2.501	0.400	3.359	6.303
11004	10.1306	75	873.0	S	2.973	1.142	0.302	2.501	0.400	3.357	6.302
				R	2.952	1.133	0.303	2.501	0.400	3.344	6.292
				V	2.946	1.131	0.302	2.480	0.403	3.344	6.279
MOLYBDENUM				H	2.929	1.124	0.303	2.480	0.403	3.333	6.272
11005	10.1155	75	973.0	S	2.927	1.123	0.303	2.480	0.403	3.332	6.271
				R	2.909	1.115	0.305	2.480	0.403	3.320	6.262

CUBIC

CODE	NAME / DENSITY	REF	TEMP(K)		E	G	SIGMA	K	BETA	VS	VP
				V	3.396	1.313	0.293	2.734	0.366	3.460	6.394
	MOLYBDENUM(93.0)-RHENIUM(7.0)			H	3.373	1.303	0.294	2.734	0.366	3.446	6.383
11006	10.9722	69	83.0	S	3.370	1.301	0.295	2.734	0.366	3.444	6.382
				R	3.345	1.290	0.296	2.734	0.366	3.429	6.372
				V	3.396	1.313	0.293	2.734	0.366	3.460	6.394
	MOLYBDENUM(93.0)-RHENIUM(7.0)			H	3.373	1.303	0.294	2.734	0.366	3.446	6.383
11007	10.9722	69	93.0	S	3.370	1.301	0.295	2.734	0.366	3.444	6.382
				R	3.345	1.290	0.296	2.734	0.366	3.429	6.372
				V	3.393	1.312	0.293	2.733	0.366	3.458	6.391
	MOLYBDENUM(93.0)-RHENIUM(7.0)			H	3.369	1.301	0.295	2.733	0.366	3.444	6.381
11008	10.9722	69	103.0	S	3.366	1.300	0.295	2.733	0.366	3.442	6.380
				R	3.341	1.289	0.296	2.733	0.366	3.427	6.369
				V	3.390	1.311	0.293	2.731	0.366	3.457	6.390
	MOLYBDENUM(93.0)-RHENIUM(7.0)			H	3.366	1.300	0.295	2.731	0.366	3.443	6.380
11009	10.9689	69	113.0	S	3.363	1.299	0.295	2.731	0.366	3.441	6.379
				R	3.338	1.288	0.296	2.731	0.366	3.426	6.368
				V	3.387	1.310	0.293	2.730	0.366	3.455	6.388
	MOLYBDENUM(93.0)-RHENIUM(7.0)			H	3.364	1.299	0.295	2.730	0.366	3.441	6.378
11010	10.9689	69	123.0	S	3.361	1.298	0.295	2.730	0.366	3.440	6.377
				R	3.336	1.287	0.296	2.730	0.366	3.425	6.366
				V	3.383	1.308	0.293	2.729	0.366	3.453	6.385
	MOLYBDENUM(93.0)-RHENIUM(7.0)			H	3.360	1.297	0.295	2.729	0.366	3.439	6.376
11011	10.9689	69	133.0	S	3.357	1.296	0.295	2.729	0.366	3.438	6.374
				R	3.333	1.285	0.296	2.729	0.366	3.423	6.364
				V	3.374	1.304	0.294	2.725	0.367	3.448	6.380
	MOLYBDENUM(93.0)-RHENIUM(7.0)			H	3.351	1.294	0.295	2.725	0.367	3.435	6.371
11012	10.9656	69	153.0	S	3.349	1.293	0.295	2.725	0.367	3.433	6.370
				R	3.325	1.282	0.297	2.725	0.367	3.419	6.359
				V	3.365	1.300	0.294	2.722	0.367	3.444	6.376
	MOLYBDENUM(93.0)-RHENIUM(7.0)			H	3.343	1.290	0.295	2.722	0.367	3.431	6.366
11013	10.9623	69	173.0	S	3.341	1.289	0.295	2.722	0.367	3.429	6.365
				R	3.317	1.279	0.297	2.722	0.367	3.416	6.355
				V	3.357	1.297	0.294	2.720	0.368	3.440	6.372
	MOLYBDENUM(93.0)-RHENIUM(7.0)			H	3.335	1.287	0.296	2.720	0.368	3.427	6.363
11014	10.9590	69	193.0	S	3.333	1.286	0.296	2.720	0.368	3.425	6.361
				R	3.310	1.276	0.297	2.720	0.368	3.412	6.352
				V	3.347	1.292	0.295	2.718	0.368	3.435	6.367
	MOLYBDENUM(93.0)-RHENIUM(7.0)			H	3.326	1.283	0.296	2.718	0.368	3.422	6.358
11015	10.9557	69	213.0	S	3.323	1.282	0.296	2.718	0.368	3.421	6.357
				R	3.301	1.272	0.298	2.718	0.368	3.407	6.347
				V	3.338	1.289	0.295	2.716	0.368	3.430	6.363
	MOLYBDENUM(93.0)-RHENIUM(7.0)			H	3.317	1.279	0.296	2.716	0.368	3.418	6.354
11016	10.9524	69	233.0	S	3.315	1.278	0.297	2.716	0.368	3.416	6.353
				R	3.293	1.269	0.298	2.716	0.368	3.403	6.343
				V	3.330	1.285	0.295	2.713	0.369	3.427	6.359
	MOLYBDENUM(93.0)-RHENIUM(7.0)			H	3.309	1.276	0.297	2.713	0.369	3.414	6.351
11017	10.9458	69	253.0	S	3.307	1.275	0.297	2.713	0.369	3.413	6.350
				R	3.286	1.266	0.298	2.713	0.369	3.400	6.340
				V	3.320	1.281	0.296	2.710	0.369	3.422	6.354
	MOLYBDENUM(93.0)-RHENIUM(7.0)			H	3.301	1.272	0.297	2.710	0.369	3.410	6.346
11018	10.9428	69	273.0	S	3.298	1.271	0.297	2.710	0.369	3.409	6.345
				R	3.277	1.262	0.298	2.710	0.369	3.396	6.336
				V	3.309	1.276	0.296	2.709	0.369	3.415	6.349
	MOLYBDENUM(93.0)-RHENIUM(7.0)			H	3.290	1.268	0.298	2.709	0.369	3.404	6.341
11019	10.9428	69	293.0	S	3.288	1.267	0.298	2.709	0.369	3.403	6.340
				R	3.268	1.258	0.299	2.709	0.369	3.391	6.331
				V	3.308	1.276	0.296	2.708	0.369	3.416	6.349
	MOLYBDENUM(93.0)-RHENIUM(7.0)			H	3.289	1.268	0.298	2.708	0.369	3.404	6.341
11020	10.9362	69	298.0	S	3.287	1.267	0.298	2.708	0.369	3.403	6.340
				R	3.267	1.258	0.299	2.708	0.369	3.391	6.332

CODE	NAME / DENSITY	REF	TEMP(K)		E	G	SIGMA	K	BETA	VS	VP
				V	3.300	1.272	0.297	2.707	0.369	3.411	6.346
	MOLYBDENUM(93.0)-RHENIUM(7.0)			H	3.282	1.264	0.298	2.707	0.369	3.400	6.338
11021	10.9329	69	313.0	S	3.280	1.263	0.298	2.707	0.369	3.399	6.337
				R	3.260	1.254	0.299	2.707	0.369	3.387	6.329
				V	3.292	1.269	0.297	2.704	0.370	3.407	6.342
	MOLYBDENUM(93.0)-RHENIUM(7.0)			H	3.273	1.261	0.298	2.704	0.370	3.396	6.334
11022	10.9296	69	333.0	S	3.271	1.260	0.298	2.704	0.370	3.395	6.333
				R	3.252	1.251	0.300	2.704	0.370	3.384	6.325
				V	3.282	1.265	0.297	2.699	0.371	3.402	6.335
	MOLYBDENUM(93.0)-RHENIUM(7.0)			H	3.264	1.257	0.298	2.699	0.371	3.392	6.328
11023	10.9263	69	353.0	S	3.262	1.256	0.299	2.699	0.371	3.391	6.327
				R	3.243	1.248	0.300	2.699	0.371	3.379	6.319
				V	3.272	1.261	0.298	2.695	0.371	3.397	6.329
	MOLYBDENUM(93.0)-RHENIUM(7.0)			H	3.254	1.253	0.299	2.695	0.371	3.387	6.322
11024	10.9230	69	373.0	S	3.252	1.252	0.299	2.695	0.371	3.386	6.321
				R	3.234	1.244	0.300	2.695	0.371	3.374	6.313
				V	3.460	1.335	0.296	2.821	0.355	3.341	6.201
	MOLYBDENUM(83.4)-RHENIUM(16.6)			H	3.453	1.332	0.296	2.821	0.355	3.337	6.198
11025	11.9670	69	83.0	S	3.453	1.332	0.296	2.821	0.355	3.336	6.198
				R	3.445	1.329	0.296	2.821	0.355	3.332	6.195
				V	3.458	1.335	0.296	2.820	0.355	3.340	6.200
	MOLYBDENUM(83.4)-RHENIUM(16.6)			H	3.451	1.332	0.296	2.820	0.355	3.336	6.197
11026	11.9670	69	93.0	S	3.451	1.331	0.296	2.820	0.355	3.335	6.197
				R	3.444	1.328	0.297	2.820	0.355	3.331	6.194
				V	3.455	1.333	0.296	2.820	0.355	3.338	6.198
	MOLYBDENUM(83.4)-RHENIUM(16.6)			H	3.448	1.330	0.296	2.820	0.355	3.334	6.195
11027	11.9670	69	103.0	S	3.447	1.330	0.296	2.820	0.355	3.333	6.195
				R	3.440	1.327	0.297	2.820	0.355	3.329	6.192
				V	3.452	1.332	0.296	2.819	0.355	3.337	6.197
	MOLYBDENUM(83.4)-RHENIUM(16.6)			H	3.446	1.329	0.296	2.819	0.355	3.333	6.195
11028	11.9634	69	113.0	S	3.445	1.329	0.296	2.819	0.355	3.333	6.195
				R	3.438	1.326	0.297	2.819	0.355	3.329	6.192
				V	3.448	1.330	0.296	2.817	0.355	3.335	6.195
	MOLYBDENUM(83.4)-RHENIUM(16.6)			H	3.442	1.327	0.296	2.817	0.355	3.331	6.192
11029	11.9634	69	123.0	S	3.441	1.327	0.296	2.817	0.355	3.331	6.192
				R	3.434	1.324	0.297	2.817	0.355	3.327	6.189
				V	3.446	1.329	0.296	2.816	0.355	3.333	6.193
	MOLYBDENUM(83.4)-RHENIUM(16.6)			H	3.439	1.326	0.296	2.816	0.355	3.330	6.190
11030	11.9634	69	133.0	S	3.439	1.326	0.296	2.816	0.355	3.330	6.190
				R	3.432	1.323	0.297	2.816	0.355	3.326	6.187
				V	3.441	1.327	0.296	2.811	0.356	3.331	6.189
	MOLYBDENUM(83.4)-RHENIUM(16.6)			H	3.434	1.325	0.296	2.811	0.356	3.328	6.187
11031	11.9598	69	153.0	S	3.434	1.324	0.296	2.811	0.356	3.328	6.186
				R	3.427	1.321	0.297	2.811	0.356	3.324	6.184
				V	3.434	1.325	0.296	2.808	0.356	3.328	6.185
	MOLYBDENUM(83.4)-RHENIUM(16.6)			H	3.428	1.322	0.297	2.808	0.356	3.325	6.183
11032	11.9562	69	173.0	S	3.427	1.322	0.297	2.808	0.356	3.325	6.182
				R	3.420	1.319	0.297	2.808	0.356	3.321	6.180
				V	3.426	1.321	0.296	2.805	0.357	3.325	6.181
	MOLYBDENUM(83.4)-RHENIUM(16.6)			H	3.420	1.319	0.297	2.805	0.357	3.321	6.179
11033	11.9526	69	193.0	S	3.419	1.318	0.297	2.805	0.357	3.321	6.179
				R	3.413	1.315	0.297	2.805	0.357	3.317	6.176
				V	3.416	1.317	0.297	2.802	0.357	3.320	6.177
	MOLYBDENUM(83.4)-RHENIUM(16.6)			H	3.411	1.315	0.297	2.802	0.357	3.317	6.174
11034	11.9490	69	213.0	S	3.410	1.314	0.297	2.802	0.357	3.317	6.174
				R	3.404	1.312	0.298	2.802	0.357	3.313	6.172
				V	3.406	1.313	0.297	2.798	0.357	3.315	6.171
	MOLYBDENUM(83.4)-RHENIUM(16.6)			H	3.401	1.311	0.297	2.798	0.357	3.312	6.169
11035	11.9454	69	233.0	S	3.400	1.310	0.297	2.798	0.357	3.312	6.169
				R	3.394	1.308	0.298	2.798	0.357	3.309	6.166

CUBIC

NAME	CODE	DENSITY	REF	TEMP(K)		E	G	SIGMA	K	BETA	VS	VP
MOLYBDENUM(83.4)-RHENIUM(16.6)	11036	11.9382	69	253.0	V	3.398	1.310	0.297	2.795	0.358	3.312	6.167
					H	3.393	1.307	0.298	2.795	0.358	3.309	6.165
					S	3.392	1.307	0.298	2.795	0.358	3.309	6.165
					R	3.387	1.304	0.298	2.795	0.358	3.306	6.163
MOLYBDENUM(83.4)-RHENIUM(16.6)	11037	11.9349	69	273.0	V	3.389	1.306	0.298	2.792	0.358	3.307	6.163
					H	3.384	1.303	0.298	2.792	0.358	3.305	6.161
					S	3.383	1.303	0.298	2.792	0.358	3.304	6.161
					R	3.378	1.301	0.298	2.792	0.358	3.301	6.158
MOLYBDENUM(83.4)-RHENIUM(16.6)	11038	11.9313	69	293.0	V	3.379	1.302	0.298	2.789	0.359	3.303	6.158
					H	3.374	1.299	0.298	2.789	0.359	3.300	6.156
					S	3.374	1.299	0.298	2.789	0.359	3.300	6.156
					R	3.369	1.297	0.299	2.789	0.359	3.297	6.154
MOLYBDENUM(83.4)-RHENIUM(16.6)	11039	11.9277	69	298.0	V	3.377	1.301	0.298	2.789	0.359	3.302	6.158
					H	3.372	1.298	0.298	2.789	0.359	3.299	6.156
					S	3.372	1.298	0.298	2.789	0.359	3.299	6.156
					R	3.367	1.296	0.299	2.789	0.359	3.296	6.154
MOLYBDENUM(83.4)-RHENIUM(16.6)	11040	11.9241	69	313.0	V	3.362	1.295	0.298	2.781	0.360	3.295	6.148
					H	3.358	1.293	0.299	2.781	0.360	3.293	6.146
					S	3.358	1.293	0.299	2.781	0.360	3.292	6.146
					R	3.353	1.291	0.299	2.781	0.360	3.290	6.144
MOLYBDENUM(83.4)-RHENIUM(16.6)	11041	11.9205	69	333.0	V	3.355	1.292	0.298	2.773	0.361	3.292	6.141
					H	3.351	1.290	0.299	2.773	0.361	3.290	6.140
					S	3.351	1.290	0.299	2.773	0.361	3.290	6.140
					R	3.346	1.288	0.299	2.773	0.361	3.287	6.138
MOLYBDENUM(83.4)-RHENIUM(16.6)	11042	11.9169	69	353.0	V	3.347	1.289	0.299	2.771	0.361	3.289	6.138
					H	3.343	1.287	0.299	2.771	0.361	3.286	6.136
					S	3.343	1.287	0.299	2.771	0.361	3.286	6.136
					R	3.339	1.285	0.299	2.771	0.361	3.284	6.134
MOLYBDENUM(83.4)-RHENIUM(16.6)	11043	11.9133	69	373.0	V	3.340	1.286	0.299	2.771	0.361	3.285	6.136
					H	3.336	1.284	0.299	2.771	0.361	3.283	6.134
					S	3.336	1.284	0.299	2.771	0.361	3.283	6.134
					R	3.332	1.282	0.300	2.771	0.361	3.280	6.132
MOLYBDENUM(73.1)-RHENIUM(26.9)	11044	13.0465	69	83.0	V	3.509	1.350	0.300	2.921	0.342	3.217	6.015
					H	3.509	1.350	0.300	2.921	0.342	3.217	6.015
					S	3.509	1.350	0.300	2.921	0.342	3.217	6.015
					R	3.509	1.350	0.300	2.921	0.342	3.217	6.015
MOLYBDENUM(73.1)-RHENIUM(26.9)	11045	13.0465	69	93.0	V	3.507	1.349	0.300	2.919	0.343	3.216	6.013
					H	3.507	1.349	0.300	2.919	0.343	3.216	6.013
					S	3.507	1.349	0.300	2.919	0.343	3.216	6.013
					R	3.507	1.349	0.300	2.919	0.343	3.216	6.013
MOLYBDENUM(73.1)-RHENIUM(26.9)	11046	13.0465	69	103.0	V	3.508	1.350	0.300	2.916	0.343	3.216	6.012
					H	3.508	1.350	0.300	2.916	0.343	3.216	6.012
					S	3.508	1.350	0.300	2.916	0.343	3.216	6.012
					R	3.508	1.350	0.300	2.916	0.343	3.216	6.012
MOLYBDENUM(73.1)-RHENIUM(26.9)	11047	13.0426	69	113.0	V	3.504	1.348	0.300	2.913	0.343	3.215	6.009
					H	3.504	1.348	0.300	2.913	0.343	3.215	6.009
					S	3.504	1.348	0.300	2.913	0.343	3.215	6.009
					R	3.504	1.348	0.300	2.913	0.343	3.215	6.009
MOLYBDENUM(73.1)-RHENIUM(26.9)	11048	13.0426	69	123.0	V	3.501	1.347	0.299	2.909	0.344	3.214	6.006
					H	3.501	1.347	0.299	2.909	0.344	3.214	6.006
					S	3.501	1.347	0.299	2.909	0.344	3.214	6.006
					R	3.501	1.347	0.299	2.909	0.344	3.214	6.006
MOLYBDENUM(73.1)-RHENIUM(26.9)	11049	13.0426	69	133.0	V	3.497	1.346	0.299	2.905	0.344	3.212	6.003
					H	3.497	1.346	0.299	2.905	0.344	3.212	6.003
					S	3.497	1.346	0.299	2.905	0.344	3.212	6.003
					R	3.497	1.346	0.299	2.905	0.344	3.212	6.003
MOLYBDENUM(73.1)-RHENIUM(26.9)	11050	13.0387	69	153.0	V	3.491	1.344	0.299	2.897	0.345	3.210	5.997
					H	3.491	1.344	0.299	2.897	0.345	3.210	5.997
					S	3.491	1.344	0.299	2.897	0.345	3.210	5.997
					R	3.491	1.344	0.299	2.897	0.345	3.210	5.997

CUBIC

NAME / CODE	DENSITY	REF	TEMP(K)		E	G	SIGMA	K	BETA	VS	VP
				V	3.485	1.341	0.299	2.889	0.346	3.208	5.990
MOLYBDENUM(73.1)-RHENIUM(26.9)				H	3.485	1.341	0.299	2.889	0.346	3.208	5.990
11051	13.0348	69	173.0	S	3.485	1.341	0.299	2.889	0.346	3.208	5.990
				R	3.485	1.341	0.299	2.889	0.346	3.208	5.990
				V	3.476	1.338	0.299	2.881	0.347	3.205	5.984
MOLYBDENUM(73.1)-RHENIUM(26.9)				H	3.476	1.338	0.299	2.881	0.347	3.205	5.984
11052	13.0309	69	193.0	S	3.476	1.338	0.299	2.881	0.347	3.205	5.984
				R	3.476	1.338	0.299	2.881	0.347	3.205	5.984
				V	3.470	1.336	0.299	2.873	0.348	3.202	5.977
MOLYBDENUM(73.1)-RHENIUM(26.9)				H	3.470	1.336	0.299	2.873	0.348	3.202	5.977
11053	13.0270	69	213.0	S	3.470	1.336	0.299	2.873	0.348	3.202	5.977
				R	3.470	1.336	0.299	2.873	0.348	3.202	5.977
				V	3.462	1.333	0.299	2.866	0.349	3.199	5.971
MOLYBDENUM(73.1)-RHENIUM(26.9)				H	3.462	1.333	0.299	2.866	0.349	3.199	5.971
11054	13.0231	69	233.0	S	3.462	1.333	0.299	2.866	0.349	3.199	5.971
				R	3.462	1.333	0.299	2.866	0.349	3.199	5.971
				V	3.454	1.330	0.299	2.859	0.350	3.197	5.966
MOLYBDENUM(73.1)-RHENIUM(26.9)				H	3.454	1.330	0.299	2.859	0.350	3.197	5.966
11055	13.0153	69	253.0	S	3.454	1.330	0.299	2.859	0.350	3.197	5.966
				R	3.454	1.330	0.299	2.859	0.350	3.197	5.966
				V	3.447	1.327	0.298	2.850	0.351	3.194	5.959
MOLYBDENUM(73.1)-RHENIUM(26.9)				H	3.447	1.327	0.298	2.850	0.351	3.194	5.959
11056	13.0114	69	273.0	S	3.447	1.327	0.298	2.850	0.351	3.194	5.959
				R	3.447	1.327	0.298	2.850	0.351	3.194	5.959
				V	3.439	1.324	0.298	2.844	0.352	3.191	5.953
MOLYBDENUM(73.1)-RHENIUM(26.9)				H	3.439	1.324	0.298	2.844	0.352	3.191	5.953
11057	13.0072	69	293.0	S	3.439	1.324	0.298	2.844	0.352	3.191	5.953
				R	3.439	1.324	0.298	2.844	0.352	3.191	5.953
				V	3.437	1.323	0.298	2.842	0.352	3.190	5.952
MOLYBDENUM(73.1)-RHENIUM(26.9)				H	3.437	1.323	0.298	2.842	0.352	3.190	5.952
11058	13.0033	69	298.0	S	3.437	1.323	0.298	2.842	0.352	3.190	5.952
				R	3.437	1.323	0.298	2.842	0.352	3.190	5.952
				V	3.430	1.321	0.298	2.837	0.353	3.188	5.947
MOLYBDENUM(73.1)-RHENIUM(26.9)				H	3.430	1.321	0.298	2.837	0.353	3.188	5.947
11059	12.9995	69	313.0	S	3.430	1.321	0.298	2.837	0.353	3.188	5.947
				R	3.430	1.321	0.298	2.837	0.353	3.188	5.947
				V	3.420	1.317	0.298	2.827	0.354	3.184	5.939
MOLYBDENUM(73.1)-RHENIUM(26.9)				H	3.420	1.317	0.298	2.827	0.354	3.184	5.939
11060	12.9956	69	333.0	S	3.420	1.317	0.298	2.827	0.354	3.184	5.939
				R	3.420	1.317	0.298	2.827	0.354	3.184	5.939
				V	3.412	1.314	0.298	2.819	0.355	3.181	5.932
MOLYBDENUM(73.1)-RHENIUM(26.9)				H	3.412	1.314	0.298	2.819	0.355	3.181	5.932
11061	12.9918	69	353.0	S	3.412	1.314	0.298	2.819	0.355	3.181	5.932
				R	3.412	1.314	0.298	2.819	0.355	3.181	5.932
				V	3.404	1.311	0.298	2.810	0.356	3.177	5.924
MOLYBDENUM(73.1)-RHENIUM(26.9)				H	3.404	1.311	0.298	2.810	0.356	3.177	5.924
11062	12.9879	69	373.0	S	3.404	1.311	0.298	2.810	0.356	3.177	5.924
				R	3.404	1.311	0.298	2.810	0.356	3.177	5.924
				V	2.063	0.784	0.316	1.867	0.536	2.968	5.720
NICKEL				H	1.952	0.736	0.326	1.867	0.536	2.876	5.657
11063	8.9000	129	ROOM	S	1.900	0.714	0.330	1.867	0.536	2.833	5.628
				R	1.768	0.659	0.342	1.867	0.536	2.720	5.553
				V	2.156	0.826	0.305	1.847	0.542	3.046	5.755
NICKEL				H	2.070	0.788	0.313	1.847	0.542	2.976	5.706
11064	8.9000	129	ROOM	S	2.038	0.774	0.316	1.847	0.542	2.950	5.688
				R	1.936	0.730	0.325	1.847	0.542	2.865	5.630
				V	2.365	0.916	0.291	1.885	0.531	3.208	5.908
NICKEL				H	2.218	0.850	0.304	1.885	0.531	3.091	5.824
11065	8.9000	129	ROOM	S	2.143	0.817	0.311	1.885	0.531	3.031	5.781
				R	1.972	0.744	0.326	1.885	0.531	2.891	5.685

CUBIC

CODE	NAME DENSITY	REF	TEMP(K)		E	G	SIGMA	K	BETA	VS	VP
				V	2.329	0.906	0.285	1.808	0.553	3.191	5.821
NICKEL				H	2.188	0.843	0.298	1.808	0.553	3.077	5.739
11066	8.9000	129	ROOM	S	2.117	0.811	0.305	1.808	0.553	3.019	5.698
				R	1.954	0.740	0.320	1.808	0.553	2.884	5.604
				V	2.392	0.929	0.288	1.883	0.531	3.230	5.922
NICKEL				H	2.247	0.864	0.301	1.883	0.531	3.115	5.839
11067	8.9000	129	ROOM	S	2.175	0.832	0.307	1.883	0.531	3.057	5.798
				R	2.006	0.758	0.322	1.883	0.531	2.919	5.702
				V	2.422	0.944	0.282	1.856	0.539	3.257	5.916
NICKEL				H	2.284	0.882	0.295	1.856	0.539	3.148	5.836
11068	8.9000	129	ROOM	S	2.218	0.852	0.301	1.856	0.539	3.095	5.799
				R	2.058	0.782	0.315	1.856	0.539	2.965	5.707
				V	2.417	0.947	0.277	1.804	0.554	3.261	5.869
NICKEL				H	2.273	0.881	0.290	1.804	0.554	3.146	5.785
11069	8.9000	129	ROOM	S	2.203	0.849	0.296	1.804	0.554	3.089	5.744
				R	2.037	0.777	0.312	1.804	0.554	2.954	5.648
				V	2.312	0.891	0.297	1.900	0.526	3.164	5.890
NICKEL				H	2.163	0.825	0.310	1.900	0.526	3.045	5.806
11070	8.9000	177	ROOM	S	2.084	0.791	0.317	1.900	0.526	2.981	5.762
				R	1.910	0.717	0.332	1.900	0.526	2.838	5.664
				V	2.571	1.011	0.272	1.876	0.533	3.359	5.999
NICKEL				H	2.432	0.947	0.284	1.876	0.533	3.251	5.919
11071	8.9600	17	0.0	S	2.369	0.918	0.290	1.876	0.533	3.201	5.882
				R	2.209	0.847	0.304	1.876	0.533	3.075	5.792
				V	2.571	1.011	0.272	1.876	0.533	3.359	5.999
NICKEL				H	2.432	0.947	0.284	1.876	0.533	3.251	5.919
11072	8.9600	17	20.0	S	2.369	0.918	0.290	1.876	0.533	3.201	5.882
				R	2.209	0.847	0.304	1.876	0.533	3.075	5.792
				V	2.569	1.010	0.272	1.875	0.533	3.357	5.997
NICKEL				H	2.429	0.946	0.284	1.875	0.533	3.249	5.917
11073	8.9600	17	40.0	S	2.366	0.917	0.290	1.875	0.533	3.199	5.880
				R	2.206	0.846	0.304	1.875	0.533	3.073	5.790
				V	2.564	1.008	0.272	1.873	0.534	3.354	5.992
NICKEL				H	2.424	0.944	0.284	1.873	0.534	3.245	5.911
11074	8.9610	17	60.0	S	2.360	0.915	0.290	1.873	0.534	3.195	5.875
				R	2.201	0.844	0.304	1.873	0.534	3.068	5.784
				V	2.556	1.004	0.272	1.872	0.534	3.348	5.986
NICKEL				H	2.416	0.940	0.285	1.872	0.534	3.239	5.906
11075	8.9600	17	80.0	S	2.353	0.912	0.290	1.872	0.534	3.190	5.870
				R	2.193	0.841	0.305	1.872	0.534	3.063	5.779
				V	2.546	1.000	0.273	1.870	0.535	3.341	5.980
NICKEL				H	2.407	0.936	0.285	1.870	0.535	3.233	5.900
11076	8.9580	17	100.0	S	2.343	0.907	0.291	1.870	0.535	3.183	5.863
				R	2.184	0.837	0.305	1.870	0.535	3.056	5.773
				V	2.535	0.995	0.274	1.866	0.536	3.334	5.971
NICKEL				H	2.396	0.932	0.286	1.866	0.536	3.225	5.891
11077	8.9550	17	120.0	S	2.332	0.903	0.292	1.866	0.536	3.175	5.855
				R	2.173	0.832	0.306	1.866	0.536	3.048	5.764
				V	2.522	0.989	0.274	1.862	0.537	3.325	5.962
NICKEL				H	2.383	0.926	0.287	1.862	0.537	3.216	5.882
11078	8.9510	17	140.0	S	2.319	0.897	0.292	1.862	0.537	3.166	5.845
				R	2.160	0.827	0.307	1.862	0.537	3.039	5.755
				V	2.508	0.984	0.275	1.858	0.538	3.316	5.952
NICKEL				H	2.370	0.920	0.287	1.858	0.538	3.207	5.872
11079	8.9470	17	160.0	S	2.306	0.892	0.293	1.858	0.538	3.157	5.836
				R	2.147	0.821	0.307	1.858	0.538	3.030	5.745
				V	2.496	0.978	0.276	1.855	0.539	3.307	5.944
NICKEL				H	2.358	0.915	0.288	1.855	0.539	3.199	5.865
11080	8.9420	17	180.0	S	2.295	0.887	0.294	1.855	0.539	3.149	5.828
				R	2.136	0.817	0.308	1.855	0.539	3.022	5.738

CUBIC

CODE	NAME	DENSITY	REF	TEMP(K)		E	G	SIGMA	K	BETA	VS	VP
11081	NICKEL	8.9360	17	200.0	V	2.483	0.972	0.277	1.852	0.540	3.299	5.936
					H	2.345	0.910	0.289	1.852	0.540	3.191	5.856
					S	2.281	0.881	0.295	1.852	0.540	3.140	5.820
					R	2.123	0.811	0.309	1.852	0.540	3.013	5.729
11082	NICKEL	8.9310	17	220.0	V	2.469	0.967	0.277	1.849	0.541	3.290	5.927
					H	2.332	0.904	0.290	1.849	0.541	3.182	5.848
					S	2.268	0.875	0.296	1.849	0.541	3.131	5.811
					R	2.110	0.806	0.310	1.849	0.541	3.003	5.721
11083	NICKEL	8.9260	17	240.0	V	2.455	0.960	0.278	1.846	0.542	3.280	5.918
					H	2.318	0.898	0.291	1.846	0.542	3.171	5.839
					S	2.253	0.869	0.297	1.846	0.542	3.120	5.802
					R	2.095	0.799	0.311	1.846	0.542	2.992	5.711
11084	NICKEL	8.9200	17	260.0	V	2.442	0.955	0.279	1.842	0.543	3.271	5.909
					H	2.304	0.892	0.291	1.842	0.543	3.163	5.830
					S	2.240	0.863	0.297	1.842	0.543	3.111	5.793
					R	2.082	0.794	0.312	1.842	0.543	2.983	5.702
11085	NICKEL	8.9120	17	280.0	V	2.428	0.948	0.280	1.839	0.544	3.262	5.901
					H	2.290	0.886	0.292	1.839	0.544	3.153	5.821
					S	2.225	0.857	0.298	1.839	0.544	3.101	5.784
					R	2.066	0.787	0.313	1.839	0.544	2.972	5.693
11086	NICKEL	8.9070	17	300.0	V	2.415	0.943	0.281	1.836	0.545	3.253	5.893
					H	2.277	0.880	0.293	1.836	0.545	3.144	5.813
					S	2.212	0.851	0.299	1.836	0.545	3.091	5.775
					R	2.053	0.782	0.314	1.836	0.545	2.962	5.684
11087	NICKEL	8.8990	17	320.0	V	2.400	0.936	0.282	1.832	0.546	3.244	5.883
					H	2.262	0.874	0.294	1.832	0.546	3.134	5.803
					S	2.197	0.845	0.300	1.832	0.546	3.081	5.765
					R	2.038	0.775	0.315	1.832	0.546	2.951	5.674
11088	NICKEL	8.8910	17	340.0	V	2.385	0.930	0.283	1.829	0.547	3.233	5.875
					H	2.247	0.867	0.295	1.829	0.547	3.123	5.795
					S	2.181	0.838	0.301	1.829	0.547	3.070	5.757
					R	2.022	0.768	0.316	1.829	0.547	2.940	5.665
11089	NICKEL	8.8830	17	360.0	V	2.370	0.923	0.284	1.825	0.548	3.224	5.866
					H	2.232	0.861	0.296	1.825	0.548	3.114	5.785
					S	2.166	0.832	0.302	1.825	0.548	3.060	5.747
					R	2.006	0.762	0.317	1.825	0.548	2.929	5.655
11090	NICKEL	8.8750	17	380.0	V	2.355	0.917	0.284	1.821	0.549	3.214	5.855
					H	2.217	0.854	0.297	1.821	0.549	3.103	5.775
					S	2.150	0.825	0.303	1.821	0.549	3.048	5.736
					R	1.990	0.755	0.318	1.821	0.549	2.917	5.644
11091	NICKEL	8.8670	17	400.0	V	2.340	0.910	0.285	1.815	0.551	3.204	5.845
					H	2.202	0.848	0.298	1.815	0.551	3.093	5.764
					S	2.134	0.818	0.304	1.815	0.551	3.038	5.725
					R	1.974	0.749	0.319	1.815	0.551	2.906	5.633
11092	NICKEL	8.8590	17	420.0	V	2.324	0.904	0.286	1.811	0.552	3.194	5.835
					H	2.186	0.842	0.299	1.811	0.552	3.082	5.754
					S	2.118	0.811	0.305	1.811	0.552	3.027	5.715
					R	1.958	0.742	0.320	1.811	0.552	2.894	5.622
11093	NICKEL	8.8510	17	440.0	V	2.310	0.897	0.287	1.806	0.554	3.184	5.824
					H	2.171	0.835	0.300	1.806	0.554	3.072	5.743
					S	2.103	0.805	0.306	1.806	0.554	3.016	5.703
					R	1.942	0.735	0.321	1.806	0.554	2.882	5.610
11094	NICKEL	8.8430	17	460.0	V	2.293	0.890	0.288	1.802	0.555	3.173	5.813
					H	2.154	0.828	0.301	1.802	0.555	3.060	5.732
					S	2.085	0.797	0.307	1.802	0.555	3.003	5.692
					R	1.924	0.728	0.322	1.802	0.555	2.868	5.599
11095	NICKEL	8.8350	17	480.0	V	2.277	0.883	0.289	1.796	0.557	3.162	5.802
					H	2.138	0.821	0.302	1.796	0.557	3.049	5.721
					S	2.068	0.791	0.308	1.796	0.557	2.991	5.680
					R	1.907	0.721	0.323	1.796	0.557	2.856	5.586

CUBIC

CODE	NAME DENSITY	REF	TEMP(K)		E	G	SIGMA	K	BETA	VS	VP
NICKEL 11096	8.8270	17	500.0	V	2.261	0.877	0.290	1.792	0.558	3.151	5.792
				H	2.122	0.815	0.303	1.792	0.558	3.038	5.710
				S	2.052	0.784	0.309	1.792	0.558	2.980	5.669
				R	1.891	0.714	0.324	1.792	0.558	2.844	5.575
NICKEL 11097	8.8190	17	520.0	V	2.245	0.870	0.290	1.785	0.560	3.141	5.779
				H	2.106	0.808	0.303	1.785	0.560	3.026	5.697
				S	2.034	0.776	0.310	1.785	0.560	2.967	5.655
				R	1.872	0.706	0.325	1.785	0.560	2.830	5.561
NICKEL 11098	8.8110	17	540.0	V	2.227	0.862	0.292	1.781	0.561	3.128	5.767
				H	2.087	0.800	0.305	1.781	0.561	3.013	5.685
				S	2.015	0.768	0.311	1.781	0.561	2.953	5.643
				R	1.853	0.698	0.327	1.781	0.561	2.815	5.548
NICKEL 11099	8.8030	17	560.0	V	2.209	0.855	0.292	1.773	0.564	3.116	5.752
				H	2.068	0.792	0.306	1.773	0.564	3.000	5.669
				S	1.995	0.760	0.313	1.773	0.564	2.938	5.626
				R	1.831	0.690	0.328	1.773	0.564	2.799	5.531
NICKEL 11100	8.7950	17	580.0	V	2.190	0.847	0.293	1.766	0.566	3.103	5.738
				H	2.050	0.784	0.307	1.766	0.566	2.986	5.654
				S	1.976	0.752	0.314	1.766	0.566	2.924	5.611
				R	1.811	0.681	0.329	1.766	0.566	2.784	5.515
NICKEL 11101	8.7870	17	600.0	V	2.173	0.840	0.294	1.759	0.568	3.091	5.724
				H	2.032	0.777	0.308	1.759	0.568	2.974	5.640
				S	1.957	0.744	0.315	1.759	0.568	2.910	5.596
				R	1.792	0.674	0.330	1.759	0.568	2.769	5.499
NICKEL 11102	8.7790	17	620.0	V	2.154	0.831	0.296	1.756	0.570	3.077	5.712
				H	2.012	0.769	0.309	1.756	0.570	2.959	5.628
				S	1.936	0.735	0.316	1.756	0.570	2.894	5.583
				R	1.770	0.665	0.332	1.756	0.570	2.751	5.486
NICKEL 11103	8.7710	17	640.0	V	2.135	0.823	0.297	1.751	0.571	3.064	5.699
				H	1.992	0.760	0.310	1.751	0.571	2.944	5.614
				S	1.915	0.727	0.318	1.751	0.571	2.878	5.568
				R	1.748	0.655	0.334	1.751	0.571	2.734	5.470
NICKEL 11104	8.7640	17	660.0	V	2.119	0.817	0.298	1.746	0.573	3.052	5.687
				H	1.976	0.753	0.311	1.746	0.573	2.932	5.602
				S	1.897	0.719	0.319	1.746	0.573	2.865	5.555
				R	1.729	0.648	0.335	1.746	0.573	2.718	5.457
NICKEL 11105	8.7560	17	680.0	V	2.105	0.810	0.298	1.741	0.574	3.042	5.676
				H	1.961	0.747	0.312	1.741	0.574	2.921	5.591
				S	1.881	0.713	0.320	1.741	0.574	2.853	5.543
				R	1.713	0.641	0.336	1.741	0.574	2.706	5.444
NICKEL 11106	8.7480	17	700.0	V	2.092	0.805	0.299	1.736	0.576	3.033	5.667
				H	1.948	0.742	0.313	1.736	0.576	2.912	5.581
				S	1.869	0.708	0.321	1.736	0.576	2.844	5.534
				R	1.701	0.636	0.337	1.736	0.576	2.697	5.435
NICKEL 11107	8.7410	17	720.0	V	2.078	0.799	0.300	1.730	0.578	3.024	5.655
				H	1.935	0.736	0.314	1.730	0.578	2.902	5.570
				S	1.855	0.702	0.321	1.730	0.578	2.834	5.522
				R	1.687	0.631	0.337	1.730	0.578	2.686	5.423
NICKEL 11108	8.7330	17	740.0	V	2.065	0.794	0.301	1.725	0.580	3.015	5.646
				H	1.922	0.731	0.314	1.725	0.580	2.894	5.560
				S	1.843	0.697	0.322	1.725	0.580	2.825	5.513
				R	1.676	0.626	0.338	1.725	0.580	2.678	5.414
NICKEL 11109	8.7220	17	760.0	V	2.052	0.788	0.301	1.720	0.581	3.007	5.637
				H	1.910	0.726	0.315	1.720	0.581	2.886	5.552
				S	1.830	0.692	0.323	1.720	0.581	2.817	5.504
				R	1.664	0.622	0.339	1.720	0.581	2.670	5.406
NICKEL 11110	8.9100	85	298.0	V	2.395	0.932	0.285	1.860	0.538	3.234	5.900
				H	2.238	0.861	0.299	1.860	0.538	3.109	5.810
				S	2.154	0.824	0.307	1.860	0.538	3.041	5.762
				R	1.973	0.745	0.323	1.860	0.538	2.892	5.659

CUBIC

CODE	NAME	DENSITY	REF	TEMP(K)		E	G	SIGMA	K	BETA	VS	VP
11111	NICKEL	8.9100	264	298.0	V	2.420	0.950	0.274	1.783	0.561	3.265	5.851
					H	2.287	0.889	0.286	1.783	0.561	3.159	5.772
					S	2.225	0.861	0.292	1.783	0.561	3.109	5.736
					R	2.073	0.793	0.306	1.783	0.561	2.984	5.647
11112	NICKEL(90.4)-ALUMINUM(9.6)	8.3800	264	298.0	V	2.522	0.990	0.274	1.860	0.538	3.437	6.160
					H	2.357	0.914	0.289	1.860	0.538	3.303	6.062
					S	2.270	0.875	0.297	1.860	0.538	3.232	6.010
					R	2.081	0.792	0.314	1.860	0.538	3.074	5.899
11113	NICKEL(49.4)-ALUMINUM(50.6)	5.9050	273	298.0	V	2.088	0.809	0.290	1.660	0.602	3.702	6.811
					H	1.906	0.728	0.309	1.660	0.602	3.512	6.675
					S	1.782	0.675	0.321	1.660	0.602	3.380	6.584
					R	1.573	0.586	0.342	1.660	0.602	3.150	6.430
11114	NICKEL(92.7)-COPPER(7.3)	8.9183	85	298.0	V	2.319	0.901	0.287	1.812	0.552	3.178	5.812
					H	2.167	0.833	0.301	1.812	0.552	3.056	5.724
					S	2.086	0.797	0.308	1.812	0.552	2.990	5.678
					R	1.910	0.721	0.324	1.812	0.552	2.844	5.576
11115	NICKEL(82.2)-COPPER(17.8)	8.9301	85	298.0	V	2.208	0.856	0.290	1.755	0.570	3.095	5.694
					H	2.058	0.789	0.305	1.755	0.570	2.972	5.606
					S	1.975	0.753	0.312	1.755	0.570	2.903	5.558
					R	1.801	0.678	0.329	1.755	0.570	2.754	5.456
11116	NICKEL(80.0)-COPPER(20.0)	8.9330	85	298.0	V	2.181	0.843	0.294	1.762	0.568	3.072	5.684
					H	2.024	0.774	0.309	1.762	0.568	2.943	5.592
					S	1.934	0.734	0.317	1.762	0.568	2.867	5.540
					R	1.752	0.657	0.334	1.762	0.568	2.711	5.434
11117	NICKEL(77.2)-COPPER(22.8)	8.9360	85	298.0	V	2.147	0.829	0.295	1.742	0.574	3.047	5.646
					H	1.999	0.764	0.309	1.742	0.574	2.923	5.558
					S	1.916	0.728	0.317	1.742	0.574	2.854	5.510
					R	1.744	0.654	0.333	1.742	0.574	2.705	5.409
11118	NICKEL(65.5)-COPPER(34.5)	8.9495	85	298.0	V	2.025	0.778	0.301	1.697	0.589	2.949	5.528
					H	1.866	0.709	0.317	1.697	0.589	2.814	5.433
					S	1.767	0.666	0.326	1.697	0.589	2.728	5.375
					R	1.582	0.588	0.345	1.697	0.589	2.564	5.266
11119	NICKEL(53.8)-COPPER(46.2)	8.9620	85	298.0	V	1.924	0.737	0.305	1.647	0.607	2.868	5.417
					H	1.770	0.670	0.321	1.647	0.607	2.734	5.324
					S	1.671	0.628	0.331	1.647	0.607	2.647	5.265
					R	1.490	0.552	0.349	1.647	0.607	2.483	5.157
11120	NICKEL(31.1)-COPPER(68.9)	8.9586	85	298.0	V	1.710	0.652	0.311	1.510	0.662	2.698	5.153
					H	1.569	0.591	0.327	1.510	0.662	2.569	5.065
					S	1.476	0.552	0.337	1.510	0.662	2.482	5.007
					R	1.311	0.484	0.355	1.510	0.662	2.323	4.904
11121	NICKEL(73.8)-IRON(26.2)	8.5390	82	300.0	V	2.273	0.887	0.281	1.731	0.578	3.223	5.841
					H	2.115	0.816	0.296	1.731	0.578	3.091	5.745
					S	2.027	0.777	0.305	1.731	0.578	3.016	5.692
					R	1.845	0.698	0.322	1.731	0.578	2.858	5.582
11122	NIOBIUM	8.5780	38	300.0	V	1.104	0.396	0.393	1.718	0.582	2.150	5.118
					H	1.062	0.380	0.397	1.718	0.582	2.105	5.093
					S	1.048	0.375	0.398	1.718	0.582	2.090	5.085
					R	1.001	0.357	0.403	1.718	0.582	2.039	5.057
11123	NIOBIUM (130 PPM TA)	8.5780	272	298.0	V	1.101	0.398	0.384	1.587	0.630	2.153	4.968
					H	1.055	0.380	0.389	1.587	0.630	2.104	4.940
					S	1.038	0.373	0.391	1.587	0.630	2.085	4.930
					R	0.988	0.354	0.396	1.587	0.630	2.030	4.899
11124	NIOBIUM (1175 PPM TA)	8.5780	272	298.0	V	1.100	0.397	0.385	1.595	0.627	2.152	4.977
					H	1.055	0.379	0.390	1.595	0.627	2.103	4.949
					S	1.038	0.373	0.391	1.595	0.627	2.085	4.939
					R	0.988	0.354	0.397	1.595	0.627	2.031	4.908
11125	NIOBIUM	8.5938	47	4.2	V	1.178	0.425	0.387	1.730	0.578	2.223	5.170
					H	1.135	0.408	0.391	1.730	0.578	2.179	5.144
					S	1.120	0.402	0.392	1.730	0.578	2.163	5.136
					R	1.072	0.384	0.397	1.730	0.578	2.113	5.108

CUBIC

CODE	NAME DENSITY	REF	TEMP(K)		E	G	SIGMA	K	BETA	VS	VP
NIOBIUM 11126	8.5938	47	20.0	V	1.174	0.423	0.387	1.738	0.575	2.219	5.175
				H	1.132	0.407	0.391	1.738	0.575	2.175	5.151
				S	1.117	0.401	0.393	1.738	0.575	2.160	5.142
				R	1.070	0.383	0.397	1.738	0.575	2.111	5.115
NIOBIUM 11127	8.5912	47	40.0	V	1.172	0.422	0.388	1.740	0.575	2.217	5.177
				H	1.129	0.406	0.392	1.740	0.575	2.173	5.152
				S	1.115	0.400	0.393	1.740	0.575	2.158	5.144
				R	1.068	0.382	0.398	1.740	0.575	2.109	5.116
NIOBIUM 11128	8.5912	47	60.0	V	1.166	0.420	0.388	1.736	0.576	2.211	5.170
				H	1.124	0.404	0.392	1.736	0.576	2.167	5.145
				S	1.110	0.398	0.393	1.736	0.576	2.153	5.137
				R	1.063	0.380	0.398	1.736	0.576	2.103	5.110
NIOBIUM 11129	8.5886	47	80.0	V	1.153	0.415	0.389	1.738	0.575	2.198	5.165
				H	1.112	0.399	0.393	1.738	0.575	2.156	5.141
				S	1.099	0.394	0.395	1.738	0.575	2.142	5.134
				R	1.054	0.377	0.399	1.738	0.575	2.094	5.107
NIOBIUM 11130	8.5861	47	100.0	V	1.142	0.411	0.391	1.740	0.575	2.187	5.162
				H	1.103	0.395	0.394	1.740	0.575	2.146	5.139
				S	1.090	0.390	0.396	1.740	0.575	2.133	5.132
				R	1.046	0.373	0.400	1.740	0.575	2.086	5.106
NIOBIUM 11131	8.5835	47	120.0	V	1.138	0.409	0.391	1.734	0.577	2.184	5.154
				H	1.098	0.394	0.394	1.734	0.577	2.142	5.130
				S	1.085	0.389	0.396	1.734	0.577	2.128	5.123
				R	1.039	0.371	0.400	1.734	0.577	2.080	5.096
NIOBIUM 11132	8.5809	47	140.0	V	1.136	0.409	0.390	1.728	0.579	2.182	5.146
				H	1.095	0.393	0.394	1.728	0.579	2.139	5.122
				S	1.081	0.387	0.396	1.728	0.579	2.124	5.114
				R	1.035	0.370	0.400	1.728	0.579	2.075	5.087
NIOBIUM 11133	8.5784	47	160.0	V	1.135	0.408	0.390	1.721	0.581	2.182	5.139
				H	1.093	0.392	0.394	1.721	0.581	2.138	5.114
				S	1.079	0.387	0.396	1.721	0.581	2.123	5.106
				R	1.032	0.369	0.400	1.721	0.581	2.073	5.079
NIOBIUM 11134	8.5732	47	180.0	V	1.134	0.408	0.390	1.714	0.584	2.182	5.132
				H	1.092	0.392	0.394	1.714	0.584	2.137	5.107
				S	1.077	0.386	0.395	1.714	0.584	2.122	5.098
				R	1.030	0.368	0.400	1.714	0.584	2.071	5.070
NIOBIUM 11135	8.5707	47	200.0	V	1.133	0.408	0.390	1.709	0.585	2.181	5.126
				H	1.089	0.391	0.394	1.709	0.585	2.135	5.101
				S	1.075	0.385	0.395	1.709	0.585	2.120	5.092
				R	1.027	0.367	0.400	1.709	0.585	2.068	5.064
NIOBIUM 11136	8.5681	47	220.0	V	1.128	0.406	0.390	1.709	0.585	2.176	5.124
				H	1.085	0.389	0.394	1.709	0.585	2.131	5.099
				S	1.071	0.384	0.396	1.709	0.585	2.116	5.090
				R	1.023	0.365	0.400	1.709	0.585	2.065	5.062
NIOBIUM 11137	8.5655	47	240.0	V	1.119	0.402	0.391	1.714	0.584	2.167	5.125
				H	1.078	0.386	0.395	1.714	0.584	2.124	5.101
				S	1.064	0.381	0.396	1.714	0.584	2.109	5.093
				R	1.019	0.364	0.401	1.714	0.584	2.060	5.066
NIOBIUM 11138	8.5630	47	260.0	V	1.109	0.398	0.393	1.721	0.581	2.157	5.129
				H	1.070	0.383	0.396	1.721	0.581	2.116	5.106
				S	1.058	0.378	0.398	1.721	0.581	2.102	5.098
				R	1.014	0.362	0.402	1.721	0.581	2.055	5.073
NIOBIUM 11139	8.5604	47	280.0	V	1.098	0.394	0.394	1.732	0.577	2.145	5.135
				H	1.062	0.380	0.398	1.732	0.577	2.106	5.114
				S	1.050	0.375	0.399	1.732	0.577	2.094	5.107
				R	1.009	0.360	0.403	1.732	0.577	2.050	5.083
NIOBIUM 11140	8.5553	47	300.0	V	1.088	0.390	0.396	1.743	0.574	2.134	5.143
				H	1.054	0.377	0.399	1.743	0.574	2.098	5.123
				S	1.043	0.372	0.400	1.743	0.574	2.086	5.117
				R	1.004	0.358	0.404	1.743	0.574	2.045	5.094

CUBIC

CODE	NAME DENSITY	REF	TEMP(K)		E	G	SIGMA	K	BETA	VS	VP
				V	1.150	0.416	0.381	1.607	0.622	2.202	5.019
NIOBIUM				H	1.102	0.398	0.386	1.607	0.622	2.152	4.990
11141	8.5821	22	103.0	S	1.084	0.391	0.388	1.607	0.622	2.134	4.980
				R	1.032	0.371	0.393	1.607	0.622	2.078	4.948
				V	1.146	0.415	0.382	1.613	0.620	2.199	5.024
NIOBIUM				H	1.098	0.396	0.386	1.613	0.620	2.148	4.995
11142	8.5815	22	113.0	S	1.081	0.389	0.388	1.613	0.620	2.130	4.984
				R	1.029	0.369	0.394	1.613	0.620	2.074	4.953
				V	1.143	0.413	0.382	1.619	0.618	2.195	5.029
NIOBIUM				H	1.095	0.395	0.387	1.619	0.618	2.145	5.000
11143	8.5809	22	123.0	S	1.078	0.388	0.389	1.619	0.618	2.126	4.989
				R	1.025	0.368	0.394	1.619	0.618	2.070	4.958
				V	1.140	0.412	0.383	1.626	0.615	2.192	5.036
NIOBIUM				H	1.092	0.393	0.388	1.626	0.615	2.141	5.007
11144	8.5802	22	133.0	S	1.075	0.387	0.390	1.626	0.615	2.123	4.996
				R	1.022	0.366	0.395	1.626	0.615	2.067	4.965
				V	1.137	0.411	0.383	1.621	0.617	2.188	5.028
NIOBIUM				H	1.089	0.392	0.388	1.621	0.617	2.138	4.999
11145	8.5796	22	143.0	S	1.071	0.385	0.390	1.621	0.617	2.119	4.988
				R	1.019	0.365	0.395	1.621	0.617	2.063	4.957
				V	1.134	0.410	0.383	1.614	0.620	2.186	5.018
NIOBIUM				H	1.086	0.391	0.388	1.614	0.620	2.136	4.989
11146	8.5790	22	153.0	S	1.068	0.384	0.390	1.614	0.620	2.117	4.978
				R	1.016	0.364	0.395	1.614	0.620	2.060	4.947
				V	1.132	0.409	0.383	1.611	0.621	2.184	5.014
NIOBIUM				H	1.083	0.390	0.388	1.611	0.621	2.133	4.985
11147	8.5783	22	163.0	S	1.066	0.383	0.390	1.611	0.621	2.114	4.974
				R	1.013	0.363	0.395	1.611	0.621	2.057	4.942
				V	1.129	0.408	0.383	1.609	0.622	2.181	5.010
NIOBIUM				H	1.080	0.389	0.388	1.609	0.622	2.130	4.980
11148	8.5777	22	173.0	S	1.063	0.382	0.390	1.609	0.622	2.111	4.969
				R	1.010	0.362	0.395	1.609	0.622	2.054	4.937
				V	1.126	0.407	0.383	1.607	0.622	2.179	5.006
NIOBIUM				H	1.078	0.388	0.388	1.607	0.622	2.128	4.977
11149	8.5770	22	183.0	S	1.060	0.381	0.390	1.607	0.622	2.109	4.966
				R	1.008	0.361	0.395	1.607	0.622	2.052	4.934
				V	1.122	0.405	0.384	1.609	0.622	2.174	5.006
NIOBIUM				H	1.074	0.387	0.389	1.609	0.622	2.123	4.977
11150	8.5758	22	203.0	S	1.056	0.380	0.391	1.609	0.622	2.104	4.966
				R	1.003	0.359	0.396	1.609	0.622	2.047	4.934
				V	1.120	0.404	0.384	1.614	0.620	2.172	5.011
NIOBIUM				H	1.072	0.386	0.389	1.614	0.620	2.121	4.982
11151	8.5751	22	213.0	S	1.054	0.379	0.391	1.614	0.620	2.102	4.971
				R	1.002	0.359	0.397	1.614	0.620	2.045	4.940
				V	1.118	0.404	0.385	1.614	0.620	2.170	5.010
NIOBIUM				H	1.070	0.385	0.390	1.614	0.620	2.119	4.981
11152	8.5745	22	223.0	S	1.052	0.378	0.391	1.614	0.620	2.100	4.970
				R	1.000	0.358	0.397	1.614	0.620	2.043	4.938
				V	1.116	0.403	0.384	1.605	0.623	2.168	4.999
NIOBIUM				H	1.068	0.384	0.389	1.605	0.623	2.117	4.970
11153	8.5738	22	233.0	S	1.050	0.377	0.391	1.605	0.623	2.098	4.959
				R	0.998	0.357	0.396	1.605	0.623	2.041	4.927
				V	1.114	0.402	0.385	1.612	0.620	2.166	5.006
NIOBIUM				H	1.066	0.384	0.390	1.612	0.620	2.115	4.977
11154	8.5732	22	243.0	S	1.049	0.377	0.392	1.612	0.620	2.096	4.966
				R	0.996	0.357	0.397	1.612	0.620	2.040	4.935
				V	1.111	0.401	0.385	1.611	0.621	2.163	5.002
NIOBIUM				H	1.063	0.382	0.390	1.611	0.621	2.112	4.974
11155	8.5719	22	263.0	S	1.045	0.376	0.392	1.611	0.621	2.093	4.963
				R	0.994	0.356	0.397	1.611	0.621	2.037	4.932

CUBIC

CODE (NAME)	DENSITY	REF	TEMP(K)		E	G	SIGMA	K	BETA	VS	VP
				V	1.109	0.400	0.385	1.613	0.620	2.161	5.004
NIOBIUM				H	1.062	0.382	0.390	1.613	0.620	2.111	4.976
11156	8.5713	22	273.0	S	1.044	0.375	0.392	1.613	0.620	2.092	4.965
				R	0.993	0.355	0.397	1.613	0.620	2.035	4.934
				V	1.108	0.400	0.386	1.615	0.619	2.159	5.006
NIOBIUM				H	1.061	0.381	0.391	1.615	0.619	2.109	4.978
11157	8.5706	22	283.0	S	1.043	0.375	0.392	1.615	0.619	2.091	4.967
				R	0.992	0.355	0.398	1.615	0.619	2.034	4.936
				V	1.106	0.399	0.387	1.638	0.611	2.156	5.031
NIOBIUM				H	1.059	0.380	0.392	1.638	0.611	2.107	5.003
11158	8.5700	22	293.0	S	1.042	0.374	0.394	1.638	0.611	2.088	4.993
				R	0.991	0.354	0.399	1.638	0.611	2.033	4.962
				V	1.105	0.399	0.386	1.617	0.618	2.157	5.007
NIOBIUM				H	1.058	0.380	0.391	1.617	0.618	2.107	4.979
11159	8.5694	22	303.0	S	1.041	0.374	0.393	1.617	0.618	2.089	4.968
				R	0.990	0.354	0.398	1.617	0.618	2.033	4.937
				V	1.104	0.398	0.386	1.615	0.619	2.156	5.005
NIOBIUM				H	1.058	0.380	0.391	1.615	0.619	2.106	4.977
11160	8.5687	22	313.0	S	1.040	0.374	0.393	1.615	0.619	2.088	4.966
				R	0.990	0.354	0.398	1.615	0.619	2.032	4.935
				V	1.102	0.397	0.386	1.614	0.620	2.154	5.002
NIOBIUM				H	1.056	0.380	0.391	1.614	0.620	2.105	4.974
11161	8.5674	22	333.0	S	1.039	0.373	0.393	1.614	0.620	2.087	4.964
				R	0.989	0.354	0.398	1.614	0.620	2.032	4.934
				V	1.101	0.397	0.386	1.609	0.622	2.153	4.996
NIOBIUM				H	1.055	0.379	0.391	1.609	0.622	2.104	4.968
11162	8.5668	22	343.0	S	1.039	0.373	0.392	1.609	0.622	2.087	4.958
				R	0.989	0.354	0.398	1.609	0.622	2.032	4.928
				V	1.100	0.397	0.386	1.604	0.623	2.152	4.990
NIOBIUM				H	1.054	0.379	0.390	1.604	0.623	2.104	4.963
11163	8.5662	22	353.0	S	1.038	0.373	0.392	1.604	0.623	2.086	4.953
				R	0.988	0.354	0.397	1.604	0.623	2.032	4.922
				V	1.099	0.397	0.385	1.598	0.626	2.152	4.983
NIOBIUM				H	1.054	0.379	0.390	1.598	0.626	2.104	4.956
11164	8.5655	22	363.0	S	1.038	0.373	0.392	1.598	0.626	2.086	4.946
				R	0.988	0.354	0.397	1.598	0.626	2.032	4.916
				V	1.098	0.396	0.385	1.592	0.628	2.151	4.975
NIOBIUM				H	1.053	0.379	0.390	1.592	0.628	2.103	4.948
11165	8.5649	22	373.0	S	1.037	0.373	0.391	1.592	0.628	2.086	4.938
				R	0.988	0.354	0.397	1.592	0.628	2.032	4.908
				V	1.096	0.396	0.385	1.583	0.632	2.150	4.965
NIOBIUM				H	1.052	0.379	0.389	1.583	0.632	2.103	4.938
11166	8.5636	22	393.0	S	1.036	0.372	0.391	1.583	0.632	2.086	4.928
				R	0.988	0.354	0.396	1.583	0.632	2.032	4.899
				V	1.096	0.396	0.384	1.580	0.633	2.150	4.961
NIOBIUM				H	1.052	0.379	0.389	1.580	0.633	2.103	4.934
11167	8.5630	22	403.0	S	1.036	0.372	0.391	1.580	0.633	2.086	4.924
				R	0.988	0.354	0.396	1.580	0.633	2.033	4.895
				V	1.095	0.395	0.384	1.573	0.636	2.149	4.952
NIOBIUM				H	1.051	0.378	0.389	1.573	0.636	2.102	4.925
11168	8.5623	22	413.0	S	1.036	0.373	0.390	1.573	0.636	2.086	4.916
				R	0.988	0.354	0.395	1.573	0.636	2.033	4.887
				V	1.096	0.396	0.383	1.567	0.638	2.151	4.946
NIOBIUM				H	1.052	0.379	0.388	1.567	0.638	2.103	4.919
11169	8.5617	22	423.0	S	1.036	0.373	0.390	1.567	0.638	2.087	4.910
				R	0.988	0.354	0.395	1.567	0.638	2.034	4.880
				V	1.093	0.395	0.384	1.566	0.639	2.148	4.944
NIOBIUM				H	1.050	0.378	0.388	1.566	0.639	2.102	4.918
11170	8.5610	22	433.0	S	1.035	0.373	0.390	1.566	0.639	2.086	4.909
				R	0.988	0.354	0.395	1.566	0.639	2.034	4.880

CUBIC

CODE	NAME DENSITY	REF	TEMP(K)		E	G	SIGMA	K	BETA	VS	VP
				V	1.093	0.395	0.384	1.565	0.639	2.148	4.942
NIOBIUM				H	1.050	0.378	0.388	1.565	0.639	2.102	4.916
11171	8.5604	22	443.0	S	1.035	0.373	0.390	1.565	0.639	2.086	4.907
				R	0.989	0.354	0.395	1.565	0.639	2.035	4.878
				V	1.092	0.395	0.384	1.564	0.639	2.148	4.942
NIOBIUM				H	1.050	0.378	0.388	1.564	0.639	2.102	4.916
11172	8.5598	22	453.0	S	1.036	0.373	0.390	1.564	0.639	2.086	4.907
				R	0.989	0.355	0.395	1.564	0.639	2.035	4.878
				V	1.092	0.395	0.384	1.564	0.639	2.147	4.941
NIOBIUM				H	1.050	0.378	0.388	1.564	0.639	2.102	4.916
11173	8.5591	22	463.0	S	1.036	0.373	0.390	1.564	0.639	2.087	4.907
				R	0.990	0.355	0.395	1.564	0.639	2.036	4.878
				V	1.092	0.395	0.384	1.563	0.640	2.147	4.941
NIOBIUM				H	1.050	0.378	0.388	1.563	0.640	2.103	4.915
11174	8.5585	22	473.0	S	1.036	0.373	0.390	1.563	0.640	2.087	4.906
				R	0.991	0.355	0.394	1.563	0.640	2.037	4.878
				V	1.091	0.394	0.384	1.564	0.639	2.147	4.942
NIOBIUM				H	1.050	0.378	0.388	1.564	0.639	2.103	4.916
11175	8.5579	22	483.0	S	1.036	0.373	0.390	1.564	0.639	2.087	4.908
				R	0.991	0.355	0.394	1.564	0.639	2.038	4.880
				V	1.091	0.394	0.384	1.566	0.639	2.147	4.944
NIOBIUM				H	1.051	0.379	0.388	1.566	0.639	2.103	4.919
11176	8.5566	22	503.0	S	1.037	0.373	0.390	1.566	0.639	2.089	4.911
				R	0.993	0.356	0.394	1.566	0.639	2.040	4.883
				V	1.091	0.394	0.384	1.567	0.638	2.146	4.946
NIOBIUM				H	1.051	0.379	0.388	1.567	0.638	2.104	4.921
11177	8.5559	22	513.0	S	1.038	0.373	0.390	1.567	0.638	2.089	4.913
				R	0.994	0.356	0.394	1.567	0.638	2.041	4.886
				V	1.091	0.394	0.384	1.566	0.639	2.146	4.944
NIOBIUM				H	1.051	0.379	0.388	1.566	0.639	2.104	4.920
11178	8.5553	22	523.0	S	1.038	0.374	0.390	1.566	0.639	2.090	4.912
				R	0.994	0.357	0.394	1.566	0.639	2.042	4.885
				V	1.091	0.394	0.384	1.569	0.637	2.146	4.948
NIOBIUM				H	1.052	0.379	0.388	1.569	0.637	2.104	4.924
11179	8.5547	22	533.0	S	1.039	0.374	0.390	1.569	0.637	2.090	4.916
				R	0.995	0.357	0.394	1.569	0.637	2.043	4.889
				V	1.091	0.394	0.384	1.571	0.636	2.146	4.951
NIOBIUM				H	1.052	0.379	0.388	1.571	0.636	2.105	4.927
11180	8.5540	22	543.0	S	1.039	0.374	0.390	1.571	0.636	2.091	4.919
				R	0.996	0.357	0.394	1.571	0.636	2.044	4.893
				V	1.091	0.394	0.385	1.574	0.635	2.146	4.954
NIOBIUM				H	1.053	0.379	0.389	1.574	0.635	2.105	4.931
11181	8.5534	22	553.0	S	1.040	0.374	0.390	1.574	0.635	2.091	4.923
				R	0.997	0.358	0.394	1.574	0.635	2.045	4.897
				V	1.090	0.394	0.385	1.577	0.634	2.145	4.957
NIOBIUM				H	1.052	0.379	0.389	1.577	0.634	2.104	4.933
11182	8.5527	22	563.0	S	1.039	0.374	0.390	1.577	0.634	2.091	4.926
				R	0.997	0.357	0.395	1.577	0.634	2.044	4.900
				V	1.091	0.394	0.385	1.578	0.634	2.146	4.960
NIOBIUM				H	1.053	0.379	0.389	1.578	0.634	2.106	4.937
11183	8.5521	22	573.0	S	1.041	0.375	0.390	1.578	0.634	2.093	4.929
				R	1.000	0.358	0.394	1.578	0.634	2.047	4.904
				V	1.091	0.394	0.385	1.581	0.633	2.146	4.963
NIOBIUM				H	1.054	0.379	0.389	1.581	0.633	2.106	4.940
11184	8.5515	22	583.0	S	1.042	0.375	0.390	1.581	0.633	2.093	4.933
				R	1.001	0.359	0.395	1.581	0.633	2.048	4.907
				V	1.091	0.394	0.385	1.582	0.632	2.146	4.964
NIOBIUM				H	1.054	0.379	0.389	1.582	0.632	2.107	4.942
11185	8.5508	22	593.0	S	1.043	0.375	0.390	1.582	0.632	2.094	4.935
				R	1.002	0.359	0.394	1.582	0.632	2.049	4.910

CUBIC

CODE	NAME DENSITY	REF	TEMP(K)		E	G	SIGMA	K	BETA	VS	VP
				V	1.091	0.394	0.385	1.583	0.632	2.146	4.966
NIOBIUM				H	1.055	0.380	0.389	1.583	0.632	2.107	4.944
11186	8.5502	22	603.0	S	1.043	0.375	0.390	1.583	0.632	2.095	4.937
				R	1.003	0.360	0.394	1.583	0.632	2.051	4.912
				V	1.091	0.394	0.385	1.584	0.632	2.146	4.966
NIOBIUM				H	1.055	0.380	0.389	1.583	0.632	2.108	4.944
11187	8.5495	22	613.0	S	1.044	0.375	0.390	1.583	0.632	2.096	4.937
				R	1.004	0.360	0.394	1.584	0.632	2.052	4.913
				V	1.091	0.394	0.385	1.586	0.630	2.146	4.969
NIOBIUM				H	1.056	0.380	0.389	1.586	0.630	2.108	4.948
11188	8.5489	22	623.0	S	1.045	0.376	0.390	1.586	0.630	2.096	4.941
				R	1.005	0.360	0.394	1.586	0.630	2.053	4.917
				V	1.091	0.394	0.385	1.588	0.630	2.146	4.971
NIOBIUM				H	1.056	0.380	0.389	1.588	0.630	2.109	4.950
11189	8.5483	22	633.0	S	1.045	0.376	0.390	1.588	0.630	2.097	4.943
				R	1.006	0.361	0.394	1.588	0.630	2.054	4.919
				V	1.091	0.394	0.385	1.588	0.630	2.147	4.972
NIOBIUM				H	1.057	0.380	0.389	1.588	0.630	2.109	4.951
11190	8.5476	22	643.0	S	1.046	0.376	0.390	1.588	0.630	2.098	4.944
				R	1.007	0.361	0.394	1.588	0.630	2.056	4.920
				V	1.091	0.394	0.386	1.590	0.629	2.147	4.975
NIOBIUM				H	1.057	0.381	0.389	1.590	0.629	2.110	4.954
11191	8.5470	22	653.0	S	1.047	0.376	0.390	1.590	0.629	2.099	4.948
				R	1.009	0.362	0.394	1.590	0.629	2.057	4.924
				V	1.092	0.394	0.386	1.592	0.628	2.147	4.977
NIOBIUM				H	1.058	0.381	0.389	1.592	0.628	2.111	4.956
11192	8.5463	22	663.0	S	1.048	0.377	0.390	1.592	0.628	2.100	4.950
				R	1.010	0.362	0.394	1.592	0.628	2.058	4.927
				V	1.092	0.394	0.386	1.595	0.627	2.147	4.981
NIOBIUM				H	1.058	0.381	0.389	1.595	0.627	2.111	4.960
11193	8.5457	22	673.0	S	1.048	0.377	0.390	1.595	0.627	2.100	4.954
				R	1.011	0.363	0.394	1.595	0.627	2.060	4.931
				V	1.092	0.394	0.386	1.593	0.628	2.147	4.980
NIOBIUM				H	1.059	0.381	0.389	1.593	0.628	2.112	4.959
11194	8.5451	22	683.0	S	1.049	0.377	0.390	1.593	0.628	2.101	4.953
				R	1.012	0.363	0.394	1.594	0.628	2.061	4.931
				V	1.092	0.394	0.386	1.595	0.627	2.148	4.982
NIOBIUM				H	1.060	0.381	0.389	1.595	0.627	2.113	4.962
11195	8.5444	22	693.0	S	1.050	0.378	0.390	1.595	0.627	2.102	4.956
				R	1.013	0.363	0.394	1.595	0.627	2.062	4.934
				V	1.092	0.394	0.386	1.596	0.627	2.148	4.983
NIOBIUM				H	1.060	0.382	0.389	1.596	0.627	2.113	4.963
11196	8.5438	22	703.0	S	1.051	0.378	0.390	1.596	0.627	2.103	4.958
				R	1.015	0.364	0.394	1.596	0.627	2.064	4.936
				V	1.093	0.394	0.386	1.600	0.625	2.148	4.988
NIOBIUM				H	1.061	0.382	0.389	1.600	0.625	2.114	4.969
11197	8.5431	22	713.0	S	1.051	0.378	0.390	1.600	0.625	2.104	4.963
				R	1.016	0.364	0.394	1.600	0.625	2.065	4.941
				V	1.093	0.394	0.386	1.597	0.626	2.148	4.985
NIOBIUM				H	1.062	0.382	0.389	1.597	0.626	2.115	4.965
11198	8.5425	22	723.0	S	1.052	0.379	0.390	1.597	0.626	2.105	4.960
				R	1.017	0.365	0.394	1.597	0.626	2.066	4.938
				V	1.093	0.394	0.386	1.598	0.626	2.149	4.987
NIOBIUM				H	1.062	0.382	0.389	1.598	0.626	2.116	4.968
11199	8.5419	22	733.0	S	1.053	0.379	0.390	1.598	0.626	2.106	4.962
				R	1.018	0.365	0.394	1.598	0.626	2.068	4.941
				V	1.093	0.394	0.386	1.597	0.626	2.149	4.986
NIOBIUM				H	1.063	0.383	0.389	1.597	0.626	2.116	4.967
11200	8.5412	22	743.0	S	1.054	0.379	0.390	1.597	0.626	2.107	4.962
				R	1.019	0.366	0.394	1.597	0.626	2.069	4.941

CUBIC

CODE	NAME DENSITY	REF	TEMP(K)		E	G	SIGMA	K	BETA	VS	VP
				V	1.094	0.395	0.386	1.601	0.625	2.149	4.990
NIOBIUM				H	1.064	0.383	0.389	1.601	0.625	2.117	4.972
11201	8.5406	22	753.0	S	1.055	0.379	0.390	1.601	0.625	2.108	4.967
				R	1.021	0.366	0.394	1.601	0.625	2.071	4.946
				V	1.094	0.395	0.386	1.604	0.624	2.150	4.994
NIOBIUM				H	1.064	0.383	0.389	1.604	0.624	2.118	4.976
11202	8.5400	22	763.0	S	1.056	0.380	0.390	1.604	0.624	2.109	4.971
				R	1.022	0.367	0.394	1.604	0.624	2.072	4.950
				V	1.094	0.395	0.386	1.603	0.624	2.150	4.994
NIOBIUM				H	1.065	0.383	0.389	1.603	0.624	2.119	4.976
11203	8.5393	22	773.0	S	1.057	0.380	0.390	1.603	0.624	2.110	4.971
				R	1.023	0.367	0.394	1.603	0.624	2.074	4.950
				V	1.408	0.520	0.353	1.595	0.627	2.427	5.090
NIOBIUM(75.5)-TIN(24.5)				H	1.366	0.503	0.357	1.595	0.627	2.387	5.065
11204	8.8340	141	300.0	S	1.354	0.498	0.359	1.595	0.627	2.375	5.058
				R	1.309	0.480	0.363	1.595	0.627	2.332	5.031
				V	0.494	0.170	0.450	1.651	0.606	1.386	4.604
NIOBIUM(75.5)-TIN(24.5)				H	0.405	0.139	0.459	1.651	0.606	1.251	4.552
11205	8.8600	142	35.0	S	0.214	0.072	0.478	1.651	0.606	0.903	4.441
				R	0.140	0.047	0.486	1.651	0.606	0.728	4.398
				V	1.408	0.520	0.353	1.595	0.627	2.424	5.083
NIOBIUM(75.5)-TIN(24.5)				H	1.366	0.503	0.357	1.595	0.627	2.384	5.058
11206	8.8600	142	300.0	S	1.355	0.499	0.358	1.595	0.627	2.372	5.051
				R	1.310	0.481	0.363	1.595	0.627	2.329	5.024
				V	1.491	0.543	0.373	1.955	0.512	2.121	4.711
PALLADIUM				H	1.406	0.510	0.380	1.955	0.512	2.055	4.672
11207	12.0690	209	0.0	S	1.362	0.492	0.384	1.955	0.512	2.019	4.651
				R	1.253	0.450	0.393	1.955	0.512	1.931	4.601
				V	1.489	0.542	0.373	1.955	0.511	2.120	4.711
PALLADIUM				H	1.405	0.509	0.380	1.955	0.511	2.053	4.672
11208	12.0690	209	20.0	S	1.360	0.491	0.384	1.955	0.511	2.018	4.651
				R	1.251	0.449	0.393	1.955	0.511	1.929	4.600
				V	1.482	0.539	0.374	1.957	0.511	2.114	4.709
PALLADIUM				H	1.397	0.506	0.381	1.957	0.511	2.047	4.670
11209	12.0680	209	40.0	S	1.351	0.488	0.385	1.957	0.511	2.011	4.649
				R	1.242	0.445	0.394	1.957	0.511	1.921	4.598
				V	1.474	0.536	0.375	1.958	0.511	2.108	4.706
PALLADIUM				H	1.388	0.502	0.382	1.958	0.511	2.040	4.666
11210	12.0670	209	60.0	S	1.341	0.484	0.386	1.958	0.511	2.003	4.645
				R	1.231	0.441	0.395	1.958	0.511	1.912	4.593
				V	1.466	0.533	0.375	1.956	0.511	2.102	4.701
PALLADIUM				H	1.379	0.499	0.382	1.956	0.511	2.033	4.661
11211	12.0660	209	80.0	S	1.332	0.480	0.386	1.956	0.511	1.995	4.639
				R	1.221	0.437	0.396	1.956	0.511	1.904	4.587
				V	1.460	0.531	0.375	1.951	0.513	2.098	4.695
PALLADIUM				H	1.373	0.496	0.383	1.951	0.513	2.028	4.654
11212	12.0650	209	100.0	S	1.324	0.477	0.387	1.951	0.513	1.989	4.631
				R	1.211	0.434	0.397	1.951	0.513	1.896	4.579
				V	1.456	0.529	0.375	1.943	0.515	2.095	4.687
PALLADIUM				H	1.367	0.494	0.383	1.943	0.515	2.025	4.645
11213	12.0623	209	120.0	S	1.317	0.475	0.387	1.943	0.515	1.984	4.622
				R	1.203	0.431	0.397	1.943	0.515	1.890	4.569
				V	1.454	0.529	0.375	1.936	0.517	2.094	4.680
PALLADIUM				H	1.363	0.493	0.383	1.936	0.517	2.022	4.637
11214	12.0596	209	140.0	S	1.312	0.473	0.387	1.936	0.517	1.980	4.613
				R	1.196	0.428	0.397	1.936	0.517	1.884	4.559
				V	1.453	0.528	0.374	1.929	0.519	2.093	4.673
PALLADIUM				H	1.361	0.492	0.382	1.929	0.519	2.020	4.630
11215	12.0569	209	160.0	S	1.308	0.471	0.387	1.929	0.519	1.977	4.605
				R	1.190	0.426	0.397	1.929	0.519	1.879	4.550

CUBIC

CODE	NAME / DENSITY	REF	TEMP(K)		E	G	SIGMA	K	BETA	VS	VP
				V	1.452	0.528	0.374	1.924	0.520	2.094	4.670
	PALLADIUM			H	1.359	0.492	0.382	1.924	0.520	2.019	4.626
11216	12.0542	209	180.0	S	1.304	0.470	0.387	1.924	0.520	1.975	4.600
				R	1.185	0.424	0.397	1.924	0.520	1.875	4.544
				V	1.453	0.529	0.374	1.919	0.521	2.095	4.666
	PALLADIUM			H	1.358	0.491	0.382	1.919	0.521	2.019	4.621
11217	12.0515	209	200.0	S	1.302	0.469	0.387	1.919	0.521	1.973	4.595
				R	1.180	0.422	0.397	1.919	0.521	1.872	4.538
				V	1.454	0.529	0.374	1.916	0.522	2.096	4.665
	PALLADIUM			H	1.358	0.491	0.382	1.916	0.522	2.019	4.619
11218	12.0488	209	220.0	S	1.300	0.469	0.387	1.916	0.522	1.972	4.592
				R	1.177	0.421	0.398	1.916	0.522	1.869	4.534
				V	1.456	0.530	0.373	1.915	0.522	2.098	4.665
	PALLADIUM			H	1.358	0.491	0.382	1.915	0.522	2.019	4.619
11219	12.0461	209	240.0	S	1.298	0.468	0.387	1.915	0.522	1.971	4.591
				R	1.173	0.420	0.398	1.915	0.522	1.866	4.533
				V	1.458	0.531	0.373	1.917	0.522	2.100	4.669
	PALLADIUM			H	1.358	0.492	0.382	1.917	0.522	2.020	4.622
11220	12.0434	209	260.0	S	1.297	0.468	0.387	1.917	0.522	1.971	4.593
				R	1.170	0.418	0.398	1.917	0.522	1.864	4.534
				V	1.460	0.532	0.373	1.924	0.520	2.101	4.676
	PALLADIUM			H	1.359	0.492	0.382	1.924	0.520	2.021	4.628
11221	12.0407	209	280.0	S	1.296	0.467	0.388	1.924	0.520	1.970	4.599
				R	1.167	0.417	0.399	1.924	0.520	1.862	4.538
				V	1.463	0.532	0.374	1.931	0.518	2.103	4.684
	PALLADIUM			H	1.360	0.492	0.383	1.931	0.518	2.021	4.635
11222	12.0380	209	300.0	S	1.296	0.467	0.388	1.931	0.518	1.969	4.605
				R	1.165	0.416	0.399	1.931	0.518	1.859	4.544
				V	1.874	0.673	0.392	2.884	0.347	1.765	4.182
	PLATINUM			H	1.849	0.663	0.393	2.884	0.347	1.752	4.175
11223	21.6200	158	0.0	S	1.843	0.661	0.393	2.884	0.347	1.749	4.174
				R	1.810	0.649	0.395	2.884	0.347	1.732	4.164
				V	1.814	0.651	0.393	2.827	0.354	1.740	4.146
	PLATINUM			H	1.780	0.638	0.395	2.827	0.354	1.723	4.136
11224	21.5000	158	300.0	S	1.772	0.635	0.396	2.827	0.354	1.718	4.134
				R	1.729	0.618	0.398	2.827	0.354	1.696	4.121
				V	0.045	0.017	0.291	0.036	27.900	1.450	2.669
	POTASSIUM			H	0.038	0.014	0.322	0.036	27.900	1.320	2.577
11225	0.8300	128	ROOM	S	0.030	0.011	0.359	0.036	27.900	1.157	2.471
				R	0.023	0.008	0.391	0.036	27.900	1.006	2.381
				V	0.046	0.017	0.310	0.040	24.896	1.424	2.716
	POTASSIUM			H	0.039	0.014	0.339	0.040	24.896	1.298	2.630
11226	0.8600	138	83.0	S	0.031	0.011	0.373	0.040	24.896	1.137	2.529
				R	0.024	0.008	0.402	0.040	24.896	0.988	2.444
				V	0.048	0.019	0.282	0.037	27.322	1.438	2.609
	POTASSIUM			H	0.040	0.015	0.318	0.037	27.322	1.295	2.508
11227	0.9030	172	4.2	S	0.029	0.011	0.366	0.037	27.322	1.092	2.376
				R	0.022	0.008	0.400	0.037	27.322	0.931	2.283
				V	0.045	0.017	0.293	0.036	27.855	1.389	2.565
	POTASSIUM			H	0.037	0.014	0.327	0.036	27.855	1.251	2.469
11228	0.8960	172	78.0	S	0.027	0.010	0.373	0.036	27.855	1.053	2.342
				R	0.020	0.007	0.406	0.036	27.855	0.897	2.254
				V	0.044	0.017	0.302	0.037	27.100	1.372	2.574
	POTASSIUM			H	0.037	0.014	0.334	0.037	27.100	1.239	2.483
11229	0.8960	172	78.0	S	0.027	0.010	0.377	0.037	27.100	1.049	2.363
				R	0.020	0.007	0.409	0.037	27.100	0.895	2.278
				V	0.042	0.016	0.308	0.036	27.624	1.338	2.540
	POTASSIUM			H	0.035	0.013	0.339	0.036	27.624	1.210	2.453
11230	0.8910	172	120.0	S	0.026	0.009	0.381	0.036	27.624	1.027	2.338
				R	0.019	0.007	0.411	0.036	27.624	0.877	2.256

CUBIC

CODE	NAME DENSITY	REF	TEMP(K)		E	G	SIGMA	K	BETA	VS	VP
POTASSIUM				V	0.039	0.015	0.315	0.036	28.143	1.304	2.508
11231	0.8830	172	160.0	H	0.033	0.012	0.344	0.036	28.143	1.182	2.426
				S	0.025	0.009	0.383	0.036	28.143	1.010	2.320
				R	0.019	0.007	0.412	0.036	28.142	0.865	2.241
POTASSIUM				V	0.038	0.014	0.319	0.035	28.819	1.277	2.477
11232	0.8760	172	195.0	H	0.032	0.012	0.347	0.035	28.818	1.160	2.399
				S	0.024	0.009	0.384	0.035	28.818	0.999	2.300
				R	0.018	0.006	0.412	0.035	28.817	0.859	2.224
POTASSIUM				V	0.033	0.012	0.334	0.033	30.060	1.207	2.419
11233	0.8510	239	295.0	H	0.028	0.010	0.360	0.033	30.060	1.100	2.350
				S	0.021	0.008	0.393	0.033	30.060	0.951	2.262
				R	0.016	0.006	0.419	0.033	30.060	0.820	2.192
POTASSIUM	BROMIDE			V	0.222	0.089	0.254	0.151	6.623	1.796	3.129
11234	2.7500	129	ROOM	H	0.204	0.080	0.275	0.151	6.623	1.705	3.060
				S	0.195	0.076	0.285	0.151	6.623	1.660	3.027
				R	0.179	0.069	0.303	0.151	6.622	1.580	2.970
POTASSIUM	BROMIDE			V	0.222	0.088	0.260	0.154	6.494	1.788	3.140
11235	2.7500	129	ROOM	H	0.203	0.079	0.280	0.154	6.494	1.698	3.073
				S	0.194	0.075	0.290	0.154	6.494	1.654	3.041
				R	0.178	0.068	0.307	0.154	6.493	1.575	2.985
POTASSIUM	BROMIDE			V	0.255	0.102	0.245	0.167	6.000	1.930	3.320
11236	2.7500	129	ROOM	H	0.238	0.094	0.262	0.167	6.000	1.852	3.261
				S	0.231	0.091	0.269	0.167	6.000	1.819	3.236
				R	0.216	0.084	0.284	0.167	6.000	1.750	3.185
POTASSIUM	BROMIDE			V	0.224	0.089	0.259	0.155	6.452	1.797	3.153
11237	2.7505	4	ROOM	H	0.205	0.080	0.280	0.155	6.452	1.707	3.085
				S	0.196	0.076	0.289	0.155	6.452	1.662	3.053
				R	0.180	0.069	0.306	0.155	6.452	1.583	2.996
POTASSIUM	BROMIDE			V	0.219	0.087	0.258	0.151	6.608	1.784	3.125
11238	2.7400	4	ROOM	H	0.201	0.079	0.279	0.151	6.608	1.694	3.058
				S	0.192	0.075	0.288	0.151	6.608	1.650	3.025
				R	0.176	0.068	0.306	0.151	6.608	1.570	2.968
POTASSIUM	BROMIDE			V	0.224	0.090	0.239	0.143	6.990	1.814	3.097
11239	2.7500	194	293.0	H	0.205	0.081	0.261	0.143	6.990	1.719	3.024
				S	0.196	0.077	0.272	0.143	6.990	1.672	2.988
				R	0.180	0.070	0.291	0.143	6.990	1.590	2.928
POTASSIUM	BROMIDE			V	0.210	0.084	0.256	0.144	6.960	1.750	3.056
11240	2.7340	194	373.0	H	0.195	0.076	0.274	0.144	6.960	1.672	2.997
				S	0.188	0.073	0.282	0.144	6.960	1.636	2.970
				R	0.174	0.067	0.298	0.144	6.960	1.566	2.919
POTASSIUM	BROMIDE			V	0.194	0.075	0.295	0.158	6.330	1.670	3.097
11241	2.6900	194	473.0	H	0.183	0.070	0.307	0.158	6.330	1.611	3.055
				S	0.178	0.068	0.312	0.158	6.330	1.587	3.039
				R	0.167	0.063	0.324	0.158	6.330	1.532	3.000
POTASSIUM	BROMIDE			V	0.179	0.068	0.315	0.161	6.210	1.603	3.082
11242	2.6510	194	573.0	H	0.171	0.064	0.324	0.161	6.210	1.559	3.052
				S	0.167	0.063	0.327	0.161	6.210	1.543	3.041
				R	0.159	0.060	0.335	0.161	6.210	1.499	3.012
POTASSIUM	BROMIDE			V	0.164	0.063	0.309	0.143	6.990	1.545	2.939
11243	2.6210	194	673.0	H	0.158	0.060	0.316	0.143	6.990	1.512	2.917
				S	0.156	0.059	0.318	0.143	6.990	1.502	2.910
				R	0.150	0.057	0.325	0.143	6.990	1.469	2.887
POTASSIUM	BROMIDE			V	0.147	0.056	0.311	0.130	7.710	1.476	2.815
11244	2.5820	194	773.0	H	0.144	0.055	0.315	0.130	7.710	1.455	2.801
				S	0.143	0.054	0.316	0.130	7.710	1.451	2.798
				R	0.139	0.053	0.321	0.130	7.710	1.430	2.784
POTASSIUM	BROMIDE			V	0.131	0.051	0.301	0.110	9.090	1.409	2.641
11245	2.5430	194	873.0	H	0.130	0.050	0.303	0.110	9.090	1.399	2.633
				S	0.129	0.050	0.304	0.110	9.090	1.397	2.632
				R	0.128	0.049	0.307	0.110	9.090	1.386	2.624

CUBIC

CODE	NAME	DENSITY	REF	TEMP(K)		E	G	SIGMA	K	BETA	VS	VP
11246	POTASSIUM BROMIDE	2.5030	194	973.0	V	0.117	0.047	0.260	0.081	12.300	1.364	2.393
					H	0.117	0.046	0.261	0.081	12.300	1.359	2.389
					S	0.116	0.046	0.261	0.081	12.300	1.358	2.389
					R	0.116	0.046	0.263	0.081	12.300	1.352	2.385
11247	POTASSIUM BROMIDE	2.4650	194	1003.0	V	0.113	0.044	0.281	0.086	11.610	1.337	2.425
					H	0.113	0.044	0.282	0.086	11.610	1.335	2.423
					S	0.113	0.044	0.282	0.086	11.610	1.335	2.423
					R	0.112	0.044	0.283	0.086	11.610	1.333	2.421
11248	POTASSIUM BROMIDE	2.7600	181	293.0	V	0.223	0.088	0.265	0.158	6.329	1.785	3.158
					H	0.204	0.079	0.285	0.158	6.329	1.696	3.092
					S	0.195	0.075	0.294	0.158	6.329	1.652	3.060
					R	0.179	0.068	0.311	0.158	6.329	1.573	3.004
11249	POTASSIUM BROMIDE	2.7440	234	298.0	V	0.222	0.088	0.260	0.154	6.482	1.793	3.147
					H	0.204	0.080	0.280	0.154	6.482	1.703	3.080
					S	0.195	0.076	0.290	0.154	6.482	1.659	3.048
					R	0.179	0.068	0.307	0.154	6.482	1.580	2.992
11250	POTASSIUM BROMIDE	2.7533	213	300.0	V	0.221	0.088	0.252	0.149	6.728	1.792	3.112
					H	0.203	0.080	0.272	0.149	6.728	1.702	3.043
					S	0.194	0.076	0.282	0.149	6.728	1.658	3.010
					R	0.178	0.069	0.300	0.149	6.728	1.579	2.953
11251	POTASSIUM CHLORIDE	1.9840	128	ROOM	V	0.256	0.108	0.185	0.135	7.398	2.333	3.751
					H	0.237	0.098	0.208	0.135	7.398	2.221	3.659
					S	0.228	0.094	0.219	0.135	7.398	2.171	3.619
					R	0.212	0.086	0.238	0.135	7.398	2.079	3.546
11252	POTASSIUM CHLORIDE	1.9840	128	ROOM	V	0.263	0.105	0.253	0.177	5.640	2.299	3.998
					H	0.255	0.101	0.261	0.177	5.640	2.256	3.965
					S	0.252	0.100	0.263	0.177	5.640	2.244	3.956
					R	0.244	0.096	0.270	0.177	5.640	2.202	3.924
11253	POTASSIUM CHLORIDE	1.9880	129	ROOM	V	0.262	0.105	0.249	0.174	5.747	2.295	3.972
					H	0.242	0.095	0.268	0.174	5.747	2.190	3.892
					S	0.233	0.091	0.277	0.174	5.747	2.142	3.856
					R	0.216	0.083	0.293	0.174	5.747	2.049	3.788
11254	POTASSIUM CHLORIDE	1.9840	129	ROOM	V	0.295	0.118	0.246	0.193	5.172	2.443	4.207
					H	0.279	0.111	0.259	0.193	5.172	2.364	4.147
					S	0.273	0.108	0.264	0.193	5.172	2.335	4.125
					R	0.259	0.102	0.277	0.193	5.172	2.262	4.070
11255	POTASSIUM CHLORIDE	1.9900	4	ROOM	V	0.261	0.104	0.256	0.178	5.607	2.286	3.991
					H	0.242	0.095	0.274	0.178	5.607	2.186	3.916
					S	0.234	0.091	0.282	0.178	5.607	2.141	3.882
					R	0.217	0.084	0.297	0.178	5.607	2.051	3.817
11256	POTASSIUM CHLORIDE	2.0253	80	80.0	V	0.309	0.124	0.241	0.199	5.028	2.478	4.243
					H	0.279	0.110	0.267	0.199	5.028	2.330	4.130
					S	0.262	0.102	0.280	0.199	5.028	2.250	4.070
					R	0.238	0.091	0.301	0.199	5.028	2.125	3.980
11257	POTASSIUM CHLORIDE	2.0240	80	90.0	V	0.307	0.123	0.244	0.200	5.010	2.470	4.242
					H	0.277	0.109	0.268	0.200	5.010	2.324	4.131
					S	0.262	0.102	0.282	0.200	5.010	2.246	4.073
					R	0.237	0.091	0.302	0.200	5.010	2.123	3.984
11258	POTASSIUM CHLORIDE	2.0181	80	130.0	V	0.300	0.120	0.246	0.196	5.094	2.441	4.204
					H	0.272	0.107	0.269	0.196	5.094	2.303	4.099
					S	0.257	0.100	0.281	0.196	5.094	2.231	4.045
					R	0.235	0.090	0.301	0.196	5.094	2.114	3.960
11259	POTASSIUM CHLORIDE	2.0166	80	140.0	V	0.297	0.119	0.248	0.196	5.091	2.431	4.198
					H	0.270	0.106	0.271	0.196	5.091	2.297	4.096
					S	0.256	0.100	0.282	0.196	5.091	2.226	4.043
					R	0.234	0.090	0.302	0.196	5.091	2.111	3.960
11260	POTASSIUM CHLORIDE	2.0150	80	150.0	V	0.295	0.119	0.246	0.194	5.151	2.425	4.181
					H	0.269	0.106	0.269	0.194	5.151	2.292	4.079
					S	0.255	0.100	0.281	0.194	5.151	2.223	4.028
					R	0.233	0.090	0.300	0.194	5.151	2.108	3.945

CUBIC

CODE / NAME	DENSITY	REF	TEMP(K)		E	G	SIGMA	K	BETA	VS	VP
				V	0.293	0.118	0.245	0.192	5.214	2.419	4.163
POTASSIUM CHLORIDE				H	0.267	0.105	0.268	0.192	5.214	2.287	4.062
11261	2.0133	80	160.0	S	0.254	0.099	0.280	0.192	5.214	2.219	4.011
				R	0.232	0.089	0.299	0.192	5.214	2.106	3.929
				V	0.291	0.117	0.247	0.192	5.217	2.409	4.155
POTASSIUM CHLORIDE				H	0.265	0.105	0.269	0.192	5.217	2.280	4.057
11262	2.0116	80	170.0	S	0.252	0.099	0.280	0.192	5.217	2.214	4.008
				R	0.231	0.089	0.299	0.192	5.217	2.102	3.927
				V	0.289	0.116	0.246	0.189	5.280	2.402	4.138
POTASSIUM CHLORIDE				H	0.264	0.104	0.268	0.189	5.280	2.274	4.040
11263	2.0099	80	180.0	S	0.251	0.098	0.279	0.189	5.280	2.210	3.992
				R	0.230	0.089	0.298	0.189	5.280	2.099	3.911
				V	0.287	0.115	0.244	0.187	5.346	2.396	4.119
POTASSIUM CHLORIDE				H	0.262	0.103	0.267	0.187	5.346	2.269	4.022
11264	2.0082	80	190.0	S	0.250	0.098	0.278	0.187	5.346	2.205	3.975
				R	0.229	0.088	0.296	0.187	5.346	2.096	3.895
				V	0.285	0.114	0.246	0.187	5.352	2.386	4.112
POTASSIUM CHLORIDE				H	0.260	0.103	0.268	0.187	5.352	2.262	4.017
11265	2.0063	80	200.0	S	0.248	0.097	0.278	0.187	5.352	2.200	3.971
				R	0.228	0.088	0.297	0.187	5.352	2.093	3.893
				V	0.283	0.114	0.245	0.185	5.418	2.380	4.094
POTASSIUM CHLORIDE				H	0.259	0.102	0.266	0.185	5.418	2.257	4.000
11266	2.0045	80	210.0	S	0.247	0.097	0.277	0.185	5.418	2.196	3.955
				R	0.227	0.088	0.295	0.185	5.418	2.090	3.877
				V	0.280	0.113	0.246	0.184	5.427	2.370	4.086
POTASSIUM CHLORIDE				H	0.257	0.101	0.268	0.184	5.427	2.250	3.994
11267	2.0026	80	220.0	S	0.246	0.096	0.278	0.184	5.427	2.191	3.950
				R	0.226	0.087	0.296	0.184	5.427	2.086	3.874
				V	0.278	0.112	0.245	0.182	5.493	2.364	4.068
POTASSIUM CHLORIDE				H	0.255	0.101	0.266	0.182	5.493	2.245	3.977
11268	2.0007	80	230.0	S	0.244	0.096	0.276	0.182	5.493	2.187	3.934
				R	0.225	0.087	0.294	0.182	5.493	2.083	3.858
				V	0.276	0.111	0.247	0.182	5.502	2.355	4.060
POTASSIUM CHLORIDE				H	0.254	0.100	0.267	0.182	5.502	2.238	3.971
11269	1.9987	80	240.0	S	0.243	0.095	0.277	0.182	5.502	2.181	3.929
				R	0.224	0.086	0.295	0.182	5.502	2.079	3.855
				V	0.274	0.110	0.248	0.181	5.511	2.346	4.053
POTASSIUM CHLORIDE				H	0.252	0.099	0.268	0.181	5.511	2.231	3.965
11270	1.9967	80	250.0	S	0.242	0.095	0.278	0.181	5.511	2.176	3.925
				R	0.223	0.086	0.295	0.181	5.511	2.076	3.851
				V	0.272	0.109	0.247	0.179	5.580	2.339	4.035
POTASSIUM CHLORIDE				H	0.250	0.099	0.267	0.179	5.580	2.226	3.949
11271	1.9947	80	260.0	S	0.240	0.094	0.277	0.179	5.580	2.172	3.908
				R	0.222	0.086	0.294	0.179	5.580	2.073	3.836
				V	0.270	0.108	0.248	0.179	5.589	2.331	4.028
POTASSIUM CHLORIDE				H	0.249	0.098	0.268	0.179	5.589	2.219	3.943
11272	1.9926	80	270.0	S	0.239	0.094	0.277	0.179	5.589	2.167	3.904
				R	0.221	0.085	0.294	0.179	5.589	2.069	3.833
				V	0.268	0.108	0.247	0.177	5.658	2.325	4.011
POTASSIUM CHLORIDE				H	0.247	0.098	0.267	0.177	5.658	2.214	3.927
11273	1.9905	80	280.0	S	0.238	0.093	0.276	0.177	5.658	2.163	3.888
				R	0.220	0.085	0.293	0.177	5.658	2.066	3.817
				V	0.266	0.106	0.259	0.184	5.448	2.306	4.041
POTASSIUM CHLORIDE (P= 0KB)				H	0.246	0.096	0.277	0.184	5.448	2.201	3.962
11274	1.9860	149	298.0	S	0.236	0.092	0.285	0.184	5.448	2.152	3.927
				R	0.219	0.084	0.301	0.184	5.448	2.058	3.859
				V	0.276	0.109	0.263	0.194	5.152	2.331	4.112
POTASSIUM CHLORIDE (P= 2KB)				H	0.252	0.098	0.284	0.194	5.152	2.212	4.023
11275	2.0082	149	298.0	S	0.241	0.093	0.293	0.194	5.152	2.152	3.980
				R	0.220	0.084	0.311	0.194	5.152	2.047	3.905

CUBIC

CODE	NAME / DENSITY	REF	TEMP(K)		E	G	SIGMA	K	BETA	VS	VP
	POTASSIUM CHLORIDE (P= 4KB)			V	0.286	0.113	0.266	0.203	4.915	2.360	4.178
11276	2.0298	149	298.C	H	0.259	0.100	0.288	0.203	4.915	2.224	4.077
				S	0.244	0.094	0.300	0.203	4.915	2.150	4.024
				R	0.221	0.084	0.319	0.203	4.915	2.032	3.941
	POTASSIUM CHLORIDE (P= 6KB)			V	0.296	0.117	0.268	0.213	4.703	2.386	4.238
11277	2.C5C5	149	298.0	H	0.265	0.103	0.292	0.213	4.703	2.237	4.128
				S	0.248	0.095	0.306	0.213	4.703	2.150	4.066
				R	0.222	0.084	0.326	0.213	4.703	2.021	3.977
	POTASSIUM CHLORIDE (P= 8KB)			V	0.305	0.120	0.266	0.217	4.605	2.412	4.272
11278	2.C705	149	298.0	H	0.270	0.105	0.292	0.217	4.605	2.248	4.150
				S	0.250	0.095	0.308	0.217	4.605	2.146	4.078
				R	0.222	0.083	0.330	0.217	4.605	2.007	3.982
	POTASSIUM CHLORIDE (P=10KB)			V	0.318	0.125	0.271	0.232	4.319	2.446	4.365
11279	2.C896	149	298.0	H	0.278	0.107	0.300	0.232	4.319	2.264	4.232
				S	0.253	0.096	0.318	0.232	4.319	2.143	4.147
				R	0.222	0.083	0.340	0.232	4.319	1.991	4.046
	POTASSIUM CHLORIDE			V	0.311	0.126	0.237	0.197	5.074	2.488	4.239
11280	2.029C	196	4.2	H	0.280	0.111	0.263	0.197	5.074	2.336	4.122
				S	0.263	0.103	0.278	0.197	5.074	2.252	4.059
				R	0.238	0.092	0.299	0.197	5.074	2.125	3.967
	POTASSIUM CHLORIDE			V	0.310	0.125	0.237	0.196	5.092	2.483	4.231
11281	2.0290	196	20.0	H	0.279	0.110	0.263	0.196	5.093	2.332	4.114
				S	0.262	0.103	0.278	0.196	5.093	2.248	4.052
				R	0.237	0.091	0.299	0.196	5.092	2.122	3.960
	POTASSIUM CHLORIDE			V	0.308	0.124	0.238	0.196	5.102	2.474	4.222
11282	2.0290	196	40.C	H	0.277	0.110	0.264	0.196	5.102	2.325	4.107
				S	0.261	0.102	0.278	0.196	5.102	2.243	4.046
				R	0.237	0.091	0.299	0.196	5.102	2.118	3.955
	POTASSIUM CHLORIDE			V	0.306	0.123	0.239	0.195	5.134	2.466	4.209
11283	2.0275	196	60.0	H	0.276	0.109	0.264	0.195	5.134	2.319	4.096
				S	0.260	0.102	0.278	0.195	5.134	2.239	4.036
				R	0.236	0.091	0.298	0.195	5.134	2.115	3.946
	POTASSIUM CHLORIDE			V	0.303	0.122	0.239	0.193	5.175	2.458	4.195
11284	2.0253	196	80.0	H	0.274	0.108	0.264	0.193	5.175	2.313	4.083
				S	0.258	0.101	0.277	0.193	5.175	2.235	4.025
				R	0.235	0.090	0.298	0.193	5.175	2.113	3.936
	POTASSIUM CHLORIDE			V	0.300	0.121	0.241	0.193	5.187	2.444	4.183
11285	2.0226	196	100.0	H	0.271	0.107	0.265	0.193	5.187	2.303	4.075
				S	0.257	0.100	0.278	0.193	5.187	2.228	4.019
				R	0.233	0.090	0.298	0.193	5.187	2.108	3.932
	POTASSIUM CHLORIDE			V	0.296	0.119	0.242	0.191	5.225	2.429	4.165
11286	2.0196	196	120.0	H	0.269	0.106	0.266	0.191	5.225	2.292	4.059
				S	0.254	0.099	0.278	0.191	5.225	2.219	4.006
				R	0.232	0.089	0.298	0.191	5.225	2.102	3.921
	POTASSIUM CHLORIDE			V	0.292	0.117	0.243	0.190	5.272	2.413	4.143
11287	2.0166	196	140.0	H	0.266	0.105	0.267	0.190	5.272	2.280	4.042
				S	0.252	0.099	0.278	0.190	5.272	2.211	3.991
				R	0.230	0.089	0.298	0.190	5.272	2.097	3.908
	POTASSIUM CHLORIDE			V	0.289	0.116	0.244	0.188	5.326	2.402	4.125
11288	2.0133	196	160.0	H	0.263	0.104	0.266	0.188	5.326	2.272	4.026
				S	0.250	0.098	0.278	0.188	5.326	2.205	3.976
				R	0.229	0.088	0.297	0.188	5.326	2.094	3.895
	POTASSIUM CHLORIDE			V	0.285	0.115	0.245	0.186	5.363	2.388	4.109
11289	2.0099	196	180.0	H	0.261	0.103	0.267	0.186	5.363	2.262	4.012
				S	0.248	0.097	0.278	0.186	5.363	2.198	3.965
				R	0.228	0.088	0.297	0.186	5.363	2.089	3.886
	POTASSIUM CHLORIDE			V	0.282	0.113	0.246	0.185	5.404	2.373	4.091
11290	2.0063	196	200.0	H	0.258	0.102	0.268	0.185	5.404	2.251	3.997
				S	0.246	0.096	0.278	0.185	5.404	2.190	3.952
				R	0.226	0.087	0.297	0.185	5.404	2.084	3.874

CUBIC

CODE	NAME / DENSITY	REF	TEMP(K)		E	G	SIGMA	K	BETA	VS	VP
				V	0.278	0.111	0.248	0.184	5.441	2.359	4.074
POTASSIUM	CHLORIDE			H	0.255	0.101	0.269	0.184	5.441	2.240	3.984
11291	2.0026	196	220.0	S	0.244	0.095	0.279	0.184	5.441	2.182	3.940
				R	0.224	0.086	0.297	0.184	5.441	2.078	3.865
				V	0.275	0.110	0.248	0.182	5.496	2.349	4.057
POTASSIUM	CHLORIDE			H	0.253	0.100	0.269	0.182	5.497	2.232	3.968
11292	1.9987	196	240.0	S	0.242	0.095	0.278	0.182	5.497	2.176	3.926
				R	0.223	0.086	0.296	0.182	5.497	2.074	3.852
				V	0.272	0.109	0.249	0.181	5.537	2.334	4.039
POTASSIUM	CHLORIDE			H	0.250	0.098	0.269	0.181	5.537	2.221	3.954
11293	1.9947	196	260.0	S	0.240	0.094	0.279	0.181	5.537	2.167	3.914
				R	0.221	0.085	0.296	0.181	5.537	2.069	3.842
				V	0.267	0.107	0.252	0.180	5.561	2.317	4.024
POTASSIUM	CHLORIDE			H	0.247	0.097	0.271	0.180	5.561	2.208	3.942
11294	1.9905	196	280.0	S	0.237	0.093	0.280	0.180	5.561	2.158	3.904
				R	0.219	0.085	0.297	0.180	5.561	2.062	3.835
				V	0.264	0.105	0.254	0.178	5.605	2.301	4.006
POTASSIUM	CHLORIDE			H	0.244	0.096	0.272	0.178	5.605	2.197	3.927
11295	1.9860	196	300.0	S	0.235	0.092	0.281	0.178	5.605	2.149	3.891
				R	0.218	0.084	0.297	0.178	5.605	2.056	3.824
				V	0.265	0.106	0.252	0.178	5.621	2.306	4.005
POTASSIUM	CHLORIDE			H	0.245	0.096	0.271	0.178	5.621	2.202	3.926
11296	1.9870	87	298.0	S	0.236	0.092	0.279	0.178	5.621	2.154	3.891
				R	0.219	0.084	0.295	0.178	5.621	2.061	3.824
				V	0.264	0.107	0.232	0.164	6.090	2.323	3.932
POTASSIUM	CHLORIDE			H	0.243	0.097	0.253	0.164	6.090	2.212	3.846
11297	1.9860	84	298.0	S	0.234	0.093	0.263	0.164	6.090	2.160	3.806
				R	0.217	0.085	0.280	0.164	6.090	2.064	3.734
				V	0.260	0.105	0.236	0.164	6.099	2.305	3.920
POTASSIUM	CHLORIDE			H	0.240	0.096	0.256	0.164	6.099	2.198	3.837
11298	1.9806	84	323.0	S	0.231	0.091	0.265	0.164	6.099	2.149	3.800
				R	0.215	0.084	0.282	0.164	6.099	2.056	3.730
				V	0.256	0.103	0.239	0.163	6.120	2.287	3.904
POTASSIUM	CHLORIDE			H	0.237	0.094	0.258	0.163	6.120	2.184	3.825
11299	1.9753	84	348.0	S	0.229	0.090	0.267	0.163	6.120	2.138	3.790
				R	0.212	0.083	0.283	0.163	6.120	2.047	3.723
				V	0.251	0.101	0.242	0.162	6.156	2.267	3.886
POTASSIUM	CHLORIDE			H	0.234	0.093	0.260	0.162	6.156	2.169	3.811
11300	1.9694	84	373.0	S	0.226	0.089	0.268	0.162	6.156	2.126	3.778
				R	0.210	0.082	0.284	0.162	6.156	2.039	3.714
				V	0.247	0.099	0.245	0.161	6.210	2.246	3.863
POTASSIUM	CHLORIDE			H	0.230	0.091	0.262	0.161	6.210	2.153	3.792
11301	1.9641	84	398.0	S	0.223	0.088	0.270	0.161	6.210	2.113	3.762
				R	0.208	0.081	0.285	0.161	6.210	2.030	3.700
				V	0.242	0.097	0.248	0.160	6.252	2.225	3.843
POTASSIUM	CHLORIDE			H	0.226	0.089	0.264	0.160	6.252	2.137	3.776
11302	1.9583	84	423.0	S	0.220	0.086	0.271	0.160	6.252	2.100	3.748
				R	0.206	0.080	0.286	0.160	6.252	2.021	3.690
				V	0.238	0.095	0.250	0.159	6.309	2.206	3.822
POTASSIUM	CHLORIDE			H	0.223	0.088	0.266	0.159	6.309	2.122	3.758
11303	1.9526	84	448.0	S	0.217	0.085	0.272	0.159	6.309	2.088	3.732
				R	0.203	0.079	0.286	0.159	6.309	2.012	3.676
				V	0.233	0.093	0.253	0.157	6.372	2.185	3.799
POTASSIUM	CHLORIDE			H	0.219	0.086	0.268	0.157	6.372	2.106	3.739
11304	1.9460	84	473.0	S	0.213	0.084	0.273	0.157	6.372	2.075	3.716
				R	0.201	0.078	0.287	0.157	6.372	2.003	3.663
				V	0.228	0.091	0.255	0.155	6.441	2.165	3.775
POTASSIUM	CHLORIDE			H	0.215	0.085	0.269	0.155	6.441	2.091	3.719
11305	1.9402	84	498.0	S	0.210	0.083	0.274	0.155	6.441	2.062	3.698
				R	0.199	0.077	0.287	0.155	6.441	1.994	3.648

CUBIC

CODE	NAME / DENSITY	REF	TEMP(K)		E	G	SIGMA	K	BETA	VS	VP
				V	0.224	0.089	0.257	0.154	6.507	2.145	3.752
POTASSIUM	CHLORIDE			H	0.212	0.083	0.270	0.154	6.507	2.075	3.700
11306	1.9337	84	523.0	S	0.207	0.081	0.275	0.154	6.507	2.050	3.681
				R	0.196	0.076	0.287	0.154	6.507	1.985	3.634
				V	0.219	0.087	0.260	0.152	6.582	2.124	3.728
POTASSIUM	CHLORIDE			H	0.208	0.082	0.272	0.152	6.582	2.060	3.679
11307	1.9281	84	548.0	S	0.204	0.080	0.276	0.152	6.582	2.036	3.662
				R	0.194	0.075	0.287	0.152	6.582	1.976	3.617
				V	0.215	0.085	0.262	0.150	6.660	2.105	3.704
POTASSIUM	CHLORIDE			H	0.204	0.080	0.273	0.150	6.660	2.044	3.658
11308	1.9217	84	573.0	S	0.201	0.079	0.277	0.150	6.660	2.023	3.643
				R	0.191	0.074	0.288	0.150	6.660	1.966	3.601
				V	0.210	0.083	0.264	0.149	6.729	2.085	3.682
POTASSIUM	CHLORIDE			H	0.201	0.079	0.275	0.149	6.729	2.029	3.640
11309	1.9149	84	598.0	S	0.198	0.077	0.278	0.149	6.729	2.010	3.626
				R	0.189	0.073	0.288	0.149	6.729	1.956	3.587
				V	0.205	0.081	0.266	0.147	6.825	2.062	3.653
POTASSIUM	CHLORIDE			H	0.197	0.077	0.276	0.147	6.825	2.010	3.615
11310	1.9081	84	623.0	S	0.194	0.076	0.279	0.147	6.825	1.994	3.603
				R	0.186	0.072	0.288	0.147	6.825	1.945	3.567
				V	0.201	0.079	0.269	0.145	6.900	2.041	3.629
POTASSIUM	CHLORIDE			H	0.193	0.076	0.278	0.145	6.900	1.994	3.594
11311	1.9021	84	648.0	S	0.191	0.075	0.280	0.145	6.900	1.980	3.584
				R	0.184	0.071	0.289	0.145	6.900	1.935	3.551
				V	0.197	0.077	0.271	0.143	6.984	2.020	3.605
POTASSIUM	CHLORIDE			H	0.190	0.074	0.279	0.143	6.984	1.977	3.573
11312	1.8952	84	673.0	S	0.188	0.073	0.282	0.143	6.984	1.965	3.564
				R	0.181	0.070	0.289	0.143	6.984	1.924	3.534
				V	0.192	0.075	0.273	0.141	7.098	1.999	3.576
POTASSIUM	CHLORIDE			H	0.186	0.073	0.280	0.141	7.098	1.961	3.548
11313	1.8884	84	698.0	S	0.184	0.072	0.282	0.141	7.098	1.950	3.540
				R	0.178	0.069	0.289	0.141	7.098	1.912	3.512
				V	0.188	0.074	0.275	0.139	7.197	1.979	3.550
POTASSIUM	CHLORIDE			H	0.182	0.071	0.281	0.139	7.197	1.944	3.525
11314	1.8818	84	723.0	S	0.181	0.070	0.283	0.139	7.197	1.935	3.518
				R	0.175	0.068	0.290	0.139	7.197	1.901	3.493
				V	0.183	0.072	0.275	0.136	7.344	1.959	3.518
POTASSIUM	CHLORIDE			H	0.179	0.070	0.281	0.136	7.344	1.928	3.495
11315	1.8752	84	748.0	S	0.177	0.069	0.283	0.136	7.344	1.920	3.489
				R	0.172	0.067	0.289	0.136	7.344	1.889	3.467
				V	0.179	0.070	0.276	0.133	7.500	1.938	3.485
POTASSIUM	CHLORIDE			H	0.175	0.068	0.282	0.133	7.500	1.910	3.465
11316	1.8681	84	773.0	S	0.174	0.068	0.283	0.133	7.500	1.904	3.460
				R	0.170	0.066	0.288	0.133	7.500	1.877	3.440
				V	0.175	0.068	0.278	0.131	7.635	1.917	3.455
POTASSIUM	CHLORIDE			H	0.171	0.067	0.282	0.131	7.635	1.893	3.438
11317	1.8609	84	798.0	S	0.170	0.066	0.283	0.131	7.635	1.888	3.434
				R	0.167	0.065	0.288	0.131	7.635	1.864	3.416
				V	0.171	0.067	0.279	0.129	7.764	1.898	3.428
POTASSIUM	CHLORIDE			H	0.168	0.065	0.283	0.129	7.764	1.877	3.413
11318	1.8533	84	823.0	S	0.167	0.065	0.284	0.129	7.764	1.873	3.410
				R	0.164	0.064	0.288	0.129	7.764	1.852	3.395
				V	0.167	0.065	0.281	0.127	7.887	1.879	3.402
POTASSIUM	CHLORIDE			H	0.164	0.064	0.284	0.127	7.887	1.861	3.389
11319	1.8462	84	848.0	S	0.164	0.064	0.285	0.127	7.887	1.857	3.386
				R	0.161	0.062	0.288	0.127	7.887	1.839	3.373
				V	0.163	0.064	0.282	0.125	8.016	1.859	3.376
POTASSIUM	CHLORIDE			H	0.161	0.063	0.285	0.125	8.016	1.844	3.365
11320	1.8386	84	873.0	S	0.160	0.062	0.286	0.125	8.016	1.842	3.363
				R	0.158	0.061	0.289	0.125	8.016	1.826	3.351

CUBIC

CODE	NAME / DENSITY	REF	TEMP(K)		E	G	SIGMA	K	BETA	VS	VP
				V	0.159	0.062	0.283	0.122	8.166	1.841	3.347
POTASSIUM	CHLORIDE			H	0.157	0.061	0.286	0.122	8.166	1.828	3.338
11321	1.8317	84	898.0	S	0.157	0.061	0.286	0.122	8.166	1.826	3.336
				R	0.155	0.060	0.289	0.122	8.166	1.813	3.327
				V	0.156	0.061	0.283	0.120	8.352	1.824	3.316
POTASSIUM	CHLORIDE			H	0.154	0.060	0.285	0.120	8.352	1.813	3.309
11322	1.8242	84	923.0	S	0.154	0.060	0.286	0.120	8.352	1.812	3.308
				R	0.152	0.059	0.288	0.120	8.352	1.801	3.300
				V	0.152	0.059	0.284	0.117	8.514	1.806	3.288
POTASSIUM	CHLORIDE			H	0.151	0.059	0.286	0.117	8.514	1.797	3.282
11323	1.8168	84	948.0	S	0.151	0.059	0.286	0.117	8.514	1.796	3.281
				R	0.149	0.058	0.288	0.117	8.514	1.787	3.275
				V	0.149	0.058	0.284	0.115	8.700	1.789	3.259
POTASSIUM	CHLORIDE			H	0.148	0.057	0.286	0.115	8.700	1.782	3.254
11324	1.8089	84	973.0	S	0.148	0.057	0.286	0.115	8.700	1.782	3.254
				R	0.147	0.057	0.287	0.115	8.700	1.774	3.248
				V	0.145	0.057	0.284	0.112	8.904	1.773	3.229
POTASSIUM	CHLORIDE			H	0.145	0.056	0.285	0.112	8.904	1.767	3.225
11325	1.8015	84	998.0	S	0.145	0.056	0.285	0.112	8.904	1.767	3.224
				R	0.144	0.056	0.287	0.112	8.904	1.761	3.220
				V	0.142	0.055	0.284	0.109	9.135	1.754	3.195
POTASSIUM	CHLORIDE			H	0.141	0.055	0.285	0.109	9.135	1.750	3.192
11326	1.7932	84	1023.0	S	0.141	0.055	0.285	0.109	9.135	1.750	3.192
				R	0.141	0.055	0.286	0.109	9.135	1.746	3.189
				V	0.141	0.055	0.284	0.109	9.204	1.748	3.184
POTASSIUM	CHLORIDE			H	0.140	0.055	0.285	0.109	9.204	1.744	3.181
11327	1.7916	84	1028.0	S	0.140	0.055	0.285	0.109	9.204	1.744	3.181
				R	0.140	0.054	0.286	0.109	9.204	1.741	3.179
				V	0.140	0.054	0.288	0.110	9.120	1.741	3.189
POTASSIUM	CHLORIDE			H	0.139	0.054	0.288	0.110	9.120	1.739	3.187
11328	1.7900	84	1033.0	S	0.139	0.054	0.288	0.110	9.120	1.738	3.187
				R	0.139	0.054	0.289	0.110	9.120	1.735	3.184
				V	0.283	0.113	0.252	0.190	5.269	2.371	4.116
POTASSIUM	CHLORIDE			H	0.259	0.102	0.273	0.190	5.269	2.249	4.024
11329	2.0089	23	195.0	S	0.247	0.096	0.283	0.190	5.269	2.187	3.978
				R	0.226	0.087	0.301	0.190	5.269	2.081	3.901
				V	0.264	0.105	0.258	0.182	5.509	2.297	4.021
POTASSIUM	CHLORIDE			H	0.244	0.096	0.276	0.182	5.509	2.194	3.944
11330	1.9874	23	295.0	S	0.235	0.092	0.284	0.182	5.509	2.147	3.909
				R	0.218	0.084	0.300	0.182	5.509	2.055	3.843
				V	0.264	0.105	0.259	0.183	5.463	2.301	4.035
POTASSIUM	CHLORIDE			H	0.245	0.096	0.277	0.183	5.463	2.198	3.958
11331	1.9840	234	298.0	S	0.236	0.092	0.285	0.183	5.463	2.151	3.923
				R	0.219	0.084	0.301	0.183	5.463	2.059	3.857
				V	0.265	0.105	0.259	0.183	5.456	2.303	4.037
POTASSIUM	CHLORIDE			H	0.245	0.096	0.277	0.183	5.457	2.197	3.958
11332	1.9865	79	ROOM	S	0.236	0.092	0.286	0.183	5.457	2.148	3.922
				R	0.218	0.084	0.302	0.183	5.457	2.054	3.854
				V	0.255	0.101	0.260	0.177	5.659	2.180	3.826
POTASSIUM	CHLORIDE(83.2)-BROMIDE(16.8)			H	0.236	0.092	0.278	0.177	5.659	2.082	3.752
11333	2.1290	234	298.0	S	0.227	0.088	0.286	0.177	5.659	2.037	3.719
				R	0.211	0.081	0.301	0.177	5.659	1.950	3.656
				V	0.255	0.101	0.260	0.177	5.656	2.181	3.828
POTASSIUM	CHLORIDE(82.9)-BROMIDE(17.1)			H	0.236	0.092	0.278	0.177	5.656	2.084	3.756
11334	2.1260	234	298.0	S	0.227	0.088	0.286	0.177	5.656	2.039	3.723
				R	0.211	0.081	0.301	0.177	5.656	1.952	3.660
				V	0.244	0.097	0.260	0.169	5.906	2.053	3.602
POTASSIUM	CHLORIDE(61.8)-BROMIDE(38.2)			H	0.226	0.088	0.278	0.169	5.906	1.960	3.533
11335	2.3020	234	298.0	S	0.218	0.085	0.286	0.169	5.906	1.918	3.501
				R	0.202	0.078	0.301	0.169	5.905	1.835	3.442

CUBIC

CODE / NAME	DENSITY	REF	TEMP(K)		E	G	SIGMA	K	BETA	VS	VP
				V	0.244	0.097	0.260	0.169	5.903	2.052	3.603
POTASSIUM CHLORIDE(61.3)-BROMIDE(38.7)				H	0.226	0.088	0.278	0.169	5.903	1.959	3.533
11336	2.3000	234	298.0	S	0.217	0.085	0.286	0.169	5.903	1.917	3.502
				R	0.201	0.077	0.302	0.169	5.903	1.834	3.443
				V	0.237	0.094	0.260	0.164	6.099	1.956	3.433
POTASSIUM CHLORIDE(42.2)-BROMIDE(57.8)				H	0.218	0.085	0.278	0.164	6.099	1.866	3.365
11337	2.4530	234	298.0	S	0.210	0.082	0.287	0.164	6.099	1.824	3.334
				R	0.194	0.075	0.303	0.164	6.099	1.743	3.276
				V	0.235	0.093	0.260	0.163	6.135	1.940	3.408
POTASSIUM CHLORIDE(40.2)-BROMIDE(59.8)				H	0.217	0.085	0.279	0.163	6.135	1.851	3.340
11338	2.4730	234	298.0	S	0.208	0.081	0.287	0.163	6.135	1.809	3.309
				R	0.193	0.074	0.303	0.163	6.135	1.728	3.252
				V	0.228	0.090	0.261	0.159	6.284	1.860	3.271
POTASSIUM CHLORIDE(20.5)-BROMIDE(79.5)				H	0.210	0.082	0.280	0.159	6.284	1.771	3.205
11339	2.6130	234	298.0	S	0.201	0.078	0.289	0.159	6.284	1.729	3.174
				R	0.186	0.071	0.305	0.159	6.284	1.650	3.118
				V	0.228	0.091	0.260	0.158	6.309	1.863	3.270
POTASSIUM CHLORIDE(20.0)-BROMIDE(80.0)				H	0.210	0.082	0.279	0.158	6.309	1.774	3.203
11340	2.6130	234	298.0	S	0.202	0.078	0.288	0.158	6.309	1.731	3.172
				R	0.186	0.071	0.304	0.159	6.309	1.652	3.115
				V	0.069	0.024	0.420	0.143	6.977	1.248	3.363
POTASSIUM CYANIDE				H	0.064	0.023	0.425	0.143	6.977	1.204	3.341
11341	1.5530	4	ROOM	S	0.062	0.022	0.428	0.143	6.977	1.180	3.329
				R	0.057	0.020	0.434	0.143	6.977	1.129	3.306
				V	0.451	0.179	0.264	0.319	3.138	2.659	4.695
POTASSIUM FLUORIDE				H	0.433	0.170	0.273	0.319	3.138	2.596	4.648
11342	2.5257	4	ROOM	S	0.428	0.168	0.276	0.319	3.138	2.576	4.633
				R	0.411	0.160	0.285	0.319	3.138	2.515	4.588
				V	0.449	0.178	0.260	0.311	3.212	2.679	4.704
POTASSIUM FLUORIDE				H	0.431	0.170	0.269	0.311	3.212	2.616	4.656
11343	2.4800	145	298.0	S	0.426	0.167	0.272	0.311	3.212	2.597	4.642
				R	0.409	0.160	0.281	0.311	3.212	2.537	4.597
				V	0.512	0.205	0.251	0.342	2.921	2.843	4.931
POTASSIUM FLUORIDE				H	0.482	0.191	0.265	0.342	2.921	2.744	4.855
11344	2.5300	152	4.2	S	0.471	0.185	0.271	0.342	2.921	2.706	4.826
				R	0.444	0.173	0.284	0.342	2.921	2.616	4.760
				V	0.440	0.173	0.269	0.318	3.148	2.644	4.704
POTASSIUM FLUORIDE				H	0.422	0.165	0.279	0.318	3.148	2.578	4.655
11345	2.4800	152	295.0	S	0.416	0.162	0.282	0.318	3.148	2.557	4.640
				R	0.398	0.154	0.291	0.318	3.148	2.495	4.594
				V	0.504	0.200	0.261	0.351	2.849	2.780	4.887
POTASSIUM FLUORIDE				H	0.474	0.186	0.275	0.351	2.849	2.680	4.812
11346	2.5858	175	4.2	S	0.462	0.180	0.281	0.351	2.849	2.641	4.783
				R	0.435	0.168	0.294	0.351	2.849	2.550	4.716
				V	0.501	0.198	0.265	0.355	2.818	2.769	4.894
POTASSIUM FLUORIDE				H	0.472	0.185	0.278	0.355	2.818	2.673	4.822
11347	2.5852	175	20.0	S	0.461	0.180	0.284	0.355	2.818	2.635	4.795
				R	0.434	0.168	0.296	0.355	2.818	2.546	4.730
				V	0.476	0.188	0.268	0.342	2.923	2.706	4.808
POTASSIUM FLUORIDE				H	0.453	0.177	0.280	0.342	2.923	2.627	4.750
11348	2.5618	175	140.0	S	0.444	0.173	0.284	0.342	2.923	2.600	4.729
				R	0.423	0.163	0.294	0.342	2.923	2.525	4.675
				V	0.471	0.186	0.268	0.339	2.948	2.696	4.791
POTASSIUM FLUORIDE				H	0.449	0.175	0.279	0.339	2.948	2.620	4.735
11349	2.5570	175	160.0	S	0.441	0.172	0.283	0.339	2.948	2.593	4.715
				R	0.420	0.163	0.293	0.339	2.948	2.521	4.663
				V	0.467	0.184	0.268	0.336	2.977	2.686	4.773
POTASSIUM FLUORIDE				H	0.445	0.174	0.279	0.336	2.977	2.612	4.718
11350	2.5524	175	180.0	S	0.438	0.171	0.283	0.336	2.977	2.587	4.699
				R	0.418	0.162	0.293	0.336	2.977	2.517	4.648

CUBIC

CODE	NAME / DENSITY	REF	TEMP(K)		E	G	SIGMA	K	BETA	VS	VP
	POTASSIUM FLUORIDE			V	0.462	0.182	0.269	0.334	2.996	2.674	4.758
11351	2.5476	175	200.0	H	0.442	0.173	0.279	0.334	2.996	2.603	4.705
				S	0.435	0.169	0.283	0.334	2.996	2.579	4.687
				R	0.415	0.161	0.293	0.334	2.996	2.511	4.638
	POTASSIUM FLUORIDE			V	0.457	0.180	0.270	0.332	3.013	2.660	4.742
11352	2.5427	175	220.0	H	0.437	0.171	0.280	0.332	3.013	2.592	4.691
				S	0.431	0.168	0.284	0.332	3.013	2.569	4.675
				R	0.412	0.159	0.293	0.332	3.013	2.504	4.628
	POTASSIUM FLUORIDE			V	0.453	0.178	0.270	0.329	3.042	2.652	4.726
11353	2.5377	175	240.0	H	0.434	0.170	0.280	0.329	3.042	2.585	4.676
				S	0.428	0.167	0.283	0.329	3.042	2.564	4.660
				R	0.410	0.159	0.292	0.329	3.042	2.500	4.614
	POTASSIUM FLUORIDE			V	0.447	0.175	0.274	0.329	3.037	2.632	4.716
11354	2.5329	175	260.0	H	0.429	0.167	0.283	0.329	3.037	2.570	4.670
				S	0.424	0.165	0.286	0.329	3.037	2.551	4.656
				R	0.407	0.157	0.294	0.329	3.037	2.491	4.612
	POTASSIUM FLUORIDE			V	0.444	0.174	0.272	0.324	3.084	2.627	4.694
11355	2.5280	175	280.0	H	0.426	0.166	0.281	0.324	3.084	2.566	4.648
				S	0.421	0.164	0.284	0.324	3.084	2.547	4.634
				R	0.404	0.156	0.292	0.324	3.084	2.488	4.591
	POTASSIUM FLUORIDE			V	0.440	0.173	0.273	0.323	3.099	2.615	4.679
11356	2.5256	175	300.0	H	0.423	0.165	0.282	0.323	3.099	2.556	4.635
				S	0.418	0.163	0.284	0.323	3.099	2.538	4.622
				R	0.402	0.155	0.292	0.323	3.099	2.481	4.581
	POTASSIUM IODIDE			V	0.175	0.070	0.251	0.117	8.520	1.496	2.595
11357	3.1300	128	ROOM	H	0.162	0.064	0.270	0.117	8.520	1.429	2.544
				S	0.156	0.061	0.278	0.117	8.520	1.398	2.521
				R	0.145	0.056	0.294	0.117	8.520	1.338	2.477
	POTASSIUM IODIDE			V	0.173	0.068	0.265	0.123	8.119	1.480	2.619
11358	3.1279	4	ROOM	H	0.158	0.061	0.287	0.123	8.119	1.400	2.559
				S	0.150	0.058	0.298	0.123	8.119	1.358	2.529
				R	0.136	0.052	0.315	0.123	8.119	1.287	2.479
	POTASSIUM IODIDE			V	0.170	0.067	0.262	0.119	8.380	1.467	2.585
11359	3.1300	4	ROOM	H	0.154	0.060	0.284	0.119	8.380	1.386	2.525
				S	0.146	0.056	0.296	0.119	8.380	1.343	2.493
				R	0.133	0.051	0.314	0.119	8.380	1.272	2.443
	POTASSIUM IODIDE			V	0.170	0.067	0.265	0.120	8.310	1.464	2.588
11360	3.1300	4	ROOM	H	0.154	0.060	0.287	0.120	8.310	1.380	2.527
				S	0.145	0.056	0.299	0.120	8.310	1.335	2.494
				R	0.132	0.050	0.318	0.120	8.310	1.263	2.444
	POTASSIUM IODIDE			V	0.209	0.085	0.226	0.127	7.853	1.640	2.758
11361	3.1690	196	4.2	H	0.180	0.071	0.264	0.127	7.853	1.500	2.649
				S	0.161	0.062	0.290	0.127	7.853	1.402	2.577
				R	0.140	0.053	0.317	0.127	7.853	1.294	2.500
	POTASSIUM IODIDE			V	0.208	0.084	0.230	0.128	7.813	1.633	2.756
11362	3.1690	196	20.0	H	0.179	0.071	0.266	0.128	7.813	1.495	2.650
				S	0.160	0.062	0.291	0.128	7.813	1.400	2.579
				R	0.140	0.053	0.318	0.128	7.812	1.293	2.504
	POTASSIUM IODIDE			V	0.205	0.083	0.233	0.128	7.813	1.621	2.747
11363	3.1680	196	40.0	H	0.178	0.070	0.268	0.128	7.813	1.488	2.644
				S	0.160	0.062	0.292	0.128	7.813	1.396	2.577
				R	0.139	0.053	0.319	0.128	7.813	1.292	2.503
	POTASSIUM IODIDE			V	0.203	0.082	0.236	0.128	7.812	1.611	2.739
11364	3.1660	196	60.0	H	0.176	0.069	0.270	0.128	7.813	1.481	2.640
				S	0.159	0.062	0.293	0.128	7.813	1.394	2.576
				R	0.139	0.053	0.319	0.128	7.812	1.292	2.504
	POTASSIUM IODIDE			V	0.200	0.081	0.239	0.128	7.833	1.597	2.727
11365	3.1640	196	80.0	H	0.175	0.069	0.272	0.128	7.833	1.472	2.632
				S	0.158	0.061	0.294	0.128	7.833	1.390	2.571
				R	0.139	0.053	0.319	0.128	7.833	1.290	2.501

CUBIC

CODE	NAME / DENSITY	REF	TEMP(K)		E	G	SIGMA	K	BETA	VS	VP
				V	0.198	0.080	0.242	0.128	7.833	1.586	2.719
POTASSIUM IODIDE				H	0.173	0.068	0.274	0.128	7.833	1.465	2.627
11366	3.1620	196	100.0	S	0.157	0.061	0.295	0.128	7.833	1.386	2.569
				R	0.139	0.053	0.319	0.128	7.833	1.289	2.500
				V	0.195	0.078	0.244	0.127	7.895	1.575	2.705
POTASSIUM IODIDE				H	0.171	0.067	0.275	0.127	7.895	1.457	2.616
11367	3.1590	196	120.0	S	0.156	0.060	0.294	0.127	7.895	1.382	2.561
				R	0.138	0.052	0.318	0.127	7.895	1.287	2.494
				V	0.192	0.077	0.247	0.127	7.895	1.563	2.697
POTASSIUM IODIDE				H	0.169	0.066	0.277	0.127	7.895	1.450	2.611
11368	3.1560	196	140.0	S	0.155	0.060	0.296	0.127	7.895	1.378	2.559
				R	0.138	0.052	0.319	0.127	7.895	1.286	2.493
				V	0.189	0.076	0.250	0.126	7.916	1.549	2.685
POTASSIUM IODIDE				H	0.167	0.065	0.279	0.126	7.916	1.440	2.602
11369	3.1530	196	160.0	S	0.154	0.059	0.297	0.126	7.916	1.372	2.553
				R	0.137	0.052	0.320	0.126	7.916	1.282	2.490
				V	0.187	0.075	0.252	0.126	7.958	1.540	2.674
POTASSIUM IODIDE				H	0.166	0.065	0.280	0.126	7.958	1.434	2.594
11370	3.1500	196	180.0	S	0.153	0.059	0.297	0.126	7.958	1.369	2.547
				R	0.136	0.052	0.319	0.126	7.958	1.281	2.485
				V	0.184	0.073	0.254	0.125	8.021	1.528	2.660
POTASSIUM IODIDE				H	0.164	0.064	0.281	0.125	8.021	1.426	2.583
11371	3.1470	196	200.0	S	0.152	0.059	0.297	0.125	8.021	1.365	2.539
				R	0.136	0.052	0.318	0.125	8.021	1.279	2.479
				V	0.182	0.073	0.254	0.123	8.108	1.520	2.646
POTASSIUM IODIDE				H	0.162	0.063	0.281	0.123	8.108	1.420	2.571
11372	3.1440	196	220.0	S	0.151	0.058	0.296	0.123	8.108	1.361	2.528
				R	0.135	0.051	0.317	0.123	8.108	1.277	2.469
				V	0.179	0.071	0.257	0.123	8.130	1.506	2.635
POTASSIUM IODIDE				H	0.160	0.062	0.283	0.123	8.130	1.411	2.563
11373	3.1400	196	240.0	S	0.150	0.058	0.297	0.123	8.130	1.355	2.523
				R	0.134	0.051	0.318	0.123	8.130	1.275	2.466
				V	0.176	0.070	0.259	0.122	8.197	1.494	2.620
POTASSIUM IODIDE				H	0.158	0.062	0.284	0.122	8.197	1.402	2.552
11374	3.1360	196	260.0	S	0.148	0.057	0.297	0.122	8.197	1.350	2.514
				R	0.134	0.051	0.318	0.122	8.197	1.271	2.459
				V	0.173	0.068	0.263	0.121	8.242	1.477	2.604
POTASSIUM IODIDE				H	0.156	0.061	0.286	0.121	8.242	1.390	2.540
11375	3.1320	196	280.0	S	0.147	0.056	0.299	0.121	8.242	1.342	2.505
				R	0.133	0.050	0.318	0.121	8.242	1.267	2.452
				V	0.170	0.067	0.265	0.120	8.310	1.464	2.589
POTASSIUM IODIDE				H	0.154	0.060	0.287	0.120	8.310	1.381	2.528
11376	3.1280	196	300.0	S	0.145	0.056	0.299	0.120	8.310	1.336	2.495
				R	0.132	0.050	0.318	0.120	8.310	1.264	2.445
				V	0.182	0.073	0.256	0.125	8.010	1.524	2.663
POTASSIUM IODIDE				H	0.167	0.065	0.277	0.125	8.010	1.447	2.605
11377	3.1260	212	303.0	S	0.160	0.062	0.287	0.125	8.010	1.410	2.577
				R	0.147	0.056	0.304	0.125	8.010	1.342	2.529
				V	0.170	0.068	0.256	0.116	8.603	1.470	2.569
POTASSIUM IODIDE				H	0.154	0.060	0.279	0.116	8.603	1.388	2.507
11378	3.1257	213	300.0	S	0.146	0.056	0.291	0.116	8.603	1.343	2.475
				R	0.132	0.051	0.310	0.116	8.603	1.272	2.424
				V	1.165	0.476	0.224	0.704	1.420	3.886	6.518
POTASSIUM MAGNESIUM FLUORIDE				H	1.165	0.476	0.224	0.704	1.420	3.886	6.518
11379	3.1500	219	290.0	S	1.165	0.476	0.224	0.704	1.420	3.886	6.518
				R	1.164	0.476	0.224	0.704	1.420	3.885	6.517
				V	0.755	0.289	0.306	0.649	1.541	2.898	5.483
POTASSIUM MANGANESE FLUORIDE				H	0.752	0.288	0.307	0.649	1.541	2.892	5.479
11380	3.4397	9	200.0	S	0.752	0.288	0.307	0.649	1.541	2.892	5.479
				R	0.749	0.286	0.308	0.649	1.541	2.885	5.474

CUBIC

CODE	NAME DENSITY	REF	TEMP(K)		E	G	SIGMA	K	BETA	VS	VP
				V	0.806	0.311	0.293	0.649	1.541	3.010	5.564
POTASSIUM MANGANESE FLUORIDE				H	0.799	0.308	0.295	0.649	1.541	2.995	5.553
11381	3.4377	9	210.0	S	0.798	0.308	0.295	0.649	1.541	2.993	5.552
				R	0.791	0.305	0.297	0.649	1.541	2.977	5.541
				V	0.826	0.321	0.288	0.649	1.541	3.056	5.598
POTASSIUM MANGANESE FLUORIDE				H	0.817	0.317	0.290	0.649	1.541	3.035	5.583
11382	3.4357	9	220.0	S	0.815	0.316	0.291	0.649	1.541	3.032	5.581
				R	0.806	0.312	0.293	0.649	1.541	3.011	5.566
				V	0.836	0.325	0.285	0.649	1.541	3.077	5.614
POTASSIUM MANGANESE FLUORIDE				H	0.825	0.320	0.288	0.649	1.541	3.054	5.597
11383	3.4337	9	230.0	S	0.823	0.319	0.289	0.649	1.541	3.050	5.595
				R	0.812	0.314	0.291	0.649	1.541	3.026	5.577
				V	0.840	0.327	0.284	0.649	1.541	3.088	5.623
POTASSIUM MANGANESE FLUORIDE				H	0.829	0.322	0.287	0.649	1.541	3.063	5.605
11384	3.4318	9	240.0	S	0.827	0.321	0.288	0.649	1.541	3.059	5.602
				R	0.815	0.316	0.291	0.649	1.541	3.033	5.584
				V	0.843	0.329	0.283	0.649	1.541	3.095	5.630
POTASSIUM MANGANESE FLUORIDE				H	0.831	0.323	0.286	0.649	1.541	3.069	5.611
11385	3.4298	9	250.0	S	0.829	0.322	0.287	0.649	1.541	3.065	5.608
				R	0.817	0.317	0.290	0.649	1.541	3.038	5.588
				V	0.845	0.329	0.283	0.649	1.541	3.100	5.634
POTASSIUM MANGANESE FLUORIDE				H	0.833	0.324	0.286	0.649	1.541	3.074	5.615
11386	3.4278	9	260.0	S	0.831	0.323	0.287	0.649	1.541	3.069	5.612
				R	0.818	0.317	0.290	0.649	1.541	3.042	5.592
				V	0.847	0.330	0.283	0.649	1.541	3.104	5.638
POTASSIUM MANGANESE FLUORIDE				H	0.834	0.324	0.286	0.649	1.541	3.077	5.619
11387	3.4259	9	270.0	S	0.832	0.323	0.286	0.649	1.541	3.073	5.615
				R	0.819	0.318	0.290	0.649	1.541	3.045	5.595
				V	0.848	0.331	0.282	0.649	1.541	3.108	5.642
POTASSIUM MANGANESE FLUORIDE				H	0.836	0.325	0.285	0.649	1.541	3.081	5.622
11388	3.4239	9	280.0	S	0.833	0.324	0.286	0.649	1.541	3.076	5.619
				R	0.820	0.318	0.289	0.649	1.541	3.048	5.598
				V	0.850	0.331	0.282	0.649	1.541	3.112	5.646
POTASSIUM MANGANESE FLUORIDE				H	0.837	0.326	0.285	0.649	1.541	3.085	5.626
11389	3.4220	9	290.0	S	0.835	0.325	0.286	0.649	1.541	3.080	5.622
				R	0.821	0.319	0.289	0.649	1.541	3.051	5.601
				V	0.851	0.332	0.281	0.649	1.541	3.116	5.650
POTASSIUM MANGANESE FLUORIDE				H	0.838	0.326	0.285	0.649	1.541	3.088	5.629
11390	3.4200	9	300.0	S	0.836	0.325	0.285	0.649	1.541	3.083	5.626
				R	0.822	0.319	0.289	0.649	1.541	3.054	5.605
				V	2.818	1.453	-0.030	0.887	1.128	5.380	7.501
PYRITE				H	2.727	1.381	-0.013	0.887	1.128	5.246	7.373
11391	5.0180	128	ROOM	S	2.702	1.362	-0.008	0.887	1.128	5.210	7.338
				R	2.632	1.309	0.005	0.887	1.128	5.107	7.242
				V	2.827	1.434	-0.014	0.916	1.092	5.346	7.507
PYRITE				H	2.734	1.364	0.002	0.916	1.092	5.213	7.381
11392	5.0180	128	ROOM	S	2.708	1.344	0.007	0.916	1.092	5.176	7.346
				R	2.636	1.292	0.020	0.916	1.092	5.074	7.251
				V	2.825	1.448	-0.025	0.897	1.114	5.376	7.513
PYRITE				H	2.733	1.377	-0.008	0.897	1.114	5.243	7.386
11393	5.0100	129	ROOM	S	2.708	1.358	-0.003	0.897	1.114	5.206	7.352
				R	2.637	1.305	0.010	0.897	1.114	5.104	7.256
				V	2.856	1.462	-0.023	0.909	1.100	5.403	7.554
PYRITE				H	2.772	1.398	-0.008	0.909	1.100	5.282	7.439
11394	5.0100	129	ROOM	S	2.751	1.381	-0.004	0.909	1.100	5.250	7.410
				R	2.685	1.332	0.008	0.909	1.100	5.157	7.322
				V	2.756	1.444	-0.045	0.842	1.187	5.368	7.432
PYRITE				H	2.663	1.368	-0.027	0.842	1.187	5.226	7.295
11395	5.0100	129	ROOM	S	2.636	1.347	-0.022	0.842	1.187	5.185	7.257
				R	2.564	1.292	-0.007	0.842	1.187	5.078	7.155

CUBIC

CODE	NAME DENSITY	REF	TEMP(K)		E	G	SIGMA	K	BETA	VS	VP
				V	2.836	1.454	-0.025	0.900	1.111	5.388	7.528
PYRITE				H	2.756	1.392	-0.010	0.900	1.111	5.271	7.417
11396	5.0100	129	ROOM	S	2.735	1.376	-0.007	0.900	1.111	5.242	7.389
				R	2.673	1.330	0.005	0.900	1.111	5.151	7.304
				V	2.866	1.456	-0.015	0.927	1.079	5.390	7.566
PYRITE				H	2.753	1.370	0.005	0.927	1.079	5.229	7.413
11397	5.0100	129	ROOM	S	2.718	1.344	0.011	0.927	1.079	5.179	7.367
				R	2.631	1.281	0.027	0.927	1.079	5.057	7.252
				V	3.090	1.344	0.150	1.470	0.680	5.178	8.067
PYRITE				H	3.034	1.312	0.156	1.470	0.680	5.116	8.014
11398	5.0130	4	ROOM	S	3.023	1.306	0.157	1.470	0.680	5.105	8.004
				R	2.973	1.278	0.163	1.470	0.680	5.050	7.957
				V	2.921	1.279	0.142	1.360	0.735	5.050	7.819
PYRITE				H	2.884	1.258	0.147	1.360	0.735	5.009	7.784
11399	5.0130	8	ROOM	S	2.879	1.255	0.147	1.360	0.735	5.003	7.778
				R	2.846	1.236	0.151	1.360	0.735	4.966	7.746
				V	2.977	1.289	0.155	1.437	0.696	5.071	7.934
PYRITE				H	2.925	1.260	0.161	1.437	0.696	5.014	7.885
11400	5.0130	8	ROOM	S	2.916	1.255	0.162	1.437	0.696	5.003	7.876
				R	2.869	1.229	0.167	1.437	0.696	4.952	7.833
				V	3.119	1.358	0.149	1.479	0.676	5.203	8.099
PYRITE				H	3.060	1.324	0.155	1.479	0.676	5.138	8.043
11401	5.0160	233	ROOM	S	3.048	1.318	0.157	1.479	0.676	5.125	8.032
				R	2.995	1.288	0.163	1.479	0.676	5.067	7.983
				V	3.101	1.350	0.148	1.470	0.680	5.233	8.145
PYRITE				H	3.042	1.317	0.155	1.470	0.680	5.168	8.089
11402	4.9290	233	ROOM	S	3.030	1.310	0.156	1.470	0.680	5.155	8.078
				R	2.977	1.280	0.162	1.470	0.680	5.097	8.029
				V	2.967	1.286	0.154	1.427	0.701	5.108	7.984
PYRITE				H	2.917	1.258	0.159	1.427	0.701	5.051	7.936
11403	4.9290	233	ROOM	S	2.907	1.253	0.161	1.427	0.701	5.041	7.927
				R	2.862	1.227	0.166	1.427	0.701	4.990	7.884
				V	0.028	0.011	0.321	0.026	38.265	0.814	1.584
RUBIDIUM				H	0.024	0.009	0.347	0.026	38.265	0.744	1.537
11404	1.6070	217	80.0	S	0.019	0.007	0.379	0.026	38.265	0.655	1.483
				R	0.015	0.005	0.406	0.026	38.264	0.570	1.435
				V	0.037	0.014	0.297	0.031	32.680	0.941	1.752
RUBIDIUM				H	0.031	0.012	0.332	0.031	32.680	0.845	1.686
11405	1.6200	122	4.2	S	0.022	0.008	0.380	0.031	32.680	0.701	1.595
				R	0.016	0.006	0.412	0.031	32.679	0.593	1.536
				V	0.034	0.013	0.306	0.029	34.443	0.896	1.695
RUBIDIUM				H	0.028	0.011	0.338	0.029	34.443	0.809	1.636
11406	1.6100	122	78.0	S	0.021	0.007	0.381	0.029	34.443	0.682	1.557
				R	0.015	0.005	0.412	0.029	34.442	0.581	1.501
				V	0.032	0.012	0.308	0.028	35.629	0.878	1.668
RUBIDIUM				H	0.027	0.010	0.340	0.028	35.629	0.793	1.610
11407	1.6000	122	110.0	S	0.020	0.007	0.382	0.028	35.629	0.670	1.534
				R	0.015	0.005	0.413	0.028	35.629	0.571	1.479
				V	0.031	0.012	0.312	0.027	36.585	0.861	1.645
RUBIDIUM				H	0.026	0.010	0.342	0.027	36.585	0.779	1.590
11408	1.5900	122	140.0	S	0.019	0.007	0.383	0.027	36.585	0.662	1.517
				R	0.014	0.005	0.412	0.027	36.584	0.566	1.465
				V	0.029	0.011	0.315	0.027	37.689	0.841	1.619
RUBIDIUM				H	0.025	0.009	0.345	0.027	37.688	0.761	1.566
11409	1.5800	122	170.0	S	0.018	0.007	0.386	0.027	37.688	0.645	1.494
				R	0.014	0.005	0.415	0.027	37.688	0.550	1.443
				V	0.195	0.077	0.264	0.138	7.238	1.518	2.682
RUBIDIUM BROMIDE				H	0.173	0.067	0.291	0.138	7.238	1.416	2.607
11410	3.3510	4	ROOM	S	0.161	0.061	0.306	0.138	7.238	1.354	2.563
				R	0.143	0.054	0.327	0.138	7.238	1.269	2.504

CUBIC

CODE	NAME / DENSITY	REF	TEMP(K)		E	G	SIGMA	K	BETA	VS	VP
				V	0.196	0.078	0.255	0.134	7.481	1.529	2.666
RUBIDIUM	BROMIDE			H	0.174	0.068	0.283	0.134	7.481	1.425	2.588
11411	3.3490	41	300.0	S	0.161	0.062	0.299	0.134	7.481	1.362	2.542
				R	0.144	0.054	0.321	0.134	7.481	1.275	2.482
				V	0.193	0.076	0.267	0.138	7.245	1.505	2.670
RUBIDIUM	BROMIDE			H	0.171	0.066	0.293	0.138	7.245	1.404	2.596
11412	3.3590	213	300.0	S	0.159	0.061	0.308	0.138	7.245	1.343	2.553
				R	0.141	0.053	0.329	0.138	7.245	1.259	2.494
				V	0.232	0.092	0.259	0.160	6.236	1.639	2.873
RUBIDIUM	BROMIDE			H	0.201	0.078	0.291	0.160	6.236	1.506	2.774
11413	3.4340	152	4.0	S	0.180	0.069	0.313	0.160	6.236	1.414	2.708
				R	0.157	0.059	0.337	0.160	6.236	1.307	2.636
				V	0.232	0.092	0.259	0.160	6.240	1.639	2.872
RUBIDIUM	BROMIDE			H	0.201	0.078	0.291	0.160	6.240	1.506	2.773
11414	3.4340	152	10.0	S	0.180	0.069	0.313	0.160	6.240	1.414	2.708
				R	0.157	0.059	0.337	0.160	6.240	1.307	2.635
				V	0.232	0.092	0.259	0.160	6.249	1.637	2.870
RUBIDIUM	BROMIDE			H	0.201	0.078	0.291	0.160	6.249	1.505	2.771
11415	3.4340	152	20.0	S	0.180	0.069	0.313	0.160	6.249	1.413	2.706
				R	0.157	0.059	0.337	0.160	6.249	1.306	2.634
				V	0.231	0.092	0.259	0.160	6.267	1.635	2.866
RUBIDIUM	BROMIDE			H	0.200	0.078	0.291	0.160	6.267	1.503	2.768
11416	3.4330	152	30.0	S	0.180	0.068	0.312	0.160	6.267	1.412	2.703
				R	0.156	0.059	0.337	0.160	6.267	1.306	2.631
				V	0.230	0.091	0.259	0.159	6.292	1.632	2.860
RUBIDIUM	BROMIDE			H	0.200	0.077	0.291	0.159	6.292	1.501	2.763
11417	3.4320	152	40.0	S	0.179	0.068	0.312	0.159	6.292	1.410	2.699
				R	0.156	0.058	0.336	0.159	6.292	1.305	2.627
				V	0.229	0.091	0.259	0.158	6.320	1.628	2.855
RUBIDIUM	BROMIDE			H	0.199	0.077	0.291	0.158	6.320	1.498	2.758
11418	3.4300	152	50.0	S	0.179	0.068	0.312	0.158	6.320	1.409	2.695
				R	0.156	0.058	0.336	0.158	6.320	1.304	2.623
				V	0.228	0.090	0.259	0.157	6.352	1.624	2.848
RUBIDIUM	BROMIDE			H	0.198	0.077	0.291	0.157	6.352	1.495	2.752
11419	3.4280	152	60.0	S	0.178	0.068	0.311	0.157	6.352	1.407	2.690
				R	0.156	0.058	0.335	0.157	6.352	1.303	2.619
				V	0.226	0.090	0.259	0.157	6.384	1.620	2.841
RUBIDIUM	BROMIDE			H	0.197	0.076	0.290	0.157	6.384	1.492	2.746
11420	3.4260	152	70.0	S	0.177	0.068	0.311	0.157	6.384	1.405	2.684
				R	0.155	0.058	0.335	0.157	6.384	1.302	2.614
				V	0.225	0.089	0.259	0.156	6.420	1.615	2.833
RUBIDIUM	BROMIDE			H	0.196	0.076	0.290	0.156	6.420	1.489	2.739
11421	3.4240	152	80.0	S	0.177	0.067	0.311	0.156	6.420	1.403	2.678
				R	0.155	0.058	0.335	0.156	6.420	1.300	2.608
				V	0.191	0.075	0.271	0.139	7.213	1.497	2.669
RUBIDIUM	BROMIDE			H	0.170	0.066	0.296	0.139	7.213	1.398	2.597
11422	3.3520	152	290.0	S	0.157	0.060	0.311	0.139	7.213	1.339	2.554
				R	0.140	0.053	0.331	0.139	7.213	1.255	2.497
				V	0.189	0.074	0.271	0.138	7.252	1.490	2.661
RUBIDIUM	BROMIDE			H	0.169	0.065	0.296	0.138	7.252	1.393	2.590
11423	3.3490	152	300.0	S	0.157	0.060	0.311	0.138	7.252	1.335	2.548
				R	0.140	0.053	0.331	0.138	7.252	1.252	2.492
				V	0.218	0.087	0.257	0.149	6.689	1.595	2.790
RUBIDIUM	BROMIDE			H	0.190	0.074	0.289	0.149	6.689	1.470	2.697
11424	3.4040	110	140.0	S	0.171	0.065	0.309	0.149	6.689	1.386	2.637
				R	0.150	0.056	0.333	0.150	6.689	1.285	2.567
				V	0.215	0.085	0.258	0.148	6.751	1.584	2.776
RUBIDIUM	BROMIDE			H	0.187	0.073	0.289	0.148	6.751	1.462	2.685
11425	3.3980	110	160.0	S	0.170	0.065	0.309	0.148	6.751	1.380	2.627
				R	0.149	0.056	0.333	0.148	6.751	1.281	2.559

CUBIC

CODE	NAME DENSITY	REF	TEMP(K)		E	G	SIGMA	K	BETA	VS	VP
				V	0.212	0.084	0.260	0.147	6.809	1.573	2.762
RUBIDIUM BROMIDE				H	0.185	0.072	0.290	0.147	6.809	1.455	2.674
11426	3.3920	110	180.0	S	0.168	0.064	0.309	0.147	6.809	1.376	2.618
				R	0.148	0.055	0.332	0.147	6.809	1.279	2.552
				V	0.208	0.083	0.261	0.145	6.881	1.562	2.747
RUBIDIUM BROMIDE				H	0.183	0.071	0.290	0.145	6.881	1.446	2.661
11427	3.3850	110	200.0	S	0.166	0.064	0.309	0.145	6.881	1.371	2.607
				R	0.147	0.055	0.332	0.145	6.881	1.275	2.542
				V	0.205	0.081	0.263	0.144	6.946	1.551	2.733
RUBIDIUM BROMIDE				H	0.180	0.070	0.291	0.144	6.946	1.438	2.650
11428	3.3780	110	220.0	S	0.165	0.063	0.309	0.144	6.946	1.366	2.598
				R	0.146	0.055	0.331	0.144	6.946	1.273	2.534
				V	0.202	0.080	0.264	0.143	7.014	1.539	2.718
RUBIDIUM BROMIDE				H	0.178	0.069	0.292	0.143	7.014	1.430	2.637
11429	3.3710	110	240.0	S	0.163	0.062	0.309	0.143	7.014	1.361	2.588
				R	0.145	0.054	0.331	0.143	7.014	1.270	2.526
				V	0.199	0.078	0.265	0.141	7.086	1.528	2.703
RUBIDIUM BROMIDE				H	0.176	0.068	0.293	0.141	7.085	1.421	2.625
11430	3.3640	110	260.0	S	0.162	0.062	0.309	0.141	7.085	1.355	2.577
				R	0.143	0.054	0.331	0.141	7.086	1.266	2.516
				V	0.195	0.077	0.267	0.140	7.156	1.516	2.688
RUBIDIUM BROMIDE				H	0.173	0.067	0.293	0.140	7.156	1.413	2.612
11431	3.3570	110	280.0	S	0.160	0.061	0.309	0.140	7.156	1.350	2.568
				R	0.143	0.054	0.330	0.140	7.156	1.264	2.508
				V	0.192	0.076	0.268	0.138	7.225	1.505	2.674
RUBIDIUM BROMIDE				H	0.171	0.066	0.294	0.138	7.225	1.405	2.601
11432	3.3500	110	300.0	S	0.159	0.061	0.309	0.138	7.225	1.344	2.558
				R	0.141	0.053	0.330	0.138	7.225	1.260	2.500
				V	0.212	0.084	0.260	0.147	6.809	1.573	2.762
RUBIDIUM BROMIDE (P=0.001KB)				H	0.185	0.072	0.290	0.147	6.809	1.455	2.674
11433	3.3920	110	180.0	S	0.168	0.064	0.309	0.147	6.809	1.376	2.618
				R	0.148	0.055	0.332	0.147	6.809	1.279	2.552
				V	0.213	0.084	0.260	0.148	6.761	1.576	2.769
RUBIDIUM BROMIDE (P=0.200KB)				H	0.186	0.072	0.291	0.148	6.761	1.456	2.679
11434	3.3971	110	180.0	S	0.168	0.064	0.310	0.148	6.761	1.375	2.622
				R	0.148	0.055	0.333	0.148	6.761	1.277	2.555
				V	0.214	0.085	0.261	0.149	6.717	1.579	2.775
RUBIDIUM BROMIDE (P=0.400KB)				H	0.186	0.072	0.291	0.149	6.717	1.456	2.684
11435	3.4013	110	180.0	S	0.168	0.064	0.311	0.149	6.717	1.374	2.626
				R	0.147	0.055	0.335	0.149	6.717	1.274	2.558
				V	0.215	0.085	0.261	0.150	6.677	1.581	2.780
RUBIDIUM BROMIDE (P=0.600KB)				H	0.187	0.072	0.292	0.150	6.677	1.458	2.689
11436	3.4063	110	180.0	S	0.169	0.064	0.312	0.150	6.677	1.373	2.629
				R	0.147	0.055	0.336	0.150	6.677	1.273	2.561
				V	0.216	0.086	0.261	0.151	6.634	1.584	2.787
RUBIDIUM BROMIDE (P=0.800KB)				H	0.188	0.073	0.293	0.151	6.634	1.459	2.694
11437	3.4103	110	180.0	S	0.169	0.064	0.313	0.151	6.634	1.373	2.633
				R	0.147	0.055	0.337	0.151	6.634	1.271	2.564
				V	0.217	0.086	0.261	0.152	6.598	1.587	2.792
RUBIDIUM BROMIDE (P=1.000KB)				H	0.188	0.073	0.293	0.152	6.598	1.460	2.698
11438	3.4154	110	180.0	S	0.169	0.064	0.314	0.152	6.598	1.372	2.636
				R	0.147	0.055	0.338	0.152	6.598	1.270	2.566
				V	0.218	0.086	0.262	0.152	6.559	1.590	2.798
RUBIDIUM BROMIDE (P=1.200KB)				H	0.189	0.073	0.293	0.152	6.559	1.461	2.703
11439	3.4196	110	180.0	S	0.169	0.064	0.315	0.152	6.559	1.372	2.640
				R	0.147	0.055	0.339	0.152	6.559	1.268	2.570
				V	0.219	0.087	0.262	0.154	6.512	1.592	2.804
RUBIDIUM BROMIDE (P=1.400KB)				H	0.189	0.073	0.294	0.154	6.512	1.462	2.708
11440	3.4247	110	180.0	S	0.169	0.064	0.316	0.154	6.512	1.370	2.643
				R	0.147	0.055	0.341	0.154	6.512	1.265	2.572

CUBIC

CODE	NAME DENSITY	REF	TEMP(K)		E	G	SIGMA	K	BETA	VS	VP
				V	0.220	0.087	0.262	0.154	6.475	1.595	2.810
RUBIDIUM BROMIDE (P=1.600KB)				H	0.190	0.073	0.295	0.154	6.475	1.463	2.713
11441	3.4288	110	180.0	S	0.169	0.064	0.317	0.154	6.475	1.369	2.646
				R	0.147	0.055	0.342	0.154	6.475	1.263	2.575
				V	0.221	0.088	0.263	0.156	6.424	1.598	2.817
RUBIDIUM BRCMIDE(P=1.800KB)				H	0.191	0.074	0.296	0.156	6.424	1.464	2.719
11442	3.4339	110	180.0	S	0.170	0.064	0.318	0.156	6.424	1.368	2.651
				R	0.147	0.055	0.343	0.156	6.424	1.262	2.580
				V	0.223	0.088	0.264	0.157	6.372	1.600	2.825
RUBIDIUM BRCMIDE (P=2.000KB)				H	0.191	0.074	0.297	0.157	6.372	1.465	2.725
11443	3.4380	110	180.0	S	0.170	0.064	0.320	0.157	6.372	1.368	2.657
				R	0.147	0.055	0.344	0.157	6.372	1.260	2.585
				V	0.224	0.088	0.263	0.158	6.345	1.603	2.829
RUBIDIUM BRCMIDE (P=2.200KB)				H	0.192	0.074	0.297	0.158	6.345	1.467	2.729
11444	3.4431	110	180.0	S	0.170	0.064	0.320	0.158	6.345	1.367	2.659
				R	0.147	0.055	0.345	0.158	6.345	1.258	2.586
				V	0.225	0.089	0.264	0.159	6.289	1.606	2.838
RUBIDIUM BRCMIDE (P=2.400KB)				H	0.193	0.074	0.298	0.159	6.289	1.468	2.736
11445	3.4472	110	180.0	S	0.170	0.064	0.322	0.159	6.289	1.366	2.665
				R	0.147	0.054	0.346	0.159	6.289	1.257	2.592
				V	0.205	0.081	0.262	0.144	6.959	1.550	2.731
RUBIDIUM BROMIDE (P=0.001KB)				H	0.180	0.070	0.291	0.144	6.959	1.437	2.647
11446	3.3780	110	220.0	S	0.165	0.063	0.309	0.144	6.959	1.365	2.596
				R	0.145	0.055	0.331	0.144	6.959	1.272	2.532
				V	0.206	0.082	0.263	0.145	6.904	1.553	2.738
RUBIDIUM BROMIDE (P=0.200KB)				H	0.181	0.070	0.292	0.145	6.904	1.439	2.653
11447	3.3831	110	220.0	S	0.165	0.063	0.310	0.145	6.904	1.364	2.601
				R	0.145	0.055	0.333	0.145	6.904	1.270	2.536
				V	0.207	0.082	0.264	0.146	6.849	1.555	2.745
RUBIDIUM BROMIDE (P=0.400KB)				H	0.182	0.070	0.293	0.146	6.849	1.440	2.659
11448	3.3882	110	220.0	S	0.165	0.063	0.311	0.146	6.849	1.364	2.606
				R	0.145	0.055	0.334	0.146	6.849	1.269	2.541
				V	0.208	0.082	0.264	0.147	6.798	1.559	2.753
RUBIDIUM BRCMIDE (P=0.600KB)				H	0.182	0.070	0.293	0.147	6.798	1.442	2.666
11449	3.3921	110	220.0	S	0.166	0.063	0.312	0.147	6.798	1.364	2.611
				R	0.145	0.054	0.335	0.147	6.798	1.267	2.545
				V	0.210	0.083	0.264	0.148	6.746	1.562	2.760
RUBIDIUM BRCMIDE (P=0.800KB)				H	0.183	0.071	0.294	0.148	6.746	1.443	2.672
11450	3.3972	110	220.0	S	0.166	0.063	0.314	0.148	6.746	1.363	2.616
				R	0.145	0.054	0.336	0.148	6.746	1.266	2.549
				V	0.211	0.083	0.265	0.149	6.692	1.565	2.767
RUBIDIUM BROMIDE (P=1.000KB)				H	0.184	0.071	0.295	0.149	6.692	1.444	2.678
11451	3.4023	110	220.0	S	0.166	0.063	0.315	0.149	6.692	1.362	2.620
				R	0.145	0.054	0.338	0.149	6.692	1.263	2.553
				V	0.212	0.084	0.266	0.151	6.640	1.568	2.774
RUBIDIUM BROMIDE (P=1.200KB)				H	0.184	0.071	0.296	0.151	6.640	1.445	2.684
11452	3.4065	110	220.0	S	0.166	0.063	0.316	0.151	6.640	1.361	2.625
				R	0.145	0.054	0.339	0.151	6.640	1.261	2.558
				V	0.213	0.084	0.266	0.152	6.591	1.571	2.782
RUBIDIUM BRCMIDE (P=1.400KB)				H	0.185	0.071	0.297	0.152	6.591	1.447	2.690
11453	3.4115	110	220.0	S	0.166	0.063	0.317	0.152	6.591	1.361	2.630
				R	0.145	0.054	0.341	0.152	6.590	1.260	2.562
				V	0.214	0.085	0.266	0.153	6.545	1.574	2.789
RUBIDIUM BRCMIDE (P=1.600KB)				H	0.186	0.072	0.297	0.153	6.545	1.448	2.696
11454	3.4156	110	220.0	S	0.167	0.063	0.318	0.153	6.545	1.360	2.635
				R	0.145	0.054	0.342	0.153	6.544	1.258	2.566
				V	0.215	0.085	0.267	0.154	6.493	1.577	2.796
RUBIDIUM BROMIDE (P=1.800KB)				H	0.187	0.072	0.298	0.154	6.494	1.449	2.702
11455	3.4207	110	220.0	S	0.167	0.063	0.319	0.154	6.494	1.360	2.639
				R	0.145	0.054	0.343	0.154	6.493	1.256	2.570

CUBIC

CODE	NAME DENSITY	REF	TEMP(K)		E	G	SIGMA	K	BETA	VS	VP
				V	0.216	0.085	0.267	0.155	6.449	1.579	2.802
RUBIDIUM BROMIDE (P=2.000KB)				H	0.187	0.072	0.299	0.155	6.449	1.450	2.707
11456	3.4248	110	220.0	S	0.167	0.063	0.321	0.155	6.449	1.358	2.643
				R	0.145	0.054	0.344	0.155	6.449	1.253	2.573
				V	0.218	0.086	0.268	0.156	6.405	1.582	2.809
RUBIDIUM BROMIDE (P=2.200KB)				H	0.188	0.072	0.300	0.156	6.405	1.451	2.713
11457	3.4299	110	220.0	S	0.167	0.063	0.322	0.156	6.405	1.357	2.647
				R	0.145	0.054	0.346	0.156	6.405	1.252	2.577
				V	0.219	0.086	0.268	0.157	6.361	1.585	2.816
RUBIDIUM BROMIDE (P=2.400KB)				H	0.188	0.072	0.300	0.157	6.361	1.453	2.719
11458	3.4341	110	220.0	S	0.167	0.063	0.323	0.157	6.361	1.357	2.652
				R	0.145	0.054	0.347	0.157	6.361	1.250	2.581
				V	0.220	0.087	0.268	0.158	6.316	1.588	2.822
RUBIDIUM BROMIDE (P=2.615KB)				H	0.189	0.073	0.301	0.158	6.316	1.454	2.724
11459	3.4392	110	220.0	S	0.167	0.063	0.324	0.158	6.316	1.356	2.656
				R	0.145	0.054	0.348	0.158	6.316	1.248	2.585
				V	0.221	0.087	0.268	0.159	6.284	1.591	2.828
RUBIDIUM BROMIDE (P=2.800KB)				H	0.190	0.073	0.301	0.159	6.284	1.455	2.729
11460	3.4433	110	220.0	S	0.168	0.063	0.324	0.159	6.284	1.356	2.659
				R	0.144	0.054	0.349	0.159	6.284	1.247	2.587
				V	0.222	0.088	0.269	0.160	6.246	1.594	2.834
RUBIDIUM BROMIDE (P=3.000KB)				H	0.190	0.073	0.302	0.160	6.246	1.456	2.733
11461	3.4474	110	220.0	S	0.168	0.063	0.326	0.160	6.246	1.354	2.662
				R	0.144	0.053	0.350	0.160	6.246	1.244	2.590
				V	0.223	0.088	0.269	0.161	6.209	1.597	2.840
RUBIDIUM BROMIDE (P=3.200KB)				H	0.191	0.073	0.302	0.161	6.209	1.458	2.739
11462	3.4515	110	220.0	S	0.168	0.063	0.326	0.161	6.209	1.353	2.666
				R	0.144	0.053	0.351	0.161	6.209	1.242	2.593
				V	0.224	0.088	0.269	0.162	6.167	1.599	2.846
RUBIDIUM BROMIDE (P=3.415KB)				H	0.192	0.074	0.303	0.162	6.166	1.458	2.744
11463	3.4566	110	220.0	S	0.168	0.063	0.328	0.162	6.166	1.352	2.670
				R	0.144	0.053	0.352	0.162	6.166	1.240	2.597
				V	0.226	0.089	0.269	0.163	6.130	1.603	2.853
RUBIDIUM BROMIDE (P=3.610KB)				H	0.192	0.074	0.304	0.163	6.130	1.460	2.749
11464	3.4608	110	220.0	S	0.168	0.063	0.328	0.163	6.130	1.352	2.674
				R	0.144	0.053	0.353	0.163	6.130	1.239	2.600
				V	0.227	0.089	0.270	0.164	6.093	1.605	2.859
RUBIDIUM BROMIDE (P=3.810KB)				H	0.193	0.074	0.304	0.164	6.093	1.461	2.754
11465	3.4649	110	220.0	S	0.168	0.063	0.330	0.164	6.093	1.350	2.677
				R	0.143	0.053	0.354	0.164	6.093	1.236	2.603
				V	0.199	0.078	0.265	0.141	7.107	1.527	2.701
RUBIDIUM BROMIDE (P=0.001KB)				H	0.176	0.068	0.292	0.141	7.107	1.421	2.622
11466	3.3640	110	260.0	S	0.162	0.062	0.309	0.141	7.107	1.355	2.575
				R	0.143	0.054	0.330	0.141	7.107	1.266	2.514
				V	0.200	0.079	0.265	0.142	7.046	1.531	2.709
RUBIDIUM BROMIDE (P=0.210KB)				H	0.176	0.068	0.293	0.142	7.046	1.423	2.629
11467	3.3691	110	260.0	S	0.162	0.062	0.310	0.142	7.046	1.355	2.581
				R	0.143	0.054	0.332	0.142	7.046	1.264	2.519
				V	0.201	0.079	0.266	0.143	7.000	1.533	2.714
RUBIDIUM BROMIDE (P=0.400KB)				H	0.177	0.068	0.294	0.143	7.000	1.424	2.634
11468	3.3742	110	260.0	S	0.162	0.062	0.311	0.143	7.000	1.354	2.584
				R	0.143	0.054	0.333	0.143	7.000	1.263	2.522
				V	0.202	0.080	0.267	0.144	6.935	1.536	2.722
RUBIDIUM BROMIDE (P=0.600KB)				H	0.178	0.069	0.295	0.144	6.935	1.425	2.641
11469	3.3792	110	260.0	S	0.162	0.062	0.312	0.144	6.935	1.353	2.590
				R	0.143	0.054	0.334	0.144	6.935	1.261	2.528
				V	0.203	0.080	0.267	0.145	6.885	1.539	2.730
RUBIDIUM BROMIDE (P=0.800KB)				H	0.178	0.069	0.295	0.145	6.885	1.426	2.647
11470	3.3832	110	260.0	S	0.163	0.062	0.313	0.145	6.885	1.353	2.595
				R	0.143	0.054	0.335	0.145	6.885	1.260	2.532

CUBIC

CODE / NAME	DENSITY	REF	TEMP(K)		E	G	SIGMA	K	BETA	VS	VP
				V	0.204	0.081	0.268	0.146	6.829	1.542	2.737
RUBIDIUM BROMIDE (P=1.000KB)				H	0.179	0.069	0.296	0.146	6.829	1.427	2.653
11471	3.3883	110	260.0	S	0.163	0.062	0.315	0.146	6.829	1.352	2.600
				R	0.143	0.054	0.337	0.146	6.829	1.257	2.535
				V	0.205	0.081	0.268	0.148	6.780	1.544	2.744
RUBIDIUM BROMIDE (P=1.200KB)				H	0.180	0.069	0.297	0.147	6.780	1.428	2.658
11472	3.3934	110	260.0	S	0.163	0.062	0.316	0.147	6.780	1.351	2.604
				R	0.143	0.053	0.338	0.148	6.780	1.255	2.539
				V	0.207	0.081	0.268	0.149	6.726	1.548	2.751
RUBIDIUM BROMIDE (P=1.400KB)				H	0.180	0.069	0.298	0.149	6.726	1.430	2.665
11473	3.3984	110	260.0	S	0.163	0.062	0.317	0.149	6.726	1.351	2.609
				R	0.143	0.053	0.340	0.149	6.726	1.254	2.544
				V	0.208	0.082	0.269	0.150	6.677	1.551	2.758
RUBIDIUM BROMIDE (P=1.600KB)				H	0.181	0.070	0.299	0.150	6.677	1.431	2.671
11474	3.4025	110	260.0	S	0.164	0.062	0.318	0.150	6.677	1.350	2.614
				R	0.143	0.053	0.341	0.150	6.677	1.252	2.548
				V	0.209	0.082	0.269	0.151	6.630	1.553	2.765
RUBIDIUM BROMIDE (P=1.800KB)				H	0.182	0.070	0.299	0.151	6.630	1.433	2.676
11475	3.4076	110	260.0	S	0.164	0.062	0.319	0.151	6.630	1.350	2.618
				R	0.143	0.053	0.342	0.151	6.630	1.251	2.552
				V	0.210	0.083	0.270	0.152	6.578	1.556	2.772
RUBIDIUM BROMIDE (P=2.000KB)				H	0.182	0.070	0.300	0.152	6.578	1.433	2.682
11476	3.4126	110	260.0	S	0.164	0.062	0.320	0.152	6.578	1.348	2.623
				R	0.143	0.053	0.344	0.152	6.578	1.248	2.555
				V	0.211	0.083	0.270	0.153	6.530	1.559	2.779
RUBIDIUM BROMIDE (P=2.200KB)				H	0.183	0.070	0.301	0.153	6.530	1.434	2.688
11477	3.4168	110	260.0	S	0.164	0.062	0.322	0.153	6.530	1.348	2.627
				R	0.143	0.053	0.345	0.153	6.530	1.246	2.560
				V	0.212	0.083	0.271	0.154	6.485	1.561	2.785
RUBIDIUM BROMIDE (P=2.400KB)				H	0.184	0.071	0.302	0.154	6.485	1.436	2.694
11478	3.4207	110	260.0	S	0.164	0.062	0.323	0.154	6.485	1.347	2.632
				R	0.143	0.053	0.346	0.154	6.485	1.245	2.564
				V	0.213	0.084	0.271	0.155	6.438	1.564	2.792
RUBIDIUM BROMIDE (P=2.600KB)				H	0.184	0.071	0.302	0.155	6.438	1.437	2.699
11479	3.4259	110	260.0	S	0.164	0.062	0.324	0.155	6.438	1.346	2.636
				R	0.143	0.053	0.347	0.155	6.438	1.243	2.568
				V	0.214	0.084	0.272	0.156	6.393	1.567	2.799
RUBIDIUM BROMIDE (P=2.800KB)				H	0.185	0.071	0.303	0.156	6.393	1.438	2.705
11480	3.4310	110	260.0	S	0.165	0.062	0.325	0.156	6.393	1.346	2.641
				R	0.143	0.053	0.348	0.156	6.393	1.241	2.572
				V	0.215	0.085	0.272	0.158	6.349	1.569	2.805
RUBIDIUM BROMIDE (P=3.000KB)				H	0.185	0.071	0.304	0.157	6.349	1.439	2.710
11481	3.4351	110	260.0	S	0.164	0.062	0.326	0.157	6.349	1.344	2.644
				R	0.142	0.053	0.350	0.157	6.349	1.238	2.575
				V	0.216	0.085	0.273	0.159	6.309	1.571	2.811
RUBIDIUM BROMIDE (P=3.200KB)				H	0.186	0.071	0.305	0.158	6.309	1.439	2.715
11482	3.4393	110	260.0	S	0.165	0.062	0.327	0.158	6.309	1.343	2.648
				R	0.142	0.053	0.351	0.158	6.309	1.236	2.578
				V	0.217	0.085	0.273	0.160	6.266	1.574	2.817
RUBIDIUM BROMIDE (P=3.400KB)				H	0.187	0.071	0.305	0.160	6.266	1.440	2.720
11483	3.4444	110	260.0	S	0.165	0.062	0.328	0.160	6.266	1.342	2.652
				R	0.142	0.053	0.352	0.160	6.266	1.235	2.582
				V	0.218	0.086	0.274	0.161	6.223	1.577	2.824
RUBIDIUM BROMIDE (P=3.600KB)				H	0.187	0.072	0.306	0.161	6.223	1.442	2.726
11484	3.4485	110	260.0	S	0.165	0.062	0.329	0.161	6.223	1.341	2.657
				R	0.142	0.052	0.353	0.161	6.223	1.233	2.586
				V	0.219	0.086	0.274	0.162	6.183	1.579	2.830
RUBIDIUM BROMIDE (P=3.800KB)				H	0.188	0.072	0.307	0.162	6.183	1.442	2.731
11485	3.4526	110	260.0	S	0.165	0.062	0.330	0.162	6.183	1.340	2.660
				R	0.141	0.052	0.354	0.162	6.183	1.230	2.589

CUBIC

CODE	NAME / DENSITY	REF	TEMP(K)		E	G	SIGMA	K	BETA	VS	VP
				V	0.192	0.076	0.268	0.138	7.227	1.504	2.674
RUBIDIUM BROMIDE (P=0.001KB)				H	0.171	0.066	0.294	0.138	7.227	1.405	2.600
11486	3.3500	110	300.0	S	0.159	0.061	0.309	0.138	7.227	1.344	2.557
				R	0.141	0.053	0.330	0.138	7.227	1.260	2.499
				V	0.196	0.077	0.270	0.142	7.059	1.514	2.696
RUBIDIUM BROMIDE (P=0.600KB)				H	0.173	0.067	0.296	0.142	7.059	1.409	2.619
11487	3.3650	110	300.0	S	0.159	0.061	0.312	0.142	7.059	1.344	2.573
				R	0.141	0.053	0.334	0.142	7.059	1.256	2.512
				V	0.199	0.078	0.271	0.145	6.897	1.523	2.717
RUBIDIUM BROMIDE (P=1.200KB)				H	0.175	0.067	0.299	0.145	6.897	1.413	2.637
11488	3.3803	110	300.0	S	0.160	0.061	0.316	0.145	6.897	1.342	2.586
				R	0.141	0.053	0.338	0.145	6.897	1.250	2.524
				V	0.203	0.080	0.272	0.148	6.736	1.533	2.740
RUBIDIUM BROMIDE (P=1.800KB)				H	0.177	0.068	0.301	0.148	6.736	1.418	2.656
11489	3.3944	110	300.0	S	0.161	0.061	0.319	0.148	6.736	1.341	2.602
				R	0.141	0.053	0.341	0.148	6.736	1.245	2.538
				V	0.206	0.081	0.273	0.152	6.593	1.541	2.760
RUBIDIUM BROMIDE (P=2.400KB)				H	0.179	0.069	0.303	0.152	6.593	1.421	2.672
11490	3.4086	110	300.0	S	0.161	0.061	0.323	0.152	6.593	1.338	2.615
				R	0.141	0.052	0.345	0.152	6.593	1.239	2.549
				V	0.207	0.081	0.274	0.153	6.545	1.545	2.767
RUBIDIUM BROMIDE (P=2.600KB)				H	0.180	0.069	0.304	0.153	6.545	1.423	2.679
11491	3.4127	110	300.0	S	0.162	0.061	0.324	0.153	6.545	1.338	2.620
				R	0.141	0.052	0.346	0.153	6.544	1.238	2.553
				V	0.209	0.082	0.274	0.154	6.501	1.548	2.774
RUBIDIUM BROMIDE (P=2.800KB)				H	0.181	0.069	0.304	0.154	6.501	1.424	2.684
11492	3.4177	110	300.0	S	0.162	0.061	0.325	0.154	6.501	1.337	2.624
				R	0.141	0.052	0.348	0.154	6.501	1.236	2.557
				V	0.210	0.082	0.275	0.155	6.452	1.550	2.780
RUBIDIUM BROMIDE (P=3.015KB)				H	0.181	0.069	0.305	0.155	6.452	1.424	2.689
11493	3.4229	110	300.0	S	0.162	0.061	0.326	0.155	6.452	1.335	2.628
				R	0.140	0.052	0.349	0.155	6.452	1.233	2.560
				V	0.211	0.083	0.275	0.156	6.408	1.552	2.787
RUBIDIUM BROMIDE (P=3.220KB)				H	0.182	0.070	0.306	0.156	6.408	1.425	2.695
11494	3.4270	110	300.0	S	0.162	0.061	0.327	0.156	6.408	1.334	2.632
				R	0.140	0.052	0.350	0.156	6.408	1.231	2.564
				V	0.212	0.083	0.276	0.157	6.360	1.555	2.793
RUBIDIUM BROMIDE (P=3.400KB)				H	0.182	0.070	0.307	0.157	6.360	1.426	2.701
11495	3.4321	110	300.0	S	0.162	0.061	0.328	0.157	6.360	1.334	2.637
				R	0.140	0.052	0.351	0.157	6.360	1.229	2.568
				V	0.213	0.083	0.276	0.158	6.318	1.558	2.801
RUBIDIUM BROMIDE (P=3.610KB)				H	0.183	0.070	0.307	0.158	6.318	1.428	2.707
11496	3.4362	110	300.0	S	0.162	0.061	0.329	0.158	6.318	1.333	2.641
				R	0.140	0.052	0.352	0.158	6.318	1.228	2.572
				V	0.214	0.084	0.276	0.159	6.276	1.561	2.807
RUBIDIUM BROMIDE (P=3.800KB)				H	0.184	0.070	0.308	0.159	6.276	1.429	2.712
11497	3.4402	110	300.0	S	0.163	0.061	0.330	0.159	6.276	1.333	2.646
				R	0.140	0.052	0.353	0.159	6.276	1.226	2.576
				V	0.215	0.084	0.277	0.160	6.233	1.563	2.813
RUBIDIUM BROMIDE (P=4.010KB)				H	0.184	0.070	0.309	0.160	6.233	1.429	2.717
11498	3.4454	110	300.0	S	0.162	0.061	0.331	0.160	6.233	1.331	2.649
				R	0.140	0.052	0.355	0.160	6.233	1.223	2.579
				V	0.226	0.089	0.268	0.162	6.167	1.785	3.169
RUBIDIUM CHLORIDE				H	0.204	0.079	0.290	0.162	6.166	1.680	3.091
11499	2.7990	4	ROOM	S	0.192	0.074	0.303	0.162	6.166	1.621	3.049
				R	0.173	0.065	0.322	0.162	6.166	1.530	2.986
				V	0.220	0.087	0.269	0.159	6.303	1.773	3.153
RUBIDIUM CHLORIDE				H	0.198	0.077	0.292	0.159	6.303	1.668	3.075
11500	2.7600	4	ROOM	S	0.186	0.071	0.304	0.159	6.303	1.608	3.033
				R	0.168	0.064	0.323	0.159	6.302	1.517	2.969

CUBIC

CODE	NAME / DENSITY	REF	TEMP(K)		E	G	SIGMA	K	BETA	VS	VP
				V	0.270	0.106	0.269	0.195	5.127	1.926	3.429
RUBIDIUM CHLORIDE				H	0.236	0.091	0.298	0.195	5.127	1.783	3.324
11501	2.8641	176	4.2	S	0.215	0.082	0.316	0.195	5.127	1.689	3.258
				R	0.189	0.071	0.339	0.195	5.127	1.570	3.177
				V	0.268	0.106	0.270	0.194	5.147	1.921	3.421
RUBIDIUM CHLORIDE				H	0.235	0.091	0.298	0.194	5.147	1.779	3.318
11502	2.8639	176	30.0	S	0.214	0.081	0.316	0.194	5.147	1.686	3.252
				R	0.188	0.070	0.338	0.194	5.147	1.568	3.172
				V	0.264	0.104	0.270	0.191	5.223	1.905	3.396
RUBIDIUM CHLORIDE				H	0.232	0.089	0.298	0.191	5.223	1.768	3.295
11503	2.8618	176	60.0	S	0.212	0.081	0.315	0.191	5.223	1.678	3.232
				R	0.187	0.070	0.337	0.191	5.223	1.563	3.154
				V	0.259	0.102	0.271	0.188	5.310	1.889	3.369
RUBIDIUM CHLORIDE				H	0.229	0.088	0.298	0.188	5.310	1.756	3.271
11504	2.8572	176	90.0	S	0.210	0.080	0.314	0.188	5.310	1.671	3.211
				R	0.185	0.069	0.336	0.188	5.310	1.559	3.135
				V	0.254	0.100	0.271	0.185	5.398	1.873	3.343
RUBIDIUM CHLORIDE				H	0.225	0.087	0.298	0.185	5.398	1.744	3.249
11505	2.8507	176	120.0	S	0.207	0.079	0.314	0.185	5.398	1.664	3.192
				R	0.184	0.069	0.335	0.185	5.398	1.555	3.118
				V	0.249	0.098	0.272	0.182	5.486	1.857	3.318
RUBIDIUM CHLORIDE				H	0.222	0.085	0.297	0.182	5.486	1.733	3.227
11506	2.8430	176	150.0	S	0.205	0.078	0.313	0.182	5.486	1.657	3.173
				R	0.182	0.068	0.333	0.182	5.486	1.551	3.101
				V	0.245	0.096	0.272	0.179	5.579	1.842	3.293
RUBIDIUM CHLORIDE				H	0.218	0.084	0.297	0.179	5.579	1.722	3.206
11507	2.8345	176	180.0	S	0.202	0.077	0.312	0.179	5.579	1.650	3.155
				R	0.181	0.068	0.332	0.179	5.579	1.547	3.085
				V	0.240	0.094	0.273	0.176	5.683	1.826	3.267
RUBIDIUM CHLORIDE				H	0.215	0.083	0.297	0.176	5.683	1.711	3.183
11508	2.8256	176	210.0	S	0.200	0.076	0.311	0.176	5.683	1.643	3.135
				R	0.179	0.067	0.330	0.176	5.683	1.544	3.067
				V	0.235	0.092	0.273	0.172	5.800	1.811	3.240
RUBIDIUM CHLORIDE				H	0.211	0.081	0.296	0.172	5.800	1.700	3.158
11509	2.8163	176	240.0	S	0.197	0.075	0.309	0.172	5.800	1.636	3.113
				R	0.177	0.067	0.328	0.172	5.800	1.540	3.047
				V	0.231	0.091	0.272	0.168	5.936	1.797	3.211
RUBIDIUM CHLORIDE				H	0.208	0.080	0.295	0.168	5.936	1.690	3.132
11510	2.8068	176	270.0	S	0.195	0.075	0.307	0.168	5.936	1.630	3.089
				R	0.176	0.066	0.326	0.168	5.936	1.537	3.025
				V	0.226	0.089	0.271	0.165	6.069	1.782	3.182
RUBIDIUM CHLORIDE				H	0.204	0.079	0.294	0.165	6.069	1.679	3.107
11511	2.7972	176	300.0	S	0.192	0.074	0.305	0.165	6.069	1.623	3.066
				R	0.174	0.066	0.324	0.165	6.069	1.533	3.004
				V	0.260	0.103	0.268	0.187	5.362	1.907	3.386
RUBIDIUM CHLORIDE				H	0.229	0.088	0.295	0.186	5.362	1.771	3.286
11512	2.8200	152	4.2	S	0.210	0.080	0.312	0.186	5.362	1.685	3.225
				R	0.186	0.070	0.334	0.187	5.362	1.571	3.147
				V	0.224	0.088	0.273	0.165	6.073	1.786	3.197
RUBIDIUM CHLORIDE				H	0.202	0.078	0.296	0.165	6.073	1.681	3.120
11513	2.7600	152	295.0	S	0.190	0.073	0.308	0.165	6.073	1.621	3.078
				R	0.171	0.065	0.327	0.165	6.073	1.529	3.014
				V	2.589	1.030	0.257	1.779	0.562	6.007	10.509
RUBIDIUM CHLORIDE				H	2.274	0.883	0.287	1.779	0.562	5.564	10.178
11514	2.8538	110	100.0	S	2.075	0.795	0.306	1.779	0.562	5.277	9.973
				R	1.830	0.689	0.329	1.779	0.562	4.913	9.722
				V	2.557	1.016	0.258	1.761	0.568	5.971	10.456
RUBIDIUM CHLORIDE				H	2.251	0.874	0.287	1.761	0.568	5.539	10.133
11515	2.8504	110	120.0	S	2.060	0.789	0.305	1.761	0.568	5.262	9.934
				R	1.821	0.686	0.328	1.761	0.568	4.905	9.688

CUBIC

CODE	NAME DENSITY	REF	TEMP(K)		E	G	SIGMA	K	BETA	VS	VP
				V	2.524	1.002	0.259	1.744	0.573	5.935	10.405
RUBIDIUM CHLORIDE				H	2.226	0.865	0.287	1.744	0.573	5.512	10.090
11516	2.8458	110	140.0	S	2.043	0.783	0.305	1.744	0.573	5.245	9.898
				R	1.810	0.682	0.327	1.744	0.573	4.895	9.656
				V	2.494	0.990	0.260	1.731	0.578	5.903	10.364
RUBIDIUM CHLORIDE				H	2.204	0.856	0.288	1.731	0.578	5.489	10.056
11517	2.8404	110	160.0	S	2.028	0.777	0.305	1.731	0.578	5.230	9.871
				R	1.799	0.678	0.327	1.731	0.578	4.886	9.632
				V	2.458	0.974	0.262	1.719	0.582	5.862	10.318
RUBIDIUM CHLORIDE				H	2.178	0.845	0.289	1.719	0.582	5.460	10.019
11518	2.8346	110	180.0	S	2.010	0.770	0.305	1.719	0.582	5.212	9.841
				R	1.787	0.674	0.327	1.719	0.582	4.875	9.608
				V	2.423	0.959	0.263	1.702	0.587	5.823	10.267
RUBIDIUM CHLORIDE				H	2.152	0.835	0.289	1.702	0.587	5.432	9.976
11519	2.8285	110	200.0	S	1.991	0.763	0.305	1.702	0.587	5.194	9.806
				R	1.775	0.669	0.326	1.702	0.587	4.864	9.578
				V	2.387	0.944	0.264	1.688	0.593	5.783	10.217
RUBIDIUM CHLORIDE				H	2.125	0.824	0.290	1.688	0.593	5.403	9.935
11520	2.8223	110	220.0	S	1.972	0.756	0.305	1.688	0.593	5.174	9.772
				R	1.762	0.664	0.326	1.688	0.593	4.852	9.549
				V	2.350	0.929	0.265	1.670	0.599	5.743	10.163
RUBIDIUM CHLORIDE				H	2.099	0.813	0.291	1.670	0.599	5.374	9.890
11521	2.8160	110	240.0	S	1.953	0.748	0.305	1.670	0.599	5.155	9.734
				R	1.749	0.660	0.325	1.670	0.599	4.841	9.516
				V	2.312	0.913	0.267	1.653	0.605	5.699	10.106
RUBIDIUM CHLORIDE				H	2.070	0.802	0.291	1.653	0.605	5.342	9.842
11522	2.8097	110	260.0	S	1.933	0.741	0.305	1.653	0.605	5.134	9.694
				R	1.736	0.655	0.325	1.653	0.605	4.828	9.482
				V	2.278	0.899	0.268	1.633	0.612	5.661	10.050
RUBIDIUM CHLORIDE				H	2.045	0.792	0.291	1.633	0.612	5.314	9.794
11523	2.8033	110	280.0	S	1.914	0.734	0.305	1.633	0.612	5.116	9.652
				R	1.723	0.650	0.324	1.633	0.612	4.817	9.445
				V	2.241	0.883	0.269	1.616	0.619	5.619	9.994
RUBIDIUM CHLORIDE				H	2.018	0.781	0.292	1.616	0.619	5.284	9.748
11524	2.7969	110	300.0	S	1.895	0.726	0.305	1.616	0.619	5.096	9.613
				R	1.710	0.646	0.324	1.616	0.619	4.806	9.412
				V	0.245	0.097	0.262	0.172	5.824	1.852	3.261
RUBIDIUM CHLORIDE (P=0.001KB)				H	0.218	0.084	0.289	0.172	5.824	1.725	3.166
11525	2.8346	110	181.0	S	0.201	0.077	0.305	0.172	5.824	1.647	3.110
				R	0.179	0.067	0.327	0.172	5.824	1.541	3.037
				V	0.247	0.098	0.262	0.174	5.763	1.857	3.272
RUBIDIUM CHLORIDE (P=0.350KB)				H	0.219	0.085	0.290	0.174	5.763	1.728	3.176
11526	2.8406	110	181.0	S	0.201	0.077	0.307	0.174	5.763	1.646	3.118
				R	0.179	0.067	0.329	0.174	5.763	1.538	3.043
				V	0.249	0.099	0.263	0.176	5.697	1.862	3.285
RUBIDIUM CHLORIDE (P=0.695KB)				H	0.220	0.085	0.291	0.176	5.697	1.730	3.187
11527	2.8466	110	181.0	S	0.202	0.077	0.309	0.176	5.697	1.645	3.127
				R	0.178	0.067	0.331	0.176	5.697	1.535	3.051
				V	0.251	0.099	0.264	0.177	5.644	1.867	3.296
RUBIDIUM CHLORIDE (P=0.982KB)				H	0.221	0.085	0.292	0.177	5.644	1.731	3.196
11528	2.8508	110	181.0	S	0.202	0.077	0.310	0.177	5.644	1.644	3.134
				R	0.178	0.067	0.332	0.177	5.644	1.532	3.057
				V	0.253	0.100	0.264	0.179	5.587	1.872	3.307
RUBIDIUM CHLORIDE (P=1.326KB)				H	0.222	0.086	0.293	0.179	5.587	1.734	3.205
11529	2.8568	110	181.0	S	0.202	0.077	0.312	0.179	5.587	1.644	3.141
				R	0.178	0.067	0.334	0.179	5.587	1.529	3.063
				V	0.255	0.101	0.265	0.181	5.519	1.877	3.321
RUBIDIUM CHLORIDE (P=1.730KB)				H	0.223	0.086	0.295	0.181	5.519	1.736	3.216
11530	2.8636	110	181.0	S	0.203	0.077	0.314	0.181	5.519	1.641	3.150
				R	0.178	0.067	0.336	0.181	5.519	1.525	3.071

CODE	NAME / DENSITY	REF	TEMP(K)		E	G	SIGMA	K	BETA	VS	VP
				V	0.257	0.102	0.266	0.183	5.470	1.881	3.331
	RUBIDIUM CHLORIDE (P=2.020KB)			H	0.224	0.087	0.296	0.183	5.470	1.737	3.225
11531	2.8679	110	181.0	S	0.203	0.077	0.315	0.183	5.470	1.640	3.156
				R	0.178	0.066	0.338	0.183	5.470	1.522	3.076
				V	0.260	0.103	0.266	0.185	5.396	1.889	3.347
	RUBIDIUM CHLORIDE (P=2.490KB)			H	0.226	0.087	0.297	0.185	5.396	1.741	3.238
11532	2.8756	110	181.0	S	0.204	0.077	0.317	0.185	5.396	1.640	3.167
				R	0.178	0.066	0.340	0.185	5.396	1.519	3.086
				V	0.238	0.094	0.265	0.169	5.932	1.825	3.227
	RUBIDIUM CHLORIDE (P=0.001KB)			H	0.212	0.082	0.290	0.169	5.932	1.706	3.139
11533	2.8222	110	222.0	S	0.197	0.075	0.305	0.169	5.932	1.634	3.088
				R	0.176	0.066	0.326	0.169	5.932	1.533	3.018
				V	0.240	0.095	0.265	0.170	5.875	1.830	3.238
	RUBIDIUM CHLORIDE (P=0.325KB)			H	0.213	0.082	0.291	0.170	5.875	1.708	3.147
11534	2.8282	110	222.0	S	0.197	0.075	0.307	0.170	5.875	1.633	3.094
				R	0.176	0.066	0.328	0.170	5.875	1.530	3.023
				V	0.242	0.095	0.266	0.172	5.806	1.835	3.251
	RUBIDIUM CHLORIDE (P=0.690KB)			H	0.214	0.083	0.293	0.172	5.806	1.710	3.159
11535	2.8342	110	222.0	S	0.198	0.076	0.309	0.172	5.806	1.633	3.104
				R	0.176	0.066	0.330	0.172	5.806	1.528	3.031
				V	0.244	0.096	0.267	0.174	5.745	1.841	3.263
	RUBIDIUM CHLORIDE (P=1.010KB)			H	0.216	0.083	0.294	0.174	5.745	1.713	3.169
11536	2.8392	110	222.0	S	0.198	0.076	0.310	0.174	5.745	1.633	3.112
				R	0.176	0.066	0.332	0.174	5.745	1.525	3.038
				V	0.247	0.097	0.268	0.177	5.657	1.848	3.280
	RUBIDIUM CHLORIDE (P=1.490KB)			H	0.217	0.084	0.295	0.177	5.657	1.716	3.183
11537	2.8477	110	222.0	S	0.199	0.076	0.312	0.177	5.657	1.631	3.123
				R	0.176	0.066	0.334	0.177	5.657	1.521	3.048
				V	0.249	0.098	0.269	0.180	5.570	1.855	3.297
	RUBIDIUM CHLORIDE (P=2.015KB)			H	0.219	0.084	0.297	0.180	5.570	1.718	3.197
11538	2.8562	110	222.0	S	0.199	0.076	0.315	0.180	5.570	1.628	3.134
				R	0.175	0.066	0.337	0.180	5.570	1.515	3.057
				V	0.252	0.099	0.269	0.182	5.486	1.862	3.314
	RUBIDIUM CHLORIDE (P=2.515KB)			H	0.220	0.085	0.298	0.182	5.486	1.721	3.211
11539	2.8648	110	222.0	S	0.200	0.076	0.317	0.182	5.486	1.627	3.145
				R	0.175	0.065	0.340	0.182	5.486	1.510	3.067
				V	0.255	0.100	0.270	0.185	5.403	1.869	3.331
	RUBIDIUM CHLORIDE (P=3.035KB)			H	0.222	0.085	0.300	0.185	5.403	1.724	3.226
11540	2.8732	110	222.0	S	0.200	0.076	0.320	0.185	5.403	1.625	3.156
				R	0.175	0.065	0.342	0.185	5.403	1.506	3.076
				V	0.258	0.101	0.271	0.188	5.330	1.876	3.347
	RUBIDIUM CHLORIDE (P=3.530KB)			H	0.224	0.086	0.301	0.188	5.330	1.727	3.239
11541	2.8810	110	222.0	S	0.201	0.076	0.322	0.188	5.330	1.623	3.166
				R	0.175	0.065	0.345	0.188	5.330	1.501	3.085
				V	0.231	0.091	0.268	0.166	6.040	1.800	3.196
	RUBIDIUM CHLORIDE (P=0.001KB)			H	0.207	0.080	0.292	0.166	6.040	1.688	3.113
11542	2.8097	110	259.0	S	0.193	0.074	0.306	0.166	6.040	1.622	3.066
				R	0.173	0.065	0.325	0.166	6.040	1.526	3.000
				V	0.233	0.092	0.268	0.167	5.981	1.805	3.207
	RUBIDIUM CHLORIDE(P=0.320KB)			H	0.208	0.080	0.293	0.167	5.981	1.690	3.122
11543	2.8157	110	259.0	S	0.194	0.074	0.307	0.167	5.981	1.622	3.073
				R	0.173	0.065	0.327	0.167	5.981	1.523	3.005
				V	0.234	0.092	0.269	0.169	5.927	1.809	3.217
	RUBIDIUM CHLORIDE (P=0.602KB)			H	0.209	0.081	0.294	0.169	5.927	1.692	3.130
11544	2.8208	110	259.0	S	0.194	0.074	0.308	0.169	5.927	1.622	3.080
				R	0.174	0.065	0.329	0.169	5.926	1.522	3.012
				V	0.237	0.093	0.270	0.171	5.845	1.815	3.232
	RUBIDIUM CHLORIDE (P=0.998KB)			H	0.210	0.081	0.295	0.171	5.845	1.694	3.143
11545	2.8274	110	259.0	S	0.195	0.074	0.310	0.171	5.845	1.620	3.091
				R	0.173	0.065	0.331	0.171	5.845	1.518	3.020

CUBIC

CODE / NAME	DENSITY	REF	TEMP(K)		E	G	SIGMA	K	BETA	VS	VP
				V	0.240	0.094	0.270	0.174	5.754	1.824	3.250
RUBIDIUM CHLORIDE (P=1.515KB)				H	0.212	0.082	0.297	0.174	5.754	1.698	3.158
11546	2.8359	110	259.0	S	0.195	0.074	0.313	0.174	5.754	1.620	3.103
				R	0.173	0.065	0.334	0.174	5.754	1.514	3.030
				V	0.242	0.095	0.271	0.177	5.664	1.830	3.267
RUBIDIUM CHLORIDE (P=1.995KB)				H	0.214	0.082	0.298	0.177	5.664	1.701	3.172
11547	2.8444	110	259.0	S	0.196	0.074	0.315	0.177	5.664	1.618	3.114
				R	0.173	0.065	0.337	0.177	5.664	1.509	3.040
				V	0.245	0.096	0.272	0.179	5.581	1.837	3.284
RUBIDIUM CHLORIDE (P=2.485KB)				H	0.215	0.083	0.300	0.179	5.581	1.704	3.186
11548	2.8520	110	259.0	S	0.196	0.075	0.317	0.179	5.581	1.617	3.125
				R	0.173	0.065	0.339	0.179	5.581	1.505	3.050
				V	0.248	0.097	0.273	0.182	5.501	1.844	3.300
RUBIDIUM CHLORIDE (P=3.015KB)				H	0.217	0.083	0.301	0.182	5.501	1.706	3.199
11549	2.8613	110	259.0	S	0.197	0.075	0.320	0.182	5.501	1.614	3.135
				R	0.173	0.064	0.342	0.182	5.501	1.499	3.058
				V	0.251	0.098	0.274	0.185	5.408	1.852	3.319
RUBIDIUM CHLORIDE (P=3.540KB)				H	0.219	0.084	0.303	0.185	5.408	1.710	3.216
11550	2.8699	110	259.0	S	0.197	0.075	0.322	0.185	5.408	1.613	3.148
				R	0.173	0.064	0.344	0.185	5.408	1.496	3.070
				V	0.253	0.099	0.275	0.188	5.326	1.856	3.335
RUBIDIUM CHLORIDE (P=4.035KB)				H	0.220	0.084	0.305	0.188	5.326	1.711	3.229
11551	2.8777	110	259.0	S	0.197	0.075	0.325	0.188	5.326	1.609	3.159
				R	0.172	0.064	0.347	0.188	5.326	1.490	3.079
				V	0.223	0.088	0.270	0.162	6.187	1.771	3.156
RUBIDIUM CHLORIDE (P=0.001KB)				H	0.201	0.078	0.293	0.162	6.187	1.666	3.079
11552	2.7966	110	304.0	S	0.189	0.072	0.305	0.162	6.187	1.607	3.037
				R	0.170	0.064	0.324	0.162	6.187	1.516	2.974
				V	0.224	0.088	0.271	0.163	6.130	1.775	3.166
RUBIDIUM CHLORIDE (P=0.295KB)				H	0.202	0.078	0.294	0.163	6.130	1.668	3.088
11553	2.8017	110	304.0	S	0.189	0.072	0.307	0.163	6.130	1.607	3.044
				R	0.170	0.064	0.326	0.163	6.130	1.515	2.980
				V	0.226	0.089	0.271	0.165	6.072	1.779	3.176
RUBIDIUM CHLORIDE (P=0.575KB)				H	0.203	0.078	0.295	0.165	6.072	1.670	3.096
11554	2.8068	110	304.0	S	0.190	0.072	0.308	0.165	6.072	1.607	3.051
				R	0.170	0.064	0.328	0.165	6.072	1.512	2.986
				V	0.228	0.090	0.272	0.167	5.992	1.786	3.192
RUBIDIUM CHLORIDE (P=0.975KB)				H	0.204	0.079	0.296	0.167	5.992	1.673	3.109
11555	2.8143	110	304.0	S	0.190	0.073	0.310	0.167	5.992	1.607	3.062
				R	0.171	0.064	0.330	0.167	5.992	1.510	2.995
				V	0.231	0.091	0.273	0.170	5.879	1.794	3.212
RUBIDIUM CHLORIDE (P=1.500KB)				H	0.206	0.079	0.298	0.170	5.879	1.676	3.126
11556	2.8235	110	304.0	S	0.191	0.073	0.313	0.170	5.879	1.605	3.076
				R	0.170	0.064	0.333	0.170	5.879	1.504	3.007
				V	0.234	0.092	0.275	0.173	5.773	1.801	3.231
RUBIDIUM CHLORIDE (P=2.030KB)				H	0.208	0.080	0.300	0.173	5.773	1.679	3.142
11557	2.8328	110	304.0	S	0.192	0.073	0.316	0.173	5.773	1.604	3.089
				R	0.170	0.064	0.336	0.173	5.773	1.500	3.019
				V	0.237	0.093	0.275	0.176	5.687	1.808	3.248
RUBIDIUM CHLORIDE (P=2.490KB)				H	0.209	0.080	0.302	0.176	5.687	1.682	3.157
11558	2.8404	110	304.0	S	0.192	0.073	0.318	0.176	5.687	1.602	3.101
				R	0.170	0.064	0.339	0.176	5.687	1.496	3.029
				V	0.240	0.094	0.274	0.177	5.650	1.820	3.260
RUBIDIUM CHLORIDE (P=2.995KB)				H	0.211	0.081	0.301	0.177	5.650	1.688	3.164
11559	2.8489	110	304.0	S	0.193	0.073	0.318	0.177	5.650	1.602	3.104
				R	0.170	0.063	0.340	0.177	5.650	1.492	3.030
				V	0.242	0.095	0.278	0.182	5.507	1.822	3.284
RUBIDIUM CHLORIDE (P=3.525KB)				H	0.212	0.081	0.305	0.182	5.507	1.688	3.187
11560	2.8574	110	304.0	S	0.193	0.073	0.323	0.182	5.507	1.599	3.125
				R	0.170	0.063	0.344	0.182	5.507	1.487	3.050

CUBIC

CODE	NAME / DENSITY	REF	TEMP(K)		E	G	SIGMA	K	BETA	VS	VP
				V	0.367	0.144	0.276	0.273	3.659	2.233	4.017
RUBIDIUM FLUORIDE				H	0.343	0.133	0.291	0.273	3.659	2.148	3.955
11561	2.8800	4	ROOM	S	0.333	0.129	0.297	0.273	3.659	2.113	3.930
				R	0.312	0.119	0.310	0.273	3.659	2.034	3.874
				V	0.155	0.061	0.268	0.111	8.996	1.311	2.328
RUBIDIUM IODIDE				H	0.135	0.052	0.297	0.111	8.996	1.211	2.255
11562	3.5540	4	ROOM	S	0.122	0.046	0.316	0.111	8.996	1.144	2.207
				R	0.107	0.040	0.339	0.111	8.996	1.061	2.152
				V	0.156	0.062	0.255	0.106	9.434	1.324	2.307
RUBIDIUM IODIDE				H	0.136	0.053	0.286	0.106	9.434	1.223	2.231
11563	3.5500	41	300.0	S	0.124	0.047	0.305	0.106	9.434	1.155	2.183
				R	0.109	0.041	0.329	0.106	9.434	1.073	2.126
				V	0.188	0.075	0.261	0.131	7.634	1.425	2.506
RUBIDIUM IODIDE				H	0.159	0.061	0.298	0.131	7.634	1.291	2.407
11564	3.6680	152	4.0	S	0.136	0.051	0.326	0.131	7.634	1.184	2.333
				R	0.116	0.043	0.353	0.131	7.634	1.080	2.264
				V	0.155	0.061	0.262	0.108	9.217	1.314	2.314
RUBIDIUM IODIDE				H	0.135	0.052	0.293	0.108	9.217	1.211	2.238
11565	3.5530	152	290.0	S	0.122	0.046	0.313	0.108	9.217	1.141	2.189
				R	0.106	0.040	0.337	0.108	9.217	1.057	2.132
				V	0.154	0.061	0.262	0.108	9.288	1.309	2.306
RUBIDIUM IODIDE				H	0.134	0.052	0.293	0.108	9.288	1.208	2.231
11566	3.5490	152	300.0	S	0.121	0.046	0.313	0.108	9.288	1.139	2.183
				R	0.106	0.040	0.336	0.108	9.288	1.056	2.126
				V	0.177	0.071	0.255	0.121	8.296	1.397	2.434
RUBIDIUM IODIDE				H	0.151	0.058	0.291	0.121	8.296	1.269	2.340
11567	3.6250	110	120.0	S	0.131	0.050	0.318	0.121	8.296	1.173	2.271
				R	0.113	0.042	0.344	0.121	8.296	1.074	2.205
				V	0.175	0.070	0.256	0.119	8.371	1.389	2.423
RUBIDIUM IODIDE				H	0.149	0.058	0.292	0.119	8.371	1.264	2.331
11568	3.6180	110	140.0	S	0.131	0.050	0.318	0.119	8.371	1.170	2.264
				R	0.112	0.042	0.344	0.119	8.371	1.073	2.199
				V	0.172	0.068	0.258	0.118	8.453	1.376	2.409
RUBIDIUM IODIDE				H	0.147	0.057	0.293	0.118	8.453	1.255	2.319
11569	3.6100	110	160.0	S	0.129	0.049	0.318	0.118	8.453	1.165	2.256
				R	0.111	0.041	0.343	0.118	8.453	1.070	2.192
				V	0.170	0.067	0.259	0.117	8.523	1.367	2.398
RUBIDIUM IODIDE				H	0.145	0.056	0.294	0.117	8.523	1.249	2.310
11570	3.6020	110	180.0	S	0.128	0.049	0.318	0.117	8.523	1.162	2.249
				R	0.111	0.041	0.343	0.117	8.523	1.069	2.186
				V	0.167	0.066	0.261	0.116	8.616	1.356	2.384
RUBIDIUM IODIDE				H	0.143	0.055	0.294	0.116	8.616	1.241	2.299
11571	3.5940	110	200.0	S	0.127	0.048	0.318	0.116	8.616	1.158	2.240
				R	0.110	0.041	0.342	0.116	8.616	1.067	2.179
				V	0.164	0.065	0.263	0.115	8.698	1.345	2.370
RUBIDIUM IODIDE				H	0.141	0.055	0.295	0.115	8.698	1.233	2.288
11572	3.5850	110	220.0	S	0.126	0.048	0.318	0.115	8.698	1.154	2.232
				R	0.109	0.041	0.342	0.115	8.698	1.065	2.172
				V	0.161	0.064	0.264	0.114	8.785	1.334	2.357
RUBIDIUM IODIDE				H	0.139	0.054	0.296	0.114	8.785	1.226	2.277
11573	3.5770	110	240.0	S	0.125	0.047	0.318	0.114	8.785	1.150	2.224
				R	0.108	0.040	0.341	0.114	8.785	1.062	2.165
				V	0.158	0.063	0.266	0.113	8.876	1.324	2.344
RUBIDIUM IODIDE				H	0.137	0.053	0.297	0.113	8.876	1.219	2.267
11574	3.5690	110	260.0	S	0.123	0.047	0.317	0.113	8.876	1.146	2.215
				R	0.108	0.040	0.341	0.113	8.876	1.060	2.158
				V	0.153	0.060	0.270	0.111	9.036	1.302	2.318
RUBIDIUM IODIDE				H	0.134	0.051	0.299	0.111	9.036	1.203	2.246
11575	3.5510	110	300.0	S	0.121	0.046	0.318	0.111	9.036	1.137	2.200
				R	0.106	0.040	0.340	0.111	9.036	1.056	2.145

CUBIC

CODE	NAME / DENSITY	REF	TEMP(K)		E	G	SIGMA	K	BETA	VS	VP
				V	0.170	0.067	0.259	0.117	8.518	1.367	2.398
RUBIDIUM IODIDE (P=0.001KB)				H	0.145	0.056	0.294	0.117	8.518	1.249	2.311
11576	3.6020	110	178.0	S	0.128	0.049	0.318	0.117	8.518	1.162	2.249
				R	0.111	0.041	0.343	0.117	8.518	1.069	2.187
				V	0.171	0.068	0.260	0.118	8.439	1.371	2.406
RUBIDIUM IODIDE (P=0.200KB)				H	0.146	0.056	0.294	0.118	8.439	1.251	2.317
11577	3.6086	110	178.0	S	0.128	0.049	0.319	0.118	8.439	1.161	2.254
				R	0.110	0.041	0.345	0.118	8.439	1.067	2.191
				V	0.172	0.068	0.261	0.120	8.354	1.374	2.414
RUBIDIUM IODIDE (P=0.400KB)				H	0.147	0.057	0.296	0.120	8.354	1.252	2.324
11578	3.6149	110	178.0	S	0.129	0.049	0.321	0.120	8.354	1.161	2.260
				R	0.110	0.041	0.346	0.120	8.354	1.065	2.196
				V	0.173	0.069	0.261	0.121	8.280	1.377	2.422
RUBIDIUM IODIDE (P=0.600KB)				H	0.148	0.057	0.296	0.121	8.280	1.254	2.331
11579	3.6203	110	178.0	S	0.129	0.049	0.322	0.121	8.280	1.160	2.265
				R	0.110	0.041	0.348	0.121	8.280	1.063	2.201
				V	0.174	0.069	0.262	0.122	8.206	1.380	2.429
RUBIDIUM IODIDE (P=0.800KB)				H	0.148	0.057	0.297	0.122	8.206	1.255	2.337
11580	3.6269	110	178.0	S	0.129	0.049	0.324	0.122	8.206	1.159	2.270
				R	0.110	0.041	0.349	0.122	8.206	1.061	2.205
				V	0.176	0.070	0.262	0.123	8.128	1.384	2.437
RUBIDIUM IODIDE (P=1.000KB)				H	0.149	0.057	0.298	0.123	8.128	1.256	2.343
11581	3.6334	110	178.0	S	0.129	0.049	0.325	0.123	8.128	1.158	2.275
				R	0.110	0.041	0.351	0.123	8.128	1.059	2.210
				V	0.177	0.070	0.263	0.124	8.034	1.387	2.446
RUBIDIUM IODIDE (P=1.250KB)				H	0.150	0.058	0.300	0.124	8.034	1.258	2.351
11582	3.6411	110	178.0	S	0.129	0.049	0.327	0.124	8.034	1.156	2.280
				R	0.110	0.041	0.353	0.124	8.034	1.055	2.214
				V	0.178	0.071	0.264	0.126	7.949	1.391	2.456
RUBIDIUM IODIDE (P=1.500KB)				H	0.151	0.058	0.301	0.126	7.949	1.260	2.359
11583	3.6476	110	178.0	S	0.129	0.049	0.329	0.126	7.949	1.155	2.287
				R	0.110	0.040	0.355	0.126	7.949	1.054	2.220
				V	0.180	0.071	0.264	0.127	7.864	1.395	2.465
RUBIDIUM IODIDE (P=1.750KB)				H	0.151	0.058	0.302	0.127	7.864	1.261	2.367
11584	3.6551	110	178.0	S	0.130	0.049	0.330	0.127	7.864	1.154	2.293
				R	0.110	0.040	0.356	0.127	7.864	1.051	2.226
				V	0.181	0.072	0.264	0.128	7.813	1.399	2.471
RUBIDIUM IODIDE (P=2.000KB)				H	0.152	0.058	0.302	0.128	7.813	1.263	2.371
11585	3.6629	110	178.0	S	0.130	0.049	0.331	0.128	7.813	1.153	2.295
				R	0.109	0.040	0.357	0.128	7.813	1.049	2.228
				V	0.183	0.072	0.265	0.129	7.730	1.402	2.479
RUBIDIUM IODIDE (P=2.250KB)				H	0.153	0.059	0.303	0.129	7.730	1.264	2.378
11586	3.6705	110	178.0	S	0.130	0.049	0.333	0.129	7.730	1.151	2.300
				R	0.109	0.040	0.359	0.129	7.730	1.046	2.232
				V	0.184	0.073	0.265	0.131	7.649	1.406	2.489
RUBIDIUM IODIDE (P=2.500KB)				H	0.154	0.059	0.304	0.131	7.649	1.266	2.386
11587	3.6771	110	178.0	S	0.130	0.049	0.335	0.131	7.649	1.150	2.306
				R	0.109	0.040	0.361	0.131	7.649	1.044	2.238
				V	0.185	0.073	0.266	0.132	7.580	1.410	2.496
RUBIDIUM IODIDE (P=2.750KB)				H	0.155	0.059	0.305	0.132	7.580	1.268	2.393
11588	3.6848	110	178.0	S	0.130	0.049	0.336	0.132	7.580	1.149	2.311
				R	0.109	0.040	0.362	0.132	7.580	1.041	2.242
				V	0.187	0.074	0.266	0.133	7.506	1.414	2.505
RUBIDIUM IODIDE (P=3.000KB)				H	0.155	0.059	0.306	0.133	7.506	1.269	2.399
11589	3.6913	110	178.0	S	0.130	0.049	0.338	0.133	7.506	1.146	2.316
				R	0.108	0.040	0.364	0.133	7.506	1.038	2.246
				V	0.164	0.065	0.263	0.115	8.693	1.345	2.371
RUBIDIUM IODIDE (P=0.001KB)				H	0.141	0.055	0.295	0.115	8.693	1.234	2.289
11590	3.5850	110	220.0	S	0.126	0.048	0.318	0.115	8.693	1.154	2.233
				R	0.109	0.041	0.342	0.115	8.693	1.065	2.173

CUBIC

NAME / CODE	DENSITY	REF	TEMP(K)		E	G	SIGMA	K	BETA	VS	VP
				V	0.165	0.065	0.263	0.116	8.631	1.348	2.377
RUBIDIUM IODIDE (P=0.175KB)				H	0.142	0.055	0.296	0.116	8.631	1.235	2.294
11591	3.5904	110	220.0	S	0.126	0.048	0.319	0.116	8.631	1.153	2.236
				R	0.109	0.041	0.343	0.116	8.631	1.063	2.176
				V	0.166	0.066	0.263	0.117	8.535	1.352	2.386
RUBIDIUM IODIDE (P=0.410KB)				H	0.143	0.055	0.297	0.117	8.535	1.237	2.301
11592	3.5979	110	220.0	S	0.126	0.048	0.320	0.117	8.535	1.153	2.242
				R	0.109	0.040	0.345	0.117	8.535	1.061	2.181
				V	0.167	0.066	0.264	0.118	8.463	1.355	2.393
RUBIDIUM IODIDE (P=0.600KB)				H	0.143	0.055	0.298	0.118	8.463	1.238	2.307
11593	3.6045	110	220.0	S	0.126	0.048	0.322	0.118	8.463	1.152	2.247
				R	0.109	0.040	0.346	0.118	8.463	1.059	2.185
				V	0.169	0.067	0.264	0.119	8.385	1.359	2.401
RUBIDIUM IODIDE (P=0.800KB)				H	0.144	0.055	0.299	0.119	8.385	1.240	2.313
11594	3.6108	110	220.0	S	0.127	0.048	0.323	0.119	8.385	1.151	2.252
				R	0.109	0.040	0.348	0.119	8.385	1.057	2.189
				V	0.170	0.067	0.265	0.120	8.313	1.361	2.407
RUBIDIUM IODIDE (P=1.000KB)				H	0.145	0.056	0.300	0.120	8.313	1.240	2.319
11595	3.6174	110	220.0	S	0.126	0.048	0.325	0.120	8.313	1.149	2.255
				R	0.108	0.040	0.350	0.120	8.313	1.053	2.192
				V	0.171	0.067	0.265	0.121	8.244	1.365	2.415
RUBIDIUM IODIDE (P=1.200KB)				H	0.145	0.056	0.300	0.121	8.244	1.243	2.325
11596	3.6227	110	220.0	S	0.127	0.048	0.326	0.121	8.244	1.150	2.261
				R	0.109	0.040	0.351	0.121	8.244	1.053	2.197
				V	0.172	0.068	0.266	0.122	8.165	1.368	2.423
RUBIDIUM IODIDE (P=1.400KB)				H	0.146	0.056	0.301	0.122	8.165	1.243	2.331
11597	3.6293	110	220.0	S	0.127	0.048	0.327	0.122	8.165	1.147	2.265
				R	0.108	0.040	0.353	0.122	8.165	1.050	2.201
				V	0.173	0.068	0.266	0.123	8.106	1.371	2.429
RUBIDIUM IODIDE (P=1.600KB)				H	0.147	0.056	0.302	0.123	8.106	1.245	2.337
11598	3.6347	110	220.0	S	0.127	0.048	0.329	0.123	8.106	1.146	2.269
				R	0.108	0.040	0.354	0.123	8.106	1.048	2.204
				V	0.174	0.069	0.267	0.125	8.028	1.374	2.437
RUBIDIUM IODIDE (P=1.800KB)				H	0.147	0.057	0.303	0.125	8.028	1.246	2.343
11599	3.6411	110	220.0	S	0.127	0.048	0.330	0.125	8.028	1.145	2.274
				R	0.108	0.040	0.356	0.125	8.028	1.046	2.209
				V	0.175	0.069	0.267	0.126	7.966	1.377	2.443
RUBIDIUM IODIDE (P=2.000KB)				H	0.148	0.057	0.304	0.126	7.966	1.247	2.348
11600	3.6477	110	220.0	S	0.127	0.048	0.331	0.126	7.966	1.144	2.278
				R	0.108	0.040	0.357	0.126	7.966	1.044	2.212
				V	0.176	0.070	0.268	0.126	7.909	1.380	2.449
RUBIDIUM IODIDE (P=2.200KB)				H	0.149	0.057	0.304	0.126	7.909	1.249	2.354
11601	3.6532	110	220.0	S	0.127	0.048	0.332	0.126	7.909	1.143	2.281
				R	0.108	0.040	0.358	0.126	7.909	1.042	2.215
				V	0.178	0.070	0.268	0.128	7.833	1.383	2.457
RUBIDIUM IODIDE (P=2.400KB)				H	0.149	0.057	0.305	0.128	7.833	1.250	2.360
11602	3.6597	110	220.0	S	0.127	0.048	0.334	0.128	7.833	1.142	2.287
				R	0.108	0.040	0.360	0.128	7.833	1.040	2.220
				V	0.179	0.070	0.268	0.129	7.778	1.386	2.464
RUBIDIUM IODIDE (P=2.600KB)				H	0.150	0.057	0.306	0.129	7.778	1.251	2.366
11603	3.6651	110	220.0	S	0.127	0.048	0.335	0.129	7.778	1.141	2.290
				R	0.107	0.039	0.361	0.129	7.778	1.038	2.223
				V	0.180	0.071	0.269	0.130	7.702	1.389	2.472
RUBIDIUM IODIDE (P=2.800KB)				H	0.151	0.058	0.307	0.130	7.702	1.253	2.373
11604	3.6706	110	220.0	S	0.128	0.048	0.336	0.130	7.702	1.140	2.296
				R	0.107	0.039	0.362	0.130	7.702	1.036	2.229
				V	0.181	0.071	0.269	0.131	7.659	1.392	2.477
RUBIDIUM IODIDE (P=3.000KB)				H	0.151	0.058	0.307	0.131	7.659	1.254	2.376
11605	3.6772	110	220.0	S	0.128	0.048	0.337	0.131	7.659	1.139	2.298
				R	0.107	0.039	0.363	0.131	7.659	1.034	2.231

CUBIC

CODE	NAME	DENSITY	REF	TEMP(K)		E	G	SIGMA	K	BETA	VS	VP
11606	RUBIDIUM IODIDE (P=0.001KB)	3.5690	110	260.0	V	0.158	0.062	0.267	0.113	8.860	1.322	2.344
					H	0.137	0.053	0.297	0.113	8.860	1.217	2.266
					S	0.123	0.047	0.318	0.113	8.860	1.144	2.215
					R	0.107	0.040	0.342	0.113	8.860	1.058	2.158
11607	RUBIDIUM IODIDE (P=0.200KB)	3.5753	110	260.0	V	0.159	0.063	0.267	0.114	8.785	1.326	2.351
					H	0.138	0.053	0.298	0.114	8.785	1.219	2.272
					S	0.123	0.047	0.319	0.114	8.785	1.143	2.220
					R	0.107	0.040	0.343	0.114	8.785	1.057	2.162
11608	RUBIDIUM IODIDE (P=0.400KB)	3.5819	110	260.0	V	0.160	0.063	0.267	0.115	8.703	1.329	2.359
					H	0.139	0.053	0.299	0.115	8.703	1.220	2.279
					S	0.124	0.047	0.321	0.115	8.703	1.143	2.225
					R	0.107	0.040	0.345	0.115	8.703	1.055	2.166
11609	RUBIDIUM IODIDE (P=0.600KB)	3.5882	110	260.0	V	0.162	0.064	0.268	0.116	8.621	1.333	2.367
					H	0.139	0.054	0.300	0.116	8.621	1.222	2.286
					S	0.124	0.047	0.322	0.116	8.621	1.142	2.230
					R	0.107	0.040	0.346	0.116	8.621	1.053	2.171
11610	RUBIDIUM IODIDE (P=0.800KB)	3.5948	110	260.0	V	0.163	0.064	0.268	0.117	8.537	1.337	2.375
					H	0.140	0.054	0.301	0.117	8.537	1.224	2.292
					S	0.124	0.047	0.324	0.117	8.537	1.142	2.235
					R	0.107	0.040	0.348	0.117	8.537	1.051	2.175
11611	RUBIDIUM IODIDE (P=1.000KB)	3.6012	110	260.0	V	0.164	0.065	0.269	0.118	8.458	1.340	2.383
					H	0.141	0.054	0.301	0.118	8.458	1.226	2.299
					S	0.124	0.047	0.325	0.118	8.458	1.141	2.240
					R	0.107	0.040	0.349	0.118	8.458	1.049	2.180
11612	RUBIDIUM IODIDE (P=1.250KB)	3.6099	110	260.0	V	0.166	0.065	0.269	0.120	8.366	1.344	2.392
					H	0.142	0.054	0.302	0.120	8.366	1.228	2.307
					S	0.125	0.047	0.326	0.120	8.366	1.140	2.246
					R	0.107	0.040	0.351	0.120	8.366	1.047	2.185
11613	RUBIDIUM IODIDE (P=1.500KB)	3.6174	110	260.0	V	0.167	0.066	0.270	0.121	8.267	1.349	2.402
					H	0.143	0.055	0.304	0.121	8.267	1.229	2.315
					S	0.125	0.047	0.328	0.121	8.267	1.139	2.252
					R	0.107	0.039	0.353	0.121	8.267	1.044	2.190
11614	RUBIDIUM IODIDE (P=1.750KB)	3.6251	110	260.0	V	0.169	0.066	0.270	0.122	8.183	1.353	2.411
					H	0.143	0.055	0.304	0.122	8.183	1.231	2.322
					S	0.125	0.047	0.330	0.122	8.183	1.138	2.258
					R	0.107	0.039	0.355	0.122	8.183	1.042	2.195
11615	RUBIDIUM IODIDE (P=2.000KB)	3.6326	110	260.0	V	0.170	0.067	0.271	0.124	8.088	1.357	2.421
					H	0.144	0.055	0.305	0.124	8.088	1.233	2.331
					S	0.125	0.047	0.331	0.124	8.088	1.137	2.264
					R	0.107	0.039	0.356	0.124	8.088	1.040	2.201
11616	RUBIDIUM IODIDE (P=2.250KB)	3.6402	110	260.0	V	0.172	0.067	0.271	0.125	8.006	1.362	2.430
					H	0.145	0.056	0.306	0.125	8.006	1.236	2.338
					S	0.125	0.047	0.333	0.125	8.006	1.137	2.270
					R	0.107	0.039	0.358	0.125	8.006	1.038	2.206
11617	RUBIDIUM IODIDE (P=2.500KB)	3.6477	110	260.0	V	0.173	0.068	0.272	0.126	7.924	1.365	2.438
					H	0.146	0.056	0.307	0.126	7.924	1.237	2.345
					S	0.125	0.047	0.335	0.126	7.924	1.134	2.275
					R	0.106	0.039	0.360	0.126	7.924	1.034	2.210
11618	RUBIDIUM IODIDE (P=2.750KB)	3.6554	110	260.0	V	0.174	0.069	0.272	0.127	7.843	1.370	2.447
					H	0.147	0.056	0.308	0.127	7.843	1.239	2.353
					S	0.125	0.047	0.336	0.127	7.843	1.134	2.281
					R	0.106	0.039	0.361	0.127	7.843	1.032	2.215
11619	RUBIDIUM IODIDE (P=2.990KB)	3.6619	110	260.0	V	0.176	0.069	0.273	0.129	7.764	1.373	2.456
					H	0.148	0.056	0.309	0.129	7.764	1.241	2.360
					S	0.126	0.047	0.337	0.129	7.764	1.133	2.286
					R	0.106	0.039	0.363	0.129	7.764	1.030	2.221
11620	RUBIDIUM IODIDE (P=3.250KB)	3.6706	110	260.0	V	0.177	0.070	0.273	0.130	7.684	1.377	2.464
					H	0.148	0.057	0.310	0.130	7.684	1.242	2.367
					S	0.126	0.047	0.339	0.130	7.684	1.130	2.291
					R	0.106	0.039	0.365	0.130	7.684	1.026	2.225

CUBIC

CODE	NAME / DENSITY	REF	TEMP(K)		E	G	SIGMA	K	BETA	VS	VP
				V	0.179	0.070	0.274	0.131	7.608	1.381	2.473
RUBIDIUM IODIDE (P=3.500KB)				H	0.149	0.057	0.311	0.131	7.608	1.244	2.374
11621	3.6771	110	260.0	S	0.126	0.047	0.341	0.131	7.608	1.129	2.297
				R	0.105	0.039	0.366	0.131	7.608	1.024	2.230
				V	0.180	0.071	0.274	0.133	7.538	1.384	2.481
RUBIDIUM IODIDE (P=3.750KB)				H	0.150	0.057	0.312	0.133	7.538	1.245	2.380
11622	3.6848	110	260.0	S	0.126	0.047	0.342	0.133	7.538	1.127	2.301
				R	0.105	0.038	0.368	0.133	7.538	1.020	2.233
				V	0.153	0.060	0.269	0.110	9.061	1.302	2.317
RUBIDIUM IODIDE (P=0.001KB)				H	0.133	0.051	0.298	0.110	9.061	1.203	2.244
11623	3.5510	110	300.0	S	0.121	0.046	0.318	0.110	9.061	1.136	2.198
				R	0.106	0.039	0.340	0.110	9.061	1.054	2.143
				V	0.154	0.061	0.270	0.111	8.977	1.306	2.325
RUBIDIUM IODIDE (P=0.200KB)				H	0.134	0.052	0.299	0.111	8.977	1.205	2.251
11624	3.5573	110	300.0	S	0.121	0.046	0.319	0.111	8.977	1.136	2.203
				R	0.106	0.039	0.342	0.111	8.977	1.053	2.147
				V	0.155	0.061	0.270	0.113	8.886	1.309	2.333
RUBIDIUM IODIDE (P=0.400KB)				H	0.135	0.052	0.300	0.113	8.886	1.206	2.258
11625	3.5648	110	300.0	S	0.121	0.046	0.320	0.113	8.886	1.135	2.208
				R	0.106	0.039	0.343	0.113	8.886	1.051	2.152
				V	0.156	0.062	0.271	0.114	8.803	1.313	2.341
RUBIDIUM IODIDE (P=0.600KB)				H	0.136	0.052	0.301	0.114	8.803	1.208	2.264
11626	3.5713	110	300.0	S	0.122	0.046	0.322	0.114	8.803	1.135	2.213
				R	0.106	0.039	0.345	0.114	8.803	1.049	2.156
				V	0.158	0.062	0.271	0.115	8.718	1.316	2.349
RUBIDIUM IODIDE (P=0.800KB)				H	0.136	0.052	0.302	0.115	8.718	1.210	2.271
11627	3.5777	110	300.0	S	0.122	0.046	0.323	0.115	8.718	1.134	2.218
				R	0.106	0.039	0.346	0.115	8.718	1.047	2.161
				V	0.159	0.062	0.271	0.116	8.633	1.320	2.357
RUBIDIUM IODIDE (P=1.000KB)				H	0.137	0.053	0.303	0.116	8.633	1.212	2.278
11628	3.5841	110	300.0	S	0.122	0.046	0.324	0.116	8.633	1.134	2.224
				R	0.106	0.039	0.348	0.116	8.633	1.046	2.165
				V	0.160	0.063	0.272	0.117	8.562	1.323	2.364
RUBIDIUM IODIDE (P=1.200KB)				H	0.138	0.053	0.304	0.117	8.562	1.213	2.284
11629	3.5905	110	300.0	S	0.122	0.046	0.326	0.117	8.562	1.133	2.228
				R	0.106	0.039	0.349	0.117	8.562	1.044	2.169
				V	0.161	0.063	0.272	0.118	8.477	1.326	2.372
RUBIDIUM IODIDE (P=1.400KB)				H	0.138	0.053	0.305	0.118	8.477	1.214	2.290
11630	3.5970	110	300.0	S	0.122	0.046	0.327	0.118	8.477	1.132	2.233
				R	0.105	0.039	0.351	0.118	8.477	1.042	2.174
				V	0.162	0.064	0.273	0.119	8.403	1.330	2.379
RUBIDIUM IODIDE (P=1.600KB)				H	0.139	0.053	0.305	0.119	8.403	1.216	2.296
11631	3.6035	110	300.0	S	0.123	0.046	0.328	0.119	8.403	1.131	2.238
				R	0.105	0.039	0.352	0.119	8.403	1.040	2.178
				V	0.163	0.064	0.273	0.120	8.324	1.333	2.387
RUBIDIUM IODIDE (P=1.800KB)				H	0.140	0.053	0.306	0.120	8.324	1.217	2.303
11632	3.6098	110	300.0	S	0.123	0.046	0.330	0.120	8.324	1.131	2.243
				R	0.105	0.039	0.354	0.120	8.324	1.038	2.183
				V	0.165	0.065	0.274	0.121	8.249	1.336	2.394
RUBIDIUM IODIDE (P=2.000KB)				H	0.140	0.054	0.307	0.121	8.249	1.219	2.309
11633	3.6164	110	300.0	S	0.123	0.046	0.331	0.121	8.249	1.130	2.248
				R	0.105	0.039	0.355	0.121	8.249	1.036	2.187
				V	0.166	0.065	0.274	0.122	8.181	1.339	2.401
RUBIDIUM IODIDE (P=2.200KB)				H	0.141	0.054	0.308	0.122	8.181	1.220	2.315
11634	3.6229	110	300.0	S	0.123	0.046	0.332	0.122	8.181	1.129	2.252
				R	0.105	0.039	0.357	0.122	8.181	1.034	2.191
				V	0.167	0.065	0.275	0.123	8.110	1.343	2.409
RUBIDIUM IODIDE (P=2.400KB)				H	0.142	0.054	0.308	0.123	8.110	1.222	2.321
11635	3.6283	110	300.0	S	0.123	0.046	0.334	0.123	8.110	1.128	2.257
				R	0.105	0.039	0.358	0.123	8.110	1.032	2.195

CUBIC

CODE	NAME	DENSITY	REF	TEMP(K)		E	G	SIGMA	K	BETA	VS	VP
11636	RUBIDIUM IODIDE (P=2.600KB)	3.6347	110	300.0	V	0.168	0.066	0.275	0.124	8.043	1.346	2.416
					H	0.142	0.054	0.309	0.124	8.043	1.223	2.327
					S	0.123	0.046	0.335	0.124	8.043	1.127	2.261
					R	0.105	0.039	0.359	0.124	8.043	1.030	2.199
11637	RUBIDIUM IODIDE (P=2.800KB)	3.6412	110	300.0	V	0.169	0.066	0.275	0.125	7.974	1.349	2.423
					H	0.143	0.055	0.310	0.125	7.974	1.224	2.333
					S	0.123	0.046	0.336	0.125	7.974	1.126	2.266
					R	0.105	0.038	0.361	0.125	7.974	1.028	2.203
11638	RUBIDIUM IODIDE (P=3.000KB)	3.6467	110	300.0	V	0.170	0.067	0.276	0.126	7.905	1.352	2.430
					H	0.144	0.055	0.311	0.126	7.905	1.226	2.339
					S	0.123	0.046	0.337	0.126	7.905	1.125	2.271
					R	0.105	0.038	0.362	0.127	7.905	1.026	2.207
11639	RUBIDIUM IODIDE (P=3.200KB)	3.6532	110	300.0	V	0.171	0.067	0.276	0.127	7.845	1.355	2.437
					H	0.144	0.055	0.311	0.127	7.845	1.227	2.344
					S	0.124	0.046	0.338	0.127	7.845	1.124	2.275
					R	0.104	0.038	0.363	0.127	7.845	1.024	2.211
11640	RUBIDIUM IODIDE (P=3.400KB)	3.6586	110	300.0	V	0.172	0.067	0.277	0.128	7.782	1.358	2.443
					H	0.145	0.055	0.312	0.128	7.782	1.228	2.350
					S	0.124	0.046	0.340	0.128	7.782	1.123	2.279
					R	0.104	0.038	0.365	0.128	7.782	1.022	2.215
11641	RUBIDIUM IODIDE (P=3.600KB)	3.6650	110	300.0	V	0.173	0.068	0.277	0.130	7.716	1.360	2.450
					H	0.145	0.055	0.313	0.130	7.716	1.229	2.356
					S	0.123	0.046	0.341	0.130	7.716	1.120	2.282
					R	0.104	0.038	0.366	0.130	7.716	1.018	2.218
11642	RUBIDIUM IODIDE (P=3.800KB)	3.6706	110	300.0	V	0.174	0.068	0.277	0.131	7.657	1.364	2.457
					H	0.146	0.056	0.314	0.131	7.657	1.231	2.362
					S	0.124	0.046	0.342	0.131	7.657	1.121	2.288
					R	0.104	0.038	0.367	0.131	7.657	1.018	2.223
11643	SILICON	2.3400	129	293.0	V	1.659	0.681	0.217	0.978	1.022	5.395	8.979
					H	1.631	0.667	0.222	0.978	1.022	5.340	8.934
					S	1.625	0.664	0.223	0.978	1.022	5.328	8.925
					R	1.595	0.650	0.228	0.978	1.022	5.268	8.878
11644	SILICON	2.3400	129	ROOM	V	1.718	0.708	0.214	1.000	1.000	5.501	9.115
					H	1.683	0.690	0.219	1.000	1.000	5.431	9.059
					S	1.675	0.686	0.221	1.000	1.000	5.415	9.046
					R	1.638	0.667	0.227	1.000	1.000	5.340	8.987
11645	SILICON	2.3310	164	298.0	V	1.659	0.681	0.217	0.979	1.022	5.407	8.998
					H	1.631	0.667	0.222	0.979	1.022	5.351	8.954
					S	1.626	0.665	0.223	0.979	1.022	5.339	8.945
					R	1.596	0.650	0.228	0.979	1.022	5.280	8.897
11646	SILICON (P=30,000PSI)	2.3360	164	298.0	V	1.664	0.682	0.219	0.988	1.013	5.405	9.013
					H	1.636	0.668	0.224	0.988	1.013	5.349	8.968
					S	1.630	0.665	0.225	0.988	1.013	5.337	8.959
					R	1.600	0.650	0.230	0.988	1.013	5.277	8.911
11647	SILICON	2.3320	164	77.2	V	1.676	0.688	0.219	0.992	1.008	5.430	9.048
					H	1.648	0.673	0.223	0.992	1.008	5.374	9.003
					S	1.642	0.671	0.224	0.992	1.008	5.362	8.994
					R	1.612	0.656	0.229	0.992	1.008	5.302	8.946
11648	SILICON (P=30,000PSI)	2.3370	164	77.2	V	1.680	0.689	0.220	1.001	0.999	5.428	9.062
					H	1.652	0.674	0.225	1.001	0.999	5.372	9.017
					S	1.646	0.671	0.226	1.001	0.999	5.360	9.007
					R	1.616	0.656	0.231	1.001	0.999	5.299	8.959
11649	SILVER	10.5000	128	ROOM	V	0.873	0.322	0.354	0.998	1.002	1.752	3.688
					H	0.809	0.296	0.365	0.998	1.002	1.680	3.642
					S	0.768	0.280	0.372	0.998	1.002	1.633	3.614
					R	0.688	0.248	0.385	0.998	1.002	1.538	3.558
11650	SILVER	10.5300	129	ROOM	V	0.914	0.338	0.353	1.036	0.965	1.791	3.757
					H	0.843	0.309	0.364	1.036	0.965	1.713	3.708
					S	0.796	0.290	0.372	1.036	0.965	1.660	3.676
					R	0.708	0.255	0.386	1.036	0.965	1.557	3.616

CUBIC

CODE	NAME DENSITY	REF	TEMP(K)		E	G	SIGMA	K	BETA	VS	VP
				V	0.862	0.318	0.357	1.003	0.997	1.740	3.686
SILVER				H	0.804	0.294	0.366	1.003	0.997	1.673	3.645
11651	10.5000	4	ROOM	S	0.769	0.280	0.372	1.003	0.997	1.633	3.620
				R	0.696	0.251	0.384	1.003	0.997	1.547	3.569
				V	1.009	0.375	0.345	1.087	0.920	1.878	3.863
SILVER				H	0.931	0.343	0.357	1.087	0.920	1.796	3.811
11652	10.6330	191	0.0	S	0.880	0.322	0.365	1.087	0.920	1.741	3.777
				R	0.785	0.284	0.380	1.087	0.920	1.635	3.714
				V	1.008	0.375	0.345	1.087	0.920	1.877	3.863
SILVER				H	0.930	0.343	0.357	1.087	0.920	1.795	3.810
11653	10.6330	191	10.0	S	0.880	0.322	0.365	1.087	0.920	1.741	3.777
				R	0.784	0.284	0.380	1.087	0.920	1.635	3.713
				V	1.007	0.374	0.346	1.087	0.920	1.876	3.862
SILVER				H	0.929	0.342	0.358	1.087	0.920	1.794	3.809
11654	10.6330	191	25.0	S	0.878	0.322	0.365	1.087	0.920	1.739	3.776
				R	0.783	0.284	0.380	1.087	0.920	1.634	3.712
				V	1.002	0.372	0.346	1.085	0.921	1.871	3.857
SILVER				H	0.924	0.340	0.358	1.085	0.921	1.789	3.805
11655	10.6300	191	50.0	S	0.874	0.320	0.366	1.085	0.921	1.735	3.771
				R	0.779	0.282	0.380	1.085	0.921	1.629	3.708
				V	0.993	0.369	0.347	1.082	0.924	1.863	3.849
SILVER				H	0.916	0.337	0.359	1.082	0.924	1.782	3.797
11656	10.6210	191	75.0	S	0.866	0.317	0.367	1.082	0.924	1.727	3.764
				R	0.771	0.279	0.381	1.082	0.924	1.621	3.700
				V	0.984	0.365	0.348	1.077	0.928	1.855	3.839
SILVER				H	0.908	0.334	0.360	1.077	0.928	1.774	3.788
11657	10.6110	191	100.0	S	0.858	0.314	0.367	1.077	0.928	1.719	3.754
				R	0.764	0.276	0.382	1.077	0.928	1.614	3.691
				V	0.976	0.362	0.348	1.072	0.933	1.848	3.830
SILVER				H	0.900	0.331	0.360	1.072	0.933	1.766	3.778
11658	10.5980	191	125.0	S	0.850	0.311	0.368	1.072	0.933	1.712	3.745
				R	0.757	0.274	0.382	1.072	0.933	1.607	3.682
				V	0.967	0.358	0.349	1.067	0.937	1.840	3.820
SILVER				H	0.891	0.327	0.361	1.067	0.937	1.759	3.769
11659	10.5850	191	150.0	S	0.842	0.307	0.369	1.067	0.937	1.704	3.736
				R	0.749	0.271	0.383	1.067	0.937	1.599	3.673
				V	0.958	0.355	0.350	1.062	0.941	1.832	3.811
SILVER				H	0.883	0.324	0.361	1.062	0.941	1.751	3.760
11660	10.5730	191	175.0	S	0.834	0.304	0.369	1.062	0.941	1.697	3.726
				R	0.741	0.268	0.384	1.062	0.941	1.592	3.664
				V	0.949	0.351	0.350	1.057	0.946	1.824	3.801
SILVER				H	0.875	0.321	0.362	1.057	0.946	1.744	3.751
11661	10.5600	191	200.0	S	0.826	0.301	0.370	1.057	0.946	1.689	3.717
				R	0.734	0.265	0.384	1.057	0.946	1.585	3.655
				V	0.940	0.348	0.351	1.053	0.950	1.816	3.792
SILVER				H	0.866	0.318	0.363	1.053	0.950	1.736	3.742
11662	10.5440	191	225.0	S	0.817	0.298	0.371	1.053	0.950	1.682	3.709
				R	0.727	0.262	0.385	1.053	0.950	1.577	3.647
				V	0.931	0.344	0.352	1.048	0.954	1.808	3.783
SILVER				H	0.858	0.315	0.364	1.048	0.954	1.728	3.733
11663	10.5280	191	250.0	S	0.809	0.295	0.371	1.048	0.954	1.674	3.700
				R	0.719	0.259	0.386	1.048	0.954	1.570	3.638
				V	0.922	0.341	0.353	1.043	0.959	1.800	3.773
SILVER				H	0.849	0.311	0.364	1.043	0.959	1.721	3.723
11664	10.5160	191	275.0	S	0.801	0.292	0.372	1.043	0.959	1.666	3.690
				R	0.712	0.257	0.386	1.043	0.959	1.562	3.629
				V	0.913	0.337	0.353	1.038	0.964	1.792	3.764
SILVER				H	0.841	0.308	0.365	1.038	0.964	1.713	3.714
11665	10.5000	191	300.0	S	0.793	0.289	0.373	1.038	0.964	1.659	3.681
				R	0.704	0.254	0.387	1.038	0.964	1.555	3.620

CUBIC

CODE	NAME DENSITY	REF	TEMP(K)		E	G	SIGMA	K	BETA	VS	VP
				V	0.917	0.339	0.353	1.040	0.962	1.797	3.770
SILVER				H	0.843	0.309	0.365	1.040	0.962	1.715	3.719
11666	10.4970	51	300.0	S	0.793	0.289	0.373	1.040	0.962	1.658	3.684
				R	0.701	0.253	0.388	1.040	0.962	1.552	3.622
				V	0.899	0.332	0.354	1.028	0.972	1.781	3.749
SILVER				H	0.827	0.303	0.366	1.028	0.972	1.701	3.699
11667	10.4680	51	350.0	S	0.778	0.283	0.374	1.028	0.972	1.645	3.665
				R	0.689	0.248	0.388	1.028	0.972	1.540	3.603
				V	0.880	0.325	0.356	1.015	0.985	1.764	3.724
SILVER				H	0.808	0.295	0.367	1.015	0.985	1.683	3.674
11668	10.4370	51	400.0	S	0.758	0.276	0.375	1.015	0.985	1.625	3.640
				R	0.670	0.241	0.390	1.015	0.985	1.519	3.578
				V	0.862	0.318	0.357	1.003	0.997	1.747	3.703
SILVER				H	0.792	0.289	0.368	1.003	0.997	1.667	3.654
11669	10.4060	51	450.0	S	0.744	0.270	0.376	1.003	0.997	1.611	3.620
				R	0.657	0.236	0.391	1.003	0.997	1.507	3.559
				V	0.841	0.310	0.358	0.990	1.010	1.727	3.677
SILVER				H	0.772	0.282	0.370	0.990	1.010	1.648	3.628
11670	10.3750	51	500.0	S	0.723	0.262	0.378	0.990	1.010	1.590	3.594
				R	0.637	0.229	0.393	0.990	1.010	1.485	3.533
				V	0.824	0.303	0.360	0.978	1.022	1.712	3.656
SILVER				H	0.757	0.276	0.371	0.978	1.022	1.633	3.607
11671	10.3450	51	550.0	S	0.709	0.257	0.379	0.978	1.022	1.577	3.574
				R	0.625	0.224	0.393	0.978	1.022	1.473	3.514
				V	0.804	0.295	0.361	0.965	1.036	1.692	3.629
SILVER				H	0.736	0.268	0.373	0.965	1.036	1.613	3.581
11672	10.3140	51	600.0	S	0.689	0.249	0.381	0.965	1.036	1.555	3.547
				R	0.605	0.217	0.395	0.965	1.036	1.450	3.487
				V	0.787	0.289	0.362	0.953	1.049	1.676	3.608
SILVER				H	0.721	0.263	0.374	0.953	1.049	1.598	3.561
11673	10.2790	51	650.0	S	0.675	0.244	0.382	0.953	1.049	1.541	3.527
				R	0.593	0.212	0.396	0.953	1.049	1.438	3.468
				V	0.766	0.281	0.364	0.940	1.064	1.656	3.582
SILVER				H	0.701	0.255	0.376	0.940	1.064	1.577	3.535
11674	10.2430	51	700.0	S	0.654	0.236	0.384	0.940	1.064	1.519	3.500
				R	0.573	0.205	0.398	0.940	1.064	1.414	3.442
				V	0.747	0.273	0.366	0.932	1.073	1.637	3.564
SILVER				H	0.682	0.248	0.378	0.932	1.073	1.557	3.516
11675	10.2060	51	750.0	S	0.634	0.229	0.387	0.932	1.073	1.497	3.481
				R	0.554	0.198	0.401	0.932	1.073	1.391	3.422
				V	0.726	0.265	0.368	0.918	1.089	1.615	3.536
SILVER				H	0.662	0.240	0.380	0.918	1.089	1.535	3.489
11676	10.1730	51	800.0	S	0.613	0.221	0.389	0.918	1.089	1.473	3.453
				R	0.533	0.190	0.403	0.918	1.089	1.367	3.394
				V	0.912	0.337	0.351	1.019	0.981	1.793	3.742
SILVER				H	0.840	0.308	0.363	1.019	0.981	1.714	3.693
11677	10.4900	207	294.0	S	0.793	0.289	0.370	1.019	0.981	1.661	3.660
				R	0.705	0.255	0.385	1.019	0.981	1.558	3.599
				V	0.906	0.335	0.351	1.012	0.988	1.787	3.728
SILVER				H	0.840	0.308	0.362	1.012	0.988	1.713	3.681
11678	10.5000	133	ROOM	S	0.798	0.292	0.369	1.012	0.988	1.666	3.652
				R	0.716	0.259	0.382	1.012	0.988	1.571	3.595
				V	0.253	0.090	0.396	0.406	2.463	1.182	2.852
SILVER BROMIDE				H	0.248	0.089	0.398	0.406	2.463	1.169	2.845
11679	6.4730	129	ROOM	S	0.246	0.088	0.399	0.406	2.463	1.166	2.844
				R	0.240	0.086	0.401	0.406	2.463	1.151	2.835
				V	0.251	0.090	0.397	0.408	2.453	1.178	2.854
SILVER BROMIDE				H	0.246	0.088	0.400	0.408	2.453	1.165	2.847
11680	6.4730	138	ROOM	S	0.244	0.087	0.400	0.408	2.453	1.161	2.845
				R	0.238	0.085	0.403	0.408	2.453	1.146	2.837

CUBIC

NAME CODE	DENSITY	REF	TEMP(K)		E	G	SIGMA	K	BETA	VS	VP
				V	0.240	0.085	0.409	0.442	2.264	1.235	3.152
SILVER CHLORIDE				H	0.232	0.082	0.413	0.442	2.264	1.212	3.140
11681	5.5890	129	ROOM	S	0.229	0.081	0.414	0.442	2.264	1.203	3.136
				R	0.219	0.077	0.417	0.442	2.264	1.176	3.122
				V	0.311	0.111	0.398	0.507	1.973	1.399	3.394
SILVER CHLORIDE				H	0.291	0.103	0.404	0.507	1.973	1.348	3.366
11682	5.6900	257	0.0	S	0.280	0.100	0.408	0.507	1.973	1.323	3.353
				R	0.259	0.091	0.415	0.507	1.973	1.267	3.324
				V	0.311	0.111	0.398	0.506	1.975	1.398	3.392
SILVER CHLORIDE				H	0.290	0.103	0.405	0.506	1.975	1.347	3.364
11683	5.6900	257	20.0	S	0.280	0.099	0.408	0.506	1.975	1.322	3.351
				R	0.258	0.091	0.415	0.506	1.975	1.266	3.322
				V	0.309	0.110	0.398	0.504	1.985	1.394	3.383
SILVER CHLORIDE				H	0.289	0.103	0.405	0.504	1.985	1.344	3.356
11684	5.6890	257	40.0	S	0.279	0.099	0.408	0.504	1.985	1.319	3.343
				R	0.257	0.091	0.415	0.504	1.985	1.264	3.315
				V	0.305	0.109	0.398	0.500	1.998	1.385	3.370
SILVER CHLORIDE				H	0.286	0.102	0.405	0.500	1.998	1.337	3.344
11685	5.6870	257	60.0	S	0.276	0.098	0.408	0.500	1.998	1.313	3.331
				R	0.255	0.090	0.415	0.500	1.998	1.259	3.304
				V	0.300	0.107	0.399	0.496	2.018	1.374	3.353
SILVER CHLORIDE				H	0.281	0.100	0.405	0.496	2.018	1.327	3.327
11686	5.6810	257	80.0	S	0.272	0.097	0.408	0.496	2.018	1.304	3.315
				R	0.252	0.089	0.415	0.496	2.018	1.252	3.289
				V	0.294	0.105	0.400	0.491	2.038	1.360	3.334
SILVER CHLORIDE				H	0.276	0.098	0.406	0.491	2.038	1.316	3.310
11687	5.6730	257	100.0	S	0.268	0.095	0.409	0.491	2.038	1.295	3.299
				R	0.249	0.088	0.415	0.491	2.038	1.246	3.274
				V	0.287	0.102	0.401	0.485	2.063	1.345	3.312
SILVER CHLORIDE				H	0.271	0.096	0.407	0.485	2.063	1.304	3.290
11688	5.6640	257	120.0	S	0.263	0.093	0.409	0.485	2.063	1.284	3.280
				R	0.246	0.087	0.416	0.485	2.063	1.238	3.256
				V	0.281	0.100	0.402	0.479	2.086	1.331	3.292
SILVER CHLORIDE				H	0.266	0.094	0.408	0.479	2.086	1.292	3.271
11689	5.6560	257	140.0	S	0.259	0.092	0.410	0.479	2.086	1.274	3.262
				R	0.242	0.086	0.416	0.479	2.086	1.230	3.240
				V	0.275	0.098	0.403	0.474	2.111	1.317	3.271
SILVER CHLORIDE				H	0.261	0.093	0.408	0.474	2.111	1.280	3.252
11690	5.6470	257	160.0	S	0.255	0.090	0.410	0.474	2.111	1.265	3.244
				R	0.239	0.084	0.416	0.474	2.111	1.223	3.223
				V	0.269	0.096	0.404	0.468	2.136	1.302	3.251
SILVER CHLORIDE				H	0.256	0.091	0.409	0.468	2.136	1.268	3.233
11691	5.6360	257	180.0	S	0.250	0.089	0.411	0.468	2.136	1.254	3.226
				R	0.236	0.083	0.416	0.468	2.136	1.215	3.206
				V	0.262	0.093	0.406	0.463	2.159	1.288	3.232
SILVER CHLORIDE				H	0.251	0.089	0.410	0.463	2.159	1.257	3.216
11692	5.6260	257	200.0	S	0.246	0.087	0.412	0.463	2.159	1.244	3.209
				R	0.232	0.082	0.416	0.463	2.159	1.208	3.190
				V	0.257	0.091	0.406	0.457	2.188	1.275	3.211
SILVER CHLORIDE				H	0.246	0.087	0.410	0.457	2.188	1.246	3.195
11693	5.6130	257	220.0	S	0.241	0.086	0.412	0.457	2.188	1.234	3.190
				R	0.229	0.081	0.416	0.457	2.188	1.201	3.172
				V	0.251	0.089	0.407	0.451	2.218	1.261	3.188
SILVER CHLORIDE				H	0.241	0.085	0.411	0.451	2.218	1.234	3.174
11694	5.6040	257	240.0	S	0.237	0.084	0.412	0.451	2.218	1.224	3.169
				R	0.226	0.080	0.417	0.451	2.218	1.192	3.153
				V	0.245	0.087	0.408	0.445	2.249	1.247	3.166
SILVER CHLORIDE				H	0.236	0.084	0.412	0.445	2.249	1.223	3.153
11695	5.5930	257	260.0	S	0.233	0.082	0.413	0.445	2.249	1.214	3.149
				R	0.222	0.078	0.417	0.445	2.249	1.185	3.134

CUBIC

CODE	NAME / DENSITY	REF	TEMP(K)		E	G	SIGMA	K	BETA	VS	VP
				V	0.239	0.085	0.409	0.439	2.276	1.233	3.146
SILVER CHLORIDE				H	0.231	0.082	0.412	0.439	2.276	1.211	3.134
11696	5.5820	257	280.0	S	0.228	0.081	0.413	0.439	2.276	1.203	3.130
				R	0.219	0.077	0.417	0.439	2.276	1.176	3.117
				V	0.234	0.083	0.410	0.434	2.303	1.220	3.127
SILVER CHLORIDE				H	0.226	0.080	0.413	0.434	2.303	1.199	3.117
11697	5.5690	257	295.0	S	0.224	0.079	0.414	0.434	2.303	1.193	3.113
				R	0.215	0.076	0.417	0.434	2.303	1.168	3.101
				V	0.321	0.115	0.396	0.514	1.947	1.421	3.421
SILVER CHLORIDE				H	0.297	0.106	0.404	0.514	1.947	1.363	3.390
11698	5.6990	132	0.0	S	0.285	0.101	0.408	0.514	1.947	1.332	3.373
				R	0.260	0.092	0.416	0.514	1.947	1.270	3.341
				V	0.321	0.115	0.396	0.513	1.948	1.420	3.420
SILVER CHLORIDE				H	0.297	0.106	0.404	0.513	1.948	1.363	3.389
11699	5.6990	132	10.0	S	0.284	0.101	0.408	0.513	1.948	1.332	3.372
				R	0.260	0.092	0.416	0.513	1.948	1.270	3.340
				V	0.320	0.114	0.396	0.513	1.951	1.417	3.417
SILVER CHLORIDE				H	0.296	0.105	0.404	0.513	1.951	1.361	3.386
11700	5.6980	132	20.0	S	0.284	0.101	0.408	0.513	1.951	1.330	3.370
				R	0.260	0.092	0.416	0.513	1.951	1.269	3.338
				V	0.318	0.114	0.396	0.512	1.954	1.414	3.413
SILVER CHLORIDE				H	0.295	0.105	0.404	0.512	1.954	1.358	3.383
11701	5.6970	132	30.0	S	0.283	0.100	0.408	0.512	1.954	1.328	3.367
				R	0.259	0.091	0.416	0.512	1.954	1.267	3.335
				V	0.316	0.113	0.397	0.511	1.958	1.410	3.408
SILVER CHLORIDE				H	0.294	0.105	0.404	0.511	1.958	1.355	3.378
11702	5.6950	132	40.0	S	0.282	0.100	0.408	0.511	1.958	1.325	3.363
				R	0.258	0.091	0.416	0.511	1.958	1.265	3.332
				V	0.314	0.112	0.397	0.509	1.964	1.404	3.401
SILVER CHLORIDE				H	0.291	0.104	0.405	0.509	1.965	1.350	3.372
11703	5.6930	132	50.0	S	0.280	0.099	0.408	0.509	1.965	1.321	3.357
				R	0.257	0.091	0.416	0.509	1.964	1.262	3.326
				V	0.311	0.111	0.398	0.507	1.972	1.398	3.394
SILVER CHLORIDE				H	0.289	0.103	0.405	0.507	1.972	1.345	3.365
11704	5.6900	132	60.0	S	0.278	0.099	0.409	0.507	1.972	1.317	3.350
				R	0.255	0.090	0.416	0.507	1.972	1.259	3.321
				V	0.307	0.110	0.399	0.505	1.981	1.390	3.385
SILVER CHLORIDE				H	0.286	0.102	0.405	0.505	1.981	1.339	3.357
11705	5.6860	132	70.0	S	0.276	0.098	0.409	0.505	1.981	1.312	3.343
				R	0.254	0.090	0.416	0.505	1.981	1.255	3.314
				V	0.303	0.108	0.399	0.502	1.991	1.380	3.373
SILVER CHLORIDE				H	0.283	0.101	0.406	0.502	1.991	1.331	3.347
11706	5.6830	132	80.0	S	0.273	0.097	0.409	0.502	1.991	1.306	3.333
				R	0.252	0.089	0.416	0.502	1.991	1.251	3.305
				V	0.243	0.086	0.413	0.464	2.156	1.233	3.198
SILVER CHLORIDE				H	0.236	0.083	0.415	0.464	2.156	1.214	3.189
11707	5.6560	132	150.0	S	0.234	0.083	0.416	0.464	2.156	1.209	3.186
				R	0.226	0.080	0.419	0.464	2.156	1.187	3.175
				V	0.274	0.097	0.405	0.482	2.075	1.313	3.290
SILVER CHLORIDE				H	0.260	0.092	0.410	0.482	2.075	1.277	3.271
11708	5.6520	132	160.0	S	0.254	0.090	0.412	0.482	2.075	1.261	3.263
				R	0.238	0.084	0.418	0.482	2.075	1.220	3.242
				V	0.271	0.096	0.406	0.479	2.086	1.307	3.281
SILVER CHLORIDE				H	0.258	0.091	0.410	0.479	2.086	1.271	3.262
11709	5.6480	132	170.0	S	0.252	0.089	0.412	0.479	2.086	1.256	3.254
				R	0.237	0.083	0.418	0.479	2.086	1.216	3.234
				V	0.253	0.090	0.409	0.463	2.161	1.265	3.220
SILVER CHLORIDE				H	0.243	0.086	0.413	0.463	2.161	1.236	3.205
11710	5.6200	132	230.0	S	0.238	0.084	0.414	0.463	2.161	1.225	3.199
				R	0.226	0.080	0.419	0.463	2.161	1.191	3.182

CUBIC

CODE	NAME	DENSITY	REF	TEMP(K)		E	G	SIGMA	K	BETA	VS	VP
11711	SILVER CHLORIDE	5.6160	132	240.0	V	0.250	0.089	0.409	0.460	2.175	1.258	3.209
					H	0.240	0.085	0.413	0.460	2.175	1.230	3.194
					S	0.236	0.084	0.414	0.460	2.175	1.220	3.189
					R	0.225	0.079	0.419	0.460	2.175	1.187	3.173
11712	SILVER CHLORIDE	5.6110	132	250.0	V	0.247	0.088	0.410	0.456	2.191	1.250	3.197
					H	0.238	0.084	0.413	0.456	2.191	1.224	3.183
					S	0.234	0.083	0.415	0.456	2.191	1.214	3.178
					R	0.223	0.078	0.419	0.456	2.191	1.183	3.162
11713	SILVER CHLORIDE	5.6060	132	260.0	V	0.245	0.087	0.410	0.453	2.207	1.244	3.185
					H	0.235	0.083	0.413	0.453	2.208	1.219	3.172
					S	0.232	0.082	0.415	0.453	2.208	1.210	3.167
					R	0.221	0.078	0.419	0.453	2.208	1.180	3.152
11714	SILVER CHLORIDE	5.6010	132	270.0	V	0.242	0.086	0.410	0.450	2.224	1.238	3.174
					H	0.233	0.082	0.414	0.450	2.224	1.213	3.161
					S	0.230	0.081	0.415	0.450	2.224	1.204	3.156
					R	0.220	0.077	0.419	0.450	2.224	1.175	3.142
11715	SILVER CHLORIDE	5.5960	132	280.0	V	0.239	0.085	0.411	0.446	2.240	1.231	3.162
					H	0.231	0.082	0.414	0.446	2.240	1.207	3.150
					S	0.228	0.080	0.415	0.446	2.240	1.199	3.146
					R	0.218	0.077	0.419	0.446	2.240	1.172	3.132
11716	SILVER CHLORIDE	5.5910	132	290.0	V	0.236	0.084	0.411	0.443	2.256	1.224	3.150
					H	0.228	0.081	0.414	0.443	2.256	1.201	3.139
					S	0.226	0.080	0.415	0.443	2.256	1.194	3.135
					R	0.216	0.076	0.419	0.443	2.256	1.167	3.121
11717	SILVER CHLORIDE	5.5890	132	293.0	V	0.236	0.083	0.411	0.442	2.263	1.222	3.146
					H	0.228	0.080	0.414	0.442	2.263	1.200	3.135
					S	0.225	0.079	0.415	0.442	2.263	1.192	3.131
					R	0.216	0.076	0.419	0.442	2.263	1.166	3.118
11718	SILVER CHLORIDE	5.5850	132	300.0	V	0.233	0.083	0.412	0.440	2.273	1.216	3.138
					H	0.226	0.080	0.415	0.440	2.273	1.195	3.128
					S	0.223	0.079	0.416	0.440	2.273	1.188	3.124
					R	0.214	0.075	0.419	0.440	2.273	1.163	3.111
11719	SILVER(98.4)-ALUMINUM(1.6)	10.3700	207	294.0	V	0.919	0.338	0.359	1.082	0.924	1.806	3.845
					H	0.844	0.308	0.370	1.082	0.924	1.723	3.794
					S	0.792	0.287	0.378	1.082	0.924	1.665	3.759
					R	0.700	0.251	0.392	1.082	0.924	1.557	3.697
11720	SILVER(96.1)-ALUMINUM(3.9)	10.2100	207	294.0	V	0.911	0.336	0.355	1.045	0.957	1.814	3.824
					H	0.835	0.306	0.367	1.045	0.957	1.730	3.772
					S	0.783	0.285	0.375	1.045	0.957	1.670	3.735
					R	0.691	0.248	0.390	1.045	0.957	1.560	3.671
11721	SILVER(94.8)-ALUMINUM(5.2)	10.1200	207	294.0	V	0.919	0.340	0.352	1.038	0.963	1.833	3.839
					H	0.839	0.307	0.365	1.038	0.963	1.743	3.783
					S	0.781	0.284	0.375	1.038	0.963	1.675	3.742
					R	0.683	0.245	0.390	1.038	0.963	1.557	3.674
11722	SILVER(98.66)-CADMIUM(1.34)	10.4890	129	ROOM	V	0.912	0.337	0.352	1.026	0.975	1.793	3.751
					H	0.840	0.308	0.364	1.026	0.975	1.713	3.701
					S	0.792	0.289	0.371	1.026	0.975	1.659	3.668
					R	0.703	0.254	0.386	1.026	0.975	1.555	3.606
11723	SILVER(98.08)-CADMIUM(1.92)	10.4830	129	ROOM	V	0.908	0.336	0.351	1.014	0.986	1.790	3.734
					H	0.836	0.307	0.363	1.014	0.986	1.711	3.685
					S	0.789	0.288	0.370	1.014	0.986	1.657	3.652
					R	0.702	0.253	0.385	1.014	0.986	1.555	3.591
11724	SILVER(98.0)-GOLD(2.0)	10.6600	207	294.0	V	0.927	0.343	0.351	1.035	0.966	1.794	3.742
					H	0.854	0.313	0.362	1.035	0.966	1.715	3.692
					S	0.806	0.294	0.370	1.035	0.966	1.662	3.659
					R	0.717	0.259	0.385	1.035	0.966	1.559	3.599
11725	SILVER(96.0)-GOLD(4.0)	10.8500	207	294.0	V	0.936	0.347	0.349	1.032	0.969	1.788	3.711
					H	0.864	0.317	0.361	1.032	0.969	1.710	3.662
					S	0.816	0.298	0.368	1.032	0.969	1.658	3.630
					R	0.728	0.263	0.382	1.032	0.969	1.557	3.570

CUBIC

NAME	CODE	DENSITY	REF	TEMP(K)		E	G	SIGMA	K	BETA	VS	VP
SILVER(91.64)-INDIUM((8.36)	11726	10.3540	129	ROOM	V	0.879	0.325	0.351	0.982	1.018	1.772	3.698
					H	0.803	0.294	0.364	0.982	1.018	1.686	3.644
					S	0.749	0.273	0.373	0.982	1.018	1.623	3.605
					R	0.656	0.236	0.389	0.982	1.018	1.511	3.539
SILVER(96.93)-MAGNESIUM(3.07)	11727	10.2360	129	ROOM	V	0.906	0.336	0.349	0.998	1.002	1.812	3.759
					H	0.834	0.306	0.361	0.998	1.002	1.730	3.707
					S	0.786	0.287	0.369	0.998	1.002	1.674	3.673
					R	0.697	0.252	0.384	0.998	1.002	1.568	3.610
SILVER(92.67)-MAGNESIUM(7.33)	11728	9.8620	129	ROOM	V	0.888	0.330	0.346	0.964	1.038	1.829	3.772
					H	0.817	0.300	0.359	0.964	1.038	1.746	3.719
					S	0.769	0.281	0.367	0.964	1.038	1.688	3.684
					R	0.681	0.246	0.382	0.964	1.038	1.581	3.620
SILVER(50.0)-MAGNESIUM(50.0) (BETA')	11729	6.1110	52	77.0	V	0.932	0.366	0.274	0.686	1.458	2.447	4.383
					H	0.839	0.324	0.296	0.686	1.458	2.301	4.276
					S	0.769	0.293	0.313	0.686	1.458	2.188	4.196
					R	0.665	0.248	0.339	0.686	1.458	2.016	4.080
SILVER(50.0)-MAGNESIUM(50.0) (BETA')	11730	6.1040	52	100.0	V	0.926	0.363	0.274	0.683	1.464	2.439	4.373
					H	0.834	0.321	0.297	0.683	1.464	2.295	4.267
					S	0.765	0.291	0.313	0.683	1.464	2.184	4.189
					R	0.662	0.247	0.338	0.683	1.464	2.013	4.073
SILVER(50.0)-MAGNESIUM(50.0) (BETA')	11731	6.0890	52	150.0	V	0.912	0.358	0.275	0.676	1.479	2.423	4.351
					H	0.823	0.317	0.297	0.676	1.479	2.282	4.248
					S	0.757	0.288	0.313	0.676	1.479	2.175	4.172
					R	0.657	0.245	0.338	0.676	1.479	2.008	4.059
SILVER(50.0)-MAGNESIUM(50.0) (BETA')	11732	6.0700	52	200.0	V	0.898	0.352	0.276	0.669	1.495	2.407	4.330
					H	0.812	0.313	0.298	0.669	1.495	2.270	4.229
					S	0.749	0.285	0.313	0.669	1.495	2.167	4.157
					R	0.652	0.244	0.338	0.669	1.495	2.004	4.046
SILVER(50.0)-MAGNESIUM(50.0) (BETA')	11733	6.0500	52	250.0	V	0.884	0.346	0.278	0.662	1.511	2.391	4.308
					H	0.800	0.308	0.299	0.662	1.511	2.256	4.210
					S	0.740	0.282	0.314	0.662	1.511	2.157	4.141
					R	0.646	0.241	0.337	0.662	1.511	1.998	4.032
SILVER(50.0)-MAGNESIUM(50.0) (BETA')	11734	6.0330	52	300.0	V	0.870	0.340	0.279	0.655	1.527	2.375	4.287
					H	0.789	0.304	0.299	0.655	1.527	2.243	4.191
					S	0.732	0.278	0.314	0.655	1.527	2.148	4.124
					R	0.640	0.239	0.337	0.655	1.527	1.992	4.019
SILVER(50.0)-MAGNESIUM(50.0) (BETA')	11735	6.0130	52	350.0	V	0.856	0.334	0.280	0.648	1.544	2.358	4.265
					H	0.778	0.299	0.300	0.648	1.544	2.231	4.172
					S	0.723	0.275	0.314	0.648	1.544	2.140	4.108
					R	0.635	0.238	0.337	0.648	1.544	1.988	4.005
SILVER(50.0)-MAGNESIUM(50.0) (BETA')	11736	5.9930	52	400.0	V	0.842	0.328	0.281	0.641	1.560	2.341	4.243
					H	0.766	0.295	0.301	0.641	1.560	2.217	4.153
					S	0.715	0.272	0.314	0.641	1.560	2.131	4.092
					R	0.630	0.236	0.336	0.641	1.560	1.983	3.992
SILVER(50.0)-MAGNESIUM(50.0) (BETA')	11737	5.9750	52	450.0	V	0.828	0.323	0.282	0.634	1.577	2.324	4.221
					H	0.755	0.290	0.301	0.634	1.577	2.203	4.133
					S	0.706	0.269	0.314	0.634	1.577	2.120	4.075
					R	0.623	0.233	0.336	0.634	1.577	1.976	3.977
SILVER(50.0)-MAGNESIUM(50.0) (BETA')	11738	5.9550	52	500.0	V	0.814	0.317	0.284	0.627	1.595	2.307	4.198
					H	0.744	0.286	0.302	0.627	1.595	2.190	4.114
					S	0.698	0.265	0.314	0.627	1.595	2.111	4.058
					R	0.618	0.231	0.336	0.627	1.595	1.971	3.963
SILVER(50.0)-MAGNESIUM(50.0) (BETA')	11739	5.9360	52	550.0	V	0.800	0.311	0.285	0.620	1.613	2.289	4.175
					H	0.732	0.281	0.303	0.620	1.613	2.176	4.094
					S	0.689	0.262	0.315	0.620	1.613	2.101	4.041
					R	0.612	0.229	0.335	0.620	1.613	1.965	3.949
SILVER(93.78)-PALLADIUM(6.22)	11740	10.6060	129	ROOM	V	0.952	0.352	0.351	1.064	0.940	1.823	3.803
					H	0.878	0.322	0.363	1.064	0.940	1.743	3.753
					S	0.829	0.302	0.370	1.064	0.940	1.689	3.720
					R	0.737	0.266	0.385	1.064	0.940	1.585	3.658

CUBIC

CODE	NAME / DENSITY	REF	TEMP(K)		E	G	SIGMA	K	BETA	VS	VP
				V	0.899	0.332	0.353	1.018	0.982	1.785	3.742
	SILVER(96.83)-TIN(3.17)			H	0.824	0.302	0.365	1.018	0.982	1.701	3.690
11741	10.4340	129	ROOM	S	0.772	0.281	0.374	1.018	0.982	1.641	3.653
				R	0.679	0.245	0.389	1.018	0.982	1.531	3.589
				V	0.901	0.333	0.352	1.014	0.987	1.787	3.737
	SILVER(97.60)-ZINC(2.40)			H	0.828	0.303	0.364	1.014	0.987	1.705	3.686
11742	10.4390	129	ROOM	S	0.777	0.283	0.372	1.014	0.987	1.647	3.651
				R	0.687	0.247	0.387	1.014	0.987	1.540	3.588
				V	0.909	0.336	0.353	1.032	0.969	1.797	3.771
	SILVER(96.47)-ZINC(3.53)			H	0.836	0.306	0.365	1.032	0.969	1.715	3.720
11743	10.4080	129	ROOM	S	0.785	0.286	0.373	1.032	0.969	1.658	3.685
				R	0.695	0.250	0.388	1.032	0.969	1.551	3.622
				V	0.069	0.028	0.240	0.044	22.500	1.697	2.901
	SODIUM			H	0.058	0.023	0.281	0.044	22.500	1.531	2.775
11744	0.9712	128	ROOM	S	0.045	0.017	0.331	0.044	22.500	1.318	2.626
				R	0.034	0.013	0.371	0.044	22.500	1.137	2.510
				V	0.104	0.040	0.292	0.083	11.986	2.002	3.691
	SODIUM			H	0.087	0.033	0.326	0.083	11.986	1.807	3.554
11745	1.0080	4	90.0	S	0.065	0.024	0.371	0.083	11.986	1.532	3.377
				R	0.048	0.017	0.403	0.083	11.986	1.309	3.250
				V	0.078	0.032	0.221	0.047	21.352	1.802	3.013
	SODIUM			H	0.065	0.026	0.267	0.047	21.352	1.617	2.869
11746	0.9870	4	210.0	S	0.049	0.018	0.325	0.047	21.352	1.369	2.691
				R	0.037	0.014	0.368	0.047	21.352	1.172	2.564
				V	0.092	0.038	0.214	0.054	18.668	1.976	3.274
	SODIUM			H	0.075	0.030	0.267	0.054	18.668	1.746	3.096
11747	0.9710	4	ROOM	S	0.050	0.019	0.344	0.054	18.668	1.384	2.841
				R	0.036	0.013	0.388	0.054	18.668	1.154	2.700
				V	0.072	0.027	0.317	0.066	15.129	1.682	3.253
	SODIUM			H	0.061	0.023	0.347	0.066	15.129	1.525	3.148
11748	0.9710	67	299.0	S	0.046	0.016	0.385	0.066	15.129	1.302	3.011
				R	0.034	0.012	0.414	0.066	15.129	1.116	2.910
				V	0.096	0.037	0.281	0.073	13.717	1.924	3.486
	SODIUM			H	0.079	0.030	0.319	0.073	13.717	1.724	3.344
11749	1.0100	76	78.0	S	0.056	0.020	0.372	0.073	13.717	1.424	3.150
				R	0.041	0.015	0.406	0.073	13.717	1.203	3.025
				V	0.092	0.036	0.285	0.071	13.999	1.887	3.443
	SODIUM			H	0.076	0.029	0.322	0.071	13.999	1.693	3.305
11750	1.0060	76	115.0	S	0.054	0.020	0.374	0.071	13.999	1.400	3.117
				R	0.040	0.014	0.407	0.071	13.999	1.183	2.995
				V	0.088	0.034	0.289	0.070	14.361	1.851	3.397
	SODIUM			H	0.073	0.028	0.325	0.070	14.361	1.663	3.264
11751	0.9990	76	155.0	S	0.052	0.019	0.374	0.070	14.361	1.383	3.085
				R	0.039	0.014	0.408	0.070	14.361	1.172	2.967
				V	0.084	0.033	0.294	0.068	14.663	1.811	3.355
	SODIUM			H	0.070	0.026	0.329	0.068	14.663	1.630	3.229
11752	0.9910	76	195.0	S	0.051	0.018	0.376	0.068	14.663	1.362	3.059
				R	0.037	0.013	0.409	0.068	14.663	1.156	2.943
				V	0.433	0.171	0.266	0.309	3.240	2.261	4.008
	SODIUM BROMATE			H	0.430	0.169	0.268	0.309	3.240	2.253	4.001
11753	3.3390	4	ROOM	S	0.429	0.169	0.268	0.309	3.240	2.252	4.000
				R	0.426	0.168	0.270	0.309	3.240	2.243	3.994
				V	0.411	0.161	0.278	0.309	3.236	2.194	3.959
	SODIUM BROMATE			H	0.410	0.160	0.279	0.309	3.236	2.191	3.957
11754	3.3390	4	ROOM	S	0.410	0.160	0.279	0.309	3.236	2.191	3.957
				R	0.409	0.160	0.280	0.309	3.236	2.187	3.954
				V	0.291	0.116	0.250	0.194	5.163	1.905	3.299
	SODIUM BROMIDE			H	0.286	0.114	0.254	0.194	5.164	1.888	3.286
11755	3.2030	4	ROOM	S	0.285	0.114	0.254	0.194	5.164	1.885	3.284
				R	0.281	0.112	0.258	0.194	5.163	1.868	3.271

CUBIC

CODE	NAME DENSITY	REF	TEMP(K)		E	G	SIGMA	K	BETA	VS	VP
				V	0.296	0.117	0.266	0.211	4.747	1.910	3.384
SODIUM BROMIDE				H	0.292	0.115	0.269	0.211	4.747	1.896	3.373
11756	3.2000	4	ROOM	S	0.291	0.115	0.269	0.211	4.747	1.894	3.371
				R	0.288	0.113	0.272	0.211	4.747	1.879	3.360
				V	0.219	0.086	0.270	0.159	6.293	1.642	2.925
SODIUM BROMIDE				H	0.216	0.085	0.273	0.159	6.293	1.629	2.915
11757	3.2030	4	ROOM	S	0.216	0.085	0.274	0.159	6.293	1.627	2.914
				R	0.213	0.083	0.277	0.159	6.293	1.613	2.903
				V	0.296	0.118	0.249	0.196	5.100	1.923	3.324
SODIUM BROMIDE				H	0.293	0.117	0.251	0.196	5.100	1.913	3.317
11758	3.2030	44	ROOM	S	0.293	0.117	0.251	0.196	5.100	1.911	3.316
				R	0.290	0.116	0.253	0.196	5.100	1.900	3.307
				V	0.303	0.121	0.250	0.202	4.948	1.947	3.371
SODIUM BROMIDE				H	0.299	0.119	0.253	0.202	4.948	1.930	3.358
11759	3.2030	145	298.0	S	0.298	0.119	0.254	0.202	4.948	1.927	3.356
				R	0.294	0.117	0.258	0.202	4.948	1.910	3.343
				V	0.349	0.140	0.242	0.226	4.430	2.064	3.538
SODIUM BROMIDE				H	0.339	0.136	0.250	0.226	4.430	2.027	3.510
11760	3.2990	152	4.0	S	0.336	0.134	0.252	0.226	4.430	2.018	3.503
				R	0.327	0.130	0.259	0.226	4.430	1.983	3.477
				V	0.340	0.137	0.243	0.221	4.531	2.040	3.502
SODIUM BROMIDE				H	0.331	0.132	0.250	0.221	4.531	2.008	3.477
11761	3.2860	152	80.0	S	0.329	0.131	0.252	0.221	4.531	2.000	3.471
				R	0.321	0.127	0.258	0.221	4.531	1.969	3.448
				V	0.310	0.124	0.248	0.205	4.884	1.962	3.389
SODIUM BROMIDE				H	0.304	0.122	0.252	0.205	4.884	1.942	3.373
11762	3.2250	152	240.0	S	0.303	0.121	0.253	0.205	4.884	1.938	3.370
				R	0.298	0.119	0.257	0.205	4.884	1.918	3.354
				V	0.308	0.123	0.248	0.204	4.907	1.957	3.381
SODIUM BROMIDE				H	0.303	0.121	0.252	0.204	4.907	1.937	3.366
11763	3.2210	152	250.0	S	0.302	0.120	0.253	0.204	4.907	1.933	3.363
				R	0.297	0.118	0.257	0.204	4.907	1.914	3.348
				V	0.306	0.123	0.249	0.203	4.929	1.952	3.374
SODIUM BROMIDE				H	0.301	0.120	0.253	0.203	4.929	1.932	3.359
11764	3.2180	152	260.0	S	0.300	0.120	0.253	0.203	4.929	1.929	3.356
				R	0.295	0.117	0.257	0.203	4.929	1.910	3.342
				V	0.304	0.122	0.249	0.202	4.952	1.946	3.367
SODIUM BROMIDE				H	0.299	0.119	0.253	0.202	4.952	1.928	3.352
11765	3.2140	152	270.0	S	0.298	0.119	0.254	0.202	4.952	1.924	3.350
				R	0.294	0.117	0.258	0.202	4.952	1.906	3.336
				V	0.302	0.121	0.249	0.201	4.976	1.941	3.359
SODIUM BROMIDE				H	0.298	0.119	0.253	0.201	4.976	1.923	3.345
11766	3.2100	152	280.0	S	0.297	0.118	0.254	0.201	4.976	1.920	3.343
				R	0.292	0.116	0.258	0.201	4.976	1.902	3.330
				V	0.300	0.120	0.250	0.200	5.000	1.936	3.352
SODIUM BROMIDE				H	0.296	0.118	0.253	0.200	5.000	1.918	3.338
11767	3.2060	152	290.0	S	0.295	0.118	0.254	0.200	5.000	1.915	3.336
				R	0.291	0.116	0.258	0.200	5.000	1.898	3.323
				V	0.298	0.119	0.250	0.199	5.023	1.930	3.344
SODIUM BROMIDE				H	0.294	0.117	0.254	0.199	5.023	1.913	3.331
11768	3.2020	152	300.0	S	0.293	0.117	0.255	0.199	5.023	1.910	3.329
				R	0.289	0.115	0.258	0.199	5.023	1.894	3.317
				V	0.327	0.237	-0.311	0.067	14.880	3.086	3.923
SODIUM CHLORATE				H	0.303	0.202	-0.251	0.067	14.880	2.851	3.679
11769	2.4900	128	ROOM	S	0.289	0.185	-0.218	0.067	14.880	2.726	3.550
				R	0.274	0.167	-0.180	0.067	14.880	2.591	3.413
				V	0.362	0.142	0.279	0.273	3.663	2.385	4.307
SODIUM CHLORATE				H	0.357	0.139	0.282	0.273	3.663	2.364	4.291
11770	2.4900	4	ROOM	S	0.356	0.139	0.283	0.273	3.663	2.360	4.288
				R	0.350	0.136	0.286	0.273	3.663	2.338	4.273

CUBIC

CODE	NAME DENSITY	REF	TEMP(K)		E	G	SIGMA	K	BETA	VS	VP
				V	0.360	0.142	0.270	0.260	3.841	2.386	4.248
SODIUM CHLORATE				H	0.354	0.139	0.273	0.260	3.841	2.363	4.231
11771	2.4900	4	ROOM	S	0.353	0.139	0.274	0.260	3.841	2.359	4.228
				R	0.347	0.136	0.278	0.260	3.841	2.335	4.210
				V	0.356	0.140	0.268	0.256	3.911	2.373	4.216
SODIUM CHLORATE				H	0.350	0.138	0.272	0.256	3.911	2.352	4.200
11772	2.4900	4	ROOM	S	0.349	0.137	0.272	0.256	3.911	2.348	4.198
				R	0.344	0.135	0.276	0.256	3.911	2.327	4.182
				V	0.355	0.140	0.271	0.259	3.866	2.370	4.231
SODIUM CHLORATE				H	0.349	0.137	0.275	0.259	3.866	2.349	4.215
11773	2.4850	265	ROOM	S	0.348	0.137	0.275	0.259	3.866	2.345	4.212
				R	0.343	0.134	0.279	0.259	3.866	2.323	4.196
				V	0.360	0.144	0.249	0.239	4.182	2.581	4.464
SODIUM CHLORIDE				H	0.358	0.143	0.251	0.239	4.182	2.570	4.455
11774	2.1650	128	ROOM	S	0.357	0.143	0.251	0.239	4.182	2.568	4.454
				R	0.355	0.142	0.253	0.239	4.182	2.557	4.446
				V	0.373	0.150	0.243	0.242	4.140	2.633	4.517
SODIUM CHLORIDE				H	0.369	0.148	0.246	0.242	4.140	2.614	4.502
11775	2.1650	128	ROOM	S	0.368	0.148	0.246	0.242	4.140	2.611	4.500
				R	0.364	0.146	0.249	0.242	4.140	2.593	4.486
				V	0.370	0.148	0.248	0.245	4.082	2.616	4.521
SODIUM CHLORIDE				H	0.365	0.146	0.251	0.245	4.082	2.597	4.507
11776	2.1650	129	ROOM	S	0.365	0.146	0.252	0.245	4.082	2.594	4.504
				R	0.360	0.144	0.255	0.245	4.082	2.575	4.490
				V	0.375	0.150	0.250	0.250	3.995	2.634	4.562
SODIUM CHLORIDE				H	0.371	0.148	0.253	0.250	3.995	2.613	4.546
11777	2.1650	129	ROOM	S	0.370	0.147	0.254	0.250	3.995	2.610	4.544
				R	0.365	0.145	0.257	0.250	3.995	2.590	4.528
				V	0.370	0.148	0.247	0.244	4.104	2.617	4.515
SODIUM CHLORIDE				H	0.366	0.146	0.250	0.244	4.104	2.599	4.501
11778	2.1650	129	ROOM	S	0.365	0.146	0.250	0.244	4.104	2.596	4.499
				R	0.361	0.144	0.253	0.244	4.104	2.577	4.485
				V	0.371	0.149	0.249	0.246	4.065	2.620	4.529
SODIUM CHLORIDE				H	0.367	0.147	0.251	0.246	4.065	2.603	4.516
11779	2.1650	129	290.0	S	0.366	0.146	0.252	0.246	4.065	2.600	4.514
				R	0.362	0.144	0.255	0.246	4.065	2.583	4.501
				V	0.378	0.152	0.246	0.248	4.032	2.648	4.561
SODIUM CHLORIDE				H	0.374	0.150	0.249	0.248	4.032	2.628	4.546
11780	2.1650	4	ROOM	S	0.373	0.149	0.249	0.248	4.032	2.625	4.544
				R	0.368	0.147	0.252	0.248	4.032	2.606	4.529
				V	0.373	0.148	0.257	0.256	3.900	2.618	4.580
SODIUM CHLORIDE				H	0.369	0.147	0.260	0.256	3.900	2.602	4.568
11781	2.1650	44	ROOM	S	0.369	0.146	0.260	0.256	3.900	2.600	4.567
				R	0.365	0.145	0.263	0.256	3.900	2.584	4.554
				V	0.425	0.172	0.238	0.270	3.704	2.737	4.751
SODIUM CHLORIDE				H	0.414	0.166	0.244	0.270	3.704	2.743	4.717
11782	2.2108	218	80.0	S	0.411	0.165	0.246	0.270	3.704	2.733	4.708
				R	0.401	0.160	0.252	0.270	3.704	2.691	4.676
				V	0.423	0.171	0.239	0.270	3.708	2.781	4.745
SODIUM CHLORIDE				H	0.412	0.166	0.245	0.270	3.708	2.738	4.712
11783	2.2088	218	90.0	S	0.410	0.164	0.247	0.270	3.708	2.728	4.704
				R	0.400	0.160	0.253	0.270	3.708	2.687	4.673
				V	0.412	0.166	0.242	0.267	3.750	2.747	4.711
SODIUM CHLORIDE				H	0.403	0.161	0.248	0.267	3.750	2.710	4.682
11784	2.1984	218	140.0	S	0.401	0.161	0.249	0.267	3.750	2.702	4.676
				R	0.392	0.156	0.255	0.267	3.750	2.667	4.649
				V	0.410	0.165	0.243	0.266	3.759	2.738	4.702
SODIUM CHLORIDE				H	0.401	0.160	0.249	0.266	3.759	2.703	4.675
11785	2.1962	218	150.0	S	0.399	0.160	0.250	0.266	3.759	2.696	4.669
				R	0.391	0.156	0.255	0.266	3.759	2.662	4.643

CUBIC

CODE	NAME DENSITY	REF	TEMP(K)		E	G	SIGMA	K	BETA	VS	VP
				V	0.407	0.164	0.244	0.265	3.769	2.730	4.694
SODIUM CHLORIDE				H	0.399	0.160	0.250	0.265	3.769	2.697	4.668
11786	2.1941	218	160.0	S	0.397	0.159	0.251	0.265	3.769	2.690	4.662
				R	0.389	0.155	0.256	0.265	3.769	2.657	4.637
				V	0.405	0.163	0.245	0.264	3.788	2.724	4.684
SODIUM CHLORIDE				H	0.397	0.159	0.250	0.264	3.788	2.691	4.658
11787	2.1919	218	170.0	S	0.395	0.158	0.251	0.264	3.788	2.684	4.653
				R	0.387	0.154	0.256	0.264	3.788	2.652	4.629
				V	0.402	0.161	0.246	0.263	3.797	2.715	4.675
SODIUM CHLORIDE				H	0.394	0.158	0.250	0.263	3.797	2.684	4.651
11788	2.1896	218	180.0	S	0.393	0.157	0.251	0.263	3.797	2.678	4.646
				R	0.385	0.153	0.256	0.263	3.797	2.647	4.623
				V	0.400	0.160	0.247	0.263	3.802	2.708	4.669
SODIUM CHLORIDE				H	0.393	0.157	0.251	0.263	3.802	2.678	4.646
11789	2.1874	218	190.0	S	0.391	0.156	0.252	0.263	3.802	2.672	4.641
				R	0.384	0.153	0.257	0.263	3.802	2.642	4.619
				V	0.398	0.159	0.247	0.262	3.822	2.701	4.659
SODIUM CHLORIDE				H	0.390	0.156	0.251	0.262	3.822	2.672	4.636
11790	2.1851	218	200.0	S	0.389	0.155	0.252	0.262	3.822	2.666	4.632
				R	0.382	0.152	0.257	0.262	3.822	2.638	4.610
				V	0.395	0.158	0.248	0.261	3.831	2.693	4.650
SODIUM CHLORIDE				H	0.388	0.155	0.252	0.261	3.831	2.665	4.629
11791	2.1829	218	210.0	S	0.387	0.154	0.253	0.261	3.831	2.660	4.625
				R	0.380	0.151	0.257	0.261	3.831	2.632	4.604
				V	0.392	0.157	0.248	0.260	3.851	2.685	4.639
SODIUM CHLORIDE				H	0.386	0.154	0.252	0.260	3.851	2.658	4.618
11792	2.1806	218	220.0	S	0.385	0.154	0.253	0.260	3.851	2.653	4.615
				R	0.378	0.150	0.257	0.260	3.851	2.627	4.594
				V	0.390	0.156	0.249	0.259	3.856	2.678	4.634
SODIUM CHLORIDE				H	0.384	0.153	0.253	0.259	3.856	2.652	4.614
11793	2.1782	218	230.0	S	0.383	0.153	0.254	0.259	3.856	2.648	4.610
				R	0.377	0.150	0.258	0.259	3.856	2.622	4.590
				V	0.388	0.155	0.249	0.258	3.876	2.671	4.623
SODIUM CHLORIDE				H	0.382	0.152	0.253	0.258	3.876	2.646	4.604
11794	2.1759	218	240.0	S	0.381	0.152	0.254	0.258	3.876	2.642	4.600
				R	0.375	0.149	0.258	0.258	3.876	2.617	4.581
				V	0.385	0.154	0.250	0.257	3.886	2.663	4.614
SODIUM CHLORIDE				H	0.380	0.151	0.254	0.257	3.886	2.639	4.596
11795	2.1735	218	250.0	S	0.379	0.151	0.255	0.257	3.886	2.635	4.593
				R	0.373	0.148	0.258	0.257	3.886	2.611	4.575
				V	0.383	0.153	0.251	0.256	3.906	2.656	4.604
SODIUM CHLORIDE				H	0.378	0.151	0.254	0.256	3.906	2.633	4.586
11796	2.1711	218	260.0	S	0.377	0.150	0.255	0.256	3.906	2.629	4.583
				R	0.371	0.147	0.258	0.256	3.906	2.606	4.566
				V	0.380	0.152	0.252	0.255	3.916	2.646	4.595
SODIUM CHLORIDE				H	0.375	0.149	0.255	0.255	3.916	2.625	4.578
11797	2.1687	80	270.0	S	0.374	0.149	0.256	0.255	3.916	2.621	4.575
				R	0.369	0.147	0.259	0.255	3.916	2.600	4.559
				V	0.378	0.151	0.252	0.254	3.932	2.640	4.586
SODIUM CHLORIDE				H	0.373	0.149	0.255	0.254	3.932	2.619	4.570
11798	2.1664	80	280.0	S	0.372	0.148	0.256	0.254	3.932	2.616	4.568
				R	0.367	0.146	0.259	0.254	3.932	2.595	4.552
				V	0.376	0.150	0.253	0.254	3.942	2.632	4.578
SODIUM CHLORIDE				H	0.371	0.148	0.256	0.254	3.942	2.612	4.563
11799	2.1639	80	290.0	S	0.370	0.147	0.257	0.254	3.942	2.609	4.560
				R	0.365	0.145	0.260	0.254	3.942	2.589	4.545
				V	0.374	0.149	0.253	0.253	3.958	2.626	4.570
SODIUM CHLORIDE				H	0.369	0.147	0.256	0.253	3.958	2.607	4.556
11800	2.1613	80	300.0	S	0.368	0.147	0.257	0.253	3.958	2.604	4.553
				R	0.364	0.144	0.260	0.253	3.958	2.585	4.539

CUBIC

CODE	NAME / DENSITY	REF	TEMP(K)		E	G	SIGMA	K	BETA	VS	VP
				V	0.371	0.148	0.255	0.252	3.968	2.618	4.562
	SODIUM CHLORIDE			H	0.367	0.146	0.257	0.252	3.968	2.600	4.548
11801	2.1586	80	310.0	S	0.366	0.146	0.258	0.252	3.968	2.597	4.546
				R	0.362	0.144	0.261	0.252	3.968	2.579	4.532
				V	0.369	0.147	0.255	0.251	3.989	2.610	4.551
	SODIUM CHLORIDE			H	0.365	0.145	0.258	0.251	3.989	2.593	4.538
11802	2.1558	80	320.0	S	0.364	0.145	0.258	0.251	3.989	2.590	4.536
				R	0.360	0.143	0.261	0.251	3.989	2.573	4.523
				V	0.366	0.146	0.255	0.249	4.011	2.603	4.541
	SODIUM CHLORIDE			H	0.362	0.144	0.258	0.249	4.011	2.587	4.528
11803	2.1528	80	330.0	S	0.362	0.144	0.258	0.249	4.011	2.585	4.526
				R	0.358	0.142	0.261	0.249	4.011	2.568	4.514
				V	0.364	0.145	0.256	0.249	4.016	2.596	4.535
	SODIUM CHLORIDE			H	0.360	0.143	0.259	0.249	4.016	2.581	4.524
11804	2.1497	80	340.0	S	0.360	0.143	0.259	0.249	4.016	2.579	4.522
				R	0.356	0.141	0.261	0.249	4.016	2.563	4.510
				V	0.362	0.144	0.257	0.248	4.038	2.589	4.525
	SODIUM CHLORIDE			H	0.358	0.142	0.259	0.248	4.038	2.574	4.514
11805	2.1464	80	350.0	S	0.358	0.142	0.259	0.248	4.038	2.572	4.512
				R	0.354	0.140	0.262	0.248	4.038	2.558	4.501
				V	0.359	0.143	0.258	0.247	4.049	2.580	4.517
	SODIUM CHLORIDE			H	0.356	0.141	0.260	0.247	4.049	2.567	4.507
11806	2.1431	80	360.0	S	0.355	0.141	0.260	0.247	4.049	2.565	4.505
				R	0.352	0.140	0.262	0.247	4.049	2.551	4.495
				V	0.356	0.142	0.258	0.246	4.071	2.573	4.506
	SODIUM CHLORIDE			H	0.353	0.140	0.260	0.246	4.071	2.560	4.497
11807	2.1396	80	370.0	S	0.353	0.140	0.260	0.246	4.071	2.559	4.496
				R	0.350	0.139	0.262	0.246	4.071	2.546	4.486
				V	0.354	0.141	0.259	0.245	4.076	2.566	4.502
	SODIUM CHLORIDE			H	0.352	0.139	0.261	0.245	4.076	2.554	4.493
11808	2.1360	80	380.0	S	0.351	0.139	0.261	0.245	4.076	2.553	4.492
				R	0.348	0.138	0.263	0.245	4.076	2.541	4.483
				V	0.352	0.140	0.260	0.244	4.098	2.559	4.492
	SODIUM CHLORIDE			H	0.349	0.138	0.261	0.244	4.098	2.548	4.484
11809	2.1322	80	390.0	S	0.349	0.138	0.262	0.244	4.098	2.547	4.483
				R	0.346	0.137	0.263	0.244	4.098	2.536	4.474
				V	0.349	0.138	0.261	0.243	4.110	2.550	4.484
	SODIUM CHLORIDE			H	0.347	0.137	0.262	0.243	4.110	2.540	4.476
11810	2.1283	80	400.0	S	0.347	0.137	0.263	0.243	4.110	2.539	4.475
				R	0.344	0.136	0.264	0.243	4.110	2.529	4.468
				V	0.347	0.138	0.261	0.242	4.127	2.546	4.478
	SODIUM CHLORIDE			H	0.345	0.137	0.263	0.242	4.127	2.536	4.470
11811	2.1243	80	410.0	S	0.345	0.137	0.263	0.242	4.127	2.535	4.469
				R	0.343	0.135	0.264	0.242	4.127	2.525	4.462
				V	0.345	0.137	0.262	0.241	4.149	2.538	4.467
	SODIUM CHLORIDE			H	0.343	0.136	0.263	0.241	4.149	2.530	4.461
11812	2.1202	80	420.0	S	0.342	0.136	0.263	0.241	4.149	2.529	4.460
				R	0.340	0.135	0.265	0.241	4.149	2.520	4.453
				V	0.342	0.135	0.263	0.240	4.161	2.529	4.460
	SODIUM CHLORIDE			H	0.340	0.135	0.264	0.240	4.161	2.522	4.454
11813	2.1159	80	430.0	S	0.340	0.134	0.264	0.240	4.161	2.521	4.453
				R	0.338	0.134	0.266	0.240	4.161	2.513	4.447
				V	0.339	0.134	0.263	0.239	4.184	2.522	4.450
	SODIUM CHLORIDE			H	0.338	0.134	0.264	0.239	4.184	2.515	4.444
11814	2.1115	80	440.0	S	0.338	0.133	0.265	0.239	4.184	2.514	4.444
				R	0.336	0.133	0.266	0.239	4.184	2.507	4.438
				V	0.337	0.133	0.265	0.238	4.196	2.513	4.442
	SODIUM CHLORIDE			H	0.335	0.132	0.266	0.238	4.196	2.507	4.438
11815	2.1069	80	450.0	S	0.335	0.132	0.266	0.238	4.196	2.506	4.437
				R	0.334	0.132	0.267	0.238	4.196	2.500	4.432

CUBIC

CODE	NAME	DENSITY	REF	TEMP(K)		E	G	SIGMA	K	BETA	VS	VP
11816	SODIUM CHLORIDE	2.1023	80	460.0	V	0.335	0.132	0.265	0.237	4.213	2.508	4.436
					H	0.333	0.132	0.266	0.237	4.213	2.502	4.431
					S	0.333	0.132	0.266	0.237	4.213	2.502	4.431
					R	0.332	0.131	0.267	0.237	4.213	2.495	4.426
11817	SODIUM CHLORIDE	2.0975	80	470.0	V	0.332	0.131	0.266	0.237	4.225	2.499	4.429
					H	0.331	0.130	0.267	0.237	4.225	2.494	4.425
					S	0.331	0.130	0.267	0.237	4.225	2.494	4.424
					R	0.329	0.130	0.268	0.237	4.225	2.488	4.420
11818	SODIUM CHLORIDE	2.0926	80	480.0	V	0.329	0.130	0.267	0.235	4.249	2.492	4.419
					H	0.328	0.129	0.268	0.235	4.249	2.488	4.415
					S	0.328	0.129	0.268	0.235	4.249	2.487	4.415
					R	0.327	0.129	0.268	0.235	4.249	2.482	4.412
11819	SODIUM CHLORIDE (P= 0KB)	2.1620	149	298.0	V	0.375	0.151	0.245	0.245	4.077	2.639	4.542
					H	0.371	0.148	0.248	0.245	4.077	2.620	4.528
					S	0.370	0.148	0.249	0.245	4.077	2.617	4.525
					R	0.365	0.146	0.252	0.245	4.077	2.598	4.510
11820	SODIUM CHLORIDE (P= 2KB)	2.1804	149	298.0	V	0.388	0.155	0.250	0.259	3.864	2.667	4.621
					H	0.382	0.152	0.254	0.259	3.864	2.642	4.602
					S	0.381	0.152	0.255	0.259	3.864	2.637	4.598
					R	0.375	0.149	0.259	0.259	3.864	2.613	4.580
11821	SODIUM CHLORIDE (P= 4KB)	2.1982	149	298.0	V	0.398	0.159	0.252	0.267	3.742	2.688	4.668
					H	0.390	0.155	0.256	0.267	3.742	2.659	4.646
					S	0.389	0.155	0.257	0.267	3.742	2.653	4.641
					R	0.382	0.151	0.262	0.267	3.742	2.624	4.619
11822	SODIUM CHLORIDE (P= 6KB)	2.2153	149	298.0	V	0.410	0.163	0.256	0.279	3.581	2.714	4.735
					H	0.401	0.159	0.261	0.279	3.581	2.678	4.709
					S	0.399	0.158	0.262	0.279	3.581	2.671	4.703
					R	0.390	0.154	0.267	0.279	3.581	2.636	4.677
11823	SODIUM CHLORIDE (P= 8KB)	2.2319	149	298.0	V	0.421	0.167	0.260	0.293	3.410	2.737	4.809
					H	0.411	0.162	0.266	0.293	3.410	2.696	4.778
					S	0.408	0.161	0.268	0.293	3.410	2.686	4.771
					R	0.398	0.156	0.274	0.293	3.410	2.646	4.740
11824	SODIUM CHLORIDE (P=10KB)	2.2477	149	298.0	V	0.433	0.171	0.265	0.306	3.264	2.759	4.877
					H	0.421	0.165	0.271	0.306	3.264	2.713	4.842
					S	0.417	0.164	0.273	0.306	3.264	2.701	4.833
					R	0.406	0.158	0.279	0.306	3.264	2.655	4.799
11825	SODIUM CHLORIDE	2.1891	23	195.0	V	0.396	0.159	0.244	0.258	3.877	2.696	4.634
					H	0.389	0.156	0.249	0.258	3.877	2.667	4.612
					S	0.388	0.155	0.250	0.258	3.877	2.662	4.608
					R	0.381	0.152	0.254	0.258	3.877	2.634	4.586
11826	SODIUM CHLORIDE	2.1644	23	295.0	V	0.373	0.149	0.249	0.247	4.047	2.625	4.539
					H	0.368	0.147	0.252	0.247	4.047	2.607	4.525
					S	0.367	0.147	0.252	0.247	4.047	2.604	4.523
					R	0.363	0.145	0.255	0.247	4.047	2.585	4.509
11827	SODIUM CHLORIDE	2.1640	251	300.0	V	0.375	0.150	0.252	0.252	3.963	2.631	4.570
					H	0.371	0.148	0.255	0.252	3.963	2.612	4.556
					S	0.370	0.147	0.256	0.252	3.963	2.610	4.554
					R	0.366	0.145	0.258	0.252	3.963	2.591	4.540
11828	SODIUM CHLORIDE	2.2159	97	4.2	V	0.429	0.173	0.239	0.274	3.651	2.795	4.772
					H	0.417	0.168	0.246	0.274	3.651	2.750	4.737
					S	0.415	0.166	0.248	0.274	3.651	2.739	4.729
					R	0.404	0.161	0.254	0.274	3.651	2.696	4.696
11829	SODIUM CHLORIDE	2.2123	97	77.3	V	0.424	0.171	0.240	0.271	3.684	2.780	4.751
					H	0.413	0.166	0.246	0.271	3.684	2.738	4.719
					S	0.411	0.165	0.248	0.271	3.684	2.728	4.711
					R	0.401	0.160	0.254	0.271	3.684	2.688	4.680
11830	SODIUM CHLORIDE	2.1630	234	298.0	V	0.376	0.150	0.250	0.251	3.988	2.637	4.568
					H	0.372	0.148	0.253	0.251	3.988	2.619	4.554
					S	0.371	0.148	0.253	0.251	3.988	2.616	4.552
					R	0.367	0.146	0.256	0.251	3.988	2.598	4.538

CODE	NAME DENSITY	REF	TEMP(K)		E	G	SIGMA	K	BETA	VS	VP
				V	0.376	0.150	0.252	0.253	3.958	2.636	4.576
SODIUM	CHLORIDE			H	0.372	0.148	0.255	0.253	3.958	2.617	4.561
11831	2.1644	113	295.0	S	0.371	0.148	0.255	0.253	3.958	2.614	4.559
				R	0.367	0.146	0.258	0.253	3.958	2.595	4.544
				V	0.425	0.172	0.234	0.266	3.760	2.786	4.727
SODIUM	CHLORIDE			H	0.413	0.167	0.241	0.266	3.760	2.741	4.692
11832	2.2170	152	4.0	S	0.411	0.165	0.243	0.266	3.760	2.730	4.684
				R	0.400	0.160	0.249	0.266	3.760	2.688	4.651
				V	0.420	0.170	0.235	0.264	3.782	2.772	4.711
SODIUM	CHLORIDE			H	0.410	0.165	0.242	0.264	3.782	2.730	4.679
11833	2.2120	152	80.0	S	0.407	0.164	0.243	0.264	3.782	2.721	4.672
				R	0.397	0.159	0.249	0.264	3.782	2.682	4.641
				V	0.384	0.154	0.250	0.256	3.913	2.656	4.597
SODIUM	CHLORIDE			H	0.379	0.151	0.253	0.256	3.913	2.635	4.582
11834	2.1780	152	230.0	S	0.378	0.151	0.253	0.256	3.913	2.632	4.579
				R	0.373	0.149	0.257	0.256	3.913	2.611	4.563
				V	0.382	0.153	0.250	0.255	3.925	2.648	4.589
SODIUM	CHLORIDE			H	0.377	0.150	0.253	0.255	3.925	2.629	4.574
11835	2.1760	152	240.0	S	0.376	0.150	0.254	0.255	3.925	2.625	4.571
				R	0.371	0.148	0.257	0.255	3.925	2.606	4.556
				V	0.379	0.152	0.251	0.254	3.938	2.641	4.580
SODIUM	CHLORIDE			H	0.375	0.149	0.254	0.254	3.938	2.622	4.566
11836	2.1740	152	250.0	S	0.374	0.149	0.254	0.254	3.938	2.619	4.563
				R	0.370	0.147	0.257	0.254	3.938	2.600	4.549
				V	0.377	0.151	0.252	0.253	3.950	2.635	4.573
SODIUM	CHLORIDE			H	0.373	0.149	0.255	0.253	3.950	2.616	4.559
11837	2.1710	152	260.0	S	0.372	0.148	0.255	0.253	3.950	2.613	4.557
				R	0.368	0.146	0.258	0.253	3.950	2.594	4.543
				V	0.375	0.150	0.252	0.252	3.964	2.628	4.565
SODIUM	CHLORIDE			H	0.371	0.148	0.255	0.252	3.964	2.610	4.551
11838	2.1690	152	270.0	S	0.370	0.147	0.255	0.252	3.964	2.607	4.549
				R	0.366	0.145	0.258	0.252	3.964	2.589	4.535
				V	0.373	0.149	0.253	0.251	3.977	2.622	4.558
SODIUM	CHLORIDE			H	0.369	0.147	0.256	0.251	3.977	2.604	4.544
11839	2.1660	152	280.0	S	0.368	0.147	0.256	0.251	3.977	2.601	4.542
				R	0.364	0.145	0.259	0.251	3.977	2.584	4.529
				V	0.371	0.148	0.253	0.251	3.991	2.615	4.549
SODIUM	CHLORIDE			H	0.367	0.146	0.256	0.251	3.991	2.598	4.536
11840	2.1640	152	290.0	S	0.366	0.146	0.256	0.251	3.991	2.595	4.534
				R	0.362	0.144	0.259	0.251	3.991	2.578	4.521
				V	0.369	0.147	0.254	0.250	4.004	2.609	4.542
SODIUM	CHLORIDE			H	0.365	0.145	0.256	0.250	4.004	2.592	4.529
11841	2.1620	152	300.0	S	0.364	0.145	0.257	0.250	4.004	2.590	4.527
				R	0.360	0.143	0.259	0.250	4.004	2.573	4.514
				V	0.375	0.150	0.252	0.252	3.963	2.631	4.570
SODIUM	CHLORIDE			H	0.371	0.148	0.255	0.252	3.963	2.612	4.556
11842	2.1640	250	ROOM	S	0.370	0.147	0.256	0.252	3.963	2.610	4.554
				R	0.366	0.145	0.258	0.252	3.963	2.591	4.540
				V	0.428	0.172	0.240	0.274	3.647	2.821	4.824
SODIUM	CHLORIDE			H	0.416	0.167	0.247	0.274	3.647	2.775	4.787
11843	2.1669	201	4.2	S	0.413	0.165	0.249	0.274	3.647	2.763	4.779
				R	0.402	0.160	0.256	0.274	3.647	2.719	4.744
				V	0.375	0.150	0.249	0.249	4.011	2.636	4.559
SODIUM	CHLORIDE			H	0.371	0.148	0.252	0.249	4.011	2.617	4.544
11844	2.1640	79	ROOM	S	0.370	0.148	0.252	0.249	4.011	2.614	4.542
				R	0.366	0.146	0.255	0.249	4.011	2.595	4.528
				V	0.771	0.311	0.240	0.494	2.024	3.338	5.706
SODIUM	FLUORIDE			H	0.767	0.309	0.241	0.494	2.024	3.326	5.697
11845	2.7900	4	ROOM	S	0.766	0.309	0.242	0.494	2.024	3.325	5.696
				R	0.762	0.306	0.243	0.494	2.024	3.314	5.688

CUBIC

CODE	NAME / DENSITY	REF	TEMP(K)		E	G	SIGMA	K	BETA	VS	VP
				V	0.774	0.314	0.234	0.486	2.059	3.344	5.676
	SODIUM FLUORIDE			H	0.769	0.311	0.236	0.486	2.059	3.331	5.666
11846	2.8050	4	ROOM	S	0.769	0.311	0.236	0.486	2.059	3.329	5.665
				R	0.764	0.308	0.238	0.486	2.059	3.316	5.655
				V	0.777	0.316	0.231	0.482	2.075	3.356	5.675
	SODIUM FLUORIDE			H	0.773	0.313	0.233	0.482	2.075	3.343	5.665
11847	2.8040	185	300.0	S	0.772	0.313	0.233	0.482	2.075	3.341	5.663
				R	0.767	0.311	0.235	0.482	2.075	3.328	5.653
				V	0.862	0.353	0.222	0.517	1.933	3.514	5.881
	SODIUM FLUORIDE			H	0.850	0.347	0.226	0.517	1.933	3.483	5.856
11848	2.8560	260	0.0	S	0.848	0.345	0.227	0.517	1.933	3.478	5.852
				R	0.836	0.340	0.231	0.517	1.933	3.448	5.828
				V	0.777	0.316	0.231	0.481	2.080	3.342	5.647
	SODIUM FLUORIDE			H	0.772	0.313	0.232	0.481	2.080	3.329	5.637
11849	2.8277	260	ROOM	S	0.772	0.313	0.232	0.481	2.080	3.328	5.636
				R	0.767	0.311	0.234	0.481	2.080	3.315	5.626
				V	0.846	0.345	0.226	0.514	1.944	3.479	5.847
	SODIUM FLUORIDE			H	0.835	0.339	0.230	0.514	1.944	3.450	5.823
11850	2.8510	152	4.0	S	0.833	0.338	0.230	0.514	1.944	3.445	5.820
				R	0.821	0.333	0.234	0.514	1.944	3.417	5.797
				V	0.838	0.342	0.227	0.511	1.956	3.465	5.828
	SODIUM FLUORIDE			H	0.828	0.337	0.230	0.511	1.956	3.438	5.807
11851	2.8470	152	80.0	S	0.826	0.336	0.231	0.511	1.956	3.434	5.803
				R	0.816	0.331	0.234	0.511	1.956	3.408	5.783
				V	0.784	0.318	0.234	0.491	2.038	3.364	5.706
	SODIUM FLUORIDE			H	0.779	0.315	0.236	0.491	2.038	3.350	5.695
11852	2.8080	152	260.0	S	0.778	0.315	0.236	0.491	2.038	3.348	5.694
				R	0.772	0.312	0.238	0.491	2.038	3.334	5.682
				V	0.780	0.316	0.234	0.489	2.044	3.357	5.699
	SODIUM FLUORIDE			H	0.775	0.314	0.236	0.489	2.044	3.344	5.688
11853	2.8050	152	270.0	S	0.775	0.313	0.236	0.489	2.044	3.342	5.687
				R	0.769	0.311	0.238	0.489	2.044	3.328	5.676
				V	0.777	0.315	0.235	0.488	2.050	3.350	5.690
	SODIUM FLUORIDE			H	0.772	0.312	0.236	0.488	2.050	3.337	5.680
11854	2.8020	152	280.0	S	0.771	0.312	0.237	0.488	2.050	3.336	5.679
				R	0.766	0.309	0.238	0.488	2.050	3.323	5.669
				V	0.773	0.313	0.235	0.486	2.056	3.343	5.682
	SODIUM FLUORIDE			H	0.768	0.310	0.237	0.486	2.056	3.331	5.672
11855	2.7990	152	290.0	S	0.768	0.310	0.237	0.486	2.056	3.329	5.671
				R	0.763	0.308	0.239	0.486	2.056	3.317	5.661
				V	0.769	0.311	0.236	0.485	2.062	3.335	5.673
	SODIUM FLUORIDE			H	0.764	0.309	0.237	0.485	2.062	3.324	5.663
11856	2.7960	152	300.0	S	0.764	0.309	0.237	0.485	2.062	3.322	5.662
				R	0.759	0.306	0.239	0.485	2.062	3.311	5.653
				V	0.219	0.086	0.274	0.161	6.205	1.530	2.740
	SODIUM IODIDE			H	0.216	0.084	0.277	0.161	6.205	1.517	2.731
11857	3.6714	4	ROOM	S	0.215	0.084	0.277	0.161	6.205	1.515	2.729
				R	0.212	0.083	0.281	0.161	6.205	1.502	2.720
				V	0.219	0.087	0.256	0.149	6.696	1.540	2.689
	SODIUM IODIDE			H	0.216	0.086	0.259	0.149	6.696	1.528	2.680
11858	3.6710	87	298.0	S	0.215	0.086	0.260	0.149	6.696	1.526	2.678
				R	0.213	0.084	0.263	0.149	6.696	1.514	2.669
				V	0.241	0.096	0.260	0.168	5.969	1.614	2.835
	SODIUM IODIDE			H	0.235	0.093	0.266	0.168	5.969	1.591	2.817
11859	3.6714	63	180.0	S	0.234	0.092	0.267	0.168	5.969	1.586	2.813
				R	0.228	0.090	0.273	0.168	5.969	1.563	2.796
				V	0.237	0.094	0.265	0.168	5.959	1.597	2.823
	SODIUM IODIDE			H	0.232	0.091	0.270	0.168	5.959	1.576	2.808
11860	3.6714	63	200.0	S	0.231	0.091	0.271	0.168	5.959	1.572	2.804
				R	0.226	0.088	0.276	0.168	5.959	1.552	2.789

CODE	NAME / DENSITY	REF	TEMP(K)		E	G	SIGMA	K	BETA	VS	VP
				V	0.233	0.092	0.268	0.167	5.976	1.581	2.809
SODIUM IODIDE				H	0.228	0.090	0.273	0.167	5.976	1.563	2.795
11861	3.6714	63	220.0	S	0.227	0.089	0.274	0.167	5.976	1.559	2.793
				R	0.223	0.087	0.278	0.167	5.976	1.541	2.779
				V	0.229	0.090	0.270	0.166	6.018	1.567	2.793
SODIUM IODIDE				H	0.225	0.088	0.274	0.166	6.018	1.550	2.780
11862	3.6714	63	240.0	S	0.224	0.088	0.275	0.166	6.018	1.547	2.778
				R	0.220	0.086	0.279	0.166	6.018	1.531	2.766
				V	0.225	0.089	0.271	0.164	6.086	1.554	2.774
SODIUM IODIDE				H	0.222	0.087	0.275	0.164	6.086	1.539	2.763
11863	3.6714	63	260.0	S	0.221	0.087	0.276	0.164	6.086	1.537	2.761
				R	0.218	0.085	0.279	0.164	6.086	1.522	2.750
				V	0.222	0.087	0.271	0.162	6.180	1.544	2.754
SODIUM IODIDE				H	0.219	0.086	0.274	0.162	6.180	1.530	2.744
11864	3.6714	63	280.0	S	0.218	0.086	0.275	0.162	6.180	1.528	2.742
				R	0.215	0.084	0.278	0.162	6.180	1.514	2.732
				V	0.219	0.086	0.270	0.159	6.293	1.534	2.732
SODIUM IODIDE				H	0.216	0.085	0.273	0.159	6.293	1.521	2.723
11865	3.6714	63	300.0	S	0.216	0.085	0.274	0.159	6.293	1.519	2.721
				R	0.213	0.083	0.277	0.159	6.293	1.507	2.712
				V	0.266	0.106	0.252	0.179	5.601	1.679	2.917
SODIUM IODIDE				H	0.256	0.102	0.261	0.179	5.601	1.644	2.889
11866	3.7620	59	4.2	S	0.254	0.100	0.263	0.179	5.601	1.634	2.882
				R	0.245	0.096	0.271	0.179	5.601	1.600	2.856
				V	0.266	0.106	0.252	0.178	5.603	1.679	2.916
SODIUM IODIDE				H	0.256	0.102	0.261	0.178	5.603	1.644	2.889
11867	3.7620	59	10.0	S	0.254	0.100	0.263	0.178	5.603	1.634	2.881
				R	0.245	0.096	0.271	0.178	5.603	1.600	2.856
				V	0.265	0.106	0.252	0.178	5.611	1.677	2.914
SODIUM IODIDE				H	0.256	0.101	0.261	0.178	5.611	1.642	2.887
11868	3.7619	59	20.0	S	0.253	0.100	0.263	0.178	5.611	1.633	2.879
				R	0.245	0.096	0.271	0.178	5.611	1.599	2.854
				V	0.265	0.106	0.252	0.178	5.622	1.676	2.911
SODIUM IODIDE				H	0.255	0.101	0.261	0.178	5.622	1.641	2.885
11869	3.7613	59	30.0	S	0.253	0.100	0.263	0.178	5.622	1.632	2.877
				R	0.244	0.096	0.271	0.178	5.622	1.599	2.852
				V	0.264	0.105	0.253	0.178	5.627	1.672	2.908
SODIUM IODIDE				H	0.255	0.101	0.261	0.178	5.627	1.639	2.882
11870	3.7606	59	40.0	S	0.252	0.100	0.263	0.178	5.627	1.629	2.875
				R	0.244	0.096	0.271	0.178	5.627	1.597	2.851
				V	0.262	0.105	0.253	0.177	5.638	1.669	2.904
SODIUM IODIDE				H	0.254	0.101	0.262	0.177	5.638	1.636	2.878
11871	3.7595	59	50.0	S	0.251	0.099	0.264	0.177	5.638	1.627	2.871
				R	0.243	0.096	0.271	0.177	5.638	1.595	2.848
				V	0.224	0.088	0.271	0.163	6.150	1.545	2.755
SODIUM IODIDE				H	0.221	0.087	0.274	0.163	6.150	1.533	2.746
11872	3.6884	59	260.0	S	0.220	0.086	0.274	0.163	6.150	1.531	2.745
				R	0.217	0.085	0.277	0.163	6.150	1.518	2.735
				V	0.221	0.087	0.272	0.161	6.193	1.537	2.746
SODIUM IODIDE				H	0.218	0.086	0.275	0.161	6.193	1.526	2.737
11873	3.6789	59	280.0	S	0.218	0.085	0.275	0.161	6.193	1.524	2.736
				R	0.215	0.084	0.278	0.161	6.193	1.513	2.728
				V	0.219	0.086	0.274	0.161	6.210	1.530	2.740
SODIUM IODIDE				H	0.216	0.085	0.276	0.161	6.210	1.519	2.733
11874	3.6689	59	300.0	S	0.216	0.085	0.277	0.161	6.210	1.518	2.731
				R	0.213	0.083	0.279	0.161	6.210	1.507	2.723
				V	3.101	1.245	0.245	2.026	0.494	5.857	10.078
SPINEL (MAGNESIA-3.5ALUMINA)				H	2.963	1.179	0.256	2.026	0.494	5.700	9.957
11875	3.6300	262	ROOM	S	2.911	1.154	0.261	2.026	0.494	5.639	9.911
				R	2.758	1.083	0.273	2.026	0.493	5.462	9.778

CUBIC

CODE	NAME / DENSITY	REF	TEMP(K)		E	G	SIGMA	K	BETA	VS	VP
				V	2.925	1.170	0.250	1.950	0.513	5.716	9.900
SPINEL (MAGNESIUM ALUMINATE)				H	2.762	1.092	0.264	1.950	0.513	5.523	9.753
11876	3.5810	153	ROOM	S	2.687	1.058	0.270	1.950	0.513	5.434	9.687
				R	2.504	0.974	0.286	1.950	0.513	5.214	9.524
				V	3.078	1.235	0.246	2.020	0.495	5.842	10.066
SPINEL (MAGNESIA-2.6ALUMINA) (P=1BAR)				H	2.940	1.169	0.257	2.020	0.495	5.683	9.944
11877	3.6193	227	298.0	S	2.886	1.144	0.262	2.020	0.495	5.621	9.896
				R	2.732	1.072	0.275	2.020	0.495	5.442	9.762
				V	3.083	1.237	0.247	2.028	0.493	5.842	10.075
SPINEL (MAGNESIA-2.6ALUMINA) (P=2KB)				H	2.945	1.170	0.258	2.028	0.493	5.684	9.953
11878	3.6229	227	298.0	S	2.891	1.145	0.262	2.028	0.493	5.622	9.906
				R	2.737	1.073	0.275	2.029	0.493	5.442	9.772
				V	0.932	0.361	0.292	0.746	1.341	2.890	5.328
STRONTIUM FLUORIDE				H	0.929	0.359	0.292	0.746	1.341	2.883	5.323
11879	4.3210	105	4.2	S	0.928	0.359	0.292	0.746	1.341	2.883	5.323
				R	0.924	0.357	0.293	0.746	1.341	2.876	5.318
				V	0.933	0.361	0.292	0.746	1.341	2.891	5.330
STRONTIUM FLUORIDE				H	0.929	0.359	0.292	0.746	1.341	2.884	5.325
11880	4.3210	105	20.0	S	0.929	0.359	0.292	0.746	1.341	2.884	5.324
				R	0.925	0.357	0.293	0.746	1.341	2.876	5.319
				V	0.932	0.361	0.291	0.744	1.343	2.890	5.326
STRONTIUM FLUORIDE				H	0.928	0.359	0.292	0.744	1.343	2.883	5.321
11881	4.3200	105	40.0	S	0.928	0.359	0.292	0.744	1.343	2.882	5.320
				R	0.924	0.357	0.293	0.744	1.343	2.875	5.315
				V	0.931	0.360	0.291	0.743	1.346	2.889	5.322
STRONTIUM FLUORIDE				H	0.927	0.359	0.292	0.743	1.346	2.882	5.316
11882	4.3200	105	60.0	S	0.927	0.359	0.292	0.743	1.346	2.881	5.316
				R	0.922	0.357	0.293	0.743	1.346	2.874	5.311
				V	0.929	0.360	0.291	0.740	1.351	2.887	5.315
STRONTIUM FLUORIDE				H	0.925	0.358	0.292	0.740	1.351	2.880	5.310
11883	4.3190	105	80.0	S	0.925	0.358	0.292	0.740	1.351	2.879	5.310
				R	0.921	0.356	0.293	0.740	1.351	2.872	5.304
				V	0.927	0.359	0.290	0.737	1.356	2.885	5.309
STRONTIUM FLUORIDE				H	0.923	0.357	0.291	0.737	1.356	2.878	5.303
11884	4.3160	105	100.0	S	0.923	0.357	0.291	0.737	1.356	2.877	5.303
				R	0.919	0.355	0.292	0.737	1.356	2.870	5.297
				V	0.924	0.358	0.290	0.733	1.364	2.882	5.299
STRONTIUM FLUORIDE				H	0.920	0.356	0.291	0.733	1.364	2.874	5.293
11885	4.3140	105	120.0	S	0.920	0.356	0.291	0.733	1.364	2.874	5.293
				R	0.915	0.354	0.292	0.733	1.364	2.866	5.287
				V	0.922	0.357	0.290	0.730	1.370	2.879	5.290
STRONTIUM FLUORIDE				H	0.917	0.355	0.291	0.730	1.370	2.871	5.284
11886	4.3110	105	140.0	S	0.917	0.355	0.291	0.730	1.370	2.871	5.283
				R	0.912	0.353	0.292	0.730	1.370	2.862	5.277
				V	0.918	0.356	0.289	0.726	1.378	2.876	5.279
STRONTIUM FLUORIDE				H	0.914	0.354	0.290	0.726	1.378	2.867	5.273
11887	4.3080	105	160.0	S	0.914	0.354	0.290	0.726	1.378	2.867	5.273
				R	0.909	0.352	0.291	0.726	1.378	2.858	5.266
				V	0.915	0.355	0.289	0.722	1.385	2.872	5.270
STRONTIUM FLUORIDE				H	0.911	0.353	0.290	0.722	1.385	2.864	5.264
11888	4.3040	105	180.0	S	0.910	0.353	0.290	0.722	1.385	2.863	5.264
				R	0.905	0.351	0.291	0.722	1.385	2.854	5.257
				V	0.913	0.354	0.288	0.718	1.392	2.870	5.262
STRONTIUM FLUORIDE				H	0.908	0.352	0.289	0.718	1.392	2.861	5.256
11889	4.3000	105	200.0	S	0.907	0.352	0.289	0.718	1.392	2.861	5.255
				R	0.902	0.350	0.291	0.718	1.392	2.851	5.248
				V	0.909	0.353	0.288	0.714	1.400	2.867	5.252
STRONTIUM FLUORIDE				H	0.905	0.351	0.289	0.714	1.400	2.859	5.246
11890	4.2950	105	220.0	S	0.904	0.351	0.289	0.714	1.400	2.858	5.245
				R	0.899	0.348	0.290	0.714	1.400	2.848	5.239

CUBIC

CODE	NAME DENSITY	REF	TEMP(K)		E	G	SIGMA	K	BETA	VS	VP
				V	0.903	0.350	0.290	0.717	1.395	2.856	5.252
STRONTIUM	FLUORIDE			H	0.898	0.348	0.291	0.717	1.395	2.848	5.246
11891	4.2900	105	240.0	S	0.898	0.348	0.291	0.717	1.395	2.847	5.245
				R	0.893	0.346	0.292	0.717	1.395	2.839	5.239
				V	0.903	0.351	0.287	0.706	1.416	2.861	5.234
STRONTIUM	FLUORIDE			H	0.898	0.349	0.288	0.706	1.416	2.852	5.228
11892	4.2860	105	260.0	S	0.898	0.348	0.288	0.706	1.416	2.851	5.227
				R	0.892	0.346	0.289	0.706	1.416	2.841	5.220
				V	0.900	0.350	0.287	0.702	1.424	2.858	5.225
STRONTIUM	FLUORIDE			H	0.894	0.347	0.288	0.702	1.424	2.848	5.218
11893	4.2810	105	280.0	S	0.894	0.347	0.288	0.702	1.424	2.847	5.217
				R	0.888	0.345	0.289	0.702	1.424	2.837	5.210
				V	0.897	0.349	0.286	0.699	1.431	2.855	5.216
STRONTIUM	FLUORIDE			H	0.891	0.346	0.287	0.699	1.431	2.845	5.208
11894	4.2770	105	300.0	S	0.891	0.346	0.288	0.699	1.431	2.844	5.208
				R	0.885	0.343	0.289	0.699	1.431	2.833	5.200
				V	0.361	0.139	0.302	0.303	3.300	2.154	4.042
STRONTIUM	NITRATE			H	0.360	0.138	0.302	0.303	3.300	2.152	4.040
11895	2.9860	129	ROOM	S	0.360	0.138	0.302	0.303	3.300	2.152	4.040
				R	0.359	0.138	0.302	0.303	3.300	2.150	4.038
				V	0.327	0.122	0.338	0.337	2.971	2.021	4.087
STRONTIUM	NITRATE			H	0.309	0.115	0.347	0.337	2.971	1.959	4.047
11896	2.9890	124	293.0	S	0.301	0.111	0.351	0.337	2.971	1.929	4.027
				R	0.279	0.102	0.362	0.337	2.971	1.850	3.978
				V	1.542	0.639	0.207	0.877	1.141	3.589	5.903
STRONTIUM	OXIDE			H	1.538	0.637	0.208	0.877	1.141	3.584	5.899
11897	4.9600	139	80.0	S	1.538	0.637	0.208	0.877	1.141	3.583	5.899
				R	1.535	0.635	0.208	0.877	1.141	3.578	5.894
				V	1.537	0.637	0.207	0.874	1.144	3.582	5.893
STRONTIUM	OXIDE			H	1.534	0.635	0.208	0.874	1.144	3.577	5.889
11898	4.9630	139	100.0	S	1.534	0.635	0.208	0.874	1.144	3.577	5.889
				R	1.531	0.633	0.208	0.874	1.144	3.572	5.885
				V	1.532	0.634	0.207	0.871	1.148	3.575	5.881
STRONTIUM	OXIDE			H	1.529	0.633	0.208	0.871	1.148	3.570	5.878
11899	4.9650	139	120.0	S	1.528	0.633	0.208	0.871	1.148	3.570	5.877
				R	1.526	0.631	0.208	0.871	1.148	3.566	5.874
				V	1.523	0.631	0.208	0.868	1.152	3.563	5.866
STRONTIUM	OXIDE			H	1.521	0.630	0.208	0.868	1.152	3.560	5.863
11900	4.9680	139	140.0	S	1.521	0.629	0.208	0.868	1.152	3.559	5.863
				R	1.518	0.628	0.209	0.868	1.152	3.556	5.860
				V	1.514	0.627	0.208	0.865	1.157	3.551	5.849
STRONTIUM	OXIDE			H	1.512	0.626	0.208	0.865	1.157	3.548	5.846
11901	4.9710	139	160.0	S	1.512	0.626	0.209	0.865	1.157	3.548	5.846
				R	1.510	0.625	0.209	0.865	1.157	3.545	5.844
				V	1.506	0.623	0.208	0.860	1.162	3.539	5.831
STRONTIUM	OXIDE			H	1.504	0.622	0.209	0.860	1.162	3.537	5.829
11902	4.9740	139	180.0	S	1.504	0.622	0.209	0.860	1.162	3.537	5.829
				R	1.502	0.621	0.209	0.860	1.162	3.534	5.827
				V	1.495	0.618	0.209	0.855	1.169	3.525	5.810
STRONTIUM	OXIDE			H	1.493	0.618	0.209	0.855	1.169	3.523	5.808
11903	4.9760	139	200.0	S	1.493	0.618	0.209	0.855	1.169	3.523	5.808
				R	1.492	0.617	0.209	0.855	1.169	3.521	5.807
				V	1.482	0.613	0.209	0.850	1.176	3.508	5.787
STRONTIUM	OXIDE			H	1.481	0.612	0.210	0.850	1.176	3.507	5.786
11904	4.9790	139	220.0	S	1.481	0.612	0.210	0.850	1.176	3.507	5.786
				R	1.481	0.612	0.210	0.850	1.176	3.506	5.785
				V	1.469	0.607	0.210	0.844	1.185	3.491	5.761
STRONTIUM	OXIDE			H	1.469	0.607	0.210	0.844	1.185	3.490	5.761
11905	4.9820	139	240.0	S	1.469	0.607	0.210	0.844	1.185	3.490	5.761
				R	1.468	0.607	0.210	0.844	1.185	3.490	5.760

CUBIC

CODE	NAME DENSITY	REF	TEMP(K)		E	G	SIGMA	K	BETA	VS	VP
				V	1.455	0.601	0.211	0.838	1.194	3.473	5.735
	STRONTIUM OXIDE			H	1.455	0.601	0.211	0.838	1.194	3.472	5.734
11906	4.9840	139	260.0	S	1.455	0.601	0.211	0.838	1.194	3.472	5.734
				R	1.455	0.601	0.211	0.838	1.194	3.472	5.734
				V	1.440	0.594	0.211	0.831	1.203	3.452	5.706
	STRONTIUM OXIDE			H	1.440	0.594	0.211	0.831	1.203	3.452	5.706
11907	4.9870	139	280.0	S	1.440	0.594	0.211	0.831	1.203	3.452	5.706
				R	1.440	0.594	0.211	0.831	1.203	3.452	5.706
				V	1.423	0.587	0.212	0.824	1.214	3.430	5.674
	STRONTIUM OXIDE			H	1.423	0.587	0.212	0.824	1.214	3.430	5.674
11908	4.9900	139	300.0	S	1.423	0.587	0.212	0.824	1.214	3.430	5.674
				R	1.423	0.587	0.212	0.824	1.214	3.430	5.674
				V	6.093	3.222	-0.054	1.832	0.546	7.931	10.937
	STRONTIUM TITANATE			H	5.565	2.801	-0.006	1.832	0.546	7.394	10.423
11909	5.1230	202	ROOM	S	5.205	2.536	0.026	1.832	0.546	7.035	10.087
				R	4.708	2.197	0.072	1.832	0.546	6.548	9.640
				V	1.893	0.707	0.339	1.961	0.510	2.059	4.172
	TANTALUM			H	1.861	0.693	0.342	1.961	0.510	2.039	4.160
11910	16.6780	38	300.0	S	1.854	0.690	0.342	1.961	0.510	2.035	4.157
				R	1.815	0.674	0.346	1.961	0.510	2.011	4.141
				V	1.891	0.706	0.340	1.965	0.509	2.057	4.174
	TANTALUM			H	1.858	0.692	0.342	1.965	0.509	2.037	4.161
11911	16.6780	38	300.0	S	1.851	0.689	0.343	1.965	0.509	2.033	4.159
				R	1.812	0.673	0.346	1.965	0.509	2.008	4.143
				V	1.971	0.740	0.331	1.942	0.515	2.105	4.187
	TANTALUM			H	1.933	0.724	0.334	1.942	0.515	2.082	4.171
11912	16.7129	89	0.0	S	1.923	0.720	0.335	1.942	0.515	2.076	4.167
				R	1.877	0.701	0.339	1.942	0.515	2.048	4.149
				V	1.970	0.740	0.331	1.943	0.515	2.104	4.186
	TANTALUM			H	1.931	0.724	0.334	1.943	0.515	2.081	4.171
11913	16.7129	89	20.0	S	1.922	0.720	0.335	1.943	0.515	2.075	4.167
				R	1.876	0.701	0.339	1.943	0.515	2.047	4.149
				V	1.966	0.738	0.331	1.944	0.514	2.102	4.187
	TANTALUM			H	1.928	0.722	0.335	1.944	0.514	2.079	4.172
11914	16.7079	89	40.0	S	1.919	0.719	0.335	1.944	0.514	2.074	4.168
				R	1.873	0.699	0.339	1.944	0.514	2.046	4.149
				V	1.962	0.736	0.332	1.946	0.514	2.100	4.187
	TANTALUM			H	1.924	0.721	0.335	1.946	0.514	2.077	4.172
11915	16.7029	89	60.0	S	1.915	0.717	0.336	1.946	0.514	2.072	4.168
				R	1.870	0.698	0.340	1.946	0.514	2.044	4.150
				V	1.956	0.734	0.333	1.948	0.513	2.097	4.187
	TANTALUM			H	1.919	0.718	0.336	1.948	0.513	2.074	4.172
11916	16.6979	89	80.0	S	1.910	0.714	0.337	1.948	0.513	2.068	4.168
				R	1.865	0.696	0.340	1.948	0.513	2.041	4.150
				V	1.949	0.731	0.333	1.949	0.513	2.093	4.185
	TANTALUM			H	1.912	0.715	0.336	1.949	0.513	2.070	4.170
11917	16.6928	89	100.0	S	1.904	0.712	0.337	1.949	0.513	2.065	4.166
				R	1.859	0.693	0.341	1.949	0.513	2.038	4.149
				V	1.941	0.728	0.334	1.947	0.514	2.088	4.181
	TANTALUM			H	1.905	0.712	0.337	1.947	0.514	2.066	4.167
11918	16.6879	89	120.0	S	1.896	0.709	0.338	1.947	0.514	2.061	4.163
				R	1.852	0.690	0.341	1.947	0.514	2.034	4.145
				V	1.934	0.725	0.334	1.945	0.514	2.084	4.178
	TANTALUM			H	1.897	0.709	0.337	1.945	0.514	2.062	4.163
11919	16.6829	89	140.0	S	1.889	0.706	0.338	1.945	0.514	2.057	4.160
				R	1.845	0.688	0.342	1.945	0.514	2.030	4.142
				V	1.926	0.721	0.335	1.943	0.515	2.080	4.174
	TANTALUM			H	1.890	0.706	0.338	1.943	0.515	2.058	4.159
11920	16.6778	89	160.0	S	1.882	0.703	0.339	1.943	0.515	2.053	4.156
				R	1.839	0.685	0.342	1.943	0.515	2.026	4.138

CUBIC

CODE	NAME DENSITY	REF	TEMP(K)		E	G	SIGMA	K	BETA	VS	VP
				V	1.918	0.718	0.335	1.941	0.515	2.075	4.169
TANTALUM				H	1.882	0.703	0.338	1.941	0.515	2.054	4.155
11921	16.6728	89	180.0	S	1.874	0.700	0.339	1.941	0.515	2.049	4.151
				R	1.831	0.682	0.343	1.941	0.515	2.022	4.134
				V	1.909	0.715	0.336	1.938	0.516	2.071	4.164
TANTALUM				H	1.874	0.700	0.339	1.938	0.516	2.049	4.150
11922	16.6679	89	200.0	S	1.866	0.697	0.340	1.938	0.516	2.044	4.147
				R	1.824	0.679	0.343	1.938	0.516	2.018	4.130
				V	1.901	0.711	0.336	1.934	0.517	2.066	4.160
TANTALUM				H	1.866	0.697	0.339	1.934	0.517	2.045	4.146
11923	16.6577	89	220.0	S	1.858	0.693	0.340	1.934	0.517	2.040	4.143
				R	1.816	0.676	0.344	1.934	0.517	2.014	4.126
				V	1.892	0.708	0.337	1.931	0.518	2.062	4.155
TANTALUM				H	1.858	0.693	0.340	1.931	0.518	2.041	4.141
11924	16.6529	89	240.0	S	1.850	0.690	0.340	1.931	0.518	2.036	4.138
				R	1.808	0.673	0.344	1.931	0.518	2.010	4.121
				V	1.884	0.704	0.337	1.927	0.519	2.057	4.149
TANTALUM				H	1.850	0.690	0.340	1.927	0.519	2.036	4.136
11925	16.6478	89	260.0	S	1.842	0.687	0.341	1.927	0.519	2.031	4.132
				R	1.801	0.670	0.344	1.927	0.519	2.006	4.116
				V	1.876	0.701	0.337	1.923	0.520	2.053	4.144
TANTALUM				H	1.842	0.687	0.340	1.923	0.520	2.032	4.130
11926	16.6413	89	280.0	S	1.834	0.684	0.341	1.923	0.520	2.027	4.127
				R	1.793	0.667	0.345	1.923	0.520	2.002	4.111
				V	1.867	0.698	0.338	1.919	0.521	2.048	4.139
TANTALUM				H	1.834	0.684	0.341	1.919	0.521	2.028	4.126
11927	16.6330	89	300.0	S	1.826	0.681	0.341	1.919	0.521	2.023	4.123
				R	1.786	0.664	0.345	1.919	0.521	1.998	4.106
				V	1.886	0.707	0.334	1.897	0.527	2.062	4.133
TANTALUM				H	1.854	0.693	0.337	1.897	0.527	2.042	4.120
11928	16.6260	241	298.0	S	1.847	0.690	0.338	1.897	0.527	2.038	4.117
				R	1.808	0.674	0.341	1.897	0.527	2.014	4.101
				V	1.881	0.705	0.335	1.895	0.528	2.059	4.130
TANTALUM				H	1.849	0.691	0.337	1.895	0.528	2.040	4.117
11929	16.6209	241	313.0	S	1.842	0.688	0.338	1.895	0.528	2.035	4.114
				R	1.804	0.672	0.341	1.895	0.528	2.011	4.098
				V	1.875	0.702	0.335	1.893	0.528	2.056	4.126
TANTALUM				H	1.843	0.689	0.338	1.893	0.528	2.036	4.113
11930	16.6158	241	333.0	S	1.836	0.686	0.338	1.893	0.528	2.032	4.111
				R	1.798	0.670	0.342	1.893	0.528	2.008	4.095
				V	1.868	0.700	0.335	1.890	0.529	2.052	4.122
TANTALUM				H	1.837	0.687	0.338	1.890	0.529	2.033	4.109
11931	16.6110	241	353.0	S	1.830	0.684	0.339	1.890	0.529	2.029	4.107
				R	1.793	0.668	0.342	1.890	0.529	2.005	4.091
				V	1.863	0.697	0.336	1.889	0.530	2.049	4.120
TANTALUM				H	1.831	0.684	0.338	1.889	0.530	2.030	4.108
11932	16.6008	241	373.0	S	1.825	0.681	0.339	1.889	0.530	2.026	4.105
				R	1.787	0.666	0.342	1.889	0.529	2.003	4.089
				V	1.856	0.695	0.336	1.886	0.530	2.046	4.117
TANTALUM				H	1.825	0.682	0.339	1.886	0.530	2.027	4.104
11933	16.5960	241	393.0	S	1.819	0.679	0.339	1.886	0.530	2.023	4.101
				R	1.781	0.663	0.343	1.886	0.530	1.999	4.086
				V	1.850	0.692	0.336	1.884	0.531	2.042	4.113
TANTALUM				H	1.819	0.679	0.339	1.884	0.531	2.023	4.101
11934	16.5909	241	413.0	S	1.813	0.677	0.340	1.884	0.531	2.019	4.098
				R	1.776	0.661	0.343	1.884	0.531	1.996	4.083
				V	1.844	0.690	0.337	1.882	0.531	2.039	4.110
TANTALUM				H	1.814	0.677	0.339	1.882	0.531	2.021	4.098
11935	16.5861	241	433.0	S	1.807	0.674	0.340	1.882	0.531	2.016	4.095
				R	1.770	0.659	0.343	1.882	0.531	1.993	4.080

CUBIC

CODE	NAME DENSITY	REF	TEMP(K)		E	G	SIGMA	K	BETA	VS	VP
				V	1.839	0.688	0.337	1.880	0.532	2.037	4.108
TANTALUM				H	1.808	0.675	0.340	1.880	0.532	2.018	4.095
11936	16.5761	241	453.0	S	1.802	0.672	0.340	1.880	0.532	2.014	4.092
				R	1.765	0.657	0.344	1.880	0.532	1.991	4.077
				V	1.833	0.685	0.337	1.878	0.533	2.033	4.104
TANTALUM				H	1.803	0.673	0.340	1.878	0.532	2.015	4.092
11937	16.5711	241	473.0	S	1.796	0.670	0.341	1.878	0.532	2.011	4.089
				R	1.760	0.655	0.344	1.878	0.532	1.988	4.074
				V	1.827	0.683	0.338	1.875	0.533	2.030	4.100
TANTALUM				H	1.797	0.670	0.340	1.875	0.533	2.012	4.088
11938	16.5661	241	493.0	S	1.790	0.668	0.341	1.875	0.533	2.008	4.085
				R	1.754	0.653	0.344	1.875	0.533	1.985	4.071
				V	1.821	0.681	0.338	1.872	0.534	2.028	4.097
TANTALUM				H	1.791	0.668	0.340	1.872	0.534	2.009	4.085
11939	16.5562	241	513.0	S	1.785	0.665	0.341	1.872	0.534	2.005	4.082
				R	1.749	0.650	0.344	1.872	0.534	1.982	4.067
				V	1.816	0.678	0.338	1.871	0.535	2.025	4.095
TANTALUM				H	1.786	0.666	0.341	1.871	0.535	2.006	4.082
11940	16.5512	241	533.0	S	1.780	0.663	0.341	1.871	0.535	2.002	4.080
				R	1.744	0.648	0.345	1.871	0.535	1.979	4.065
				V	1.810	0.676	0.339	1.868	0.535	2.022	4.092
TANTALUM				H	1.780	0.664	0.341	1.868	0.535	2.003	4.079
11941	16.5463	241	553.0	S	1.774	0.661	0.342	1.868	0.535	1.999	4.077
				R	1.738	0.646	0.345	1.868	0.535	1.976	4.062
				V	1.805	0.674	0.339	1.866	0.536	2.018	4.088
TANTALUM				H	1.775	0.662	0.341	1.866	0.536	2.000	4.076
11942	16.5413	241	573.0	S	1.769	0.659	0.342	1.866	0.536	1.996	4.074
				R	1.733	0.644	0.345	1.866	0.536	1.973	4.059
				V	1.799	0.672	0.339	1.863	0.537	2.016	4.085
TANTALUM				H	1.770	0.659	0.342	1.863	0.537	1.997	4.073
11943	16.5314	241	593.0	S	1.763	0.657	0.342	1.863	0.537	1.993	4.070
				R	1.727	0.642	0.345	1.863	0.537	1.971	4.056
				V	1.794	0.670	0.339	1.862	0.537	2.013	4.083
TANTALUM				H	1.764	0.657	0.342	1.862	0.537	1.994	4.071
11944	16.5265	241	613.0	S	1.758	0.655	0.343	1.862	0.537	1.990	4.068
				R	1.722	0.640	0.346	1.862	0.537	1.968	4.053
				V	1.788	0.667	0.340	1.859	0.538	2.010	4.079
TANTALUM				H	1.759	0.655	0.342	1.859	0.538	1.991	4.067
11945	16.5216	241	633.0	S	1.752	0.652	0.343	1.859	0.538	1.987	4.064
				R	1.717	0.638	0.346	1.859	0.538	1.965	4.050
				V	1.783	0.665	0.340	1.857	0.539	2.007	4.077
TANTALUM				H	1.754	0.653	0.343	1.857	0.539	1.989	4.064
11946	16.5117	241	653.0	S	1.747	0.650	0.343	1.857	0.539	1.985	4.062
				R	1.712	0.636	0.346	1.857	0.538	1.962	4.047
				V	1.777	0.663	0.340	1.855	0.539	2.004	4.073
TANTALUM				H	1.748	0.651	0.343	1.855	0.539	1.986	4.061
11947	16.5066	241	673.0	S	1.742	0.648	0.344	1.855	0.539	1.982	4.059
				R	1.706	0.633	0.347	1.855	0.539	1.959	4.044
				V	1.772	0.661	0.341	1.852	0.540	2.001	4.070
TANTALUM				H	1.743	0.649	0.343	1.852	0.540	1.983	4.058
11948	16.5018	241	693.0	S	1.736	0.646	0.344	1.852	0.540	1.979	4.055
				R	1.701	0.631	0.347	1.852	0.540	1.956	4.040
				V	1.767	0.659	0.341	1.849	0.541	1.999	4.066
TANTALUM				H	1.738	0.647	0.343	1.849	0.541	1.980	4.054
11949	16.4968	241	713.0	S	1.731	0.644	0.344	1.849	0.541	1.976	4.051
				R	1.696	0.629	0.347	1.849	0.541	1.953	4.037
				V	1.762	0.657	0.341	1.847	0.541	1.996	4.063
TANTALUM				H	1.733	0.645	0.344	1.847	0.542	1.977	4.051
11950	16.4919	241	733.0	S	1.726	0.642	0.344	1.847	0.542	1.973	4.048
				R	1.691	0.627	0.347	1.847	0.541	1.950	4.034

CUBIC

CODE	NAME / DENSITY	REF	TEMP(K)		E	G	SIGMA	K	BETA	VS	VP
				V	1.757	0.655	0.341	1.844	0.542	1.993	4.060
TANTALUM				H	1.728	0.643	0.344	1.844	0.542	1.975	4.048
11951	16.4821	241	753.0	S	1.721	0.640	0.344	1.844	0.542	1.971	4.045
				R	1.685	0.625	0.348	1.844	0.542	1.948	4.030
				V	1.752	0.653	0.342	1.843	0.543	1.991	4.059
TANTALUM				H	1.723	0.641	0.344	1.843	0.543	1.972	4.046
11952	16.4722	241	773.0	S	1.716	0.638	0.345	1.843	0.543	1.968	4.044
				R	1.680	0.623	0.348	1.843	0.543	1.945	4.029
				V	3.330	1.338	0.244	2.170	0.461	3.022	5.195
TANTALUM MONOCARBIDE				H	3.071	1.215	0.264	2.170	0.461	2.880	5.086
11953	14.6500	24	ROOM	S	2.952	1.159	0.273	2.170	0.461	2.813	5.036
				R	2.732	1.059	0.290	2.170	0.461	2.688	4.944
				V	0.241	0.091	0.321	0.225	4.451	1.099	2.141
THALLIUM BROMIDE				H	0.237	0.090	0.324	0.225	4.451	1.089	2.134
11954	7.5600	125	293.0	S	0.237	0.089	0.324	0.225	4.451	1.087	2.133
				R	0.233	0.088	0.327	0.225	4.451	1.076	2.125
				V	0.296	0.112	0.320	0.274	3.645	1.203	2.339
THALLIUM BROMIDE				H	0.294	0.111	0.321	0.274	3.645	1.199	2.336
11955	7.7490	259	4.2	S	0.294	0.111	0.321	0.274	3.645	1.199	2.336
				R	0.292	0.111	0.322	0.274	3.645	1.194	2.333
				V	0.295	0.112	0.319	0.272	3.671	1.202	2.334
THALLIUM BROMIDE				H	0.293	0.111	0.321	0.272	3.671	1.198	2.331
11956	7.7390	259	30.0	S	0.293	0.111	0.321	0.272	3.671	1.198	2.331
				R	0.291	0.110	0.322	0.272	3.671	1.193	2.328
				V	0.292	0.111	0.318	0.267	3.739	1.197	2.317
THALLIUM BROMIDE				H	0.290	0.110	0.319	0.267	3.739	1.192	2.314
11957	7.7350	259	60.0	S	0.290	0.110	0.319	0.267	3.739	1.192	2.313
				R	0.288	0.109	0.321	0.267	3.739	1.187	2.310
				V	0.287	0.109	0.317	0.262	3.824	1.187	2.295
THALLIUM BROMIDE				H	0.284	0.108	0.319	0.262	3.824	1.182	2.291
11958	7.7200	259	90.0	S	0.284	0.108	0.319	0.262	3.824	1.181	2.291
				R	0.282	0.107	0.320	0.262	3.824	1.176	2.287
				V	0.279	0.106	0.318	0.256	3.910	1.174	2.273
THALLIUM BROMIDE				H	0.277	0.105	0.319	0.256	3.910	1.169	2.269
11959	7.6870	259	120.0	S	0.277	0.105	0.320	0.256	3.910	1.168	2.269
				R	0.274	0.104	0.321	0.256	3.910	1.162	2.264
				V	0.272	0.103	0.319	0.250	3.995	1.161	2.252
THALLIUM BROMIDE				H	0.269	0.102	0.321	0.250	3.995	1.155	2.248
11960	7.6440	259	150.0	S	0.269	0.102	0.321	0.250	3.995	1.154	2.247
				R	0.266	0.101	0.323	0.250	3.995	1.147	2.243
				V	0.265	0.100	0.320	0.245	4.083	1.147	2.230
THALLIUM BROMIDE				H	0.262	0.099	0.322	0.245	4.083	1.141	2.225
11961	7.6150	259	180.0	S	0.261	0.099	0.322	0.245	4.083	1.140	2.224
				R	0.259	0.098	0.324	0.245	4.083	1.132	2.220
				V	0.258	0.097	0.321	0.240	4.167	1.134	2.209
THALLIUM BROMIDE				H	0.255	0.096	0.323	0.240	4.167	1.127	2.204
11962	7.5830	259	210.0	S	0.254	0.096	0.323	0.240	4.167	1.125	2.203
				R	0.251	0.095	0.326	0.240	4.167	1.118	2.198
				V	0.250	0.095	0.322	0.234	4.267	1.120	2.187
THALLIUM BROMIDE				H	0.247	0.093	0.324	0.234	4.267	1.113	2.181
11963	7.5400	259	240.0	S	0.247	0.093	0.325	0.234	4.267	1.111	2.181
				R	0.243	0.092	0.327	0.234	4.267	1.103	2.175
				V	0.243	0.092	0.323	0.229	4.365	1.107	2.165
THALLIUM BROMIDE				H	0.240	0.090	0.325	0.229	4.365	1.098	2.160
11964	7.5000	259	270.0	S	0.239	0.090	0.326	0.229	4.365	1.097	2.158
				R	0.236	0.089	0.328	0.229	4.365	1.088	2.152
				V	0.237	0.090	0.324	0.224	4.458	1.096	2.146
THALLIUM BROMIDE				H	0.234	0.088	0.326	0.224	4.458	1.087	2.140
11965	7.4620	259	300.0	S	0.233	0.088	0.327	0.224	4.458	1.085	2.139
				R	0.229	0.086	0.330	0.224	4.458	1.075	2.132

CUBIC

CODE	NAME DENSITY	REF	TEMP(K)		E	G	SIGMA	K	BETA	VS	VP
				V	0.310	0.120	0.299	0.257	3.887	1.241	2.316
THALLIUM BROMIDE				H	0.309	0.119	0.300	0.257	3.887	1.237	2.314
11966	7.7653	187	0.0	S	0.309	0.119	0.300	0.257	3.887	1.236	2.313
				R	0.307	0.118	0.301	0.257	3.887	1.232	2.310
				V	0.295	0.113	0.304	0.251	3.991	1.214	2.288
THALLIUM BROMIDE				H	0.292	0.112	0.306	0.251	3.991	1.209	2.284
11967	7.6623	187	100.0	S	0.292	0.112	0.306	0.251	3.991	1.208	2.284
				R	0.289	0.111	0.307	0.251	3.991	1.202	2.279
				V	0.279	0.107	0.308	0.243	4.115	1.185	2.250
THALLIUM BROMIDE				H	0.277	0.106	0.310	0.243	4.115	1.178	2.246
11968	7.6115	187	150.0	S	0.276	0.105	0.310	0.243	4.115	1.177	2.245
				R	0.274	0.104	0.312	0.243	4.115	1.171	2.240
				V	0.266	0.101	0.312	0.236	4.233	1.158	2.217
THALLIUM BROMIDE				H	0.263	0.100	0.314	0.236	4.233	1.150	2.211
11969	7.5588	187	200.0	S	0.263	0.100	0.315	0.236	4.233	1.149	2.211
				R	0.259	0.098	0.317	0.236	4.233	1.141	2.205
				V	0.253	0.096	0.316	0.229	4.372	1.132	2.181
THALLIUM BROMIDE				H	0.249	0.095	0.318	0.229	4.372	1.123	2.175
11970	7.5043	187	250.0	S	0.249	0.094	0.319	0.229	4.372	1.121	2.174
				R	0.245	0.093	0.321	0.229	4.372	1.112	2.167
				V	0.241	0.091	0.319	0.223	4.494	1.108	2.150
THALLIUM BROMIDE				H	0.237	0.090	0.322	0.223	4.494	1.098	2.143
11971	7.4529	187	298.0	S	0.237	0.089	0.323	0.223	4.494	1.096	2.142
				R	0.233	0.088	0.326	0.223	4.494	1.085	2.134
				V	0.229	0.087	0.322	0.215	4.660	1.083	2.113
THALLIUM BROMIDE				H	0.225	0.085	0.325	0.215	4.660	1.071	2.105
11972	7.3948	187	350.0	S	0.224	0.084	0.326	0.215	4.660	1.068	2.103
				R	0.219	0.082	0.330	0.215	4.660	1.055	2.094
				V	0.220	0.083	0.322	0.207	4.842	1.065	2.080
THALLIUM BROMIDE				H	0.215	0.081	0.326	0.207	4.842	1.051	2.071
11973	7.3377	187	400.0	S	0.214	0.081	0.327	0.207	4.842	1.048	2.069
				R	0.209	0.078	0.332	0.207	4.842	1.034	2.059
				V	0.209	0.079	0.325	0.199	5.033	1.041	2.043
THALLIUM BROMIDE				H	0.203	0.076	0.329	0.199	5.033	1.025	2.032
11974	7.2811	187	450.0	S	0.202	0.076	0.331	0.199	5.033	1.021	2.029
				R	0.196	0.073	0.336	0.199	5.033	1.004	2.018
				V	0.198	0.075	0.326	0.190	5.260	1.018	2.004
THALLIUM BROMIDE				H	0.192	0.072	0.332	0.190	5.260	1.000	1.992
11975	7.2143	187	500.0	S	0.190	0.071	0.333	0.190	5.260	0.995	1.989
				R	0.184	0.069	0.339	0.190	5.260	0.977	1.977
				V	0.188	0.071	0.327	0.181	5.519	0.995	1.963
THALLIUM BROMIDE				H	0.181	0.068	0.333	0.181	5.519	0.975	1.950
11976	7.1484	187	550.0	S	0.179	0.067	0.335	0.181	5.519	0.968	1.946
				R	0.172	0.064	0.342	0.181	5.519	0.947	1.932
				V	0.175	0.066	0.331	0.173	5.796	0.962	1.916
THALLIUM BROMIDE				H	0.167	0.062	0.339	0.173	5.796	0.939	1.900
11977	7.0831	187	600.0	S	0.165	0.061	0.341	0.173	5.796	0.931	1.895
				R	0.157	0.058	0.348	0.173	5.796	0.907	1.880
				V	0.165	0.062	0.331	0.163	6.139	0.940	1.870
THALLIUM BROMIDE				H	0.157	0.059	0.339	0.163	6.139	0.915	1.854
11978	7.0188	187	650.0	S	0.155	0.058	0.342	0.163	6.139	0.906	1.848
				R	0.147	0.054	0.350	0.163	6.139	0.881	1.832
				V	0.153	0.057	0.334	0.154	6.513	0.906	1.817
THALLIUM BROMIDE				H	0.144	0.053	0.344	0.154	6.513	0.877	1.798
11979	6.9552	187	700.0	S	0.140	0.052	0.348	0.154	6.513	0.865	1.790
				R	0.132	0.048	0.357	0.154	6.513	0.835	1.771
				V	0.241	0.091	0.319	0.223	4.494	1.108	2.150
THALLIUM BROMIDE (P=0KB)				H	0.237	0.090	0.322	0.223	4.494	1.098	2.143
11980	7.4529	187	ROOM	S	0.237	0.089	0.323	0.223	4.494	1.096	2.142
				R	0.233	0.088	0.326	0.223	4.494	1.085	2.134

CUBIC

CODE	NAME / DENSITY	REF	TEMP(K)		E	G	SIGMA	K	BETA	VS	VP
				V	0.250	0.095	0.318	0.229	4.362	1.125	2.179
THALLIUM BROMIDE (P=0.981KB)				H	0.246	0.093	0.321	0.229	4.362	1.115	2.172
11981	7.4888	187	ROOM	S	0.245	0.093	0.322	0.229	4.362	1.113	2.171
				R	0.241	0.091	0.325	0.229	4.362	1.103	2.164
				V	0.255	0.097	0.321	0.237	4.224	1.133	2.204
THALLIUM BROMIDE (P=1.962KB)				H	0.252	0.095	0.323	0.237	4.224	1.125	2.198
11982	7.5236	187	ROOM	S	0.251	0.095	0.323	0.237	4.224	1.123	2.198
				R	0.248	0.093	0.326	0.237	4.224	1.115	2.192
				V	0.262	0.099	0.320	0.243	4.120	1.145	2.227
THALLIUM BROMIDE (P=2.943KB)				H	0.259	0.098	0.322	0.243	4.120	1.138	2.222
11983	7.5579	187	ROOM	S	0.258	0.098	0.323	0.243	4.120	1.137	2.222
				R	0.256	0.096	0.325	0.243	4.120	1.130	2.217
				V	0.267	0.101	0.321	0.249	4.017	1.153	2.248
THALLIUM BROMIDE (P=3.924KB)				H	0.264	0.100	0.323	0.249	4.017	1.147	2.244
11984	7.5918	187	ROOM	S	0.264	0.100	0.323	0.249	4.017	1.146	2.243
				R	0.261	0.099	0.325	0.249	4.017	1.140	2.239
				V	0.271	0.102	0.323	0.256	3.909	1.159	2.269
THALLIUM BROMIDE (P=4.905KB)				H	0.269	0.102	0.325	0.256	3.909	1.154	2.265
11985	7.6244	187	ROOM	S	0.269	0.101	0.325	0.256	3.909	1.154	2.265
				R	0.267	0.101	0.326	0.256	3.909	1.148	2.261
				V	0.242	0.091	0.323	0.228	4.392	1.120	2.189
THALLIUM BROMIDE-CHLORIDE				H	0.237	0.089	0.326	0.228	4.392	1.107	2.181
11986	7.2920	129	ROOM	S	0.236	0.089	0.327	0.228	4.392	1.104	2.179
				R	0.231	0.087	0.331	0.228	4.392	1.091	2.170
				V	0.199	0.075	0.333	0.198	5.042	1.003	2.004
THALLIUM BROMIDE-IODIDE				H	0.194	0.072	0.337	0.198	5.042	0.988	1.995
11987	7.4110	129	ROOM	S	0.192	0.072	0.338	0.198	5.042	0.984	1.992
				R	0.187	0.070	0.343	0.198	5.042	0.969	1.982
				V	0.203	0.075	0.346	0.220	4.545	1.006	2.075
THALLIUM BROMIDE(0.6)-IODIDE(0.4)				H	0.196	0.072	0.352	0.220	4.545	0.986	2.063
11988	7.4380	129	ROOM	S	0.193	0.071	0.353	0.220	4.545	0.980	2.059
				R	0.186	0.068	0.359	0.220	4.545	0.959	2.045
				V	0.252	0.095	0.322	0.236	4.243	1.166	2.276
THALLIUM CHLORIDE				H	0.246	0.093	0.326	0.236	4.243	1.152	2.266
11989	7.0000	125	293.0	S	0.245	0.092	0.327	0.236	4.243	1.149	2.264
				R	0.239	0.090	0.331	0.236	4.243	1.133	2.254
				V	0.906	0.366	0.240	0.580	1.723	1.762	3.011
THORIUM				H	0.822	0.325	0.264	0.580	1.723	1.662	2.934
11990	11.7790	21	80.0	S	0.764	0.299	0.280	0.580	1.723	1.592	2.882
				R	0.673	0.257	0.307	0.580	1.723	1.478	2.800
				V	0.904	0.364	0.240	0.580	1.725	1.759	3.009
THORIUM				H	0.820	0.324	0.264	0.580	1.725	1.660	2.932
11991	11.7720	21	100.0	S	0.762	0.297	0.281	0.580	1.725	1.589	2.880
				R	0.670	0.256	0.307	0.580	1.725	1.475	2.798
				V	0.901	0.363	0.241	0.580	1.725	1.756	3.006
THORIUM				H	0.816	0.323	0.265	0.580	1.725	1.656	2.930
11992	11.7650	21	120.0	S	0.758	0.296	0.282	0.580	1.725	1.586	2.877
				R	0.666	0.255	0.308	0.580	1.725	1.471	2.795
				V	0.896	0.361	0.242	0.579	1.726	1.752	3.003
THORIUM				H	0.812	0.321	0.266	0.579	1.726	1.652	2.926
11993	11.7580	21	140.0	S	0.754	0.294	0.283	0.579	1.726	1.581	2.874
				R	0.662	0.253	0.310	0.579	1.726	1.466	2.792
				V	0.892	0.358	0.243	0.579	1.727	1.747	2.999
THORIUM				H	0.808	0.319	0.268	0.579	1.727	1.647	2.923
11994	11.7500	21	160.0	S	0.749	0.292	0.284	0.579	1.727	1.575	2.870
				R	0.657	0.251	0.311	0.579	1.727	1.461	2.788
				V	0.886	0.356	0.245	0.578	1.729	1.742	2.995
THORIUM				H	0.803	0.316	0.269	0.578	1.729	1.641	2.919
11995	11.7400	21	180.0	S	0.744	0.289	0.286	0.578	1.729	1.570	2.866
				R	0.652	0.248	0.312	0.579	1.729	1.455	2.784

CUBIC

CODE	NAME / DENSITY	REF	TEMP(K)		E	G	SIGMA	K	BETA	VS	VP
				V	0.882	0.354	0.246	0.578	1.730	1.736	2.991
THORIUM				H	0.798	0.314	0.270	0.578	1.730	1.636	2.915
11996	11.7330	21	200.0	S	0.739	0.287	0.287	0.578	1.730	1.564	2.862
				R	0.648	0.247	0.313	0.578	1.730	1.449	2.780
				V	0.876	0.351	0.247	0.578	1.731	1.730	2.986
THORIUM				H	0.792	0.311	0.272	0.578	1.731	1.630	2.910
11997	11.7260	21	220.0	S	0.732	0.284	0.289	0.578	1.731	1.556	2.856
				R	0.640	0.243	0.315	0.578	1.731	1.440	2.774
				V	0.870	0.348	0.249	0.577	1.732	1.724	2.982
THORIUM				H	0.787	0.309	0.273	0.577	1.732	1.624	2.906
11998	11.7190	21	240.0	S	0.727	0.282	0.290	0.577	1.732	1.551	2.852
				R	0.636	0.241	0.317	0.577	1.732	1.435	2.770
				V	0.864	0.345	0.251	0.577	1.732	1.718	2.977
THORIUM				H	0.781	0.306	0.275	0.577	1.732	1.617	2.901
11999	11.7120	21	260.0	S	0.721	0.279	0.292	0.577	1.732	1.544	2.847
				R	0.630	0.239	0.318	0.577	1.732	1.428	2.765
				V	0.858	0.343	0.252	0.577	1.732	1.711	2.973
THORIUM				H	0.775	0.304	0.276	0.577	1.732	1.611	2.897
12000	11.7030	21	280.0	S	0.715	0.277	0.293	0.577	1.732	1.537	2.843
				R	0.624	0.236	0.320	0.577	1.732	1.421	2.762
				V	0.852	0.340	0.254	0.577	1.733	1.704	2.968
THORIUM				H	0.769	0.301	0.278	0.577	1.733	1.604	2.892
12001	11.6940	21	300.0	S	0.708	0.273	0.295	0.577	1.733	1.529	2.837
				R	0.617	0.233	0.322	0.577	1.733	1.413	2.756
				V	0.846	0.337	0.256	0.577	1.732	1.698	2.964
THORIUM				H	0.763	0.298	0.280	0.577	1.732	1.597	2.888
12002	11.6870	21	320.0	S	0.702	0.271	0.297	0.577	1.732	1.522	2.833
				R	0.611	0.231	0.324	0.577	1.732	1.405	2.752
				V	0.839	0.334	0.258	0.577	1.733	1.690	2.958
THORIUM				H	0.756	0.295	0.282	0.577	1.733	1.589	2.882
12003	11.6800	21	340.0	S	0.695	0.267	0.299	0.577	1.733	1.513	2.827
				R	0.603	0.228	0.326	0.577	1.733	1.396	2.746
				V	2.559	1.000	0.279	1.930	0.518	3.164	5.715
THORIUM DIOXIDE				H	2.503	0.975	0.284	1.930	0.518	3.124	5.686
12004	9.9910	157	298.0	S	2.491	0.969	0.285	1.930	0.518	3.115	5.679
				R	2.435	0.944	0.290	1.930	0.518	3.074	5.649
				V	0.783	0.321	0.218	0.463	2.161	2.227	3.709
TIN TELLURIDE				H	0.662	0.263	0.261	0.463	2.161	2.013	3.543
12005	6.4760	28	0.0	S	0.574	0.222	0.293	0.463	2.161	1.851	3.422
				R	0.491	0.186	0.323	0.463	2.161	1.693	3.311
				V	0.782	0.321	0.218	0.462	2.163	2.226	3.708
TIN TELLURIDE				H	0.662	0.262	0.261	0.462	2.163	2.013	3.541
12006	6.4760	28	10.0	S	0.574	0.222	0.293	0.462	2.163	1.851	3.421
				R	0.491	0.186	0.323	0.462	2.163	1.693	3.311
				V	0.781	0.320	0.219	0.462	2.163	2.224	3.706
TIN TELLURIDE				H	0.661	0.262	0.262	0.462	2.163	2.011	3.540
12007	6.4760	28	20.0	S	0.573	0.221	0.294	0.462	2.163	1.849	3.420
				R	0.490	0.185	0.323	0.462	2.163	1.691	3.310
				V	0.777	0.319	0.219	0.461	2.169	2.218	3.699
TIN TELLURIDE				H	0.658	0.261	0.262	0.461	2.169	2.006	3.534
12008	6.4750	28	40.0	S	0.570	0.220	0.294	0.461	2.169	1.845	3.414
				R	0.488	0.184	0.323	0.461	2.169	1.688	3.304
				V	0.772	0.316	0.220	0.460	2.174	2.210	3.690
TIN TELLURIDE				H	0.654	0.259	0.263	0.460	2.174	1.999	3.526
12009	6.4740	28	60.0	S	0.567	0.219	0.295	0.460	2.174	1.839	3.408
				R	0.485	0.183	0.324	0.460	2.174	1.682	3.298
				V	0.765	0.313	0.221	0.457	2.188	2.200	3.676
TIN TELLURIDE				H	0.648	0.256	0.264	0.457	2.188	1.990	3.513
12010	6.4720	28	80.0	S	0.562	0.217	0.295	0.457	2.188	1.831	3.396
				R	0.481	0.182	0.325	0.457	2.188	1.675	3.287

CUBIC

CODE	NAME / DENSITY	REF	TEMP(K)		E	G	SIGMA	K	BETA	VS	VP
				V	0.758	0.310	0.222	0.455	2.199	2.189	3.663
	TIN TELLURIDE			H	0.642	0.254	0.265	0.455	2.199	1.981	3.502
12011	6.4700	28	100.0	S	0.558	0.215	0.296	0.455	2.199	1.824	3.386
				R	0.478	0.180	0.325	0.455	2.199	1.669	3.278
				V	0.751	0.307	0.223	0.452	2.214	2.179	3.649
	TIN TELLURIDE			H	0.637	0.252	0.265	0.452	2.214	1.973	3.489
12012	6.4670	28	125.0	S	0.553	0.213	0.296	0.452	2.214	1.817	3.374
				R	0.474	0.179	0.325	0.452	2.214	1.663	3.267
				V	0.741	0.303	0.224	0.447	2.237	2.163	3.627
	TIN TELLURIDE			H	0.629	0.248	0.266	0.447	2.237	1.960	3.470
12013	6.4640	28	150.0	S	0.547	0.211	0.296	0.447	2.237	1.807	3.357
				R	0.469	0.177	0.325	0.447	2.237	1.655	3.251
				V	0.724	0.295	0.226	0.440	2.273	2.138	3.593
	TIN TELLURIDE			H	0.616	0.243	0.267	0.440	2.273	1.940	3.440
12014	6.4580	28	200.0	S	0.539	0.208	0.296	0.440	2.273	1.794	3.332
				R	0.463	0.175	0.325	0.440	2.273	1.645	3.228
				V	0.707	0.288	0.228	0.433	2.309	2.112	3.558
	TIN TELLURIDE			H	0.603	0.238	0.268	0.433	2.309	1.920	3.410
12015	6.4510	28	250.0	S	0.530	0.204	0.296	0.433	2.309	1.780	3.307
				R	0.456	0.172	0.324	0.433	2.309	1.635	3.205
				V	0.689	0.280	0.230	0.425	2.353	2.086	3.520
	TIN TELLURIDE			H	0.590	0.233	0.268	0.425	2.353	1.900	3.378
12016	6.4450	28	300.0	S	0.521	0.201	0.296	0.425	2.353	1.766	3.279
				R	0.450	0.170	0.324	0.425	2.353	1.624	3.180
				V	4.373	1.824	0.199	2.420	0.413	6.076	9.911
	TITANIUM CARBIDE			H	4.369	1.822	0.199	2.420	0.413	6.073	9.908
12017	4.9400	112	ROOM	S	4.369	1.822	0.199	2.420	0.413	6.073	9.907
				R	4.365	1.820	0.199	2.420	0.413	6.069	9.905
				V	4.090	1.911	0.070	1.586	0.631	6.231	9.164
	TITANIUM CARBIDE			H	4.081	1.905	0.071	1.586	0.631	6.221	9.155
12018	4.9220	73	ROOM	S	4.080	1.904	0.071	1.586	0.631	6.220	9.155
				R	4.071	1.899	0.072	1.586	0.631	6.211	9.146
				V	4.499	1.890	0.190	2.422	0.413	6.207	10.037
	TITANIUM-0.91CARBIDE			H	4.492	1.886	0.191	2.422	0.413	6.201	10.032
12019	4.9050	49	298.0	S	4.491	1.886	0.191	2.422	0.413	6.200	10.031
				R	4.484	1.882	0.191	2.422	0.413	6.194	10.026
				V	3.937	1.530	0.287	3.081	0.325	2.815	5.151
	TUNGSTEN			H	3.937	1.530	0.287	3.081	0.325	2.815	5.151
12020	19.3000	128	ROOM	S	3.937	1.530	0.287	3.081	0.325	2.815	5.151
				R	3.937	1.530	0.287	3.081	0.325	2.815	5.151
				V	3.903	1.522	0.282	2.990	0.334	2.808	5.099
	TUNGSTEN			H	3.903	1.522	0.282	2.990	0.335	2.808	5.099
12021	19.3000	128	ROOM	S	3.903	1.522	0.282	2.990	0.335	2.808	5.099
				R	3.903	1.522	0.282	2.990	0.334	2.808	5.099
				V	4.165	1.629	0.278	3.129	0.320	2.904	5.239
	TUNGSTEN			H	4.165	1.629	0.278	3.129	0.320	2.904	5.239
12022	19.3160	39	77.0	S	4.165	1.629	0.278	3.129	0.320	2.904	5.239
				R	4.165	1.629	0.278	3.129	0.320	2.904	5.239
				V	4.164	1.629	0.278	3.128	0.320	2.904	5.238
	TUNGSTEN			H	4.164	1.629	0.278	3.128	0.320	2.904	5.238
12023	19.3150	39	80.0	S	4.164	1.629	0.278	3.128	0.320	2.904	5.238
				R	4.164	1.629	0.278	3.128	0.320	2.904	5.238
				V	4.159	1.627	0.278	3.121	0.320	2.903	5.234
	TUNGSTEN			H	4.159	1.627	0.278	3.121	0.320	2.903	5.234
12024	19.3110	39	100.0	S	4.159	1.627	0.278	3.121	0.320	2.903	5.234
				R	4.159	1.627	0.278	3.121	0.320	2.903	5.234
				V	4.154	1.626	0.278	3.115	0.321	2.902	5.231
	TUNGSTEN			H	4.154	1.626	0.278	3.115	0.321	2.902	5.231
12025	19.3070	39	120.0	S	4.154	1.626	0.278	3.115	0.321	2.902	5.231
				R	4.154	1.626	0.278	3.115	0.321	2.902	5.231

CUBIC

CODE	NAME DENSITY	REF	TEMP(K)		E	G	SIGMA	K	BETA	VS	VP
				V	4.149	1.624	0.278	3.111	0.321	2.900	5.228
TUNGSTEN				H	4.149	1.624	0.278	3.111	0.321	2.900	5.228
12026	19.3040	39	140.0	S	4.149	1.624	0.278	3.111	0.321	2.900	5.228
				R	4.149	1.624	0.278	3.111	0.321	2.900	5.228
				V	4.144	1.622	0.278	3.106	0.322	2.899	5.224
TUNGSTEN				H	4.144	1.622	0.278	3.106	0.322	2.899	5.224
12027	19.3000	39	160.0	S	4.144	1.622	0.278	3.106	0.322	2.899	5.224
				R	4.144	1.622	0.278	3.106	0.322	2.899	5.224
				V	4.138	1.619	0.278	3.102	0.322	2.897	5.222
TUNGSTEN				H	4.138	1.619	0.278	3.102	0.322	2.897	5.221
12028	19.2950	39	180.0	S	4.138	1.619	0.278	3.102	0.322	2.897	5.221
				R	4.138	1.619	0.278	3.102	0.322	2.897	5.221
				V	4.130	1.616	0.278	3.098	0.323	2.894	5.218
TUNGSTEN				H	4.130	1.616	0.278	3.098	0.323	2.894	5.218
12029	19.2910	39	200.0	S	4.130	1.616	0.278	3.098	0.323	2.894	5.218
				R	4.130	1.616	0.278	3.098	0.323	2.894	5.218
				V	4.123	1.613	0.278	3.094	0.323	2.892	5.215
TUNGSTEN				H	4.123	1.613	0.278	3.094	0.323	2.892	5.215
12030	19.2860	39	220.0	S	4.123	1.613	0.278	3.094	0.323	2.892	5.215
				R	4.123	1.613	0.278	3.094	0.323	2.892	5.215
				V	4.116	1.610	0.278	3.092	0.323	2.890	5.213
TUNGSTEN				H	4.116	1.610	0.278	3.092	0.323	2.890	5.213
12031	19.2800	39	240.0	S	4.116	1.610	0.278	3.092	0.323	2.890	5.213
				R	4.116	1.610	0.278	3.092	0.323	2.890	5.213
				V	4.109	1.607	0.278	3.089	0.324	2.888	5.210
TUNGSTEN				H	4.109	1.607	0.278	3.089	0.324	2.888	5.210
12032	19.2760	39	260.0	S	4.109	1.607	0.278	3.089	0.324	2.888	5.210
				R	4.109	1.607	0.278	3.089	0.324	2.888	5.210
				V	4.102	1.604	0.279	3.087	0.324	2.885	5.207
TUNGSTEN				H	4.102	1.604	0.279	3.087	0.324	2.885	5.207
12033	19.2710	39	280.0	S	4.102	1.604	0.279	3.087	0.324	2.885	5.207
				R	4.102	1.604	0.279	3.087	0.324	2.885	5.207
				V	4.096	1.602	0.279	3.083	0.324	2.883	5.205
TUNGSTEN				H	4.096	1.602	0.279	3.083	0.324	2.883	5.205
12034	19.2650	39	300.0	S	4.096	1.602	0.279	3.083	0.324	2.883	5.205
				R	4.096	1.602	0.279	3.083	0.324	2.883	5.205
				V	4.079	1.595	0.279	3.077	0.325	2.878	5.198
TUNGSTEN				H	4.079	1.595	0.279	3.077	0.325	2.878	5.198
12035	19.2540	39	340.0	S	4.079	1.595	0.279	3.077	0.325	2.878	5.198
				R	4.079	1.595	0.279	3.077	0.325	2.878	5.198
				V	4.064	1.588	0.279	3.071	0.326	2.873	5.192
TUNGSTEN				H	4.064	1.588	0.279	3.071	0.326	2.873	5.192
12036	19.2460	39	380.0	S	4.064	1.588	0.279	3.071	0.326	2.873	5.192
				R	4.064	1.588	0.279	3.071	0.326	2.873	5.192
				V	4.048	1.581	0.280	3.065	0.326	2.867	5.186
TUNGSTEN				H	4.048	1.581	0.280	3.065	0.326	2.867	5.186
12037	19.2340	39	420.0	S	4.048	1.581	0.280	3.065	0.326	2.867	5.186
				R	4.048	1.581	0.280	3.065	0.326	2.867	5.186
				V	4.031	1.574	0.280	3.059	0.327	2.862	5.180
TUNGSTEN				H	4.031	1.574	0.280	3.059	0.327	2.861	5.179
12038	19.2250	39	460.0	S	4.031	1.574	0.280	3.059	0.327	2.861	5.179
				R	4.031	1.574	0.280	3.059	0.327	2.861	5.179
				V	4.015	1.567	0.281	3.053	0.328	2.856	5.173
TUNGSTEN				H	4.014	1.567	0.281	3.053	0.328	2.856	5.173
12039	19.2160	39	500.0	S	4.014	1.567	0.281	3.053	0.328	2.856	5.173
				R	4.014	1.567	0.281	3.053	0.328	2.856	5.173
				V	4.178	1.634	0.278	3.142	0.318	2.908	5.248
TUNGSTEN				H	4.178	1.634	0.278	3.141	0.318	2.908	5.248
12040	19.3204	89	0.0	S	4.178	1.634	0.278	3.141	0.318	2.908	5.248
				R	4.178	1.634	0.278	3.142	0.318	2.908	5.248

CUBIC

CODE	NAME / DENSITY	REF	TEMP(K)		E	G	SIGMA	K	BETA	VS	VP
				V	4.178	1.634	0.278	3.142	0.318	2.908	5.248
	TUNGSTEN			H	4.178	1.634	0.278	3.141	0.318	2.908	5.248
12041	19.3204	89	20.0	S	4.178	1.634	0.278	3.141	0.318	2.908	5.248
				R	4.178	1.634	0.278	3.142	0.318	2.908	5.248
				V	4.178	1.634	0.278	3.141	0.318	2.908	5.247
	TUNGSTEN			H	4.178	1.634	0.278	3.141	0.318	2.908	5.247
12042	19.3204	89	40.0	S	4.178	1.634	0.278	3.141	0.318	2.908	5.247
				R	4.178	1.634	0.278	3.141	0.318	2.908	5.247
				V	4.177	1.634	0.278	3.142	0.318	2.908	5.248
	TUNGSTEN			H	4.177	1.634	0.278	3.142	0.318	2.908	5.248
12043	19.3146	89	60.0	S	4.177	1.634	0.278	3.142	0.318	2.908	5.248
				R	4.177	1.634	0.278	3.142	0.318	2.908	5.248
				V	4.177	1.634	0.278	3.141	0.318	2.909	5.249
	TUNGSTEN			H	4.177	1.634	0.278	3.141	0.318	2.909	5.249
12044	19.3088	89	80.0	S	4.177	1.634	0.278	3.141	0.318	2.909	5.249
				R	4.177	1.634	0.278	3.141	0.318	2.909	5.249
				V	4.176	1.633	0.278	3.142	0.318	2.909	5.249
	TUNGSTEN			H	4.176	1.633	0.278	3.141	0.318	2.909	5.249
12045	19.3031	89	100.0	S	4.176	1.633	0.278	3.141	0.318	2.909	5.249
				R	4.176	1.633	0.278	3.142	0.318	2.909	5.249
				V	4.173	1.632	0.279	3.142	0.318	2.908	5.249
	TUNGSTEN			H	4.173	1.632	0.279	3.142	0.318	2.908	5.249
12046	19.2973	89	120.0	S	4.173	1.632	0.279	3.142	0.318	2.908	5.249
				R	4.173	1.632	0.279	3.142	0.318	2.908	5.249
				V	4.167	1.629	0.279	3.143	0.318	2.906	5.249
	TUNGSTEN			H	4.167	1.629	0.279	3.143	0.318	2.906	5.249
12047	19.2915	89	140.0	S	4.167	1.629	0.279	3.143	0.318	2.906	5.249
				R	4.167	1.629	0.279	3.143	0.318	2.906	5.249
				V	4.159	1.625	0.280	3.144	0.318	2.902	5.247
	TUNGSTEN			H	4.159	1.625	0.280	3.144	0.318	2.902	5.247
12048	19.2915	89	160.0	S	4.159	1.625	0.280	3.144	0.318	2.902	5.247
				R	4.159	1.625	0.280	3.144	0.318	2.902	5.247
				V	4.150	1.621	0.280	3.142	0.318	2.899	5.244
	TUNGSTEN			H	4.150	1.621	0.280	3.142	0.318	2.899	5.244
12049	19.2857	89	180.0	S	4.150	1.621	0.280	3.142	0.318	2.899	5.244
				R	4.150	1.621	0.280	3.142	0.318	2.899	5.244
				V	4.141	1.618	0.280	3.137	0.319	2.897	5.240
	TUNGSTEN			H	4.141	1.618	0.280	3.137	0.319	2.897	5.240
12050	19.2799	89	200.0	S	4.141	1.618	0.280	3.137	0.319	2.897	5.240
				R	4.141	1.618	0.280	3.137	0.319	2.897	5.240
				V	4.132	1.614	0.280	3.132	0.319	2.894	5.236
	TUNGSTEN			H	4.132	1.614	0.280	3.132	0.319	2.894	5.236
12051	19.2742	89	220.0	S	4.132	1.614	0.280	3.132	0.319	2.894	5.236
				R	4.132	1.614	0.280	3.132	0.319	2.894	5.236
				V	4.124	1.611	0.280	3.126	0.320	2.891	5.232
	TUNGSTEN			H	4.124	1.611	0.280	3.126	0.320	2.891	5.232
12052	19.2684	89	240.0	S	4.124	1.611	0.280	3.126	0.320	2.891	5.232
				R	4.124	1.611	0.280	3.126	0.320	2.891	5.232
				V	4.116	1.608	0.280	3.120	0.320	2.889	5.227
	TUNGSTEN			H	4.116	1.608	0.280	3.120	0.320	2.889	5.227
12053	19.2684	89	260.0	S	4.116	1.608	0.280	3.120	0.320	2.889	5.227
				R	4.116	1.608	0.280	3.120	0.320	2.889	5.227
				V	4.108	1.605	0.280	3.114	0.321	2.886	5.222
	TUNGSTEN			H	4.108	1.605	0.280	3.114	0.321	2.886	5.222
12054	19.2626	89	280.0	S	4.108	1.605	0.280	3.114	0.321	2.886	5.222
				R	4.108	1.605	0.280	3.114	0.321	2.886	5.222
				V	4.101	1.602	0.280	3.108	0.322	2.884	5.218
	TUNGSTEN			H	4.101	1.602	0.280	3.108	0.322	2.884	5.218
12055	19.2568	89	300.0	S	4.101	1.602	0.280	3.108	0.322	2.884	5.218
				R	4.101	1.602	0.280	3.108	0.322	2.884	5.218

CUBIC

CODE	NAME DENSITY	REF	TEMP(K)		E	G	SIGMA	K	BETA	VS	VP
				V	4.098	1.601	0.280	3.104	0.322	2.883	5.216
TUNGSTEN				H	4.098	1.601	0.280	3.104	0.322	2.883	5.216
12056	19.2570	156	297.0	S	4.098	1.601	0.280	3.104	0.322	2.883	5.216
				R	4.098	1.601	0.280	3.104	0.322	2.883	5.216
				V	4.068	1.588	0.281	3.090	0.324	2.873	5.204
TUNGSTEN				H	4.067	1.588	0.281	3.090	0.324	2.873	5.203
12057	19.2339	156	373.0	S	4.067	1.588	0.281	3.090	0.324	2.873	5.203
				R	4.067	1.588	0.281	3.090	0.324	2.873	5.203
				V	4.030	1.572	0.281	3.073	0.325	2.861	5.187
TUNGSTEN				H	4.030	1.572	0.281	3.073	0.325	2.860	5.187
12058	19.2165	156	473.0	S	4.030	1.572	0.281	3.073	0.325	2.860	5.187
				R	4.030	1.572	0.281	3.073	0.325	2.860	5.187
				V	3.991	1.556	0.282	3.056	0.327	2.848	5.170
TUNGSTEN				H	3.991	1.556	0.282	3.056	0.327	2.847	5.170
12059	19.1934	156	573.0	S	3.991	1.556	0.282	3.056	0.327	2.847	5.170
				R	3.990	1.556	0.282	3.056	0.327	2.847	5.170
				V	3.951	1.539	0.283	3.038	0.329	2.834	5.153
TUNGSTEN				H	3.950	1.539	0.283	3.038	0.329	2.834	5.153
12060	19.1705	156	673.0	S	3.950	1.539	0.283	3.038	0.329	2.834	5.153
				R	3.950	1.539	0.283	3.038	0.329	2.833	5.153
				V	3.910	1.522	0.284	3.021	0.331	2.820	5.137
TUNGSTEN				H	3.909	1.522	0.284	3.021	0.331	2.820	5.136
12061	19.1418	156	773.0	S	3.909	1.522	0.284	3.021	0.331	2.820	5.136
				R	3.908	1.521	0.284	3.021	0.331	2.819	5.136
				V	3.867	1.504	0.285	3.004	0.333	2.805	5.119
TUNGSTEN				H	3.866	1.504	0.286	3.004	0.333	2.805	5.119
12062	19.1131	156	873.0	S	3.866	1.504	0.286	3.004	0.333	2.805	5.119
				R	3.864	1.503	0.286	3.004	0.333	2.804	5.119
				V	3.823	1.486	0.287	2.987	0.335	2.790	5.101
TUNGSTEN				H	3.821	1.485	0.287	2.986	0.335	2.789	5.100
12063	19.0903	156	973.0	S	3.821	1.485	0.287	2.986	0.335	2.789	5.100
				R	3.819	1.484	0.287	2.987	0.335	2.788	5.100
				V	3.777	1.466	0.288	2.969	0.337	2.773	5.082
TUNGSTEN				H	3.775	1.465	0.288	2.969	0.337	2.772	5.081
12064	19.0674	156	1073.0	S	3.775	1.465	0.288	2.969	0.337	2.772	5.081
				R	3.772	1.464	0.288	2.969	0.337	2.771	5.081
				V	3.731	1.447	0.289	2.952	0.339	2.757	5.064
TUNGSTEN				H	3.728	1.445	0.290	2.952	0.339	2.756	5.063
12065	19.0332	156	1173.0	S	3.727	1.445	0.290	2.952	0.339	2.756	5.063
				R	3.724	1.444	0.290	2.952	0.339	2.754	5.062
				V	3.682	1.426	0.291	2.936	0.341	2.739	5.044
TUNGSTEN				H	3.679	1.425	0.291	2.936	0.341	2.737	5.043
12066	19.0106	156	1273.0	S	3.678	1.424	0.291	2.936	0.341	2.737	5.043
				R	3.674	1.423	0.291	2.936	0.341	2.736	5.042
				V	3.633	1.405	0.293	2.919	0.343	2.721	5.025
TUNGSTEN				H	3.628	1.403	0.293	2.919	0.343	2.719	5.023
12067	18.9822	156	1373.0	S	3.628	1.403	0.293	2.919	0.343	2.719	5.023
				R	3.623	1.401	0.293	2.919	0.343	2.716	5.021
				V	3.582	1.384	0.294	2.902	0.345	2.702	5.004
TUNGSTEN				H	3.576	1.381	0.295	2.902	0.345	2.699	5.003
12068	18.9539	156	1473.0	S	3.576	1.381	0.295	2.902	0.345	2.699	5.002
				R	3.569	1.378	0.295	2.902	0.345	2.696	5.000
				V	3.530	1.362	0.296	2.885	0.347	2.683	4.985
TUNGSTEN				H	3.522	1.358	0.297	2.885	0.347	2.680	4.983
12069	18.9144	156	1573.0	S	3.522	1.358	0.297	2.885	0.347	2.680	4.983
				R	3.514	1.355	0.297	2.885	0.347	2.676	4.980
				V	3.476	1.339	0.298	2.868	0.349	2.663	4.965
TUNGSTEN				H	3.467	1.335	0.299	2.868	0.349	2.659	4.962
12070	18.8805	156	1673.0	S	3.467	1.335	0.299	2.868	0.349	2.659	4.962
				R	3.457	1.331	0.299	2.868	0.349	2.655	4.959

CUBIC

CODE	NAME DENSITY	REF	TEMP(K)		E	G	SIGMA	K	BETA	VS	VP
				V	3.421	1.316	0.300	2.852	0.351	2.642	4.944
TUNGSTEN				H	3.410	1.311	0.301	2.852	0.351	2.637	4.940
12071	18.8469	156	1773.0	S	3.410	1.311	0.301	2.852	0.351	2.637	4.940
				R	3.398	1.306	0.301	2.852	0.351	2.632	4.936
				V	3.364	1.292	0.302	2.835	0.353	2.620	4.922
TUNGSTEN				H	3.352	1.286	0.303	2.835	0.353	2.615	4.918
12072	18.8132	156	1873.0	S	3.351	1.286	0.303	2.835	0.353	2.614	4.918
				R	3.337	1.280	0.304	2.835	0.353	2.608	4.913
				V	3.306	1.267	0.305	2.819	0.355	2.598	4.900
TUNGSTEN				H	3.292	1.261	0.305	2.819	0.355	2.591	4.895
12073	18.7796	156	1973.0	S	3.290	1.260	0.305	2.819	0.355	2.590	4.895
				R	3.274	1.253	0.306	2.819	0.355	2.583	4.889
				V	3.246	1.242	0.307	2.802	0.357	2.574	4.878
TUNGSTEN				H	3.230	1.235	0.308	2.802	0.357	2.567	4.872
12074	18.7406	156	2073.0	S	3.228	1.234	0.308	2.802	0.357	2.566	4.872
				R	3.209	1.226	0.309	2.802	0.357	2.557	4.866
				V	2.191	0.858	0.276	1.633	0.612	2.509	4.514
URANIUM CARBIDE				H	2.123	0.827	0.283	1.633	0.612	2.463	4.480
12075	13.6300	118	300.0	S	2.105	0.819	0.285	1.633	0.612	2.451	4.471
				R	2.038	0.789	0.292	1.633	0.612	2.406	4.438
				V	2.446	0.935	0.308	2.127	0.470	2.919	5.545
URANIUM DIOXIDE				H	2.327	0.883	0.318	2.127	0.470	2.837	5.488
12076	10.9700	269	298.0	S	2.284	0.864	0.321	2.127	0.470	2.807	5.467
				R	2.168	0.815	0.330	2.127	0.470	2.726	5.412
				V	1.728	0.642	0.346	1.870	0.535	2.419	4.985
URANIUM-2.23OXIDE				H	1.620	0.598	0.356	1.870	0.535	2.334	4.931
12077	10.9690	71	ROOM	S	1.573	0.578	0.360	1.870	0.535	2.296	4.907
				R	1.466	0.535	0.369	1.870	0.535	2.209	4.854
				V	1.422	0.528	0.346	1.540	0.649	2.976	6.136
VANADIUM				H	1.360	0.503	0.353	1.540	0.649	2.905	6.090
12078	5.9600	4	ROOM	S	1.335	0.493	0.355	1.540	0.649	2.875	6.071
				R	1.260	0.462	0.364	1.540	0.649	2.784	6.014
				V	1.360	0.502	0.356	1.570	0.637	2.884	6.092
VANADIUM				H	1.355	0.499	0.356	1.570	0.637	2.877	6.088
12079	6.0340	14	0.0	S	1.354	0.499	0.356	1.570	0.637	2.877	6.088
				R	1.348	0.497	0.357	1.570	0.637	2.869	6.083
				V	1.359	0.501	0.356	1.571	0.637	2.882	6.092
VANADIUM				H	1.353	0.499	0.356	1.571	0.637	2.876	6.088
12080	6.0330	14	25.0	S	1.353	0.499	0.356	1.571	0.637	2.875	6.087
				R	1.346	0.496	0.357	1.571	0.637	2.867	6.082
				V	1.354	0.499	0.356	1.569	0.637	2.877	6.087
VANADIUM				H	1.348	0.497	0.357	1.569	0.637	2.870	6.083
12081	6.0320	14	50.0	S	1.348	0.497	0.357	1.569	0.637	2.870	6.082
				R	1.341	0.494	0.358	1.569	0.637	2.862	6.077
				V	1.347	0.497	0.357	1.567	0.638	2.869	6.080
VANADIUM				H	1.341	0.494	0.357	1.567	0.638	2.862	6.075
12082	6.0310	14	75.0	S	1.340	0.494	0.357	1.567	0.638	2.861	6.075
				R	1.333	0.491	0.358	1.567	0.638	2.853	6.069
				V	1.340	0.493	0.357	1.565	0.639	2.861	6.071
VANADIUM				H	1.333	0.491	0.358	1.565	0.639	2.853	6.067
12083	6.0300	14	100.0	S	1.333	0.491	0.358	1.565	0.639	2.852	6.066
				R	1.325	0.488	0.359	1.565	0.639	2.844	6.061
				V	1.332	0.490	0.358	1.564	0.639	2.852	6.065
VANADIUM				H	1.325	0.488	0.359	1.564	0.639	2.844	6.060
12084	6.0290	14	125.0	S	1.324	0.487	0.359	1.564	0.639	2.843	6.059
				R	1.317	0.484	0.360	1.564	0.639	2.834	6.054
				V	1.325	0.488	0.359	1.562	0.640	2.845	6.058
VANADIUM				H	1.319	0.485	0.359	1.562	0.640	2.837	6.053
12085	6.0280	14	150.0	S	1.318	0.485	0.359	1.562	0.640	2.836	6.052
				R	1.310	0.482	0.360	1.562	0.640	2.826	6.047

CUBIC

CODE	NAME / DENSITY	REF	TEMP(K)		E	G	SIGMA	K	BETA	VS	VP
				V	1.318	0.485	0.359	1.561	0.641	2.836	6.052
	VANADIUM			H	1.311	0.482	0.360	1.561	0.641	2.828	6.046
12086	6.0270	14	175.0	S	1.310	0.482	0.360	1.561	0.641	2.827	6.046
				R	1.302	0.478	0.361	1.561	0.641	2.817	6.040
				V	1.311	0.482	0.360	1.559	0.641	2.828	6.045
	VANADIUM			H	1.304	0.479	0.361	1.559	0.641	2.820	6.039
12087	6.0260	14	200.0	S	1.303	0.479	0.361	1.559	0.641	2.819	6.039
				R	1.295	0.476	0.362	1.559	0.641	2.809	6.033
				V	1.304	0.479	0.360	1.557	0.642	2.820	6.038
	VANADIUM			H	1.297	0.476	0.361	1.557	0.642	2.812	6.033
12088	6.0250	14	225.0	S	1.296	0.476	0.361	1.557	0.642	2.811	6.032
				R	1.288	0.473	0.362	1.558	0.642	2.801	6.026
				V	1.299	0.477	0.361	1.555	0.643	2.815	6.032
	VANADIUM			H	1.292	0.474	0.362	1.556	0.643	2.806	6.027
12089	6.0240	14	250.0	S	1.291	0.474	0.362	1.556	0.643	2.805	6.026
				R	1.283	0.471	0.363	1.556	0.643	2.795	6.020
				V	1.294	0.475	0.361	1.553	0.644	2.810	6.026
	VANADIUM			H	1.287	0.472	0.362	1.553	0.644	2.801	6.021
12090	6.0230	14	275.0	S	1.286	0.472	0.362	1.553	0.644	2.800	6.020
				R	1.277	0.469	0.363	1.553	0.644	2.789	6.013
				V	1.290	0.474	0.361	1.551	0.645	2.805	6.021
	VANADIUM			H	1.282	0.471	0.362	1.551	0.645	2.796	6.015
12091	6.0220	14	300.0	S	1.281	0.470	0.362	1.551	0.645	2.795	6.014
				R	1.273	0.467	0.363	1.551	0.645	2.784	6.008
				V	1.302	0.478	0.361	1.556	0.643	2.818	6.035
	VANADIUM			H	1.294	0.475	0.361	1.556	0.643	2.810	6.030
12092	6.0220	38	300.0	S	1.294	0.475	0.361	1.556	0.643	2.809	6.029
				R	1.285	0.472	0.362	1.556	0.643	2.799	6.023
				V	0.833	0.329	0.267	0.595	1.680	2.499	4.432
	ZINC SELENIDE (CONSTANT E)			H	0.776	0.302	0.283	0.595	1.680	2.397	4.356
12093	5.2620	32	298.0	S	0.745	0.288	0.291	0.595	1.680	2.341	4.315
				R	0.680	0.260	0.310	0.595	1.680	2.222	4.230
				V	0.870	0.337	0.291	0.694	1.440	2.873	5.292
	ZINC SULFIDE			H	0.825	0.317	0.302	0.694	1.440	2.785	5.230
12094	4.0830	128	ROOM	S	0.805	0.308	0.307	0.694	1.440	2.746	5.202
				R	0.752	0.285	0.320	0.694	1.440	2.642	5.130
				V	0.849	0.319	0.332	0.841	1.189	2.793	5.568
	ZINC SULFIDE			H	0.805	0.300	0.340	0.841	1.189	2.713	5.514
12095	4.0830	128	ROOM	S	0.786	0.292	0.344	0.841	1.189	2.675	5.490
				R	0.733	0.270	0.355	0.841	1.189	2.574	5.425
				V	0.734	0.274	0.340	0.767	1.304	2.589	5.263
	ZINC SULFIDE			H	0.709	0.263	0.346	0.767	1.304	2.539	5.230
12096	4.0870	204	ROOM	S	0.700	0.260	0.348	0.767	1.304	2.521	5.218
				R	0.669	0.247	0.355	0.767	1.304	2.458	5.178
				V	0.898	0.348	0.292	0.719	1.391	2.920	5.384
	ZINC SULFIDE			H	0.852	0.327	0.303	0.719	1.391	2.831	5.320
12097	4.0790	83	302.0	S	0.831	0.318	0.307	0.719	1.391	2.791	5.292
				R	0.776	0.294	0.320	0.719	1.391	2.684	5.218
				V	0.940	0.360	0.304	0.801	1.248	2.965	5.592
	ZINC SULFIDE (CONSTANT E)			H	0.892	0.339	0.314	0.801	1.248	2.878	5.531
12098	4.0990	32	77.0	S	0.871	0.330	0.319	0.801	1.248	2.839	5.504
				R	0.815	0.306	0.331	0.801	1.248	2.733	5.432
				V	0.926	0.355	0.303	0.784	1.275	2.948	5.547
	ZINC SULFIDE (CONSTANT E)			H	0.877	0.334	0.314	0.784	1.275	2.858	5.484
12099	4.0880	32	298.0	S	0.855	0.324	0.318	0.784	1.275	2.816	5.455
				R	0.797	0.300	0.331	0.784	1.275	2.707	5.381
				V	0.939	0.361	0.301	0.788	1.270	2.969	5.567
	ZINC SULFIDE			H	0.890	0.339	0.312	0.788	1.270	2.880	5.504
12100	4.0940	281	93.0	S	0.868	0.330	0.316	0.788	1.270	2.839	5.476
				R	0.811	0.305	0.328	0.788	1.270	2.730	5.402

CUBIC

CODE	NAME DENSITY	REF	TEMP(K)		E	G	SIGMA	K	BETA	VS	VP
				V	0.938	0.361	0.301	0.787	1.271	2.968	5.565
ZINC SULFIDE				H	0.890	0.339	0.312	0.787	1.271	2.878	5.502
12101	4.0940	281	113.0	S	0.867	0.330	0.316	0.787	1.271	2.837	5.473
				R	0.810	0.305	0.328	0.787	1.271	2.729	5.399
				V	0.937	0.360	0.301	0.786	1.272	2.966	5.562
ZINC SULFIDE				H	0.888	0.338	0.312	0.786	1.272	2.876	5.498
12102	4.0928	281	133.0	S	0.866	0.329	0.316	0.786	1.272	2.834	5.469
				R	0.808	0.304	0.329	0.786	1.272	2.725	5.395
				V	0.936	0.360	0.301	0.785	1.274	2.965	5.559
ZINC SULFIDE				H	0.887	0.338	0.312	0.785	1.274	2.875	5.495
12103	4.0916	281	153.0	S	0.865	0.328	0.316	0.785	1.274	2.833	5.466
				R	0.807	0.304	0.329	0.785	1.274	2.724	5.391
				V	0.934	0.359	0.301	0.784	1.276	2.962	5.554
ZINC SULFIDE				H	0.885	0.337	0.312	0.784	1.276	2.871	5.490
12104	4.0916	281	173.0	S	0.862	0.327	0.317	0.784	1.276	2.829	5.461
				R	0.804	0.303	0.329	0.784	1.276	2.719	5.386
				V	0.933	0.359	0.301	0.783	1.278	2.961	5.552
ZINC SULFIDE				H	0.884	0.337	0.312	0.783	1.278	2.869	5.488
12105	4.0903	281	193.0	S	0.861	0.327	0.317	0.783	1.278	2.827	5.458
				R	0.802	0.302	0.329	0.783	1.278	2.717	5.383
				V	0.931	0.358	0.301	0.781	1.280	2.957	5.547
ZINC SULFIDE				H	0.881	0.336	0.312	0.781	1.280	2.865	5.483
12106	4.0891	281	213.0	S	0.858	0.326	0.317	0.781	1.280	2.823	5.453
				R	0.799	0.301	0.329	0.781	1.280	2.712	5.377
				V	0.930	0.357	0.301	0.780	1.281	2.956	5.544
ZINC SULFIDE				H	0.880	0.335	0.312	0.780	1.282	2.863	5.479
12107	4.0879	281	233.0	S	0.856	0.325	0.317	0.780	1.282	2.820	5.449
				R	0.797	0.300	0.330	0.780	1.282	2.708	5.373
				V	0.927	0.356	0.302	0.779	1.284	2.953	5.539
ZINC SULFIDE				H	0.877	0.334	0.312	0.779	1.284	2.860	5.474
12108	4.0854	281	253.0	S	0.853	0.324	0.317	0.779	1.284	2.816	5.443
				R	0.794	0.299	0.330	0.779	1.284	2.703	5.367
				V	0.925	0.355	0.302	0.777	1.286	2.949	5.534
ZINC SULFIDE				H	0.874	0.333	0.313	0.777	1.286	2.855	5.468
12109	4.0842	281	273.0	S	0.850	0.323	0.318	0.777	1.286	2.811	5.437
				R	0.791	0.297	0.330	0.777	1.286	2.697	5.360
				V	0.922	0.354	0.302	0.776	1.289	2.945	5.528
ZINC SULFIDE				H	0.871	0.332	0.313	0.776	1.289	2.851	5.462
12110	4.0830	281	293.0	S	0.847	0.321	0.318	0.776	1.289	2.806	5.431
				R	0.788	0.296	0.331	0.776	1.289	2.692	5.354
				V	0.640	0.248	0.291	0.510	1.962	2.098	3.861
ZINC TELLURIDE (CONSTANT E)				H	0.615	0.237	0.299	0.510	1.962	2.050	3.827
12111	5.6360	32	298.0	S	0.606	0.233	0.302	0.510	1.962	2.032	3.815
				R	0.577	0.220	0.311	0.510	1.962	1.976	3.775
				V	4.072	1.702	0.196	2.231	0.448	5.076	8.255
ZIRCONIUM CARBIDE				H	4.062	1.697	0.197	2.231	0.448	5.069	8.248
12112	6.6060	49	298.0	S	4.062	1.697	0.197	2.231	0.448	5.069	8.248
				R	4.052	1.692	0.197	2.231	0.448	5.061	8.242
				V	3.718	1.643	0.132	1.682	0.595	5.003	7.681
ZIRCONIUM CARBIDE				H	3.694	1.629	0.134	1.682	0.595	4.982	7.663
12113	6.5640	46	300.0	S	3.692	1.628	0.134	1.682	0.595	4.980	7.661
				R	3.670	1.615	0.136	1.682	0.595	4.960	7.644
				V	2.015	0.784	0.285	1.566	0.639	3.069	5.600
ZIRCONIUM-COBALT INTERMETALLIC				H	1.996	0.775	0.288	1.566	0.639	3.051	5.588
12114	8.3240	229	0.0	S	1.993	0.774	0.288	1.566	0.639	3.049	5.586
				R	1.971	0.764	0.290	1.566	0.639	3.029	5.572
				V	2.015	0.784	0.285	1.566	0.639	3.069	5.600
ZIRCONIUM-COBALT INTERMETALLIC				H	1.996	0.775	0.288	1.566	0.639	3.051	5.588
12115	8.3240	229	77.0	S	1.993	0.774	0.288	1.566	0.639	3.049	5.586
				R	1.971	0.764	0.290	1.566	0.639	3.029	5.572

CUBIC

CODE	NAME DENSITY	REF	TEMP(K)		E	G	SIGMA	K	BETA	VS	VP
				V	2.015	0.784	0.285	1.565	0.639	3.069	5.600
ZIRCONIUM-COBALT INTERMETALLIC				H	1.996	0.775	0.287	1.565	0.639	3.052	5.587
12116	8.3220	229	100.0	S	1.993	0.774	0.288	1.565	0.639	3.049	5.585
				R	1.971	0.764	0.290	1.565	0.639	3.030	5.571
				V	2.000	0.777	0.287	1.562	0.640	3.058	5.590
ZIRCONIUM-COBALT INTERMETALLIC				H	1.982	0.769	0.288	1.562	0.640	3.042	5.579
12117	8.3140	229	150.0	S	1.980	0.768	0.289	1.562	0.640	3.039	5.577
				R	1.959	0.759	0.291	1.562	0.640	3.021	5.564
				V	1.973	0.766	0.288	1.552	0.644	3.037	5.567
ZIRCONIUM-COBALT INTERMETALLIC				H	1.955	0.758	0.290	1.552	0.644	3.021	5.556
12118	8.3020	229	200.0	S	1.953	0.757	0.290	1.552	0.644	3.019	5.554
				R	1.933	0.748	0.293	1.552	0.644	3.001	5.541
				V	1.946	0.755	0.289	1.539	0.650	3.018	5.542
ZIRCONIUM-COBALT INTERMETALLIC				H	1.928	0.747	0.291	1.539	0.650	3.001	5.530
12119	8.2890	229	250.0	S	1.926	0.745	0.292	1.539	0.650	2.999	5.528
				R	1.905	0.736	0.294	1.539	0.650	2.980	5.515
				V	1.918	0.743	0.291	1.528	0.654	2.996	5.516
ZIRCONIUM-COBALT INTERMETALLIC				H	1.899	0.734	0.293	1.528	0.654	2.979	5.504
12120	8.2770	229	300.0	S	1.896	0.733	0.293	1.528	0.654	2.976	5.502
				R	1.875	0.724	0.296	1.528	0.654	2.957	5.489

CUBIC

CODE / NAME	DENSITY	REF	TEMP(K)		E	G	SIGMA	K	BETA	VS	VP
					PRESSURE DERIVATIVES						
ALUMINUM				V	4.862	1.840	-1.644	4.417	-7.675	8.872	15.637
12121	2.6970	254	298.0	R	4.751	1.793	-1.478	4.417	-7.675	8.657	15.493
BARIUM FLUORIDE				V	1.506	0.348	11.382	4.900	-13.672	-0.388	8.772
12122	4.9120	278	195.0	R	1.472	0.333	11.472	4.900	-13.672	-0.452	8.728
BARIUM FLUORIDE				V	1.640	0.396	11.817	5.053	-14.829	-0.181	9.413
12123	4.8860	278	298.0	R	1.621	0.388	11.871	5.053	-14.829	-0.219	9.387
CALCIUM FLUORIDE				V	2.789	0.778	8.645	5.477	-7.043	1.029	10.962
12124	3.2085	134	77.4	R	3.301	1.049	6.872	5.477	-7.043	2.367	12.063
CALCIUM FLUORIDE				V	3.289	1.026	7.353	5.320	-7.070	2.085	11.498
12125	3.1963	134	194.5	R	3.595	1.199	5.932	5.320	-7.070	3.066	12.323
CALCIUM FLUORIDE				V	3.271	0.958	9.240	5.977	-8.235	1.822	12.943
12126	3.1835	134	273.1	R	3.716	1.204	7.360	5.977	-8.235	3.137	14.047
CALCIUM FLUORIDE				V	3.216	0.906	10.165	6.260	-8.727	1.606	13.485
12127	3.1795	134	295.5	R	3.811	1.224	7.915	6.260	-8.727	3.237	14.834
CALCIUM FLUORIDE				V	3.350	1.102	5.936	4.757	-6.440	2.380	10.445
12128	3.1980	278	195.0	R	3.691	1.284	4.506	4.757	-6.440	3.419	11.292
CALCIUM FLUORIDE				V	3.431	1.126	6.299	4.917	-6.959	2.526	11.053
12129	3.1810	278	298.0	R	3.781	1.314	4.745	4.917	-6.959	3.629	11.956
CESIUM BROMIDE				V	5.727	2.180	12.029	4.786	-148.750	11.477	24.372
12130	4.6124	213	77.0	R	6.194	2.404	7.267	4.786	-148.751	13.167	25.667
CESIUM BROMIDE				V	5.787	2.208	11.800	4.850	-191.603	12.989	27.098
12131	4.4560	213	300.0	R	6.902	2.732	-2.282	4.850	-191.604	17.654	30.623
CESIUM BROMIDE				V	6.497	2.482	13.683	5.387	-216.600	15.011	31.104
12132	4.4560	55	286.0	R	7.764	3.081	-2.207	5.387	-216.599	20.254	35.087
CESIUM CHLORIDE				V	6.573	2.490	14.542	5.640	-169.708	14.810	31.736
12133	3.9880	55	286.0	R	8.170	3.243	-4.215	5.640	-169.709	21.652	36.921
CESIUM IODIDE				V	6.704	2.566	18.510	5.457	-343.708	16.969	35.074
12134	4.5250	55	286.0	R	7.740	3.062	2.415	5.457	-343.707	21.664	38.681
COPPER				V	4.529	1.642	1.640	5.587	-2.977	2.824	7.221
12135	8.9300	68	298.0	R	3.061	1.070	1.676	5.587	-2.977	2.069	6.829
COPPER				V	4.716	1.728	1.342	5.440	-2.972	2.999	7.198
12136	8.9300	133	ROOM	R	2.563	0.874	2.204	5.440	-2.972	1.545	6.362
GALLIUM ANTIMONIDE				V	2.660	0.670	14.066	4.747	-14.949	0.141	7.875
12137	5.6137	170	298.0	R	2.108	0.467	14.262	4.747	-14.949	-0.419	7.587
GALLIUM ARSENIDE				V	2.693	0.702	9.863	4.490	-7.897	0.176	6.527
12138	5.3169	169	298.0	R	2.044	0.452	10.209	4.490	-7.897	-0.456	6.165
GARNET (ALMANDITE-PYROPE)				V	4.141	1.400	3.045	5.433	-1.734	2.188	7.874
12139	4.1602	242	298.0	R	4.132	1.396	3.051	5.433	-1.734	2.179	7.867
GOLD				V	3.521	1.249	-0.234	6.430	-2.159	2.182	5.333
12140	19.3000	68	298.0	R	2.373	0.824	0.195	6.430	-2.158	1.587	5.045
GOLD				V	2.992	1.064	-0.362	5.213	-1.732	1.821	4.194
12141	19.3000	133	ROOM	R	2.035	0.709	0.019	5.213	-1.732	1.332	3.952
GOLD(100.0)				V	3.594	1.280	-0.446	6.150	-2.060	2.246	5.147
12142	19.2830	114	296.5	R	2.396	0.835	0.063	6.150	-2.060	1.611	4.833
GOLD(97.05)-NICKEL(2.95)				V	3.642	1.296	-0.319	6.360	-2.135	2.257	5.344
12143	19.0600	114	296.5	R	2.417	0.841	0.174	6.360	-2.135	1.612	5.027
GOLD(91.28)-NICKEL(9.72)				V	3.633	1.295	-0.231	6.210	-2.114	2.217	5.280
12144	18.5330	114	296.5	R	2.412	0.840	0.245	6.209	-2.114	1.579	4.963
GOLD(75.80)-NICKEL(24.20)				V	3.866	1.388	-0.130	5.990	-2.056	2.266	5.279
12145	17.3410	114	296.5	R	2.658	0.931	0.339	5.990	-2.056	1.647	4.956

CUBIC

CODE	NAME / DENSITY	REF	TEMP(K)		E	G	SIGMA	K	BETA	VS	VP
							PRESSURE DERIVATIVES				
GOLD(57.58)-NICKEL(42.42)				V	4.017	1.446	0.391	5.970	-2.054	2.194	5.421
12146	15.7120	114	296.5	R	2.606	0.907	0.870	5.970	-2.054	1.476	5.042
IRON				V	5.681	2.060	2.441	5.973	-2.144	2.879	7.387
12147	7.8720	220	300.0	R	4.889	1.754	2.067	5.973	-2.144	2.701	7.337
IRON				V	5.374	1.982	1.840	5.293	-1.911	2.733	6.569
12148	7.8672	121	300.0	R	4.552	1.652	1.596	5.293	-1.911	2.503	6.464
LEAD				V	3.598	1.294	0.067	5.447	-0.254	1.430	3.471
12149	11.4420	184	195.0	R	1.790	0.615	0.267	5.447	-0.254	0.768	3.143
LEAD				V	3.762	1.356	-0.062	5.530	-0.276	1.630	3.693
12150	11.3440	184	296.0	R	1.883	0.651	0.182	5.530	-0.276	0.907	3.329
LITHIUM FLUORIDE				V	6.203	2.276	7.041	5.143	-10.606	6.598	16.214
12151	2.6390	185	300.0	R	8.159	3.301	0.634	5.143	-10.606	11.889	20.537
LITHIUM FLUORIDE				V	4.879	1.716	7.261	4.460	-9.162	4.200	12.470
12152	2.6380	79	ROOM	R	6.385	2.519	2.184	4.460	-9.162	8.358	15.899
MAGNESIUM OXIDE				V	6.475	2.420	2.886	4.769	-1.761	3.575	8.324
12153	3.5995	20	77.4	R	7.545	3.017	1.530	4.769	-1.761	5.079	9.598
MAGNESIUM OXIDE				V	5.926	2.193	2.861	4.487	-1.705	3.124	7.601
12154	3.5833	20	296.0	R	7.170	2.882	1.267	4.487	-1.705	4.890	9.092
MAGNESIUM OXIDE				V	5.735	2.140	2.613	4.267	-1.620	3.006	7.189
12155	3.5790	54	298.0	R	6.961	2.816	1.056	4.267	-1.620	4.740	8.651
MAGNESIUM OXIDE				V	5.503	2.110	1.999	3.847	-1.451	2.935	6.520
12156	3.5833	243	300.0	R	6.742	2.787	0.461	3.847	-1.451	4.668	7.973
MAGNESIUM OXIDE				V	5.514	2.150	1.718	3.743	-1.586	3.242	6.792
12157	3.5206	243	800.0	R	7.193	3.043	-0.554	3.743	-1.586	5.742	8.857
MAGNESIUM OXIDE (P=8KB)				V	5.549	2.130	1.952	3.880	-1.412	2.964	6.547
12158	3.6010	243	300.0	R	6.735	2.776	0.521	3.880	-1.412	4.593	7.912
MAGNESIUM OXIDE (P=8KB)				V	5.568	2.176	1.640	3.767	-1.535	3.284	6.812
12159	3.5396	243	800.0	R	7.184	3.035	-0.481	3.767	-1.535	5.643	8.761
POTASSIUM BROMIDE				V	5.666	2.077	26.360	5.381	-243.558	15.013	37.097
12160	2.7533	213	300.0	R	1.201	0.151	58.966	5.381	-243.558	-3.579	24.386
POTASSIUM CHLORIDE				V	5.463	1.984	21.959	5.347	-148.418	14.582	37.482
12161	2.0089	23	195.0	R	0.905	0.026	48.052	5.347	-148.420	-5.166	24.056
POTASSIUM CHLORIDE				V	5.493	2.010	20.801	5.340	-162.040	15.691	39.105
12162	1.9874	23	95.0	R	1.190	0.148	48.015	5.340	-162.040	-3.855	25.664
POTASSIUM CHLORIDE				V	5.374	1.952	21.837	5.373	-159.982	15.053	38.717
12163	1.9865	79	ROOM	R	0.757	-0.040	51.292	5.373	-159.983	-6.089	24.237
POTASSIUM FLUORIDE				V	4.851	1.745	12.757	5.020	-51.802	8.826	23.935
12164	2.4800	145	298.0	R	1.798	0.382	25.678	5.020	-51.801	-1.039	16.865
POTASSIUM IODIDE				V	6.055	2.169	44.688	6.277	-464.638	17.273	46.037
12165	3.1257	213	300.0	R	1.333	0.187	83.455	6.277	-464.636	-3.121	32.645
RUBIDIUM BROMIDE				V	5.436	1.877	44.611	6.539	-343.215	13.120	40.743
12166	3.3590	213	300.0	R	0.548	-0.091	74.291	6.539	-343.216	-5.637	29.265
RUBIDIUM BROMIDE				V	5.505	2.060	18.715	4.965	-230.200	13.942	31.748
12167	3.3920	110	180.0	R	-0.380	-0.397	61.029	4.965	-230.197	-8.929	16.941
RUBIDIUM BROMIDE				V	5.759	2.142	21.477	5.345	-257.869	15.065	34.931
12168	3.3780	110	220.0	R	-0.256	-0.366	65.620	5.345	-257.867	-8.678	19.562
RUBIDIUM BROMIDE				V	5.530	2.019	26.758	5.538	-278.032	14.233	35.677
12169	3.3640	110	60.0	R	-0.190	-0.350	68.732	5.538	-278.031	-8.595	21.038
RUBIDIUM BROMIDE				V	5.645	2.080	24.316	5.513	-287.835	15.195	36.593
12170	3.3500	110	0.0	R	-0.064	-0.299	68.613	5.513	-287.834	-8.088	21.514

CODE	NAME DENSITY	REF	TEMP(K)		E	G	SIGMA PRESSURE	K DERIVATIVES	BETA	VS	VP
RUBIDIUM CHLORIDE				V	5.810	2.148	20.063	5.509	-186.565	15.044	35.774
12171	2.8346	110	80.0	R	-0.143	-0.343	56.926	5.509	-186.566	-8.407	20.494
RUBIDIUM CHLORIDE				V	5.657	2.088	19.907	5.427	-190.528	14.817	35.455
12172	2.8223	110	222.0	R	-0.150	-0.344	57.400	5.427	-190.529	-8.517	20.205
RUBIDIUM CHLORIDE				V	5.555	2.044	20.627	5.433	-198.928	14.731	35.764
12173	2.8097	110	260.0	R	-0.069	-0.314	58.202	5.433	-198.927	-8.273	20.700
RUBIDIUM CHLORIDE				V	5.703	2.112	19.489	5.475	-209.641	15.747	37.116
12174	2.7969	110	304.0	R	0.110	-0.245	58.629	5.475	-209.642	-7.580	21.716
RUBIDIUM IODIDE				V	5.749	2.145	25.825	5.236	-380.331	15.950	36.648
12175	3.6020	110	180.0	R	-0.446	-0.400	76.383	5.236	-380.329	-9.751	20.539
RUBIDIUM IODIDE				V	5.689	2.122	25.411	5.225	-395.341	16.162	37.088
12176	3.5850	110	220.0	R	-0.403	-0.385	77.679	5.225	-395.340	-9.677	20.808
RUBIDIUM IODIDE				V	5.823	2.170	26.293	5.407	-425.983	17.091	39.214
12177	3.5690	110	260.0	R	-0.281	-0.346	80.539	5.407	-425.980	-9.273	22.540
RUBIDIUM IODIDE				V	5.653	2.103	25.796	5.335	-435.590	16.874	38.956
12178	3.5510	110	300.0	R	-0.208	-0.315	80.166	5.335	-435.585	-8.965	22.566
SILICON				V	2.262	0.468	8.212	4.230	-4.337	-0.884	6.970
12179	2.3320	164	77.2	R	1.896	0.328	8.366	4.230	-4.337	-1.341	6.707
SILICON				V	2.356	0.508	8.215	4.237	-4.422	-0.746	7.118
12180	2.3310	164	298.0	R	1.986	0.365	8.380	4.237	-4.422	-1.213	6.844
SILVER				V	4.541	1.642	1.469	6.180	-5.758	3.492	8.775
12181	10.5000	68	298.0	R	3.234	1.137	1.588	6.180	-5.758	2.721	8.376
SILVER				V	5.538	2.126	-3.056	4.113	-4.016	4.781	7.033
12182	10.5000	133	ROOM	R	3.820	1.410	-1.500	4.113	-4.016	3.500	6.162
SODIUM BROMIDE				V	5.859	2.218	13.004	4.953	-121.273	12.968	28.294
12183	3.2030	145	298.0	R	4.348	1.510	23.556	4.953	-121.273	7.613	24.261
SODIUM CHLORIDE				V	5.667	2.096	14.203	5.123	-77.009	12.529	30.043
12184	2.1891	23	195.0	R	4.030	1.330	22.841	5.123	-77.007	6.430	25.459
SODIUM CHLORIDE				V	5.777	2.138	14.657	5.273	-86.365	13.502	32.159
12185	2.1644	23	295.0	R	4.407	1.496	22.538	5.273	-86.364	8.137	28.116
SODIUM CHLORIDE				V	5.867	2.174	14.106	5.383	-84.547	13.880	32.812
12186	2.1640	250	ROOM	R	4.472	1.523	21.999	5.383	-84.547	8.443	28.732
SODIUM CHLORIDE				V	5.543	2.068	12.544	4.927	-79.248	12.844	29.802
12187	2.1640	79	ROOM	R	3.999	1.342	21.597	4.927	-79.247	6.745	25.192
SODIUM FLUORIDE				V	5.619	2.039	9.478	5.183	-22.311	7.355	18.944
12188	2.8040	185	300.0	R	4.568	1.529	12.731	5.183	-22.311	4.740	16.916

CUBIC

CODE	NAME DENSITY	REF	TEMP(K)		E	G	SIGMA	K	BETA	VS	VP
					TEMPERATURE DERIVATIVES						
ALUMINUM				V	-0.362	-0.144	0.465	-0.162	0.281	-0.745	-0.797
12189	2.6970	254	298.0	R	-0.366	-0.145	0.475	-0.162	0.281	-0.756	-0.803
BARIUM FLUORIDE				V	-0.148	-0.055	-0.070	-0.152	0.424	-0.188	-0.418
12190	4.9120	278	195.0	R	-0.147	-0.055	-0.074	-0.152	0.424	-0.186	-0.416
BARIUM FLUORIDE				V	-0.156	-0.059	-0.058	-0.155	0.454	-0.200	-0.428
12191	4.8860	278	298.0	R	-0.156	-0.058	-0.061	-0.155	0.454	-0.198	-0.427
CALCIUM FLUORIDE				V	-0.263	-0.107	0.100	-0.157	0.213	-0.356	-0.536
12192	3.1980	278	195.0	R	-0.292	-0.119	0.181	-0.157	0.213	-0.430	-0.587
CALCIUM FLUORIDE				V	-0.287	-0.116	0.100	-0.176	0.249	-0.382	-0.583
12193	3.1810	278	298.0	R	-0.314	-0.127	0.184	-0.176	0.249	-0.457	-0.636
GARNET (ALMANDITE-PYROPE)				V	-0.271	-0.106	-0.002	-0.201	0.064	-0.217	-0.391
12194	4.1602	242	298.0	R	-0.271	-0.106	-0.002	-0.201	0.064	-0.217	-0.391
MAGNESIUM OXIDE				V	-0.429	-0.198	0.136	-0.153	0.058	-0.348	-0.431
12195	3.5833	243	300.0	R	-0.550	-0.261	0.271	-0.153	0.058	-0.508	-0.561
MAGNESIUM OXIDE				V	-0.462	-0.204	0.090	-0.200	0.085	-0.363	-0.500
12196	3.5206	243	800.0	R	-0.603	-0.276	0.268	-0.200	0.085	-0.566	-0.663
MAGNESIUM OXIDE (P=8KB)				V	-0.426	-0.197	0.138	-0.150	0.055	-0.344	-0.423
12197	3.6010	243	300.0	R	-0.540	-0.256	0.261	-0.150	0.055	-0.492	-0.543
MAGNESIUM OXIDE (P=8KB)				V	-0.463	-0.205	0.095	-0.198	0.081	-0.362	-0.494
12198	3.5396	243	800.0	R	-0.596	-0.273	0.258	-0.198	0.081	-0.550	-0.646
RUBIDIUM BROMIDE				V	-0.162	-0.068	0.771	-0.070	3.668	-0.600	-0.762
12199	3.3500	110	300.0	R	-0.055	-0.020	-0.204	-0.070	3.668	-0.169	-0.447
RUBIDIUM CHLORIDE				V	-0.176	-0.074	0.670	-0.080	3.063	-0.643	-0.831
12200	2.7969	110	300.0	R	-0.064	-0.023	-0.210	-0.080	3.063	-0.187	-0.497
RUBIDIUM IODIDE				V	-0.138	-0.058	0.852	-0.059	4.790	-0.556	-0.697
12201	3.5510	110	300.0	R	-0.040	-0.014	-0.251	-0.059	4.790	-0.127	-0.387
SODIUM CHLORIDE				V	0.139	0.050	0.493	0.144	-2.256	0.451	1.088
12202	2.1640	250	ROOM	R	0.072	0.018	0.897	0.144	-2.256	0.180	0.883

TETRAGONAL 1

NAME / CODE	DENSITY	REF	TEMP(K)		E	G	SIGMA	K	BETA	VS	VP
AMMONIUM DIHYDROGEN ARSENATE				V	0.340	0.135	0.257	0.233	4.287	2.420	4.231
22203	2.3110	1	298.0	R	0.225	0.084	0.336	0.229	4.370	1.909	3.842
AMMONIUM DIHYDROGEN PHOSPHATE				V	0.339	0.136	0.246	0.222	4.500	2.746	4.730
22204	1.8030	129	293.0	R	0.253	0.097	0.305	0.216	4.628	2.321	4.378
AMMONIUM DIHYDROGEN PHOSPHATE				V	0.344	0.132	0.298	0.283	3.531	2.711	5.050
22205	1.8030	4	ROOM	R	0.253	0.094	0.343	0.269	3.722	2.286	4.676
AMMONIUM DIHYDROGEN PHOSPHATE				V	0.333	0.128	0.307	0.288	3.475	2.660	5.039
22206	1.8030	4	ROOM	R	0.251	0.093	0.348	0.274	3.646	2.270	4.699
AMMONIUM DIHYDROGEN PHOSPHATE				V	0.331	0.127	0.305	0.283	3.535	2.654	5.008
22207	1.8030	4	ROOM	R	0.251	0.093	0.345	0.270	3.707	2.273	4.675
AMMONIUM DIHYDROGEN PHOSPHATE (D=0)				V	0.341	0.131	0.304	0.290	3.452	2.692	5.072
22208	1.8030	138	ROOM	R	0.262	0.098	0.341	0.275	3.630	2.329	4.745
AMMONIUM DIHYDROGEN PHOSPHATE				V	0.313	0.119	0.311	0.276	3.626	2.572	4.911
22209	1.8030	138	ROOM	R	0.244	0.091	0.346	0.265	3.775	2.243	4.626
AMMONIUM DIHYDROGEN PHOSPHATE				V	0.394	0.167	0.180	0.205	4.875	3.041	4.869
22210	1.8030	138	ROOM	R	0.250	0.097	0.293	0.202	4.948	2.317	4.286
AMMONIUM DIHYDROGEN PHOSPHATE				V	0.336	0.129	0.306	0.290	3.452	2.672	5.058
22211	1.8030	1	298.0	R	0.253	0.094	0.347	0.275	3.630	2.281	4.713
BARIUM TITANATE (E=0)				V	1.623	0.599	0.355	1.866	0.536	3.160	6.665
22212	5.9988	31	298.0	R	1.300	0.476	0.367	1.629	0.614	2.815	6.142
BARIUM TITANATE (D=0)				V	1.941	0.731	0.327	1.869	0.535	3.491	6.886
22213	5.9988	31	298.0	R	1.663	0.623	0.335	1.678	0.596	3.222	6.466
INDIUM				V	0.170	0.059	0.432	0.416	2.404	0.901	2.604
22214	7.3000	138	ROOM	R	0.108	0.037	0.457	0.416	2.405	0.714	2.525
MAGNESIUM FLUORIDE				V	1.422	0.579	0.228	0.872	1.146	4.269	7.195
22215	3.1770	62	ROOM	R	1.216	0.481	0.266	0.865	1.156	3.889	6.884
NICKEL SULPHATE HEXAHYDRATE				V	0.301	0.126	0.195	0.165	6.074	2.467	4.008
22216	2.0700	177	ROOM	R	0.241	0.097	0.237	0.153	6.550	2.168	3.693
POTASSIUM DIHYDROGEN ARSENATE				V	0.351	0.164	0.071	0.137	7.317	2.392	3.520
22217	2.8670	129	ROOM	R	0.291	0.128	0.138	0.134	7.452	2.113	3.260
POTASSIUM DIHYDROGEN PHOSPHATE				V	0.430	0.163	0.321	0.400	2.500	2.609	5.082
22218	2.3880	129	280.0	R	0.345	0.127	0.356	0.400	2.501	2.309	4.884
POTASSIUM DIHYDROGEN PHOSPHATE				V	0.423	0.167	0.265	0.301	3.326	2.646	4.683
22219	2.3880	4	ROOM	R	0.315	0.119	0.322	0.295	3.389	2.233	4.359
POTASSIUM DIHYDROGEN PHOSPHATE				V	0.439	0.177	0.241	0.283	3.539	2.723	4.661
22220	2.3880	4	ROOM	R	0.329	0.126	0.305	0.282	3.550	2.297	4.340
POTASSIUM DIHYDROGEN PHOSPHATE				V	0.442	0.180	0.227	0.270	3.702	2.746	4.622
22221	2.3880	4	ROOM	R	0.330	0.127	0.296	0.270	3.706	2.311	4.292
POTASSIUM DIHYDROGEN PHOSPHATE				V	0.445	0.182	0.223	0.268	3.738	2.761	4.623
22222	2.3880	138	ROOM	R	0.332	0.128	0.293	0.267	3.739	2.317	4.284
POTASSIUM DIHYDROGEN PHOSPHATE				V	0.394	0.144	0.369	0.502	1.991	2.456	5.392
22223	2.3880	138	ROOM	R	0.327	0.118	0.391	0.502	1.992	2.220	5.252
RUBIDIUM DIHYDROGEN PHOSPHATE				V	0.394	0.158	0.245	0.258	3.879	2.349	4.044
22224	2.8660	186	293.0	R	0.242	0.090	0.343	0.257	3.891	1.773	3.628
RUBIDIUM DIHYDROGEN PHOSPHATE				V	0.360	0.157	0.148	0.170	5.867	2.366	3.681
22225	2.8000	2	298.0	R	0.235	0.092	0.270	0.170	5.870	1.817	3.238
RUTILE				V	2.840	1.112	0.277	2.120	0.472	5.109	9.196
22226	4.2600	263	ROOM	R	1.864	0.691	0.349	2.062	0.485	4.027	8.368
RUTILE				V	3.171	1.259	0.260	2.198	0.455	5.433	9.534
22227	4.2640	35	ROOM	R	2.617	1.012	0.293	2.106	0.475	4.872	9.002

CODE	NAME DENSITY	REF	TEMP(K)		E	G	SIGMA	K	BETA	VS	VP
RUTILE				V	3.117	1.244	0.253	2.104	0.475	5.409	9.408
22228	4.2500	267	ROOM	R	2.556	0.991	0.290	2.030	0.493	4.828	8.879
RUTILE				V	3.161	1.254	0.261	2.201	0.454	5.425	9.535
22229	4.2600	171	298.0	R	2.579	0.995	0.296	2.109	0.474	4.833	8.981
TELLURIUM OXIDE				V	0.715	0.291	0.230	0.441	2.268	2.203	3.719
22230	5.9900	256	ROOM	R	0.250	0.089	0.404	0.434	2.306	1.220	3.037
TIN (WHITE)				V	1.046	0.442	0.183	0.550	1.818	2.459	3.948
22231	7.3100	129	ROOM	R	0.980	0.410	0.195	0.535	1.868	2.368	3.847
TIN (WHITE)				V	0.512	0.191	0.338	0.527	1.898	1.621	3.278
22232	7.2800	138	ROOM	R	0.410	0.150	0.370	0.526	1.900	1.434	3.158
TIN (WHITE)				V	0.601	0.236	0.270	0.436	2.294	1.802	3.213
22233	7.2800	138	ROOM	R	0.596	0.235	0.270	0.431	2.318	1.796	3.198
TIN (WHITE)				V	0.676	0.259	0.305	0.579	1.727	1.873	3.539
22234	7.3800	210	4.2	R	0.606	0.229	0.325	0.579	1.727	1.760	3.461
TIN (WHITE)				V	0.654	0.249	0.313	0.583	1.715	1.838	3.524
22235	7.3670	210	77.0	R	0.585	0.219	0.333	0.583	1.715	1.726	3.448
TIN (WHITE)				V	0.545	0.204	0.335	0.550	1.818	1.675	3.361
22236	7.2790	210	300.0	R	0.424	0.154	0.371	0.549	1.821	1.456	3.221
TIN (WHITE)				V	0.634	0.241	0.317	0.578	1.730	1.817	3.511
22237	7.2900	135	93.0	R	0.565	0.211	0.337	0.578	1.730	1.703	3.435
TIN (WHITE)				V	0.629	0.238	0.322	0.588	1.700	1.806	3.524
22238	7.2900	135	108.0	R	0.558	0.208	0.342	0.588	1.700	1.688	3.445
TIN (WHITE)				V	0.612	0.231	0.323	0.575	1.740	1.781	3.480
22239	7.2900	135	138.0	R	0.538	0.200	0.344	0.575	1.740	1.657	3.398
TIN (WHITE)				V	0.597	0.225	0.326	0.571	1.750	1.758	3.458
22240	7.2900	135	168.0	R	0.520	0.193	0.348	0.571	1.750	1.627	3.372
TIN (WHITE)				V	0.584	0.219	0.333	0.581	1.720	1.734	3.462
22241	7.2900	135	198.0	R	0.502	0.185	0.356	0.581	1.720	1.593	3.370
TIN (WHITE)				V	0.570	0.213	0.339	0.588	1.700	1.709	3.459
22242	7.2900	135	228.0	R	0.481	0.176	0.364	0.588	1.700	1.555	3.360
TIN (WHITE)				V	0.551	0.206	0.340	0.575	1.739	1.679	3.413
22243	7.2900	135	258.0	R	0.456	0.167	0.368	0.575	1.740	1.513	3.307
TIN (WHITE)				V	0.537	0.199	0.352	0.606	1.650	1.651	3.457
22244	7.2900	135	288.0	R	0.433	0.157	0.381	0.606	1.650	1.466	3.343
TIN (WHITE)				V	0.519	0.192	0.353	0.589	1.698	1.622	3.403
22245	7.2900	135	318.0	R	0.406	0.147	0.385	0.588	1.700	1.418	3.279
TIN (WHITE)				V	0.503	0.184	0.365	0.622	1.607	1.589	3.450
22246	7.2900	135	348.0	R	0.380	0.136	0.398	0.621	1.610	1.366	3.318
TIN (WHITE)				V	0.483	0.176	0.374	0.639	1.564	1.553	3.462
22247	7.2900	135	378.0	R	0.350	0.124	0.408	0.637	1.570	1.305	3.318
TIN (WHITE)				V	0.464	0.166	0.393	0.721	1.387	1.511	3.597
22248	7.2900	135	408.0	R	0.315	0.111	0.426	0.714	1.400	1.231	3.438
TIN (WHITE)				V	0.444	0.157	0.410	0.825	1.212	1.469	3.768
22249	7.2900	135	438.0	R	0.276	0.096	0.443	0.813	1.230	1.145	3.592
TIN (WHITE)				V	0.420	0.149	0.408	0.757	1.321	1.431	3.622
22250	7.2900	135	468.0	R	0.234	0.081	0.447	0.730	1.370	1.053	3.390
ZIRCON				V	0.485	0.217	0.116	0.210	4.752	2.182	3.311
22251	4.5600	129	ROOM	R	0.417	0.184	0.135	0.191	5.248	2.007	3.090
ZIRCON				V	2.102	0.789	0.332	2.081	0.481	4.098	8.164
22252	4.7000	223	ROOM	R	1.873	0.695	0.347	2.045	0.489	3.845	7.951

TETRAGONAL 1

CODE	NAME DENSITY	REF	TEMP(K)		E	G	SIGMA	K	BETA	VS	VP
					PRESSURE DERIVATIVES						
RUTILE				V	5.476	1.869	3.050	6.618	-1.366	2.811	9.048
22253	4.2600	171	298.0	R	0.477	-0.305	6.364	6.975	-1.440	-1.886	6.456

TETRAGONAL 1

CODE	NAME DENSITY	REF	TEMP(K)		E	G	SIGMA	K	BETA	VS	VP
					TEMPERATURE DERIVATIVES						
RUTILE				V	-0.725	-0.289	0.017	-0.489	0.101	-0.626	-1.077
22254	4.2600	171	298.0	R	-0.297	-0.096	-0.241	-0.493	0.102	-0.233	-0.811

CODE	NAME DENSITY	REF	TEMP(K)		E	G	SIGMA	K	BETA	VS	VP
CALCIUM MCLYBDATE (+Z=-Z OF REF 266?)				V	1.051	0.409	0.286	0.817	1.224	3.099	5.657
72255	4.2550	19	298.0	R	1.000	0.387	0.293	0.805	1.242	3.015	5.572
CALCIUM MCLYBDATE				V	1.054	0.410	0.286	0.822	1.216	3.103	5.671
72256	4.2550	266	298.0	R	1.003	0.387	0.295	0.815	1.226	3.017	5.595

ORTHORHOMBIC

CODE / NAME	DENSITY	REF	TEMP(K)		E	G	SIGMA	K	BETA	VS	VP
ACENAPHTHENE				V	0.084	0.033	0.287	0.066	15.205	1.636	2.993
32257	1.2200	183	293.0	R	0.075	0.029	0.309	0.066	15.212	1.534	2.920
AMMONIUM SULFATE				V	0.238	0.091	0.317	0.217	4.616	2.259	4.361
32258	1.7740	126	293.0	R	0.234	0.089	0.319	0.215	4.659	2.235	4.331
ARAGONITE				V	0.951	0.404	0.176	0.490	2.043	3.714	5.924
32259	2.9300	128	ROOM	R	0.864	0.367	0.178	0.447	2.235	3.537	5.653
BARITE				V	0.653	0.253	0.293	0.525	1.903	2.396	4.427
32260	4.4000	128	ROOM	R	0.565	0.214	0.320	0.524	1.910	2.204	4.287
BARITE				V	0.647	0.246	0.317	0.589	1.697	2.337	4.514
32261	4.5000	129	ROOM	R	0.566	0.211	0.340	0.588	1.699	2.166	4.397
BARITE				V	0.628	0.241	0.302	0.530	1.886	2.315	4.351
32262	4.5000	138	ROOM	R	0.544	0.205	0.329	0.529	1.890	2.133	4.222
BARITE				V	0.640	0.244	0.309	0.560	1.786	2.326	4.428
32263	4.5170	214	308.0	R	0.562	0.212	0.329	0.546	1.830	2.164	4.283
BARITE				V	0.630	0.238	0.323	0.594	1.682	2.297	4.494
32264	4.5170	214	338.0	R	0.562	0.211	0.334	0.565	1.769	2.160	4.328
BARITE				V	0.644	0.235	0.369	0.818	1.223	2.281	5.004
32265	4.5170	214	378.0	R	0.563	0.205	0.377	0.764	1.309	2.128	4.791
BARITE				V	0.795	0.274	0.450	2.646	0.378	2.464	8.165
32266	4.5170	214	413.0	R	0.571	0.199	0.438	1.527	0.655	2.097	6.298
BARITE				V	0.753	0.261	0.443	2.220	0.450	2.403	7.540
32267	4.5170	214	438.0	R	0.565	0.197	0.435	1.445	0.692	2.089	6.149
BARITE				V	1.303	0.441	0.478	9.804	0.102	3.124	15.168
32268	4.5170	214	458.0	R	0.566	0.194	0.457	2.203	0.454	2.073	7.382
BENZENE				V	0.075	0.029	0.288	0.058	17.107	1.651	3.024
32269	1.0610	130	170.0	R	0.062	0.023	0.323	0.058	17.199	1.484	2.901
BENZENE				V	0.071	0.028	0.293	0.057	17.408	1.613	2.981
32270	1.0610	130	180.0	R	0.059	0.022	0.327	0.057	17.502	1.449	2.861
BENZENE				V	0.069	0.026	0.297	0.056	17.804	1.578	2.935
32271	1.0610	130	190.0	R	0.056	0.021	0.332	0.056	17.883	1.413	2.817
BENZENE				V	0.066	0.025	0.299	0.055	18.326	1.544	2.885
32272	1.0610	130	200.0	R	0.054	0.020	0.335	0.054	18.398	1.380	2.768
BENZENE				V	0.063	0.024	0.303	0.053	18.801	1.507	2.836
32273	1.0610	130	210.0	R	0.051	0.019	0.339	0.053	18.868	1.344	2.721
BENZENE				V	0.060	0.023	0.306	0.051	19.422	1.471	2.782
32274	1.0610	130	220.0	R	0.049	0.018	0.340	0.051	19.467	1.316	2.674
BENZENE				V	0.056	0.021	0.313	0.050	19.969	1.422	2.723
32275	1.0610	130	230.0	R	0.045	0.017	0.348	0.050	19.993	1.261	2.614
BENZENE				V	0.054	0.020	0.313	0.048	20.887	1.388	2.661
32276	1.0610	130	240.0	R	0.044	0.016	0.347	0.048	20.903	1.240	2.561
BENZENE				V	0.051	0.019	0.316	0.046	21.744	1.348	2.600
32277	1.0610	130	250.0	R	0.042	0.016	0.347	0.046	21.754	1.215	2.510
BENZOPHENONE				V	0.067	0.026	0.294	0.054	18.518	1.455	2.693
32278	1.2190	4	ROOM	R	0.059	0.023	0.299	0.049	20.334	1.369	2.556
BRONZITE				V	1.534	0.613	0.251	1.027	0.974	4.259	7.387
32279	3.3800	223	ROOM	R	1.507	0.601	0.253	1.017	0.983	4.218	7.335
BRONZITE (84.5% ENSTATITE)				V	1.838	0.759	0.211	1.061	0.942	4.769	7.883
32280	3.3350	147	ROOM	R	1.804	0.745	0.210	1.038	0.963	4.728	7.806
CALCIUM SULFATE				V	0.883	0.355	0.244	0.575	1.738	3.462	5.950
32281	2.9620	228	ROOM	R	0.604	0.231	0.307	0.522	1.915	2.794	5.295

CODE	NAME DENSITY	REF	TEMP(K)		E	G	SIGMA	K	BETA	VS	VP
CELESTITE				V	0.628	0.229	0.372	0.821	1.219	2.404	5.332
32282	3.9600	129	ROOM	R	0.555	0.200	0.387	0.820	1.220	2.248	5.238
CESIUM SULFATE				V	0.326	0.126	0.294	0.263	3.797	1.723	3.188
32283	4.2430	126	293.0	R	0.324	0.125	0.294	0.261	3.825	1.717	3.177
FORSTERITE				V	2.046	0.826	0.239	1.307	0.765	5.061	8.643
32284	3.2240	148	298.0	R	1.966	0.792	0.241	1.265	0.791	4.958	8.485
FORSTERITE				V	2.060	0.832	0.238	1.312	0.762	5.081	8.669
32285	3.2217	116	298.0	R	1.985	0.801	0.239	1.269	0.788	4.986	8.516
IODIC ACID				V	0.421	0.181	0.161	0.207	4.823	1.980	3.115
32286	4.6300	177	ROOM	R	0.396	0.169	0.172	0.201	4.970	1.911	3.036
LITHIUM AMMONIUM TARTRATE				V	0.351	0.139	0.260	0.244	4.099	2.854	5.013
32287	1.7100	129	ROOM	R	0.298	0.117	0.279	0.225	4.440	2.612	4.719
MAGNESIUM SULFATE HEPTAHYDRATE				V	0.488	0.185	0.315	0.440	2.273	3.366	6.480
32288	1.6360	129	ROOM	R	0.424	0.159	0.336	0.430	2.324	3.113	6.263
MAGNESIUM SULFATE HEPTAHYDRATE				V	0.244	0.090	0.360	0.291	3.435	2.310	4.944
32289	1.6800	4	ROOM	R	0.218	0.079	0.374	0.289	3.464	2.170	4.844
MAGNESIUM SULFATE HEPTAHYDRATE				V	0.240	0.091	0.321	0.223	4.491	2.356	4.584
32290	1.6360	11	ROOM	R	0.218	0.081	0.336	0.222	4.507	2.231	4.494
NATROLITE				V	0.744	0.298	0.248	0.493	2.029	3.640	6.291
32291	2.2500	223	ROOM	R	0.658	0.263	0.252	0.443	2.259	3.417	5.936
NICKEL SULFATE HEPTAHYDRATE				V	0.262	0.099	0.322	0.244	4.093	2.254	4.395
32292	1.9480	11	ROOM	R	0.234	0.087	0.340	0.244	4.106	2.117	4.299
OLIVINE (91.7% FORSTERITE)				V	2.051	0.824	0.244	1.337	0.748	4.979	8.559
32293	3.3240	262	ROOM	R	1.975	0.793	0.245	1.289	0.776	4.886	8.403
OLIVINE (93.0% FORSTERITE)				V	2.006	0.805	0.246	1.315	0.760	4.932	8.494
32294	3.3110	148	298.0	R	1.931	0.774	0.247	1.272	0.786	4.835	8.343
POTASSIUM PENTABORATE				V	0.246	0.091	0.350	0.274	3.653	2.286	4.765
32295	1.7400	138	ROOM	R	0.171	0.062	0.381	0.241	4.154	1.888	4.311
POTASSIUM SULFATE				V	0.443	0.175	0.268	0.319	3.136	2.560	4.551
32296	2.6650	126	293.0	R	0.439	0.173	0.271	0.319	3.136	2.544	4.538
ROCHELLE SALT				V	0.361	0.123	0.463	1.626	0.615	2.625	10.001
32297	1.7900	128	ROOM	R	0.223	0.077	0.452	0.769	1.300	2.071	6.978
ROCHELLE SALT				V	0.255	0.092	0.386	0.374	2.677	2.267	5.265
32298	1.7900	128	ROOM	R	0.174	0.062	0.399	0.286	3.500	1.863	4.538
ROCHELLE SALT				V	0.237	0.082	0.447	0.752	1.330	2.140	6.936
32299	1.7900	128	ROOM	R	0.186	0.064	0.453	0.667	1.500	1.892	6.482
ROCHELLE SALT				V	0.250	0.090	0.395	0.396	2.523	2.246	5.391
32300	1.7750	128	ROOM	R	0.193	0.068	0.414	0.375	2.670	1.963	5.122
ROCHELLE SALT				V	0.242	0.093	0.300	0.201	4.964	2.280	4.265
32301	1.7900	138	ROOM	R	0.191	0.072	0.334	0.193	5.189	2.002	4.014
RUBIDIUM SULFATE				V	0.390	0.152	0.280	0.296	3.375	2.052	3.714
32302	3.6210	126	293.0	R	0.389	0.152	0.281	0.296	3.378	2.048	3.711
SODIUM AMMONIUM TARTRATE				V	0.219	0.078	0.400	0.365	2.738	2.220	5.440
32303	1.5870	129	ROOM	R	0.179	0.063	0.412	0.339	2.950	1.996	5.165
SODIUM TARTRATE				V	0.272	0.098	0.386	0.399	2.508	2.340	5.434
32304	1.7940	129	ROOM	R	0.208	0.074	0.409	0.382	2.617	2.027	5.175
STAUROLITE				V	1.865	0.772	0.208	1.063	0.941	4.787	7.881
32305	3.3690	4	ROOM	R	1.638	0.692	0.183	0.862	1.160	4.533	7.280
STRONTIUM FORMATE DIHYDRATE				V	0.340	0.168	0.011	0.116	8.632	2.734	3.888
32306	2.2500	129	ROOM	R	0.307	0.149	0.028	0.108	9.240	2.575	3.695

ORTHORHOMBIC

CODE	NAME / DENSITY	REF	TEMP(K)		E	G	SIGMA	K	BETA	VS	VP
SULPHUR				V	0.194	0.072	0.343	0.206	4.854	1.868	3.821
32307	2.0700	129	ROOM	R	0.166	0.062	0.343	0.176	5.695	1.726	3.529
TERPIN MONOHYDRATE				V	0.084	0.032	0.309	0.073	13.657	1.697	3.231
32308	1.1100	4	ROOM	R	0.078	0.030	0.311	0.069	14.460	1.640	3.133
THALLIUM SULFATE				V	0.254	0.094	0.355	0.293	3.418	1.176	2.482
32309	6.7760	126	293.0	R	0.246	0.090	0.360	0.292	3.424	1.155	2.467
TOPAZ				V	2.835	1.161	0.221	1.691	0.591	5.744	9.594
32310	3.5200	128	ROOM	R	2.777	1.136	0.222	1.667	0.600	5.681	9.507
URANIUM (ALPHA)				V	2.287	0.976	0.171	1.160	0.862	2.251	3.575
32311	19.2560	166	73.0	R	2.121	0.900	0.178	1.097	0.912	2.162	3.454
URANIUM (ALPHA)				V	2.284	0.974	0.173	1.164	0.859	2.249	3.576
32312	19.2472	166	83.0	R	2.119	0.898	0.180	1.105	0.905	2.160	3.458
URANIUM (ALPHA)				V	2.280	0.971	0.174	1.166	0.858	2.246	3.576
32313	19.2383	166	93.0	R	2.116	0.895	0.182	1.111	0.900	2.157	3.460
URANIUM (ALPHA)				V	2.275	0.968	0.175	1.167	0.857	2.243	3.575
32314	19.2294	166	103.0	R	2.111	0.891	0.184	1.114	0.897	2.153	3.461
URANIUM (ALPHA)				V	2.268	0.964	0.176	1.167	0.857	2.240	3.573
32315	19.2204	166	113.0	R	2.106	0.888	0.186	1.117	0.895	2.150	3.460
URANIUM (ALPHA)				V	2.262	0.961	0.177	1.167	0.857	2.236	3.570
32316	19.2114	166	123.0	R	2.100	0.884	0.187	1.119	0.894	2.146	3.459
URANIUM (ALPHA)				V	2.254	0.957	0.178	1.167	0.857	2.232	3.566
32317	19.2023	166	133.0	R	2.093	0.881	0.188	1.120	0.893	2.142	3.456
URANIUM (ALPHA)				V	2.246	0.953	0.179	1.166	0.858	2.228	3.563
32318	19.1931	166	143.0	R	2.086	0.877	0.190	1.121	0.892	2.137	3.454
URANIUM (ALPHA)				V	2.238	0.949	0.180	1.165	0.858	2.224	3.559
32319	19.1838	166	153.0	R	2.079	0.873	0.191	1.121	0.892	2.133	3.451
URANIUM (ALPHA)				V	2.230	0.944	0.181	1.164	0.859	2.219	3.555
32320	19.1745	166	163.0	R	2.071	0.868	0.192	1.121	0.892	2.128	3.448
URANIUM (ALPHA)				V	2.221	0.940	0.182	1.164	0.859	2.214	3.551
32321	19.1651	166	173.0	R	2.063	0.864	0.193	1.121	0.892	2.123	3.444
URANIUM (ALPHA)				V	2.212	0.935	0.183	1.162	0.860	2.210	3.547
32322	19.1556	166	183.0	R	2.054	0.860	0.195	1.121	0.892	2.119	3.441
URANIUM (ALPHA)				V	2.203	0.931	0.184	1.161	0.861	2.205	3.542
32323	19.1461	166	193.0	R	2.046	0.855	0.196	1.121	0.892	2.114	3.437
URANIUM (ALPHA)				V	2.195	0.926	0.185	1.160	0.862	2.200	3.538
32324	19.1364	166	203.0	R	2.038	0.851	0.197	1.121	0.892	2.109	3.433
URANIUM (ALPHA)				V	2.185	0.922	0.186	1.159	0.863	2.195	3.533
32325	19.1267	166	213.0	R	2.029	0.847	0.198	1.120	0.893	2.104	3.429
URANIUM (ALPHA)				V	2.176	0.917	0.187	1.157	0.864	2.190	3.528
32326	19.1169	166	223.0	R	2.020	0.842	0.199	1.120	0.893	2.099	3.425
URANIUM (ALPHA)				V	2.167	0.912	0.188	1.156	0.865	2.185	3.524
32327	19.1069	166	233.0	R	2.011	0.838	0.200	1.119	0.894	2.094	3.421
URANIUM (ALPHA)				V	2.158	0.908	0.189	1.154	0.866	2.180	3.519
32328	19.0969	166	243.0	R	2.002	0.833	0.202	1.118	0.894	2.089	3.417
URANIUM (ALPHA)				V	2.148	0.903	0.189	1.153	0.867	2.175	3.514
32329	19.0868	166	253.0	R	1.993	0.829	0.203	1.117	0.895	2.083	3.412
URANIUM (ALPHA)				V	2.139	0.898	0.190	1.151	0.868	2.170	3.509
32330	19.0766	166	263.0	R	1.984	0.824	0.204	1.117	0.896	2.078	3.408
URANIUM (ALPHA)				V	2.129	0.893	0.191	1.150	0.870	2.165	3.504
32331	19.0663	166	273.0	R	1.974	0.819	0.205	1.116	0.896	2.073	3.403

CODE	NAME / DENSITY	REF	TEMP(K)		E	G	SIGMA	K	BETA	VS	VP
URANIUM	(ALPHA)			V	2.119	0.888	0.193	1.149	0.871	2.159	3.499
32332	19.0558	166	283.0	R	1.964	0.814	0.206	1.115	0.897	2.067	3.398
URANIUM	(ALPHA)			V	2.109	0.883	0.194	1.147	0.872	2.154	3.494
32333	19.0453	166	293.0	R	1.955	0.809	0.208	1.114	0.898	2.061	3.394
URANIUM	(ALPHA)			V	2.099	0.878	0.195	1.146	0.873	2.148	3.489
32334	19.0347	166	303.0	R	1.945	0.804	0.209	1.113	0.898	2.056	3.389
URANIUM	(ALPHA)			V	2.088	0.873	0.196	1.144	0.874	2.142	3.483
32335	19.0239	166	313.0	R	1.935	0.799	0.210	1.112	0.899	2.050	3.384
URANIUM	(ALPHA)			V	2.078	0.868	0.197	1.142	0.876	2.137	3.478
32336	19.0130	166	323.0	R	1.925	0.795	0.211	1.110	0.901	2.044	3.378
URANIUM	(ALPHA)			V	2.203	0.944	0.167	1.104	0.906	2.214	3.503
32337	19.2539	91	44.0	R	1.984	0.860	0.154	0.955	1.048	2.113	3.304
URANIUM	(ALPHA)			V	2.230	0.955	0.167	1.115	0.897	2.228	3.523
32338	19.2524	91	46.0	R	2.028	0.876	0.158	0.988	1.012	2.133	3.346
URANIUM	(ALPHA)			V	2.245	0.961	0.168	1.126	0.888	2.235	3.537
32339	19.2510	91	48.0	R	2.053	0.883	0.162	1.014	0.987	2.142	3.374
URANIUM	(ALPHA)			V	2.255	0.965	0.168	1.133	0.882	2.239	3.546
32340	19.2495	91	50.0	R	2.070	0.888	0.166	1.032	0.969	2.147	3.392
URANIUM	(ALPHA)			V	2.279	0.973	0.171	1.153	0.868	2.249	3.569
32341	19.2424	91	60.0	R	2.106	0.897	0.174	1.077	0.928	2.159	3.437
URANIUM	(ALPHA)			V	2.287	0.976	0.171	1.160	0.862	2.253	3.578
32342	19.2327	91	73.0	R	2.121	0.900	0.178	1.097	0.912	2.164	3.456
URANIUM	(ALPHA)			V	2.277	0.969	0.175	1.166	0.857	2.246	3.577
32343	19.2142	91	98.0	R	2.113	0.893	0.183	1.113	0.899	2.156	3.462
URANIUM	(ALPHA)			V	2.262	0.961	0.177	1.167	0.857	2.237	3.571
32344	19.1952	91	123.0	R	2.100	0.885	0.187	1.119	0.894	2.147	3.460
URANIUM	(ALPHA)			V	2.242	0.951	0.179	1.166	0.858	2.227	3.562
32345	19.1759	91	148.0	R	2.083	0.875	0.190	1.121	0.892	2.136	3.454
URANIUM	(ALPHA)			V	2.221	0.940	0.182	1.163	0.860	2.215	3.552
32346	19.1561	91	173.0	R	2.063	0.864	0.193	1.121	0.892	2.124	3.445
URANIUM	(ALPHA)			V	2.199	0.928	0.184	1.161	0.861	2.203	3.540
32347	19.1359	91	198.0	R	2.042	0.853	0.196	1.121	0.892	2.112	3.436
URANIUM	(ALPHA)			V	2.176	0.917	0.187	1.158	0.864	2.190	3.529
32348	19.1153	91	223.0	R	2.020	0.842	0.199	1.120	0.893	2.099	3.426
URANIUM	(ALPHA)			V	2.153	0.906	0.189	1.154	0.867	2.178	3.517
32349	19.0942	91	248.0	R	1.998	0.831	0.202	1.118	0.894	2.086	3.414
URANIUM	(ALPHA)			V	2.129	0.893	0.191	1.150	0.870	2.164	3.503
32350	19.0727	91	273.0	R	1.974	0.819	0.205	1.116	0.896	2.072	3.402
URANIUM	(ALPHA)			V	2.103	0.881	0.194	1.147	0.872	2.150	3.491
32351	19.0507	91	298.0	R	1.949	0.807	0.208	1.114	0.897	2.058	3.390
URANIUM	(ALPHA)			V	2.078	0.868	0.197	1.144	0.874	2.136	3.478
32352	19.0283	91	323.0	R	1.924	0.794	0.212	1.112	0.899	2.043	3.378
URANIUM	(ALPHA)			V	2.052	0.855	0.200	1.140	0.877	2.121	3.463
32353	19.0054	91	348.0	R	1.898	0.781	0.215	1.110	0.901	2.027	3.364
URANIUM	(ALPHA)			V	2.026	0.843	0.203	1.135	0.881	2.107	3.450
32354	18.9821	91	373.0	R	1.872	0.768	0.218	1.106	0.904	2.012	3.351
URANIUM	(ALPHA)			V	2.000	0.830	0.205	1.131	0.884	2.092	3.435
32355	18.9582	91	398.0	R	1.846	0.756	0.221	1.103	0.907	1.997	3.337
URANIUM	(ALPHA)			V	1.974	0.817	0.207	1.125	0.889	2.078	3.420
32356	18.9338	91	423.0	R	1.820	0.744	0.224	1.098	0.911	1.982	3.322

ORTHORHOMBIC

CODE	NAME / DENSITY	REF	TEMP(K)		E	G	SIGMA	K	BETA	VS	VP
URANIUM (ALPHA)				V	1.946	0.804	0.210	1.118	0.894	2.062	3.404
32357	18.9090	91	448.0	R	1.792	0.730	0.227	1.093	0.915	1.965	3.306
URANIUM (ALPHA)				V	1.917	0.790	0.213	1.113	0.898	2.045	3.387
32358	18.8836	91	473.0	R	1.763	0.717	0.230	1.089	0.918	1.948	3.291
URANIUM (ALPHA)				V	1.887	0.776	0.216	1.107	0.903	2.028	3.370
32359	18.8577	91	498.0	R	1.734	0.703	0.234	1.085	0.922	1.930	3.274
URANIUM (ALPHA)				V	1.854	0.760	0.220	1.102	0.907	2.009	3.352
32360	18.8313	91	523.0	R	1.701	0.687	0.238	1.081	0.925	1.910	3.256
URANIUM (ALPHA)				V	1.823	0.746	0.223	1.096	0.913	1.991	3.334
32361	18.8043	91	548.0	R	1.669	0.672	0.241	1.076	0.930	1.891	3.239
URANIUM (ALPHA)				V	1.791	0.730	0.226	1.090	0.917	1.972	3.316
32362	18.7768	91	573.0	R	1.637	0.657	0.245	1.071	0.934	1.871	3.220
URANIUM (ALPHA)				V	1.757	0.714	0.230	1.084	0.923	1.952	3.296
32363	18.7488	91	598.0	R	1.602	0.641	0.250	1.066	0.938	1.849	3.201
URANIUM (ALPHA)				V	1.723	0.698	0.234	1.078	0.927	1.931	3.276
32364	18.7202	91	623.0	R	1.566	0.624	0.254	1.061	0.942	1.826	3.180
URANIUM (ALPHA)				V	1.688	0.682	0.238	1.073	0.932	1.910	3.257
32365	18.6911	91	648.0	R	1.529	0.607	0.259	1.057	0.946	1.803	3.160
URANIUM (ALPHA)				V	1.652	0.665	0.242	1.067	0.937	1.888	3.236
32366	18.6614	91	673.0	R	1.490	0.590	0.264	1.051	0.951	1.777	3.138
URANIUM (ALPHA)				V	1.617	0.649	0.246	1.061	0.942	1.866	3.215
32367	18.6311	91	698.0	R	1.451	0.572	0.269	1.046	0.956	1.752	3.115
URANIUM (ALPHA)				V	1.580	0.632	0.250	1.055	0.947	1.843	3.194
32368	18.6002	91	723.0	R	1.411	0.554	0.274	1.041	0.961	1.725	3.093
URANIUM (ALPHA)				V	1.541	0.614	0.255	1.049	0.953	1.818	3.172
32369	18.5688	91	748.0	R	1.367	0.534	0.280	1.035	0.966	1.696	3.067
URANIUM (ALPHA)				V	1.500	0.595	0.261	1.045	0.957	1.791	3.149
32370	18.5368	91	773.0	R	1.320	0.513	0.287	1.032	0.969	1.664	3.042
URANIUM (ALPHA)				V	1.457	0.575	0.266	1.040	0.962	1.763	3.124
32371	18.5043	91	798.0	R	1.271	0.491	0.294	1.027	0.973	1.629	3.015
URANIUM (ALPHA)				V	1.411	0.554	0.273	1.035	0.966	1.733	3.099
32372	18.4711	91	823.0	R	1.217	0.468	0.302	1.023	0.977	1.591	2.986
URANIUM (ALPHA)				V	1.367	0.534	0.279	1.029	0.972	1.702	3.073
32373	18.4374	91	848.0	R	1.165	0.445	0.309	1.018	0.982	1.553	2.956
URANIUM (ALPHA)				V	1.320	0.514	0.285	1.023	0.978	1.671	3.046
32374	18.4031	91	873.0	R	1.111	0.422	0.317	1.012	0.988	1.514	2.925
URANIUM (ALPHA)				V	1.274	0.493	0.291	1.017	0.983	1.639	3.020
32375	18.3682	91	898.0	R	1.057	0.399	0.325	1.007	0.993	1.473	2.895
URANIUM (ALPHA)				V	1.227	0.473	0.298	1.011	0.989	1.606	2.992
32376	18.3327	91	923.0	R	1.000	0.375	0.334	1.001	0.999	1.430	2.861
URANIUM (ALPHA)				V	2.131	0.912	0.168	1.071	0.934	2.172	3.440
32377	19.3189	92	4.2	R	1.828	0.804	0.136	0.837	1.195	2.041	3.144
ZINC SULFATE HEPTAHYDRATE				V	0.234	0.088	0.332	0.232	4.317	2.111	4.207
32378	1.9700	11	ROOM	R	0.210	0.078	0.348	0.230	4.347	1.988	4.117

ORTHORHOMBIC

CODE	NAME DENSITY	REF	TEMP(K)		E	G	SIGMA	K	BETA	VS	VP
					PRESSURE DERIVATIVES						
FORSTERITE				V	5.085	1.762	4.349	5.427	-3.176	3.464	10.647
32379	3.2240	148	298.0	R	5.228	1.852	3.981	5.306	-3.105	3.835	10.857
FORSTERITE				V	4.992	1.778	3.538	4.953	-2.877	3.494	9.808
32380	3.2217	116	298.0	R	5.079	1.843	3.187	4.798	-2.787	3.772	9.865
OLIVINE (93.0% FORSTERITE)				V	4.943	1.741	3.766	5.188	-3.000	3.455	10.120
32381	3.3110	148	298.0	R	5.126	1.845	3.387	5.082	-2.938	3.861	10.372

ORTHORHOMBIC

CODE	NAME DENSITY	REF	TEMP(K)		E	G	SIGMA	K	BETA	VS	VP
					TEMPERATURE DERIVATIVES						
AMMONIUM SULFATE				V	-0.064	-0.019	-0.833	-0.160	3.302	-0.107	-0.943
32382	1.7779	126	273.0	R	-0.041	-0.009	-0.975	-0.156	3.234	0.011	-0.849
CESIUM SULFATE				V	-0.197	-0.077	0.090	-0.147	2.078	-0.427	-0.740
32383	4.2523	126	273.0	R	-0.202	-0.079	0.069	-0.154	2.169	-0.440	-0.775
FORSTERITE				V	-0.301	-0.127	0.082	-0.151	0.088	-0.325	-0.466
32384	3.2240	148	298.0	R	-0.311	-0.132	0.104	-0.149	0.087	-0.351	-0.489
FORSTERITE				V	-0.318	-0.132	0.058	-0.173	0.101	-0.341	-0.519
32385	3.2078	116	473.0	R	-0.322	-0.134	0.068	-0.173	0.101	-0.357	-0.537
OLIVINE (93.0% FORSTERITE)				V	-0.303	-0.127	0.081	-0.157	0.091	-0.327	-0.474
32386	3.3110	148	298.0	R	-0.313	-0.132	0.103	-0.155	0.089	-0.352	-0.494
POTASSIUM SULFATE				V	-0.229	-0.090	-0.029	-0.169	1.625	-0.514	-0.933
32387	2.6708	126	273.0	R	-0.212	-0.082	-0.090	-0.167	1.607	-0.463	-0.888
RUBIDIUM SULFATE				V	-0.221	-0.087	0.027	-0.164	1.826	-0.464	-0.824
32388	3.6289	126	273.0	R	-0.221	-0.086	0.009	-0.167	1.859	-0.464	-0.836
THALLIUM SULFATE				V	-0.196	-0.072	0.036	-0.218	2.469	-0.379	-0.776
32389	6.7909	126	273.0	R	-0.189	-0.070	0.038	-0.217	2.457	-0.372	-0.769

NAME / CODE	DENSITY	REF	TEMP(K)		E	G	SIGMA	K	BETA	VS	VP
AEGIRITE				V	1.530	0.602	0.272	1.117	0.895	4.146	7.405
42390	3.5000	7	ROOM	R	1.455	0.571	0.274	1.072	0.932	4.040	7.239
AEGIRITE-AUGITE				V	1.244	0.477	0.305	1.061	0.943	3.734	7.042
42391	3.4200	10	ROOM	R	1.142	0.441	0.294	0.923	1.084	3.593	6.647
ANTHRACENE				V	0.084	0.032	0.316	0.076	13.200	1.589	3.064
42392	1.2580	3	ROOM	R	0.062	0.023	0.333	0.062	16.100	1.364	2.723
AUGITE				V	1.503	0.600	0.252	1.011	0.990	4.252	7.385
42393	3.3200	10	ROOM	R	1.391	0.559	0.243	0.902	1.108	4.105	7.046
COBALT SULFATE HEPTAHYDRATE				V	0.219	0.081	0.346	0.236	4.233	2.043	4.206
42394	1.9480	11	ROOM	R	0.184	0.067	0.367	0.230	4.353	1.858	4.049
DIALLAGE				V	1.489	0.616	0.208	0.849	1.178	4.322	7.115
42395	3.3000	10	ROOM	R	1.414	0.584	0.210	0.812	1.231	4.207	6.944
DIBENZYL				V	0.063	0.024	0.295	0.051	19.418	1.490	2.761
42396	1.1040	253	ROOM	R	0.054	0.021	0.298	0.045	22.367	1.376	2.564
DIOPSIDE (BAIKALITE)				V	1.663	0.658	0.264	1.176	0.850	4.457	7.876
42397	3.3100	10	ROOM	R	1.544	0.615	0.255	1.051	0.951	4.310	7.519
DIPOTASSIUM TARTRATE				V	0.281	0.107	0.314	0.251	3.977	2.329	4.472
42398	1.9700	129	ROOM	R	0.252	0.096	0.317	0.229	4.360	2.204	4.257
DIPOTASSIUM TARTRATE				V	0.318	0.121	0.317	0.289	3.465	2.474	4.776
42399	1.9700	129	ROOM	R	0.280	0.107	0.314	0.251	3.980	2.326	4.469
EPIDOTE				V	1.622	0.650	0.248	1.074	0.931	4.372	7.555
42400	3.4000	223	ROOM	R	1.457	0.574	0.269	1.049	0.953	4.109	7.305
ETHYLENE DIAMINE TARTRATE				V	0.285	0.101	0.416	0.563	1.778	2.557	6.730
42401	1.5380	129	ROOM	R	0.164	0.059	0.399	0.270	3.700	1.951	4.759
ETHYLENE DIAMINE TARTRATE				V	0.237	0.090	0.316	0.215	4.654	2.418	4.666
42402	1.5380	129	ROOM	R	0.165	0.063	0.313	0.147	6.790	2.023	3.877
FELDSPAR (LABRADORITE)				V	0.947	0.367	0.291	0.757	1.320	3.700	6.820
42403	2.6800	6	ROOM	R	0.816	0.310	0.318	0.746	1.340	3.400	6.577
FELDSPAR (MICROCLINE)				V	0.832	0.328	0.267	0.596	1.677	3.582	6.356
42404	2.5600	6	ROOM	R	0.630	0.245	0.289	0.497	2.012	3.091	5.670
FELDSPAR (OLIGOCLASE)				V	0.832	0.319	0.303	0.704	1.420	3.477	6.541
42405	2.6400	6	ROOM	R	0.622	0.234	0.329	0.607	1.648	2.978	5.900
FELDSPAR (PLAGIOCLASE-9%AN)				V	0.805	0.314	0.282	0.616	1.624	3.467	6.295
42406	2.6100	221	ROOM	R	0.690	0.269	0.283	0.531	1.884	3.210	5.837
FELDSPAR (PLAGIOCLASE-24%AN)				V	0.843	0.328	0.285	0.653	1.531	3.525	6.427
42407	2.6400	221	ROOM	R	0.728	0.282	0.291	0.579	1.726	3.269	6.016
FELDSPAR (PLAGIOCLASE-29%AN)				V	0.863	0.336	0.284	0.665	1.503	3.569	6.495
42408	2.6400	221	ROOM	R	0.748	0.290	0.291	0.596	1.679	3.313	6.099
FELDSPAR (PLAGIOCLASE-53%AN)				V	0.926	0.359	0.291	0.739	1.354	3.658	6.738
42409	2.6800	221	ROOM	R	0.811	0.312	0.301	0.680	1.470	3.411	6.395
FELDSPAR (PLAGIOCLASE-56%AN)				V	0.953	0.370	0.289	0.753	1.328	3.706	6.806
42410	2.6900	221	ROOM	R	0.825	0.317	0.300	0.688	1.454	3.434	6.426
FELDSPAR (53.5%OR-34.6%AB-9.15%AN)				V	0.748	0.286	0.305	0.639	1.565	3.332	6.290
42411	2.5800	222	ROOM	R	0.523	0.197	0.329	0.509	1.963	2.762	5.470
FELDSPAR (60.7%OR-35.6%AB-1.64%AN)				V	0.782	0.309	0.267	0.560	1.786	3.466	6.148
42412	2.5700	222	ROOM	R	0.552	0.218	0.264	0.390	2.566	2.915	5.147
FELDSPAR (66.6%OR-28.6%AB)				V	0.702	0.274	0.282	0.535	1.868	3.283	5.954
42413	2.5400	222	ROOM	R	0.516	0.199	0.298	0.425	2.354	2.798	5.212
FELDSPAR (74%OR-18.9%AB-1.95%AN)				V	0.738	0.286	0.289	0.583	1.716	3.338	6.126
42414	2.5700	222	ROOM	R	0.555	0.212	0.307	0.479	2.088	2.875	5.446

MONOCLINIC

NAME CODE	DENSITY	REF	TEMP(K)		E	G	SIGMA	K	BETA	VS	VP
FELDSPAR (75%OR-22%AB)				V	0.707	0.276	0.279	0.532	1.879	3.299	5.955
42415	2.5400	222	ROOM	R	0.519	0.202	0.287	0.407	2.458	2.817	5.157
FELDSPAR (78.5%OR-19.4%AB-2.1%AN)				V	0.818	0.319	0.283	0.627	1.595	3.529	6.411
42416	2.5600	222	ROOM	R	0.581	0.226	0.284	0.448	2.231	2.972	5.412
FELDSPAR (64.9%OR-26.6%AB-3.60%AN)				V	0.747	0.290	0.286	0.582	1.718	3.361	6.141
42417	2.5700	222	ROOM	R	0.557	0.216	0.292	0.447	2.239	2.897	5.344
FERROUS SULPHATE HEPTAHYDRATE				V	0.213	0.079	0.354	0.244	4.104	2.035	4.285
42418	1.8980	11	ROOM	R	0.201	0.074	0.359	0.237	4.211	1.977	4.210
HORNBLENDE				V	1.155	0.449	0.286	0.901	1.110	3.794	6.933
42419	3.1200	7	ROOM	R	1.067	0.414	0.289	0.842	1.188	3.643	6.684
HORNBLENDE				V	1.309	0.512	0.277	0.978	1.022	4.033	7.262
42420	3.1500	7	ROOM	R	1.066	0.407	0.308	0.925	1.081	3.596	6.826
LITHIUM SULPHATE MONOHYDRATE				V	0.466	0.194	0.201	0.259	3.857	2.956	4.829
42421	2.2210	4	ROOM	R	0.425	0.174	0.221	0.254	3.937	2.801	4.679
LITHIUM SULPHATE MONOHYDRATE				V	0.537	0.219	0.229	0.330	3.032	3.137	5.289
42422	2.2210	4	ROOM	R	0.469	0.188	0.250	0.313	3.197	2.907	5.035
L-RHAMNOSE MONOHYDRATE				V	0.173	0.065	0.340	0.181	5.526	2.096	4.261
42423	1.4710	4	ROOM	R	0.153	0.058	0.326	0.147	6.800	1.984	3.904
NAPHTHALENE				V	0.081	0.032	0.258	0.055	18.036	1.687	2.952
42424	1.1270	13	ROOM	R	0.022	0.008	0.432	0.053	18.869	0.819	2.366
POTASSIUM TARTRATE				V	0.278	0.106	0.315	0.251	3.986	2.308	4.442
42425	1.9870	4	ROOM	R	0.245	0.093	0.319	0.226	4.421	2.163	4.198
POTASSIUM TARTRATE				V	0.330	0.124	0.325	0.313	3.193	2.503	4.911
42426	1.9870	4	ROOM	R	0.279	0.105	0.332	0.276	3.619	2.297	4.576
POTASSIUM TARTRATE				V	0.280	0.107	0.310	0.245	4.085	2.318	4.414
42427	1.9870	4	ROOM	R	0.251	0.095	0.314	0.224	4.455	2.191	4.207
SODIUM THIOSULPHATE				V	0.219	0.081	0.348	0.241	4.153	2.154	4.466
42428	1.7499	129	ROOM	R	0.110	0.039	0.403	0.189	5.278	1.498	3.717
STILBENE				V	0.079	0.030	0.305	0.068	14.778	1.376	2.599
42429	1.6000	253	ROOM	R	0.055	0.020	0.359	0.066	15.225	1.128	2.409
TARTARIC ACID				V	0.304	0.113	0.348	0.334	2.996	2.529	5.243
42430	1.7600	4	ROOM	R	0.205	0.079	0.298	0.170	5.890	2.120	3.955
TAURINE				V	0.225	0.086	0.315	0.202	4.941	2.236	4.300
42431	1.7120	127	293.0	R	0.166	0.064	0.301	0.139	7.203	1.930	3.616
TOLANE				V	0.074	0.030	0.218	0.043	22.989	1.636	2.725
42432	1.1280	253	ROOM	R	0.060	0.024	0.242	0.039	25.652	1.466	2.514
TRIGLYCINE SULFATE				V	0.227	0.084	0.345	0.244	4.100	2.240	4.605
42433	1.6800	4	ROOM	R	0.178	0.065	0.368	0.224	4.458	1.970	4.304

HEXAGONAL

CODE / NAME	DENSITY	REF	TEMP(K)		E	G	SIGMA	K	BETA	VS	VP
APATITE				V	1.530	0.638	0.199	0.846	1.182	4.453	7.261
52434	3.2180	129	ROOM	R	1.404	0.575	0.221	0.839	1.193	4.228	7.063
BERYL				V	2.005	0.798	0.257	1.373	0.728	5.476	9.572
52435	2.6600	128	ROOM	R	1.958	0.777	0.260	1.359	0.736	5.405	9.488
BERYL				V	2.065	0.822	0.257	1.417	0.706	5.557	9.718
52436	2.6600	129	ROOM	R	2.009	0.796	0.261	1.403	0.713	5.472	9.626
BERYL				V	2.257	0.902	0.251	1.511	0.662	5.845	10.139
52437	2.6400	129	ROOM	R	2.214	0.883	0.254	1.498	0.667	5.783	10.067
BERYL				V	2.078	0.828	0.255	1.416	0.706	5.557	9.695
52438	2.6800	129	ROOM	R	2.017	0.800	0.260	1.401	0.714	5.464	9.595
BERYL				V	2.133	0.850	0.255	1.451	0.689	5.590	9.748
52439	2.7200	129	ROOM	R	2.086	0.830	0.257	1.432	0.699	5.523	9.659
BERYLLIUM				V	3.077	1.377	0.117	1.339	0.747	8.747	13.282
52440	1.8000	129	ROOM	R	2.788	1.252	0.114	1.203	0.831	8.339	12.631
BERYLLIUM				V	3.186	1.531	0.041	1.156	0.865	9.102	13.154
52441	1.8477	237	0.0	R	3.164	1.517	0.043	1.154	0.866	9.061	13.112
BERYLLIUM				V	3.186	1.531	0.041	1.156	0.865	9.102	13.154
52442	1.8477	237	75.0	R	3.164	1.517	0.043	1.154	0.866	9.061	13.112
BERYLLIUM				V	3.185	1.530	0.041	1.155	0.866	9.101	13.151
52443	1.8477	237	100.0	R	3.163	1.516	0.043	1.154	0.867	9.059	13.109
BERYLLIUM				V	3.185	1.528	0.042	1.159	0.863	9.094	13.153
52444	1.8477	237	125.0	R	3.163	1.514	0.045	1.157	0.864	9.053	13.111
BERYLLIUM				V	3.181	1.526	0.042	1.158	0.863	9.087	13.146
52445	1.8477	237	150.0	R	3.159	1.512	0.045	1.157	0.865	9.046	13.104
BERYLLIUM				V	3.175	1.523	0.042	1.156	0.865	9.080	13.134
52446	1.8477	237	175.0	R	3.154	1.509	0.045	1.154	0.866	9.039	13.092
BERYLLIUM				V	3.170	1.518	0.044	1.159	0.863	9.064	13.125
52447	1.8477	237	200.0	R	3.148	1.504	0.047	1.157	0.864	9.023	13.083
BERYLLIUM				V	3.161	1.514	0.044	1.156	0.865	9.052	13.107
52448	1.8477	237	225.0	R	3.139	1.500	0.046	1.154	0.867	9.011	13.065
BERYLLIUM				V	3.151	1.507	0.045	1.155	0.866	9.032	13.088
52449	1.8477	237	250.0	R	3.130	1.494	0.048	1.153	0.867	8.991	13.046
BERYLLIUM				V	3.138	1.501	0.045	1.150	0.869	9.012	13.060
52450	1.8477	237	275.0	R	3.116	1.487	0.048	1.148	0.871	8.971	13.017
BERYLLIUM				V	3.122	1.493	0.045	1.145	0.873	8.989	13.027
52451	1.8477	237	300.0	R	3.100	1.480	0.048	1.143	0.875	8.949	12.985
BERYLLIUM OXIDE				V	3.953	1.599	0.236	2.496	0.401	7.289	12.400
52452	3.0100	30	ROOM	R	3.925	1.585	0.238	2.494	0.401	7.258	12.373
BERYLLIUM OXIDE				V	4.004	1.664	0.203	2.244	0.446	7.436	12.177
52453	3.0100	60	298.0	R	3.961	1.642	0.206	2.244	0.446	7.386	12.137
BIOTITE				V	1.028	0.424	0.213	0.597	1.675	3.727	6.172
52454	3.0500	5	ROOM	R	0.339	0.124	0.363	0.412	2.430	2.020	4.351
CADMIUM				V	0.609	0.232	0.315	0.549	1.821	1.637	3.151
52455	8.6420	4	ROOM	R	0.492	0.187	0.315	0.444	2.250	1.471	2.834
CADMIUM				V	0.666	0.251	0.324	0.629	1.589	1.706	3.341
52456	8.6420	4	ROOM	R	0.545	0.206	0.319	0.502	1.991	1.546	3.000
CADMIUM				V	0.605	0.230	0.315	0.544	1.837	1.632	3.138
52457	8.6420	4	ROOM	R	0.490	0.186	0.316	0.444	2.254	1.467	2.829
CADMIUM				V	0.655	0.247	0.324	0.622	1.607	1.692	3.319
52458	8.6420	177	ROOM	R	0.495	0.188	0.315	0.445	2.247	1.475	2.838

HEXAGONAL

CODE / NAME	DENSITY	REF	TEMP(K)		E	G	SIGMA	K	BETA	VS	VP
CADMIUM				V	0.808	0.315	0.283	0.621	1.610	1.887	3.431
52459	8.8415	101	0.0	R	0.685	0.267	0.283	0.525	1.904	1.738	3.157
CADMIUM				V	0.807	0.314	0.283	0.620	1.612	1.886	3.429
52460	8.8415	101	20.0	R	0.684	0.266	0.283	0.524	1.907	1.736	3.154
CADMIUM				V	0.803	0.313	0.284	0.618	1.618	1.881	3.422
52461	8.8361	101	40.0	R	0.679	0.265	0.283	0.522	1.917	1.731	3.146
CADMIUM				V	0.796	0.310	0.285	0.617	1.622	1.874	3.416
52462	8.8254	101	60.0	R	0.672	0.262	0.284	0.519	1.927	1.722	3.136
CADMIUM				V	0.789	0.307	0.286	0.614	1.627	1.866	3.408
52463	8.8121	101	80.0	R	0.665	0.259	0.285	0.516	1.936	1.713	3.126
CADMIUM				V	0.781	0.303	0.287	0.612	1.634	1.857	3.399
52464	8.7988	101	100.0	R	0.657	0.255	0.287	0.513	1.948	1.703	3.115
CADMIUM				V	0.772	0.300	0.288	0.609	1.643	1.847	3.387
52465	8.7856	101	120.0	R	0.649	0.252	0.288	0.510	1.959	1.693	3.104
CADMIUM				V	0.763	0.296	0.290	0.605	1.652	1.837	3.377
52466	8.7697	101	140.0	R	0.641	0.248	0.290	0.508	1.970	1.683	3.093
CADMIUM				V	0.754	0.292	0.291	0.602	1.660	1.827	3.366
52467	8.7539	101	160.0	R	0.632	0.245	0.291	0.505	1.981	1.672	3.081
CADMIUM				V	0.745	0.288	0.293	0.599	1.669	1.816	3.355
52468	8.7381	101	180.0	R	0.624	0.241	0.293	0.502	1.993	1.662	3.070
CADMIUM				V	0.735	0.284	0.294	0.596	1.678	1.805	3.343
52469	8.7248	101	200.0	R	0.615	0.238	0.294	0.499	2.004	1.651	3.058
CADMIUM				V	0.725	0.280	0.296	0.592	1.688	1.793	3.330
52470	8.7065	101	220.0	R	0.606	0.234	0.296	0.496	2.016	1.639	3.046
CADMIUM				V	0.714	0.275	0.298	0.588	1.700	1.780	3.315
52471	8.6934	101	240.0	R	0.597	0.230	0.298	0.493	2.029	1.627	3.033
CADMIUM				V	0.703	0.271	0.299	0.583	1.714	1.766	3.299
52472	8.6752	101	260.0	R	0.588	0.226	0.300	0.489	2.043	1.615	3.020
CADMIUM				V	0.691	0.266	0.301	0.579	1.727	1.752	3.283
52473	8.6595	101	280.0	R	0.578	0.222	0.302	0.486	2.057	1.601	3.005
CADMIUM				V	0.679	0.261	0.303	0.574	1.741	1.736	3.265
52474	8.6440	101	300.0	R	0.568	0.218	0.304	0.482	2.073	1.588	2.990
CADMIUM				V	0.681	0.261	0.303	0.576	1.736	1.738	3.269
52475	8.6520	50	300.0	R	0.570	0.219	0.303	0.484	2.068	1.590	2.994
CADMIUM				V	0.667	0.255	0.306	0.572	1.748	1.720	3.251
52476	8.6330	50	320.0	R	0.557	0.213	0.307	0.481	2.079	1.571	2.977
CADMIUM				V	0.653	0.249	0.309	0.568	1.760	1.701	3.234
52477	8.6141	50	340.0	R	0.544	0.208	0.310	0.478	2.091	1.552	2.961
CADMIUM				V	0.638	0.243	0.311	0.564	1.774	1.682	3.214
52478	8.5953	50	360.0	R	0.531	0.202	0.313	0.475	2.106	1.534	2.943
CADMIUM				V	0.622	0.237	0.315	0.559	1.789	1.661	3.193
52479	8.5766	50	380.0	R	0.518	0.197	0.317	0.471	2.122	1.514	2.924
CADMIUM				V	0.605	0.230	0.318	0.554	1.805	1.638	3.170
52480	8.5579	50	400.0	R	0.504	0.191	0.321	0.468	2.137	1.492	2.905
CADMIUM				V	0.588	0.223	0.322	0.549	1.821	1.614	3.147
52481	8.5394	50	420.0	R	0.489	0.185	0.324	0.464	2.154	1.471	2.885
CADMIUM				V	0.570	0.215	0.325	0.544	1.839	1.590	3.123
52482	8.5188	50	440.0	R	0.475	0.179	0.328	0.460	2.172	1.449	2.864
CADMIUM				V	0.552	0.207	0.329	0.538	1.857	1.562	3.097
52483	8.4984	50	460.0	R	0.459	0.172	0.333	0.457	2.190	1.423	2.842

NAME / CODE	DENSITY	REF	TEMP(K)		E	G	SIGMA	K	BETA	VS	VP
CADMIUM				V	0.532	0.200	0.333	0.533	1.877	1.534	3.070
52484	8.4780	50	480.0	R	0.444	0.166	0.337	0.453	2.209	1.399	2.819
CADMIUM				V	0.512	0.191	0.338	0.527	1.897	1.503	3.041
52485	8.4578	50	500.0	R	0.426	0.159	0.342	0.449	2.228	1.371	2.795
CADMIUM				V	0.490	0.182	0.343	0.521	1.919	1.470	3.010
52486	8.4377	50	520.0	R	0.409	0.152	0.347	0.445	2.247	1.341	2.770
CADMIUM				V	0.467	0.173	0.349	0.515	1.942	1.434	2.977
52487	8.4176	50	540.0	R	0.389	0.144	0.353	0.441	2.266	1.308	2.743
CADMIUM				V	0.444	0.164	0.354	0.508	1.970	1.397	2.940
52488	8.3977	50	560.0	R	0.370	0.136	0.359	0.437	2.288	1.273	2.714
CADMIUM				V	0.426	0.157	0.358	0.501	1.994	1.367	2.911
52489	8.3828	50	575.0	R	0.355	0.130	0.364	0.434	2.307	1.246	2.691
CADMIUM SELENIDE (CONSTANT E)				V	0.420	0.154	0.368	0.532	1.878	1.648	3.610
52490	5.6550	32	298.0	R	0.407	0.148	0.372	0.532	1.879	1.620	3.593
CADMIUM SELENIDE (HEXAGONAL)				V	0.423	0.155	0.369	0.537	1.861	1.649	3.616
52491	5.6840	60	298.0	R	0.409	0.149	0.373	0.537	1.862	1.618	3.597
CADMIUM SULFIDE				V	0.428	0.155	0.380	0.594	1.684	1.794	4.076
52492	4.8200	4	ROOM	R	0.426	0.154	0.380	0.594	1.684	1.790	4.073
CADMIUM SULFIDE				V	0.425	0.154	0.380	0.591	1.692	1.786	4.062
52493	4.8250	123	ROOM	R	0.423	0.153	0.381	0.591	1.692	1.783	4.060
CADMIUM SULFIDE				V	0.468	0.170	0.373	0.614	1.629	1.878	4.174
52494	4.8250	43	300.0	R	0.456	0.166	0.376	0.613	1.630	1.853	4.159
CADMIUM SULFIDE (CONSTANT E)				V	0.469	0.170	0.382	0.661	1.512	1.875	4.288
52495	4.8250	32	298.0	R	0.460	0.166	0.384	0.661	1.513	1.856	4.277
CADMIUM SULFIDE				V	1.095	0.455	0.204	0.618	1.619	3.071	5.039
52496	4.8200	61	296.0	R	0.694	0.264	0.313	0.618	1.619	2.342	4.487
CADMIUM SULFIDE				V	0.463	0.169	0.373	0.608	1.644	1.869	4.155
52497	4.8240	107	298.0	R	0.451	0.164	0.376	0.608	1.645	1.844	4.140
CALCIUM(33.33)-MAGNESIUM(66.67)				V	0.524	0.214	0.222	0.315	3.178	3.444	5.765
52498	1.8064	248	100.0	R	0.520	0.212	0.225	0.315	3.178	3.428	5.752
CALCIUM(33.33)-MAGNESIUM(66.67)				V	0.517	0.211	0.224	0.312	3.201	3.426	5.745
52499	1.8000	248	125.0	R	0.514	0.209	0.226	0.312	3.202	3.411	5.733
CALCIUM(33.33)-MAGNESIUM(66.67)				V	0.515	0.211	0.222	0.309	3.235	3.426	5.735
52500	1.7939	248	150.0	R	0.511	0.209	0.224	0.309	3.236	3.411	5.722
CALCIUM(33.33)-MAGNESIUM(66.67)				V	0.511	0.209	0.222	0.306	3.264	3.418	5.720
52501	1.7877	248	175.0	R	0.507	0.207	0.224	0.306	3.266	3.404	5.707
CALCIUM(33.33)-MAGNESIUM(66.67)				V	0.506	0.207	0.223	0.304	3.290	3.408	5.705
52502	1.7814	248	200.0	R	0.502	0.205	0.224	0.304	3.292	3.393	5.693
CALCIUM(33.33)-MAGNESIUM(66.67)				V	0.501	0.205	0.223	0.302	3.315	3.397	5.690
52503	1.7753	248	225.0	R	0.498	0.203	0.225	0.301	3.317	3.383	5.678
CALCIUM(33.33)-MAGNESIUM(66.67)				V	0.496	0.203	0.224	0.299	3.341	3.386	5.675
52504	1.7691	248	250.0	R	0.493	0.201	0.225	0.299	3.344	3.373	5.663
CALCIUM(33.33)-MAGNESIUM(66.67)				V	0.491	0.200	0.225	0.297	3.362	3.372	5.659
52505	1.7630	248	275.0	R	0.488	0.199	0.226	0.297	3.366	3.359	5.648
CALCIUM(33.33)-MAGNESIUM(66.67)				V	0.486	0.198	0.226	0.295	3.384	3.357	5.643
52506	1.7570	248	300.0	R	0.483	0.197	0.227	0.295	3.388	3.345	5.632
CANCRINITE				V	0.562	0.241	0.167	0.282	3.552	3.128	4.949
52507	2.4600	4	ROOM	R	0.546	0.236	0.159	0.267	3.750	3.095	4.859
COBALT				V	1.573	0.580	0.357	1.828	0.547	2.552	5.406
52508	8.9000	129	ROOM	R	1.547	0.569	0.359	1.828	0.547	2.529	5.391

HEXAGONAL

CODE	NAME DENSITY	REF	TEMP(K)		E	G	SIGMA	K	BETA	VS	VP
COBALT				V	2.209	0.845	0.307	1.903	0.525	3.093	5.856
52509	8.8360	160	298.0	R	2.109	0.802	0.315	1.903	0.525	3.012	5.800
COBALT				V	1.767	0.660	0.339	1.824	0.548	2.729	5.524
52510	8.8630	178	293.0	R	1.748	0.652	0.340	1.824	0.548	2.713	5.513
COBALT				V	2.365	0.911	0.298	1.948	0.513	3.194	5.950
52511	8.9332	94	4.0	R	2.266	0.868	0.306	1.948	0.513	3.117	5.895
COBALT				V	2.364	0.911	0.298	1.947	0.513	3.193	5.949
52512	8.9332	94	23.0	R	2.265	0.867	0.306	1.947	0.514	3.116	5.894
COBALT				V	2.352	0.906	0.298	1.944	0.514	3.186	5.944
52513	8.9208	94	73.0	R	2.252	0.862	0.307	1.944	0.514	3.108	5.888
COBALT				V	2.325	0.894	0.300	1.936	0.517	3.170	5.928
52514	8.9023	94	123.0	R	2.227	0.851	0.308	1.936	0.517	3.092	5.873
COBALT				V	2.294	0.881	0.302	1.927	0.519	3.150	5.909
52515	8.8837	94	173.0	R	2.194	0.837	0.310	1.926	0.519	3.070	5.853
COBALT				V	2.259	0.866	0.304	1.916	0.522	3.126	5.886
52516	8.8643	94	223.0	R	2.158	0.822	0.312	1.916	0.522	3.046	5.829
COBALT				V	2.222	0.851	0.306	1.904	0.525	3.101	5.861
52517	8.8466	94	273.0	R	2.120	0.807	0.314	1.904	0.525	3.020	5.803
COBALT				V	2.203	0.843	0.306	1.898	0.527	3.089	5.848
52518	8.8360	94	298.0	R	2.101	0.799	0.315	1.897	0.527	3.007	5.790
COBALT				V	2.185	0.835	0.307	1.891	0.529	3.077	5.835
52519	8.8254	94	323.0	R	2.082	0.791	0.316	1.891	0.529	2.994	5.777
COBALT				V	2.150	0.821	0.309	1.877	0.533	3.054	5.809
52520	8.8068	94	373.0	R	2.047	0.776	0.318	1.877	0.533	2.969	5.750
COBALT				V	2.113	0.806	0.311	1.863	0.537	3.027	5.780
52521	8.7909	94	423.0	R	2.009	0.761	0.320	1.862	0.537	2.942	5.720
COBALT				V	2.071	0.788	0.313	1.847	0.541	2.998	5.748
52522	8.7724	94	473.0	R	1.966	0.743	0.323	1.847	0.542	2.911	5.688
COBALT				V	2.029	0.771	0.315	1.828	0.547	2.968	5.712
52523	8.7565	94	523.0	R	1.924	0.726	0.325	1.828	0.547	2.880	5.651
DYSPROSIUM				V	0.624	0.250	0.247	0.411	2.435	1.710	2.949
52524	8.5600	94	298.0	R	0.622	0.250	0.247	0.411	2.435	1.707	2.947
DYSPROSIUM				V	0.620	0.248	0.248	0.410	2.442	1.707	2.947
52525	8.5275	94	323.0	R	0.618	0.248	0.248	0.409	2.442	1.704	2.945
DYSPROSIUM				V	0.612	0.245	0.250	0.407	2.457	1.700	2.943
52526	8.4649	94	373.0	R	0.609	0.244	0.250	0.407	2.457	1.697	2.940
DYSPROSIUM				V	0.603	0.241	0.252	0.405	2.471	1.693	2.939
52527	8.4018	94	423.0	R	0.600	0.240	0.253	0.405	2.471	1.688	2.935
DYSPROSIUM				V	0.593	0.237	0.254	0.402	2.488	1.684	2.933
52528	8.3394	94	473.0	R	0.590	0.235	0.255	0.402	2.488	1.679	2.929
DYSPROSIUM				V	0.584	0.232	0.256	0.399	2.503	1.676	2.927
52529	8.2779	94	523.0	R	0.581	0.231	0.258	0.399	2.503	1.670	2.923
DYSPROSIUM				V	0.575	0.228	0.259	0.397	2.520	1.667	2.921
52530	8.2173	94	573.0	R	0.571	0.227	0.260	0.397	2.520	1.660	2.916
DYSPROSIUM				V	0.565	0.224	0.261	0.394	2.536	1.658	2.915
52531	8.1576	94	623.0	R	0.561	0.222	0.263	0.394	2.536	1.651	2.910
DYSPROSIUM				V	0.556	0.220	0.263	0.392	2.552	1.649	2.909
52532	8.0986	94	673.0	R	0.552	0.218	0.265	0.392	2.553	1.641	2.903
DYSPROSIUM				V	0.551	0.218	0.265	0.390	2.563	1.644	2.905
52533	8.0694	94	698.0	R	0.547	0.216	0.266	0.390	2.563	1.636	2.899

HEXAGONAL

CODE	NAME DENSITY	REF	TEMP(K)		E	G	SIGMA	K	BETA	VS	VP
ERBIUM				V	0.757	0.307	0.230	0.468	2.137	1.837	3.104
52534	9.1093	94	81.0	R	0.754	0.306	0.231	0.466	2.144	1.834	3.099
ERBIUM				V	0.756	0.307	0.231	0.469	2.133	1.836	3.105
52535	9.1093	94	83.0	R	0.754	0.306	0.231	0.467	2.140	1.833	3.100
ERBIUM				V	0.755	0.307	0.231	0.468	2.135	1.835	3.104
52536	9.1066	94	98.0	R	0.752	0.305	0.232	0.467	2.141	1.831	3.099
ERBIUM				V	0.752	0.305	0.232	0.467	2.141	1.831	3.099
52537	9.1012	94	123.0	R	0.749	0.304	0.232	0.466	2.146	1.827	3.094
ERBIUM				V	0.747	0.303	0.233	0.466	2.147	1.826	3.093
52538	9.0948	94	148.0	R	0.745	0.302	0.233	0.465	2.152	1.822	3.088
ERBIUM				V	0.743	0.301	0.233	0.464	2.154	1.820	3.086
52539	9.0894	94	173.0	R	0.740	0.300	0.234	0.463	2.159	1.817	3.082
ERBIUM				V	0.733	0.297	0.235	0.460	2.172	1.808	3.071
52540	9.0794	94	223.0	R	0.730	0.296	0.235	0.460	2.176	1.804	3.066
ERBIUM				V	0.722	0.292	0.237	0.457	2.187	1.794	3.055
52541	9.0694	94	273.0	R	0.720	0.291	0.237	0.456	2.191	1.791	3.051
ERBIUM				V	0.717	0.290	0.238	0.456	2.195	1.788	3.048
52542	9.0640	94	298.0	R	0.714	0.288	0.238	0.455	2.199	1.784	3.043
FLUORAPATITE				V	1.196	0.470	0.272	0.874	1.145	3.824	6.832
52543	3.2147	280	ROOM	R	1.178	0.464	0.271	0.856	1.168	3.798	6.772
GADOLINIUM				V	0.615	0.246	0.247	0.405	2.467	1.768	3.050
52544	7.8880	93	4.0	R	0.610	0.244	0.249	0.405	2.468	1.759	3.043
GADOLINIUM				V	0.612	0.245	0.248	0.404	2.475	1.763	3.044
52545	7.8880	93	50.0	R	0.607	0.243	0.250	0.404	2.476	1.754	3.037
GADOLINIUM				V	0.604	0.242	0.248	0.400	2.499	1.751	3.027
52546	7.8880	93	100.0	R	0.599	0.239	0.250	0.400	2.500	1.742	3.020
GADOLINIUM				V	0.595	0.238	0.251	0.398	2.513	1.736	3.010
52547	7.8880	93	150.0	R	0.590	0.235	0.253	0.398	2.514	1.727	3.003
GADOLINIUM				V	0.583	0.233	0.253	0.393	2.542	1.718	2.987
52548	7.8880	93	200.0	R	0.579	0.231	0.255	0.393	2.543	1.710	2.981
GADOLINIUM				V	0.578	0.230	0.254	0.391	2.557	1.709	2.975
52549	7.8880	93	218.0	R	0.574	0.228	0.256	0.391	2.558	1.702	2.969
GADOLINIUM				V	0.577	0.230	0.253	0.389	2.570	1.708	2.970
52550	7.8880	93	225.0	R	0.572	0.228	0.255	0.389	2.570	1.700	2.964
GADOLINIUM				V	0.567	0.226	0.254	0.385	2.599	1.693	2.950
52551	7.8880	93	250.0	R	0.563	0.224	0.256	0.385	2.599	1.686	2.944
GADOLINIUM				V	0.547	0.218	0.255	0.372	2.689	1.663	2.898
52552	7.8880	93	286.0	R	0.545	0.217	0.256	0.372	2.689	1.658	2.895
GADOLINIUM				V	0.547	0.217	0.258	0.377	2.652	1.660	2.907
52553	7.8880	93	289.0	R	0.544	0.216	0.259	0.377	2.652	1.655	2.904
GADOLINIUM				V	0.545	0.216	0.260	0.378	2.643	1.656	2.907
52554	7.8880	93	298.0	R	0.542	0.215	0.261	0.378	2.643	1.651	2.903
HAFNIUM				V	1.542	0.608	0.268	1.106	0.904	2.182	3.874
52555	12.7730	95	4.2	R	1.537	0.606	0.268	1.106	0.904	2.178	3.871
HAFNIUM				V	1.541	0.608	0.268	1.106	0.904	2.182	3.874
52556	12.7690	95	23.0	R	1.536	0.605	0.268	1.106	0.905	2.177	3.870
HAFNIUM				V	1.530	0.603	0.269	1.105	0.905	2.173	3.867
52557	12.7610	95	73.0	R	1.525	0.600	0.270	1.105	0.905	2.169	3.864
HAFNIUM				V	1.511	0.594	0.271	1.101	0.908	2.159	3.853
52558	12.7540	95	123.0	R	1.506	0.592	0.272	1.101	0.908	2.154	3.850

HEXAGONAL

CODE / NAME	DENSITY	REF	TEMP(K)		E	G	SIGMA	K	BETA	VS	VP
HAFNIUM				V	1.490	0.585	0.274	1.098	0.911	2.142	3.838
52559	12.7460	95	173.0	R	1.484	0.582	0.275	1.098	0.911	2.137	3.834
HAFNIUM				V	1.469	0.575	0.276	1.095	0.914	2.125	3.823
52560	12.7380	95	223.0	R	1.462	0.572	0.277	1.094	0.914	2.120	3.819
HAFNIUM				V	1.446	0.565	0.279	1.089	0.918	2.107	3.805
52561	12.7310	95	273.0	R	1.438	0.562	0.280	1.089	0.918	2.101	3.800
HAFNIUM				V	1.433	0.560	0.280	1.087	0.920	2.097	3.795
52562	12.7270	95	298.0	R	1.426	0.556	0.281	1.086	0.920	2.091	3.790
ICE				V	0.094	0.036	0.324	0.089	11.240	1.828	3.580
52563	1.0640	129	257.0	R	0.093	0.035	0.327	0.089	11.240	1.810	3.568
ICE				V	0.096	0.037	0.304	0.081	12.275	1.919	3.614
52564	1.0000	4	257.0	R	0.095	0.036	0.306	0.081	12.275	1.904	3.603
ICE				V	0.094	0.036	0.303	0.080	12.535	1.900	3.577
52565	1.0000	4	257.0	R	0.092	0.035	0.308	0.080	12.542	1.871	3.556
ICE				V	0.096	0.037	0.300	0.080	12.553	2.000	3.742
52566	0.9188	244	248.0	R	0.093	0.036	0.305	0.080	12.560	1.969	3.720
ICE				V	0.095	0.036	0.302	0.080	12.530	1.990	3.738
52567	0.9183	244	253.0	R	0.092	0.035	0.307	0.080	12.540	1.959	3.715
ICE				V	0.094	0.036	0.303	0.079	12.587	1.984	3.729
52568	0.9178	244	258.0	R	0.091	0.035	0.308	0.079	12.600	1.952	3.705
ICE				V	0.093	0.036	0.301	0.078	12.797	1.977	3.705
52569	0.9173	244	263.0	R	0.091	0.035	0.306	0.078	12.810	1.945	3.682
ICE				V	0.092	0.036	0.300	0.077	12.977	1.969	3.685
52570	0.9168	244	268.0	R	0.090	0.034	0.305	0.077	12.990	1.938	3.661
ICE				V	0.092	0.034	0.344	0.098	10.196	1.946	3.990
52571	0.9020	37	258.0	R	0.091	0.034	0.342	0.096	10.401	1.940	3.959
ICE				V	0.082	0.031	0.349	0.091	10.972	1.841	3.825
52572	0.9017	37	263.0	R	0.081	0.030	0.349	0.089	11.205	1.826	3.787
ICE				V	0.069	0.025	0.344	0.073	13.645	1.681	3.449
52573	0.9017	37	273.0	R	0.064	0.023	0.355	0.073	13.274	1.612	3.398
ICE				V	0.112	0.042	0.326	0.107	9.350	2.127	4.182
52574	0.9333	205	50.0	R	0.110	0.041	0.329	0.107	9.350	2.101	4.164
ICE				V	0.109	0.041	0.325	0.104	9.600	2.100	4.127
52575	0.9333	205	100.0	R	0.107	0.040	0.329	0.104	9.600	2.074	4.110
ICE				V	0.105	0.040	0.326	0.100	10.000	2.061	4.055
52576	0.9322	205	150.0	R	0.103	0.039	0.329	0.100	10.000	2.037	4.039
ICE				V	0.100	0.038	0.327	0.096	10.420	2.015	3.972
52577	0.9261	205	200.0	R	0.098	0.037	0.330	0.096	10.420	1.992	3.957
ICE				V	0.093	0.035	0.328	0.091	11.040	1.956	3.866
52578	0.9200	205	250.0	R	0.091	0.034	0.332	0.091	11.050	1.930	3.848
INDIUM				V	0.193	0.067	0.431	0.464	2.156	0.950	2.722
52579	7.4713	48	4.2	R	0.159	0.055	0.443	0.461	2.168	0.859	2.676
INDIUM				V	0.173	0.060	0.438	0.462	2.167	0.900	2.701
52580	7.4244	48	77.0	R	0.151	0.052	0.445	0.461	2.170	0.838	2.673
INDIUM				V	0.116	0.040	0.454	0.424	2.356	0.740	2.562
52581	7.2788	48	300.0	R	0.087	0.030	0.466	0.424	2.358	0.639	2.524
MAGNESIUM				V	0.446	0.175	0.270	0.323	3.099	3.176	5.656
52582	1.7400	128	ROOM	R	0.443	0.174	0.271	0.323	3.100	3.164	5.646
MAGNESIUM				V	0.448	0.174	0.284	0.346	2.890	3.165	5.766
52583	1.7400	128	ROOM	R	0.445	0.173	0.286	0.346	2.890	3.155	5.758

HEXAGONAL

CODE / NAME	DENSITY	REF	TEMP(K)		E	G	SIGMA	K	BETA	VS	VP
MAGNESIUM				V	0.495	0.194	0.276	0.369	2.710	3.305	5.946
52584	1.7752	236	0.0	R	0.492	0.193	0.278	0.369	2.710	3.295	5.938
MAGNESIUM				V	0.494	0.194	0.277	0.369	2.712	3.303	5.943
52585	1.7752	236	20.0	R	0.492	0.192	0.278	0.369	2.712	3.293	5.935
MAGNESIUM				V	0.493	0.193	0.277	0.368	2.714	3.300	5.939
52586	1.7746	236	40.0	R	0.491	0.192	0.278	0.368	2.715	3.289	5.932
MAGNESIUM				V	0.492	0.192	0.277	0.368	2.719	3.294	5.933
52587	1.7741	236	60.0	R	0.489	0.191	0.278	0.368	2.719	3.284	5.925
MAGNESIUM				V	0.490	0.192	0.278	0.367	2.723	3.289	5.928
52588	1.7723	236	80.0	R	0.487	0.190	0.279	0.367	2.723	3.278	5.920
MAGNESIUM				V	0.487	0.191	0.278	0.366	2.730	3.280	5.918
52589	1.7714	236	100.0	R	0.485	0.189	0.279	0.366	2.730	3.269	5.910
MAGNESIUM				V	0.484	0.189	0.279	0.365	2.738	3.271	5.909
52590	1.7690	236	120.0	R	0.482	0.188	0.280	0.365	2.738	3.261	5.901
MAGNESIUM				V	0.481	0.188	0.280	0.364	2.747	3.259	5.896
52591	1.7676	236	140.0	R	0.478	0.187	0.281	0.364	2.747	3.249	5.888
MAGNESIUM				V	0.477	0.186	0.281	0.363	2.756	3.248	5.884
52592	1.7655	236	160.0	R	0.475	0.185	0.282	0.363	2.756	3.238	5.876
MAGNESIUM				V	0.473	0.184	0.282	0.361	2.767	3.235	5.870
52593	1.7627	236	180.0	R	0.470	0.183	0.283	0.361	2.767	3.225	5.863
MAGNESIUM				V	0.469	0.183	0.283	0.360	2.777	3.221	5.856
52594	1.7606	236	200.0	R	0.466	0.182	0.284	0.360	2.778	3.211	5.848
MAGNESIUM				V	0.465	0.181	0.284	0.359	2.789	3.209	5.842
52595	1.7580	236	220.0	R	0.462	0.180	0.285	0.359	2.789	3.199	5.834
MAGNESIUM				V	0.461	0.179	0.285	0.357	2.800	3.196	5.828
52596	1.7552	236	240.0	R	0.458	0.178	0.286	0.357	2.800	3.186	5.820
MAGNESIUM				V	0.457	0.177	0.286	0.356	2.812	3.182	5.813
52597	1.7528	236	260.0	R	0.454	0.176	0.287	0.356	2.812	3.172	5.805
MAGNESIUM				V	0.452	0.176	0.287	0.354	2.824	3.168	5.798
52598	1.7501	236	280.0	R	0.450	0.175	0.288	0.354	2.825	3.158	5.790
MAGNESIUM				V	0.448	0.174	0.288	0.353	2.836	3.156	5.784
52599	1.7474	236	300.0	R	0.446	0.173	0.289	0.353	2.836	3.145	5.776
MAGNESIUM				V	0.448	0.174	0.287	0.351	2.846	3.166	5.796
52600	1.7370	225	294.0	R	0.446	0.173	0.289	0.351	2.846	3.155	5.788
MAGNESIUM				V	0.446	0.173	0.291	0.356	2.813	3.153	5.808
52601	1.7364	275	298.0	R	0.443	0.172	0.292	0.355	2.813	3.143	5.800
MAGNESIUM (PURE)				V	0.489	0.191	0.279	0.368	2.715	3.293	5.946
52602	1.7620	88	78.0	R	0.486	0.190	0.280	0.368	2.716	3.282	5.938
MAGNESIUM (PURE)				V	0.445	0.173	0.290	0.353	2.830	3.151	5.794
52603	1.7380	88	298.0	R	0.443	0.171	0.291	0.353	2.831	3.140	5.786
MAGNESIUM(98.98)-INDIUM(1.02)				V	0.488	0.190	0.282	0.373	2.678	3.228	5.859
52604	1.8272	88	78.0	R	0.485	0.189	0.283	0.373	2.679	3.217	5.850
MAGNESIUM(98.98)-INDIUM(1.02)				V	0.445	0.172	0.294	0.359	2.783	3.088	5.714
52605	1.8017	88	298.0	R	0.442	0.171	0.295	0.359	2.783	3.077	5.706
MAGNESIUM(98.65)-INDIUM(1.35)				V	0.492	0.192	0.280	0.373	2.684	3.225	5.831
52606	1.8498	88	4.0	R	0.490	0.191	0.281	0.372	2.685	3.214	5.823
MAGNESIUM(98.65)-INDIUM(1.35)				V	0.488	0.190	0.281	0.371	2.697	3.209	5.814
52607	1.8481	88	78.0	R	0.485	0.189	0.282	0.371	2.698	3.199	5.806
MAGNESIUM(98.65)-INDIUM(1.35)				V	0.444	0.172	0.293	0.357	2.803	3.071	5.670
52608	1.8223	88	298.0	R	0.442	0.171	0.294	0.357	2.803	3.060	5.662

CODE	NAME / DENSITY	REF	TEMP(K)		E	G	SIGMA	K	BETA	VS	VP
MAGNESIUM(98.04)-INDIUM(1.96)				V	0.486	0.190	0.283	0.373	2.683	3.169	5.757
52609	1.8866	88	78.0	R	0.483	0.188	0.284	0.373	2.684	3.159	5.749
MAGNESIUM(98.04)-INDIUM(1.96)				V	0.442	0.171	0.294	0.358	2.791	3.030	5.613
52610	1.8603	88	298.0	R	0.440	0.170	0.296	0.358	2.791	3.020	5.605
MAGNESIUM(98.49)-LITHIUM(1.51)				V	0.444	0.172	0.291	0.354	2.823	3.159	5.822
52611	1.7204	275	298.0	R	0.441	0.171	0.292	0.354	2.824	3.149	5.814
MAGNESIUM(96.98)-LITHIUM(3.02)				V	0.441	0.171	0.291	0.352	2.838	3.166	5.835
52612	1.7032	275	298.0	R	0.438	0.170	0.293	0.352	2.839	3.155	5.827
MAGNESIUM(94.90)-LITHIUM(5.10)				V	0.438	0.169	0.292	0.350	2.854	3.173	5.854
52613	1.6815	275	298.0	R	0.435	0.168	0.293	0.350	2.854	3.163	5.846
MAGNESIUM(93.00)-LITHIUM(7.00)				V	0.434	0.168	0.291	0.346	2.888	3.172	5.845
52614	1.6693	275	298.0	R	0.431	0.167	0.292	0.346	2.888	3.162	5.837
MAGNESIUM(90.00)-LITHIUM(10.00)				V	0.428	0.166	0.291	0.342	2.922	3.171	5.844
52615	1.6497	275	298.0	R	0.426	0.165	0.293	0.342	2.923	3.160	5.836
MAGNESIUM(87.95)-LITHIUM(12.05)				V	0.425	0.165	0.291	0.339	2.953	3.191	5.876
52616	1.6167	275	298.0	R	0.423	0.164	0.292	0.339	2.954	3.181	5.868
MAGNESIUM(85.85)-LITHIUM(14.15)				V	0.421	0.163	0.291	0.336	2.973	3.209	5.914
52617	1.5830	275	298.0	R	0.419	0.162	0.293	0.336	2.974	3.198	5.906
MAGNESIUM(84.06)-LITHIUM(15.94)				V	0.417	0.162	0.290	0.331	3.024	3.224	5.927
52618	1.5542	275	298.0	R	0.414	0.160	0.291	0.331	3.025	3.213	5.919
MAGNESIUM(99.93)-SILVER(0.07)				V	0.448	0.173	0.292	0.358	2.790	3.154	5.817
52619	1.7425	154	298.0	R	0.445	0.172	0.293	0.358	2.790	3.142	5.808
MAGNESIUM(99.74)-SILVER(0.26)				V	0.450	0.174	0.292	0.360	2.775	3.152	5.812
52620	1.7547	154	298.0	R	0.448	0.173	0.293	0.360	2.777	3.142	5.804
MAGNESIUM(99.63)-SILVER(0.37)				V	0.448	0.173	0.290	0.356	2.811	3.138	5.773
52621	1.7618	154	298.0	R	0.445	0.172	0.291	0.356	2.812	3.127	5.764
MAGNESIUM(99.74)-SILVER(0.26)				V	0.492	0.192	0.279	0.371	2.696	3.291	5.943
52622	1.7764	88	78.0	R	0.489	0.191	0.280	0.371	2.697	3.281	5.935
MAGNESIUM(99.74)-SILVER(0.26)				V	0.448	0.174	0.290	0.356	2.811	3.147	5.786
52623	1.7547	88	298.0	R	0.446	0.173	0.291	0.356	2.811	3.136	5.778
MAGNESIUM(99.63)-SILVER(0.37)				V	0.493	0.193	0.279	0.372	2.686	3.285	5.935
52624	1.7867	88	78.0	R	0.490	0.192	0.280	0.372	2.687	3.274	5.926
MAGNESIUM(99.63)-SILVER(0.37)				V	0.449	0.174	0.290	0.357	2.803	3.144	5.782
52625	1.7618	88	298.0	R	0.447	0.173	0.291	0.357	2.804	3.133	5.773
MAGNESIUM(99.79)-TIN(0.21)				V	0.446	0.173	0.293	0.359	2.783	3.139	5.801
52626	1.7516	154	298.0	R	0.444	0.171	0.294	0.359	2.784	3.128	5.793
MAGNESIUM(99.73)-TIN(0.27)				V	0.446	0.173	0.291	0.357	2.803	3.138	5.784
52627	1.7554	154	298.0	R	0.444	0.172	0.293	0.357	2.803	3.127	5.776
MAGNESIUM(99.57)-TIN(0.43)				V	0.447	0.173	0.293	0.360	2.775	3.127	5.783
52628	1.7658	154	298.0	R	0.444	0.172	0.295	0.360	2.775	3.116	5.775
MAGNESIUM(99.48)-TIN(0.52)				V	0.444	0.172	0.293	0.357	2.800	3.114	5.752
52629	1.7716	154	298.0	R	0.441	0.171	0.294	0.357	2.801	3.103	5.744
MAGNESIUM(99.28)-TIN(0.72)				V	0.444	0.172	0.293	0.358	2.792	3.102	5.736
52630	1.7845	154	298.0	R	0.442	0.171	0.294	0.358	2.793	3.091	5.727
MAGNESIUM(99.00)-TIN(1.00)				V	0.441	0.170	0.295	0.359	2.782	3.074	5.704
52631	1.8025	154	298.0	R	0.438	0.169	0.297	0.359	2.784	3.063	5.695
MAGNESIUM(99.79)-TIN(0.21)				V	0.489	0.191	0.279	0.370	2.704	3.281	5.931
52632	1.7764	88	78.0	R	0.487	0.190	0.281	0.370	2.705	3.270	5.922
MAGNESIUM(99.79)-TIN(0.21)				V	0.446	0.173	0.291	0.355	2.816	3.139	5.780
52633	1.7516	88	298.0	R	0.443	0.171	0.292	0.355	2.817	3.128	5.772

HEXAGONAL

NAME / CODE	DENSITY	REF	TEMP(K)		E	G	SIGMA	K	BETA	VS	VP
MAGNESIUM(99.54)-TIN(0.46)				V	0.487	0.190	0.281	0.370	2.702	3.256	5.898
52634	1.7927	88	78.0	R	0.484	0.189	0.282	0.370	2.703	3.245	5.889
MAGNESIUM(99.54)-TIN(0.46)				V	0.444	0.172	0.292	0.355	2.816	3.117	5.748
52635	1.7677	88	298.0	R	0.441	0.171	0.293	0.355	2.817	3.106	5.740
MAGNESIUM(99.33)-TIN(0.67)				V	0.489	0.191	0.281	0.371	2.693	3.251	5.886
52636	1.8065	88	78.0	R	0.486	0.190	0.282	0.371	2.695	3.240	5.877
MAGNESIUM(99.33)-TIN(0.67)				V	0.445	0.172	0.292	0.356	2.805	3.110	5.737
52637	1.7813	88	298.0	R	0.443	0.171	0.293	0.356	2.807	3.100	5.728
MAGNESIUM(99.06)-TIN(0.94)				V	0.486	0.190	0.282	0.372	2.685	3.225	5.855
52638	1.8241	88	78.0	R	0.484	0.188	0.284	0.372	2.685	3.214	5.847
MAGNESIUM(99.06)-TIN(0.94)				V	0.443	0.171	0.294	0.358	2.794	3.084	5.708
52639	1.7987	88	298.0	R	0.440	0.170	0.295	0.358	2.795	3.073	5.700
MUSCOVITE				V	1.008	0.411	0.227	0.615	1.625	3.837	6.456
52640	2.7900	5	ROOM	R	0.568	0.222	0.279	0.429	2.331	2.822	5.099
PHLOGOPITE				V	0.965	0.386	0.249	0.642	1.558	3.714	6.427
52641	2.8000	5	ROOM	R	0.325	0.118	0.381	0.454	2.205	2.051	4.670
PHLOGOPITE				V	0.987	0.405	0.220	0.587	1.704	3.788	6.320
52642	2.8200	5	ROOM	R	0.366	0.135	0.350	0.407	2.456	2.191	4.565
QUARTZ (BETA)				V	1.006	0.418	0.203	0.565	1.771	4.064	6.657
52643	2.5330	129	873.0	R	0.987	0.408	0.208	0.564	1.774	4.015	6.614
RHENIUM				V	4.888	1.912	0.279	3.678	0.272	3.015	5.442
52644	21.0240	230	4.2	R	4.803	1.873	0.282	3.678	0.272	2.985	5.420
RHENIUM				V	4.888	1.912	0.278	3.668	0.273	3.016	5.439
52645	21.0240	230	20.0	R	4.800	1.872	0.282	3.668	0.273	2.984	5.415
RHENIUM				V	4.884	1.912	0.277	3.656	0.273	3.015	5.433
52646	21.0240	230	40.0	R	4.794	1.870	0.281	3.656	0.273	2.983	5.409
RHENIUM				V	4.875	1.908	0.277	3.646	0.274	3.013	5.426
52647	21.0240	230	60.0	R	4.782	1.866	0.281	3.646	0.274	2.979	5.402
RHENIUM				V	4.871	1.908	0.277	3.634	0.275	3.012	5.421
52648	21.0240	230	78.0	R	4.776	1.864	0.281	3.634	0.275	2.978	5.395
RHENIUM				V	4.848	1.898	0.277	3.630	0.275	3.004	5.413
52649	21.0240	230	100.0	R	4.752	1.854	0.282	3.630	0.275	2.969	5.387
RHENIUM				V	4.828	1.889	0.278	3.625	0.276	2.997	5.406
52650	21.0240	230	120.0	R	4.729	1.844	0.283	3.625	0.276	2.961	5.379
RHENIUM				V	4.811	1.881	0.279	3.620	0.276	2.992	5.399
52651	21.0240	230	140.0	R	4.712	1.836	0.283	3.620	0.276	2.955	5.372
RHENIUM				V	4.790	1.872	0.280	3.622	0.276	2.984	5.394
52652	21.0240	230	160.0	R	4.691	1.827	0.284	3.622	0.276	2.948	5.368
RHENIUM				V	4.772	1.863	0.280	3.619	0.276	2.977	5.388
52653	21.0240	230	180.0	R	4.672	1.818	0.285	3.619	0.276	2.941	5.361
RHENIUM				V	4.753	1.855	0.281	3.619	0.276	2.971	5.383
52654	21.0240	230	200.0	R	4.655	1.810	0.286	3.619	0.276	2.934	5.357
RHENIUM				V	4.732	1.845	0.282	3.622	0.276	2.963	5.379
52655	21.0240	230	220.0	R	4.636	1.802	0.287	3.622	0.276	2.927	5.353
RHENIUM				V	4.714	1.837	0.283	3.621	0.276	2.956	5.374
52656	21.0240	230	240.0	R	4.620	1.794	0.287	3.621	0.276	2.921	5.348
RHENIUM				V	4.694	1.828	0.284	3.625	0.276	2.949	5.370
52657	21.0240	230	260.0	R	4.603	1.786	0.288	3.625	0.276	2.915	5.345
RHENIUM				V	4.673	1.818	0.285	3.630	0.275	2.940	5.366
52658	21.0240	230	280.0	R	4.587	1.779	0.289	3.630	0.275	2.909	5.343

HEXAGONAL

CODE / NAME	DENSITY	REF	TEMP(K)		E	G	SIGMA	K	BETA	VS	VP
RHENIUM				V	4.657	1.810	0.287	3.635	0.275	2.934	5.364
52659	21.0240	230	298.0	R	4.576	1.773	0.290	3.635	0.275	2.904	5.342
RHENIUM				V	4.942	1.933	0.278	3.715	0.269	3.025	5.458
52660	21.1207	94	4.0	R	4.830	1.882	0.283	3.715	0.269	2.985	5.429
RHENIUM				V	4.939	1.932	0.278	3.714	0.269	3.024	5.457
52661	21.1207	94	23.0	R	4.828	1.881	0.283	3.714	0.269	2.984	5.428
RHENIUM				V	4.912	1.920	0.279	3.707	0.270	3.015	5.447
52662	21.1207	94	73.0	R	4.804	1.871	0.284	3.707	0.270	2.976	5.419
RHENIUM				V	4.859	1.896	0.281	3.700	0.270	2.997	5.432
52663	21.1102	94	123.0	R	4.757	1.850	0.286	3.700	0.270	2.960	5.405
RHENIUM				V	4.801	1.871	0.283	3.691	0.271	2.979	5.416
52664	21.0850	94	173.0	R	4.705	1.827	0.288	3.691	0.271	2.944	5.391
RHENIUM				V	4.741	1.844	0.285	3.682	0.272	2.959	5.400
52665	21.0576	94	223.0	R	4.649	1.803	0.290	3.682	0.272	2.926	5.376
RHENIUM				V	4.681	1.818	0.288	3.672	0.272	2.940	5.383
52666	21.0345	94	273.0	R	4.593	1.778	0.292	3.672	0.272	2.907	5.360
RHENIUM				V	4.651	1.805	0.289	3.669	0.273	2.930	5.375
52667	21.0240	94	298.0	R	4.565	1.766	0.293	3.669	0.273	2.898	5.352
RHENIUM				V	4.622	1.792	0.290	3.662	0.273	2.920	5.366
52668	21.0111	94	323.0	R	4.537	1.754	0.293	3.662	0.273	2.889	5.344
RHENIUM				V	4.567	1.768	0.291	3.650	0.274	2.902	5.350
52669	20.9923	94	373.0	R	4.485	1.731	0.295	3.650	0.274	2.872	5.328
RHENIUM				V	4.514	1.745	0.293	3.638	0.275	2.885	5.333
52670	20.9735	94	423.0	R	4.434	1.710	0.297	3.638	0.275	2.855	5.312
RHENIUM				V	4.460	1.722	0.295	3.624	0.276	2.867	5.315
52671	20.9546	94	473.0	R	4.382	1.687	0.299	3.624	0.276	2.838	5.294
RHENIUM				V	4.407	1.699	0.297	3.612	0.277	2.849	5.299
52672	20.9357	94	523.0	R	4.330	1.665	0.300	3.612	0.277	2.820	5.278
RHENIUM				V	4.353	1.676	0.298	3.599	0.278	2.831	5.281
52673	20.9170	94	573.0	R	4.278	1.643	0.302	3.599	0.278	2.803	5.261
RHENIUM				V	4.303	1.655	0.300	3.588	0.279	2.814	5.266
52674	20.8983	94	623.0	R	4.228	1.622	0.304	3.588	0.279	2.786	5.246
RHENIUM				V	4.254	1.634	0.302	3.580	0.279	2.797	5.252
52675	20.8732	94	673.0	R	4.180	1.601	0.305	3.580	0.279	2.769	5.232
RHENIUM				V	4.204	1.613	0.304	3.570	0.280	2.781	5.238
52676	20.8482	94	723.0	R	4.131	1.580	0.307	3.570	0.280	2.753	5.218
RHENIUM				V	4.157	1.592	0.305	3.560	0.281	2.765	5.223
52677	20.8296	94	773.0	R	4.083	1.560	0.309	3.560	0.281	2.737	5.203
RHENIUM				V	4.107	1.571	0.307	3.549	0.282	2.748	5.208
52678	20.8046	94	823.0	R	4.033	1.539	0.311	3.549	0.282	2.720	5.189
RHENIUM				V	4.060	1.551	0.309	3.538	0.283	2.732	5.193
52679	20.7861	94	873.0	R	3.987	1.519	0.312	3.538	0.283	2.704	5.174
RHENIUM				V	4.014	1.532	0.310	3.526	0.284	2.716	5.179
52680	20.7612	94	923.0	R	3.942	1.500	0.314	3.526	0.284	2.688	5.159
RUTHENIUM				V	4.938	1.993	0.239	3.152	0.317	4.011	6.848
52681	12.3875	94	4.0	R	4.915	1.982	0.240	3.149	0.318	4.000	6.838
RUTHENIUM				V	4.937	1.993	0.239	3.151	0.317	4.011	6.847
52682	12.3875	94	23.0	R	4.914	1.982	0.240	3.148	0.318	4.000	6.837
RUTHENIUM				V	4.920	1.984	0.240	3.150	0.317	4.002	6.840
52683	12.3875	94	73.0	R	4.897	1.973	0.241	3.147	0.318	3.991	6.830

HEXAGONAL

CODE	NAME / DENSITY	REF	TEMP(K)		E	G	SIGMA	K	BETA	VS	VP
RUTHENIUM				V	4.894	1.973	0.240	3.142	0.318	3.991	6.827
52684	12.3862	94	123.0	R	4.870	1.962	0.241	3.139	0.319	3.980	6.816
RUTHENIUM				V	4.860	1.957	0.241	3.133	0.319	3.976	6.811
52685	12.3800	94	173.0	R	4.837	1.946	0.242	3.130	0.320	3.965	6.800
RUTHENIUM				V	4.821	1.940	0.243	3.123	0.320	3.960	6.794
52686	12.3714	94	223.0	R	4.798	1.929	0.244	3.120	0.320	3.948	6.783
RUTHENIUM				V	4.783	1.922	0.244	3.114	0.321	3.943	6.776
52687	12.3627	94	273.0	R	4.759	1.911	0.245	3.111	0.321	3.932	6.766
RUTHENIUM				V	4.764	1.914	0.245	3.109	0.322	3.935	6.767
52688	12.3615	94	298.0	R	4.741	1.903	0.246	3.106	0.322	3.923	6.756
RUTHENIUM				V	4.744	1.905	0.245	3.103	0.322	3.927	6.759
52689	12.3504	94	323.0	R	4.720	1.894	0.246	3.100	0.323	3.916	6.749
RUTHENIUM				V	4.705	1.887	0.246	3.092	0.323	3.911	6.742
52690	12.3392	94	373.0	R	4.681	1.876	0.247	3.089	0.324	3.899	6.731
RUTHENIUM				V	4.663	1.869	0.248	3.080	0.325	3.893	6.723
52691	12.3273	94	423.0	R	4.639	1.857	0.249	3.078	0.325	3.882	6.712
RUTHENIUM				V	4.620	1.849	0.249	3.070	0.326	3.875	6.704
52692	12.3176	94	473.0	R	4.595	1.838	0.250	3.067	0.326	3.863	6.693
RUTHENIUM				V	4.574	1.829	0.251	3.057	0.327	3.855	6.682
52693	12.3065	94	523.0	R	4.549	1.817	0.252	3.054	0.327	3.843	6.671
RUTHENIUM				V	4.532	1.810	0.252	3.045	0.328	3.837	6.663
52694	12.2954	94	573.0	R	4.507	1.798	0.253	3.043	0.329	3.824	6.652
RUTHENIUM				V	4.485	1.789	0.254	3.034	0.330	3.816	6.642
52695	12.2843	94	623.0	R	4.460	1.777	0.255	3.032	0.330	3.803	6.631
RUTHENIUM				V	4.440	1.769	0.255	3.022	0.331	3.796	6.621
52696	12.2725	94	673.0	R	4.415	1.757	0.256	3.020	0.331	3.784	6.610
RUTHENIUM				V	4.395	1.749	0.256	3.008	0.332	3.777	6.600
52697	12.2579	94	723.0	R	4.369	1.737	0.258	3.006	0.333	3.764	6.589
RUTHENIUM				V	4.350	1.729	0.258	2.995	0.334	3.758	6.580
52698	12.2434	94	773.0	R	4.324	1.717	0.259	2.993	0.334	3.745	6.568
RUTHENIUM				V	4.306	1.709	0.259	2.982	0.335	3.739	6.559
52699	12.2289	94	823.0	R	4.279	1.697	0.261	2.980	0.336	3.725	6.548
RUTHENIUM				V	4.260	1.689	0.261	2.971	0.337	3.718	6.539
52700	12.2145	94	873.0	R	4.232	1.676	0.262	2.969	0.337	3.704	6.527
RUTHENIUM				V	4.213	1.668	0.263	2.958	0.338	3.698	6.518
52701	12.1985	94	923.0	R	4.185	1.656	0.264	2.956	0.338	3.684	6.506
SILVER(66.67)-ALUMINUM(33.33)				V	0.978	0.363	0.349	1.078	0.928	2.096	4.348
52702	8.2596	53	77.0	R	0.961	0.356	0.351	1.073	0.932	2.075	4.328
SILVER(66.67)-ALUMINUM(33.33)				V	0.970	0.359	0.349	1.073	0.932	2.087	4.336
52703	8.2533	53	100.0	R	0.952	0.352	0.351	1.068	0.937	2.066	4.316
SILVER(66.67)-ALUMINUM(33.33)				V	0.955	0.354	0.350	1.062	0.941	2.072	4.316
52704	8.2362	53	150.0	R	0.937	0.346	0.352	1.058	0.945	2.051	4.296
SILVER(66.67)-ALUMINUM(33.33)				V	0.938	0.347	0.352	1.053	0.950	2.055	4.295
52705	8.2165	53	200.0	R	0.920	0.340	0.354	1.049	0.953	2.034	4.276
SILVER(66.67)-ALUMINUM(33.33)				V	0.925	0.342	0.352	1.038	0.963	2.041	4.264
52706	8.2165	53	250.0	R	0.905	0.334	0.354	1.035	0.966	2.017	4.244
SILVER(66.67)-ALUMINUM(33.33)				V	0.914	0.338	0.351	1.021	0.979	2.035	4.245
52707	8.1700	53	298.0	R	0.892	0.329	0.354	1.019	0.981	2.008	4.225
SILVER(66.67)-ALUMINUM(33.33)				V	0.897	0.332	0.351	1.006	0.994	2.020	4.219
52708	8.1391	53	350.0	R	0.874	0.323	0.355	1.004	0.996	1.991	4.198

CODE	NAME DENSITY	REF	TEMP(K)		E	G	SIGMA	K	BETA	VS	VP
SILVER(66.67)-ALUMINUM(33.33)				V	0.875	0.323	0.354	0.998	1.002	1.996	4.196
52709	8.1137	53	400.0	R	0.853	0.314	0.357	0.996	1.004	1.968	4.175
SILVER(66.67)-ALUMINUM(33.33)				V	0.848	0.312	0.357	0.990	1.010	1.964	4.169
52710	8.0911	53	450.0	R	0.827	0.304	0.360	0.987	1.013	1.938	4.148
SILVER(66.67)-ALUMINUM(33.33)				V	0.816	0.300	0.362	0.984	1.017	1.927	4.140
52711	8.0711	53	500.0	R	0.798	0.293	0.364	0.979	1.021	1.904	4.119
SILVER(66.67)-ALUMINUM(33.33)				V	0.784	0.287	0.366	0.977	1.024	1.887	4.108
52712	8.0526	53	550.0	R	0.768	0.281	0.368	0.969	1.032	1.868	4.085
THALLIUM				V	0.215	0.076	0.407	0.382	2.615	0.803	2.024
52713	11.8170	90	4.2	R	0.161	0.056	0.430	0.381	2.627	0.690	1.964
THALLIUM				V	0.209	0.074	0.408	0.380	2.632	0.794	2.015
52714	11.7940	90	50.0	R	0.157	0.055	0.431	0.378	2.645	0.683	1.956
THALLIUM				V	0.205	0.073	0.410	0.378	2.648	0.785	2.007
52715	11.7740	90	75.0	R	0.154	0.054	0.432	0.376	2.662	0.675	1.949
THALLIUM				V	0.203	0.072	0.409	0.370	2.699	0.782	1.992
52716	11.7510	90	100.0	R	0.151	0.053	0.432	0.369	2.711	0.670	1.933
THALLIUM				V	0.196	0.069	0.410	0.361	2.767	0.771	1.970
52717	11.7050	90	150.0	R	0.145	0.051	0.433	0.360	2.777	0.658	1.911
THALLIUM				V	0.190	0.067	0.411	0.358	2.796	0.760	1.959
52718	11.6560	90	200.0	R	0.143	0.050	0.433	0.356	2.806	0.654	1.905
THALLIUM				V	0.175	0.062	0.418	0.357	2.802	0.731	1.949
52719	11.5600	90	300.0	R	0.130	0.045	0.439	0.356	2.810	0.625	1.897
TITANIUM				V	1.314	0.505	0.301	1.100	0.909	3.342	6.262
52720	4.5225	95	4.0	R	1.298	0.498	0.303	1.100	0.909	3.318	6.245
TITANIUM				V	1.314	0.505	0.301	1.098	0.910	3.342	6.259
52721	4.5225	95	23.0	R	1.298	0.498	0.303	1.098	0.911	3.318	6.242
TITANIUM				V	1.303	0.501	0.302	1.095	0.913	3.328	6.244
52722	4.5198	95	73.0	R	1.286	0.493	0.304	1.095	0.913	3.302	6.226
TITANIUM				V	1.279	0.490	0.304	1.090	0.918	3.295	6.213
52723	4.5171	95	123.0	R	1.260	0.482	0.307	1.089	0.918	3.267	6.193
TITANIUM				V	1.248	0.477	0.308	1.085	0.921	3.250	6.175
52724	4.5144	95	173.0	R	1.226	0.467	0.312	1.085	0.921	3.217	6.151
TITANIUM				V	1.214	0.462	0.313	1.080	0.926	3.201	6.132
52725	4.5117	95	223.0	R	1.187	0.451	0.317	1.080	0.926	3.161	6.105
TITANIUM				V	1.181	0.448	0.317	1.075	0.930	3.153	6.091
52726	4.5090	95	273.0	R	1.150	0.435	0.322	1.075	0.930	3.106	6.058
TITANIUM				V	1.163	0.441	0.319	1.073	0.932	3.128	6.070
52727	4.5063	95	298.0	R	1.129	0.426	0.325	1.073	0.932	3.076	6.034
TITANIUM				V	1.146	0.434	0.321	1.070	0.935	3.103	6.050
52728	4.5022	95	323.0	R	1.109	0.418	0.327	1.070	0.935	3.047	6.011
TITANIUM				V	1.113	0.420	0.326	1.064	0.940	3.055	6.008
52729	4.4982	95	373.0	R	1.071	0.402	0.332	1.064	0.940	2.989	5.963
TITANIUM				V	1.081	0.407	0.330	1.058	0.945	3.008	5.967
52730	4.4928	95	423.0	R	1.033	0.386	0.337	1.058	0.945	2.932	5.916
TITANIUM				V	1.048	0.393	0.334	1.051	0.951	2.959	5.925
52731	4.4874	95	473.0	R	0.993	0.370	0.343	1.051	0.951	2.870	5.866
TITANIUM				V	1.017	0.380	0.338	1.044	0.958	2.913	5.882
52732	4.4834	95	523.0	R	0.954	0.354	0.348	1.044	0.958	2.809	5.814
TITANIUM				V	0.987	0.368	0.341	1.037	0.964	2.867	5.842
52733	4.4754	95	573.0	R	0.916	0.338	0.353	1.037	0.964	2.750	5.766

HEXAGONAL

CODE	NAME DENSITY	REF	TEMP(K)		E	G	SIGMA	K	BETA	VS	VP
TITANIUM				V	0.956	0.355	0.345	1.030	0.971	2.819	5.800
52734	4.4700	95	623.0	R	0.877	0.323	0.358	1.030	0.971	2.687	5.716
TITANIUM				V	0.925	0.343	0.349	1.023	0.977	2.772	5.759
52735	4.4634	95	673.0	R	0.838	0.307	0.363	1.023	0.977	2.624	5.666
TITANIUM				V	0.895	0.331	0.353	1.018	0.982	2.724	5.721
52736	4.4568	95	723.0	R	0.800	0.292	0.369	1.018	0.982	2.561	5.620
TITANIUM				V	0.868	0.320	0.357	1.009	0.991	2.681	5.680
52737	4.4502	95	773.0	R	0.765	0.279	0.374	1.009	0.991	2.502	5.570
TITANIUM				V	0.839	0.308	0.361	1.003	0.997	2.634	5.642
52738	4.4434	95	823.0	R	0.729	0.264	0.379	1.003	0.997	2.439	5.523
TITANIUM				V	0.810	0.297	0.364	0.997	1.003	2.587	5.603
52739	4.4366	95	873.0	R	0.692	0.250	0.384	0.997	1.003	2.373	5.475
TITANIUM				V	0.782	0.286	0.368	0.990	1.010	2.540	5.563
52740	4.4300	95	923.0	R	0.656	0.236	0.390	0.990	1.010	2.308	5.426
TITANIUM				V	0.754	0.275	0.372	0.983	1.017	2.493	5.524
52741	4.4234	95	973.0	R	0.619	0.222	0.395	0.983	1.017	2.239	5.377
TITANIUM				V	0.727	0.264	0.376	0.977	1.024	2.446	5.486
52742	4.4169	95	1023.0	R	0.581	0.208	0.401	0.977	1.024	2.168	5.327
TITANIUM				V	0.701	0.254	0.380	0.971	1.030	2.401	5.450
52743	4.4089	95	1073.0	R	0.543	0.193	0.407	0.971	1.030	2.092	5.277
TITANIUM DIBORIDE				V	4.731	1.793	0.319	4.356	0.230	6.019	11.675
52744	4.9500	112	ROOM	R	4.187	1.580	0.325	3.989	0.251	5.650	11.097
YTTRIUM				V	0.692	0.282	0.228	0.423	2.362	2.503	4.215
52745	4.4990	238	4.2	R	0.687	0.280	0.228	0.421	2.373	2.494	4.202
YTTRIUM				V	0.691	0.282	0.228	0.423	2.362	2.502	4.214
52746	4.4990	238	20.0	R	0.687	0.280	0.228	0.421	2.373	2.493	4.202
YTTRIUM				V	0.689	0.280	0.229	0.424	2.361	2.496	4.210
52747	4.4980	238	50.0	R	0.684	0.278	0.229	0.421	2.373	2.487	4.198
YTTRIUM				V	0.687	0.280	0.226	0.418	2.391	2.496	4.196
52748	4.4970	238	75.0	R	0.682	0.278	0.227	0.416	2.405	2.486	4.182
YTTRIUM				V	0.683	0.278	0.227	0.418	2.394	2.488	4.189
52749	4.4950	238	100.0	R	0.677	0.276	0.228	0.415	2.407	2.477	4.174
YTTRIUM				V	0.680	0.278	0.225	0.412	2.429	2.485	4.171
52750	4.4930	238	125.0	R	0.673	0.274	0.226	0.409	2.444	2.472	4.153
YTTRIUM				V	0.674	0.275	0.226	0.410	2.440	2.475	4.159
52751	4.4900	238	150.0	R	0.667	0.272	0.227	0.407	2.454	2.461	4.141
YTTRIUM				V	0.667	0.271	0.230	0.412	2.425	2.458	4.153
52752	4.4870	238	175.0	R	0.661	0.268	0.231	0.410	2.437	2.446	4.138
YTTRIUM				V	0.661	0.268	0.231	0.410	2.438	2.446	4.139
52753	4.4840	238	200.0	R	0.655	0.266	0.233	0.408	2.449	2.434	4.124
YTTRIUM				V	0.655	0.265	0.235	0.413	2.424	2.432	4.134
52754	4.4810	238	225.0	R	0.650	0.263	0.236	0.411	2.433	2.422	4.122
YTTRIUM				V	0.647	0.261	0.240	0.415	2.412	2.413	4.126
52755	4.4780	238	250.0	R	0.643	0.259	0.241	0.413	2.419	2.405	4.116
YTTRIUM				V	0.641	0.258	0.241	0.413	2.424	2.403	4.113
52756	4.4750	238	275.0	R	0.637	0.257	0.242	0.411	2.431	2.394	4.103
YTTRIUM				V	0.635	0.255	0.245	0.415	2.408	2.387	4.109
52757	4.4720	238	300.0	R	0.631	0.253	0.246	0.414	2.414	2.380	4.101
YTTRIUM				V	0.630	0.253	0.246	0.414	2.415	2.378	4.099
52758	4.4690	238	325.0	R	0.627	0.251	0.247	0.413	2.420	2.371	4.091

HEXAGONAL

CODE / NAME	DENSITY	REF	TEMP(K)		E	G	SIGMA	K	BETA	VS	VP
YTTRIUM				V	0.626	0.251	0.248	0.413	2.421	2.370	4.091
52759	4.4660	238	350.0	R	0.622	0.249	0.248	0.412	2.425	2.362	4.083
YTTRIUM				V	0.622	0.249	0.249	0.412	2.428	2.362	4.082
52760	4.4630	238	375.0	R	0.618	0.247	0.249	0.411	2.432	2.354	4.074
YTTRIUM				V	0.617	0.247	0.250	0.411	2.434	2.353	4.074
52761	4.4590	238	400.0	R	0.613	0.245	0.251	0.410	2.438	2.345	4.066
ZINC				V	1.121	0.455	0.233	0.701	1.427	2.523	4.279
52762	7.1400	128	ROOM	R	0.889	0.357	0.246	0.583	1.714	2.236	3.852
ZINC				V	1.098	0.441	0.244	0.716	1.397	2.485	4.273
52763	7.1400	128	ROOM	R	0.861	0.350	0.230	0.532	1.880	2.214	3.740
ZINC				V	1.138	0.462	0.231	0.705	1.419	2.544	4.301
52764	7.1400	128	ROOM	R	0.856	0.340	0.259	0.592	1.690	2.182	3.825
ZINC				V	1.251	0.504	0.241	0.804	1.244	2.657	4.547
52765	7.1400	128	ROOM	R	0.925	0.360	0.283	0.711	1.406	2.247	4.086
ZINC				V	1.173	0.482	0.216	0.688	1.453	2.599	4.318
52766	7.1400	128	ROOM	R	0.894	0.365	0.226	0.543	1.840	2.260	3.797
ZINC				V	1.110	0.446	0.245	0.725	1.378	2.499	4.299
52767	7.1400	129	ROOM	R	0.856	0.340	0.259	0.591	1.692	2.182	3.824
ZINC				V	1.123	0.454	0.236	0.708	1.412	2.522	4.290
52768	7.1400	4	ROOM	R	0.890	0.356	0.248	0.590	1.696	2.234	3.861
ZINC				V	1.265	0.511	0.238	0.804	1.244	2.650	4.517
52769	7.2810	16	4.2	R	0.987	0.394	0.251	0.661	1.513	2.327	4.037
ZINC				V	1.244	0.502	0.239	0.795	1.258	2.629	4.490
52770	7.2620	16	77.0	R	0.962	0.384	0.254	0.652	1.533	2.299	4.003
ZINC				V	1.209	0.486	0.243	0.784	1.276	2.595	4.453
52771	7.2220	16	150.0	R	0.929	0.369	0.258	0.641	1.560	2.261	3.961
ZINC				V	1.181	0.474	0.246	0.775	1.290	2.567	4.423
52772	7.1920	16	200.0	R	0.905	0.359	0.262	0.633	1.580	2.233	3.931
ZINC				V	1.152	0.461	0.248	0.763	1.310	2.538	4.387
52773	7.1620	16	250.0	R	0.885	0.350	0.264	0.624	1.603	2.211	3.902
ZINC				V	1.121	0.448	0.251	0.751	1.332	2.506	4.347
52774	7.1340	16	295.0	R	0.864	0.341	0.266	0.616	1.624	2.187	3.874
ZINC				V	1.089	0.434	0.254	0.737	1.357	2.474	4.306
52775	7.0970	16	350.0	R	0.842	0.332	0.268	0.606	1.649	2.163	3.844
ZINC				V	1.055	0.419	0.257	0.723	1.382	2.437	4.261
52776	7.0640	16	400.0	R	0.816	0.321	0.272	0.598	1.673	2.131	3.810
ZINC				V	1.017	0.403	0.261	0.710	1.409	2.394	4.212
52777	7.0320	16	450.0	R	0.786	0.308	0.278	0.590	1.696	2.092	3.771
ZINC				V	0.977	0.386	0.266	0.696	1.437	2.348	4.159
52778	6.9980	16	500.0	R	0.753	0.293	0.284	0.582	1.719	2.047	3.728
ZINC				V	0.937	0.368	0.272	0.685	1.460	2.300	4.109
52779	6.9660	16	550.0	R	0.721	0.279	0.291	0.574	1.742	2.002	3.686
ZINC				V	0.898	0.351	0.279	0.677	1.477	2.250	4.064
52780	6.9310	16	600.0	R	0.690	0.266	0.297	0.567	1.765	1.960	3.646
ZINC				V	0.833	0.322	0.294	0.674	1.483	2.164	4.006
52781	6.8780	16	670.0	R	0.649	0.248	0.306	0.557	1.796	1.900	3.593
ZINC				V	1.073	0.455	0.180	0.558	1.793	2.531	4.049
52782	7.1000	189	ROOM	R	0.853	0.359	0.187	0.455	2.199	2.249	3.626
ZINC OXIDE (CONSTANT E)				V	1.242	0.458	0.356	1.437	0.696	2.840	6.005
52783	5.6760	26	298.0	R	1.230	0.453	0.357	1.435	0.697	2.826	5.994

HEXAGONAL

CODE	NAME DENSITY	REF	TEMP(K)		E	G	SIGMA	K	BETA	VS	VP
ZINC SULFIDE (CONSTANT E)				V	0.879	0.347	0.266	0.626	1.597	2.913	5.160
52784	4.0900	143	ROOM	R	0.845	0.331	0.275	0.625	1.599	2.846	5.108
ZINC SULFIDE (CONSTANT D)				V	0.881	0.337	0.308	0.765	1.307	2.870	5.448
52785	4.0900	282	ROOM	R	0.855	0.325	0.314	0.765	1.307	2.820	5.413
ZINC SULFIDE (HEXAGONAL)				V	0.881	0.337	0.309	0.768	1.303	2.870	5.455
52786	4.0890	60	298.0	R	0.856	0.326	0.314	0.768	1.303	2.822	5.421
ZINC SULFIDE (ALPHA)				V	0.887	0.336	0.320	0.822	1.217	2.867	5.572
52787	4.0890	74	ROOM	R	0.864	0.326	0.325	0.821	1.218	2.824	5.542
ZIRCONIUM				V	1.112	0.425	0.310	0.973	1.027	2.551	4.856
52788	6.5280	95	4.0	R	1.095	0.417	0.312	0.972	1.029	2.528	4.838
ZIRCONIUM				V	1.110	0.424	0.310	0.973	1.027	2.549	4.856
52789	6.5260	95	23.0	R	1.093	0.416	0.313	0.972	1.029	2.525	4.837
ZIRCONIUM				V	1.097	0.418	0.312	0.972	1.029	2.531	4.842
52790	6.5230	95	73.0	R	1.079	0.411	0.315	0.971	1.030	2.509	4.824
ZIRCONIUM				V	1.074	0.408	0.315	0.968	1.034	2.503	4.816
52791	6.5190	95	123.0	R	1.057	0.401	0.318	0.966	1.035	2.480	4.798
ZIRCONIUM				V	1.045	0.396	0.319	0.961	1.040	2.467	4.782
52792	6.5150	95	173.0	R	1.028	0.389	0.321	0.960	1.042	2.444	4.764
ZIRCONIUM				V	1.015	0.384	0.323	0.958	1.044	2.427	4.750
52793	6.5110	95	223.0	R	0.998	0.376	0.326	0.956	1.046	2.404	4.731
ZIRCONIUM				V	0.984	0.370	0.328	0.954	1.048	2.386	4.717
52794	6.5070	95	273.0	R	0.966	0.363	0.331	0.953	1.050	2.362	4.698
ZIRCONIUM				V	0.969	0.364	0.331	0.954	1.048	2.366	4.704
52795	6.5050	95	298.0	R	0.951	0.356	0.334	0.952	1.050	2.341	4.685
ZIRCONIUM				V	0.953	0.358	0.333	0.951	1.052	2.345	4.686
52796	6.5030	95	323.0	R	0.935	0.350	0.336	0.950	1.053	2.320	4.666
ZIRCONIUM				V	0.924	0.346	0.337	0.947	1.056	2.306	4.655
52797	6.4970	95	373.0	R	0.905	0.337	0.341	0.946	1.058	2.279	4.635
ZIRCONIUM				V	0.896	0.334	0.341	0.942	1.062	2.268	4.621
52798	6.4950	95	423.0	R	0.875	0.325	0.345	0.940	1.064	2.238	4.600
ZIRCONIUM				V	0.870	0.323	0.345	0.937	1.067	2.232	4.592
52799	6.4890	95	473.0	R	0.847	0.314	0.349	0.936	1.069	2.200	4.569
ZIRCONIUM				V	0.841	0.312	0.350	0.933	1.071	2.193	4.561
52800	6.4840	95	523.0	R	0.817	0.302	0.354	0.932	1.073	2.157	4.536
ZIRCONIUM				V	0.817	0.302	0.353	0.929	1.076	2.158	4.533
52801	6.4780	95	573.0	R	0.790	0.291	0.358	0.928	1.078	2.119	4.506
ZIRCONIUM				V	0.791	0.292	0.357	0.924	1.082	2.122	4.503
52802	6.4740	95	623.0	R	0.762	0.280	0.362	0.923	1.084	2.079	4.474
ZIRCONIUM				V	0.768	0.282	0.361	0.919	1.088	2.088	4.475
52803	6.4680	95	673.0	R	0.736	0.269	0.366	0.918	1.090	2.040	4.443
ZIRCONIUM				V	0.743	0.272	0.365	0.914	1.094	2.053	4.446
52804	6.4620	95	723.0	R	0.707	0.258	0.371	0.913	1.096	1.998	4.410
ZIRCONIUM				V	0.718	0.262	0.368	0.909	1.100	2.016	4.416
52805	6.4560	95	773.0	R	0.679	0.247	0.375	0.908	1.102	1.955	4.376
ZIRCONIUM				V	0.694	0.253	0.372	0.903	1.107	1.981	4.385
52806	6.4510	95	823.0	R	0.651	0.236	0.380	0.902	1.109	1.912	4.342
ZIRCONIUM				V	0.670	0.243	0.375	0.897	1.115	1.944	4.354
52807	6.4430	95	873.0	R	0.622	0.225	0.384	0.895	1.117	1.867	4.305
ZIRCONIUM				V	0.839	0.310	0.355	0.963	1.039	2.194	4.623
52808	6.4370	95	923.0	R	0.804	0.295	0.360	0.959	1.042	2.142	4.585

TRIGONAL 1

CODE	NAME DENSITY	REF	TEMP(K)		E	G	SIGMA	K	BETA	VS	VP
ALUMINUM OXIDE				V	4.690	1.997	0.174	2.400	0.417	7.092	11.293
62809	3.9700	129	ROOM	R	3.913	1.593	0.228	2.398	0.417	6.335	10.673
ALUMINUM OXIDE				V	4.773	2.030	0.176	2.454	0.407	7.150	11.401
62810	3.9700	129	ROOM	R	3.846	1.554	0.237	2.440	0.410	6.257	10.661
ALUMINUM OXIDE				V	3.265	1.275	0.281	2.480	0.403	5.660	10.248
62811	3.9800	180	300.0	R	1.965	0.718	0.368	2.480	0.403	4.248	9.293
ALUMINUM OXIDE				V	4.010	1.626	0.233	2.504	0.399	6.387	10.826
62812	3.9860	33	300.0	R	3.904	1.574	0.240	2.499	0.400	6.285	10.740
ALUMINUM OXIDE				V	4.021	1.627	0.235	2.532	0.395	6.390	10.860
62813	3.9860	33	300.0	R	3.917	1.577	0.242	2.526	0.396	6.291	10.777
ALUMINUM OXIDE				V	4.082	1.661	0.229	2.512	0.398	6.455	10.889
62814	3.9860	268	298.0	R	3.973	1.607	0.236	2.508	0.399	6.350	10.801
ALUMINUM OXIDE				V	4.141	1.688	0.227	2.524	0.396	6.502	10.935
62815	3.9929	252	0.0	R	4.038	1.638	0.233	2.520	0.397	6.405	10.853
ALUMINUM OXIDE				V	4.139	1.686	0.227	2.528	0.396	6.499	10.938
62816	3.9929	252	100.0	R	4.036	1.636	0.233	2.524	0.396	6.401	10.856
ALUMINUM OXIDE				V	4.131	1.682	0.228	2.528	0.396	6.491	10.932
62817	3.9929	252	150.0	R	4.027	1.632	0.234	2.524	0.396	6.392	10.849
ALUMINUM OXIDE				V	4.118	1.676	0.229	2.528	0.396	6.481	10.925
62818	3.9905	252	200.0	R	4.012	1.624	0.235	2.524	0.396	6.380	10.841
ALUMINUM OXIDE				V	4.102	1.668	0.229	2.525	0.396	6.468	10.913
62819	3.9881	252	250.0	R	3.994	1.616	0.236	2.521	0.397	6.365	10.827
ALUMINUM OXIDE				V	4.085	1.660	0.230	2.523	0.396	6.454	10.901
62820	3.9860	252	298.0	R	3.975	1.607	0.237	2.519	0.397	6.349	10.814
ALUMINUM OXIDE				V	4.084	1.660	0.230	2.523	0.396	6.453	10.900
62821	3.9860	252	300.0	R	3.974	1.606	0.237	2.518	0.397	6.348	10.813
ALUMINUM OXIDE				V	4.063	1.651	0.231	2.516	0.397	6.438	10.883
62822	3.9824	252	350.0	R	3.951	1.596	0.238	2.511	0.398	6.331	10.793
ALUMINUM OXIDE				V	4.042	1.641	0.232	2.510	0.398	6.422	10.867
62823	3.9788	252	400.0	R	3.928	1.585	0.239	2.506	0.399	6.312	10.775
ALUMINUM OXIDE				V	4.020	1.631	0.232	2.503	0.399	6.407	10.850
62824	3.9740	252	450.0	R	3.903	1.574	0.240	2.498	0.400	6.294	10.755
ALUMINUM OXIDE				V	3.997	1.621	0.233	2.494	0.401	6.390	10.830
62825	3.9692	252	500.0	R	3.877	1.563	0.240	2.489	0.402	6.274	10.733
ALUMINUM OXIDE				V	3.973	1.610	0.234	2.486	0.402	6.372	10.809
62826	3.9656	252	550.0	R	3.850	1.551	0.241	2.481	0.403	6.253	10.710
ALUMINUM OXIDE				V	3.949	1.600	0.234	2.476	0.404	6.356	10.788
62827	3.9608	252	600.0	R	3.823	1.539	0.242	2.470	0.405	6.234	10.686
ALUMINUM OXIDE				V	3.926	1.590	0.235	2.468	0.405	6.339	10.768
62828	3.9560	252	650.0	R	3.797	1.527	0.243	2.462	0.406	6.213	10.663
ALUMINUM OXIDE				V	3.900	1.579	0.235	2.455	0.407	6.322	10.744
62829	3.9502	252	700.0	R	3.768	1.515	0.244	2.449	0.408	6.193	10.637
ALUMINUM OXIDE				V	3.876	1.568	0.236	2.450	0.408	6.303	10.727
62830	3.9454	252	750.0	R	3.741	1.503	0.245	2.443	0.409	6.171	10.616
ALUMINUM OXIDE				V	3.850	1.556	0.237	2.440	0.410	6.284	10.704
62831	3.9408	252	800.0	R	3.712	1.490	0.246	2.433	0.411	6.149	10.590
ALUMINUM OXIDE				V	3.825	1.545	0.238	2.431	0.411	6.266	10.682
62832	3.9360	252	850.0	R	3.684	1.477	0.247	2.424	0.413	6.126	10.565
ALUMINUM OXIDE				V	3.800	1.534	0.239	2.424	0.413	6.248	10.663
62833	3.9302	252	900.0	R	3.655	1.465	0.248	2.416	0.414	6.104	10.543

CODE	NAME DENSITY	REF	TEMP(K)		E	G	SIGMA	K	BETA	VS	VP
ALUMINUM OXIDE				V	4.085	1.657	0.233	2.546	0.393	6.448	10.922
62834	3.9860	111	298.0	R	3.983	1.607	0.239	2.542	0.393	6.350	10.842
ALUMINUM PHOSPHATE				V	0.760	0.285	0.332	0.755	1.325	3.335	6.652
62835	2.5660	177	ROOM	R	0.600	0.222	0.351	0.671	1.490	2.942	6.140
ANTIMONY				V	0.622	0.251	0.237	0.394	2.535	1.938	3.302
62836	6.6910	128	ROOM	R	0.515	0.204	0.263	0.362	2.760	1.746	3.079
ANTIMONY				V	0.803	0.333	0.207	0.456	2.191	2.229	3.665
62837	6.7000	86	ROOM	R	0.552	0.219	0.262	0.387	2.581	1.807	3.184
ANTIMONY				V	0.800	0.327	0.223	0.482	2.077	2.210	3.701
62838	6.7000	72	295.0	R	0.550	0.216	0.273	0.404	2.474	1.795	3.214
BISMUTH				V	0.354	0.133	0.336	0.360	2.776	1.166	2.347
62839	9.7470	4	ROOM	R	0.309	0.115	0.347	0.337	2.970	1.084	2.241
BISMUTH				V	0.382	0.145	0.317	0.347	2.880	1.216	2.348
62840	9.8000	81	300.0	R	0.291	0.108	0.351	0.325	3.074	1.048	2.187
BISMUTH				V	0.381	0.145	0.318	0.349	2.864	1.215	2.351
62841	9.8000	72	295.0	R	0.291	0.108	0.352	0.327	3.059	1.048	2.191
CALCITE				V	0.951	0.374	0.271	0.692	1.445	3.711	6.620
62842	2.7170	128	ROOM	R	0.727	0.277	0.312	0.645	1.550	3.193	6.111
CALCITE				V	0.958	0.377	0.269	0.692	1.445	3.727	6.633
62843	2.7170	129	ROOM	R	0.743	0.284	0.309	0.649	1.542	3.232	6.148
CALCITE				V	0.956	0.376	0.271	0.695	1.438	3.720	6.636
62844	2.7170	4	ROOM	R	0.732	0.279	0.312	0.647	1.546	3.204	6.124
CALCITE				V	0.928	0.357	0.301	0.778	1.286	3.627	6.798
62845	2.7120	200	298.0	R	0.691	0.258	0.339	0.717	1.395	3.083	6.253
CALCITE (P=0KB)				V	0.952	0.367	0.296	0.778	1.285	3.679	6.837
62846	2.7120	64	298.0	R	0.717	0.269	0.333	0.716	1.398	3.149	6.293
CALCITE (P=1KB)				V	0.954	0.368	0.296	0.780	1.281	3.680	6.841
62847	2.7158	64	298.0	R	0.720	0.270	0.333	0.718	1.393	3.155	6.301
CALCITE (P=2KB)				V	0.954	0.368	0.297	0.784	1.276	3.678	6.846
62848	2.7193	64	298.0	R	0.721	0.271	0.333	0.721	1.387	3.154	6.307
CALCITE (P=3KB)				V	0.953	0.367	0.298	0.787	1.271	3.671	6.846
62849	2.7231	64	298.0	R	0.719	0.269	0.335	0.725	1.379	3.145	6.310
CALCITE (P=4KB)				V	0.952	0.366	0.300	0.795	1.258	3.663	6.858
62850	2.7267	64	298.0	R	0.714	0.267	0.337	0.732	1.366	3.128	6.316
CALCITE (P=5KB)				V	0.948	0.364	0.303	0.802	1.247	3.650	6.865
62851	2.7302	64	298.0	R	0.706	0.263	0.341	0.739	1.352	3.104	6.319
CALCITE (P=6KB)				V	0.943	0.361	0.306	0.810	1.234	3.633	6.873
62852	2.7341	64	298.0	R	0.695	0.258	0.345	0.748	1.337	3.073	6.321
CALCITE				V	0.956	0.368	0.298	0.789	1.267	3.686	6.871
62853	2.7120	65	298.0	R	0.714	0.267	0.338	0.733	1.365	3.136	6.334
CALCITE				V	0.948	0.365	0.298	0.781	1.280	3.671	6.839
62854	2.7110	65	323.0	R	0.705	0.263	0.338	0.727	1.375	3.117	6.307
CALCITE				V	0.940	0.363	0.297	0.772	1.295	3.658	6.807
62855	2.7100	65	348.0	R	0.701	0.262	0.338	0.720	1.388	3.109	6.283
CALCITE				V	0.932	0.359	0.296	0.763	1.311	3.643	6.772
62856	2.7080	65	373.0	R	0.694	0.260	0.338	0.713	1.402	3.096	6.254
CALCITE				V	0.924	0.356	0.296	0.754	1.326	3.629	6.739
62857	2.7070	65	398.0	R	0.690	0.258	0.337	0.707	1.415	3.088	6.231
CALCITE				V	0.916	0.354	0.295	0.746	1.341	3.616	6.708
62858	2.7060	65	423.0	R	0.686	0.257	0.337	0.700	1.428	3.080	6.207

TRIGONAL 1

CODE	NAME DENSITY	REF	TEMP(K)		E	G	SIGMA	K	BETA	VS	VP
CALCITE				V	0.909	0.351	0.295	0.738	1.356	3.602	6.677
62859	2.7040	65	448.0	R	0.680	0.255	0.337	0.694	1.441	3.068	6.181
CALCITE				V	0.901	0.348	0.294	0.730	1.370	3.589	6.646
62860	2.7030	65	473.0	R	0.677	0.254	0.336	0.688	1.454	3.063	6.160
CALCITE				V	0.891	0.344	0.295	0.723	1.382	3.570	6.616
62861	2.7010	65	498.0	R	0.668	0.250	0.337	0.682	1.465	3.041	6.131
CALCITE				V	0.895	0.348	0.285	0.694	1.441	3.591	6.549
62862	2.7000	65	523.0	R	0.675	0.254	0.327	0.652	1.534	3.069	6.058
CALCITE				V	0.996	0.383	0.301	0.834	1.198	3.754	7.035
62863	2.7170	66	160.0	R	0.744	0.278	0.338	0.764	1.309	3.199	6.463
CALCITE				V	0.991	0.381	0.301	0.829	1.206	3.744	7.015
62864	2.7170	66	180.0	R	0.740	0.277	0.338	0.760	1.315	3.191	6.447
CALCITE				V	0.985	0.379	0.300	0.822	1.216	3.734	6.991
62865	2.7160	66	200.0	R	0.736	0.275	0.338	0.756	1.323	3.184	6.430
CALCITE				V	0.979	0.377	0.300	0.817	1.224	3.724	6.970
62866	2.7150	66	220.0	R	0.733	0.274	0.337	0.752	1.330	3.178	6.416
CALCITE				V	0.974	0.375	0.300	0.810	1.234	3.714	6.945
62867	2.7150	66	240.0	R	0.730	0.273	0.337	0.747	1.339	3.170	6.396
CALCITE				V	0.966	0.372	0.300	0.804	1.244	3.701	6.920
62868	2.7140	66	260.0	R	0.724	0.271	0.338	0.743	1.346	3.158	6.377
CALCITE				V	0.961	0.370	0.299	0.797	1.255	3.692	6.896
62869	2.7130	66	280.0	R	0.721	0.270	0.337	0.737	1.356	3.153	6.358
CALCITE				V	0.955	0.367	0.299	0.791	1.264	3.681	6.872
62870	2.7120	66	300.0	R	0.716	0.268	0.337	0.733	1.364	3.142	6.341
DEXTROSE SODIUM BROMIDE				V	0.176	0.070	0.255	0.120	8.357	2.036	3.550
62871	1.6900	129	ROOM	R	0.174	0.069	0.254	0.118	8.490	2.023	3.525
DEXTROSE SODIUM CHLORIDE				V	0.168	0.066	0.277	0.126	7.935	2.052	3.698
62872	1.5640	129	ROOM	R	0.164	0.064	0.278	0.123	8.160	2.023	3.646
DEXTROSE SODIUM IODIDE				V	0.185	0.073	0.272	0.136	7.366	1.977	3.535
62873	1.8640	129	ROOM	R	0.172	0.068	0.275	0.127	7.860	1.903	3.414
GUANIDINE AL SELENATE HEXAHYDRATE				V	0.212	0.079	0.347	0.232	4.318	1.924	3.980
62874	2.1239	4	ROOM	R	0.158	0.060	0.307	0.136	7.348	1.686	3.194
GUANIDINE AL SULFATE HEXAHYDRATE				V	0.229	0.086	0.340	0.238	4.198	2.177	4.417
62875	1.8064	4	ROOM	R	0.178	0.069	0.294	0.143	6.972	1.949	3.606
GUANIDINE GA SELENATE HEXAHYDRATE				V	0.201	0.074	0.351	0.225	4.450	1.805	3.769
62876	2.2798	4	ROOM	R	0.149	0.056	0.316	0.135	7.408	1.574	3.037
GUANIDINE GA SULFATE HEXAHYDRATE				V	0.221	0.082	0.346	0.240	4.174	2.038	4.199
62877	1.9801	4	ROOM	R	0.168	0.064	0.306	0.145	6.918	1.805	3.412
HEMATITE				V	2.152	0.948	0.135	0.981	1.019	4.254	6.547
62878	5.2400	128	ROOM	R	2.099	0.921	0.140	0.972	1.029	4.192	6.479
LITHIUM NIOBATE (CONSTANT D)				V	2.148	0.896	0.198	1.187	0.843	4.366	7.118
62879	4.7000	271	ROOM	R	2.084	0.868	0.201	1.161	0.861	4.297	7.023
LITHIUM NIOBATE (CONSTANT E)				V	1.728	0.689	0.255	1.174	0.851	3.828	6.673
62880	4.7000	271	ROOM	R	1.681	0.669	0.257	1.154	0.867	3.771	6.597
LITHIUM TANTALATE (CONSTANT E)				V	2.240	0.927	0.208	1.278	0.782	3.527	5.808
62881	7.4540	240	ROOM	R	2.198	0.911	0.206	1.246	0.802	3.496	5.746
LITHIUM TANTALATE (CONSTANT E)				V	2.224	0.918	0.211	1.283	0.779	3.510	5.801
62882	7.4500	271	ROOM	R	2.186	0.903	0.211	1.259	0.794	3.481	5.749
LITHIUM TANTALATE (CONSTANT D)				V	2.431	1.024	0.187	1.293	0.773	3.707	5.974
62883	7.4500	271	ROOM	R	2.328	0.976	0.193	1.264	0.791	3.619	5.867

TRIGONAL 1

CODE	NAME DENSITY	REF	TEMP(K)		E	G	SIGMA	K	BETA	VS	VP
MERCURY				V	0.224	0.081	0.389	0.335	2.985	0.747	1.749
62884	14.4690	129	83.0	R	0.095	0.033	0.451	0.322	3.101	0.476	1.591
QUARTZ				V	0.997	0.467	0.067	0.384	2.603	4.198	6.164
62885	2.6500	128	ROOM	R	0.895	0.405	0.105	0.377	2.650	3.909	5.884
QUARTZ				V	1.011	0.485	0.041	0.367	2.723	4.280	6.187
62886	2.6500	128	ROOM	R	0.903	0.417	0.084	0.361	2.768	3.965	5.882
QUARTZ				V	1.011	0.479	0.056	0.379	2.635	4.246	6.192
62887	2.6540	4	ROOM	R	0.902	0.410	0.099	0.375	2.670	3.932	5.893
QUARTZ				V	1.015	0.472	0.075	0.398	2.512	4.214	6.216
62888	2.6600	138	ROOM	R	0.911	0.410	0.112	0.391	2.555	3.925	5.938
QUARTZ				V	1.012	0.476	0.063	0.386	2.592	4.232	6.195
62889	2.6600	138	ROOM	R	0.913	0.415	0.099	0.380	2.634	3.952	5.924
QUARTZ				V	1.011	0.479	0.056	0.379	2.635	4.250	6.198
62890	2.6487	144	293.0	R	0.902	0.410	0.099	0.375	2.670	3.936	5.899
QUARTZ (CONSTANT E)				V	1.014	0.479	0.058	0.382	2.618	4.249	6.202
62891	2.6540	179	293.0	R	0.906	0.413	0.099	0.376	2.657	3.942	5.908
QUARTZ (CONSTANT E)				V	1.011	0.478	0.057	0.380	2.629	4.244	6.193
62892	2.6540	29	293.0	R	0.903	0.411	0.099	0.375	2.668	3.935	5.896
QUARTZ				V	1.027	0.481	0.067	0.395	2.531	4.251	6.239
62893	2.6638	168	77.2	R	0.912	0.411	0.110	0.390	2.565	3.927	5.933
QUARTZ				V	1.010	0.478	0.056	0.379	2.639	4.249	6.195
62894	2.6485	168	298.0	R	0.901	0.410	0.098	0.374	2.673	3.936	5.897
SELENIUM				V	0.279	0.106	0.318	0.256	3.905	1.479	2.866
62895	4.8380	188	300.0	R	0.110	0.044	0.264	0.078	12.841	0.949	1.676
SODIUM NITRATE				V	0.630	0.264	0.193	0.342	2.928	3.410	5.527
62896	2.2700	129	ROOM	R	0.529	0.220	0.200	0.294	3.407	3.116	5.087
TOURMALINE				V	2.026	0.879	0.153	0.973	1.028	5.368	8.385
62897	3.0500	128	ROOM	R	1.872	0.820	0.141	0.870	1.150	5.186	8.023
TOURMALINE				V	2.076	0.891	0.165	1.032	0.969	5.362	8.464
62898	3.1000	129	ROOM	R	1.923	0.820	0.172	0.977	1.024	5.145	8.173
TOURMALINE				V	1.905	0.785	0.213	1.106	0.905	5.074	8.401
62899	3.0500	129	ROOM	R	1.756	0.723	0.215	1.026	0.975	4.868	8.077
TOURMALINE				V	2.155	0.893	0.206	1.222	0.818	5.457	8.969
62900	3.0000	129	ROOM	R	1.985	0.828	0.200	1.102	0.908	5.252	8.573

TRIGONAL 1

CODE	NAME DENSITY	REF	TEMP(K)		E	G	SIGMA	K	BETA	VS	VP
					PRESSURE DERIVATIVES						
ALUMINUM OXIDE				V	4.541	1.637	1.525	4.280	-0.661	1.918	5.277
62901	3.9860	111	298.0	R	4.939	1.845	1.147	4.269	-0.659	2.395	5.652
QUARTZ				V	4.813	0.231	44.882	5.944	-38.080	-4.361	10.912
62902	2.6638	168	77.2	R	4.613	0.676	37.885	5.758	-36.887	-1.809	13.457
QUARTZ				V	5.602	0.313	51.670	6.511	-45.337	-4.217	12.936
62903	2.6485	168	298.0	R	5.059	0.607	45.403	6.330	-44.073	-2.348	14.972

References for Tables

1. Adhav, R.S., Elastic, piezoelectric, and dielectric properties of ammonium dihydrogen arsenate (ADA). I, J. Acoust. Soc. Am., 43, 835-838, 1968.

2. Adhav, R.S., Elastic, piezoelectric, and dielectric properties of rubidium dihydrogen phosphate, J. Appl. Phys., 40, 2725-2727, 1969.

3. Afanaseva, G.K., K.S. Aleksandrov, and A.I. Kitaigorodskii, Elastic constants of anthracene, Phys. Stat. Sol., 24, K61-K63, 1967.

4. Alexandrov, K.S., and T.V. Ryzhova, The elastic properties of crystals, Sov. Phys.-Cryst., 6, 228-252, 1961.

5. Alexandrov, K.S., and T.V. Ryzhova, Elastic properties of rock-forming minerals. II. layered silicates, Bull. Acad. Sci. USSR, Geophys. Ser., English Transl., no. 12, 1165-1168, 1961.

6. Alexandrov, K.S., and T.V. Ryzhova, Elastic properties of rock-forming minerals: III feldspars, Bull. Acad. Sci. USSR, Geophys. Ser., English Transl., no. 2, 129-131, 1962.

7. Alexandrov, K.S., and T.V. Ryzhova, The elastic properties of rock-forming minerals, pyroxenes and amphiboles, Bull. Acad. Sci. USSR, Geophys. Ser., English Transl., no. 8, 871-875, 1961.

8. Alexandrov, K.S., and T.V. Ryzhova, Modulus of elasticity of pyrite, Izv., Sibirsk, Otd. Akad. Nauk, SSSR, 6, 43, 1961.

9. Aleksandrov, K.S., L.M. Reshchikova, and B.V. Beznosikov, Behaviour of the elastic constants of $KMnF_3$ single crystals near the transition of puckering, Phys. Stat. Sol., 18, K17-K20, 1966.

10. Aleksandrov, K.S., T.V. Ryzhova, and B.P. Belikov, The elastic properties of pyroxenes, Sov. Phys.-Cryst., 8, 589-591, 1964.

11. Aleksandrov, K.S., T.V. Ryzhova, and A.I. Rostuntseva, Elastic properties of some sulfate heptahydrate crystals, Sov. Phys.-Cryst., 7, 753-755, 1963.

12. Aleksandrov, K.S., L.A. Shabanova, and L.M. Reshchikova, Anomalies of elastic properties and internal friction of a CoO single crystal, Sov. Phys.-Solid State, 10, 1316-1321, 1968.

13. Aleksandrov, K.S., G.S. Belikova, A.P. Ryzhenkov, V.R. Teslenko, and A.I. Kitaigorodskii, The elastic constants of molecular crystals. The elastic constants of napthalene, Sov. Phys.-Cryst., 8, 164-166, 1963.

14. Alers, G.A., Elastic moduli of vanadium, Phys. Rev., 119, 1532-1535, 1960.

15. Alers, G.A., and J.A. Karbon, Elastic moduli of the lead-thallium alloys, J. Appl. Phys., 37, 4252-4255, 1966.

16. Alers, G.A., and J.R. Neighbours, The elastic constants of zinc between 4.2° and 670°K, J. Phys. Chem. Solids, 7, 58-64, 1958.

17. Alers, G.A., J.R. Neighbours, and H. Sato, Temperature dependent magnetic contributions to the high field elastic constants of nickel and an Fe-Ni alloy, J. Phys. Chem. Solids, 13, 40-55, 1960.

18. Alper, T., and G.A. Saunders, The elastic constants of mercury telluride, J. Phys. Chem Solids, 28, 1637-1642, 1967.

19. Alton, W.J., and A.J. Barlow, Acoustic-wave propagation in tetragonal crystals and measurements of the elastic constants of calcium molybdate, J. Appl. Phys., 38, 3817-3820, 1967.

20. Anderson, O.L. and P. Andreatch, Jr., Pressure derivatives of elastic constants of single-crystal MgO at 23° and -195.8°C, J. Amer. Ceramic Soc., 49, 404-409, 1966.

21. Armstrong, P.E., O.N. Carlson, and J.F. Smith, Elastic constants of thorium single crystals in the range 77-400°K, J. Appl. Phys., 30, 36-41, 1959.

22. Armstrong, P.E., J.M. Dickinson, and H.L. Brown, Temperature dependence of the elastic stiffness coefficients of niobium (columbium), Trans. Metal. Soc. AIME, 236, 1404-1408, 1966.

23. Bartels, R.A., and D.E. Schuele, Pressure derivatives of the elastic constants of NaCl and KCl at 295°K and 195°K, J. Phys. Chem. Solids, 26, 537-549, 1965.

24. Bartlett, R.W., and C.W. Smith, Elastic constants of tantalum monocarbide, $TaC_{0.90}$, J. Appl. Phys., 38, 5428-5429, 1967.

25. Bateman, T.B., Elastic moduli of single-crystal europium iron garnet and yttrium iron garnet, J. Appl. Phys., 37, 2194-2195, 1966.

26. Bateman, T.B., Elastic moduli of single-crystal zinc oxide, J. Appl. Phys., 33, 3309-3312, 1962.

27. Bateman, T.B., H.J. McSkimin, and J.M. Whelan, Elastic moduli of single-crystal gallium arsenide, J. Appl. Phys., 30, 544-545, 1959.

28. Beattie, A.G., Temperature dependence of the elastic constants of tin telluride, J. Appl. Phys., 40, 4818-4821, 1969.

29. Bechmann, R., Elastic and piezoelectric constants of alpha-quartz, Phys. Rev., 110, 1060-1061, 1958.

30. Bentle, G.G., Elastic constants of single-crystal BeO at room temperature, J. Amer. Ceramic Soc., 49, 125-128, 1966.

31. Berlincourt, D., and H. Jaffe, Elastic and piezoelectric coefficients of single-crystal barium titanate, Phys. Rev., 111, 143-148, 1958.

32. Berlincourt, D., H. Jaffe, and L.R. Shiozawa, Electroelastic properties of the sulfides, selenides, and tellurides of Zn and Cd, Phys. Rev., 129, 1009-1017, 1963.

33. Bernstein, B.T., Elastic constants of synthetic sapphire at 27°C, J. Appl. Phys., 34, 169-172, 1963.

34. Bhimasenachar, J., and T.S. Rao, Elastic constants of some cubic nitrates, Proc. Nat. Inst. Sci., India, 16, 235-240, 1949.

35. Birch, F., Elastic constants of rutile - a correction to a paper by R.K. Verma, 'elasticity of some high-density crystals', J. Geophys. Res., 65, 3855-3856, 1960.

36. Bogardus, E.H., Third-order elastic constants of Ge, MgO, and fused SiO_2, J. Appl. Phys., 36, 2504-2513, 1965.

37. Bogorodskii, V.V., Elastic moduli of ice crystals, Sov. Phys.-Acoust., 10, 124-126, 1964.

38. Bolef, D.I., Elastic constants of single crystals of the bcc transition elements V, Nb, and Ta, J. Appl. Phys., 32, 100-105, 1961.

39. Bolef, D.I., and J. de Klerk, Elastic constants of single-crystal Mo and W between 77° and 500°K, J. Appl. Phys., 33, 2311-2314, 1962.

40. Bolef, D.I., and J. de Klerk, Anomalies in the elastic constants and thermal expansion of chromium single crystals, Phys. Rev., 129, 1063-1067, 1963.

41. Bolef, D.I., and M. Menes, Measurement of elastic constants of RbBr, RbI, CsBr, and CsI by an ultrasonic cw resonance technique, J. Appl. Phys., 31, 1010-1017, 1960.

42. Bolef, D.I., and M. Menes, Elastic constants of single-crystal aluminum antimonide, J. Appl. Phys., 31, 1426-1427, 1960.

43. Bolef, D.I., N.T. Melamed, and M. Menes, Elastic constants of hexagonal cadmium sulfide, J. Phys. Chem. Solids, 17, 143-148, 1960.

44. Bridgman, P.W., The elastic moduli of five alkali halides, Proc. Amer. Acad. Sci., 64, 19-38, 1928.

45. Briscoe, C.V., and C.F. Squire, Elastic constants of LiF from 4.2°K to 300°K by ultrasonic methods, Phys. Rev., 106, 1175-1177, 1957.

46. Brown, H.L., and C.P. Kempter, Elastic properties of zirconium carbide, Phys. Stat. Sol., 18, K21-K23, 1966.

47. Carroll, K.J., Elastic constants of niobium from 4.2° to 300°K, J. Appl. Phys., 36, 3689-3690, 1965.

48. Chandrasekhar, B.S., and J.A. Rayne, Elastic constants of indium from 1.4° to 300°K, Phys. Rev., 124, 1011-1014, 1961.

49. Chang, R., and L.J. Graham, Low-temperature elastic properties of ZrC and TiC, J. Appl. Phys., 37, 3778-3783, 1966.

50. Chang, Y.A., and L. Himmel, Elastic constants of cadmium from 300° to 575°K, J. Appl. Phys., 37, 3787-3790, 1966.

51. Chang, Y.A., and L. Himmel, Temperature dependence of the elastic constants of Cu, Ag, and Au above room temperature, J. Appl. Phys., 37, 3567-3572, 1966.

52. Chang, Y.A., L. Himmel, and J.P. Neumann, Thermal expansion and elastic constants of β'-AgMg. II. single-crystal elastic constants from 77° to 750°K, J. Appl. Phys., 38, 649-652, 1967.

53. Chang, Y.A., and J.P. Neumann, The elastic constants of the intermetallic compound Ag_2Al from 77° to 700°K, J. Phys. Chem. Solids, 28, 2117-2122, 1967.

54. Chang, Z.P., and G.R. Barsch, Pressure dependence of the elastic constants of single-crystalline magnesium oxide, J. Geophys. Res., 74, 3291-3294, 1969.

55. Chang, Z.P., G.R. Barsch, and D.L. Miller, Pressure dependence of elastic constants of cesium halides, Phys. Stat. Sol., 23, 577-586, 1967.

56. Cheng, C.H., The elastic constants of MgAg and $MgCu_2$ single crystals, J. Phys. Chem. Solids, 28, 413-416, 1967.

57. Chernov, Y.M., and A.V. Stepanov, The temperature dependence of the elastic constants of LiF monocrystals, Sov. Phys.-Solid State, 3, 2097-2098, 1961.

57A. Chung, D.-H., Elastic moduli of single crystal and polycrystalline MgO, Phil. Mag., 8, 833-841, 1963.

58. Clark, A.E., and R.E. Strakna, Elastic constants of single-crystal YIG, J. Appl. Phys., 32, 1172-1173, 1961.

59. Claytor, R.N., and B.J. Marshall, Specific heat and elastic constants of sodium iodide at low temperatures, Phys. Rev., 120, 332-334, 1960.

60. Cline, C.F., H.L. Dunegan, and G.W. Henderson, Elastic constants of hexagonal BeO, ZnS, and CdSe, J. Appl. Phys., 38, 1944-1948, 1967.

61. Corll, J.A., Effect of pressure on the elastic parameters and structures of CdS, Phys. Rev., 157, 623-626, 1967.

62. Cutler, H.R., J.J. Gibson, and K.A. McCarthy, The elastic constants of magnesium fluoride, Solid State Commun., 6, 431-433, 1968.

63. Dalven, R., and C.W. Garland, Elastic constants of sodium iodide from 180°K to 300°K, J. Chem. Phys., 30, 346-347, 1959.

64. Dandekar, D.P., Pressure dependence of the elastic constants of calcite, Phys. Rev., 172, 873-877, 1968.

65. Dandekar, D.P., Variation in the elastic constants of calcite with temperature, J. Appl. Phys., 39, 3694-3699, 1968.

66. Dandekar, D.P., and A.L. Ruoff, Temperature dependence of the elastic constants of calcite between 160° and 300°K, J. Appl. Phys., 39, 6004-6009, 1968.

67. Daniels, W.B., Pressure variation of the elastic constants of sodium, Phys. Rev., 119, 1246-1252, 1960.

68. Daniels, W.B., and C.S. Smith, Pressure derivatives of the elastic constants of copper, silver, and gold to 10 000 bars, Phys. Rev., 111, 713-721, 1958.

69. Davidson, D.L., and F.R. Brotzen, Elastic constants of molybdenum-rich rhenium alloys in the temperature range -190°C to +100°C, J. Appl. Phys., 39, 5768-5775, 1968.

70. Davis, L.C., W.B. Whitten, and G.C. Danielson, Elastic constants and calculated lattice vibration frequencies of Mg_2Sn, J. Phys. Chem. Solids, 28, 439-447, 1967.

71. De Batist, R., Room temperature elastic constants of single crystal U_4O_9, Mat. Res. Bull., 1, 75-81, 1966.

72. De Brettville, A., Jr., E.R. Cohen, A.D. Ballato, I.N. Greenberg, and S. Epstein, Least-squares determination of the elastic constants of antimony and bismuth, Phys. Rev., 148, 575-579, 1966.

73. De Klerk, J., Elastic constants and Debye temperature of TiC using a new ultrasonic coherent pulse /cw technique, Rev. Sci. Inst., 36, 1540-1545, 1965.

74. De Klerk, J., Elastic constants of α-ZnS, J. Phys. Chem. Solids, 28, 1831-1837, 1967.

75. Dickinson, J.M., and P.E. Armstrong, Temperature dependence of the elastic constants of molybdenum, J. Appl. Phys., 38, 602-606, 1967.

76. Diederich, M.E., and J. Trivisonno, Temperature dependence of the elastic constants of sodium, J. Phys. Chem. Solids, 27, 637-642, 1966.

77. Drabble, J.R., and A.J. Brammer, The third-order elastic constants of indium antimonide, Proc. Phys. Soc., 91, 959-964, 1967.

78. Drabble, J.R., and A.J. Brammer, Third-order elastic constants of gallium arsenide, Solid State Commun., 4, 467-468, 1966.

79. Drabble, J.R., and R.E.B. Strathen, The third-order elastic constants of potassium chloride, sodium chloride, and lithium fluoride, Proc. Phys. Soc., 92, 1090-1995, 1967.

80. Durand, M.A., The temperature variation of the elastic moduli of NaCl, KCl, and MgO, Phys. Rev., 50, 449-455, 1936.

81. Eckstein, Y., A.W. Lawson, and D.H. Reneker, Elastic constants of bismuth, J. Appl. Phys., 31, 1534-1538, 1960.

82. Einspruch, N.G., and L.T. Claiborne, Elastic constants of a (73.8% Ni-26.2% Fe) ferromagnetic alloy, J. Appl. Phys., 35, 175, 1964.

83. Einspruch, N.G., and R.J. Manning, Elastic constants of compound semiconductors ZnS, PbTe, GaSb, J. Acoust. Soc. Am., 35, 215-216, 1963.

84. Enck, F.D., Behavior of the principal elastic moduli and specific heat at constant volume of KCl at elevated temperatures, Phys. Rev., 119, 1873-1877, 1960.

85. Epstein, S.G., and O.N. Carlson, The elastic constants of nickel-copper alloy single crystals, Acta Metal., 13, 487-491, 1965.

86. Epstein, S.G., and A.P. DeBretteville, Jr., Elastic constants of wave propagation in antimony and bismuth, Phys. Rev., 138, A771-A779, 1965.

87. Eros, S., and J.R. Reitz, Elastic constants by the ultrasonic pulse echo method, J. Appl. Phys., 29, 683-686, 1958.

88. Eros, S., and C.S. Smith, Low-temperature elastic constants of magnesium alloys, Acta Metal., 9, 14-22, 1961.

89. Featherstone, F.H., and J.R. Neighbours, Elastic constants of tantalum, tungsten, and molybdenum, Phys. Rev., 130, 1324-1333, 1963.

90. Ferris, R.W., M.L. Shepard, and J.F. Smith, Elastic constants of thallium single crystals in the temperature range 4.2°K - 300°K, J. Appl. Phys., 34, 768-770, 1963.

91. Fisher, E.S., Temperature dependence of the elastic moduli in alpha uranium single crystals, Part IV (298° to 923°K), J. Nucl. Mat., 18, 39-54, 1966.

92. Fisher, E.S., and D. Dever, Elastic moduli and phase transition in uranium at T<43°K, Phys. Rev., 170, 607-613, 1968.

93. Fisher, E.S., and D. Dever, Elastic moduli of Gd metal, Proc. Conf. Rare Earth Res., 6th, Gatlinburg, Tenn., 522-533, 1967.

94. Fisher, E.S., and D. Dever, Temperature dependence of elastic moduli of ruthenium, rhenium, cobalt, dysprosium, and erbium; a study of the elastic anisotropy-phase transformation relationship, Trans. Metal. Soc. AIME, 239, 48-57, 1967.

95. Fisher, E.S., and C.J. Renken, Single-crystal elastic moduli and the hcp → bcc transformation in Ti, Zr and Hf, Phys. Rev., 135, A482-A494, 1964.

96. Flinn, P.A., G.M. McManus, and J.A. Rayne, Elastic constants of ordered and disordered Cu_3Au from 4.2 to 300°K, J. Phys. Chem. Solids, 15, 189-195, 1960.

97. Fugate, R.Q., and D.E. Schuele, The elastic constants of NaCl at 77.3°K and 4.2°K, J. Phys. Chem. Solids, 27, 493-498, 1966.

98. Garland, C.W., and J.S. Jones, Elastic constants of ammonium chloride near the lambda point, J. Chem. Phys., 39, 2874-2880, 1963.

99. Garland, C.W., and K.C. Park, Low-temperature elastic constants of gallium arsenide, J. Appl. Phys., 33, 759-760, 1962.

100. Garland, C.W., and R. Renard, Order-disorder phenomena. III. effect of temperature and pressure on the elastic constants of ammonium chloride, J. Chem. Phys., 44, 1130-1139, 1966.

101. Garland, C.W., and J. Silverman, Elastic constants of cadmium from 4.2°K to 300°K, Phys. Rev., 119, 1218-1222, 1960; Errata, Phys. Rev., 127, 2287, 1962.

102. Garland, C.W., and C.F. Yarnell, Temperature and pressure dependence of the elastic constants of ammonium bromide, J. Chem. Phys., 44, 1112-1120, 1966.

103. Gerlich, D., Elastic constants of barium fluoride between 4.2 and 300°K, Phys. Rev., 135, A1331-A1333, 1964.

104. Gerlich, D., Elastic constants of single-crystal indium arsenide, J. Appl. Phys., 34, 2915, 1963.

105. Gerlich, D., Elastic constants of strontium fluoride between 4.2 and 300°K, Phys. Rev., 136, A1366-A1368, 1964.

106. Gerlich, D., Low-temperature elastic constants of indium arsenide, J. Appl. Phys., 35, 3062, 1964.

107. Gerlich, D., The elastic constants of cadmium sulfide between 4.2 - 300°K, J. Phys. Chem. Solids, 28, 2575-2579, 1967.

108. Gerlich, D., Third-order elastic moduli of barium fluoride, Phys. Rev., 168, 947-952, 1968.

109. Gerlich, D., and E.S. Fisher, The high temperature elastic moduli of aluminum, J. Phys. Chem. Solids, 30, 1197-1205, 1969.

110. Ghafelehbashi, M., D.P. Dandekar, and A.L. Ruoff, Pressure and temperature dependence of the elastic constants of RbCl, RbBr, and RbI, J. Appl. Phys., 41, 652-666, 1970.

111. Gieske, J.H., and G.R. Barsch, Pressure dependence of the elastic constants of single crystalline aluminum oxide, Phys. Stat. Sol., 29, 121-131, 1967.

112. Gilman, J.J., and B.W. Roberts, Elastic constants of TiC and TiB_2, J. Appl. Phys., 32, 1405, 1961.

113. Gluyas, M., The second- and third-order elastic constants of sodium chloride at 295°K, Brit. J. Appl., Phys., 18, 913-918, 1967.

114. Golding, B., S.C. Moss, and B.L. Averbach, Composition and pressure dependence of the elastic constants of gold-nickel alloys, Phys. Rev., 158, 637-646, 1967.

115. Good, W.A., Rigidity modulus of beta-brass single crystals, Phys. Rev., 60, 605-609, 1941.

116. Graham, E.K., Jr., and G.R. Barsch, Elastic constants of single-crystal forsterite as a function of temperature and pressure, J. Geophys. Res., 74, 5949-5960, 1969.

117. Graham, L.J., and R. Chang, Elastic moduli of single-crystal gadolinium gallium garnet, J. Appl. Phys., 41, 2247-2248, 1970.

118. Graham, L.J., H. Nadler, and R. Chang, Elastic constants of UC, J. Appl. Phys., 34, 1572-1573, 1963.

119. Graham, L.J., H. Nadler, and R. Chang, Third-order elastic constants of single-crystal and polycrystalline columbium, J. Appl. Phys., 39, 3025-3033, 1968.

120. Gsänger, M., H. Egger, and E. Lüscher, Determination of the elastic constants of argon, Physics Letters, 27A, 695-696, 1968.

121. Guinan, M.W., and D.N. Beshers, Pressure derivatives of the elastic constants of α-iron to 10 kbs, J. Phys. Chem. Solids, 29, 541-549, 1968.

122. Gutman, E.J., and J. Trivisonno, Temperature dependence of the elastic constants of rubidium, J. Phys. Chem. Solids, 28, 805-809, 1967.

123. Gutsche, E., Messung der elastischen konstanten von CdS-Einkristallen, Die Naturwissenschaften, 45, 566, 1958.

124. Haussühl, S., Das elastische verhalten von flusspat und strukturverwandten kristallen, Phys. Stat. Sol., 3, 1072-1076, 1963.

125. Haussühl, S., Elastische eigenschaften von kristallen des CsCl-typs, Acta Cryst., 13, 685-687, 1960.

126. Haussühl, S., Elastische und thermoelastische konstanten von K_2SO_4, Rb_2SO_4, Cs_2SO_4, $(NH_4)_2SO_4$, Tl_2SO_4, und $K_2Mg_2(SO_4)_3$, Acta Cryst., 18, 839-842, 1965.

127. Haussühl, S., Elastische und thermoelastische konstanten von $NH_2(CH_2)_2SO_3H$ (taurin), Z. Naturforsch, 20A, 1235-1236, 1965.

128. Hearmon, R.F.S., The elastic constants of anisotropic materials, Rev. Mod. Phys., 18, 409-440, 1946.

129. Hearmon, R.F.S., The elastic constants of anisotropic materials II, Advan. Phys., 5, 323-382, 1956.

130. Heseltine, J.C.W., D.W. Elliott, and O.B. Wilson, Jr., Elastic constants of single-crystal benzene, J. Chem. Phys., 40, 2584-2587, 1964.

131. Hickernell, F.S., and W.R. Gayton, Elastic constants of single-crystal indium phosphide, J. Appl. Phys., 37, 462, 1966.

132. Hidshaw, W., J.T. Lewis, and C.V. Briscoe, Elastic constants of silver chloride from 4.2 to 300°K, Phys. Rev., 163, 876-881, 1967.

133. Hiki, Y., and A. Granato, Anharmonicity in noble metals; higher order elastic constants, Phys. Rev., 144, 411-419, 1966.

134. Ho, P.S., and A.L. Ruoff, Pressure dependence of the elastic constants and an experimental equation of state for CaF_2, Phys. Rev., 161, 864-869, 1967.

135. House, D.G., and E.V. Vernon, Determination of the elastic moduli of tin single crystals, and their variation with temperature, Brit. J. Appl. Phys., 11, 254-259, 1960.

136. Houston, B., R.E. Strakna, and H.S. Belson, Elastic constants, thermal expansion, and Debye temperature of lead telluride, J. Appl. Phys., 39, 3913-3916, 1968.

137. Huffman, D.R., and M.H. Norwood, Specific heat and elastic constants of calcium fluoride at low temperatures, Phys. Rev., 117, 709-711, 1960.

138. Huntington, H.B., The elastic constants of crystals, in Solid State Physics, Vol.7, edited by F. Seitz and D. Turnbull, pp. 213-285, Academic Press, New York, 1958.

139. Johnston, D.L., P.H. Thrasher, and R.J. Kearney, Elastic constants of SrO, J. Appl. Phys., 41, 427-428, 1970.

140. Kamm, G.N., and G.A. Alers, Low-temperature elastic moduli of aluminum, J. Appl. Phys., 35, 327-330, 1964.

141. Keller, K.R., and J.J. Hanak, Lattice softening in single crystal Nb_3Sn, Physics Letters, 21, 263-264, 1966.

142. Keller, K.R., and J.J. Hanak, Ultrasonic measurements in single-crystal Nb_3Sn, Phys. Rev., 154, 628-631, 1967.

143. Kobyakov, I.B., Piezoelectric, elastic, and dielectric constants of hexagonal zinc sulfide, Sov. Phys.-Cryst., 11, 369-371, 1966.

144. Koga, I., M. Aruga, and Y. Yoshinaka, Theory of plane elastic waves in a piezoelectric crystalline medium and determination of elastic and piezoelectric constants of quartz, Phys. Rev., 109, 1467-1473, 1958.

145. Koliwad, K.M., P.B. Ghate, and A.L. Ruoff, Pressure derivatives of the elastic constants of NaBr and KF, Phys. Stat. Sol., 21, 507-516, 1967.

146. Kollarits, F.J., and J. Trivisonno, Single-crystal elastic constants of cesium, J. Phys. Chem. Solids, 29, 2133-2139, 1968.

147. Kumazawa, M., The elastic constants of single-crystal orthopyroxene, J. Geophys. Res., 74, 5973-5980, 1969.

148. Kumazawa, M., and O.L. Anderson, Elastic moduli, pressure derivatives, and temperature derivatives of single-crystal olivine and single-crystal forsterite, J. Geophys. Res., 74, 5961-5972, 1969.

149. Lazarus, D., The variation of the adiabatic elastic constants of KCl, NaCl, CuZn, Cu, and Al with pressure to 10,000 bars, Phys. Rev., 76, 545-553, 1949.

150. Leamy, H.J., E.D. Gibson, and F.X. Kayser, The elastic stiffness coefficients of iron-aluminum alloys - I. Experimental results and thermodynamic analysis, Acta Metal., 15, 1827-1838, 1967.

151. Leese, J., and A.E. Lord, Jr., Elastic stiffness coefficients of single-crystal iron from room temperature to 500°C, J. Appl. Phys., 39, 3986-3988, 1968.

152. Lewis, J.T., A. Lehoczky, and C.V. Briscoe, Elastic constants of the alkali halides at 4.2°K, Phys. Rev., 161, 877-887, 1967.

153. Lewis, M.F., Elastic constants of magnesium aluminate spinel, J. Acoust. Soc. Am., 40, 728-729, 1966.

154. Long, T.R., and C.S. Smith, Single-crystal elastic constants of magnesium and magnesium alloys, Acta Metal., 5, 200-207, 1957.

155. Lord, A.E., Jr., and D.N. Beshers, Elastic stiffness coefficients of iron from 77° to 673°K, J. Appl. Phys., 36, 1620-1623, 1965.

156. Lowrie, R., and A.M.Gonas, Single-crystal elastic properties of tungsten from 24° to 1800°C, J. Appl. Phys., 38, 4505-4509, 1967.

157. Macedo, P.M., W. Capps, and J.B. Wachtman, Jr., Elastic constants of single crystal ThO_2 at 25°C, J. Amer. Ceramic Soc., 47, 651, 1964.

158. MacFarlane, R.E., J.A. Rayne, and C.K. Jones, Anomalous temperature dependence of shear modulus c_{44} for platinum, Physics Letters, 18, 91-92, 1965.

159. MacFarlane, R.E., J.A. Rayne, and C.K. Jones, Temperature dependence of elastic moduli of iridium, Physics Letters, 20, 234-235, 1966.

160. McSkimin, H.J., Measurements of the elastic constants of single crystal cobalt, J. Appl. Phys., 26, 406-409, 1955.

161. McSkimin, H.J., Measurement of ultrasonic wave velocities and elastic moduli for small solid specimens at high temperatures, J. Acoust. Soc. Am., 31, 287-295, 1959.

162. McSkimin, H.J., Ultrasonic measurement technique applicable to small solid specimens, J. Acoust. Soc. Am., 22, 413-418, 1950.

163. McSkimin, H.J., and P. Andreatch, Jr., Elastic moduli of germanium versus hydrostatic pressure at 25.0°C and -195.8°C, J. Appl. Phys., 34, 651-655, 1963.

164. McSkimin, H.J., and P. Andreatch, Jr., Elastic moduli of silicon vs hydrostatic pressure at 25.0°C and -195.8°C, J. Appl. Phys., 35, 2161-2165, 1964.

165. McSkimin, H.J., and W.L. Bond, Elastic moduli of diamond, Phys. Rev., 105, 116-121, 1957.

166. McSkimin, H.J., and E.S. Fisher, Temperature dependence of the adiabatic elastic moduli of single-crystal alpha uranium, J. Appl. Phys., 31, 1627-1639, 1960.

167. McSkimin, H.J. and D.G. Thomas, Elastic moduli of cadmium telluride, J. Appl. Phys., 33, 56-59, 1962.

168. McSkimin, H.J., P. Andreatch, Jr., and R.N. Thurston, Elastic moduli of quartz versus hydrostatic pressure at 25° and -195.8°C, J. Appl. Phys., 36, 1624-1632, 1965.

169. McSkimin, H.J., A. Jayaraman, and P. Andreatch, Jr., Elastic moduli of GaAs at moderate pressures and the evaluation of compression to 250 kbar, J. Appl. Phys., 38, 2362-2364, 1967.

170. McSkimin, H.J., A. Jayaraman, P. Andreatch, Jr., and T.B. Bateman, Elastic moduli of gallium antimonide under pressure and evaluation of compression to 80 kbar, J. Appl. Phys., 39, 4127-4128, 1968.

171. Manghnani, M.H., Elastic constants of single-crystal rutile under pressures to 7.5 kilobars, J. Geophys. Res., 74, 4317-4328, 1969.

172. Marquardt, W.R., and J. Trivisonno, Low temperature elastic constants of potassium, J. Phys. Chem. Solids, 26, 273-278, 1965.

173. Marshall, B.J., Elastic constants of CsBr from 4.2°K to 300°K, Phys. Rev., 121, 72-73, 1961.

174. Marshall, B.J., and J.R. Kunkel, Heat capacity and elastic constants of CsI at low temperatures, J. Appl. Phys., 40, 5191-5192, 1969.

175. Marshall, B.J., and R.E. Miller, Elastic constants of KF from 300° to 4.2°K, J. Appl. Phys., 38, 4749-4750, 1967.

176. Marshall, B.J., D.O. Pederson, and G.G. Dorris, Elastic constants of LiCl and RbCl from 300°K to 4.2°K, J. Phys. Chem. Solids, 28, 1061-1067, 1967.

177. Mason, W.P., Piezoelectric Crystals and Their Application to Ultrasonics, 508 pp., D. Van Nostrand Co., Inc., New York, 1950.

178. Masumoto, H., H. Saito, and M. Kikuchi, Thermal expansion and temperature dependence of Young's modulus of single crystal of hexagonal cobalt, Sci. Rep. Ritu, 19, 172-183, 1967.

179. Mayer, G. and J. Gigon, Fast neutron irradiation effects on certain physical constants of crystalline quartz and vitreous silica, J. Phys. Rad., 18, 109-114, 1957.

180. Mayer, W.G., and E.A. Hiedemann, Optical methods for the ultrasonic determination of the elastic constants of sapphire, J. Acoust. Soc. Am., 30, 756-760, 1958. Correction: J. Acoust. Soc. Am., 32, 1699-1700, 1960.

181. Merkulov, L.G., Absorption of ultrasonic waves in alkali-haloid crystals, Sov. Phys. Acoust., 5, 444-450, 1959.

182. Michard, F., and A. Zarembowitch, Constantes élastiques du nitrate de plomb, C.R. Acad. Sci., Paris, 260, 1921-1922, 1965.

183. Michard, F., J. Julien, and A. Zarembowitch, Propriétés élastiques des monocristaux d'acénaphtène, C.R. Acad. Sci., Paris, 265, 565-567, 1967.

184. Miller, R.A., and D.E. Schuele, The pressure derivatives of the elastic constants of lead, J. Phys. Chem. Solids, 30, 589-600, 1969.

185. Miller, R.A., and C.S. Smith, Pressure derivatives of the elastic constants of LiF and NaF, J. Phys. Chem. Solids, 25, 1279-1292, 1964.

186. Mnatsakanyan, A.V., L.A. Shuvalov, I.S. Zheludev, and I.V. Gavrilova, The elastic parameters of RbH_2PO_4 crystals over a wide temperature range, Sov. Phys.-Cryst., 11, 412-415, 1966.

187. Morse, G.E., and A.W. Lawson, The temperature and pressure dependence of the elastic constants of thallium bromide, J. Phys. Chem. Solids, 28, 939-950, 1967.

188. Mort, J., Elastic constants of trigonal selenium, J. Appl. Phys., 38, 3414-3415, 1967.

189. Musgrave, M.J.P., Calculations relating to the propagation of elastic waves in anisotropic media, Nat. Phys. Lab. Basic Physics Div., Report No. 7, 1961.

190. Nash, H.C., and C.S. Smith, Single-crystal elastic constants of lithium, J. Phys. Chem. Solids, 9, 113-118, 1959.

191. Neighbours, J.R. and G.A. Alers, Elastic constants of silver and gold, Phys. Rev., 111, 707-712, 1958.

192. Neighbours, J.R., and C.S. Smith, The elastic constants of copper alloys, Acta Metal., 2, 591-596, 1954.

193. Nikanorov, S.P., B.K. Kardashev and N.S. Kas'kovich, Temperature dependence of elastic constants of calcium fluoride, Sov. Phys.-Sol. State, 10, 703-705, 1968.

194. Nikanorov, S.P., and A.V. Stepanov, The temperature dependence of the elastic constants of potassium bromide single crystals, J. Expt. Theoret. Phys., USSR, English transl., 37, 1280-1281, 1959.

195. Nikanorov, S.P., V.A. Tatarchenko, and A.V. Stepanov, Temperature dependence of the elastic constants of cesium bromide and iodide single crystals, Sov. Phys.-Sol. State, 5, 450-455, 1963.

196. Norwood, M.H., and C.V. Briscoe, Elastic constants of potassium iodide and potassium chloride, Phys. Rev., 112, 45-48, 1958.

197. Nouet, J., and A. Zarembowitch, Physique cristalline-constantes élastiques du chlorostannate d'ammonium, C.R. Acad. Sci., Paris, 265, 856-858, 1967.

198. Oliver, D.W., The elastic moduli of MnO, J. Appl. Phys., 40, 893, 1969.

199. Overton, W.C., Jr., and J. Gaffney, Temperature variation of the elastic constants of cubic elements. I. copper, Phys. Rev., 98, 969-977, 1955.

200. Peselnick, L., and R.A. Robie, Elastic constants of calcite, J. Appl. Phys., 34, 2494-2495, 1963.

201. Pettersson, G., J. Vallin, J.L. Calais, and K. Mansikka, Experimental and theoretical determination of the elastic constants of NaCl, Arkiv för Fysik, 34, 371-376, 1967.

202. Poindexter, E., and A.A. Giardini, Elastic constants of strontium titanate ($SrTiO_3$), Phys. Rev., 110, 1069, 1958.

203. Potter, R.F., Elastic moduli of indium antimonide Phys. Rev., 103, 47-50, 1956.

204. Prince, E.,and W.A. Wooster, The elastic constants of zincblende, determined from thermal diffuse scattering of X-rays, Acta Cryst., 4, 191, 1951.

205. Proctor, T.M., Jr., Low-temperature speed of sound in single-crystal ice, J. Acoust. Soc. Am., 39, 972-977, 1966.

206. Purwins, H.-G., H. Hieber and R. Labusch, Elastische einkristallkonstanten von iridium, Phys. Stat. Sol., 11, K63-K66, 1965.

207. Purwins, H.-G., R. Labusch and P. Haasen, Elastische einkristallkonstanten von silberlegierungen, Z. Metallkde, 57, 867-871, 1966.

208. Rayne, J.A., Elastic constants of α-brasses: room-temperature variation with solute concentration, Phys. Rev., 112, 1125-1130, 1958.

209. Rayne, J.A., Elastic constants of palladium from 4.2-300°K, Phys. Rev., 118, 1545-1549, 1960.

210. Rayne, J.A., and B.S. Chandrasekhar, Elastic constants of β tin from 4.2°K to 300°K, Phys. Rev., 120, 1658-1663, 1960.

211. Rayne, J.A., and B.S. Chandrasekhar, Elastic constants of iron from 4.2°K to 300°K, Phys. Rev., 122, 1714-1716, 1961.

212. Reddy, P.J., and J. Bhimasenachar, Variation with temperature of the elastic compliances of potassium iodide, Acta Cryst., 17, 28-30, 1964.

213. Reddy, P.J., and A.L. Ruoff, Pressure derivatives of the elastic constants in some alkali halides, in Physics of Solids at High Pressures, edited by C.T. Tomizuka and R.M. Emrick, pp. 510-523, Academic Press, New York, 1965.

214. Reddy, P.J., and S.V. Subrahmanyam, Temperature variation of the elastic constants of barytes, Proc. Indian Acad. Sci., 50, 380-384, 1959.

215. Reifenberger, R., M.J. Kerk, and J. Trivisonno, Low temperature elastic constants of indium arsenide, J. Appl. Phys., 40, 5403-5404, 1969.

216. Reinitz, K., Elastic constants of CsBr, CsI, RbBr, and RbI, Phys. Rev., 123, 1615-1619, 1961.

217. Roberts, C.A., and R. Meister, The elastic constants of rubidium, J. Phys. Chem. Solids, 27, 1401-1407, 1966.

218. Rose, F.C., The variation of the adiabatic elastic moduli of rocksalt with temperature between 80°K and 270°K, *Phys. Rev.*, *49*, 50-54, 1936.

219. Rosenberg, H.M., and J.K. Wigmore, The elastic constants of potassium magnesium fluoride, *Physics Letters*, *24A*, 317, 1967.

220. Rotter, C.A., and C.S. Smith, Ultrasonic equation of state of iron. I. low pressure, room temperature, *J. Phys. Chem. Solids*, *27*, 267-276, 1966.

221. Ryzhova, T.V., Elastic properties of plagioclase, *Bull. Acad. Sci. USSR, Geophys. Ser.*, English Transl., no. 7, 633-635, 1964.

222. Ryzhova, T.V., and K.S. Aleksandrov, The elastic properties of potassium-sodium feldspars, *Bull. Acad. Sci. USSR, Earth Physics Series*, English Transl., no. 1, 53-56, 1965.

223. Ryzhova, T.V., K.S. Aleksandrov, V.M. Korobkova, The elastic properties of rock-forming minerals V. Additional data on silicates, *Bull. Acad. Sci. USSR, Earth Physics Series*, English Transl. no. 2, 111-113, 1966.

224. Ryzhova, T.V., L.M. Reshchikova, and K.S. Aleksandrov, Elastic properties of rock-forming minerals, VI. garnets, *Bull. Acad. Sci. USSR, Earth Physics Series*, English Transl., no. 7, 447-450, 1966.

225. Scharf, H. and H.-G. Friedel, Elastische konstanten von magnesiumreichen magnesium-kadmium-legierungen, *Z. Metallke*, *58*, 729-730, 1967.

226. Schmunk, R.E., and C.S. Smith, Elastic constants of copper-nickel alloys, *Acta Metal.*, *8*, 396-401, 1960.

227. Schreiber, E., Elastic moduli of single-crystal spinel at 25°C and to 2 Kbar, *J. Appl. Phys.*, *38*, 2508-2511, 1967.

228. Schwerdtner, W.M., J.C.-M. Tou, and P.B. Hertz, Elastic properties of single crystals of anhydrite, *Canadian J. Earth Sci.*, *2*, 673-683, 1965.

229. Shannette, G.W., and J.F. Smith, Single-crystalline elastic constants of $ZrCo_2$ and $HfCo_2$ in the temperature range 4.2°-300°K, J. Appl. Phys., 40, 79-82, 1969.

230. Shepard, M.L., and J.F. Smith, Elastic constants of rhenium single crystals in the temperature range 4.2°-298°K, J. Appl. Phys., 36, 1447-1450, 1965.

231. Shepard, M.L., and J.F. Smith, Single crystalline elastic constants of lead-thallium alloys, Acta Metal., 15, 357-363, 1967.

232. Siegel, S., The variation of the principal elastic moduli of Cu_3Au with temperature, Phys. Rev., 57, 537-545, 1940.

233. Simmons, G., and F. Birch, Elastic constants of pyrite, J. Appl. Phys., 34, 2736-2738, 1963.

234. Slagle, O.D., and H.A. McKinstry, Temperature dependence of the elastic constants of the alkali halides. II and III. J. Appl. Phys., 38, 446-458, 1967.

235. Slutsky, L.J., and C.W. Garland, Elastic constants of indium antimonide from 4.2°K to 300°k, Phys. Rev., 113, 167-169, 1959.

236. Slutsky, L.J., and C.W. Garland, Elastic constants of magnesium from 4.2°K to 300°K, Phys. Rev., 107, 972-976, 1957.

237. Smith, J.F., and C.L. Arbogast, Elastic constants of single crystal beryllium, J. Appl. Phys., 31, 99-102, 1960.

238. Smith, J.F., and J.A. Gjevre, Elastic constants of yttrium single crystals in the temperature range 4.2-400°K, J. Appl. Phys., 31, 645-647, 1960.

239. Smith, P.A., and C.S. Smith, Pressure derivatives of the elastic constants of potassium, J. Phys. Chem. Solids, 26, 279-289, 1965.

240. Smith, R.T., Elastic, piezoelectric, and dielectric properties of lithium tantalate, Appl. Phys. Letters, 11, 146-148, 1967.

241. Soga, N., Comparison of measured and predicted bulk moduli of tantalum and tungsten at high temperatures, J. Appl. Phys., 37, 3416-3420, 1966.

242. Soga, N., Elastic constants of garnet under pressure and temperature, J. Geophys. Res., 72, 4227-4234, 1967.

243. Spetzler, H., Equation of state of polycrystalline and single-crystal MgO to 8 kilobars and 800°K, J. Geophys. Res., 75, 2073-2087, 1970.

244. Stephens, R.W.B., The mechanical properties of ice. II. the elastic constants and mechanical relaxation of single crystal ice, Advances in Physics, 7, 266-275, 1958.

245. Subrahmanyam, S.V., Elastic constants of thallium alum, Curr. Sci., 25, 51, 1956.

246. Suess, C., Measurement of the elastic constants of LiF and MgO as a function of temperature and pressure, J. des Recherches du C.N.R.S., 54, 23-59, 1961.

247. Sumer, A., and J.F. Smith, A comparison of the elastic constants of chromium as determined from diffuse X-ray and ultrasonic techniques, J. Appl. Phys., 34, 2691-2694, 1963.

248. Sumer, A., and J.F. Smith, Elastic constants of single-crystal $CaMg_2$, J. Appl. Phys., 33, 2283-2286, 1962.

249. Sutton, P.M., The variation of the elastic constants of crystalline aluminum with temperature between 63°K and 773°K, Phys. Rev., 91, 816-821, 1953.

250. Swartz, K.D., Anharmonicity in sodium chloride, J. Acoust. Soc. Am., 41, 1083-1092, 1967.

251. Swartz, K.D., and A.V. Granato, Experimental test of the Laval-Raman-Viswanathan Theory of Elasticity, J. Acoust. Soc. Am., 38, 824-826, 1965.

252. Tefft, W.E., Elastic constants of synthetic single crystal corundum, J. Res. Nat. Bur. Stand., 70A, 277-280, 1966.

253. Teslenko, V.F., A.P. Ryzhenkov, T.V. Ryzhova, K.S. Aleksandrov, and A.I. Kitaigorodskii, Elasticity constants of molecular crystals II. elasticity constants of dibenzyl, tolane, and stilbene, Sov. Phys.-Cryst., 10, 744-747, 1966.

254. Thomas, J.F., Jr., Third-order elastic constants of aluminum, Phys. Rev., 175, 955-962, 1968.

255. Trivisonno, J., and C.S. Smith, Elastic constants of lithium-magnesium alloys, Acta Metal., 9, 1064-1071, 1961.

256. Uchida, N., and Y. Ohmachi, Elastic and photoelastic properties of TeO_2 single crystal, J. Appl. Phys., 40, 4692-4695, 1969.

257. Vallin, J., Elastic constants of AgCl, Arkiv för Fysik, 34, 367-369, 1967.

258. Vallin, J., O. Beckman, and K. Salama, Elastic constants of CsBr and CsI from 4.2°K to room temperature, J. Appl. Phys., 35, 1222-1223, 1964.

259. Vallin, J., K. Marklund, and J.O. Sikström, Elastic constants and thermal expansion of TlBr, Arkiv för Fysik, 33, 345-356, 1966.

260. Vallin, J., K. Marklund, J.O. Sikström, and O. Beckman, Elastic properties of NaF, Arkiv för Fysik, 32, 515-527, 1966.

261. Vallin, J., M. Mongy, K. Salama, and O. Beckman, Elastic constants of aluminum, J. Appl. Phys., 35, 1825-1826, 1964.

262. Verma, R.K., Elasticity of some high-density crystals, J. Geophys. Res., 65, 757-766, 1960.

263. Vick, G.L., and L.E. Hollander, Ultrasonic measurement of the elastic moduli of rutile, J. Acoust. Soc. Am., 32, 947-949, 1960.

264. Vintaikin, E.Z., X-ray determination of the elastic constants of pure nickel and its alloy with aluminum, Sov. Phys.-Doklady, 11, 91-93, 1966.

265. Viswanathan, R., Elastic constants of sodium chlorate single crystals by pulse-echo method, J. Appl. Phys. 37, 884-886, 1966.

266. Wachtman, J.B., Jr., W.S. Brower, Jr., and E.N. Farabaugh, Elastic constants of single crystal calcium molybdate ($CaMoO_4$), J. Am. Ceramic Soc., 51, 341-344, 1968.

267. Wachtman, J.B., Jr., W.E. Tefft, and D.G. Lam, Jr., Elastic constants of rutile (TiO_2), J. Res. Nat. Bur. Stand., 66A, 465-471, 1962.

268. Wachtman, J.B., Jr., W.E. Tefft, D.G. Lam, Jr., and R.P. Stinchfield, Elastic constants of synthetic single crystal corundum at room temperature, J. Res. Nat. Bur. Stand., 64A, 213-228, 1960.

269. Wachtman, J.B., Jr., M.L. Wheat, H.J. Anderson, and J.L. Bates, Elastic constants of single crystal UO_2 at 25°C, J. Nucl. Mat., 16, 39-41, 1965.

270. Waldorf, D.L., and G.A. Alers, Low-temperature elastic moduli of lead, J. Appl. Phys., 33, 3266-3269, 1962.

271. Warner, A.W., M. Onoe, and G.A. Coquin, Determination of elastic and piezoelectric constants for crystals in class (3m), J. Acoust. Soc. Am., 42, 1223-1231, 1967.

272. Wasilewski, R.J., Elastic constants of Nb by resonance at sonic frequencies, J. Phys. Chem. Solids, 26, 1643-1647, 1965.

273. Wasilewski, R.J., Elastic constants and Young's modulus of NiAl, Trans. Metal. Soc. AIME, 236, 455-457, 1966.

274. Wasilik, J.H., and M.L. Wheat, Elastic constants of cubic lead fluoride at room temperature, J. Appl. Phys., 36, 791-793, 1965.

275. Wazzan, A.R., and L.B. Robinson, Elastic constants of magnesium-lithium alloys, Phys. Rev., 155, 586-594, 1967.

276. Weil, R., and W.O. Groves, The elastic constants of gallium phosphide, J. Appl. Phys., 39, 4049-4051, 1968.

277. Whitten, W.B., P.L. Chung, and G.C. Danielson, Elastic constants and lattice vibration frequencies of Mg_2Si, J. Phys. Chem. Solids, 26, 49-56, 1965.

278. Wong, C., and D.E. Schuele, Pressure and temperature derivatives of the elastic constants of CaF_2 and BaF_2, J. Phys. Chem. Solids, 29, 1309-1330, 1968.

279. Wong, C., and D.E. Schuele, The pressure derivatives of the elastic constants of CaF_2, J. Phys. Chem. Solids, 28, 1225-1231, 1967.

280. Yoon, H.S., and R.E. Newnham, Elastic properties of fluorapatite, Am. Mineralogist, 54, 1193-1197, 1969.

281. Zarembovitch, A., Étude des constantes élastiques de la blende et de leur variation avec la température, Le Journal de Physique, 24, 1097-1102, 1963.

282. Zubov, V.G., L.A. Sysoev, and M.M. Firsova, Growth of hexagonal zinc sulfide and investigation of some of its physical properties, Sov. Phys.-Cryst., 12, 67-70, 1967.

INDEX

NAME	SYSTEM	CRYSTAL PAGE	AGGREGATE PAGE
Argon	Cubic	12	166
Augite	Monoclinic	128	311
Barite	Orthorhombic	119	304
Barium Fluoride	Cubic	13	166
Barium Fluoride (P-derivative)	Cubic	109	295
Barium Fluoride (T-derivative	Cubic	113	298
Barium Nitrate	Cubic	13	166
Barium Titanate	Cubic	13	167
Barium Titanate	Tetragonal	114	299
Benzene	Orthorhombic	119	304
Benzophenone	Orthorhombic	120	304
Beryl	Hexagonal	131	313
Beryllium	Hexagonal	131	313
Beryllium Oxide	Hexagonal	131	313
Biotite	Hexagonal	131	313
Bismuth	Trigonal	147	329
Bronzite	Orthorhombic	120	304
Cadmium	Hexagonal	131	313
Cadmium Selenide	Hexagonal	133	315
Cadmium Sulfide	Hexagonal	133	315
Cadmium Telluride	Cubic	13	167
Calcite	Trigonal	147	329
Calcium Fluoride	Cubic	13	167
Calcium Fluoride (P-derivative)	Cubic	109	295
Calcium Fluoride (T-derivative)	Cubic	113	298

NAME	SYSTEM	CRYSTAL PAGE	AGGREGATE PAGE
Rhenium	Hexagonal	139	321
Rochelle Salt	Orthorhombic	121	305
Rubidium	Cubic	73	246
Rubidium Bromide	Cubic	73	246
Rubidium Bromide (P-derivative)	Cubic	111	296
Rubidium Bromide (T-derivative)	Cubic	113	298
Rubidium Chloride	Cubic	77	252
Rubidium Chloride (P-derivative)	Cubic	111	297
Rubidium Chloride (T-derivative)	Cubic	113	298
Rubidium Dihydrogen Phosphate	Tetragonal	114	299
Rubidium Fluoride	Cubic	81	257
Rubidium Iodide	Cubic	81	257
Rubidium Iodide (P-derivative)	Cubic	111	297
Rubidium Iodide (T-derivative)	Cubic	113	298
Rubidium Sulfate	Orthorhombic	121	305
Rubidium Sulfate (T-derivative)	Orthorhombic	127	310
Ruthenium	Hexagonal	140	322
Rutile	Tetragonal	114	299
Rutile (P-derivative)	Tetragonal	116	301
Rutile (T-derivative)	Tetragonal	117	302
Selenium	Trigonal	149	331
Silicon	Cubic	85	262
Silicon (P-derivative)	Cubic	111	297